D0165640

Elementary Quantum Chemistry

Elementary Quantum Chemistry

Second Edition

Frank L. Pilar

Professor Emeritus of Chemistry
University of New Hampshire

DOVER PUBLICATIONS, INC.
Mineola, New York

Copyright

Copyright © 1968, 1990 by Frank L. Pilar
All rights reserved.

Bibliographical Note

This Dover edition, first published in 2001, is an unabridged republication of the second edition, originally published by McGraw-Hill Publishing Company, New York, in 1990.

Library of Congress Cataloging-in-Publication Data

Pilar, Frank L.
Elementary quantum chemistry / Frank L. Pilar.—Dover ed.
p. cm.
Originally published: New York : McGraw-Hill Pub., 1990.
Includes bibliographical references and index.
ISBN-13: 978-0-486-41464-5 (pbk.)
ISBN-10: 0-486-41464-7 (pbk.)
1. Quantum chemistry. I. Title.

QD462 .P55 2001
541.2'8—dc21

00-050464

Manufactured in the United States by Courier Corporation
41464710 2019
www.doverpublications.com

ABOUT THE AUTHOR

Frank L. Pilar received B.S. and M.S. degrees from the University of Nebraska (Lincoln) in 1951 and 1953, respectively, and a Ph.D. from the University of Cincinnati in 1957. Between 1953 and 1955 he was a research chemist at the Standard Oil Company (Indiana), where he worked on Ziegler-Natta olefin polymerization. Since 1957 he taught at the University of New Hampshire and served as chairman of the chemistry department from 1982 to his retirement in 1992. Since retirement he has devoted his activities to town government and to volunteer trail work with the Appalachian Mountain Club.

Dedicated to those former undergraduate and graduate students at the University of New Hampshire who shared my interests in quantum chemistry: M. Donald Jordan, Jr., John R. Morris, III, Larry Siegel, James J. Eberhardt, Mark Springgate, James D. Quirk, Donald R. Land, John R. Sabin, Robert H. Carrier, Frank Block, and Shu-jun Su.

CONTENTS

PREFACE

"Elementary" is often used to denote a level at which one is exposed superficially to a body of knowledge which will probably never be used again. The term, as used here, implies the lowest level of completeness and sophistication necessary in order for the chemist to acquire the competence needed to begin a serious, nontrivial understanding of the research literature of the late twentieth century—a research literature in which quantum concepts are playing an ever-increasing role.

Experimental chemists have progressed well beyond the point of studying the "average" behavior of reacting species described by the Arrhenius rate equation and are beginning to probe the step-by-step behavior of individual atoms and molecules as they collide, form "transition states," and ultimately form products. Such experiments are generally assisted by sophisticated quantum mechanical calculations of potential-energy surfaces—computations which help to fill gaps in observation and which assist in the interpretation of what is observed. Similarly, organic and inorganic chemists are studying increasingly sophisticated aspects of molecular behavior (e.g., photodissociation), the understanding of which requires a much deeper insight into quantum theory than provided by the "hand-waving" treatments of the past. As Fritz Schaefer has pointed out, theory has become accepted by organic, inorganic, and physical chemists as a legitimate tool for the study of legitimate chemical problems. Although this text stops far short of describing the level of computations and concepts needed in all such studies, it does attempt to provide a suitable foundation upon which expertise in such endeavors can be built.

The author has taught the material in this text to advanced undergraduate students and to beginning graduate students. At the undergraduate level it is sometimes necessary (particularly in a one-semester course) to limit oneself to only the simpler aspects of a topic and to spend less time on details of mathematical formalism, molecular symmetry, and molecular orbital calculations. It is assumed that the student has had mathematics through calculus and at least one year of undergraduate physics taught on the basis of calculus. A background in differential equations, linear algebra, and modern (or atomic) physics is very helpful but not absolutely essential. Mathematical and physical material not necessarily assumed to be part of the students' background—and which may be useful to some as a review—is supplied in a number of appendixes. The first edition worked most of these right into the text material itself; I hope the change won't tempt some students to forgo these topics entirely.

The present edition differs from the first in two very important aspects: first, rather than being derived entirely from the published works of others, many features of atomic and molecular structure are illustrated by calculations carried out specifically for this text; second, there are a number of computer-generated diagrams (e.g., so-called three-dimensional, or surface, orbital plots) which I have also constructed personally. Furthermore, all these computations and graphics can be reproduced by students using relatively modest computational facilities and readily available programs and software, e.g., GAMESS, Gaussian 88 (and previous versions), and RS/1. Some calculations can even be done on microcomputers such as the IBM PC (or compatibles) and the Macintosh series—or even on an Apple II, II+, IIe, IIc, IIGS, or various other popular micros.

I am constantly aware of what a huge debt I owe to the sources of much of my material: the numerous books, journals, and technical reports I have read—and perhaps most important of all, the very inspiring teachers and colleagues I have had. Those who have influenced me to a special degree deserve at least my explicit thanks: Prof. Hans H. Jaffé (University of Cincinnati), my first teacher in quantum chemistry; Prof. Per-Olov Löwdin (Uppsala University and University of Florida), who influenced my philosophical approach to quantum chemistry; Prof. Ruben Pauncz (Technion, Israel Institute of Technology), who gave the clearest, most beautiful lectures I have ever heard in quantum chemistry; Prof. J. de Heer (University of Colorado), whose trenchant wit enlivened some otherwise mundane topics; and the late Charles A. Coulson (Oxford University), who was my gracious host during a pleasant year's stay at the Mathematical Institute. Also, in a very special and personal way, I thank my wife, Anita, for her years of loyal support of a project which has benefited her very little in a material way. I also thank my five children for outgrowing the crayon years; this means that the backsides of manuscript pages of the second edition escaped becoming "artwork," a fate befalling some of the pages of the first edition.

Finally, my special thanks to four colleagues who reviewed portions of the final manuscript: Ernest Davidson, Indiana University; Mark Gordon, North Dakota State University; Hans Jaffé, the University of Cincinnati; and George Petersson, Wesleyan University.

Frank L. Pilar

Elementary Quantum Chemistry

CHAPTER 1

ORIGINS OF THE QUANTUM THEORY

Toward the close of the nineteenth century, many scientists thought that physics was virtually a closed book. As Sir William Cecil Dampier wrote in *A History of Science*: "It seemed as though the main framework had been put together once and for all, and that little remained to be done but to measure physical constants to the increased accuracy represented by another decimal point."[1] Yet, the beginnings of a profound revolution were already brewing—a revolution which would change drastically how scientists and philosophers would view the structure of the universe. In just one short generation the theories of relativity and quanta changed physics and its dependent sciences more comprehensively than had ever occurred before. The present chapter summarizes some of the early work which led to modern quantum mechanics and some of its more important applications to chemistry.

[1] Quoted by O. W. Greenberg, *American Scientist*, July–August 1988, p. 361. See also comments in *Phys. Today*, April 1968, p. 56; August 1968, pp. 9, 11; and January 1969, p. 9.

1-1 THE SPECTRAL SHAPE OF BLACKBODY RADIATION

When a solid is heated to some temperature T, it emits radiation (of which visible light is one specific example). Experiments show that the radiation consists of a spread of different wavelengths, each wavelength generally appearing with a different intensity. Normally, each temperature is characterized by a given radiation wavelength whose energy density is higher than that of radiation of either higher or lower wavelengths; i.e., the energy of the radiation exhibits a maximum value for some particular frequency, and such maxima occur at different frequencies for different temperatures. Figure 1-1 shows how the energy density varies with wavelength for several different temperatures. This figure represents the radiation emitted by an idealized material known as a *blackbody*—a hypothetical material which absorbs all incident radiation and is also a perfect emitter of this radiation. For experimental purposes, an acceptable blackbody may be approximated by an enclosed cavity, the walls of which are kept at some temperature T and which have a small hole in one side. The blackbody radiation shown at 5700 °C is very close to that emitted by our sun; most of the emitted radiation falls within that portion of the electromagnetic spectrum known as *visible light* [approximately 400 to 700 nanometers (nm)].

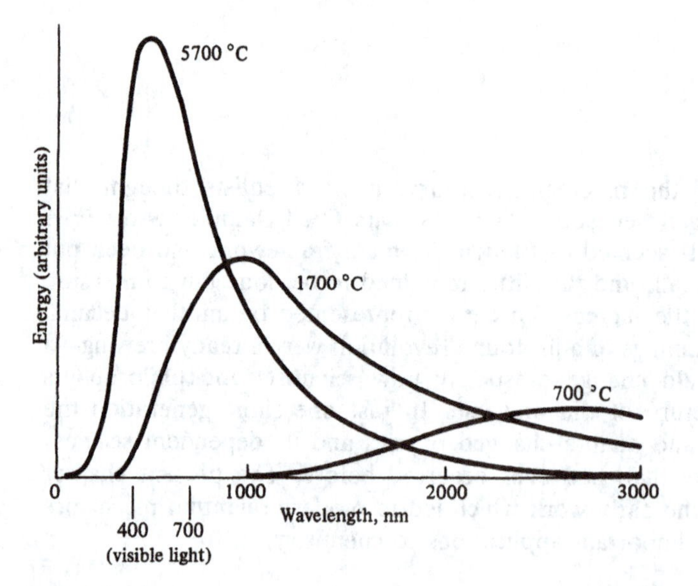

FIGURE 1-1
The spectral shape of blackbody radiation at three different temperatures. The curve at 5700 °C closely resembles the emissive behavior of the sun. The 1700 °C curve represents a body that emits primarily infrared radiation.

Although many renowned physicists tried to provide a theoretical explanation of the details of blackbody radiation, none of the attempts based on classical mechanics succeeded. Nevertheless, it was known that the emitted energy obeyed a relationship of the general form

$$\rho_\nu \, d\nu = \frac{8\pi}{c^3} \nu^3 F(x) \, d\nu \tag{1-1}$$

where $\rho_\nu \, d\nu$ = energy density of radiation having a frequency between ν and $\nu + d\nu$

c = velocity of light (in a vacuum)

$F(x)$ = some unknown function of $x = \nu/T$

The appearance of the variable x originates in the *Wien displacement law*

$$\lambda_{max} T = 2.90 \times 10^6 \text{ nm} \cdot \text{K (approx)} \tag{1-2}$$

an expression which the German physicist Wilhelm Wien managed to derive using classical methods. Here λ_{max} represents that predominant wavelength for which the energy density of the blackbody emission is a maximum. However, all attempts by Wien and his contemporaries to obtain an explicit mathematical form for the function $F(x)$ by the use of classical mechanics failed. In particular, all classical attempts to account for the spectral shape of blackbody radiation predicted what is often called the *ultraviolet catastrophe*; i.e.,

$$\lim_{\nu \to \infty} \rho(\nu) = \lim_{\lambda \to 0} \rho(\lambda) = \infty$$

This means that classical theory could not account for the appearance of a maximum in the spectral distribution.

In 1900 the German thermodynamicist Max Planck obtained an empirical form for $F(x)$:

$$F(x) = k\beta(e^{\beta x} - 1)^{-1} \tag{1-3}$$

where k is Boltzmann's constant (the ideal gas constant R divided by Avogadro's number) and β is an empirical constant (of unknown significance at this point). Planck then proceeded to derive Eq. (1-3) by making some unconventional assumptions about the nature of the blackbody emitter. Since Planck's original approach remains somewhat unclear even to this day, and since Planck himself subsequently modified his assumptions several times, only those ideas which have remained essentially unmodified to this day are given here. Basically, Planck treated the blackbody as a collection of isotropic oscillators capable of interacting with electromagnetic radiation, each oscillator having a vibrational frequency ν. Planck then proposed two new nonclassical ideas:

1. Each of the oscillators has a discrete set of possible energy values given by

$$\epsilon_n = nh\nu \tag{1-4}$$

where $n = 0, 1, 2, \ldots,$ and h is a constant independent of blackbody composition.[2] Unlike in classical mechanics (which would allow ϵ_n to have a continuum of values) Planck's formula implies that the energy of a blackbody oscillator is *quantized*, i.e., exists as packets, bundles, or *quanta* of size $h\nu$.

2. The emission and absorption of radiation are associated with transitions, or jumps, between two different energy "levels." Each emission or absorption involves loss or gain of a quantum of radiant energy of magnitude $h\nu$, ν being the frequency of the radiation absorbed or emitted.

Planck also imposed the requirement that the entropy and the energy must be related by the relationship $dS = dE/T$, where (by the second law of thermodynamics) T must be the same for all radiation frequencies. He then calculated the average energy of an oscillator, using Eq. (1-4) and classical Maxwell-Boltzmann statistics, to show that the constant β in $F(x)$ was simply h/k so that

$$F(x) = h(e^{h\nu/kT} - 1)^{-1} \tag{1-5}$$

The constant h (now called *Planck's constant*) has the dimensions of *action* (energy × time) and is sometimes called the *quantum of action*. It also has the dimensions of angular momentum. The modern numerical value of Planck's constant is 6.626196×10^{-34} J · s. Although Planck spent most of his life believing that the assumption of Eq. (1-4) was fundamentally incorrect and only fortuitously led to a successful blackbody equation, we now know that the constant h is a fundamental constant related to all dynamic discontinuities in nature, especially evident on the atomic and subatomic scale.

EXERCISES

1-1. Verify Eq. (1-5) using the following information: the average energy of an oscillator is given by Maxwell-Boltzmann statistics as

$$\bar{\epsilon} = \nu F(x) = \frac{\sum_{n=0}^{\infty} \epsilon_n e^{-\epsilon_n/kT}}{\sum_{n=0}^{\infty} e^{-\epsilon_n/kT}}$$

You will also need the relationships $(1-y)^{-1} = 1 + y + y^2 + y^3 + \cdots$ and $(1-y)^{-2} = 1 + 2y + 3y^2 + 4y^3 + \cdots$.

---

[2] In actuality, Planck did not quantize the individual oscillators but assumed that the *total* energy possessed at equilibrium by *all* oscillators in the frequency range ν to $\nu + d\nu$ equals a multiple of $h\nu$. See T. S. Kuhn, *Black-Body Radiation and the Quantum Discontinuity*, Oxford University Press, New York, 1978. It is now known that the correct formula is $\epsilon_n = (n + \frac{1}{2})h\nu$, but this is of no consequence in the present context.

1-2. Show that

$$\rho = \int_0^\infty \rho_\nu \, d\nu = aT^4$$

where a is the Stefan-Boltzmann constant given by

$$a = \frac{8\pi k^4}{c^3 h^3} \int_0^\infty \frac{x^3}{e^x - 1} \, dx \qquad \text{and} \qquad \int_0^\infty \frac{x^3}{e^x - 1} \, dx = \frac{\pi^4}{15}$$

The final result was derived classically in 1884 by Boltzmann; it shows that the *total* energy density of blackbody emission is proportional to the fourth power of the absolute temperature.

1-3. Show that at very low values of ν/T, Planck's formula reduces to

$$\rho_\nu \, d\nu = \frac{8\pi kT}{c^3} \nu^2 \, d\nu$$

This is the *Rayleigh-Jeans formula*, an unsuccessful, classically derived equation valid only for very low values of ν/T. What is $F(x)$ in the above formula? (*Note*: $e^x = 1 + x + x^2/2! + x^3/3! + \cdots$.)

1-4. Show that for high values of ν/t, Planck's formula reduces to

$$\rho_\nu \, d\nu = \frac{8\pi h}{c^3} \nu^3 e^{-h\nu/kt} \, d\nu$$

This is *Wien's formula*, another unsuccessful, classically derived equation restricted to large values of ν/T. What is $F(x)$ in Wien's formula?

1-5. According to Fig. 1-1 the surface temperature of the sun is about 5700 °C (5973 K). Estimate λ_{max} (in nm) of the sun.

1-6. The star Antares has a maximum emission wavelength of around 1160 nm. What is the surface temperature of this star, and how does its color compare with that of the sun?

1-7. Show that the Wien displacement law follows from the Planck blackbody equation. To do this, find the value of λ which represents a maximum in the distribution, i.e., solve $d\rho_\nu/d\nu = 0$. Note that $d\nu = -c\lambda^{-2} \, d\lambda$. You will find that the constant in the Wien equation is approximately $hc/5k$ if you assume that ν/T is very large.

1-2 THE PHOTOELECTRIC EFFECT

If an electropositive metal such as cesium or potassium is used as a cathode (negative potential relative to a plate), illumination of this cathode with light of a suitable frequency (usually in the visible or ultraviolet regions) causes a flow of electrons from cathode to plate—the photoelectric effect.[3] However, for each given metal, there is a characteristic wavelength (a threshold value) above which the effect ceases. It is found that the more electropositive the cathode

[3] This phenomenon was discovered accidentally by Heinrich Hertz in 1887 while he was attempting to furnish experimental support for some of the electromagnetic waves (radio waves) predicted by Maxwell's theory.

metal, the longer the wavelength one can use to produce the effect; the alkali metals respond quite well to visible light. Figure 1-2 shows an arrangement which is capable of producing a continuous photocurrent.

Although the magnitude of the photocurrent is directly proportional to the intensity of the incident light, early investigators were puzzled to discover that the *maximum* kinetic energy of the emitted electrons depended only on frequency and was *completely independent of the light intensity*. Classical mechanics failed to explain this anomalous result. According to the classical wave theory of light, radiant energy should be distributed continuously and uniformly over the entire wave front. How, then, is it that only one electron out of millions gathers together enough energy to be emitted? Furthermore, since the photocurrent begins virtually instantaneously upon illumination, there appears to be insufficient time for a single electron to accumulate enough energy from a uniform wave source.

In 1905, Albert Einstein, then a low-paid patent clerk in Switzerland, suggested an extension of Planck's quantum ideas which accounted for the details of the photoelectric effect. Einstein postulated that the radiant energy itself is quantized; i.e., radiation consists of "particles," or quanta (named *photons* by G. N. Lewis in 1926) with an energy $h\nu$ (this is just the energy units which a blackbody emits or absorbs). Assuming that the electrons in a metal behave like an "electron gas" moving freely within the metal, the photon energy $h\nu$ would be used to overcome the attraction of the electron to the bulk of the metal and, if any energy were left over, to impart a kinetic energy to the freed electron. This is expressed in mathematical form as

$$h\nu = \tfrac{1}{2}mv_{\max}^2 + \phi \tag{1-6}$$

where m = mass of the electron
$v_{\max}$ = the electron's velocity
ϕ = *work function* of the metal

The work function is the *minimum* work needed to remove an electron from the metal; that is, ϕ is an *ionization energy* of bulk metal. Einstein's model

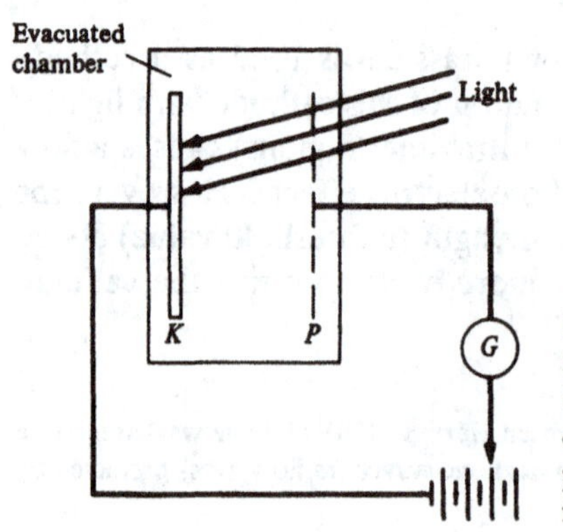

FIGURE 1-2
Schematic representation of an apparatus for observing the photoelectric effect. When light illuminates the cathode K (made of the metal to be studied), a photocurrent flows from it to the plate (anode) P and is indicated on the sensitive galvonometer G.

further implies that the work function is related to the threshold frequency ν_0 by

$$\phi = h\nu_0 \tag{1-7}$$

Because of difficulties with obtaining and maintaining scrupulously uncontaminated metal surfaces, Einstein's model remained experimentally unverified for a decade. Then in 1916, the American physicist Robert Millikan overcame the experimental difficulties by working with freshly exposed metal surfaces in vacuum atmospheres. The maximum kinetic energies of the photoelectrons were determined by measuring the minimum electrostatic potential V_0 (a retardation potential) required to prevent the flow of photocurrent. Thus, mv^2_{max} may be equated to V_0e (e is the charge on an electron—first measured by Millikan), and the Einstein equation may be rewritten

$$V_0e = h\nu - h\nu_0 \tag{1-8}$$

The correctness of Einstein's model was indicated when Millikan's measurements showed that a plot of V_0e as a function of frequency produced a straight line of slope h—the same slope resulting for each different cathode material used. Furthermore, the intercept on the frequency axis produced the threshold frequency ν_0, which differed from metal to metal. These results are illustrated in Fig. 1-3.

Although Sir Isaac Newton (1642–1726) had long ago suggested that light was corpuscular in nature, he was never able to support this idea with concrete evidence and, in fact, found that such a supposition led to certain embarrassingly incorrect predictions about how light behaved. Yet, Einstein's model also suggests a corpuscular nature for light. According to Einstein, the rest mass m_0 of a photon is zero, but since it travels with the speed of light, the requirements of *special relativity* (another "invention" of Einstein, the patent clerk) attribute to it a nonzero rest mass m. The energy of the photon can then be written

$$E = h\nu = \frac{hc}{\lambda} = mc^2 = pc \tag{1-9}$$

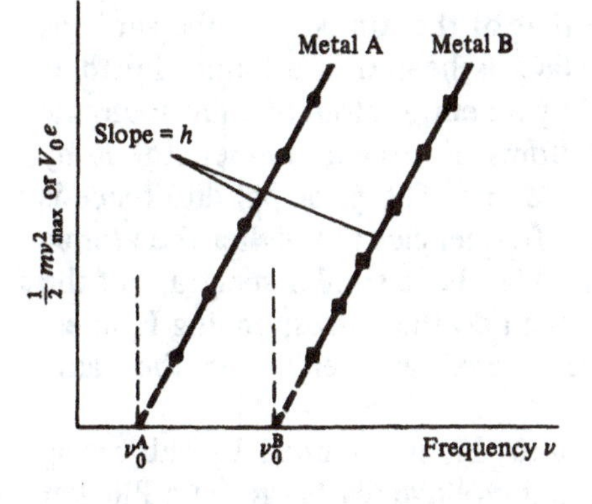

FIGURE 1-3
Millikan's verification of the Einstein photoelectric equation. The slope h is the same for all metals, but the threshold frequency, ν_0, differs from metal to metal. Metal A is the more electropositive metal of the two shown above.

where $p = mc$ is the linear momentum of the photon. Equating the third and fifth terms, one obtains

$$p = \frac{h}{\lambda} \qquad \text{or} \qquad p\lambda = h \tag{1-10}$$

This very important equation suggests that light has a dual nature; i.e., light of experimentally measurable wavelength λ sometimes behaves as if it has a particlelike momentum h/λ accompanying it.

EXERCISES

1-8. What wavelength (in nm) must a photon have in order to eject an electron from sodium metal [work function 2.28 electronvolts (eV)] with a maximum kinetic energy of 1 eV? What is the maximum (limiting) wavelength a photon can have and still eject an electron from sodium metal?

1-9. Repeat the above problem for tungsten (work function 4.5 eV).

1-10. Light of wavelength 552 nm or greater will not eject photoelectrons from a potassium surface.
(*a*) What is the threshold frequency ν_0 of potassium?
(*b*) What is the work function (in eV) of potassium?
(*c*) What is the kinetic energy [in joules (J) and eV] of photoelectrons emitted by light of wavelength 300 nm from potassium metal?
(*d*) What value of the retardation potential [in volts (V)] does the kinetic energy in part (*c*) correspond to?

1-3 LINE SPECTRA OF ATOMS

When a gaseous element is energetically excited so that it emits radiation, the emitted radiation—when passed through a prism—is found to consist of a series of well-defined lines (called the *spectrum* of the element), each associated with a different wavelength. When the excitation is carried out by heating to incandescence in a flame (arc spectrum), the spectra are found to be associated with neutral atoms, but if the excitation is more energetic, e.g., due to a high-voltage electrical discharge or spark, the resulting spectrum (spark spectrum) is found to be associated with ionized atoms. Thus, the spark spectrum of sodium vapor, assumed to be that of the ion Na^+, is the same as that observed when an ionic salt such as NaCl is heated in a flame. Furthermore, the same type of spectrum is emitted by a neutral atom of a given atomic number Z as is emitted by singly ionized atoms of atomic number $Z+1$, by doubly ionized atoms of atomic number $Z+2$, etc. The principal difference is that corresponding spectral lines have higher frequencies the higher the atomic number of the emitting species. Thus, H and He^+ have similar spectra, but the He^+ lines always occur at higher frequency than do the corresponding H lines. The spectra of He and Li^+, Be^{2+}, and B^{3+}, etc., are related in the same manner.

The line spectrum of the hydrogen atom, H, is obtained by subjecting low-pressure molecular hydrogen, H_2, to a high-voltage discharge (in a Plücker tube), as illustrated in Fig. 1-4. The discharge not only atomizes the molecular hydrogen but also excites the atoms energetically so that they emit radiation. This spectrum—the atomic hydrogen spectrum— is the simplest spectrum known.

Classical mechanics provides a very simple (but incorrect) explanation for the appearance of discrete values of ν in atomic spectra: The electrons in an atom carry out periodic motions confined to a definite region of space, and, hence, the emission frequencies should occur as integral multiples of some fundamental vibrational frequency ν_0. But even the simplest case, the hydrogen atom (with only a single electron), shows that the pattern of emission frequencies is quite different from the classical prediction.

In 1885 a German schoolmaster, J. J. Balmer, found that the positions of the spectral lines of atomic hydrogen appearing in the visible and near-ultraviolet regions could be fitted empirically by a very simple formula:

$$\lambda = \frac{m^2 A}{m^2 - 2^2} \tag{1-11}$$

where $A = 364.56$ nm, and m is 3, 4, 5, or 6. Balmer's formula was generalized subsequently by Rydberg in 1896 and by Ritz in 1908 to accommodate newly discovered spectral lines in the ultraviolet and infrared regions. The modern form, called the *Balmer-Rydberg-Ritz formula*, is

$$\tilde{\nu} = \frac{1}{\lambda} = R\left(\frac{1}{n_1^2} - \frac{1}{n_2^2}\right) \qquad n_1, n_2 = 1, 2, 3, \ldots, \text{but } n_2 > n_1 \tag{1-12}$$

where $\tilde{\nu}$ (the reciprocal wavelength) is called the *wave number* (a quantity favored by many spectroscopists in lieu of wavelength or frequency),[4] and R

[4] Both ν and $\tilde{\nu}$ (which are directly proportional to each other) are sometimes referred to as *frequencies*. Sometimes the former is called the *wave-number frequency* or the *frequency in wave numbers* (the most commonly used unit). Context generally suffices to avoid confusion between the two.

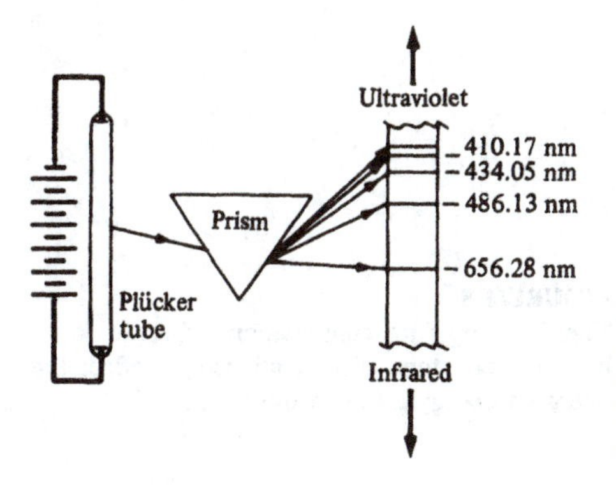

FIGURE 1-4
Schematic representation of the spectrum of atomic hydrogen as observed from a gas discharge tube. The Plücker tube contains molecular hydrogen gas, H_2, at low pressure. The lines shown belong to the Balmer series.

(equal to $4/A$) is called the *Rydberg constant* (empirical value of 109,677.8 cm^{-1}). When $n_2 = m$ and $n_1 = 2$, Eq. (1-12) reduces to the original Balmer formula. If n_1 is assigned integral values from unity upward (with n_2 also an integer but larger than n_1), one can represent all the known regions of the atomic hydrogen spectrum with high accuracy.

The Balmer-Rydberg-Ritz formula can be put into the alternative form

$$\tilde{\nu} = T_1 - T_2 \tag{1-13}$$

where T_i ($i = 1$ or 2) is called a *term* and is defined by

$$T_i = \frac{R}{n_i^2} \tag{1-14}$$

This shows that the entire spectrum of atomic hydrogen can be represented as a difference between pairs of terms. Rydberg found that the spectra of many other atoms, e.g., the alkali metals, could be approximated fairly well by differences between modified terms of the form

$$T_i = \frac{R}{(n_i + \alpha)^2} \tag{1-15}$$

where α is an empirically determined constant characteristic of a given atom. The frequencies of many spectral lines can be expressed in terms of the Rydberg-Ritz combination principle, which can be stated as follows: If $\tilde{\nu}_a = T_1 - T_4$ and $\tilde{\nu}_d = T_2 - T_3$ are two observed frequencies in the spectrum of a given atom, then $T_2 - T_4$ and $T_1 - T_3$ also correspond to two observable lines with frequencies $\tilde{\nu}_c$ and $\tilde{\nu}_b$, respectively. Thus one can write

$$\begin{aligned} \tilde{\nu}_a - \tilde{\nu}_b = \tilde{\nu}_c - \tilde{\nu}_d = T_3 - T_4 = x \\ \tilde{\nu}_a - \tilde{\nu}_c = \tilde{\nu}_b - \tilde{\nu}_d = T_1 - T_2 = y \end{aligned} \tag{1-16}$$

In these formulas $\tilde{\nu}_a > \tilde{\nu}_b > \tilde{\nu}_c > \tilde{\nu}_d$. These relationships are depicted graphically in Fig. 1-5.

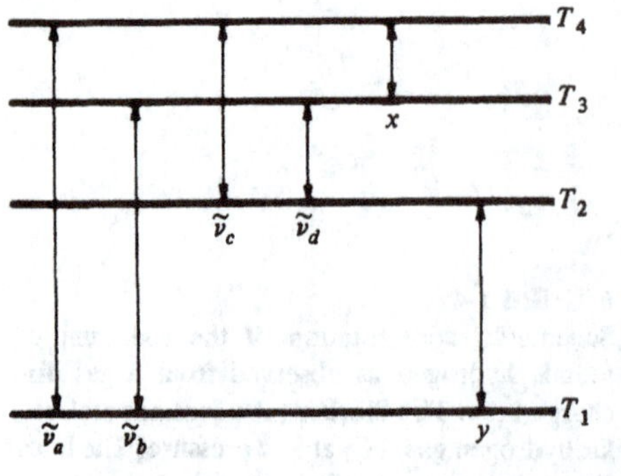

FIGURE 1-5
The Rydberg-Ritz combination principle for a hypothetical atom. The relationships reflect the quantum energy levels of the atom.

It is quite apparent that although the spectral patterns appear to be rather simple, they nevertheless do not feature integral multiples of a fundamental vibrational frequency as predicted by classical mechanics.

1-4 THE BOHR-RUTHERFORD MODEL OF THE ONE-ELECTRON ATOM

About 1910, Rutherford, Geiger, and Marsden of the University of Manchester (England) bombarded a piece of gold foil with α particles and analyzed the angular distribution of the deflected particles. Rutherford concluded that the atom was mostly empty space—a small, dense, positively charged nucleus containing most of the atom's mass, surrounded by negative electrons. The major difficulty with the model was that it appeared to be impossible for such an organization of oppositely charged particles to be stable. If, on the one hand, the electrons were motionless (relative to the nucleus), Coulomb's law predicts that the entire assembly should collapse as the unlike charges attracted each other. On the other hand, if the electrons were in orbital motion about the nucleus (as the planets orbit the sun), stability could be achieved—except that Maxwell's electrodynamic equations predicted that accelerating charges would lose energy by radiation. Hence, the orbiting electrons would quickly spiral into the nucleus, analogous to the way earth satellites respond to atmospheric drag and ultimately spiral into the earth (that's why *Skylab I* crashed into Australia).

In 1913 the Danish physicist Niels Bohr, who had recently returned to Denmark after working with Rutherford, proposed a bold, novel solution to the stability dilemma and, furthermore, provided a model for the Balmer-Rydberg-Ritz formula. Since Bohr's original approach is somewhat awkward to follow (and is incorrect by today's theories), it is presented here in what might be called a "cleaned-up" version.[5]

First, assume that Maxwell's electrodynamic laws do not apply at the subatomic level (after all, these laws were designed to describe ultraatomic phenomena) so that there is no concern about the electrodynamic stability of electrons orbiting the nucleus. In short, electrons are assumed to exist indefinitely in their nucleus-circling paths without losing energy by radiation. Next, consider only the simplest case, the hydrogen atom (or any ion with only a single electron), so that one has to deal only with the rotational motion of a single proton (or other nucleus) and a single electron moving about their mutual center of mass (see Fig. 1-6). The total energy of the hydrogen atom is

---

[5] Arguments against the use of the cleaned-up version have been given by B. L. Haendler, *J. Chem. Ed.*, **59**:372 (1982). Summaries of the various approaches Bohr actually used are given by M. Jammer (see the suggested readings at the end of this chapter).

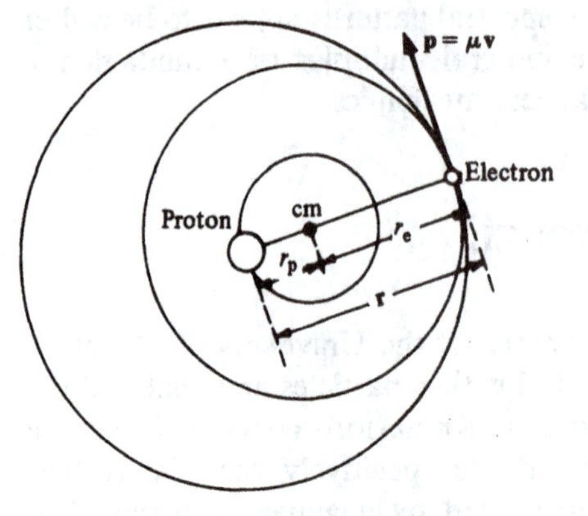

FIGURE 1-6
The Bohr model of the hydrogen atom. The angular momentum ($\mathbf{L} = \mathbf{r} \times \mathbf{p}$) is a vector quantity perpendicular to the plane defined by the vectors **r** and **p**. The center of mass is actually very close to the center of the proton since the proton is so much more massive than the electron (by a factor of over 1800). In a similar fashion, the earth and moon revolve about a point located approximately 4800 km from the earth's center.

given by the classical expression

$$E = T + V \tag{1-17}$$

where E = total energy
T = kinetic energy
V = potential energy

The rotational kinetic energy is

$$T = \frac{\mu v^2}{2} \tag{1-18}$$

where μ is the reduced mass of the system given by

$$\mu = \frac{m_e}{m_e + m_n} \tag{1-19}$$

(m_e = mass of the electron; m_n = mass of the nucleus) and v is the electron orbital velocity.[6] The coulombic potential energy is

$$V = -\frac{Ze^2K}{r} \tag{1-20}$$

where Z is the atomic number ($Z = 1$ for H, but the model is also valid for one-electron ions such as He^+, Li^{2+}, etc.), e is the charge on an electron, and r is the proton-electron distance. The quantity K represents the reciprocal of $4\pi\epsilon_0$ where ϵ_0 is the permittivity of free space; K's numerical value is unity in

[6] Strictly speaking, v is the velocity that a single particle of mass μ must have in order to have the same moment of inertia as the electron-nucleus system.

electrostatic units but 8.98755×10^9 J · m · C^{-2} in SI units. If we combine Eqs. (1-18) and (1-20), the total energy becomes

$$E = \frac{\mu v^2}{2} - \frac{Ze^2K}{r} \tag{1-21}$$

The unobservable quantities v and r can be eliminated by use of two additional relationships: one a well-known classical law and the other a novel quantum postulate. The classical relationship is that the electrostatic force (Ze^2K/r^2) between the nucleus and the electron produces a centripetal (center-seeking) acceleration, resulting in a circular path. The corresponding centripetal force $(\mu v^2/r)$ then satisfies

$$\frac{\mu v^2}{r} = \frac{Ze^2K}{r^2} \tag{1-22}$$

The quantum postulate (which Bohr did not use explicitly in his original paper) is that the orbital angular momentum of the system is *quantized*, i.e.,

$$\mu vr = \frac{nh}{2\pi} \qquad \text{with } n = 1, 2, 3, \ldots \tag{1-23}$$

where $h/2\pi$ is usually written in the compact form $\hbar$. If Eqs. (1-22) and (1-23) are solved simultaneously for r and v, one obtains

$$v = \frac{Ze^2K}{n\hbar} \qquad \text{and} \qquad r = \frac{n^2 a_H}{Z} \tag{1-24}$$

where

$$a_H = \frac{\hbar^2}{\mu e^2 K} \tag{1-25}$$

Substituting the relationships in Eq. (1-24) into Eq. (1-21) and simplifying, one obtains

$$E_n = -\frac{Z^2e^2K}{2n^2a_H} \tag{1-26}$$

where $n = 1, 2, 3, \ldots$. The subscript on E arises because each different value of n (called a *quantum number*) leads to a different energy value; i.e., the total energy of the hydrogen is predicted to be quantized. The quantity a_H is called the *first Bohr radius*; it has a numerical value of about 5.29×10^{-11} m (0.529 Å or 52.9 pm) for the H atom; for other ions one must take into account the variation due to a changed reduced mass.

The lowest possible energy of the hydrogen atom (the ground-state energy) occurs when $n = 1$. The negative of this energy, $-E_1$, should represent the ionization energy of the hydrogen atom. The calculated value of $-E_1$ is 13.605 eV (2.1798×10^{-18} J), which agrees with the experimental value.

To obtain the Balmer-Rydberg-Ritz formula, Bohr assumed that absorption or emission of radiation involves transition from one quantum state of the atom to another. For transitions between two states whose quantum numbers are n_2 and n_1 (where $n_2 > n_1$), the energy difference is assumed to satisfy a condition similar to that satisfied by the blackbody emitters of Planck, namely,

$$\Delta E = E(n_1) - E(n_2) = h\nu = hc\tilde{\nu} \tag{1-27}$$

where ν is the frequency of the emitted or absorbed radiation. This relationship implies that emission involves the production of photons of energy $h\nu$ and that absorption of a photon can occur only if the photon has an energy corresponding to a ΔE given by Eq. (1-27). Using the above along with Eq. (1-26) produces the Balmer-Rydberg-Ritz formula

$$\tilde{\nu} = \frac{\Delta E}{hc} = \frac{Ze^2K}{2a_{\mathrm{H}}\hbar c}\left(\frac{1}{n_1^2} - \frac{1}{n_2^2}\right) \tag{1-28}$$

in which the Rydberg constant is identified as

$$R = \frac{e^2K}{2a_{\mathrm{H}}\hbar c} \tag{1-29}$$

Figure 1-7 shows an energy-level diagram for the Bohr model of the hydrogen atom along with the origins of some of the spectral series.

Although the Bohr model is in very close agreement with many experimentally observed features of one-electron systems, e.g., H, He^+, Li^{2+}, Be^{3+} (and presumably U^{91+}), it does not account for the fine structure of the spectral lines. For example, what appears to be a single spectral line under low resolution turns out to be several closely spaced lines under higher resolution. Worse still, the Bohr theory is totally incapable of accounting for the spectral details of atoms with more than one electron. However, the Bohr model does suggest that the energies of all atoms are quantized and that the spectral lines

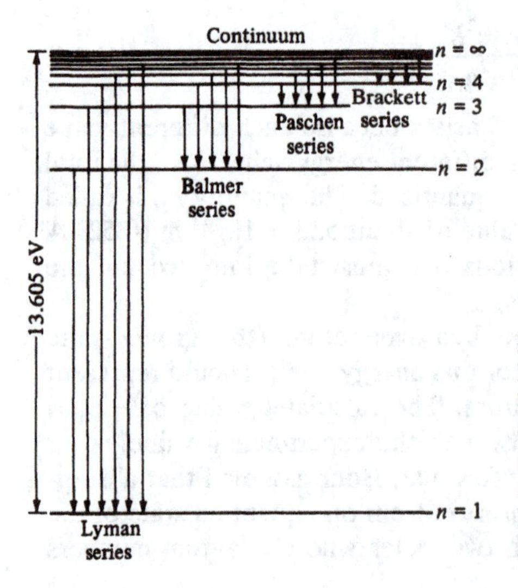

FIGURE 1-7
Energy-level diagram for the hydrogen atom based on the Bohr model. The Lyman series occurs in the ultraviolet, the Balmer in the visible and near ultraviolet, the Paschen in the infrared, and the Brackett also in the infrared. Another infrared series, the Pfund series, begins with $n_l = 5$ but is not shown.

arise from transitions governed by Eq. (1-27). The fundamental deficiency in the model is that, except for the special case of hydrogen, it cannot predict numerical values for the individual energies of the state involved in the transitions.

EXERCISES

1-11. The moment of inertia for a system of coupled masses is given by $I = \Sigma_i m_i r_i^2$, where r_i is the distance of the mass m_i from the center of mass of the system. Show for the hydrogen atom that $I = \mu r^2 = m_e r_e^2 + m_p r_p^2$, where r_e and r_p are as shown in Fig. 1-6 and μ is the reduced mass. *Hint*: $m_e r_e = m_p r_p$.

1-12. The Balmer series of the emission spectrum of the hydrogen atom includes all transitions ending up at $n = 2$. Calculate the energy (in joules and electronvolts) of the Balmer transition of wavelength 486.13 nm, and determine the n quantum number of the emitting state.

1-13. Calculate the Bohr-model value for the second ionization energy of helium, i.e., the minimum energy required for the process

$$He^+ \rightarrow He^{2+} + e^-$$

The experimental value is 54.403 eV.

1-14. The total electronic energy of an atom equals the sum of its successive ionization energies. If the first and second ionization energies of lithium are 5.39 and 75.619 eV, respectively, use the Bohr model to calculate the total electronic energy of lithium. The experimental value is −203.428 eV.

1-15. Calculate the wavelength (in nm) and frequency (in s^{-1} and cm^{-1}) of the hydrogen atom emission line arising from $n = 10$ and $n = 3$ states. This is a line in the Paschen series. Repeat for He^+.

1-16. The Brackett series of hydrogen contains all transitions having a lower state of $n = 4$. What is the longest wavelength (in nm) of emission lines in this series? (This is called the *series limit*.)

1-17. How would the Balmer series spectrum for deuterium differ from that of ordinary (light) hydrogen? The deuterium isotope was first detected through observation of this difference.

1-18. The wavelengths of the very intense yellow lines appearing in the emission spectrum of sodium vapor are 589.0 and 589.6 nm. What is the energy difference (in electronvolts and joules) between the two quantum states implied by these two lines, and what is the energy of each state relative to the ground state of sodium?

1-5 WAVE-PARTICLE DUALITY

Before 1920 the corpuscular nature of electrons appeared to be well established. The most convincing arguments for this view were based on the following empirical observations:

1. Deflection of cathode rays by electric and magnetic fields and the subsequent measurement of an e/m ratio for electrons

2. Determination of the charge (and, hence, the mass) of an electron by Millikan's oil-drop experiment
3. Demonstration that electron velocities are not uniform in a given medium
4. Observation of electron tracks in cloud chambers.

Furthermore, even the Bohr quantum model of the hydrogen atom *appeared* to support the concept of a corpuscular electron orbiting the nucleus.

In 1924, the French physicist Louis de Broglie wondered if the particlelike nature of light—as suggested by Einstein's treatment of the photoelectric effect—might not imply a reciprocal behavior on the part of corpuscular matter, i.e., did electrons (and other particles) exhibit a *wave* nature? De Broglie noted that in a broad sense, all natural phenomena involve just two entities: *matter* (described by mechanics) and *radiation* (described by electromagnetic wave theory). Since it was known that symmetry often played an important role in natural phenomena, why not expect matter and radiation to exhibit symmetry with respect to their behaviors? De Broglie began by noting that there are certain rather striking analogies between classical mechanics and geometric optics. For example, the trajectory of a particle is analogous to the path of a light ray, and the potential of a particle (a function of position) is analogous to the refractive index (also a function of position). Furthermore, there are two very important principles of mechanics and optics which are remarkably analogous. In mechanics there is the *principle of de Maupertuis* (principle of least action) which states that the integral

$$\int \sqrt{E-V}\, dS \tag{1-30}$$

over the trajectory of a particle is a minimum. Similarly, in optics there is the *principle of Fermat* (principle of least time) which states that the integral

$$\int \left(\frac{1}{n}\right)^{1/2} dS \tag{1-31}$$

(where n is the refractive index) over the path of a ray is a minimum. De Broglie then showed that waves which follow the same trajectory as the particle with which they are associated must satisfy the relationship

$$p\lambda = h \tag{1-32}$$

which is just the Einstein relationship Eq. (1-10) but now with an expanded significance. This relationship establishes a hitherto unsuspected bridge between optics and mechanics and, more importantly, accounts for certain quantum phenomena. As noted earlier, Eq. (1-32) is compatible with the corpuscular nature of light (photons); it also accounts for the quantization of angular momentum appearing in the Bohr model of the one-electron atom and, thereby, modifies the concept of a purely corpuscular electron orbiting the nucleus.

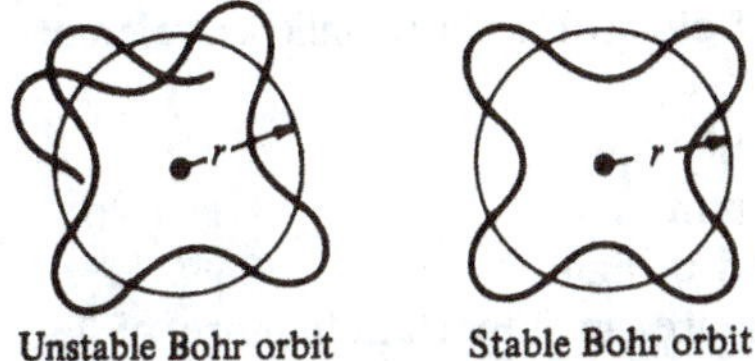
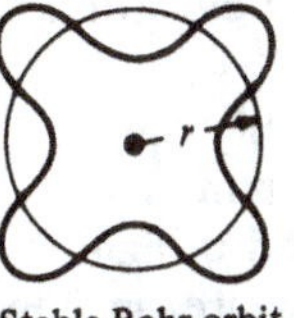

Unstable Bohr orbit Stable Bohr orbit

FIGURE 1-8
De Broglie representation of an electron in a Bohr orbit of the hydrogen atom. The wave on the left is out of phase with itself after 2π radians and will decay by destructive interference. The wave on the right contains an integral number of wavelengths and represents a stable standing wave.

As pointed out by de Broglie, the electron in a stable Bohr orbit may be regarded as a standing wave (otherwise it would destroy itself by destructive interference) and thus must consist of an integral number of wavelengths (see Fig. 1-8). This condition may be written

$$n\lambda = 2\pi r \qquad \text{with } n = 1, 2, 3, \ldots \tag{1-33}$$

Substituting for λ from Eq. (1-32) and rearranging leads to

$$pr = \frac{nh}{2\pi} = n\hbar \tag{1-34}$$

which is just the Bohr quantum condition Eq. (1-23).

Experimental verification of de Broglie's *matter waves* (as they are often called) was first provided in 1927 when two Bell Laboratories scientists, C. Davisson and L. H. Germer, bombarded the surface of a nickel crystal with a beam of electrons and found that these were diffracted like x-rays by a crystal lattice. Furthermore, using the Bragg equation and the known crystal spacings of nickel to calculate the apparent wavelength of the electrons led to a value in excellent agreement with that required by the duality relationship Eq. (1-32).

EXERCISES

1-19. Calculate the de Broglie wavelength (in nm) of an electron having an energy of

(a) 1 eV (b) 100 eV (c) 10,000 eV

Note: Kinetic energy $= mv^2/2 = p^2/2m$.

1-20. An electron is traveling at one-fourth the speed of light. Neglecting the relativistic change in mass, what is its de Broglie wavelength?

How does the result change when the relativistic change in mass is taken into account? *Note*: $m = m_0/\sqrt{1 - v^2/c^2}$, where m_0 is the rest mass.

1-21. A hydrogen atom has a kinetic energy corresponding to a temperature of 300 K. Would a beam of such atoms be diffracted by a crystal lattice with spacings on the order of 0.1 to 0.2 nm? Explain.

1-22. What accelerating voltage is needed to give an electron a de Broglie wavelength of 0.1 nm?

1-23. Compare the de Broglie wavelengths of the following:
(*a*) A 60-kg man walking at a rate of $100\ \text{m}\cdot\text{min}^{-1}$.
(*b*) An electron moving at 1 percent the speed of light.
In which case, if either, is the dual nature an important feature of the motion?

1-6 THE UNCERTAINTY PRINCIPLE

Although a discussion of the uncertainty principle is chronologically out of order at this point, a preliminary view will be of value in the study of modern quantum theory as presented in Chap. 2.

Taking wave-particle duality into account, let us consider representing a particle by a superposition of waves. Consider first the case of a single particle, free to move anywhere in space. If the particle is not localized at all, it may be represented by *one* wave with wavelength λ (Fig. 1-9). In this case the *momentum* of the particle ($p = h/\lambda$) is known precisely, but the *position* (x) of the particle is completely unknown.

Now consider the opposite extreme; if one uses a superposition of all possible wavelengths from $\lambda = 0$ to $\lambda =$ infinity, the waves can be made to reinforce at some definite value of x and be zero everywhere else (Fig. 1-9). In this case the position x is known precisely, but the momentum p is completely unknown.

In the general case, a finite number of waves of varying λ are superposed to form a pulse—a train of waves of finite length Δx. If one wants to have zero

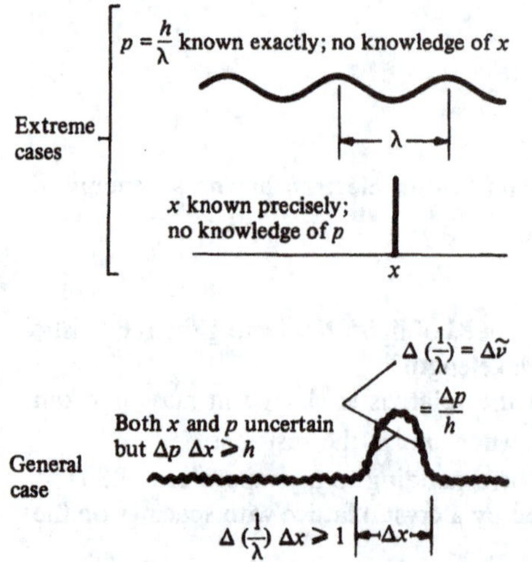

FIGURE 1-9
The uncertainty principle as a consequence of wave-particle duality.

amplitude outside the region of the pulse then it is necessary that

$$\Delta\left(\frac{1}{\lambda}\right)\Delta x \geq 1 \tag{1-35}$$

where Δx is the width of the pulse and $\Delta(1/\lambda) = \Delta\tilde{\nu}$ is the spread of $1/\lambda$ (or $\tilde{\nu}$) in the group.[7] Using the de Broglie relationship Eq. (1-32), one can write

$$\Delta\left(\frac{1}{\lambda}\right) = \frac{\Delta p}{h} \tag{1-36}$$

Substitution of this relationship into Eq. (1-35) and rearrangement produces

$$\Delta p \Delta x \geq h \tag{1-37}$$

This relationship, first proposed by the German physicist Werner Heisenberg in 1927, is called the *uncertainty principle*.[8] Although regarded by many as the hallmark of modern quantum theory, it is probably the strangest and most controversial principle of that theory.[9]

There are alternative ways of deducing the uncertainty relations. Heisenberg himself showed that they arise as an unavoidable consequence of using radiation to observe the trajectory of a moving particle; the very act of measurement disturbs a "state" or "destroys" it so that a precise, simultaneous measurement of p and x is impossible. Thus, any attempt to reduce Δx (say, by using radiation of smaller wavelength, i.e., greater resolution) will impart a more energetic "kick" to a particle, thereby making Δp larger. Conversely, using a larger wavelength to reduce Δp will ruin the resolving power and increase Δx.

A more philosophical way of viewing the uncertainty principle is as follows: The nature of the world within the submicroscopic atom is not directly observable—we deduce it on the basis of highly indirect evidence and then proceed to describe it in terms which were invented on the basis of experience with the *macroscopic* world. Consequently, it is hardly surprising that classical quantities such as position and momentum—so intuitively natural for the description of large particles which we can see—have decreased suitability for describing the physics of the subatomic world. Thus, the uncertainty principle serves as a warning device—a red flag, so to speak—telling us that if we insist on carrying macroscopic terms into the world of the atom, then we must agree on some restrictions as to how those concepts must be used—lest we are led to

---

[7] $\Delta(1/\lambda) = \Delta\tilde{\nu}$ is the uncertainty in the number of waves in the pulse whose width is Δx.

[8] The original name used by Heisenberg was the "unsharpness" principle (*Unshärfeprinzip*). Later the name was changed to the less appropriate "uncertainty" principle (*Unsicherheitsrelation*).

[9] For example, Einstein refused to accept the uncertainty principle as a legitimate, indispensable part of quantum theory. From time to time Einstein would present a new "proof" that the uncertainty principle was false, only to have his "proof" refuted by Bohr.

nonsense statements. As the British geneticist J. B. S. Haldane so aptly stated, "the universe is not only queerer than we suppose, but queerer than we can suppose. . . ."[10]

EXERCISES

1-24. The momentum of the electron in a given state of the H atom is given by $p^2 = 2mE$, where E is the energy needed to ionize the atom in that state ($E = -E_1$ for the ground state). Show that the uncertainty in momentum of the electron is $\Delta p = 2\sqrt{2mE}$, and estimate the minimum "size" of the atom in its ground state. Compare this with the value of the first Bohr orbit a_H.

1-25. What would be the effect on the way the universe behaves if the following conditions were imposed?

(*a*) Planck's constant was made very, very small, i.e., approaching zero.

(*b*) Planck's constant was made very, very large, i.e., approaching infinity.

SUGGESTED READINGS

Condon, E. U.: "60 Years of Quantum Physics," *Phys. Today*, October 1962, p. 37. An unusually interesting historical account by one of the pioneering contributors to quantum theory.

Jammer, M.: *The Conceptual Development of Quantum Mechanics*, McGraw-Hill, New York, 1966. Contains references to original articles and traces the development of quantum ideas in great detail.

---

[10] J. B. S. Haldane, *Possible Worlds*, Chatto & Windus, London, 1930.

CHAPTER 2

THE SCHRÖDINGER WAVE EQUATION

Although the Bohr theory is unable to account for the quantum states of atoms with more than one electron, Bohr and his coworkers nevertheless managed to obtain many valuable insights into some of the salient features that a more general quantum theory must possess. In particular, Bohr introduced and made extensive use of the *correspondence principle*, which recognized that the complete laws of quantum mechanics—whatever they might be—must reduce to classical laws under the limiting conditions for which the latter were known to be valid. By clever use of this principle, it is possible to predict the relative intensities of many spectral lines and to begin a classification of hitherto apparently jumbled spectra. Nevertheless, the approach is inherently unsatisfactory, since it admits of no definite methodology and depends strongly on intuition. The present chapter introduces the basic relationships and rules which permit comprehensive, methodical explanations of quantum phenomena.

The reader who is unfamiliar with operator algebra, simple vectors, complex numbers, and classical wave motion—or who wishes to review one or more of these areas— should consult the appropriate appendix or appendixes.

2-1 WAVE MECHANICS

The existence of matter waves suggests the existence of a wave equation describing them. Such a wave equation was first proposed by the Austrian physicist Erwin Schrödinger in 1926. One of the most striking features of this equation is that it leads to quantum numbers naturally; i.e., there is no need to assume them a priori as Planck and Bohr found it necessary to do.

Whereas de Broglie (Sec. 1-5) was led to wave-particle duality on the basis of a restricted analogy between optics and mechanics, Schrödinger greatly extended the scope of the analogy and developed an equation which provided a general description of quantum phenomena. Schrödinger reasoned that the most characteristic feature of waves was their interference behavior, and thus it ought to be possible to begin with the classical wave equation itself and with it construct a bridge beginning at wave optics and leading to a general wave mechanics capable of accounting for quantum phenomena. An overview of the conceptual design is shown in the block diagram of Fig. 2-1.

Restricting our treatment to one dimension for the present, the classical wave equation is

$$\frac{\partial^2 \Psi}{\partial x^2} = \frac{1}{c^2} \frac{\partial^2 \Psi}{\partial t^2} \tag{2-1}$$

where $\Psi = \Psi(x, t)$, the amplitude in classical theory, remains undefined in the

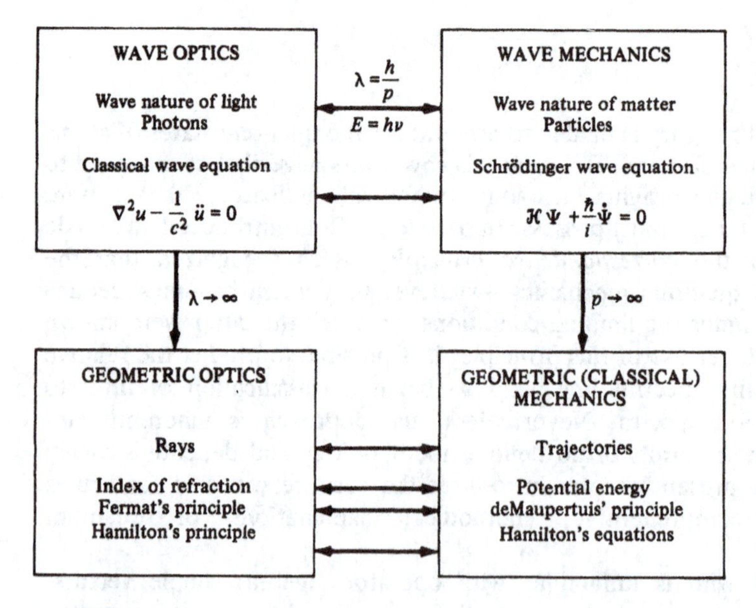

FIGURE 2-1
The Schrödinger equation as an extension of de Broglie's analogy between optics and mechanics.

present context. A solution to this equation which is general enough to include the interference characteristics of waves is

$$\Psi(x, t) = Ce^{i\alpha} \tag{2-2}$$

where C is a constant, and α is the *phase*, given by

$$\alpha = 2\pi\left(\frac{x}{\lambda} - \nu t\right) \tag{2-3}$$

Let us now seek another differential equation having solutions of the form in Eq. (2-2) but with the additional requirement that this equation reflect the two most important quantum relationships hitherto known, namely, wave-particle duality and the quantization of radiation. Using de Broglie's relationship Eq. (1-32) for λ and the Planck-Bohr quantum relationship for ν, the phase becomes

$$\alpha = \frac{xp_x - E_t}{\hbar} \tag{2-4}$$

Differentiating Ψ once with respect to t, one obtains

$$\frac{\partial\Psi}{\partial t} = i\Psi\frac{\partial\alpha}{\partial t} = i\Psi\left(\frac{-E}{\hbar}\right) \tag{2-5}$$

This may be rearranged to

$$\frac{-\hbar}{i}\frac{\partial\Psi}{\partial t} = E\Psi \tag{2-6}$$

Next, we try to replace E with an expression containing the derivative of Ψ with respect to x. Differentiating Ψ once with respect to x leads to

$$\frac{\partial\Psi}{\partial x} = i\Psi\frac{\partial\alpha}{\partial x} = i\Psi\left(\frac{p_x}{\hbar}\right) \tag{2-7}$$

which may be rearranged to

$$\frac{\hbar}{i}\frac{\partial\Psi}{\partial x} = p_x\Psi \tag{2-8}$$

The latter equation may be interpreted as an operator equation in which the x component of the linear momentum is represented by the differential operator $(\hbar/i)\partial/\partial x$ (see App. 4 for a discussion of operator algebra). For a system of mass m with a conservative potential $V(x)$ the total energy is given by

$$E = \frac{p_x^2}{2m} + V(x) \tag{2-9}$$

An operator representation of E, based on the operator representation of p_x, is

$$\hat{E} = -\frac{\hbar^2}{2m}\frac{\partial^2}{\partial x^2} + \hat{V}(x) \tag{2-10}$$

where the explicit form of $\hat{V}(x)$ depends on the particular system. Now we can rewrite Eq. (2-6) in the operator form

$$-\left[\frac{\hbar^2}{2m}\frac{\partial^2}{\partial x^2}+\hat{V}(x)\right]\Psi=-\frac{\hbar}{i}\frac{\partial}{\partial t}\Psi \tag{2-11}$$

This is the *Schrödinger wave equation* in its one-dimensional form. Its ultimate justification resides not in the admittedly vague and questionable method used here to obtain it but in how well its solutions are able to describe quantum phenomena in nature.

The Schrödinger wave equation is readily generalized to three dimensions by replacing $\partial^2/\partial x^2$ with the laplacian operator

$$\nabla^2=\frac{\partial^2}{\partial x^2}+\frac{\partial^2}{\partial y^2}+\frac{\partial^2}{\partial x^2} \tag{2-12}$$

The circumflex ˆ is customarily omitted for certain well-known operators such as ∇^2 and $V(x)$ or obvious operators involving derivatives. The obvious notational changes

$$\begin{aligned} V(x) &\rightarrow V(x, y, z) \\ \Psi(x, t) &\rightarrow \Psi(x, y, z, t) \end{aligned} \tag{2-13}$$

are also employed.

Even before Schrödinger's discovery of the wave equation (2-11), the German physicist Werner Heisenberg developed an alternative formulation of quantum theory based on the Bohr correspondence principle; this formulation is known as *matrix mechanics*.[1] Heisenberg's matrix mechanics at first escaped attention because its matrix form was unfamiliar to physicists of the time, but it was later shown to be equivalent to Schrödinger's wave mechanics.

2-2 PROBABILISTIC INTERPRETATION OF THE FUNCTION Ψ

In dealing with systems composed of a large number of particles, classical mechanics has to resort to various statistical methods, e.g., probability theory, in order to obtain useful predictive relationships. This necessarily arises from the *practical* difficulty of knowing the initial coordinates and momenta of all parts of the system. Quantum systems also require a probabilistic description—but for a fundamentally different reason: the uncertainty principle precludes predictive laws involving precisely known coordinates and momenta *at any time*.

[1] An interesting, nonmathematical account of matrix mechanics is given by Banesh Hoffmann in *The Strange Story of the Quantum*, Dover, New York, 1959.

Classical statistics makes use of a probability distribution function $P(x)$ which is positive definite and defined such that $P(x)\,dx$ is the probability that the variable x, which can take on any real value, lies between x and $x + dx$. The *mean value* (or *expectation value*) of x is given by

$$\langle x \rangle = \int_{-\infty}^{\infty} xP(x)\,dx \tag{2-14}$$

provided $P(x)$ is first *normalized* as follows:

$$\int_{-\infty}^{\infty} P(x)\,dx = 1 \tag{2-15}$$

The mean value of x is the average of the values one would expect to find after a large number of repeated measurements.[2] Normalization of $P(x)$ to unity means simply that we are using zero to unity as the numerical range of the probability; a probability of zero means no chance at all, and a probability of unity means absolute certainty.[3]

If we assume for the moment a simple one-dimensional system consisting of a single particle described by $\Psi(x, t)$, the quantum description replaces $P(x)$ not with $\Psi(x, t)$ as one might first guess, but with $\Psi^*\Psi$ (we will suppress the variables x and t for notational convenience) where Ψ^* is the *complex conjugate* of Ψ. This difference arises because, in general, Ψ is a complex-valued function and the probability density must be real. Note that $\Psi^*\Psi = \Psi\Psi^* = |\Psi|^2$. In three dimensions, $|\Psi(x, y, z, t)|^2$ becomes the *probability distribution function* (or *probability density*), and the probability of a particle's having an x coordinate between x and $x + dx$, a y coordinate between y and $y + dy$, and a z coordinate between z and $z + dz$ *at the time* t is given by

$$|\Psi|^2\,dx\,dy\,dz \tag{2-16}$$

The mean value at a given time t of some property f, which depends only on the *coordinates*, is given by

$$\langle f \rangle = \iiint_{-\infty}^{\infty} f(x, y, z)|\Psi|^2\,dx\,dy\,dz \tag{2-17}$$

where it is assumed by analogy with Eq. (2-15) that

$$\iiint_{-\infty}^{\infty} |\Psi|^2\,dx\,dy\,dz = 1 \tag{2-18}$$

In the case that f also depends on the *momenta* (which, in turn, involve

[2] If N successive measurements give $x_1, x_2, \ldots, x_N$, the mean value of x is $\Sigma_i\, x_i/N$.

[3] Sometimes $P(x)$ is normalized to 100, and one speaks of probability in *percent*; for example, a 30 percent probability means that 3 favorable occurrences out of 10 tries are expected on the average.

differential operators), the mean value (or expectation value) of f is given by

$$\langle \hat{f} \rangle = \iiint_{-\infty}^{\infty} \Psi^* \hat{f} \Psi \, dx\, dy\, dz \tag{2-19}$$

where $\hat{f}$ is the *operator representation* of the property f. All of this will be discussed more thoroughly in the sections following. Note especially that in Eq. (2-19) the operator is sandwiched between Ψ and its complex conjugate; there is no analogy to this in classical statistics.

We now consider an important posulate: For every isolated system there exists a mathematical function of the coordinates and the time. $\Psi(x, y, z, t)$, such that this function contains encoded within itself all possible meaningful information about the state of the system, including any intrinsic uncertainties. We call $\Psi(x, y, z, t)$ the *wavefunction* or *state function* of the system. The system wavefunction must, of course, be obtained as a solution to the Schrödinger wave equation (2-11).

It should be noted that, in general, the matter waves of de Broglie are distinct from the "waves" suggested by solutions to Schrödinger's equation. The matter waves of de Broglie are three-dimensional, even for a system of N particles, whereas the Schrödinger waves are $3N$-dimensional. The matter waves have a rather simple intuitive physical interpretation; they provide a *guide* for the system. By contrast, the highly abstract Schrödinger waves *are the system*. In some ways the Schrödinger waves are like $3N$-dimensional complex vibrational modes of a continuum (system), and, thereby, they do have a classical analogy. The vibrational motions of a solid consisting of N atoms can be viewed in three-dimensional space, but lagrangian mechanics (an extension of Newton's laws of motion) treats the entire motion in a coordinates space of $3N$ dimensions.

EXERCISES

2-1. The probability that a variable x has a value betweeen $-x$ and x is given by $P(x) = Ne^{-ax^2}$ where a is a positive constant. Show that N is equal to $\sqrt{a/\pi}$ when $P(x)$ is normalized (to unity). Sketch the appearance of $P(x)$ as a function of x. Where do the points of inflection occur?

2-2. Use the results of Exercise 2-1 to find the mean value of x. Does the result make intuitive sense? Repeat for the mean value of x^2. Why doesn't $\langle x^2 \rangle$ equal $\langle x \rangle^2$?

2-3. How is the expression for the average energy given in Exercise 1-1 related to Eq. (2-14)?

2-3 THE TIME-INDEPENDENT WAVE EQUATION

The Schrödinger equation is often written in the abbreviated form

$$\hat{H}\Psi = -\frac{\hbar}{i}\frac{\partial \Psi}{\partial t} \tag{2-20}$$

where $\hat{H}$ is a symbol for the operator

$$\hat{H} = -\frac{\hbar^2}{2m}\nabla^2 + \hat{V} \tag{2-21}$$

$\hat{H}$ is called the *hamiltonian operator* or, simply, the *hamiltonian*. Its name comes from Hamilton's equations of classical mechanics which employ an analogous function used to generalize Newton's laws of motion.[4] For conservative systems, the classical hamiltonian H represents the total energy of the system.

Let us investigate a partial solution of the Schrödinger equation by use of the technique of separation of variables. We seek a particular solution of Eq. (2-20) having the form

$$\Psi(x, y, z, t) = \psi(x, y, z)\phi(t)$$

or, to simplify the notation,

$$\Psi = \psi\phi \tag{2-22}$$

Substituting Eq. (2-22) into Eq. (2-20) produces

$$\hat{H}\psi\phi = -\frac{\hbar}{i}\frac{\partial(\psi\phi)}{\partial t} \tag{2-23}$$

Since $\hat{H}$ operates only on ψ (it contains x, y, and z but not t), and since $\partial/\partial t$ operates only on ϕ, we may rewrite Eq. (2-23) as

$$\phi\hat{H}\psi = \psi\left(-\frac{\hbar}{i}\frac{\partial\phi}{\partial t}\right) \tag{2-24}$$

Dividing both sides by $\psi\phi$ we obtain

$$\frac{\hat{H}\psi}{\psi} = -\frac{1}{\phi}\frac{\hbar}{i}\frac{\partial\phi}{\partial t} \tag{2-25}$$

The left-hand side of the expression is a function of x, y, and z, but it equals the right-hand side, which is a function of t alone! This is noncontradictory only if neither side is a variable, i.e., each equals a common *constant*. Letting W represent this constant (called the *separation constant*), we can replace Eq. (2-25) with the two separate equations

$$\frac{\hat{H}\psi}{\psi} = W \qquad \text{or} \qquad \hat{H}\psi = W\psi \tag{2-26}$$

$$-\frac{1}{\phi}\frac{\hbar}{i}\frac{\partial\phi}{\partial t} = W \qquad \text{or} \qquad -\frac{\hbar}{i}\frac{d\phi}{dt} = W\phi \tag{2-27}$$

Note that Eq. (2-27) is an ordinary differential equation, and thus partial differentiation notation is dropped. This equation is solved by rearrangement

[4] Hamilton's equations of motions are introduced in Sec. 2-11, Eq. (2-77).

followed by integration. One form of the solution is

$$\phi(t) = e^{-iWt/\hbar} \tag{2-28}$$

Since $|\phi|^2 = \phi^*\phi = e^{iWt/\hbar}e^{-iWt/H} = 1$, the total probability distribution function is

$$|\Psi|^2 = |\psi\phi|^2 = |\psi|^2|\phi|^2 = |\psi|^2 \tag{2-29}$$

which is time-independent. This means that the particular solution seen in Eq. (2-22) represents physical situations in which the probability density does not vary with time. Thus, we conclude that Eq. (2-26), the time-independent Schrödinger equation, represents *stationary states* of the system.

Equations (2-26) and (2-27) are of the general form

$$\hat{A}f(q) = af(q) \tag{2-30}$$

where $\hat{A}$ is an operator, a is a constant, and $f(q)$ is some function of the independent variable or variables q. Such equations are called *eigenvalue equations*; the constant a is called an *eigenvalue* of the operator $\hat{A}$, and $f(q)$ is called an *eigenfunction* of the operator $\hat{A}$. In Eq. (2-26) the separation constant W is an eigenvalue of the hamiltonian operator. Since in classical mechanics the hamiltonian H represents the total energy of a conservative system, we shall identify W with E, the total energy of the system in one of its stationary states. Consequently, we rewrite the time-independent Schrödinger equation in the compact form

$$\hat{H}\psi = E\psi \tag{2-31}$$

Equation (2-31) has a number of mathematical solutions; however, the only physically acceptable ones, corresponding to various values of the energy E, are those satisfying the following conditions:

1. ψ must be quadratically integrable; i.e., the integral $\int \ldots \int |\psi|^2 \, d\tau$ over all configuration space must equal a finite number for a finite number of particles.[5] Alternatively, one may say that ψ must vanish at the boundaries of the system.
2. ψ must be single-valued.
3. ψ must be continuous.

Wavefunctions satisfying the above conditions are said to be *well-behaved*. The well-behaved requirement has as its basis the physically reasonable expectation that $|\psi|^2$, since it represents a probability density, must lead to a finite total

---

[5] The quantity $d\tau$ represents a general volume element of integration. For a three-dimensional cartesian space, $d\tau = dx\,dy\,dz$, and a triple integration is required.

probability, must be continuous (as are classical probabilities), and must assign a single, unambiguous probability density to each point of the system.

Occasionally a fourth condition is required, namely

4. The gradient of ψ must be continuous. For the one-dimensional case, grad $\psi(x)$ is given by

$$\nabla\psi(x) = -\frac{2m}{\hbar^2}\int [E - V(x)]\psi(x)\,dx \tag{2-32}$$

   which has a higher degree of continuity than $\psi(x)$ unless $V(x)$ suddenly becomes infinite. As a general rule, potentials with this behavior are not encountered in nature, although we may occasionally employ models which have such infinite discontinuities.

The various solutions to the time-independent Schrödinger equation are designated by $\psi_n(x)$, where $n = 1, 2, 3, \ldots$ and represents different stationary states. These solutions are an example of a *complete set of functions*. A set of functions is called *complete* if an arbitrary function of the same variables found in $\{\psi_n\}$, and satisfying the same restrictive conditions (e.g., square integrability) as $\{\psi_n\}$, can be expanded as

$$f(x) = \sum_{i=1}^{\infty} c_i\psi_i(x) \tag{2-33}$$

where the coefficients $\{c_i\}$ are, in general, complex.

A general solution to the time-dependent Schrödinger equation (2-20) can be expressed as a linear combination of the stationary-state solutions [Eq. (2-22)]:

$$\Psi(x, y, z, t) = \sum_{n=1}^{\infty} a_n\psi_n(x, y, z)e^{-iE_nt/\hbar} \tag{2-34}$$

This does not, in general, represent a time-independent probability density. Note that Ψ enables one to represent the motion of a particle as a *wave packet*, i.e., a superposition of waves such that the particle is to some extent localized. However, as more and more waves are superposed (to make Δx smaller), the wavelength λ (and hence the momentum) becomes more uncertain. This is just the *uncertainty principle* discussed in Sec. 1-6.

EXERCISES

2-4. Solve the classical wave equation (2-1) by assuming the separation of variables $\Psi(x, t) = f(x)g(t)$ and letting $-\omega^2$ be the separation constant. Show that $g(t) = e^{-i\omega t}$ is a solution to the time-dependent equation.

2-5. Show that Eq. (2-28) also satisfies the time-dependent equation in Exercise 2-4.

2-6. Substitute $\phi = e^{-i\omega t}$ (see Exercise 2-4) into the classical wave equation (2-1), set $\omega = 2\pi\nu$ and $\lambda = h/p$, and show that the time-independent Schrödinger equation (2-26) results.

2-7. Show that $e^{-iWt/\hbar}$ is an eigenfunction of the operator $-(\hbar/i)\, d/dt$.

2-8. Show that e^{-ax} (a is a constant) is an eigenfunction of the operators d/dx and d^2/dx^2, and find the corresponding eigenvalues. What is the corresponding eigenvalue for the operator d^n/dx^n?

2-9. Show that $\sin nx$ and $\cos nx$ (n is an integer) are both eigenfunctions of the operator d^2/dx^2 but not of d/dx. What is the corresponding eigenvalue in the former case?

2.10. Supply the eigenvalues corresponding to the following operator-eigenfunction pairs:
(*a*) $-d^2/dx^2 + x^2$, xe^{-x^2}
(*b*) ∇^2, $\sin kx \sin ly \sin mz$
(*c*) d^2/dx^2, $Ae^{ikx} + Be^{-ikx}$
(*d*) d^2/dx^2, $A \cos \kappa x + B \sin \kappa x$
(*e*) $-(1/2r^2)\, d/dr\, (r^2)\, d/dr - 1/r$, e^{-r}

2-11. Consider the operator $d^2/dx^2 - kx^2$, where k is a constant. What value must the constant a in e^{ax^2} have in order for this function to be an eigenfunction of the above operator? What is the corresponding eigenvalue?

2-4 VECTOR INTERPRETATION OF WAVEFUNCTIONS

The Schrödinger wave equation (2-20) is a linear, homogeneous, second-order differential equation, the solutions of which have the following very important superposition property: If Ψ_1 and Ψ_2 are two solutions of the Schrödinger equation and a_1 and a_2 are arbitrary constants (generally complex), then $a_1\Psi_1 + a_2\Psi_2$ is also a solution of Schrödinger's equation. More generally, we can combine any number of particular solutions $\Psi_1, \Psi_2, \Psi_3, \ldots$ with constants $a_1, a_2, a_3, \ldots$ to obtain a new solution $\Sigma\, a_i\Psi_i$. The infinite sequence of particular solutions represented by Eq. (2-34) is one example of such a superposition.

The superposition principle implies that states can somehow be added to produce new states. This suggests that the state functions themselves are analogous to *vectors*, since vectors can be added to produce new vectors. However, the vectors one must use as an analogy must be defined in a complex space of infinite dimensions. Although it is impossible to concretely visualize vectors beyond a real three-dimensional space, much of the geometric terminology of real three-dimensional space can be carried over to complex spaces of any dimension.

Adopting a notation and terminology introduced by British physicist P. A. M. Dirac, the vectors which describe the quantum states $A, B, C, \ldots$ are called *ket vectors* (or *kets*) and are written $|A\rangle$, $|B\rangle$, $|C\rangle, \ldots$. These vectors are *representations* of the wavefunctions $\psi_A, \psi_B, \psi_C, \ldots$ describing various stationary states. Since these vectors are defined in a *complex* vector space, it is mathematically necessary to introduce a second set of vectors $\langle A|$,

$\langle B|, \langle C|, \ldots$ called *bra vectors* (or *bras*).[6] Together the bra and ket vectors form a *dual set.* The ket $|A\rangle$ and the bra $\langle A|$ are equally capable of representing the state A. The bra vector $\langle A|$ is the *complex-conjugate transpose* (or *adjoint*) of the ket vector $|A\rangle$ and vice versa.[7] Thus, one may write

$$\langle A|^{\dagger} = |A\rangle \qquad \text{and} \qquad |A\rangle^{\dagger} = \langle A|$$

The scalar product of a bra $\langle A|$ and a ket $|B\rangle$ is written $\langle A|B\rangle$; in general this is not equal to the scalar product of the bra $\langle B|$ and the ket $|A\rangle$, that is, $\langle A|B\rangle \neq \langle B|A\rangle$.[8]

Suppose we have a stationary state described by the wavefunction

$$\psi_A(\varphi) = (2\pi)^{-1/2} e^{-i\varphi}$$

where $0 \leq \varphi \leq 2\pi$ defines the configuration space. The ket vector $|A\rangle$ then *represents* $\psi_A(\varphi)$ as written above, and the bra vector $\langle A|$ represents the complex conjugate $\psi_A^*(\varphi) = (2\pi)^{-1/2} e^{i\varphi}$. The square of the norm (or modulus) of the vector $|A\rangle$ or $\langle A|$ is given by the scalar product

$$\langle A|A\rangle = \int_0^{2\pi} \psi_A^* \psi_A \, d\varphi = \frac{1}{2\pi} \int_0^{2\pi} e^{i\varphi} e^{-i\varphi} \, d\varphi = \frac{1}{2\pi} \int_0^{2\pi} d\varphi = 1$$

Note that normalization of ψ_A is equivalent to making the vector $|A\rangle$ of unit length.

Any vector $|A\rangle$, when superimposed upon itself, produces a vector which still defines the same quantum state A. This is in contrast to classical mechanics, in which such a superposition produces a new "state"; for example, superposing a vibrational mode onto itself produces a new vibrational state of increased amplitude. This property of bra and ket vectors implies that it is the *direction* of a vector which specifies a state—not its magnitude.

Bra and ket vectors which are everywhere zero imply $\psi = 0$ and, thus, represent *no state at all.* In classical mechanics, analogous functions usually represent systems at rest.

---

[6] The names "bra" and "ket" come from the word "bracket," in reference to the symbols $\langle$ and $\rangle$.

[7] The complex-conjugate transpose of a vector is a generalization of the complex conjugate of a complex number. However, the bras and kets are, in general, *complex functions* and cannot always be divided into a real and imaginary part as can ordinary complex numbers. Nevertheless, a restricted analogy with complex numbers is useful. When a bra vector is multiplied on the right by its dual ket, the result is a scalar. With every pair of real numbers a and b there are associated two complex numbers, $z = a + ib$ and $z^* = a - ib$, having the same modulus, the real number $\sqrt{a^2 + b^2}$; hence the *dual* space. The "transpose" part of the term "adjoint" comes about when vectors are represented as row or column matrices of their components; this changes (transposes) a row vector into a column vector and vice versa. The reader unfamiliar with matrix algebra, or in need of a review, should consult App. 5.

[8] This is analogous to the fact that if z_1 and z_2 are two different complex numbers, then $z_1^* z_2$ is not equal to $z_2^* z_1$, unless, of course, z_1 and z_2 are purely real or purely imaginary. Note that if the vectors $|A\rangle$ and $|B\rangle$ are real, then $\langle A|B\rangle$ (or $\langle B|A\rangle$) represents the ordinary dot product of two vectors.

EXERCISES

2-12. Show that, in general, $\langle A|B\rangle^{\dagger} = \langle B|A\rangle$ and, for the special case of real space, $\langle A|B\rangle = \langle B|A\rangle$.

2-13. Consider the column vectors

$$|r_1\rangle = \begin{bmatrix} 2 \\ 1+i \\ -3i \end{bmatrix} \qquad |r_2\rangle = \begin{bmatrix} i \\ 2-2i \\ 4 \end{bmatrix}$$

(*a*) Compute and compare the scalar products $\langle r_1|r_2\rangle$ and $\langle r_2|r_1\rangle$.
(*b*) Normalize $|r_1\rangle$ and $|r_2\rangle$.

2-14. Set the imaginary part of each component of $|r_1\rangle$ and $|r_2\rangle$ in Exercise 2-13 equal to zero and repeat (*a*) and (*b*) of the exercise.

2.5 ORTHONORMALITY OF WAVEFUNCTIONS

For convenience we let $|\psi_i\rangle$ and $|\psi_j\rangle$ represent the ket vectors for the states whose wavefunctions are ψ_i and ψ_j, respectively (the simpler notation $|i\rangle$ and $|j\rangle$ is also used). We then introduce the integral notation

$$\underset{R}{\int \cdots \int} \psi_i^* \hat{A} \psi_j \, d\tau = \langle \psi_i | \hat{A} | \psi_j \rangle \tag{2-35}$$

where $\hat{A}$ represents an arbitrary operator. Integrals of this form are the most commonly encountered in quantum chemistry.[9] Note that $\hat{A}|\psi_i\rangle = |\hat{A}\psi_i\rangle$ is also a ket vector, but it is customary to isolate the operator between the bra and ket vectors. When $\hat{A}$ is the unit operator ($\hat{1}$, or just 1), the integral is closed up and one writes simply $\langle \psi_i | \psi_j \rangle$. This particular integral notation is especially convenient for symbolic representation and algebraic manipulation of integral quantities. Since the notation suppresses the interval of integration R and the volume element $d\tau$, these must be made explicit whenever actual evaluation of the integral is to be performed. As the notation makes clear, the process of integration (of an integral of the type illustrated) is analogous to forming the scalar product of two vectors.

As shown later by Theorem 2-3 (Sec. 2-6), the state vectors of quantum mechanics are generally either automatically orthogonal or can be made orthogonal for convenience (without any loss of generality). If these vectors are also chosen to have unit moduli, we write

$$\langle \psi_i | \psi_j \rangle = \delta_{ij} \qquad i, j = 1, 2, \ldots \tag{2-36}$$

where δ_{ij} is the Kronecker delta.[10] The vectors $|\psi_i\rangle$ and $|\psi_j\rangle$ (or, equivalently,

---

[9] This integral is also called the *ij-matrix element* of the operator $\hat{A}$ and is also symbolized as A_{ij}.

[10] The Kronecker delta δ_{ij} will equal zero whenever i and j are not equal and will equal unity whenever i equals j.

$\langle \psi_i |$ and $\langle \psi_j |$) are then said to belong to an *orthonormal* set.[11] Alternatively, we say the wavefunctions $\psi_i, \psi_j, \ldots$ are orthonormal over the configuration space of the system.

Suppose we are given a wavefunction ψ which is not normalized. The scalar product of ψ with itself is

$$\langle \psi | \psi \rangle = Q^2 \tag{2-37}$$

where Q is the modulus or length of the vector $|\psi\rangle$. Consequently, the new vector $|Q^{-1}\psi\rangle$, has *unit length*. Since it is irrelevant whether Q, the length of $|\psi\rangle$, is assumed to be positive or negative, we adopt the customary choice of a positive value. Thus

$$\langle Q^{-1}\psi | Q^{-1}\psi \rangle = Q^{-2} \langle \psi | \psi \rangle = Q^{-2}Q^2 = 1 \tag{2-38}$$

To normalize a wavefunction ψ, simply multiply it by the reciprocal of its modulus.

As a simple example, let us normalize the function $\psi = \sin x$ in the interval $0 \leq x \leq \pi$.

$$\langle \psi | \psi \rangle = \int_0^\pi \sin^2 x \, dx = \frac{x}{2} - \frac{1}{4} \sin 2x \Big|_0^\pi = \frac{\pi}{2}$$

Therefore, $Q = (\pi/2)^{1/2}$, and the normalized function is

$$\psi = \left(\frac{2}{\pi}\right)^{1/2} \sin x \tag{2-39}$$

The reciprocal of the modulus of the unnormalized functions $[(2/\pi)^{1/2}$ in the case above] is called the *normalization constant*.

The concept of orthogonality of vectors is easily visualized by the example of vectors in a two- or three-dimensional cartesian space; orthogonal vectors are at right angles to each other, and thus neither one can be projected onto the axis of the other.[12] Although orthogonality is more difficult to illustrate in the general case, familiarity with these special cases is usually sufficient to make the concept acceptable. As a simple example, consider the vectors which represent the functions $\psi_1 = \sin x$ and $\psi_2 = \cos x$ (see Fig. 2-2). One readily finds by integration:

$$\langle \psi_1 | \psi_2 \rangle = \int_0^\pi \sin x \cos x \, dx = 0 \tag{2-40}$$

As is evident from Fig. 2-2, the orthogonality of ψ_1 and ψ_2 is a consequence of the fact that the integrated area between 0 and $\pi/2$ is exactly canceled

---

[11] Orthonormal = *ortho*gonal and *normal*ized.

[12] If $\mathbf{r}_1$ and $\mathbf{r}_2$ are two vectors separated by an angle θ, the dot product (analogous to $\langle r_1 | r_2 \rangle$) is $\mathbf{r}_1 \cdot \mathbf{r}_2 = |\mathbf{r}_1||\mathbf{r}_2| \cos \theta$. The projection of $\mathbf{r}_1$ onto the axis of $\mathbf{r}_2$ is $|\mathbf{r}_1| \cos \theta$; if $\theta = 90°$, $\cos \theta = 0$ and the projection vanishes.

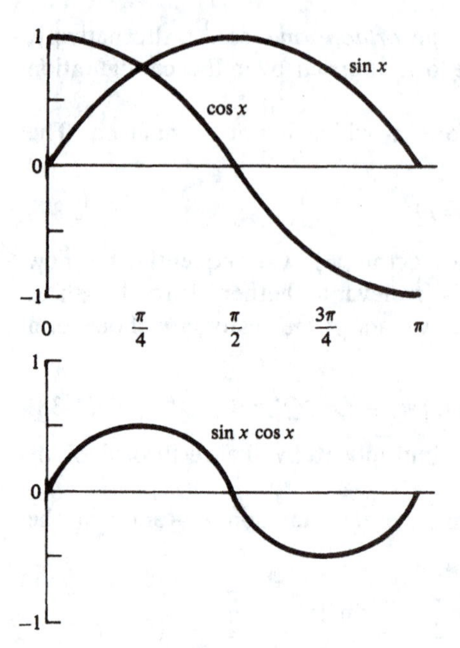

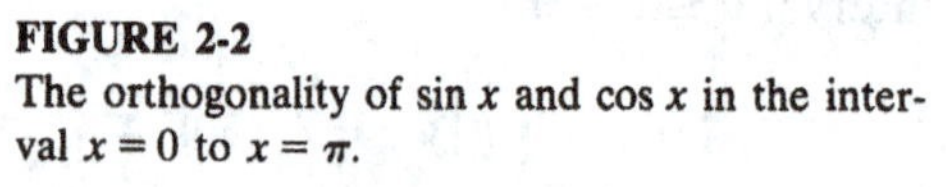

FIGURE 2-2
The orthogonality of $\sin x$ and $\cos x$ in the interval $x = 0$ to $x = \pi$.

(algebraically) by the integrated area between $\pi/2$ and π. Although not so easy to visualize as the first example, this latter type of orthogonality is *analogous* to the type of orthogonality many state vectors exhibit.

If a set of vectors $|u_1\rangle, |u_2\rangle, \ldots$ is originally not orthogonal, there exist a number of ways of transforming it to an equivalent set of orthogonal vectors. As a simple example, consider two vectors $|u_1\rangle$ and $|u_2\rangle$ (see Fig. 2-3) defined in real space and assumed not to be orthogonal. For convenience we assume the vectors are of equal length; if not, we can arbitrarily change their lengths (but not their directions) so that this is so. The diagram shows at once that two new vectors—one the sum of $|u_1\rangle$ and $|u_2\rangle$ and the other their difference—are orthogonal (i.e., at right angles to each other). Calling the new vectors $|v_1\rangle$ and $|v_2\rangle$, we may write

$$|v_1\rangle = |u_1\rangle + |u_2\rangle = |u_1 + u_2\rangle \qquad |v_2\rangle = |u_1\rangle - |u_2\rangle = |u_1 - u_2\rangle$$

$$\langle v_1 | v_2\rangle = \langle u_1 + u_2 | u_1 - u_2\rangle = \langle u_1 | u_1\rangle - \langle u_1 | u_2\rangle + \langle u_2 | u_1\rangle - \langle u_2 | u_2\rangle = 0$$

Alternatively, one could define two new vectors as follows (see Fig. 2-3 again):

$$|v_1\rangle = |u_1\rangle$$

$$|v_2\rangle = a|u_1\rangle + |u_2\rangle$$

where a is a constant chosen such that $\langle v_1 | v_2\rangle = 0$. It is left to the reader to

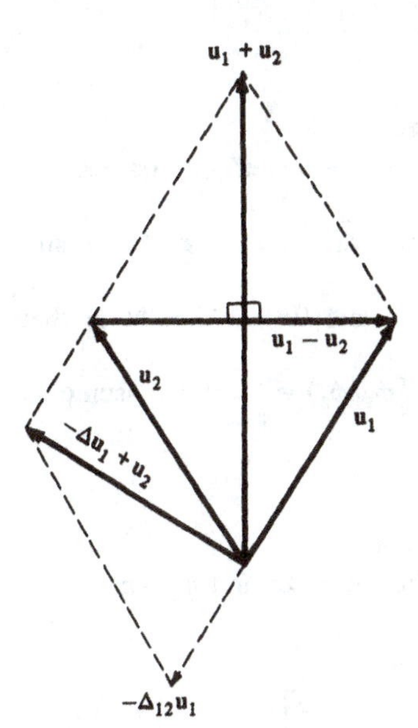

FIGURE 2-3
Two different ways of transforming two linearly independent vectors $\mathbf{u}_1$ and $\mathbf{u}_2$ to two new orthogonal vectors.

show that the desired result is obtained when a is given by

$$a = -\frac{\langle u_1 | u_2 \rangle}{\langle u_1 | u_1 \rangle}$$

or, if $|u_1\rangle$ is normalized,

$$a = -\langle u_1 | u_2 \rangle = -\Delta_{12}$$

The quantity Δ_{12} is often called an *overlap integral*; it reduces to the Kronecker delta when $|u_1\rangle$ and $|u_2\rangle$ are orthonormal. The overlap integral is a measure of the nonorthogonality, i.e., the extent to which one of the vectors can be projected onto the other.

A third vector $|v_3\rangle$ orthogonal to both the above may be defined as follows:

$$|v_2\rangle = b|u_1\rangle + c|v_2\rangle + |u_3\rangle \tag{2-41}$$

The constants b and c are determined by the orthogonality conditions

$$\langle v_1 | v_3 \rangle = 0 \qquad \langle v_2 | v_3 \rangle = 0$$

This is called the *Schmidt orthogonalization procedure*. The procedure can be continued to produce any desired number of orthogonal functions.

EXERCISES

2-15. Normalize the function $\phi = u_1 + Ku_2$ (K is a constant) for each of the following cases:

(*a*) u_1 and u_2 are orthogonal.
(*b*) $\langle u_1 | u_2 \rangle = \Delta_{12} \neq 0$.
Assume in both cases that u_1 and u_2 are normalized.

2-16. Normalize $u(r, \theta, \varphi) = e^{-ar^2}$ in the intervals $0 \leq r \leq \infty$, $0 \leq \theta \leq \pi$, and $0 \leq \varphi \leq 2\pi$. Note that $d\tau = r^2 \sin\theta \, d\theta \, d\varphi \, dr$.

2-17. Normalize the functions $u_1 = x$ and $u_2 = x^2$ in the interval $-a \leq x \leq a$. Are u_1 and u_2 orthogonal? Explain.

2-18. Normalize $u(r, \varphi) = re^{-r}$ in the intervals $0 \leq r \leq \infty$ and $0 \leq \varphi \leq 2\pi$. Note that $d\tau = r^2 \, d\varphi \, dr$.

2-19. Normalize $\psi = c_1\phi_1 + c_2\phi_2$ given that $c_1 = c_2$ and $\langle \phi_1 | \phi_2 \rangle = \Delta_{12} \neq 0$. Assume ϕ_1 and ϕ_2 are normalized. Repeat with $c_1 = -c_2$.

2-20. Normalize $u = x + 4$ in the interval $-1 \leq x \leq 1$.

2-21. Normalize $u = e^{-ix}$ in the interval $0 \leq x \leq 2\pi$.

2-22. Normalize $u = \sin x + \sin 2x$ in the interval $0 \leq x \leq \pi$.

2-23. Evaluate $\int \psi_1\psi_2 \, dx$ in the interval $-1 \leq x \leq 1$ when $\psi_1 = ax$ and $\psi_2 = b(1 - x)$. What does the result mean?

2-24. Consider the vectors

$$|r_1\rangle = \begin{bmatrix} a \\ 0 \\ 0 \end{bmatrix} \qquad |r_2\rangle = \begin{bmatrix} 0 \\ b \\ 0 \end{bmatrix} \qquad |r_3\rangle = \begin{bmatrix} 0 \\ 0 \\ c \end{bmatrix}$$

where a, b, and c are real constants. Normalize each vector and show that all three are mutually orthogonal.

2-25. Consider the vectors

$$|r_1\rangle = \begin{bmatrix} 1 \\ -3 \\ 2 \end{bmatrix} \qquad \text{and} \qquad |r_2\rangle = \begin{bmatrix} 3 \\ 2 \\ -1 \end{bmatrix}$$

(*a*) Show that $|r_1\rangle$ and $|r_2\rangle$ are not orthogonal.
(*b*) Show that two new vectors, the sum and difference of the above vectors, are orthogonal.
(*c*) Normalize the vectors $|r_1\rangle$ and $|r_2\rangle$. Also normalize the orthogonal vectors obtained in part (*b*).

2-26. Verify that Schmidt orthogonalization of $|u_1\rangle$, $|u_2\rangle$, and $|u_3\rangle$ leads to $b = -\Delta_{13}$ and $c = (\Delta_{12}\Delta_{13} - \Delta_{23})(1 - \Delta_{12}^2)^{-1}$ if b and c are defined as in Eq. (2-41). Also, normalize $|v_1\rangle$, $|v_2\rangle$, and $|v_3\rangle$. Show how to carry out the extension to a fourth vector $|v_4\rangle$.

2-27. Use Schmidt orthogonalization to orthonormalize the following vectors:

$$|r_1\rangle = \begin{bmatrix} 3 \\ 1 \\ 2 \end{bmatrix} \qquad |r_2\rangle = \begin{bmatrix} -1 \\ 2 \\ 0 \end{bmatrix} \qquad |r_3\rangle = \begin{bmatrix} 1 \\ -1 \\ 2 \end{bmatrix}$$

2-28. Construct the first few orthogonal combinations of the linearly independent functions: $1, x, x^2, x^3, x^4, \ldots$ in the interval $0 \leq x \leq 1$.

2-6 HERMITIAN OPERATORS

For every dynamical variable $a(p, q)$ of classical mechanics depending on the momenta (p) and coordinates (q) of the system, there exists a corresponding quantum mechanical operator $\hat{A}(p, q)$. In addition, there exist quantum mechanical operators for which no classical analogy exists.

The result of the precise measurement of a dynamical variable associated with the operator $\hat{A}$ is given by an eigenvalue a_i satisfying

$$\hat{A}\psi_i = a_i\psi_i \tag{2-42}$$

where ψ_i is the corresponding eigenfunction of the ith eigenstate. The mean value of a large number of measurements of the dynamical variable associated with the operator $\hat{A}$ on *initially identical* systems is given by

$$\langle \hat{A} \rangle = \frac{\langle \phi | \hat{A} | \phi \rangle}{\langle \phi | \phi \rangle} \tag{2-43}$$

where ϕ is the wavefunction of the system. This quantity is called the *expectation value* [see Eq. (2-19)] of the operator $\hat{A}$; it is analogous to the classical mean value or expectation value defined in Eq. (2-14).

The quantum mechanical operators related to measurable quantities must be linear so that the results of operating on a composite system are the same as occur when operating separately on the (independent) components and then combining the results. Furthermore, since all the measurable quantities are ultimately related to the measurements obtained by meter sticks, clocks, and balances, it is necessary that the corresponding operators have *real* expectation values. This requirement, we will see, is met by a class of operators known as *self-adjoint*, or *hermitian*, operators.

Two operators, $\hat{G}$ and $\hat{G}^\dagger$, are said (by definition) to be *adjoint* if all their respective expectation values are complex conjugates of each other. Thus if $\hat{G}$ and $\hat{G}^\dagger$ are a pair of adjoint operators (having the same domain),[13] then

$$\langle \hat{G}^\dagger \rangle = \langle \hat{G} \rangle^* \tag{2-44}$$

and, furthermore, $(\hat{G}^\dagger)^\dagger = \hat{G}$.

If an operator is to lead to real expectation values, it is necessary that $\hat{G}^\dagger = \hat{G}$; that is, the operator $\hat{G}$ must be *self-adjoint*, or *hermitian*. For self-adjoint operators, Eq. (2-44) becomes

$$\langle \hat{G} \rangle = \langle \hat{G} \rangle^* = \text{real quantity} \tag{2-45}$$

We will now prove an important theorem concerning adjoint operators which is very useful in other proofs and in manipulations of integrals involving operators.

[13] A function ϕ is said to belong to the domain of an operator $\hat{G}$ if $|\hat{G}\phi| = \sqrt{\langle \hat{G}\phi | \hat{G}\phi \rangle}$ exists.

Theorem 2-1. If $\hat{G}$ and $\hat{G}^\dagger$ are a pair of adjoint operators, and ϕ_1 and ϕ_2 are functions in their domain, then

$$\langle \phi_1|\hat{G}^\dagger|\phi_2\rangle = \langle \hat{G}\phi_1 \,|\, \phi_2\rangle$$

Proof. By definition of adjoint operators [Eq. (2-44)] we can write

$$\langle \phi|\hat{G}^\dagger\phi_2\rangle = \langle \phi|\hat{G}|\phi\rangle^* \tag{2-46}$$

where ϕ is an arbitrary function in the domain of $\hat{G}$ and $\hat{G}^\dagger$. Now consider the two special cases:

1. $\phi = \phi_1 + \phi_2$
2. $\phi = \phi_1 + i\phi_2$

When Eq. (2-46) is expanded in terms of case 1, the following relation results:

$$\langle \phi_1|\hat{G}^\dagger|\phi_2\rangle + \langle \phi_2|\hat{G}^\dagger|\phi_1\rangle = \langle \phi_1|\hat{G}|\phi_2\rangle^* + \langle \phi_2|\hat{G}|\phi_1\rangle^* \tag{2-47}$$

In like manner, expanding Eq. (2-46) in terms of case 2, simplifying, and dividing through by the common factor i, one gets

$$\langle \phi_1|\hat{G}^\dagger|\phi_2\rangle - \langle \phi_2|\hat{G}^\dagger|\phi_1\rangle = \langle \phi_2|\hat{G}|\phi_1\rangle^* - \langle \phi_1|\hat{G}|\phi_2\rangle^* \tag{2-48}$$

Adding Eqs. (2-47) and (2-48) and dividing by 2 gives

$$\langle \phi_1|\hat{G}^\dagger|\phi_2\rangle = \langle \phi_2|\hat{G}|\phi_1\rangle^*$$

Carrying out the complex-conjugate operation within the right-hand integral, one obtains[14]

$$\langle \phi_1|\hat{G}^\dagger|\phi_2\rangle = \langle \hat{G}\phi_1 \,|\, \phi_2\rangle \quad \text{QED} \tag{2-49}$$

Equation (2-49)—valid for any pair of adjoint operators—is called the *turnover rule*.[15] A special case of the turnover rule arises when the operator $\hat{G}$ is hermitian. In this case the turnover rule becomes

$$\langle \phi_1|\hat{G}|\phi_2\rangle = \langle G\phi_1 \,|\, \phi_2\rangle \tag{2-50}$$

This relationship often appears in standard integral notation as

$$\int \phi_1^* \hat{G}\phi_2 \, d\tau = \int (\hat{G}\phi_1)^* \phi_2 \, dt \tag{2-51}$$

The above equation [or its equivalent, Eq. (2-50)] is often used to provide a practical definition of a hermitian operator.

One important use of the turnover rule is in finding adjoints of given operators. As an example, consider the differential operator d/dx. We examine

---

[14] Note that the adjoint of a complex number $z = a + ib$ is just its complex conjugate. Thus, $z^\dagger = z^* = a - ib$. Also $\langle u|^\dagger = |u\rangle$ and $\langle u_1 \,|\, u_2\rangle^\dagger = \langle u_1 \,|\, u_2\rangle^* = \langle u_2 \,|\, u_1\rangle$.

[15] The name comes from the exchange of positions, or *turnover*, of $\hat{G}$ and ϕ_1.

the integral

$$I = \int \left(\frac{d\phi_1}{dx}\right)^* \phi_2 \, dx \tag{2-52}$$

which has the same form as the right-hand side of Eq. (2-51). We integrate I by parts,[16] letting

$$u = \phi_2 \qquad dv = \left(\frac{d\phi_1}{dx}\right)^* dx$$

$$du = \frac{d\phi_2}{dx}\, dx \qquad v = \phi_1^*$$

The partial integration leads to

$$I = \phi_1^* \phi_2 - \int \phi_1^* \left(\frac{d}{dx}\right) \phi_2 \, dx = -\int \phi_1^* \left(\frac{d}{dx}\right) \phi_2 \, dx = \int \phi_1^* \left(-\frac{d}{dx}\right) \phi_2 \, dx \tag{2-53}$$

where $\phi_1^* \phi_2$ vanishes at the limits of integration for well-behaved functions. Comparison of Eq. (2-53) with the left-hand side of Eq. (2-49) shows that

$$-\frac{d}{dx} = \left(\frac{d}{dx}\right)^\dagger \tag{2-54}$$

Consequently, d/dx is *not* a hermitian operator. Note, however, that the operator $(\hbar/i)\, d/dx$ (the operator for the x component of linear momentum p_x) is hermitian. This result follows from Eq. (2-54) and the fact that

$$\left(\frac{\hbar}{i}\right)^\dagger = \left(\frac{\hbar}{i}\right)^* = -\frac{\hbar}{i}$$

so that

$$\left(\frac{\hbar}{i}\frac{d}{dx}\right)^\dagger = \left(\frac{\hbar}{i}\right)^\dagger \left(\frac{d}{dx}\right)^\dagger = -\frac{\hbar}{i}\left(-\frac{d}{dx}\right) = \frac{\hbar}{i}\frac{d}{dx} \tag{2-55}$$

The turnover rule may be used to prove some relationships concerning the sums and products of operators:

1. $(\hat{F} + \hat{G})^\dagger = \hat{F}^\dagger + \hat{G}^\dagger$

Proof

$$\langle \phi_1 | (\hat{F} + \hat{G})^\dagger | \phi_2 \rangle = \langle (\hat{F} + \hat{G})\phi_1 | \phi_2 \rangle = \langle \hat{F}\phi_1 | \phi_2 \rangle + \langle \hat{G}\phi_1 | \phi_2 \rangle$$
$$= \langle \phi_1 | \hat{F}^\dagger | \phi_2 \rangle + \langle \phi_1 | \hat{G}^\dagger | \phi_2 \rangle = \langle \phi_1 | \hat{F}^\dagger + \hat{G}^\dagger | \phi_2 \rangle \quad \text{QED}$$

2. $(\hat{F}\hat{G})^\dagger = \hat{G}^\dagger \hat{F}^\dagger$

Proof

$$\langle \phi_1 | (\hat{F}\hat{G})^\dagger | \phi_2 \rangle = \langle (\hat{F}\hat{G})\phi_1 | \phi_2 \rangle = \langle \hat{G}\phi_1 | F^\dagger | \phi_2 \rangle = \langle \phi_1 | \hat{G}^\dagger \hat{F}^\dagger | \phi_2 \rangle \quad \text{QED}$$

---

[16] $\int u \, dv = uv - \int v \, du$.

Note that a trivial example of the latter relationship occurs in Eq. (2-55).

The turnover rule also helps to establish two very important properties of hermitian operators as indicated in the following two theorems:

Theorem 2-2. The eigenvalues of a hermitian operator are real.

Proof. Consider the eigenvalue equation

$$\hat{A}\psi = a\psi$$

where $\hat{A}$ is assumed to be hermitian ($\hat{A} = \hat{A}^\dagger$). Then

$$\langle \psi | \hat{A} | \psi \rangle = a \langle \psi | \psi \rangle \tag{2-56}$$

Since $\hat{A}$ is hermitian, the turnover rule Eq. (2-50) allows us to write

$$\langle \psi | \hat{A} | \psi \rangle = \langle \hat{A}\psi | \psi \rangle = a^* \langle \psi | \psi \rangle \tag{2-57}$$

Equating Eqs. (2-56) and (2-57) leads to

$$a = a^*$$

Thus, the eigenvalue a is real.[17] QED

Theorem 2-3. The nondegenerate eigenfunctions of a hermitian operator are automatically orthogonal.

Proof. Let ψ_k and ψ_l be two different eigenfunctions of the hermitian operator $\hat{A}$. The eigenvalue equations are

$$\hat{A}\psi_k = a_k \psi_k$$

$$\hat{A}\psi_l = a_l \psi_l$$

If $a_k \neq a_l$, the functions ψ_k and ψ_l are nondegenerate; otherwise, they are degenerate. We assume ψ_k and ψ_l are nondegenerate. Now consider the integral

$$\langle \psi_k | \hat{A} | \psi_l \rangle = a_l \langle \psi_k | \psi_l \rangle$$

But the original integral can be transformed differently by use of the turnover rule and Theorem 2-2

$$\langle \psi_k | \hat{A} | \psi_l \rangle = \langle \hat{A}\psi_k | \psi_l \rangle = a_k \langle \psi_k | \psi_l \rangle$$

Thus one may write

$$a_l \langle \psi_k | \psi_l \rangle = a_k \langle \psi_k | \psi_l \rangle \qquad \text{or, alternatively,} \qquad (a_k - a_l)\langle \psi_k | \psi_l \rangle = 0$$

Since by hypothesis $a_k - a_l \neq 0$, we conclude that

$$\langle \psi_k | \psi_l \rangle = 0 \quad \text{QED}$$

[17] This result also follows from Eq. (2-45) once it is recognized that an expectation value is an *average* of eigenvalues.

Since degenerate eigenfunctions are linearly independent, they can always be transformed to an equivalent set of orthogonal functions.[18] Hence, when dealing with a hermitian operator and its eigenfunctions, it is convenient and customary to assume that all are orthogonal—either automatically or by choice. If all the eigenfunctions are also normalized, we sum up all this information in the compact expression

$$\langle \psi_k | \psi_l \rangle = \delta_{kl} \qquad \text{for all } k, l$$

EXERCISES

2-29. Verify Eqs. (2-47) and (2-48).

2-30. Find the adjoints of
  (*a*) The real function $A(x, y)$
  (*b*) The complex function $W(z) = A(x, y) + iB(x, y)$, where A and B are real functions
  (*c*) The linear momentum operator $(\hbar/i)\, d/dx$
  (*d*) The operator d^2/dx^2
  (*e*) The laplacian operator ∇^2
  (*f*) The hamiltonian operator $\hat{H} = -(\hbar^2/2m)\nabla^2 + \hat{V}$ where $\hat{V}$ (or V) is a real function

2-29. How would you rationalize the fact that $\hat{x}$ (or x) is a hermitian operator?

2-31. If ψ_k and ψ_l are nondegenerate eigenfunctions of a hermitian operator $\hat{A}$, prove that the integral $\langle \psi_k | \hat{A} | \psi_l \rangle$ is zero. What can you say about this integral if ψ_k and ψ_l are degenerate?

2-32. Assume that $\hat{A}$ is a hermitian operator and that ϕ_1, ϕ_2, and ϕ_3 are three of its normalized eigenfunctions with corresponding eigenvalues $a_1 = 2$, $a_2 = 3$, and $a_3 = 2$, respectively. Examine each of the following integrals and state its numerical value if such a value unambiguously exists. If the value is indeterminate, say so.
  (*a*) $\langle \phi_1 | \hat{A} | \phi_1 \rangle$ (*f*) $\langle \phi_3 | \phi_1 \rangle$
  (*b*) $\langle \phi_2 | \hat{A} | \phi_1 \rangle$ (*g*) $\langle \phi_1 | \hat{A} | \phi_3 \rangle$
  (*c*) $\langle \phi_1 | \phi_2 \rangle$ (*h*) $\langle \phi_3 | \phi_2 \rangle$
  (*d*) $\langle \phi_1 | \phi_3 \rangle$ (*i*) $\langle \phi_1 + \phi_3 | \hat{A} | \phi_1 - \phi_3 \rangle$
  (*e*) $\langle \phi_2 | \hat{A} | \phi_3 \rangle$ (*j*) $\langle \phi_1 - \phi_3 | \hat{A} | \phi_1 - \phi_3 \rangle$

2-33. Given an arbitrary operator $\hat{F}$, prove the following:
  (*a*) $\hat{F}^\dagger \hat{F}$ is hermitian.
  (*b*) $\langle \hat{F}^\dagger \hat{F} \rangle \geq 0$. (*Hint*: Use the turnover rule and note the form of the integral.)

2-34. A pair of degenerate eigenfunctions of the operator $\hat{A}$ are ϕ_1 and ϕ_2. Show that the two new functions $u_1 = \phi_1$ and $u_2 = \phi_2 + K\phi_1$ are orthogonal, provided K is properly chosen. Show that u_1 and u_2 are also degenerate and have the same eigenvalues as ϕ_1 and ϕ_2.

---

[18] See H. Margenau and G. M. Murphy, pp. 273 and 311, in the suggested readings at the end of this chapter.

2-7 NORMAL OPERATORS

Some of the important properties of hermitian operators are most easily deduced by considering a more general class of operators defined by

$$\Lambda = \hat{A} + i\hat{B} \tag{2-58}$$

where $\hat{A}$ and $\hat{B}$ are hermitian operators (for notational simplicity we omit the circumflex above the operator Λ). Taking the adjoint of the operator $i\hat{B}$ we see that $(i\hat{B})^\dagger = -i\hat{B}$. Any operator $\hat{M}$ whose adjoint is its negative $(-\hat{M})$ is said to be *antihermitian*. Thus, the operator Λ is a sum of a hermitian operator $\hat{A}$ and an antihermitian operator $i\hat{B}$. This is analogous to the formation of a complex number $z = a + ib$ as the sum of the real number a and the imaginary number ib (with both a and b real). Just as a complex number reduces to a real number when its imaginary component is zero, so the general operator Λ reduces to a hermitian operator when its antihermitian component is zero.

Let us restrict ourselves to those operators Λ which commute with their adjoints, that is, $[\Lambda, \Lambda^\dagger] = 0$. Such operators are called *normal operators*. Obviously, normal operators are generalizations of hermitian operators, since every hermitian operator is self-commuting and, therefore, normal. However, the converse is *not* true; not all normal operators are hermitian. From the general definition we can deduce the following relationships:

$$\Lambda\Lambda^\dagger = (\hat{A} + i\hat{B})(\hat{A} - i\hat{B}) = \hat{A}^2 + \hat{B}^2 + i[\hat{B}, \hat{A}]$$

$$\Lambda^\dagger\Lambda = (\hat{A} - i\hat{B})(\hat{A} + i\hat{B}) = \hat{A}^2 + \hat{B}^2 - i[\hat{B}, \hat{A}]$$

Subtracting the two expressions gives us

$$[\Lambda, \Lambda^\dagger] = 2i[\hat{B}, \hat{A}] \tag{2-59}$$

This shows that Λ is normal only if $\hat{A}$ and $\hat{B}$ themselves commute, and Λ is hermitian only if $\hat{B}$ is zero.

We now prove several important theorems concerning normal operators, which, of course, also hold for hermitian operators.

Theorem 2-4. If a normal operator Λ has an eigenfunction ψ_k with eigenvalue λ_k, the adjoint operator $\Lambda^\dagger$ has the eigenvalue λ_k^* for the same eigenfunction ψ_k.

Proof. Let $\Lambda\psi_k = \lambda_k\psi_k$ and consider the integral

$$\langle(\Lambda^\dagger - \lambda_k^*)\psi_k|(\Lambda^\dagger - \lambda_k^*)\psi_k\rangle$$

Applying the turnover rule, this integral becomes

$$\langle\psi_k|(\Lambda - \lambda_k)(\Lambda^\dagger - \lambda_k^*)\psi_k\rangle$$

Since Λ and $\Lambda^\dagger$ commute, the central terms may be interchanged so that

$$\langle\psi_k|(\Lambda^\dagger - \lambda_k^*)(\Lambda - \lambda_k)|\psi_k\rangle = 0$$

Therefore, the original integral also vanishes, which implies

$$\Lambda^\dagger\psi_k = \lambda_k^*\psi_k \quad \text{QED}$$

This theorem—which is trivial for hermitian operators— will now be used to prove a more important theorem.

Theorem 2-5. If Λ is a normal operator which commutes with an arbitrary operator $\hat{F}$, and ψ_k and ψ_l are nondegenerate eigenfunctions of Λ, then

$$\langle \psi_k | \hat{F} | \psi_l \rangle = F_{kl} = 0$$

i.e., the nondiagonal matrix elements of the operator $\hat{F}$ in the basis $\{\psi_i\}$ vanish.

Proof. Consider

$$\lambda_l \langle \psi_k | \hat{F} | \psi_l \rangle = \langle \psi_k | \hat{F} | \lambda_l \psi_l \rangle = \langle \psi_k | \hat{F}\Lambda | \psi_l \rangle = \langle \psi_k | \Lambda \hat{F} | \psi_l \rangle$$
$$= \langle \Lambda^\dagger \psi_k | \hat{F} | \psi_l \rangle = \lambda_k \langle \psi_k | \hat{F} | \psi_l \rangle$$

Subtracting the first integral from the last produces

$$(\lambda_k - \lambda_l) \langle \psi_k | \hat{F} | \psi_l \rangle = (\lambda_k - \lambda_l) F_{kl} = 0$$

Since, by hypothesis, $\lambda_k \neq \lambda_l$, it follows that

$$F_{kl} = 0 \quad \text{QED}$$

The preceding theorem is very useful when one uses the eigenfunctions of one normal operator (such as the hamiltonian) to form the matrix representation of another operator which commutes with the first.

The next theorem reveals one of the most important properties of *commuting hermitian operators*.

Theorem 2-6. If two hermitian operators $\hat{A}$ and $\hat{B}$ commute, there exists a common set of eigenfunctions $\{\psi_k\}$ of $\hat{A}$ and $\hat{B}$ such that

$$\hat{A}\psi_k = a_k \psi_k \qquad \hat{B}\psi_k = b_k \psi_k$$

Proof. Since, by hypothesis, $[\hat{A}, \hat{B}] = 0$, the operators $\Lambda = \hat{A} + i\hat{B}$ and $\Lambda^\dagger = \hat{A} - i\hat{B}$ are normal. Thus we may write

$$\Lambda \psi_k = (\hat{A} + i\hat{B})\psi_k = (a_k + ib_k)\psi_k = \lambda_k \psi_k$$
$$\Lambda^\dagger \psi_k = (\hat{A} - i\hat{B})\psi_k = (a_k - ib_k)\psi_k = \lambda_k^* \psi_k$$

(Note the use of Theorem 2-4 in the last step.) Adding and subtracting the above two equations leads to

$$\hat{A}\psi_k = a_k \psi_k \qquad \hat{B}\psi_k = b_k \psi_k \quad \text{QED}$$

A converse of this theorem is also true: If there exists a complete set of eigenfunctions common to both $\hat{A}$ and $\hat{B}$, then $\hat{A}$ and $\hat{B}$ must commute. Note that this theorem does not preclude the possibility that $\hat{A}$ and $\hat{B}$ have some eigenfunctions which they do not have in common. Theorem 2-6 is much more awkward to prove using hermitian operators alone.[19]

[19] A proof is given in H. Eyring, J. Walter, and G. E. Kimball, *Quantum Chemistry*, Wiley, New York, 1944, pp. 34–37.

EXERCISES

2-35. If $\hat{A}$ and $\hat{B}$ are hermitian operators, under what conditions is the product $\hat{A}\hat{B}$ also hermitian?

2-36. If $\hat{A}$ and $\hat{B}$ are hermitian operators, is the operator $\hat{A}\hat{B} + \hat{B}\hat{A}$ always hermitian?

2-37. Let $\hat{A}$ be a hermitian operator and $\hat{B}$ an antihermitian operator. Under what conditions is the product $\hat{A}\hat{B}$ antihermitian?

2-38. If both $\hat{A}$ and $\hat{B}$ are antihermitian operators, under what conditions is the product $\hat{A}\hat{B}$ hermitian?

2-39. Examine the following operators—each a product of hermitian operators—and determine if they are hermitian:

$$(a)\ \frac{\hbar}{i}\,\hat{x}\,\frac{d}{dx} \qquad (b)\ (\hat{\mathbf{r}}\cdot\hat{\mathbf{p}}) \qquad (c)\ (\hat{\mathbf{r}}\cdot\nabla V)$$

2-40. Show that although the hermitian operators $(\hbar/i)\, d/dx$ and $-(\hbar^2/2m)\, d^2/dx^2$ commute, the eigenfunctions ($\sin nx$ or $\cos nx$) of the latter are not eigenfunctions of the former. What about the functions $e^{\pm inx} = \cos nx \pm i \sin nx$?

2-41. Given that $[\hat{B}, \hat{A}] = 0$ (where $\hat{H}$ is the hamiltonian of the system) and that $\hat{H}\phi_1 = a\phi_1$, $\hat{H}\phi_2 = b\phi_2$, $\hat{H}\phi_3 = b\phi_3$, and $\hat{H}\phi_4 = c\phi_4$ $(a \neq b \neq c)$, pick out the nonzero integrals in the following:

(a) $\langle \phi_1 | \phi_4 \rangle$
(b) $\langle \phi_2 | \phi_4 \rangle$
(c) $\langle \phi_2 | \phi_3 \rangle$
(d) $\langle \phi_3 | \phi_4 \rangle$
(e) $\langle \phi_1 | \hat{H} | \phi_3 \rangle$
(f) $\langle \phi_2 | \hat{H} | \phi_3 \rangle$
(g) $\langle \phi_1 | \hat{B} | \phi_4 \rangle$
(h) $\langle \phi_2 | \hat{B} | \phi_3 \rangle$
(i) $\langle \phi_2 + \phi_3 | \hat{B} | \phi_2 - \phi_3 \rangle$
(j) $\langle \phi_2 + \phi_3 | \hat{H} | \phi_2 - \phi_3 \rangle$
(k) $\langle \phi_2 + \phi_3 | \phi_2 - \phi_3 \rangle$

2-8 EXPECTATION VALUES IN QUANTUM MECHANICS

The expectation value of an operator $\hat{A}$, now assumed to be hermitian, was defined in Eq. (2-43) as

$$\langle \hat{A} \rangle = \frac{\langle \phi | \hat{A} | \phi \rangle}{\langle \phi | \phi \rangle} \tag{2-60}$$

where ϕ is the wavefunction of the system. Since the eigenfunctions of $\hat{A}$ (represented by $\psi_1, \psi_2, \ldots$) form a complete basis set, we may express the wavefunction ϕ as a linear combination of these, namely

$$\phi = \sum_{i=1}^{\infty} c_i \psi_i \tag{2-61}$$

If ϕ is normalized, then

$$\langle \phi | \phi \rangle = \sum_{i=1}^{\infty} c_i^* c_i = 1 \tag{2-62}$$

where all cross terms such as $c_i^* c_j \langle \psi_i | \psi_j \rangle$ $(i \neq j)$ vanish owing to orthogonality.

Two different situations arise depending on whether or not the operator $\hat{A}$ commutes with the hamiltonian of the system in question. In the case that $\hat{A}$ and $\hat{H}$ commute, the two operators share a common set of eigenfunctions (Theorem 2-6) so that ϕ will simply be one of the ψ_i.[20] Thus

$$\langle \hat{A} \rangle = a_i \tag{2-63}$$

This means that the system is in the stationary state ψ_i with energy E_i and that it also has the definite value a_i of the dynamical variable represented by the operator $\hat{A}$.

On the other hand, if $\hat{A}$ does not commute with $\hat{H}$, then $\hat{A}$ and $\hat{H}$ do not have common eigenfunctions and the expectation value of $\hat{A}$ has the general form

$$\langle \hat{A} \rangle = \sum_{i=1}^{\infty} c_i^* c_i a_i \tag{2-64}$$

Here, even though the system may be in a definite energy state, it may have any number of possible values for the property represented by $\hat{A}$. Any observation made on the system has the effect of forcing it into a definite state ψ_i in which the property represented by $\hat{A}$ has the value a_i. Thus $c_i^* c_i$ represents the *probability* that one gets the value a_i upon measurement, and $\langle \hat{A} \rangle$ is simply the probability-weighted mean of repeated measurements.

It is easy to verify that a necessary and sufficient condition in order for $\langle \hat{A} \rangle$ to belong to the first case is that

$$\langle \hat{A}^n \rangle = \langle \hat{A} \rangle^n \qquad n = 1, 2, 3, \ldots$$

(see Exercise 2-43).

EXERCISES

2-42. The hamiltonian operator of a given system is $\hat{H} = -(\hbar^2/2m)\, d^2/dx^2 + V$ (where V is a constant). The corresponding eigenfunctions (not normalized) are $\psi_n = e^{\pm inx}$ $(n = 1, 2, \ldots)$.

(*a*) What is the expectation value of $\hat{H}$ when the system is in its $n = 3$ stationary state?

(*b*) What is the expectation value of the x component of the linear momentum in the $n = 3$ state?

2-43. The hamiltonian operator of a given state is the same as given in the previous problem, but the eigenfunctions are now given in the form $\Psi(x, t) = A \sin(n\pi x/a) e^{-iE_1 t/\hbar}$, where A and a are constants and the interval of integration is from $x = 0$ to $x = a$.

(*a*) Evaluate and compare $\langle \hat{H} \rangle$, $\langle \hat{H}^2 \rangle$, and $\langle \hat{H} \rangle^2$.

(*b*) Evaluate and compare $\langle \hat{x} \rangle$, $\langle \hat{x}^2 \rangle$, and $\langle \hat{x} \rangle^2$.

(*c*) How do cases (*a*) and (*b*) differ? Explain.

---

[20] This means that all $c_i^* c_j$ $(j \neq i)$ are zero and that $c_i^* = 1$ for some particular value of i.

2-44. An operator $\hat{A}$ has only three eigenfunctions ψ_1, ψ_2, and ψ_3, with corresponding eigenvalues $a_1 = 1$, $a_2 = 2$, and $a_3 = 3$, respectively. There is a 50 percent chance that a measurement produces a_1 and equal chances for either a_2 or a_3.
(*a*) Calculate $\langle \hat{A} \rangle$.
(*b*) Express the wavefunction ϕ in terms of the eigenfunctions of $\hat{A}$.
(*c*) Does $\hat{A}$ commute with $\hat{H}$? Explain.

2-45. An operator $\hat{H}$ has only four eigenfunctions ψ_i $(i = 1, 2, 3, 4)$, with corresponding eigenvalues $a_1 = 1$, $a_2 = 2$, $a_3 = 1$, and $a_4 = 3$. The wavefunction of the system is $\phi = 0.500\psi_1 + 0.632\psi_2 + 0.500\psi_3 + 0.316\psi_4$. Calculate $\langle \hat{A} \rangle$ for this system.

2-46. The following illustrate in a simple way the difference between the two different types of expectation values:
(*a*) Assume you have 65 coins, 10 of which weigh 5 g each, 50 of which weigh 6 g each, and 5 of which weigh 7 g each. Compute the average weight $\langle w \rangle$ of a coin. Also compute $\langle w^2 \rangle$ and compare with $\langle w \rangle^2$.
(*b*) Assume you have 100 coins weighing 5 g each. Compute the average weight $\langle w \rangle$ and compare $\langle w^2 \rangle$ with $\langle w \rangle^2$.

2-9 CONSTRUCTION OF QUANTUM MECHANICAL OPERATORS

In time-independent states, the wavefunction $\psi(x, y, z)$ is a function of the position coordinates of a system. Thus, it is convenient to employ quantum mechanical operators which are also expressed in terms of these coordinates.[21] In such a system the operator forms of x, y, and z become simply $\hat{x}$, $\hat{y}$, and $\hat{z}$, respectively. Since $\hat{x}$ means simply "multiply by x" (with corresponding meanings for $\hat{y}$ and $\hat{z}$), it is customary to omit the circumflex, which denotes an operator; this ordinarily leads to no confusion.

On the basis of our earlier development in Sec. 2-1, the components of linear momentum $(p_x = m\dot{x},\ p_y = m\dot{y},$ and $p_z = m\dot{z})$ are represented by *differential operators* as follows:

$$\hat{p}_x = \frac{\hbar}{i}\frac{\partial}{\partial x} \qquad \hat{p}_y = \frac{\hbar}{i}\frac{\partial}{\partial y} \qquad \hat{p}_z = \frac{\hbar}{i}\frac{\partial}{\partial z} \tag{2-65}$$

The total linear momentum (a vector quantity) is represented by

$$\hat{\mathbf{p}} = \mathbf{i}\hat{p}_x + \mathbf{j}\hat{p}_y + \mathbf{k}\hat{p}_z = \frac{\hbar}{i}\left(\mathbf{i}\frac{\partial}{\partial x} + \mathbf{j}\frac{\partial}{\partial y} + \mathbf{k}\frac{\partial}{\partial z}\right) = \frac{\hbar}{i}\nabla \tag{2-66}$$

The total energy of a conservative system is given by

$$E = T + V = \frac{p^2}{2m} + V \tag{2-67}$$

[21] For some purposes it may be convenient to express ψ and the operators in terms of momenta rather than coordinates, but this situation will not arise in this text. The two alternative representations, coordinate space and momentum space, are equivalent and are related by a *Fourier transform* which reflects the uncertainty principle.

where T = kinetic energy
V = potential energy
$p^2 = \mathbf{p} \cdot \mathbf{p}$

The operator representation of p^2 is just

$$\hat{p}^2 = \frac{\hbar}{i} \nabla \cdot \frac{\hbar}{i} \nabla = -\hbar^2 \nabla^2 \tag{2-68}$$

so that the total energy operator (the hamiltonian) becomes

$$\hat{E} = \hat{H} = -\frac{\hbar^2}{2m} \nabla^2 + \hat{V} \tag{2-69}$$

The potential-energy operator $\hat{V}$ varies from system to system and cannot be given a general explicit form. As in the case of $\hat{x}$, $\hat{y}$, and $\hat{z}$, $\hat{V}$ is normally not differential in form, so the circumflex is omitted.

The construction of quantum mechanical operators for other dynamical variables can usually be carried out by writing the classical expression in terms of coordinates and momenta and then replacing these with their operator representations. As an example, consider angular momentum, the classical expression for which is

$$\mathbf{L} = \mathbf{r} \times \mathbf{p} = \begin{vmatrix} \mathbf{i} & \mathbf{j} & \mathbf{k} \\ x & y & z \\ p_x & p_y & p_z \end{vmatrix} \tag{2-70}$$

The corresponding quantum mechanical operator is

$$\mathbf{L} = \frac{\hbar}{i} \mathbf{r} \times \nabla = \frac{\hbar}{i} \begin{vmatrix} \mathbf{i} & \mathbf{j} & \mathbf{k} \\ x & y & z \\ \dfrac{\partial}{\partial x} & \dfrac{\partial}{\partial y} & \dfrac{\partial}{\partial z} \end{vmatrix} \tag{2-71}$$

where $\mathbf{r} = \mathbf{i}x + \mathbf{j}y + \mathbf{k}z$.

Situations do arise when the above procedure is inadequate to establish the correct operator, e.g., when there is ambiguity in the order of factors in the classical expression or when there is no classical analog for the operator. Except for the second case, which we will discuss later, such situations do not arise in this text.[22]

[22] The problem of a unique correlation of operators with classical dynamical variables has been discussed by H. Weyl, *Z. Physik*, **46F**:1 (1927) and by K. Grjotheim and P. C. Hemmer, *Z. Physik. Chem.* (Leipzig), **208**:378 (1958). See also the text by Tomonaga, pp. 138–140, cited in the suggested readings at the end of this chapter.

EXERCISES

2-47. Find the operator representations of the x, y, and z components of **L** (which is equal to $\mathbf{i}L_x + \mathbf{j}L_y + \mathbf{k}L_z$).

2-48. Find the operator representations of xp_x and p_xx. These quantities are equivalent in classical mechanics. Are the operator representations equivalent in quantum mechanics? Explain.

2-10 THE GENERALIZED UNCERTAINTY PRINCIPLE

The uncertainty principle introduced in Sec. 1-6 can be derived in a more general and rigorous manner.[23] If $\hat{F}$ and $\hat{G}$ are two hermitian operators representing the dynamical variables f and g, respectively, then the uncertainties in these variables satisfy the inequality

$$\Delta f \Delta g \geq \tfrac{1}{2}|\langle[\hat{F}, \hat{G}]\rangle| \tag{2-72}$$

This means that, in general, two dynamical variables f and g can be measured simultaneously and exactly only if their operators commute.

In the special case in which f and g are the position x and the linear momentum component p_x, the commutator is

$$\left[\frac{\hbar}{i}\frac{d}{dx}, x\right] = \frac{\hbar}{i} \tag{2-73}$$

Thus, the uncertainty principle for these two quantities becomes

$$\Delta x \Delta p \geq \frac{\hbar}{2} \tag{2-74}$$

This, except for an essentially irrelevant numerical factor, agrees with the previously derived expression [Eq. (1-37)]. In general, if f and g are a pair of canonically conjugate variables,[24] they will satisfy an uncertainty principle of the same form as Eq. (2-74).

EXERCISES

2-49. Obtain the correct uncertainty principle for the variables E (total energy) and t (time). This uncertainty accounts for the finite width of spectral lines and for the nonmonochromatic nature of radioactive decay. *Hint*: Consider the operators $-(\hbar/i)\,\partial/\partial t$ and $\hat{t}$ (or t).

2-50. Is it possible to measure p_y and x simultaneously and exactly? Explain.

---

[23] See Margenau and Murphy pp. 348–350, cited in the suggested readings at the end of this chapter.

[24] In classical physics, generalized coordinates q_i and generalized momenta p_i which satisfy Hamilton's equations of motion, $\partial H/\partial p_i = q_i$ and $\partial H/\partial q_i = -p_i$, are said to be *canonically conjugate* and are called *conjugate variables*.

2-11 CONSTANTS OF THE MOTION

In classical mechanics the *Poisson bracket* of two dynamical variables f and g is defined by

$$\{f, g\} = \sum_i \left(\frac{\partial f}{\partial q_i} \frac{\partial g}{\partial p_i} - \frac{\partial f}{\partial p_i} \frac{\partial g}{\partial q_i} \right) \tag{2-75}$$

where q_i and p_i are *generalized* coordinates and momenta, respectively. Now consider the time derivative of one of these variables, e.g., the variable f:

$$\frac{df}{dt} = \frac{\partial f}{\partial t} + \sum_i \left(\frac{\partial f}{\partial q_i} \frac{\partial q_i}{\partial t} + \frac{\partial f}{\partial p_i} \frac{\partial p_i}{\partial t} \right) \tag{2-76}$$

Using Hamilton's equations of motion

$$\frac{\partial q_i}{\partial t} = \dot{q}_i = \frac{\partial H}{\partial p_i} \quad \text{and} \quad \frac{\partial p_i}{\partial t} = \dot{p}_i = -\frac{\partial H}{\partial q_i} \tag{2-77}$$

we may rewrite Eq. (2-76) as

$$\frac{df}{dt} = \frac{\partial f}{\partial t} + \sum_i \left(\frac{\partial f}{\partial q_i} \frac{\partial H}{\partial p_i} - \frac{\partial f}{\partial p_i} \frac{\partial H}{\partial q_i} \right) = \frac{\partial f}{\partial t} + \{f, H\} \tag{2-78}$$

The physical meaning of this expression is: The dynamical variable f is conserved (time-independent), provided two conditions hold. These are

1. f does not depend explicitly on t (so that $\partial f/\partial t = 0$).
2. The Poisson bracket of f with the hamiltonian H is zero.

A quantum mechanical analogy to Eq. (2-78) is obtained by replacing the dynamical variables f and g with their operator representatives $\hat{F}$ and $\hat{G}$ and the Poisson bracket by the commutator.[25] Specifically $\{f, H\}$ is replaced with

$$-\frac{i}{\hbar} [\hat{F}, \hat{G}] \tag{2-79}$$

Thus, the quantum mechanical analog of Eq. (2-78) is

$$\frac{d\hat{F}}{dt} = \frac{\partial \hat{F}}{\partial t} - \frac{i}{\hbar} [\hat{F}, \hat{H}] = \frac{\partial \hat{F}}{\partial t} + \frac{i}{\hbar} [\hat{H}, \hat{F}] \tag{2-80}$$

This result is called *Heisenberg's equation of motion*; it is equivalent to Schrödinger's equation and represents an alternative formalism for the expression of quantum mechanics.[26] If the operator $\hat{F}$ commutes with the hamiltonian

[25] The Poisson brackets of a, b, and c satisfy $\{a, \{b, c\}\} + \{b, \{c, a\}\} + \{c, \{a, b\}\} = 0$. Using an analogous relationship for the corresponding commutator of the operators and the uncertainty principle leads to Eq. (2-79).

[26] Whereas the Schrödinger formalism deals with fixed dynamical variables and time-dependent states, the Heisenberg formalism deals with fixed states and time-dependent dynamical variables.

$\hat{H}$ and if $\hat{F}$ is time-independent, we say that the dynamical variable f is a *constant of the motion*, that is, f is a time-independent property. In general, it is not possible for $\hat{H}$ and $\hat{F}$ to commute at *all times* unless $\hat{H}$ is also constant, implying that $\hat{H}$ is aso a time-independent operator. This is one justification for associating $\hat{H}$ with the total energy of a conservative system.

EXERCISES

2-51. Show that Hamilton's equations of motion are special cases of Eq. (2-78). *Hint*: Replace f with the canonically conjugate variables p_i and q_i.

2-52. Apply the Heisenberg equation of motion to the expectation value of $\hat{F}$ and show by comparision with $d\langle \hat{F}\rangle / dt = (d/dt)\langle \Psi|\hat{F}|\Psi\rangle$ in expanded form that Schrödinger's Eq. (2-11) results.

2-53. Use Eq. (2-79) to obtain commutation relations for $\hat{p}_x$ and x and for $\hat{E}$ and t.

2-54. Show that:

(*a*) $m \dfrac{d\langle x\rangle}{dt} = \dfrac{\hbar}{i}\left\langle \dfrac{\partial}{\partial x}\right\rangle$

(*b*) $\hbar i \dfrac{d}{dt}\left\langle \dfrac{\partial}{\partial x}\right\rangle = -\langle \nabla V\rangle$

(*c*) These are quantum mechanical analogs of Newton's equations of motion and are known as the *Ehrenfest relations*.

2-55. The operators $\hat{A}$ and $\hat{B}$ represent constants of the motion but do not commute. Show that this implies that the energy levels of the system are generally degenerate.

2-12 THE QUANTUM MECHANICAL VIRIAL THEOREM

Consider a stationary state ψ satisfying the Schrödinger equation $\hat{H}\psi = E\psi$, and let $\hat{F}$ be any arbitrary operator depending only on the coordinates and momenta of the system. We now show that the commutator of $\hat{H}$ and $\hat{F}$ has a zero expectation value in the state ψ so that we can write

$$\frac{d\langle F\rangle}{dt} = \frac{i}{\hbar}\langle[\hat{H}, \hat{F}]\rangle \tag{2-81}$$

The proof is as follows: Expanding $\langle[\hat{H}, \hat{F}]\rangle$ and using $\hat{H}\psi = E\psi$ along with the turnover rule, we get

$$\begin{aligned}\langle[\hat{H}, \hat{F}]\rangle &= \langle\psi|\hat{H}\hat{F} - \hat{F}\hat{H}|\psi\rangle = \langle\psi|\hat{H}\hat{F}|\psi\rangle - \langle\psi|\hat{F}\hat{H}|\psi\rangle \\ &= \langle\hat{H}\psi|\hat{F}|\psi\rangle - \langle\psi\hat{F}\hat{H}|\psi\rangle = E\langle\psi|\hat{F}|\psi\rangle - E\langle\psi|\hat{F}|\psi\rangle \\ &= 0 \quad \text{QED}\end{aligned}$$

Equation (2-81) is called the *hypervirial theorem*.[27] This theorem states that the expectation values of time-independent operators do not vary with time in stationary states.

Now consider the special case where the operator $\hat{F}$ is given by

$$\hat{F} = \hat{\mathbf{r}} \cdot \hat{\mathbf{p}} = \frac{\hbar}{i}\left(x\frac{\partial}{\partial x} + y\frac{\partial}{\partial y} + z\frac{\partial}{\partial z}\right) \tag{2-82}$$

Letting $\hat{H} = -(\hbar^2/2m)\nabla^2 + V(r)$, the commutator of $\hat{H}$ and $\hat{F}$ is

$$[\hat{H}, \hat{F}] = \frac{\hbar}{i}\left(-\frac{\hbar^2}{m}\nabla^2 - \hat{\mathbf{r}}\cdot\nabla\right) = \frac{\hbar}{i}(2\hat{T} - \mathbf{r}\cdot\nabla) \tag{2-83}$$

The expectation value of the commutator then is

$$\frac{\hbar}{i}(2\langle\hat{T}\rangle - \langle\hat{\mathbf{r}}\cdot\nabla\rangle) = 0$$

which may be rewritten as

$$2\langle\hat{T}\rangle = \langle(\hat{\mathbf{r}}\cdot\nabla)\rangle \tag{2-84}$$

This latter relationship is called the *quantum mechanical virial theorem*.[28] If the potential energy is of the general form[29]

$$V(r) = kr^n \tag{2-85}$$

where k is a constant, then Eq. (2-84) becomes

$$2\langle T\rangle = n\langle V\rangle \tag{2.86}$$

This is one of the most useful forms of the virial theorem. The theorem now states that if all the coordinates of the system are multiplied by some arbitrary factor β, the potential energy will be increased by a factor βn.

In a system of coulombically interacting charged particles, all potential-energy terms contain $1/r_{ij}$ (r_{ij} is the distance between two charged particles i and j), so that $n = -1$ and the virial theorem becomes

$$2\langle\hat{T}\rangle = -\langle\hat{V}\rangle \tag{2-87}$$

This result can be used to prove a theorem which will be useful later in the text.

THEOREM 2-7. The formation of a stable molecule from the elements in their ground states is characterized by an increase in kinetic energy and a decrease in potential energy.

[27] J. O. Hirschfelder, *J. Chem. Phys.*, **33**:1472 (1960); J. O. Hirschfelder and C. A. Coulson, *J. Chem. Phys.*, **36**:941 (1962).

[28] An analogous virial theorem exists in classical mechanics.

[29] This potential is said to be *homogeneous in r of degree n*.

Proof. The change in total energy ΔE for the process of atoms forming molecules is negative for a stable molecule ($-\Delta E$ is the dissociation energy of the molecule). Thus, $\Delta E = \Delta T + \Delta V < 0$. By the virial theorem [Eq. (2-87)], $\Delta T = -\Delta V/2$, and we can write

$$-\frac{\Delta V}{2} + \Delta V = \frac{\Delta V}{2} < 0$$

Thus $\Delta V < 0$, and since $\Delta T = -\Delta V/2$, $\Delta T > 0$. QED.

The decrease in V means that the electrons are more tightly bound to many nuclei than to single nuclei. The increase in T implies that the electrons undergo more complicated motions about multinuclear positive centers than they do about individual positive centers (nuclei).

EXERCISES

2-56. Prove Eq. (2-86). Note that $\mathbf{r} \cdot \mathbf{r} = r^2 = x^2 + y^2 + z^2$.

2-57. Develop the virial theorem for a system with potential energy given by $V(r) = a/r^{12} - b/r^6$, where a and b are constants. This is called a *Lennard-Jones potential* and is used in describing intermolecular forces.

2-58. If E_0 is the energy of two isolated atoms, the energy of the two atoms a finite distance apart r may be approximated by $E = E_0 - b/r^6$, where b is a positive constant. The term "$-b/r^6$" is known as a *van der Waals potential*. Interpret the van der Waals attraction between a pair of atoms in terms of increases and decreases in the kinetic and potential energies.

2-59. How are the average kinetic and potential energies related in a one-dimensional system in which $V = kx^2/2$ (k is a constant)? The force due to this potential ($\mathbf{F} = -k\mathbf{x}$) is called a *Hooke's law force* (the restoring force is proportional to the displacement).

SUGGESTED READINGS

Courant, R., and D. Hilbert: *Methods of Mathematical Physics*, vol. 1, Interscience Publishers, New York, 1953. Contains virtually all the mathematical ideas needed to master quantum theory.

Dirac, P. A. M.: *Quantum Mechanics*, Oxford University Press, Fair Lawn, N.J., 1958. Has an excellent section on bra and ket vectors.

Margenau, H., and G. M. Murphy: *Mathematics of Physics and Chemistry*, Van Nostrand, Princeton, N.J., 1956. Has several sections dealing with quantum mechanics and the mathematical ideas needed to understand it.

Tomonaga, S.: *Quantum Mechanics*, vol. II, North-Holland, Amsterdam, 1962. Has some novel approaches to many aspects of modern quantum theory.

CHAPTER 3

THE QUANTUM MECHANICS OF SOME SIMPLE SYSTEMS

Beginning students often find it difficult to assimilate new abstract concepts when they are presented in an equally unfamiliar mathematical formalism; the struggle to master the *physics* as well as the *mathematics* simultaneously can be formidable. In the present chapter we will consider some of the most important concepts and results of quantum mechanics—all within the framework of relatively simple mathematics. Later, when we encounter some of these same concepts in more mathematically complex systems, we can draw simple analogies to these simpler systems to facilitate understanding.

Since we will frequently need to consider the transition of a system from one stationary state to another (a time-dependent process), the first section is a brief digression into some elementary aspects of such transitions. In particular, we will consider the mechanism by which atoms and molecules interact with electromagnetic radiation and the rules which determine which quantum states participate in the interaction.

3-1 INTERACTION OF MATTER AND RADIATION

Figure 3-1 represents an energy-level diagram of a system which can exist in a number of different stationary quantum states. Three of these states are given the labels k, l, and m, and the corresponding energies are arbitrarily ordered

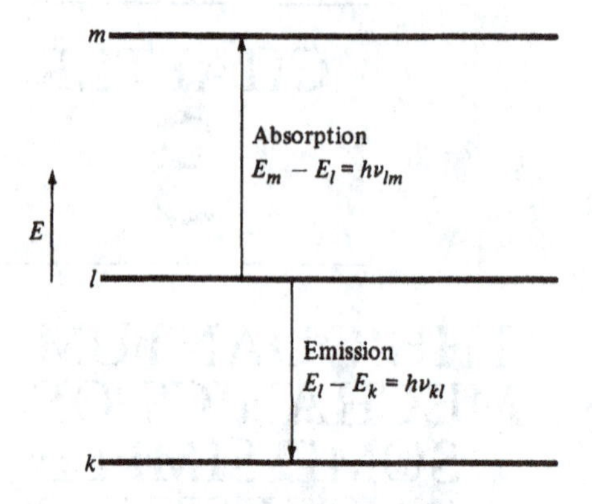

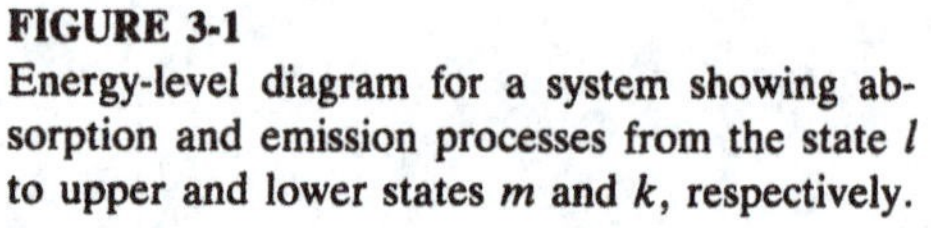
FIGURE 3-1
Energy-level diagram for a system showing absorption and emission processes from the state l to upper and lower states m and k, respectively.

$E_k < E_l < E_m$. The fundamental relationships governing transitions between these states were first developed by Einstein.[1]

Consider a collection of a large number of initially identical systems, e.g., individual atoms or molecules. Each system is assumed to be capable of acquiring energy from an external electromagnetic field so as to pass to a higher energy state, or, alternatively, of emitting energy as electromagnetic radiation so as to pass to a lower energy state. In the emission process, the frequency of the emitted radiation involving states l and k is given by the Planck-Bohr relationship

$$\nu_{kl} = \frac{E_l - E_k}{h} \tag{3-1}$$

If there are N_l systems in the state l at the time t, the time-rate of change of the population as a result of the energy emission is given by the first-order kinetics expression

$$-\frac{dN_l}{dt} = -\dot{N}_l = N_l \sum_k A_{l\to k} \tag{3-2}$$

where $A_{l\to k}$ represents the probability that a system in the state l will spontaneously emit energy and pass to a lower state k. The reciprocal of the mean lifetime τ_l of the level l is defined by

$$\frac{1}{\tau_l} = \sum_k A_{l\to k} \tag{3-3}$$

Thus, Eq. (3-2) may be rewritten as

$$-\dot{N}_l = \frac{N_l}{\tau_l} \tag{3-4}$$

[1] A. Einstein, *Verh. d. Deutsch. Phys. Ges.*, **18**:318 (1916) and *Phys. Z.*, **18**:121 (1917).

Integrating this latter equation and converting the solution to the exponential form produces

$$N_l = N_0 e^{-t/\tau_l} \tag{3-5}$$

where N_0 is the system population at $t = 0$. Note that it has been assumed that no other processes are occurring to change the population of the state l. If we restrict ourselves to a particular transition from the state l to a single particular state k, the transition rate [Eq. (3-2)] becomes simply

$$-\dot{N}_l = N_l A_{l \to k} \tag{3-6}$$

Now consider a different process: the interaction of a system in the state l with electromagnetic radiation of frequency

$$\nu_{lm} = \frac{E_m - E_l}{h} \tag{3-7}$$

which *induces* a transition to a higher state m. The rate of this induced transition is given by the second-order expression

$$-\dot{N}_l = B_{l \to m} N_l \rho(\nu_{lm}) \tag{3-8}$$

where $\rho(\nu_{lm})$ is the radiation density (units of energy $\times$ time $\times$ volume^{-1}) due to radiation of frequency ν_{lm}, and $B_{l \to m}$ is known as the *Einstein transition-probability coefficient of induced absorption.* Simultaneously, the radiation of frequency ν_{lm} also induces the *emission* of radiation as the state m undergoes a transition back to the state l.[2] The rate of depopulation of the state m due to the induced emission is

$$-\dot{N}_m = B_{m \to l} \rho(\nu_{lm}) \tag{3-9}$$

where $B_{m \to l}$ is called the *Einstein transition-probability coefficient of induced emission.*[3]

The three transition-probability coefficients, $A_{l \to k}$, $B_{l \to m}$, and $B_{m \to l}$, are fundamental measures of the interaction of matter with electromagnetic radiation and are independent of whether or not the systems are in a state of thermodynamic equilibrium. However, in the special case when thermodynamic equilibrium does exist, the following conditions must be satisfied:

---

[2] In classical physics an oscillator interacting with electromagnetic waves could either absorb energy from the field or lose energy to it, depending on the relative phases of the oscillator and waves. Thus, the probability of emission involves two parts: one independent of the radiation density and the other proportional to it.

[3] This coefficient is of interest in the field of lasers. For a short introduction to this topic, see W. F. Coleman, *J. Chem. Ed.*, **59**:441 (1982).

1. The relative numbers of systems in the various stationary states must satisfy the Maxwell-Boltzmann distribution law
$$\frac{N_i}{N_j} = \frac{g_i}{g_j} \exp\left(-\frac{E_i - E_j}{kT}\right) = \frac{g_i}{g_j} \exp\left(-\frac{h\nu_{ij}}{kT}\right) \tag{3-10}$$
where g_i and g_j are the degrees of degeneracy of the states having the energies E_i and E_j, respectively.
2. The radiation density is given by Planck's radiation equation
$$\rho_\nu \, d\nu = \frac{8\pi h\nu^3}{c^3} \left[\exp\left(\frac{h\nu}{kT}\right) - 1\right]^{-1} \tag{3-11}$$
where ν (with subscripts i and j omitted for notational clarity) is the Planck-Bohr frequency for a quantum transition between two stationary states of energies E_i and E_j.
3. The sum of the rates of the two emission processes (induced and spontaneous) must equal the rate of the induced absorption process. For transitions involving only the two states k and l, this condition becomes
$$N_l(A_{l\to k} + B_{l\to k}\rho_\nu) = N_k B_{k\to l}\rho_\nu \tag{3-12}$$

For the special cases of nondegenerate energy states ($g_i = g_j = 1$ for all i, j), the Maxwell-Boltzmann law [Eqs. (3-10) and (3-12)] lead to
$$\frac{N_k}{N_l} = \exp\left(\frac{E_l - E_k}{kT}\right) = \exp\left(-\frac{h\nu}{kT}\right) = \frac{A_{l\to k} + B_{l\to k}\rho_\nu}{B_{k\to l}\rho_\nu} \tag{3-13}$$
Solving Eq. (3-13) for the radiation density, one obtains
$$\rho_\nu = \frac{A_{l\to k}}{B_{k\to l}\exp(h\nu/kT) - B_{l\to k}} \tag{3-14}$$
In order for Eq. (3-14) to be identical with the Planck radiation law [Eq. (3-11)], it is necessary to assume the correspondences
$$B_{k\to l} = B_{l\to k} \qquad \frac{A_{l\to k}}{B_{l\to k}} = \frac{8\pi h\nu^3}{c^3} \tag{3-15}$$
Consequently, the Einstein A and B coefficients are related as follows:
$$A_{l\to k} = \frac{8\pi h\nu^3}{c^3} B_{l\to k} \tag{3-16}$$

The preceding equations show that the rate of transition from one state to another depends primarily on the Einstein coefficient $B_{k\to l}$ (which equals $B_{l\to k}$). Whenever this coefficient is identically zero, we say the transition is *forbidden*; otherwise, we say the transition is *allowed*. For allowed transitions, the magnitude of $B_{k\to l}$ determines the intensity of the emission or absorption of radiation accompanying the transition. Hence, it is apparent that a more detailed knowledge of $B_{k\to l}$ is needed if we are to describe and understand quantum transitions. Fortunately, the time-dependent Schrödinger equation allows us to describe the kinetics of quantum transitions and provides a detailed explanation of just what $B_{k \to l}$ is. For the sake of brevity we will consider a highly condensed account of this analysis; the interested reader should consult various excellent spectroscopy texts for more complete treatments.[4]

Since the coefficient for induced emission is equal to that for induced absorption, the two processes will be equal whenever the two states k and l contain the same number of systems; hence, no net absorption of energy will occur. However, we are usually concerned with the situation where lower states are much more populated than upper states so that net absorption is observed. Also, regarding spontaneous emission, this process has a much lower probability than does the induced process and is usually ignored.

Limiting our discussion to induced absorption from a lower state k to a higher state l, these two states (in the absence of electromagnetic radiation) satisfy the time-dependent Schrödinger equation

$$\hat{H}^{(0)}\Psi_j = -\frac{\hbar}{i}\frac{\partial}{\partial t}\Psi_j \qquad j = k \text{ or } l \tag{3-17}$$

where the (0) superscript on $\hat{H}$ denotes the absence of an electromagnetic field, and where

$$\Psi_j(x, y, z, t) = \psi_j(x, y, z)e^{-iE_jt/\hbar} \tag{3-18}$$

For notational simplicity we will limit the remaining discussion to the single dimension x. During the passage of the state k to the state l under the influence of radiation, the wavefunction of the system is given by the linear combination

$$\Psi(x, t) = a_k\Psi_k(x, t) + a_l\Psi_l(x, t) \tag{3-19}$$

where the coefficients a_k and a_l depend on the time. We see that when $t = 0$ (before the radiation field is turned on), $a_k = 1$ and $a_l = 0$. Similarly, after the transition is complete, $a_k = 0$ and $a_l = 1$. The wavefunction (3-19) satisfies the Schrödinger equation

$$\hat{H}\Psi(x, t) = -\frac{\hbar}{i}\frac{\partial\Psi(x, t)}{\partial t} \tag{3-20}$$

where $\hat{H}$ (without a superscript) is the hamiltonian of the system in the presence of the radiation field.[5]

Skipping over the details beyond this point, one can eventually derive an expression for the rate at which the system changes from state k to state l as a

[4] For example, the text by Barrow (see the readings at the end of this chapter), pp. 61–69, presents a fairly short but adequate treatment.

[5] $\hat{H}$ is just $\hat{H}^{(0)}$ plus a term accounting for the presence of a radiation field.

result of the absorption of radiation. In the special case that it is the *electric field* of the radiation[6] that interacts with the system, one obtains

$$\frac{d(a_l^* a_l)}{dt} = B_{k\to l}\rho_\nu \tag{3-21}$$

where

$$B_{k\to l} = \frac{2\pi}{3\hbar^2}|\langle\psi_l|e\mathbf{r}|\psi_k\rangle|^2 \tag{3-22}$$

The integral whose absolute value appears in the above is often given the symbol $\boldsymbol{\mu}_{kl}$, so that

$$\boldsymbol{\mu}_{kl} = \langle\psi_l|e\mathbf{r}|\psi_k\rangle = e\langle\psi_l|\mathbf{r}|\psi_k\rangle$$

or

$$|\boldsymbol{\mu}_{kl}|^2 = e^2(\langle\psi_1|x|\psi_k\rangle^2 + \langle\psi_l|y|\psi_k\rangle^2 + \langle\psi_l|z|\psi_k\rangle^2) \tag{3-23}$$

The quantity $|\boldsymbol{\mu}_{kl}|$ (more simply written as μ_{kl}) is called the *transition moment* or *dipole strength* of the transition; $\boldsymbol{\mu}_{kl}$ itself is a vector quantity whose direction is that of the polarization of the absorbed radiation. It is evident that the transition $k \to l$ can occur (have a nonzero rate) only if at least one component of this vector is nonzero. Thus, the transition moment is of primary importance in governing which transitions are allowed and which are forbidden. The conditions which must be met in order for the integral $\langle\psi_l|\mathbf{r}|\psi_k\rangle$ to be nonzero will be illustrated later for several specific cases; these conditions constitute what we call *selection rules*.

The interaction of matter and radiation just discussed is of a special type known as *electric dipole interaction*; this consists of a coupling between the oscillating electric field associated with the radiation and an oscillating electric dipole associated with the matter. All atoms and molecules possess a high-frequency oscillating electric dipole due to very rapid fluctuations in the negative electronic distribution relative to the positive nucleus or nuclei. Such oscillations can couple with the electric field of high-frequency electromagnetic radiation in the ultraviolet or visible regions (ν is typically larger than about 10^{14} s^{-1}). Another type of coupling arises when a molecule with a permanent dipole moment (e.g., HCl) undergoes vibrational motion; the effect of the vibrations is to set up an oscillation in the permanent dipole moment (see Fig. 3-2). Such vibrational oscillations are slower than electron density oscillations and, consequently, couple with lower-frequency radiation—in this case with radiation typically in the infrared (ν is typically around 10^{12} to 10^{13} s^{-1}). Molecules without a permanent dipole moment (provided they have at least

[6] Electromagnetic radiation has both an oscillating electric field and an oscillating magnetic field associated with it. The oscillations are at right angles to each other, and both are at right angles to the direction of propagation.

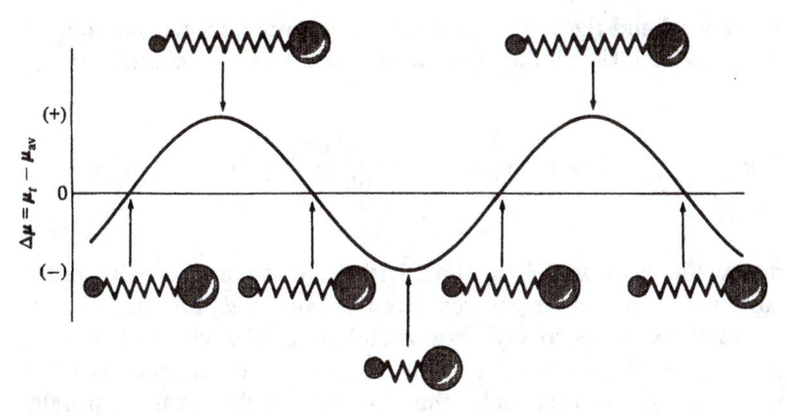

FIGURE 3-2
The oscillatory nature of the electric dipole moment of a heteronuclear diatomic molecule as a consequence of vibration.

three atoms) can also interact in this fashion. Some of the vibrational motions of such molecules create transient dipoles which average out to zero in a sufficiently long time period but which are nevertheless associated with oscillations of the right frequency to couple with infrared radiation. A third mechanism for interaction occurs when a molecule with a permanent dipole moment rotates and, thereby, behaves like an oscillating electric dipole (see Fig. 3-3). These oscillations are generally slower than those of vibration and can couple with radiation in the far infrared or microwave regions (ν is typically in the range of 10^{11} to 10^{10} s^{-1}). Later, we will examine specific examples of each of the above types of couplings.

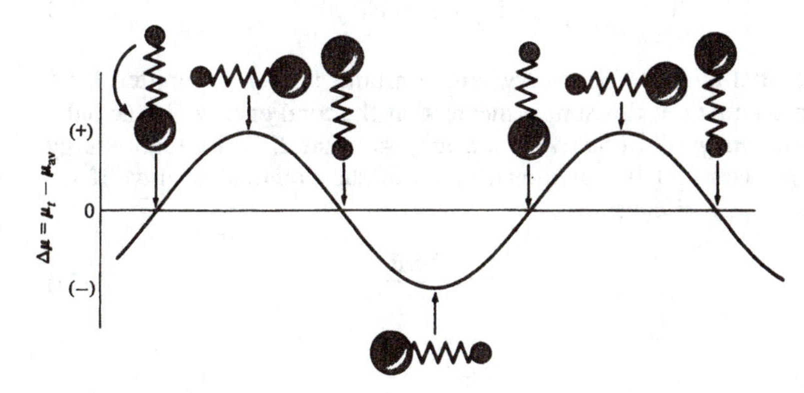

FIGURE 3-3
The oscillatory nature of the electric dipole moment of a heteronuclear diatomic molecule as a consequence of rotation. The molecule is also vibrating; the atom-atom separations shown are the average ones.

In a more general and rigorous treatment, where interaction is not limited to that involving an electric dipole mechanism, the Einstein coefficient becomes

$$B_{k\to l} = \frac{2\pi}{3\hbar^2}\left(|\boldsymbol{\mu}_{kl}|^2 + \frac{e}{2mc}\,|\langle\psi_l|\mathbf{r}\times\mathbf{p}|\psi_k\rangle|^2 + \frac{3e^2\pi^3\nu^2}{10c^2}\,|\langle\psi_l|\mathbf{r}\rangle\langle\mathbf{r}|\psi_k\rangle|^2 + \cdots\right) \tag{3-24}$$

The first term is the portion of the coefficient due to an electric dipole mechanism, and the next two terms represent *magnetic dipole* and *electric quadrupole* interactions, respectively. For radiation in the visible spectrum (400 to 700 nm), these three terms typically have relative magnitudes of $1 : 10^{-5} : 10^{-7}$, respectively, so that only the electric dipole term is usually significant. However, in systems of sufficient symmetry, μ_{kl} may vanish, and the remaining terms may become important.

EXERCISES

3-1. Verify Eqs. (3-15) and (3-16).

3-2. Show in detail how Eq. (3-23) follows from the immediately preceding equation.

3.2 THE FREE PARTICLE

A free particle is a particle moving in a conservative, uniform potential field with absolutely no coordinate restrictions on its position.[7] If we restrict ourselves to the simple one-dimensional case, the Schrödinger equation of such a particle is given by

$$\left(-\frac{\hbar^2}{2m}\frac{d^2}{dx^2} + V\right)\psi(x) = E\psi(x) \tag{3-25}$$

Since the potential energy V is everywhere constant, it may be omitted from the Schrödinger equation; this simply means that the total energy E is actually only the *kinetic energy*. Alternatively, we may say that E is the total energy *relative* to some constant but unknown value of the potential energy. If we define the positive, real constant

$$k^2 = \frac{2mE}{\hbar^2} \tag{3-26}$$

[7] A crude analogy to a one-dimensional free particle is a bead moving without friction along a wire of infinite length. A constant potential means there is no tendency for the bead to favor any particular position along the wire; i.e., there are no kinks to trap the bead or hinder its motion.

the Schrödinger equation (3-25) may be rewritten as

$$\frac{d^2\psi}{dx^2} + k^2\psi = 0 \tag{3-27}$$

This is a linear, second-order differential equation with a constant coefficient. A convenient form of the general solution is given by

$$\psi = A \cos kx + B \sin kx \tag{3-28}$$

where A and B are as yet undetermined constants. Since there are no constraints or restrictions on the constant k (other than that it be real), it may have any value whatsoever, and, thus, the energy E may have any positive value. This means that the energy of the free particle is *not* quantized—the only system for which this is true. Obviously, this situation is like that generally assumed to be true of classical particles—the energy can have any real value whatsoever.

The wavefunction of Eq. (3-28) is not quadratically integrable in the usual sense. It is readily verified that

$$\int \psi^* \psi \, dx = \infty \tag{3-29}$$

(where the integration is from $x = -\infty$ to $x = +\infty$). Nevertheless, we know that the total probability of finding the particle must be finite and, consequently, ψ must be normalizable in some fashion. Consider, then, a beam of identical, noninteracting particles having a linear density of n_0 particles per unit length L; where L is chosen to be large compared with the de Broglie wavelength of a particle. This restriction on L ensures that normalization covers a sufficiently large portion of space to be representative of the entire system and also ensures that we do not localize the particle so much that we render its momentum uncertain. Since the particle can be in any of an infinite number of different segments of length L, the above restriction does not confine the particle in any way. The normalization condition now becomes

$$\int_x^{x+L} \psi^* \psi \, dx = n_0 L \tag{3-30}$$

Upon substitution of Eq. (3-28) for ψ and integration, we obtain

$$|A|^2 + |B|^2 = 2n_0 \tag{3-31}$$

We can arbitrarily set either A or B equal to zero and choose one of the solutions

$$\psi(x) = \sqrt{2n_0} \sin kx \qquad \text{or} \qquad \sqrt{2n_0} \cos kx \tag{3-32}$$

Either of these solutions represents a superposition of two beams of equal intensity moving in opposite directions (see Exercises 3-3 and 3-4) such that the *net* current is zero.

An alternative form for the wavefunction ψ is the complex exponential

$$\psi(x) = A'e^{ikx} + B'e^{-ikx} \tag{3-33}$$

where

$$|A'|^2 + |B'|^2 = n_0 \tag{3-34}$$

The first function represents a beam of particles moving in the $+x$ direction (with a phase $\alpha = kx$), and the second represents an equal flow in the opposite direction. Again, if either A' or B' is chosen as zero, one obtains the solutions

$$\psi(x) = \sqrt{n_0}\, e^{ikx} \qquad \text{or} \qquad \sqrt{n_0}e^{-ikx} \tag{3-35}$$

EXERCISES

3-3. Using either solution in Eq. (3-32), show that the expectation value of the linear momentum is zero. Explain why this is so.

3-4. Using both the solutions in Eq. (3-35), calculate the expectation value of the linear momentum and compare it with the results of the previous exercise. Why are the results different?

3-5. Show that the probability density $|\psi|^2$ implied by either solution in Eq. (3-32) is sinusoidal whereas the densities implied by either solution in Eq. (3-35) are independent of x (i.e., constant everywhere). Explain, noting that the former is a standing wave formed by interference of oppositely directed de Broglie waves.

3-6. Show how the solutions in Eqs. (3-28) and (3-33) are related mathematically. *Note*: Use Euler's relationship $e^{\pm ix} = \cos x \pm i \sin x$.

3-7. What kind of emission spectrum (in general terms) would you expect for a system of free particles?

3-8. (*a*) Show that $\hat{p}_x$ and $\hat{H}$ commute for the free particle, and identify the common eigenfunctions.

(*b*) What are the uncertainties in x and p_x for the free particle? Do these satisfy the uncertainty principle? Explain.

3-3 THE PARTICLE IN A BOX

If the particle in the preceding section is constrained to remain in a *finite* region of space defined by $0 \le x \le L$ (where L is a finite length), then the system is referred to as a *particle in a box* (PIB). This system serves as a simple model of several real systems of physical interest: the translational motion of ideal gas molecules,[8] electrons in metals, and pi electrons in conjugated hydrocarbons and related molecules. Since the PIB model is mathematically simple, it can be used to provide illustrations of many important quantum mechanical concepts without obscuring the principles with mathematical details. No other quantum

---

[8] The free particle can also serve as a model for this system if the box has a very large length.

mechanical system is capable of providing so much information with so little mathematical manipulation.

The Schrödinger equation for the PIB is the same as for the free particle [Eq. (3-27)] if we assume that the potential energy inside the box ($0 \leq x \leq L$) is everywhere the same. To ensure that the particle remains confined to the box, we set the potential energy equal to infinity outside the box ($x < 0$ and $x > L$). Thus we can write three *boundary conditions* that the wavefunction must satisfy:

$$\begin{aligned} &\textbf{1.}\ \psi(x) = 0 \text{ for } x < 0 \text{ and } x > L \\ &\textbf{2.}\ \psi(0) = 0 \\ &\textbf{3.}\ \psi(L) = 0 \end{aligned} \tag{3-36}$$

Furthermore, in general $\psi(x) \neq 0$ for $0 < x < L$. Conditions 2 and 3 ensure that the total wavefunction is continuous in the interval $-\infty$ to $+\infty$. Condition 2 forces us to set the constant A in Eq. (3-28) equal to zero, since the cosine function does not vanish at $x = 0$. To make the remaining part of ψ (the sine function) vanish at the other end of the box ($x = L$), we require that the argument kL be equal to a multiple of 180° (π rad), since the sine function periodically vanishes at intervals of 180°. This condition may be written

$$kL = n\pi \qquad n = 1, 2, 3, \ldots \tag{3-37}$$

Then, solving for k:

$$k = \frac{n\pi}{L} \tag{3-38}$$

Thus, the well-behaved solutions of Schrödinger's equation for the PIB are given by

$$\psi_n(x) = B \sin \frac{n\pi x}{L} \tag{3-39}$$

The constant B is obtained by normalizing ψ_n:[9]

$$B = \left(\int_0^L \sin^2 \frac{n\pi x}{L}\, dx \right)^{-1/2} = \left(\frac{2}{L} \right)^{1/2} \tag{3-40}$$

Thus, the normalized PIB wavefunction is

$$\psi_n(x) = \left(\frac{2}{L} \right)^{1/2} \sin \frac{n\pi x}{L} \tag{3-41}$$

[9] The integral involving $\sin^2 u$ may be found in standard tables, or alternatively, one may use the trigonometric identity: $\cos 2u = 1 - 2\sin^2 u$. Or one may simply do two successive integrations by parts. Even more simply: $\int (\cos^2 u + \sin^2 u)\, du = \int du = b$; therefore, $\int \sin^2 u\, du = b/2$. (The interval of integration in each case is 0 to b.)

Figure 3-4 shows the first three wavefunctions and their squares (probability densities) for the PIB. Note that both ψ_n and ψ_n^2 (equal to $\psi_n^* \psi_n$, since ψ_n is real) have $n-1$ nodes (excluding those at the ends). For $n=1$ the particle has its maximum probability density at $x = L/2$. As $n \to \infty$, the probability density tends to approach a constant value and thus to be independent of x.

The condition $kL = n\pi$ also establishes the quantization of the particle's energy. If the value of k from Eq. (3-38) is substituted into Eq. (3-26), the energy of the PIB becomes

$$E_n = \frac{k^2\hbar^2}{2m} = \frac{n^2\pi^2\hbar^2}{2mL^2} = n^2 \frac{h^2}{8mL^2} \qquad (3\text{-}42)$$

The integers $n = 1, 2, 3, \ldots$ are the PIB *quantum numbers*, analogous to the quantum numbers which appear in the Bohr theory of the hydrogen atom [see Eq. (1-26)]. Note, however, that the PIB quantization does not have to be assumed a priori but arises naturally as one of the consequences of the fact that well-behaved solutions are required for the Schrödinger equation.

Equation (3-42) shows that the allowed energy levels are inversely proportional to the square of the box length. Thus, as the box becomes longer, the energies become smaller (for a given n) and they become more closely spaced. In the limit of an infinitely long box, the PIB reduces to a *free particle*, which has, of course, continuous (nonquantized) energies.[10]

[10] In statistical thermodynamic analysis of the translational motions of ideal gases, this is the justification for the argument that translational energy may be assumed to be continuous. For any macroscopic sample of gas (approximately 1 mol at STP), the size of the container is large enough so that $\Delta E \ll RT$.

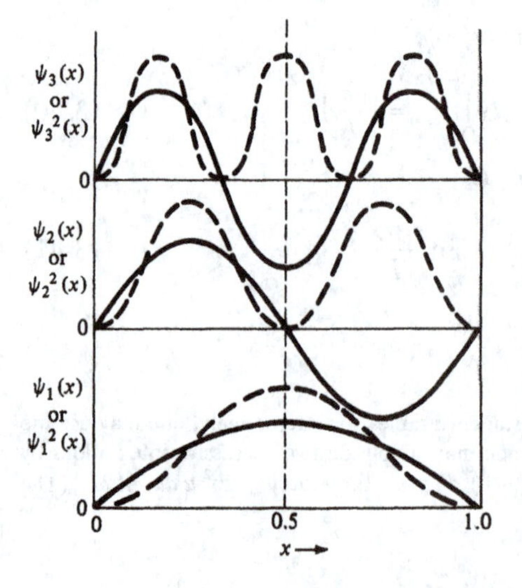

FIGURE 3-4
The first three wavefunctions of the particle in a box and their corresponding probability distribution functions. The wavefunctions are shown by solid lines; their squares are shown by dashed lines. For simplicity, L is set equal to unity.

The energy-level diagram for the PIB is shown in Fig. 3-5 for the first five quantum numbers. Note that the lowest possible energy the particle can have (the ground-state energy) is not zero but $h^2/8mL^2$. One may ask: Why would the lowest possible energy not be zero (i.e., the point at which the particle is not moving)? There are two important reasons for excluding a zero energy value. First, in order to have $E = 0$ (which means $n = 0$), it is necessary for ψ to be zero *everywhere in the box*, a trivial situation which implies *no state at all* (see Sec. 2-4). The second and nontrivial reason arises from the uncertainty principle. The best we can say about the location of the particle at any moment is that it is *someplace* in the box, that is, $\Delta x = L$. Since $E = 0$ implies $p_x = 0$, then $\Delta p_x = 0$ and the product $\Delta x \Delta p_x = 0$ violates the uncertainty principle. We see that quantum mechanics dooms the PIB never to rest![11]

The lowest possible energy, $E_1 = h^2/8mL^2$, is an example of a *zero-point energy*. Any system in thermodynamic equilibrium with its environment at a temperature approaching absolute zero would have a very high probability of being in its zero-point stationary state [see Eq. (3-10)]. Nonzero values for zero-point energies are characteristic of systems in which a particle is restricted to a finite volume of space. Note that the free particle can have $E = 0$ without violating the uncertainty principle, since $\Delta x = \infty$.

The PIB model is readily extendable to two or three dimensions. For the two-dimensional case, the Schrödinger equation is

$$\frac{-\hbar^2}{2m}\left(\frac{\partial^2}{\partial x^2} + \frac{\partial^2}{\partial y^2}\right)\psi(x, y) = E\psi(x, y) \tag{3-43}$$

Solution by the separation of variables technique is straightforward.[12] The

[11] Classical mechanics permits $E = 0$ as the lowest possible energy of the PIB; i.e., the particle can be at rest.

[12] The total energy E must be regarded as the sum of two parts: one part from each dimension.

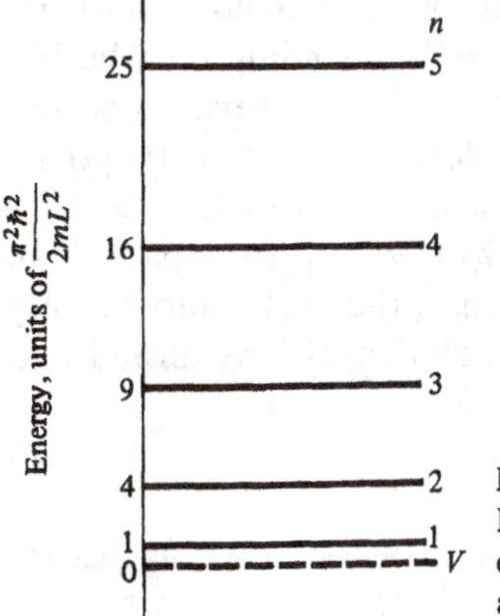

FIGURE 3-5
Energy-level diagram for the particle in a box. Note that the energy is *directly* proportional to n^2, whereas in the hydrogen atom (Bohr model) the energy is *inversely* proportional to n^2.

wavefunctions have the form

$$\psi_{n,m}(x, y) = \psi_n(x)\psi_m(y) = 2\left(\frac{1}{L_1L_2}\right)^{1/2} \sin\frac{n\pi x}{L_1} \sin\frac{m\pi y}{L_2} \qquad (3\text{-}44)$$

where L_1 and L_2 are the box lengths in the x and y directions, respectively, and n and m are the respective quantum numbers. The total energy is given by

$$E_{n,m} = E_n + E_m = \left(\frac{n^2}{L_1^2} + \frac{m^2}{L_2^2}\right)\frac{\pi^2\hbar^2}{2m} \qquad (3\text{-}45)$$

For the special case of a *square* box ($L_1 = L_2 = L$), the total energy is

$$E_{n,m} = (n^2 + m^2)\frac{\pi^2\hbar^2}{2mL^2} \qquad (3\text{-}46)$$

We can now show that degenerate solutions will arise whenever different combinations of n and m lead to the same value of the sum $n^2 + m^2$. For example,

$$E_{1,2} = E_{2,1} = (1+4)\frac{\pi^2\hbar^2}{2mL^2} = (4+1)\frac{\pi^2\hbar^2}{2mL^2} = \frac{5\pi^2\hbar^2}{2mL^2} \qquad (3\text{-}47)$$

This means that the states $\psi_1(x)\psi_2(y)$ and $\psi_2(x)\psi_1(y)$ (which are *different, distinct* states) are *doubly degenerate*. Furthermore, this degeneracy arises because of a basic *symmetry* of the system: the x and y directions are *indistinguishable*. As we will note later, symmetry and degeneracy are related in a fundamental way.[13]

The wavefunctions of the two-dimensional PIB can be represented as *surfaces*, i.e., as distortions of a rectangular planar surface as shown in Fig. 3-6. Thus the ground state ($n = m = 1$) is a "positive bulge" in the xy plane (Fig. 3-6*a*), and the $n = 2$, $m = 1$ state (Fig. 3-6*b*) is depicted as a positive bulge in one half of the plane and a corresponding negative bulge, or depression, in the other half of the plane. Note that the states $n = 1$, $m = 2$ and $n = 2$, $m = 1$ are degenerate (for the *square* box) and that their bulges are identical except for a symmetry rotation of 90°.

Figure 3-7 illustrates a simpler way of depicting the wavefunctions of the two-dimensional box. For a given state (n, m) the plane is divided into nm rectangles, and each rectangle is labeled "+" or "−" depending on whether the wavefunction is positive (positive bulge) or negative (depression) within that area. Boundaries between areas represent nodes, i.e., a locus of points along which the wavefunction changes sign and, thus, has a value of zero.

Figure 3-8 illustrates the probability density $[\psi_1(x)\psi_2(y)]^2$ for a particle in a two-dimensional box. Note that in a qualitative sense, the probability density looks just like the corresponding wavefunction with all depressions turned into

---

[13] Symmetry is a *necessary* but *insufficient* condition for degeneracy. The relationships are generally discussed in terms of *group theory*, a topic which is the subject of App. 6.

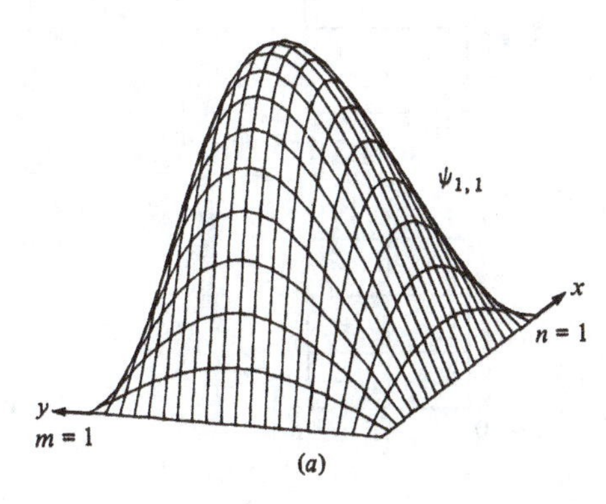

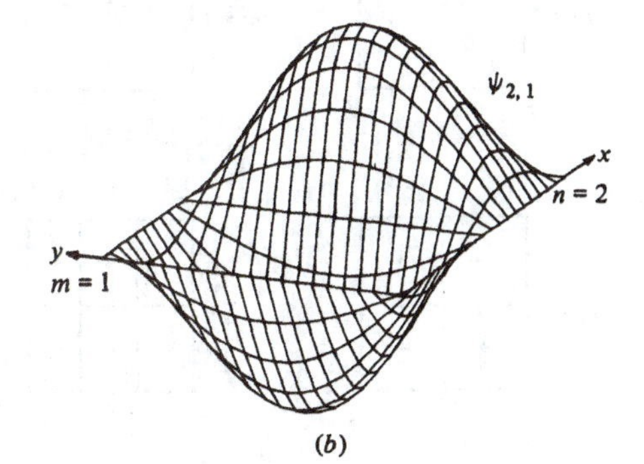

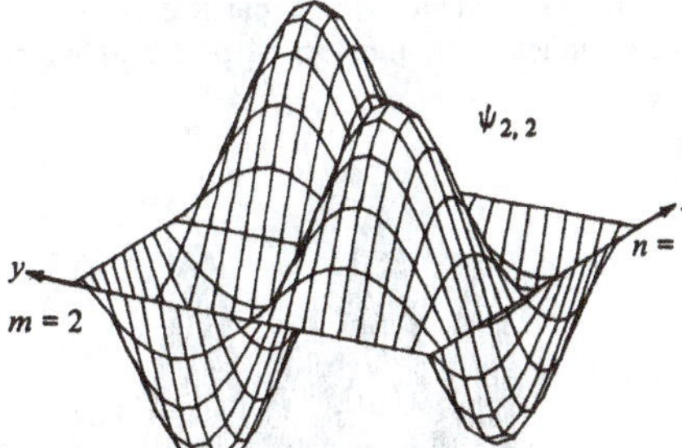

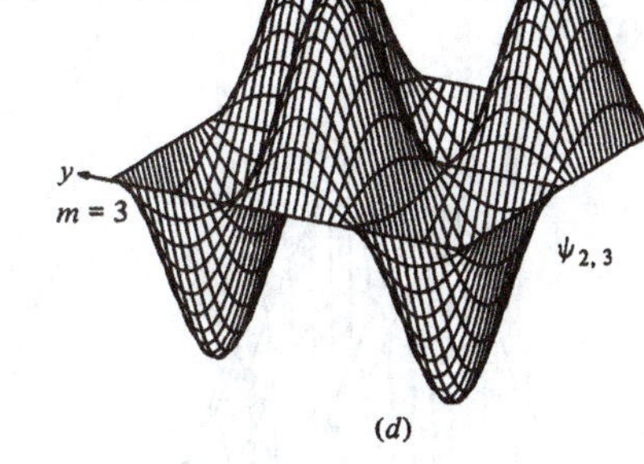

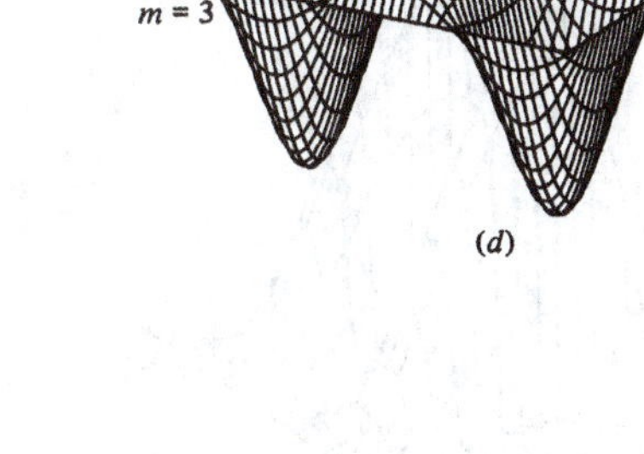

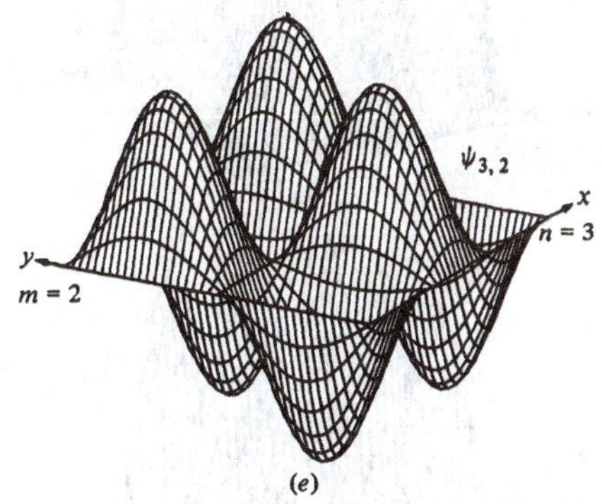

FIGURE 3-6
The wavefunctions of a particle in a two-dimensional square box. The height above the *xy* plane at which two grid lines intersect is the value of the wavefunction $\psi_{n,m}(x, y)$ at the point (x, y): (*a*) $n = m = 1$; (*b*) $n = 2$, $m = 1$; (*c*) $n = m = 2$; (*d*) $n = 2$, $m = 3$; (*e*) $n = 3$, $m = 2$.

"hills." It should also be noted that the "altitudes" of the intersection of any two grid lines above the *xy* plane represent the relative values of the probability density at a point *in the plane*. This very important point often eludes the beginning student of quantum mechanics. To represent the numerical value of a quantity associated with a system existing in n dimensions always requires $n + 1$ dimensions to illustrate it. Of course, we have only *two* dimensions

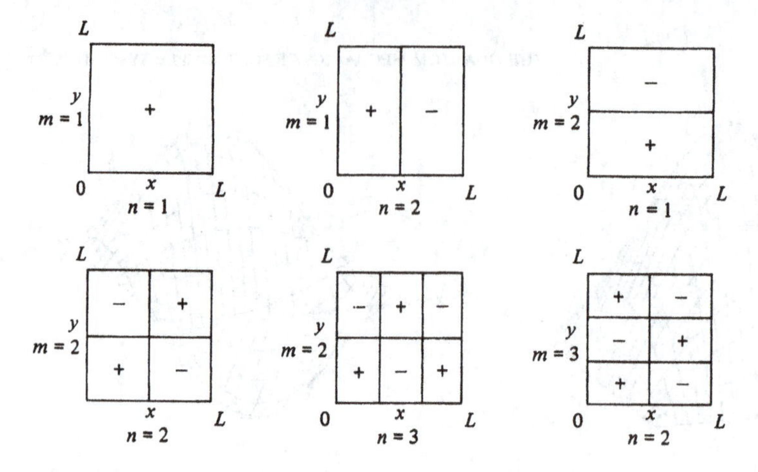

FIGURE 3-7
A simple representation of the nodal properties of the wavefunctions of a particle in a two-dimensional square box. The boundary lines represents nodes, i.e., the loci of points at which $\psi_{n,m}(x, y) = 0$.

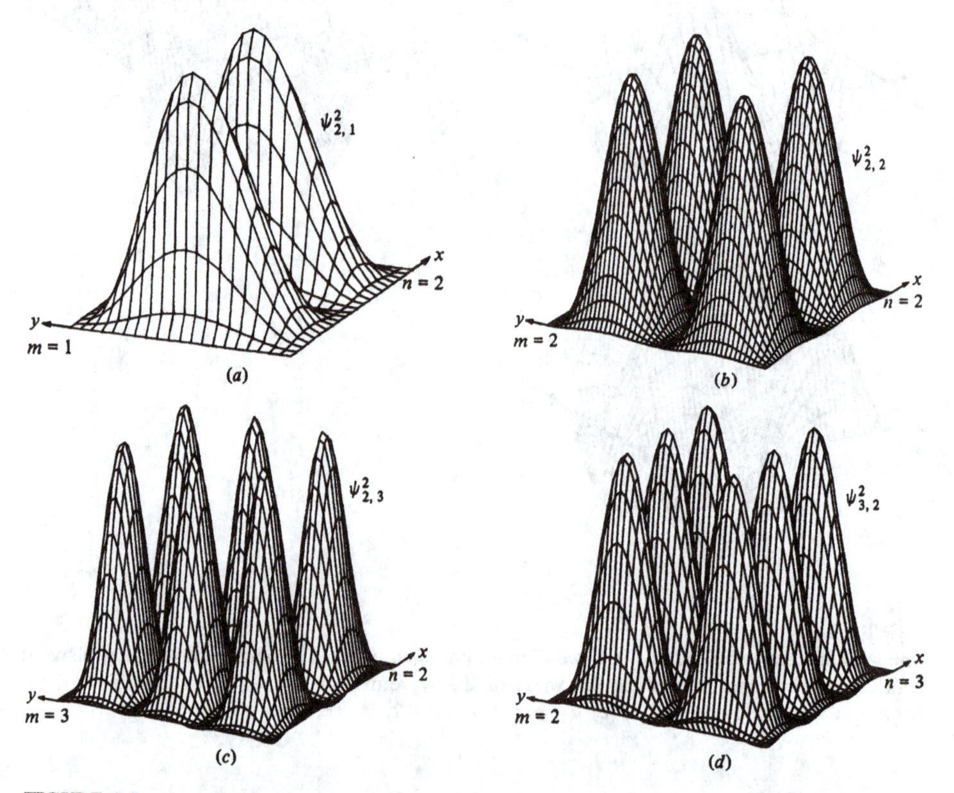

FIGURE 3-8
Probability densities for a particle in a two-dimensional square box. The height above the *xy* plane at which two grid lines intersect is the value of the probability density at the point (x, y) immediately below this intersection. For an "empty" state (no particle at all), all grid lines would be straight lines in the *xy* plane and all would intersect at 90°: (*a*) $n = 2$, $m = 1$; (*b*) $n = m = 2$; (*c*) $n = 2$, $m = 3$; (*d*) $n = 3$, $m = 2$.

available here, but a third is suggested by the use of grid lines. Thus, in attempting to illustrate probability densities for a *three-dimensional* PIB, the best one could so is to choose two-dimensional slices or cross sections of the box and show what the probability densities look like in that slice.

Referring again to the two-dimensional box, it is evident that each probability density diagram for a state (n, m) exhibits nm peaks, or hills.

EXERCISES

3-9. Show that the wavefunctions for the one-dimensional PIB are orthogonal. Is it necessary to demonstrate this explicitly?

3-10. In solving the PIB Schrödinger equation, how are the eigenfunctions and eigenvalues affected if the potential energy is not zero but rather has some finite, constant value V_0?

3-11. (*a*) Evaluate $\langle x \rangle$ and $\langle x^2 \rangle$ for the PIB.
(*b*) Why does the value for $\langle x^2 \rangle$ depend on the quantum number n?
(*c*) Why isn't $\langle x^2 \rangle$ equal to $\langle x \rangle^2$?
(*d*) Determine $\langle \hat{H} \rangle$ and $\langle \hat{H}^2 \rangle$ and compare the latter with $\langle \hat{H} \rangle^2$.
(*e*) Contrast the results in (*d*) with those in (*c*).

3-12. Classical mechanics predicts $P(x) = C$ (a constant) for the PIB. Show that this leads to $\langle x \rangle = L/2$ and $\langle x^2 \rangle = L^2/3$ for the PIB, and contrast this with the quantum theory results of Exercise 3-11*a*.

3-13. The Bohr correspondence principle states that quantum mechanics reduces to classical mechanics for large quantum numbers. Use the results of Exercises 3-11 and 3-12 to show that this is true for $\langle x^2 \rangle$.

3-14. (*a*) What is the probability of finding the PIB within each of the following intervals: 0 to $L/4$; 0 to $L/2$; $L/4$ to $L/2$?
(*b*) Repeat (*a*) for both the $n = 1$ and the $n = 2$ states.
(*c*) Repeat (*a*) for n approaching infinity.

3-15. The value of $\psi_n^2(x)$ at the point x can be regarded as a *relative* particle density. Calculate the relative particle densities for the $n = 1$ and $n = 2$ states at the points $L/4$, $L/2$, and $3L/4$.

3-16. What n quantum number would be needed to describe a particle 1 kg in mass moving with a velocity of $1\ \text{m} \cdot \text{s}^{-1}$ in a box 1 m in length? Would you be able to observe (detect) energy quantization in such a system? Explain.

3-17. A proton is in a one-dimensional box in the $n = 5$ state. When it drops down to the $n = 4$ state, it emits radiation with a wavelength of 2000 nm. What is the length of the box?

3-18. For a two-dimensional PIB, show that $\langle x^2 \rangle \neq \langle x \rangle^2$ and that $\langle y^2 \rangle \neq \langle y \rangle^2$, but that $\langle xy \rangle = \langle x \rangle \langle y \rangle$. Explain.

3-19. What are the x and y coordinates of the locations in a square box at which the probability densities are maxima for the $n = m = 2$ state?

3-20. Using Fig. 3-6 as a basis, sketch the general features of a flat *contour* diagram which represents each of the states shown.

3-21. Discuss the degeneracies possible in a three-dimensional cubic box.

3-22. Could degeneracies arise in a two-dimensional box in which L_1 and L_2 are simple multiples of each other, e.g., if $L_2 = 2L_1$? What type of symmetry is illustrated by this? (See Theorem A6-1, App. 6.)

3-23. Show that $\nabla\psi(x)$ is not continuous unless the box has a finite V at $x \le 0$ and $x \ge L$.

3-24. Redo the solution of the Schrödinger equation for the one-dimensional PIB, but now let $x = -L/2$ define the left-hand end of the box, $x = 0$ the center of the box, and $x = L/2$ the right-hand end of the box. You may find it easier to understand the results if you construct a sketch of the sine and cosine functions showing their basic symmetries about the vertical axis. It may also be illuminating to recall the trigonometric identities involving $\sin(x + y)$ and $\cos(x + y)$.

3-4 SPECTROSCOPY OF THE PARTICLE IN A BOX

Consider an electron moving along a one-dimensional, positively charged surface (a line) of length L and constant potential. The oscillatory motion of the electron relative to the center of positive charge produces an oscillating electric dipole which can undergo resonance coupling with electromagnetic radiation having a frequency comparable to that of the dipole oscillation. As shown in Sec. 3-1, this coupling leads to a transition between two stationary quantum states which we arbitrarily label k and l. The transition moment [Eq. (3-23)] for the one-dimensional PIB is

$$\mu_{kl} = e|\langle \psi_k|x|\psi_l\rangle| \tag{3-48}$$

The explicit form of the transition-moment integral is

$$\mu_{kl} = \frac{2e}{L}\int_0^L x \sin\frac{k\pi x}{L} \sin\frac{l\pi x}{L}\, dx \tag{3-49}$$

Since the PIB wavefunctions are symmetrical about the midpoint of the box ($x = L/2$), the above integral could be rewritten in terms of a new variable $x' = x - L/2$ $(-L/2 \le x' \le L/2)$. Referring to Fig. 3-9, we note that all the wavefunctions having *no* nodes at $x = L/2$ (n odd) are symmetric about this point. We say these functions have *even*, or +, *parity*. In general, if $f(-x) = f(x)$, we say the function $f(x)$ has even parity. Conversely, all the wavefunctions having nodes at $x = L/2$ have *odd*, or $-$, *parity*, that is, $f(-x) = -f(x)$. We now make use of the elementary fact that an integral over a symmetric interval $-a \le x \le a$

$$\int_{-a}^{a} f(x)\, dx$$

is zero if $f(x)$ is of odd parity and nonzero otherwise. Since the function x itself is of odd parity, the transition moment [Eq. (3-49)] will have an integrand of even parity only if the product $\psi_k(x)\psi_l(x)$ is of odd parity.[14] This can happen

[14] The parity multiplication rules are $(+)(+) = (-)(-) = (+)$ and $(+)(-) = (-)(+) = (-)$.

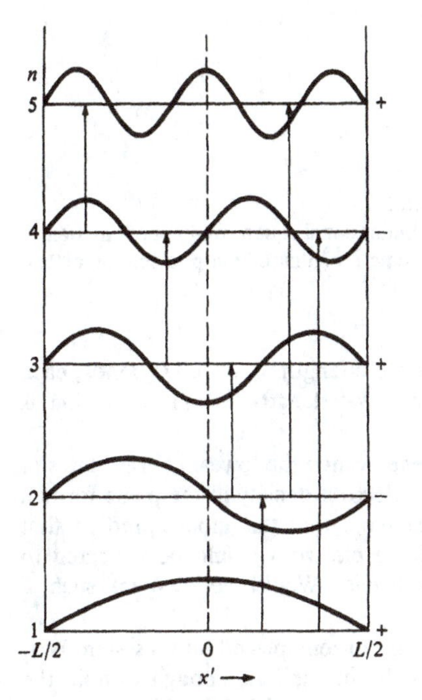

FIGURE 3-9
Symmetry properties of the wavefunctions of a particle in a one-dimensional box. The parities "+" and "−" are indicated to the right of each wavefunction. The arrows indicate stationary states between which radiative transitions can occur by an allowed electric dipole coupling mechanism.

only if $\psi_k(x)$ and $\psi_l(x)$ have *opposite* parities. Hence, the only allowed transition will be between states of *opposite* parity. This fact may be stated in the form of the *selection rules*

$$+ \rightarrow - \quad \text{and} \quad - \rightarrow +$$

$$+ \not\rightarrow + \quad \text{and} \quad - \not\rightarrow -$$

or, more succinctly

$$\Delta n = \pm\text{odd integer}$$

The interested reader may wish to refer to Exercise 3-24 and thereby verify that the same selection rules are obtained by changing explicitly to symmetric coordinates.

EXERCISES

3-25. Assume that the six pi electrons of 1,3,5-hexatriene are placed in pairs (with opposite spins) into the three lowest energy states of a one-dimensional box. This assumes that electron repulsions may be ignored and that the periodic potential of the carbon and hydrogen atoms (and their sigma electrons) may be replaced with a linear, constant potential (analogous to that shown for 1,3-butadiene in Fig. 3-10). Calculate the *longest* wavelength of light absorbed when one electron is

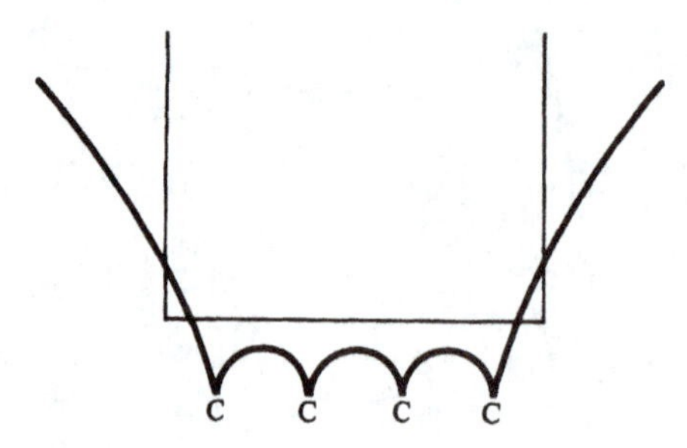

FIGURE 3-10
A one-dimensional square-well approximation of the periodic potential of the pi electrons of 1,3-butadiene.

promoted to a higher level. Assume the box has a length of five C–C bonds, each 0.14 nm long.[15] The experimentally observed wavelength for this transition is about 258 nm.

3-26. Assume the four pi electrons of cyclobutadiene occupy the lowest energy states of a square box. What is the appearance of the electron density in the plane formed by the four carbon atoms of this molecule? Note that this model predicts that cyclobutadiene is a *diradical*, since two of the electrons would be expected to occupy separate, but degenerate, energy levels. Would you expect such a molecule to be especially stable? Explain.

3-27. Carry out a treatment of the six pi electrons of benzene placed into a square box, using $L_x = L_y = L = 0.42$ nm. (This assumes the box is long enough to hold the benzene hexagon along its longest dimension plus an additional half-bond length at each end.) Work out the complete selection rules for this model, and calculate the wavelength of the lowest-energy spectroscopic transition allowed. The experimental value is about 205 nm. What are the selection rules for the particle in a square box? *Caution*: Examine this situation carefully before reaching a conclusion!

3-28. Show that for the one-dimensional PIB

$$\mu_{kl} = \frac{eL}{\pi^2}\left\{\frac{\cos[(k-1)\pi]-1}{(k-1)^2} - \frac{\cos[(k+1)]-1}{(k+1)^2}\right\}$$

Note: It is convenient to use the relationships

$$\sin a \sin b = \tfrac{1}{2}[\cos(a-b) - \cos(a+b)]$$

$$\int y \cos ay\, dy = \frac{1}{a^2}\cos ay + \frac{y}{a}\sin ay$$

3-29. Can a molecule whose electrons behave as particles in a box (as assumed in Exercises 3-25 and 3-26) ever ionize? Explain.

3-30. The parity operator $\hat{\pi}$ may be defined via the relationship $\hat{\pi}^2 = \hat{1}$ ($\hat{1}$ = unit operator). Show that:

(*a*) The eigenvalues of $\hat{\pi}$ are ± 1.

(*b*) $[\hat{\pi}, \hat{H}] = 0$ whenever V has even parity.

(*c*) $\hat{\pi}$ is self-adjoint (hermitian).

---

[15] There are various ways to determine the box length. Here we assume it is the sum of the C–C bond distances. No single method for choosing L is satisfactory for general use. See N. S. Bayliss, *Quart. Rev.*, **6**:319 (1952).

3-31. For this problem we use the notation $\psi_n(i)$ for the wavefunction of particle i in the nth quantum state of a one-dimensional box. A system consists of two indistinguishable particles in such a box.

(*a*) Write the wavefunction for the state in which one particle is in the $n = 1$ state and the other is in the $n = 2$ state. Sketch the appearance of the wavefunction.

(*b*) What is the probability density for finding both particles simultaneously at a given location x? Sketch the appearance of this probability density.

(*c*) What is the probability density (normalized to unity) of finding just *one* particle (either 1 or 2) at a given location x? Sketch the appearance of this probability density. *Hint*: Recall the difference between the probability of simultaneous occurrence of *all* of a group of independent events versus the probability of occurrence of any one of these independent events: the former is the product of the individual probabilities, and the latter is the sum of their probabilities.

(*d*) Referring to Exercise 3-26, what is the probability of finding any *one* of the four electrons in cyclobutadiene at a given location (x, y) in the square box?

3-5 THE TUNNELING EFFECT

Consider a particle in a box of length x_1, the potential barrier of which is infinite at one end ($x = 0$) and finite at the other end ($x = x_1$). Furthermore, let the finite barrier V_0 have a finite thickness of $x_2 - x_1 = a$. We now investigate the quantum behavior of a particle with energy $E < V_0$ which is initially confined to the region between $x = 0$ and $x = x_1$. The shape of the box is shown in Fig. 3-11. According to classical mechanics, such a confined particle could never escape to the region $x > x_2$. Nevertheless, we shall find that quantum theory predicts a finite probability of finding the particle beyond the barrier.

It is convenient to divide the above system into three regions:

Region I: $V = 0$ for $0 < x < x_1$
Region II: $V = V_0$ for $x_1 \leq x \leq x_2$
Region III: $V = 0$ for $x_2 < x < \infty$

In addition, we have $V = \infty$ for $x = 0$, so that the particle is strictly confined

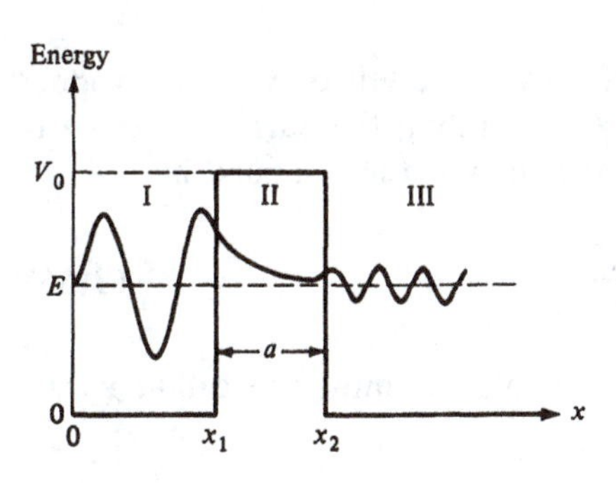

FIGURE 3-11
Particle confined behind a potential barrier of finite height and thickness. The energy of the particle is too low to permit it to penetrate the barrier classically. The wavy line represents the qualitative appearance of the wavefunction of the system.

from the left as in the PIB problem. We will now see how to calculate an approximate expression for the probability of finding the particle in region III.

First, we solve the Schrödinger equation for each of the three regions.

Region I: $V = 0$. The Schrödinger equation is

$$\frac{d^2\psi}{dx^2} + \kappa^2\psi = 0 \qquad \text{where } \kappa^2 = \frac{2mE}{\hbar^2} \tag{3-50}$$

with the solution

$$\psi_{\text{I}} = Ae^{i\kappa x} + Be^{-i\kappa x} \tag{3-51}$$

Region II: $V = V_0$. The Schrödinger equation is

$$\frac{d^2\psi}{dx^2} - k^2\psi = 0 \qquad \text{where } k^2 = \frac{2m(V_0 - E)}{\hbar^2} \tag{3-52}$$

with the solution

$$\psi_{\text{II}} = Ce^{kx} + De^{-kx} \tag{3-53}$$

Region III: $V = 0$. The Schrödinger equation is as for region I, and the solution is

$$\psi_{\text{III}} = Fe^{i\kappa x} + Ge^{-i\kappa x} \tag{3-54}$$

Recalling from previous discussions that a term such as $e^{-i\kappa x}$ refers to a particle moving in the $-x$ direction, we set $G = 0$ in Eq. (3-54). This simply states that there is no source of particles coming from the positive direction. For mathematical simplicity we now set $x_1 = 0$ and $x_2 = a$. In order for the total wavefunction to be well-behaved, it is necessary that ψ and grad ψ be continuous at the boundaries $x = 0$ and $x = a$. The boundary conditions then are

$$\begin{aligned} &\textbf{1.}\ \psi_{\text{I}}(0) = \psi_{\text{II}}(0) \\ &\textbf{2.}\ \psi_{\text{II}}(a) = \psi_{\text{III}}(a) \\ &\textbf{3.}\ \text{grad}\ \psi_{\text{I}}(0) = \text{grad}\ \psi_{\text{II}}(0) \\ &\textbf{4.}\ \text{grad}\ \psi_{\text{II}}(a) = \text{grad}\ \psi_{\text{III}}(a) \end{aligned} \tag{3-55}$$

The number of particles striking the barrier from the left of $x = 0$ is proportional to $|A|^2$, and the number of particles penetrating the barrier at $x = a$ is proportional to $|F|^2$. The transmission coefficient χ is then defined by

$$\chi = \frac{|F|^2}{|A|^2} \tag{3-56}$$

To show that penetration of the barrier is possible, we must prove that $\chi > 0$.

By application of the boundary conditions [Eqs. (3-55)] we obtain the relationships

$$\begin{aligned} A + B &= C + D \\ Ce^{ka} + De^{-ka} &= Fe^{i\kappa a} \\ i\kappa(A - B) &= k(C - D) \\ kCe^{ka} - kDe^{-ka} &= i\kappa Fe^{i\kappa a} \end{aligned} \tag{3-57}$$

These four equations may be regarded as involving five unknowns: A, B, C, D, and F. We can obtain A in terms of F by first obtaining C and D in terms of F using the second and fourth equations. We get

$$C = \frac{F}{2}\left(1 + \frac{i}{Z}\right) e^{i\kappa a - ka} \qquad D = \frac{F}{2}\left(1 - \frac{i}{Z}\right) e^{i\kappa a + ka} \tag{3-58}$$

where we have employed the convenient substitution

$$Z = \frac{k}{\kappa} = \left(\frac{V_0 - E}{V_0}\right)^{1/2} \tag{3-59}$$

If the third relationship in Eqs. (3-57) is divided by $i\kappa$ and subtracted from the first equation, we obtain

$$A = \tfrac{1}{2}[(1 - iZ)C + (1 + iZ)D] \tag{3-60}$$

If the barrier is thick and if $E \ll V_0$, we note that $|D| \gg |C|$. Thus, as a first approximation, we may neglect C in the above equation. Substituting for D in Eq. (3-60) from Eq. (3-58), we get

$$A = \frac{F}{4}(1 + iZ)\left(1 - \frac{i}{Z}\right) e^{i\kappa a + ka} \tag{3-61}$$

The transmission coefficient then becomes

$$\chi = \frac{|F|^2}{|A|^2} = \frac{16e^{-2ka}}{(1 + Z^2)(1 + Z^{-2})} \tag{3-62}$$

The factor in the denominator varies with Z^2, that is, with the relative values of E and V_0, and will simply multiply the exponential by some numerical factor. The essential part of the transmission coefficient is then expressed by

$$e^{-2ka} = \exp\left\{-\frac{2a}{\hbar}[2m(V_0 - E)]^{1/2}\right\} \tag{3-63}$$

Clearly, the transmission coefficient will not be zero unless $V_0 = \infty$ (particle in a box), or $a = \infty$, or $m = \infty$. It is apparent that for a given $V_0 - E$ and a the transmission coefficient increases as the mass of the particle decreases. Although we have resorted to several approximations in arriving at the final result, the same qualitative picture is obtained by a more rigorous treatment.

Such a penetration of a barrier by a particle whose energy is insufficient (classically) to transcend the barrier is called the *quantum mechanical tunneling effect*. It was first suggested by Gamow[16] in 1928 and independently by Gurney and Condon[17] in 1929 in order to explain the empirical observations summarized in the Geiger-Nuttall law of radioactive decay.[18] One of the perplexing problems of classical theory was to explain the spontaneous emission of α particles from the nuclei of atoms. The potential energy of an α particle close to the nucleus is depicted schematically in Fig. 3-12. It is known that the α particle is held in the nucleus by very strong nuclear binding forces, which would appear to preclude spontaneous radioactive decay. The kinetic energies of emitted α particles (which are experimentally measurable) are found to be too small to allow a classical penetration. For example, Rutherford and Royds[19] measured the kinetic energy of α particles emitted by Ra(C′) and found it to be around 7.7×10^6 eV. The binding energy of the Ra(C′) nucleus is about 10^8 to 10^9 eV, considerably greater than the α-particle kinetic energies.

A glance at Fig. 3-12 shows that the shape of the potential barrier to α decay is not square, as assumed in Fig. 3-11, but varies with distance in a smooth fashion. Nevertheless, the two barriers are at least qualitatively similar, and the square potential shape is far easier to analyze mathematically.

If we let N represent the number of collisions of the particles with the barrier per unit time, the probability of penetrating the barrier is $N\chi$. The reciprocal of this quantity is the *mean lifetime* of the particle.

Quantum mechanical tunneling is believed to be of importance in oxidation-reduction reactions and in electrode reactions, in which electrons must move from one atom or molecule to another across a phase boundary. All other factors considered equal, the small mass of the electron is conducive to a large transmission coefficient.

The so-called *umbrella inversion* of pyramidal molecules such as NH_3, PH_3, and AsH_3 may also be described as a quantum mechanical tunneling

---

[16] G. Gamow, *Z. Physik*, **51**:204 (1928).

[17] R. W. Gurney and E. U. Condon, *Phys. Rev.*, **33**:127 (1929).

[18] This law states that the average velocity of emitted α particles is inversely proportional to the half-life of the decaying nuclear species.

[19] E. Rutherford and T. Royds, *Phil. Mag.*, **17**:281 (1909).

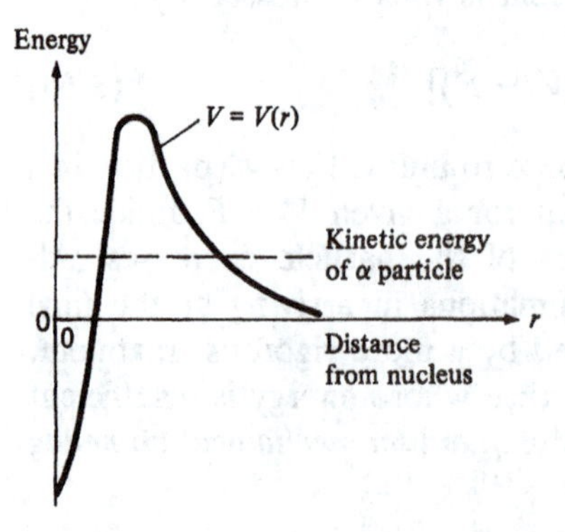

FIGURE 3-12
Potential-energy diagram for the nuclear binding forces holding an α particle in the nucleus of an atom. That portion of $V(r)$ having a positive slope represents nuclear attraction; the portion with a negative slope represents coulombic repulsion.

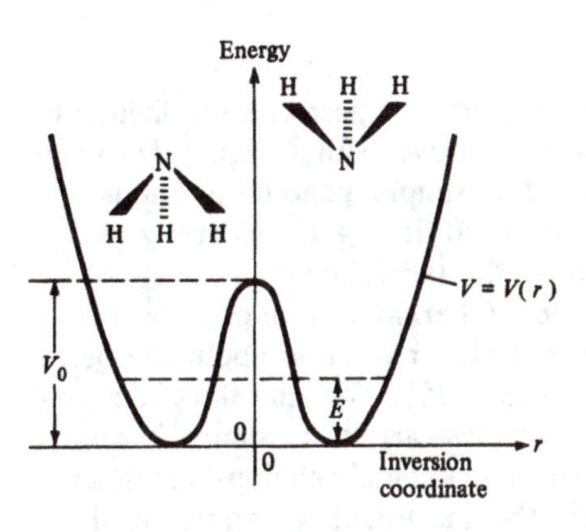

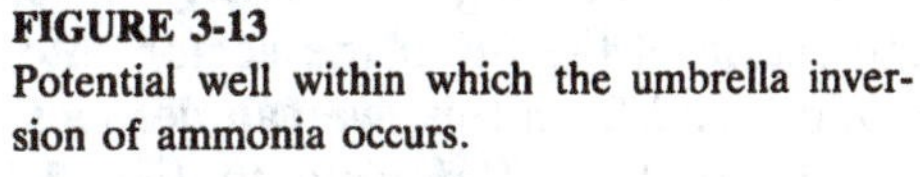

FIGURE 3-13
Potential well within which the umbrella inversion of ammonia occurs.

effect. In this case protons tunnel through the barrier maintaining the atoms in a pyramidal shape. The potential-energy diagram for such a molecule is depicted in Fig. 3-13. One would expect the transmission coefficient in such a case to decrease as the mass of the heavy atom increases.

In some cases, intermolecular electron transfer appears to occur via a quantum tunneling mechanism.[20]

EXERCISES

3-32. Referring to Fig. 3-11, show that if $E \geq V_0$, there is a finite probability that some particles will be reflected by the barrier at $x = x_1$. Classically, all particles would pass through the barrier.

3-33. Referring to Fig. 3-14, show that the reflection and transmission coefficients are of the same form as for Fig. 3-11 except for a multiplicative factor of k_1/k_0, where

$$k_0^2 = \frac{2mE}{\hbar^2} \qquad k_1^2 = \frac{2m(E - V_1)}{\hbar^2}$$

[20] John R. Miller, *Science*, **189**:221 (1975).

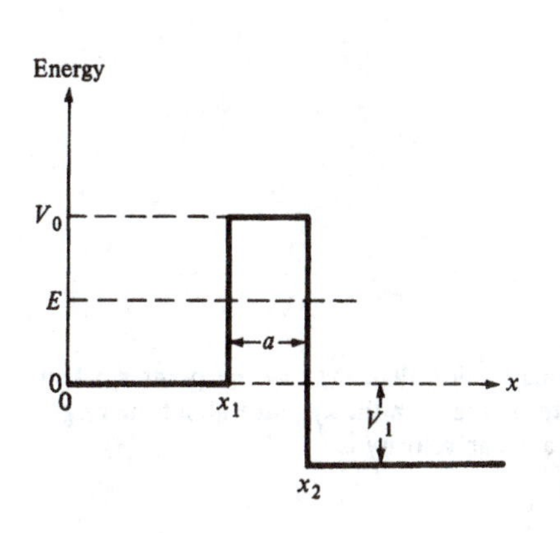

FIGURE 3-14
Particle confined behind a finite barrier such that its kinetic energy is greater outside the barrier.

3-6 THE HARMONIC OSCILLATOR

Whenever a body has a periodic motion such that the acceleration divided by the displacement is a constant, it is said to move with (simple) *harmonic motion*. Ideally, the to-and-fro motion of a simple pendulum fulfills this definition; the acceleration ÷ displacement ratio is $-g/L$, where g is the gravitational constant and L is the length of the pendulum.

As a somewhat generalized example of harmonic motion, consider a radius vector of modulus r which carries out planar rotation about the origin with a uniform angular velocity ω (see Fig. 3-15). We can show that the projection of the tip of the vector onto the x axis carries out simple harmonic motion, and from this, we can derive some fundamental relationships describing the motion. Referring to Fig. 3-15, the instantaneous value of the x component of the rotating vector at any time t is

$$x_t = r\cos\theta = r\cos\omega t \tag{3-64}$$

The instantaneous velocity of the point x_t is

$$\frac{dx_t}{dt} = \dot{x}_t = -r\omega\sin\omega t \tag{3-65}$$

and the acceleration is

$$\frac{d^2x_t}{dt^2} = \frac{d\dot{x}_t}{dt} = -r\omega^2\cos\omega t = -\omega^2 x_t \tag{3-66}$$

The latter equation may be rewritten as

$$\frac{d\dot{x}_t}{dt} + \omega^2 x_t = 0 \tag{3-67}$$

which is the differential equation of the harmonic motion. Note that acceleration ÷ displacement $= -\omega^2$ (a constant). If we associate a mass m with

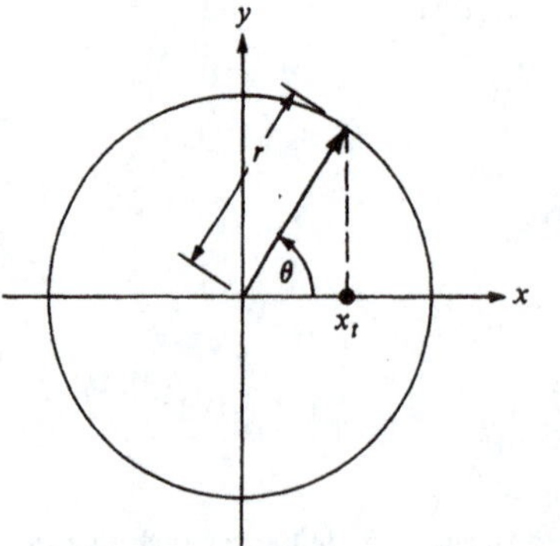

FIGURE 3-15
Simple harmonic motion described by the point x_t when the radius vector r rotates in the xy plane about the origin with constant angular velocity ω.

the point x_t, we call the system a *harmonic oscillator*; its kinetic energy is

$$T = \frac{m\dot{x}_t^2}{2} = \frac{p_x^2}{2m} \tag{3-68}$$

The potential energy may be found from the fundamental relationship

$$\mathbf{F} = -\nabla V \tag{3-69}$$

where **F** is the force acting on the system tending to restore it to its equilibrium position [see App. 4, Eq. (A4-24)]. For the one-dimensional case, we may dispense with vector notation and write

$$F = -\frac{dV}{dx} = \frac{m d\dot{x}_t}{dt} = -m\omega^2 x_t = -kx_t \tag{3-70}$$

where $k = m\omega^2$ is called the *force constant* (or Hooke's law constant).[21] If the angular velocity ω is rewritten in terms of the *frequency* ν_0 of the oscillator ($\omega = 2\pi\nu_0$), the force constant becomes

$$k = 4\pi^2 m\nu_0^2 \tag{3-71}$$

Note that the force constant is simply the *force per unit displacement* tending to restore the particle to $x_t = 0$. The potential energy is now obtained by integration of Eq. (3-70):

$$V = -\int_0^{x_t} F\,dx = k\int_0^{x_t} x\,dx = \tfrac{1}{2}kx_t^2 \tag{3-72}$$

Combining Eqs. (3-68) and (3-72), the *total energy* of the harmonic oscillator becomes

$$T + V = E \qquad \text{or} \qquad \frac{p_x^2}{2m} + \frac{kx^2}{2} = E \tag{3-73}$$

where the subscript on x_t has been dropped for convenience.

Let us now consider a system of more practical importance—one we will use to model the vibrational motion of a diatomic molecule—two masses m_1 and m_2 (same or unequal) connected by a Hooke's law spring (see Fig. 3-16).

[21] Hooke's law—long ago applied to bent beams and stretched springs—states that the deformation (through bending or stretching) is proportional to the stress producing it.

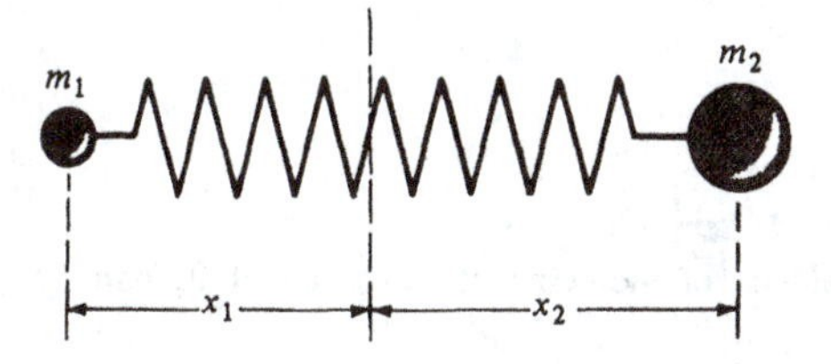

FIGURE 3-16
Two particles connected by a Hooke's law spring and undergoing vibrational motion along the line of centers.

We let x_1 and x_2 represent the instantaneous positions of the masses m_1 and m_2, respectively, relative to an arbitrary reference point. In the completely general case, the motion of this two-body system consists of three types of motion: movement of the system as a whole through space (*translation*), *rotation* of the system about its center of mass, and motion of the two masses relative to each other (*vibration*). For simplicity we will constrain the particles to move only along the line of their centers and thereby eliminate the rotational motion.[22] The total energy (translational and vibrational) of the system is

$$E = \frac{m_1\dot{x}_1^2}{2} + \frac{m_2\dot{x}_2^2}{2} + \frac{k}{2}(x_1 - x_2)^2 = \frac{p_1^2}{2m_1} + \frac{p_2^2}{2m_2} + \frac{k}{2}(x_1 + x_2)^2 \tag{3-74}$$

The Schrödinger equation for the system is

$$\left(-\frac{\hbar^2}{2m_1}\frac{\partial^2}{\partial x_1^2} - \frac{\hbar^2}{2m_2}\frac{\partial^2}{\partial x_2^2}\right)\psi(x_1, x_2) + \frac{k}{2}(x_1 - x_2)^2\psi(x_1, x_2) = E\psi(x_1, x_2) \tag{3-75}$$

This partial differential equation in two independent variables may be separated into two ordinary differential equations by introducing the change of variables

$$x = x_1 - x_2$$

$$x_{\text{cm}} = \frac{m_1x_1 + m_2x_2}{m_1 + m_2} = \frac{m_1x_1 + m_2x_2}{M} \tag{3-76}$$

We call x a *relative* coordinate and x_{cm} a *center-of-mass coordinate*.[23] Assuming the separation of variables

$$\psi(x_1, x_2) \rightarrow \psi(x, x_{\text{cm}}) = \psi(x)\phi(x_{\text{cm}}) \tag{3-77}$$

and

$$E = E_{\text{vib}} + E_{\text{trans}} \tag{3-78}$$

one readily obtains

$$-\frac{\hbar^2}{2M}\frac{d^2\phi}{dx_{\text{cm}}^2} = E_{\text{trans}}\phi \tag{3-79}$$

$$\left(-\frac{\hbar^2}{2\mu}\frac{d^2}{dx^2} + \frac{kx^2}{2}\right)\psi = E_{\text{vib}}\,\psi \tag{3-80}$$

---

[22] The rotational motion will be taken up in Chap. 4, Sec. 4-9.

[23] The center of mass x_{cm} is defined by the balance of moments of forces about it, namely $m_1(x_1 - x_{\text{cm}}) = m_2(x_{\text{cm}} - x_2)$.

where $\mu = m_1 m_2 / M$ is the reduced mass of the system. The first equation describes the translational motion of the two masses in a constant potential field; its solution is just that of the free particle. The second equation is the Schrödinger equation of the one-dimensional harmonic oscillator. Note that the use of the reduced mass μ enables us to treat the two-particle system with the same mathematical equations that describe a single particle.[24]

In solving the harmonic oscillator Schrödinger equation (3-80), it is convenient to reexpress it in terms of dimensionless variables. To do this we define the quantities

$$\alpha = \frac{\mu\omega}{\hbar} = \frac{k}{h\nu_0} \qquad \epsilon = \frac{2E}{\hbar\omega} = \frac{2E}{h\nu_0} \tag{3-81}$$

This permits us to introduce the dimensionless operators

$$\hat{\xi} = \sqrt{\alpha}\,\hat{x} \qquad \text{and} \qquad \hat{P} = \frac{1}{i}\frac{d}{d\xi} \tag{3-82}$$

and the dimensionless hamiltonian

$$\bar{H} = \frac{\hat{H}}{h\nu_0} = \frac{1}{2}\left(-\frac{d^2}{d\xi^2} + \hat{\xi}^2\right) = \frac{1}{2}(\hat{P}^2 + \hat{\xi}^2) \tag{3-83}$$

The dimensionless Schrödinger equation then becomes

$$\bar{H}\psi(\xi) = \tfrac{1}{2}\epsilon\psi(\xi) \tag{3-84}$$

Because of the way ϵ is defined in Eq. (3-81), it is apparent that $\hat{H}$ and $\bar{H}$ have the same eigenfunctions.

A very elegant and simple method of solving the Schrödinger equation for the harmonic oscillator is based on the use of ladder operators. Whenever one has an operator of the form $\hat{A}^2 + \hat{B}^2$, it is possible to factor the operator as follows:

$$\hat{A}^2 + \hat{B}^2 = (\hat{A} + i\hat{B})(\hat{A} - i\hat{B}) + i[\hat{A}, \hat{B}] = (\hat{A} - i\hat{B})(\hat{A} + i\hat{B}) - i[\hat{A}, \hat{B}] \tag{3-85}$$

The operators $\hat{A} + i\hat{B}$ and $\hat{A} - i\hat{B}$ are called *ladder operators*. In the case of the harmonic operator, the dimensionless hamiltonian [Eq. (3-83)] may be factored as follows:[25]

$$\bar{H} = \frac{\hat{\xi} + i\hat{P}}{\sqrt{2}}\frac{\hat{\xi} - i\hat{P}}{\sqrt{2}} - \frac{1}{2} \qquad \text{or} \qquad \frac{\hat{\xi} - i\hat{P}}{\sqrt{2}}\frac{\hat{\xi} + i\hat{P}}{\sqrt{2}} + \frac{1}{2} \tag{3-86}$$

---

[24] The reduced mass μ is a fictitious mass which, when at distance r from the origin, obeys the same laws of motion as two separate masses m_1 and m_2, separated by the same distance r.

[25] The solution presented here is due to Dirac. For a standard solution using differential equations, see Pauling and Wilson (in the readings at the end of this chapter). The text by Kauzmann (also in the readings at the end of this chapter) also presents the solution using differential equations and provides a wealth of physical insight.

From now on, the circumflex (^) on the operator ξ will be omitted just as was done before on x. We now introduce the pair of ladder operators

$$\hat{F} = \frac{\xi + i\hat{P}}{\sqrt{2}} \qquad \text{and} \qquad \hat{F}^\dagger = \frac{\xi - i\hat{P}}{\sqrt{2}} \tag{3-87}$$

which satisfy the commutation relation

$$[\hat{F}, \hat{F}^\dagger] = 1 \tag{3-88}$$

Thus, the dimensionless hamiltonian becomes

$$\bar{H} = \hat{F}\hat{F}^\dagger - \tfrac{1}{2} \qquad \text{or} \qquad \hat{F}^\dagger\hat{F} + \tfrac{1}{2} \tag{3-89}$$

The operators $\hat{F}\hat{F}^\dagger$ and $\hat{F}^\dagger\hat{F}$ (which are self-adjoint: see Exercise 2-33) differ from $\bar{H}$ by only an additive constant ($-\frac{1}{2}$ and $\frac{1}{2}$, respectively), and they have the same eigenfunctions as $\bar{H}$ has, i.e.,

$$\begin{aligned} \bar{H}\psi &= \tfrac{1}{2}\epsilon\psi \\ (\hat{F}\hat{F}^\dagger)\psi &= \tfrac{1}{2}(\epsilon + 1)\psi \\ (\hat{F}^\dagger\hat{F})\psi &= \tfrac{1}{2}(\epsilon - 1)\psi \end{aligned} \tag{3-90}$$

As an aid in clarifying certain algebraic manipulations, we denote the operators $\hat{F}^\dagger\hat{F}$ and $\hat{F}\hat{F}^\dagger$ by the single symbols $\hat{G}$ and $\hat{G}^\dagger$, respectively. Forming the operator $\hat{G}\hat{F}$ and expanding produces

$$\hat{G}\hat{F} = \hat{F}^\dagger\hat{F}\hat{F} = (\hat{F}\hat{F}^\dagger - 1)\hat{F} = \hat{F}\hat{F}^\dagger\hat{F} - \hat{F} = \hat{F}(\hat{F}^\dagger\hat{F} - 1) = \hat{F}(\hat{G} - 1) \tag{3-91}$$

where the commutation relation [Eq. (3-88)] has been used in the second step. Similarly, one obtains

$$\hat{G}\hat{F}^\dagger = \hat{F}^\dagger(\hat{G} + 1) \tag{3-92}$$

Now consider a particular eigenfunction ψ_v of $\hat{G}$ which satisfies the eigenvalue equation

$$\hat{G}\psi_v = \varepsilon_v \psi_v \tag{3-93}$$

where $\varepsilon_v = \frac{1}{2}(\epsilon_v - 1)$ (see Eq. 3-90). Since the expectation value of the operator $\hat{G}$ is always positive or zero (see Exercise 2-33), $\varepsilon_v \geq 0$ means that $\epsilon_v \geq 1$. The equality can hold only if

$$\hat{F}\psi_v = 0 \tag{3-94}$$

i.e., only if the operator $\hat{F}$ annihilates ψ_v. Now consider the case $\varepsilon_v > 0$ and $\hat{F}\psi_v \neq 0$. Applying the operator $\hat{G}$ to $\hat{F}\psi_v$ and using the second alternative in Eq. (3-86), we get

$$\hat{G}(\hat{F}\psi_v) = (\hat{G}\hat{F})\psi_v = \hat{F}(\hat{G} - 1)\psi_v = \hat{F}(\varepsilon_v - 1)\psi_v = (\varepsilon_v - 1)\hat{F}\psi_v \tag{3-95}$$

This shows that $\hat{F}\psi_v$ is an eigenfunction of $\hat{G}$ with the eigenvalue $\varepsilon_v - 1$. In general

$$\hat{G}(\hat{F}^n\psi_v) = (\varepsilon_v - n)\hat{F}^n\psi_v \tag{3-96}$$

Thus the operator $\hat{F}$ has the effect of operating on an eigenfunction of $\hat{G}$ with eigenvalue ε_v to produce a new eigenfunction $\hat{F}\psi_v$ with an eigenvalue *decreased* by *one* unit. Since $\varepsilon_v \geq 0$, this reduction must stop when the lowest eigenvalue $\varepsilon_v = 0$ is reached. Consequently, there must exist an eigenfunction, say ψ_0, such that $\hat{F}\psi_0 = 0$ and $\varepsilon_0 = 0$. Then from Eq. (3-95) we see that the eigenvalues ε_v are $0, 1, 2, \ldots$. For a given function ψ_v it will take v operations of $\hat{F}$ to lead to annihilation so that

$$\varepsilon_v = v = 0, 1, 2, \ldots \tag{3-97}$$

Since $\varepsilon_v = \frac{1}{2}(\epsilon_v - 1)$ and $\epsilon_v = 2E_v/h\nu_0$, we obtain

$$E_v = (v + \tfrac{1}{2})h\nu_0 \qquad v = 0, 1, 2, \ldots \tag{3-98}$$

This equation describes the quantized energy states of the harmonic oscillator. The integer v is called the *vibrational quantum number*. We see that the energy of the harmonic oscillator is quantized in units of $h\nu_0$, where ν_0 is the fundamental frequency given by

$$\nu_0 = \frac{1}{2\pi}\left(\frac{k}{\mu}\right)^{1/2} \tag{3-99}$$

Note that the zero-point energy of the harmonic oscillator is $\frac{1}{2}h\nu_0$; classical mechanics predicts zero for this quantity.

The harmonic oscillator wavefunctions are obtained by solving

$$\hat{F}\psi_0 = (\xi + i\hat{P})\psi_0 = 0 \tag{3-100}$$

or, in standard differential equation form,

$$\frac{d\psi_0}{d\xi} + \xi\psi_0 = 0 \tag{3-101}$$

A well-behaved solution to the above is

$$\psi_0(\xi) = e^{-\xi^2/2} \tag{3-102}$$

The remaining functions are found as follows: By the same method used to obtain Eq. (3-96), one gets

$$\hat{G}[(\hat{F}^\dagger)^n\psi_v] = (\varepsilon_v + n)(\hat{F}^\dagger)^n\psi_v \tag{3-103}$$

Thus the operator $\hat{F}^\dagger$ operates on an eigenfunction of $\hat{G}$ with eigenvalue ε_v to produce a new eigenfunction $\hat{F}^\dagger\psi_v$ with eigenvalue $\varepsilon_v + 1$. Similarly, n operations of $\hat{F}^\dagger$ on ψ_v produces the eigenfunction $(\hat{F}^\dagger)^n\psi_v$ with eigenvalue $\varepsilon_v + n$. The ladder operators $\hat{F}$ and $\hat{F}^\dagger$ are also called *step-down* and *step-up operators*, respectively. Thus, the higher eigenfunctions of $\hat{G}$ are obtained by operating on ψ_0 with $\hat{F}^\dagger$ an appropriate number of times, i.e.,

$$\psi_v(\xi) = \left(\xi - \frac{d}{d\xi}\right)^v e^{-\xi^2/2} \tag{3-104}$$

The eigenfunctions $\psi_v(\xi)$ can be written in the general form

$$\psi_v(\xi) = N_v H_v(\xi) e^{-\xi^2/2} \tag{3-105}$$

where N_v is a normalization constant and the $H_v(\xi)$ functions are known as *hermite polynomials*. These have the general form

$$H_v(\xi) = \begin{cases} \sum_{j=0}^{v/2} a_{2j}\xi^{2j} & v = 0, 2, 4, \ldots \\ \sum_{j=1}^{(v+1)/2} a_{2j-1}\xi^{2j-1} & v = 1, 3, 5, \ldots \end{cases} \tag{3-106}$$

Generally, the coefficient of ξ^v in H_v is chosen as 2^v, and the rest are determined by the recursion formula

$$\frac{a_{j+2}}{a_j} = \frac{2j + 1 - \epsilon_v}{(j+1)(j+2)} \tag{3-107}$$

Thus, $H_3(\xi) = a_1\xi + a_3\xi^3$, with $a_3 = 2^3 = 8$. Then, a_1 is chosen as follows:

$$\frac{a_3}{a_1} = \frac{(2)(1) + 1 - \epsilon_3}{(1+1)(1+2)} = \frac{3 - \epsilon_3}{6} = \frac{3 - 7}{6} = -\frac{2}{3}$$

Thus, $a_1 = -\frac{3}{2}a_3 = -12$ and $H_3(\xi) = 8\xi^3 - 12\xi$. The first six hermite polynomials are

$$\begin{array}{ll} H_0(\xi) = 1 & H_3(\xi) = 8\xi^3 - 12\xi \\ H_1(\xi) = 2\xi & H_4(\xi) = 16\xi^4 - 48\xi^2 + 12 \\ H_2(\xi) = 4\xi^2 - 2 & H_5(\xi) = 32\xi^5 - 160\xi^3 + 120\xi \end{array} \tag{3-108}$$

The hermite polynomials are also conveniently defined by the generating function

$$G(\xi, m) = e^{\xi^2 - (m-\xi)^2} = \sum_n \frac{H_n(\xi)m^n}{n!} \tag{3-109}$$

where m is called an *auxiliary variable*. This generating function may be used to derive the three equations[26]

$$H_n(\xi) = (-1)^n e^{\xi^2} \frac{d^n}{d\xi^n} e^{-\xi^2} \tag{3-110}$$

[26] Details of the derivation are given in Pauling and Wilson (see the readings at the end of this chapter).

$$H_n'(\xi) = \frac{dH_n(\xi)}{d\xi} = 2nH_{n-1}(\xi) \tag{3-111}$$

$$H_{n+1}(\xi) - 2\xi H_n(\xi) + 2nH_{n-1}(\xi) = 0 \tag{3-112}$$

The first equation provides a quick route to the hermite polynomials. For example,

$$H_0(\xi) = (-1)^0 e^{\xi^2} \frac{d^0}{d\xi^0} e^{-\xi^2} = (-1)^0 e^{\xi^2-\xi^2} = 1$$

$$H_1(\xi) = (-1)^1 e^{\xi^2} \frac{d}{d\xi} e^{-\xi^2} = (-1)e^{\xi^2}(-2\xi e^{-\xi^2}) = 2\xi$$

The remaining equations are called *recursion* formulas; we will use them in the next section to develop selection rules.

The generating function [Eq. (3-109)] also leads to a general expression for the normalization constant N_v for $\psi_v(\xi)$. We consider the integral

$$\int_{-\infty}^{\infty} H_v(\xi)H_w(\xi)e^{-\xi^2}\, d\xi \tag{3-113}$$

Then, using the generating function [Eq. (3-109)], we construct the integral

$$I = \int_{-\infty}^{\infty} G(\xi, m)G(\xi, n)e^{-\xi^2}\, d\xi = \sum_v \sum_w \frac{m^v n^w}{v!w!} \int_{-\infty}^{\infty} H_v(\xi)H_w(\xi)e^{-\xi^2}\, d\xi$$

$$= \int_{-\infty}^{\infty} e^{-\xi^2} e^{\xi^2-(m-\xi)^2} e^{\xi^2-(n-\xi)^2}\, d\xi = e^{2mn} \int_{-\infty}^{\infty} e^{-(\xi-m-n)^2}\, d\xi \tag{3-114}$$

If we employ the change of variable $\xi - m - n = y$, the above integral becomes

$$I = e^{2mn} \int_{-\infty}^{\infty} e^{-y^2}\, dy = \sqrt{\pi} e^{2mn} \tag{3-115}$$

Expanding e^{2mn} in a Maclaurin's series, I becomes

$$I = \sqrt{\pi}\left[1 + 2mn + \frac{(2mn)^2}{2!} + \cdots + \frac{(2mn)^v}{v!} + \cdots\right] \tag{3-116}$$

Comparison of Eqs. (3-116) and (3-115) shows that the integral given in Eq. (3-113) is just the coefficient of $m^v n^w$ in expression (3-116). This coefficient is nonzero only when $v = w$. Thus, we obtain the orthogonality relationship

$$\int_{-\infty}^{\infty} H_v(\xi)H_w(\xi)e^{-\xi^2}\, d\xi = 0 \qquad \text{when } v \neq w \tag{3-117}$$

Of course, this also follows at once from Theorem 2-3, since Eq. (3-98) shows that the eigenfunctions $\psi_v(\xi)$ are nondegenerate. The normalization constant N_v now follows by setting $v = w$ and comparing coefficients between Eqs. (3-115) and (3-113). One obtains

$$\int_{-\infty}^{\infty} H_v^2(\xi)e^{-\xi^2}\, d\xi = 2^v v! \sqrt{\pi} \tag{3-118}$$

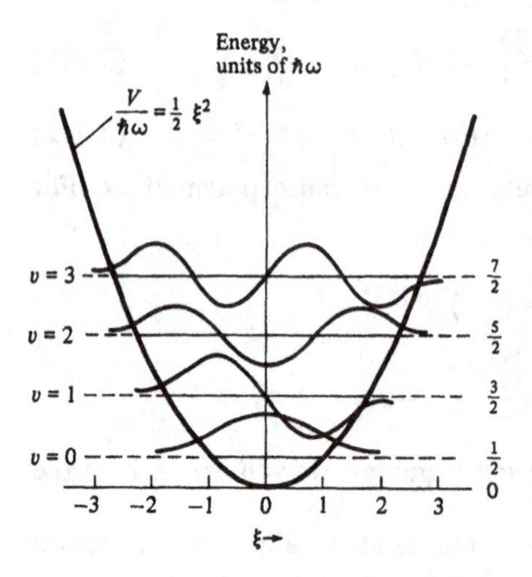

FIGURE 3-17
Eigenvalues and eigenfunctions of the dimensionless operator of the harmonic oscillator. All but the eigenfunctions are drawn to scale. The solid horizontal lines indicating allowed values of $\epsilon_v/2$ have a length equal to $2|\xi| = 2\sqrt{2v+1}$, where $|\xi|$ is the maximum value of ξ allowed by classical mechanics for a harmonic oscillator having the same total energy. Although the eigenfunctions are shown superposed onto the energy-level diagram, only the abscissa (horizontal axis) of the latter is relevant to these functions. The vertical scale of the eigenfunctions is arbitrary and is intended to be interpreted qualitatively only.

Thus

$$N_v = \left(\frac{1}{2^v v!\sqrt{\pi}}\right)^{1/2} \tag{3-119}$$

In terms of the variable x

$$\psi_v(x) = N'_v H_v(x) e^{-\alpha x^2/2} \tag{3-120}$$

where

$$N'_v = \left(\frac{1}{2^v v!\sqrt{\pi/\alpha}}\right)^{1/2} \tag{3-121}$$

Figure 3-17 summarizes the eigenvalues and eigenfunctions of the dimensionless operator $\bar{H}$ for $v = 0$, 1, 2, and 3. Note the similarity in general shapes of the harmonic oscillator wavefunctions with the PIB wavefunctions. The fact that the wavefunctions of the harmonic oscillator "leak out" of the "parabolic box" may be regarded as an example of quantum tunneling.

EXERCISES

3-34. Prove Eq. (3-85).

3-35. (*a*) Show that a general solution to the differential equation for the classical harmonic oscillator [Eq. (3-67)] is $x_t = x_0 \sin(\omega t + \delta)$, where x_0 and δ are arbitrary constants.

(*b*) If a mass m is associated with the oscillation, show that $E = T$ when $x_t = 0$, and that $E = V$ when $x_t = x_0$, and that, in general, $E = m\omega^2 x_0^2$. Note that since x_0 can have any positive value, the classical energy is not quantized.

(*c*) What restriction does quantum theory place on the value of x_0?

3-36. Calculate $\langle x \rangle$ and $\langle x^2 \rangle$ for the harmonic oscillator. What is the latter quantity related to physically?

3-37. The vibrational frequency of a harmonic oscillator is always ν_0, regardless of the quantum state it is in. What, then, accounts for the increase in energy of higher quantum states? Relate this to how the energy of a simple classical pendulum increases if one gives it a bigger swing. What happens to the pendulum's frequency at the same time?

3-38. Does the particle in a box (Sec. 3-3) exhibit simple harmonic motion? Explain.

3-39. Show that if the coordinate transformation [Eq. (3-76)] is not made, the Schrödinger equation describes a complicated mixture of translational and vibrational motions.

3-40. (*a*) Show that the energy of a three-dimensional harmonic oscillator is given by

$$E(v_1, v_2, v_3) = \sum_{i=1}^{3} (v_i + \tfrac{1}{2})h\nu_{0i}$$

where the subscripts 1, 2, and 3 denote the x, y, and z directions, respectively.

(*b*) Compare the isotropic, three-dimensional harmonic oscillator with a particle in a cubic box, and discuss the degeneracies encountered.

(*c*) Prove that the general formula for the degree of degeneracy of the isotropic, three-dimensional harmonic oscillator is given by $g = (v+1)(v+2)/2$, where v is the sum of the three quantum numbers v_1, v_2, and v_3.

3-41. Verify the hermite polynomials of Eqs. (3-108), and obtain the normalized expressions for $\psi_v(\xi)$ for $v = 0$ through 5.

3-42. Demonstrate the following relationships for the step-up and step-down operators: $\hat{F}\psi_0 = 0$, $\hat{F}^\dagger\psi_0 = \psi_1$, $\hat{F}\psi_1 = \psi_0$, and $\hat{F}^\dagger\psi_1 = \psi_2$.

3-43. Prove the commutation relation seen in Eq. (3-88).

3-44. Find the expectation values of the operators $\hat{P}$ and ξ. *Hint*: Use Eqs. (3-111) and (3-112).

3-45. Prove that a classical harmonic oscillator with total energy $\epsilon/2 = v + \frac{1}{2}$ has a maximum extension of $\sqrt{2v+1}$, as shown in Fig. 3-17. Note that $|\xi|$ is a maximum when $\dot{\xi} = 0$.

3-46. Find the commutator of $\hat{P}^2$ and ξ^2. How is this related to the fact that the wavefunctions of the harmonic oscillator do not vanish at the classical limit $|\xi| = \sqrt{2v+1}$?

3-47. Two particles of masses 3.32×10^{-27} kg and 31.5×10^{-27} kg are connected by a Hooke's law spring which requires force of 13.2×10^2 N to stretch it by 1.5 m.

(*a*) Calculate the force constant (in $\text{N} \cdot \text{m}^{-1}$) of the system.

(*b*) What is the fundamental vibrational frequency (in s^{-1} and cm^{-1}) of the system?

(*c*) What is the zero-point energy (based on quantum theory) of the system?

(*d*) What does quantum theory predict will happen to the vibrations of atoms (assumed to be harmonic oscillators) in a crystal lattice at absolute zero?

3-48. Assume that HCl and DCl ($\text{H} = {}^1\text{H}$, $\text{D} = {}^2\text{H}$, and $\text{Cl} = {}^{35}\text{Cl}$) can be treated as harmonic oscillators and that both have the same force constant (4.8×10^2 $\text{N} \cdot \text{m}^{-1}$). Calculate the fundamental vibrational frequency of each molecule.

3-49. Is the operator $\hat{F} = (\xi + i\hat{P})/\sqrt{2}$ a normal operator? Explain.

3-50. Can a diatomic molecule AB, behaving as a harmonic oscillator, ever vibrate strenuously enough to dissociate into atoms? (See Exercise 3-29).

3-51. Show that if the constant a is properly chosen, then the function e^{-ax^2} is an eigenfunction of the operator $d^2/dx^2 - qx^2$. Use this result to compute the ground-state energy of the harmonic oscillator. How could one continue to determine the excited-state energies and wavefunctions as well?

3-7 SPECTROSCOPY OF THE HARMONIC OSCILLATOR

Since the harmonic oscillator resembles the particle in a box in many respects, one might suspect that the selection rules are also very similar. This is true to an extent, but the harmonic oscillator selection rules are somewhat more restrictive.

Consider a heteronuclear diatomic molecule AB having a permanent dipole moment. Vibrational motion of the atoms A and B produces an oscillating dipole moment (relative to a fixed reference axis as shown in Fig. 3-2) which can undergo resonance coupling with the electric field oscillations of electromagnetic radiation. Typically, such oscillations correspond in frequency to the infrared region. In terms of the dimensionless variable $\xi = \sqrt{\alpha}x$, the transition moment for a transition $\psi_k(\xi) \rightarrow \psi_l(\xi)$ is given by

$$\langle \psi_l | \xi | \psi_k \rangle = N_l N_k \int_{-\infty}^{\infty} H_l(\xi) H_k(\xi) \xi e^{-\xi^2} \, d\xi \qquad (3\text{-}122)$$

Since the normalization constants N_k and N_l are irrelevant to the selection rules, we will omit them in what follows. Using the recurrence relations [Eq. (3-112)] for $H_k(\xi)\xi$, the integral in Eq. (3-122) becomes

$$\frac{1}{2} \int_{-\infty}^{\infty} H_l(\xi)[H_{k+1}(\xi) + 2kH_{k-1}(\xi)] e^{-\xi^2} \, d\xi \qquad (3\text{-}123)$$

From the orthogonality of the wavefunctions, we see that this integral is nonzero only when $l = k + 1$ or $l = k - 1$, that is, when the upper and lower states are adjacent. The selection rule then is

$$\Delta v = \pm 1 \qquad (3\text{-}124)$$

Recall that for the PIB, Δn could be any odd value $\pm n$; that is, the states did not have to be adjacent.

EXERCISES

3-52. Using data in Exercise 3-48,

(*a*) Calculate the frequency (in cm^{-1}) of the radiation absorbed when HCl in its ground state is excited to the $v = 1$ excited state.

(*b*) Repeat for DCl.

Note: Infrared spectroscopy of HCl and DCl in an inert solvent (which dampens rotational motions) shows absorption of radiation of 2886 and 2070 cm^{-1}, respectively.

3-53. Use the recurrence relation [Eq. (3-12)] to show that the transition moments for the $k \rightarrow k+1$ and $k \rightarrow k-1$ transitions are given by

$$\langle \psi_{k+1} | \xi | \psi_k \rangle = \sqrt{\frac{k+1}{2}} \qquad \text{and} \qquad \langle \psi_{k-1} | \xi | \psi_k \rangle = \sqrt{\frac{k}{2}}$$

SUGGESTED READINGS

Barrow, G. M.: *Introduction to Molecular Spectroscopy*, McGraw-Hill, New York, 1962. Contains a simplified description of how matter and radiation interact.

Kauzmann, W.: *Quantum Chemistry*, Academic, New York, 1957. One of the best texts for integrating the physics and mathematics of quantum theory.

Pauling, L., and E. B. Wilson, Jr.: *Introduction to Quantum Mechanics*, McGraw-Hill, New York, 1935. This classic text has never been equaled in its treatment of quantum theory by means of the standard differential equations approach.

CHAPTER 4

QUANTUM THEORY OF ANGULAR MOMENTUM: THE RIGID ROTATOR

Angular momentum plays an important role in many physical and chemical systems in which quantum effects are dominant. In certain systems angular momentum is a constant of the motion and thus is useful in classifying the quantum states. In this chapter we will examine the construction of quantum mechanical operators for angular momentum and its components, and we will see how to use ladder operators to deduce the eigenfunctions and eigenvalues of these operators. Finally, we will use the results to solve the Schrödinger equation of the rigid rotator: a system of two masses a fixed distance apart which rotates about its center of mass. Some of these results will be employed in the next chapter to solve the Schrödinger equation of the hydrogen atom.

4-1 REVIEW OF CLASSICAL ANGULAR MOMENTUM

Consider a particle of mass m following a circular path of radius r about the origin (see Fig. 4-1). The classical angular momentum of the particle is defined (in cartesian coordinates) by the vector product

$$\mathbf{L} = \mathbf{r} \times \mathbf{p} = \begin{vmatrix} \mathbf{i} & \mathbf{j} & \mathbf{k} \\ x & y & z \\ p_x & p_y & p_z \end{vmatrix} \tag{4-1}$$

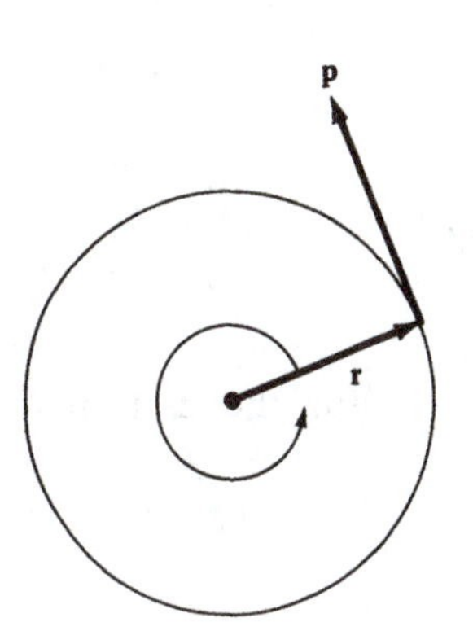

FIGURE 4-1
Classical orbital angular momentum. In this case the angular momentum, $\mathbf{L} = \mathbf{r} \times \mathbf{p}$, is a vector quantity perpendicular to the plane of the rotation and pointing toward the reader.

where $\mathbf{r} = \mathbf{i}x + \mathbf{j}y + \mathbf{k}z$ and $\mathbf{p} = \mathbf{i}p_x + \mathbf{j}p_y + \mathbf{k}p_z$ (see App. 4, Exercise A4-8). The cartesian components of $\mathbf{L}$ are

$$L_x = yp_z - zp_y \qquad L_y = zp_x - xp_z \qquad L_z = xp_y - yp_x \tag{4-2}$$

[see Eq. (A4-9) in App. 4]. (Note that the components are related by a cyclic permutation of the variables x, y, and z.) Of particular interest is the scalar square of the angular momentum:

$$\mathbf{L} \cdot \mathbf{L} = L^2 = L_x^2 + L_y^2 + L_z^2 \tag{4-3}$$

It is useful to examine the behavior of $\mathbf{L}$ as a function of time. First, consider the linear momentum $\mathbf{p}$. According to Newton's second law, the force on a single particle of mass m undergoing a constant acceleration $\mathbf{a}$ is given by

$$\mathbf{F} = m\mathbf{a} = \frac{d(m\mathbf{v})}{dt} = \frac{d\mathbf{p}}{dt} = \dot{\mathbf{p}} \tag{4-4}$$

It is apparent that if the total force on the particle is zero, then linear momentum is conserved. To analyze angular momentum in an analogous manner, consider the moment of force (torque) about the origin given by

$$\mathbf{T} = \mathbf{r} \times \mathbf{F} = \mathbf{r} \times \dot{\mathbf{p}} \tag{4-5}$$

Differentiating $\mathbf{L}$ in Eq. (4-1) with respect to time and recognizing that $\mathbf{v} \times \mathbf{v} = 0$, we get

$$\dot{\mathbf{L}} = \mathbf{r} \times \dot{\mathbf{p}} = \mathbf{T} \tag{4-6}$$

thus, angular momentum is conserved if there is no torque (e.g., a tipping force on a spinning top) acting on the particle.

The quantum mechanical operators for angular momentum and its components are readily constructed from the classical equation (4-1) and the rules illustrated in Sec. 2-9. In fact, we have already obtained the angular momentum operator in Eq. (2-71):

$$\hat{\mathbf{L}} = \frac{\hbar}{i}\begin{vmatrix} \mathbf{i} & \mathbf{j} & \mathbf{k} \\ x & y & z \\ \dfrac{\partial}{\partial x} & \dfrac{\partial}{\partial y} & \dfrac{\partial}{\partial z} \end{vmatrix} = \frac{\hbar}{i}\,\hat{\mathbf{r}} \times \hat{\nabla} \tag{4-7}$$

Use of the components in Eq. (4-2) leads to operators for the cartesian components of **L**:

$$\begin{aligned} \hat{L}_x &= \frac{\hbar}{i}\left(y\frac{\partial}{\partial z} - z\frac{\partial}{\partial y}\right) \\ \hat{L}_y &= \frac{\hbar}{i}\left(z\frac{\partial}{\partial x} - x\frac{\partial}{\partial z}\right) \\ \hat{L}_z &= \frac{\hbar}{i}\left(x\frac{\partial}{\partial y} - y\frac{\partial}{\partial x}\right) \end{aligned} \tag{4-8}$$

The order of operators such as y and $\partial/\partial z$ is immaterial, since they commute; the same is not true, of course, for operator pairs such as x and $\partial/\partial x$, y and $\partial/\partial y$, and z and $\partial/\partial z$.

The quantum mechanical operator for L^2 is

$$\hat{\mathbf{L}}\cdot\hat{\mathbf{L}} = \hat{L}^2 = \hat{L}_x^2 + \hat{L}_y^2 + \hat{L}_z^2 \tag{4-9}$$

EXERCISES

4-1. Show that $\hat{L}^2$ and its components are linear, self-adjoint operators.

4-2. Given that $\varphi = \arctan y/x$, show that $e^{i\varphi}$ is an eigenfunction of $\hat{L}_z$ [Eq. (4-8)] with the eigenvalue $\hbar$. Note that $d(\arctan u) = [1/(1+u^2)]\,du$.

4-2 COMMUTATION PROPERTIES OF THE ANGULAR MOMENTUM OPERATORS

As we have done previously in the case of the harmonic oscillator (Sec. 3-6), it is convenient to convert all operators to dimensionless form. This is done simply by dividing $\hat{L}^2$ by $\hbar^2$ and by dividing $\hat{L}_x$, $\hat{L}_y$, and $\hat{L}_z$ by $\hbar$. Note that $\hbar$ has the units of angular momentum (J · s in SI units).

Many of the important properties of angular momentum operators are readily derived as a consequence of their commutation relations. The commutation properties of the components of $\hat{\mathbf{L}}$ are obtained from Eq. (4-8) as follows:

$$\begin{aligned} \hat{L}_x\hat{L}_y &= -\left(y\frac{\partial}{\partial z} - z\frac{\partial}{\partial y}\right)\left(z\frac{\partial}{\partial x} - x\frac{\partial}{\partial z}\right) \\ &= -\left(y\frac{\partial}{\partial z}z\frac{\partial}{\partial x} - y\frac{\partial}{\partial z}x\frac{\partial}{\partial z} - z\frac{\partial}{\partial y}z\frac{\partial}{\partial x} + z\frac{\partial}{\partial y}x\frac{\partial}{\partial z}\right) \end{aligned} \tag{4-10}$$

Similarly,

$$\hat{L}_y\hat{L}_x = -\left(z\frac{\partial}{\partial x}y\frac{\partial}{\partial z} - z\frac{\partial}{\partial x}z\frac{\partial}{\partial y} - x\frac{\partial}{\partial z}y\frac{\partial}{\partial z} + x\frac{\partial}{\partial z}z\frac{\partial}{\partial y}\right) \tag{4-11}$$

Subtracting Eq. (4-11) from Eq. (4-10) and simplifying leads to

$$\hat{L}_x\hat{L}_y - \hat{L}_y\hat{L}_x = [\hat{L}_x, \hat{L}_y] = i\hat{L}_z \tag{4-12}$$

The remaining commutators follow at once by use of the cyclic symmetry of x, y, and z:

$$[\hat{L}_y, \hat{L}_z] = i\hat{L}_x \qquad [\hat{L}_z, \hat{L}_x] = i\hat{L}_y \tag{4-13}$$

Next, we determine how $\hat{L}^2$ commutes with the components of $\hat{\mathbf{L}}$. Considering $\hat{L}^2$ and $\hat{L}_x$ first, we can use Eq. (4-12) to write

$$[\hat{L}_x, \hat{L}_y]\hat{L}_y = i\hat{L}_z\hat{L}_y \tag{4-14}$$

Similarly,

$$\hat{L}_y[\hat{L}_x, \hat{L}_y] = i\hat{L}_y\hat{L}_z \tag{4-15}$$

Adding the last two equations gives

$$[\hat{L}_x, \hat{L}_y^2] = i(\hat{L}_z\hat{L}_y + \hat{L}_y\hat{L}_z) \tag{4-16}$$

Beginning with $[\hat{L}_z, \hat{L}_x]\hat{L}_z$ and repeating the procedure from Eqs. (4-14) to (4-16), we get

$$[\hat{L}_z^2, \hat{L}_x] = i(\hat{L}_z\hat{L}_y + \hat{L}_y\hat{L}_z) \tag{4-17}$$

Subtracting Eq. (4-17) from Eq. (4-16) gives

$$[\hat{L}_x, \hat{L}_y^2] - [\hat{L}_z^2, \hat{L}_x] = 0$$

or, alternatively,

$$\hat{L}_x(\hat{L}_y^2 + \hat{L}_z^2) - (\hat{L}_y^2 + \hat{L}_z^2)\hat{L}_x = 0 \tag{4-18}$$

Using Eq. (4-9) to replace $\hat{L}_y^2 + \hat{L}_z^2$ with $\hat{L}^2 - \hat{L}_x^2$ we obtain

$$\hat{L}_x(\hat{L}^2 - \hat{L}_x^2) - (\hat{L}^2 - \hat{L}_x^2)\hat{L}_x = \hat{L}_x\hat{L}^2 - \hat{L}^2\hat{L}_x = [\hat{L}_x, \hat{L}^2] = 0 \tag{4-19}$$

As before, the cyclic symmetry of x, y, and z allows us to write down at once the remaining commutators:

$$[\hat{L}_y, \hat{L}^2] = [\hat{L}_z, \hat{L}^2] = 0 \tag{4-20}$$

The physical significance of the various angular momentum commutation relations is evident from the uncertainty principle; since $\hat{L}^2$ commutes with all the components of $\hat{\mathbf{L}}$ but these components do not commute with themselves, it follows that only $\hat{L}^2$ and *one* of the components of $\hat{\mathbf{L}}$ are simultaneously measurable. Classically, this means that the circular orbital motion of the particle about the origin behaves as follows: the component of **L** about an axis normal to the plane of motion is conserved if **L** is conserved. However, the other two cartesian components (with axes at right angles to the axis of rotation) both change individually with time, although their resultant is conserved. Also, according to Theorem 2-6, the operators $\hat{L}^2$ and just one of the set ($\hat{L}_x$, $\hat{L}_y$, or $\hat{L}_z$) can have common eigenfunctions.

EXERCISES

4-3. Show that the commutators in Eqs. (4-12) and (4-13) are the cartesian components of $\hat{\mathbf{L}} \times \hat{\mathbf{L}}$.

4-4. Show that the commutation relations in Eqs. (4-12) and (4-13) are derivable from the relationship $\hat{\mathbf{L}} \times \hat{\mathbf{L}} = i\hat{\mathbf{L}}$. Why is this not zero as in ordinary vector algebra?

4-3 ANGULAR MOMENTUM IN SPHERICAL POLAR COORDINATES

The operators $\hat{L}^2$, $\hat{L}_x$, $\hat{L}_y$, and $\hat{L}_z$ are often conveniently expressed in spherical polar coordinates. The reason for this—as we will note later—is that the simultaneous eigenfunctions of $\hat{L}^2$ and one of the components of $\hat{\mathbf{L}}$ ($\hat{L}_x$, $\hat{L}_y$, or $\hat{L}_z$) are also eigenfunctions of certain hamiltonian operators, particularly the hamiltonians of systems with spherically symmetric potential energies, namely, the rigid rotator (Sec. 4-9) and the hydrogen atom (Chap. 5).

The transformation equations from the cartesian coordinates (x, y, z) to the spherical coordinates (r, θ, φ) are

$$x = r \sin\theta \cos\varphi \qquad y = r \sin\theta \sin\varphi \qquad z = r\cos\theta \tag{4-21}$$

The inverse relations are

$$r = \sqrt{x^2 + y^2 + z^2} \qquad \theta = \cos^{-1}\frac{z}{r} \qquad \varphi = \tan^{-1}\frac{y}{x} \tag{4-22}$$

The coordinates are defined in the intervals

$$0 \le r \le \infty \qquad 0 \le \theta \le \pi \qquad 0 \le \varphi \le 2\pi \qquad -\infty \le x, y, z \le \infty \tag{4-23}$$

The geometrical relationships between the two coordinate systems are illustrated in Fig. 4-2.[1]

[1] Examination of Fig. 4-2 should lead one to anticipate that $\hat{L}_z$ will turn out to be the component of $\hat{\mathbf{L}}$ which shares eigenfunctions of $\hat{L}^2$. This is merely the result of the arbitrary labeling of axes; one could just as well label the axis about which planar rotation occurs as either x or y.

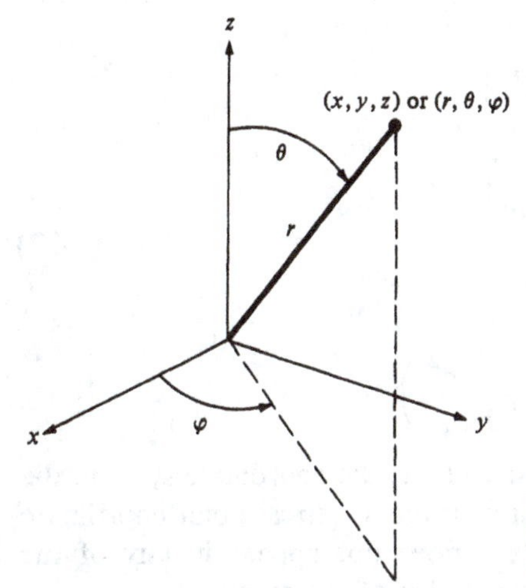

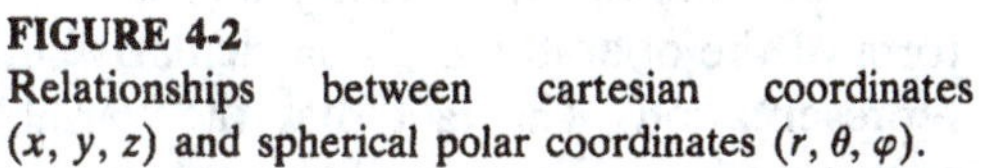
FIGURE 4-2
Relationships between cartesian coordinates (x, y, z) and spherical polar coordinates (r, θ, φ).

Two other transformation equations are of importance. If we let q represent one of the coordinates x, y, or z, the first such relationship is

$$\frac{\partial}{\partial q} = \frac{\partial r}{\partial q}\frac{\partial}{\partial r} + \frac{\partial \theta}{\partial q}\frac{\partial}{\partial \theta} + \frac{\partial \varphi}{\partial q}\frac{\partial}{\partial \varphi} \tag{4-24}$$

This relationship allows one to transform differential operators in (x, y, z) to differential operators in (r, θ, φ). [Note that for higher derivatives, $\partial^n/\partial q^n = (\partial/\partial q)^n$.] The second important transformation equation is needed to transform integrals in (x, y, z) to integrals in (r, θ, φ):

$$\iiint_{-\infty}^{\infty} f(x, y, z)\, dx\, dy\, dz = \int_0^{\infty}\int_0^{2\pi}\int_0^{\pi} g(r, \theta, \varphi) \left|\frac{\partial(z, y, z)}{\partial(r, \theta, \varphi)}\right| d\theta\, d\varphi\, dr \tag{4-25}$$

where

$$\left|\frac{\partial(x, y, z)}{\partial(r, \theta, \varphi)}\right| = \begin{vmatrix} x_r & x_\theta & x_\varphi \\ y_r & y_\theta & y_\varphi \\ z_r & z_\theta & z_\varphi \end{vmatrix} \tag{4-26}$$

and where $x_r = \partial x/\partial r$, $x_\theta = \partial x/\partial \theta$, etc. The determinant (4-26) is called the *jacobian* of the transformation. Note that the second integral in Eq. (4-25) is to be integrated using the *inside-out* convention; $d\theta$ is integrated from 0 to π, $d\varphi$ from 0 to 2π, and dr from 0 to ∞ (the order of integration is immaterial, however).

By the use of Eq. (4-24) it is a straightforward (but tedious) operation to show that

$$\hat{L}_x = i\left(\sin\varphi\,\frac{\partial}{\partial\theta} + \cot\theta\cos\varphi\,\frac{\partial}{\partial\varphi}\right)$$

$$\hat{L}_y = i\left(\cot\theta\sin\varphi\,\frac{\partial}{\partial\varphi} - \cos\varphi\,\frac{\partial}{\partial\theta}\right)$$

$$\hat{L}_z = \frac{1}{i}\frac{\partial}{\partial\varphi} = -i\,\frac{\partial}{\partial\varphi} \tag{4-27}$$

$$\hat{L}^2 = -\left(\frac{1}{\sin\theta}\frac{\partial}{\partial\theta}\sin\theta\,\frac{\partial}{\partial\theta} + \frac{1}{\sin^2\theta}\frac{\partial^2}{\partial\varphi^2}\right)$$

Because of the arbitrary way we chose to label the coordinate system, the form of the operator $\hat{L}_z$ is particularly simple in the spherical polar coordinate representation. The fact that the variable r does not appear in any of the operators is a reflection of the spherical symmetry of the rotation.

EXERCISES

4-5. Show that the jacobian determinant [Eq. (4-26)] is equal to $r^2 \sin\theta$.

4-6. Generalize Eq. (4-25) to the transformation from the coordinate system $(x_1, x_2, \ldots, x_n)$ to $(x'_1, x'_2, \ldots, x'_n)$.

4-7. Verify the form of $\hat{L}_z$ in Eqs. (4-27).

4-8. Show that $e^{in\varphi}$ (n is a constant) is an eigenfunction of $\hat{L}_z$ (see Exercise 4-2). Is $e^{in\varphi}$ also an eigenfunction of $\hat{L}_x$, $\hat{L}_y$, or $\hat{L}^2$? Explain.

4-4 LADDER OPERATORS FOR ANGULAR MOMENTUM

Analogous to the solution of the harmonic oscillator Schrödinger equation by the ladder operator technique (Sec. 3-6), the eigenvalues and eigenfunctions of $\hat{L}^2$ and $\hat{L}_z$ may be deduced in a similar fashion. We consider the operator

$$\hat{L}^2 - \hat{L}_z^2 = \hat{L}_x^2 + \hat{L}_y^2 \tag{4-28}$$

Using Eq. (3-85) and noting that the commutator of $\hat{L}_x$ and $\hat{L}_y$ is $i\hat{L}_z$, we can write

$$\hat{L}^2 - \hat{L}_z^2 = (\hat{L}_x + i\hat{L}_y)(\hat{L}_x - i\hat{L}_y) - \hat{L}_z = (\hat{L}_x - i\hat{L}_y)(\hat{L}_x + i\hat{L}_y) + \hat{L}_z \tag{4-29}$$

The ladder operators in the above are given the compact representations

$$\hat{L}_x + i\hat{L}_y = \hat{L}_+$$

$$\hat{L}_x - i\hat{L}_y = \hat{L}_- \tag{4-30}$$

Thus Eq. (4-29) becomes

$$\hat{L}^2 - \hat{L}_z^2 = \hat{L}_+\hat{L}_- - \hat{L}_z = \hat{L}_-\hat{L}_+ + \hat{L}_z \tag{4-31}$$

We now prove a useful relationship:

$$\hat{L}_z\hat{L}_+ = \hat{L}_+(\hat{L}_z + 1) \tag{4-32}$$

Proof. Replace $\hat{L}_+$ by its definition in Eq. (4-30) and expand the left-hand side of Eq. (4-32):

$$\hat{L}_z(\hat{L}_x + i\hat{L}_y) = \hat{L}_z\hat{L}_x + i\hat{L}_z\hat{L}_y$$

Using the commutation relations involving $\hat{L}_z\hat{L}_x$ and $\hat{L}_z\hat{L}_y$, we rewrite the above as

$$\begin{aligned}\hat{L}_z\hat{L}_+ &= \hat{L}_x\hat{L}_z + i\hat{L}_y + i(\hat{L}_y\hat{L}_z - i\hat{L}_x) = (\hat{L}_x + i\hat{L}_y)\hat{L}_z + (\hat{L}_x + i\hat{L}_y)\\ &= \hat{L}_+\hat{L}_z + \hat{L}_+ = \hat{L}_+(\hat{L}_z + 1) \quad \text{QED}\end{aligned}$$

Similarly, one can show

$$\hat{L}_z\hat{L}_- = \hat{L}_-(\hat{L}_z - 1) \tag{4-33}$$

The relationships in Eqs. (4-32) and (4-33) may be used to demonstrate that $\hat{L}_+$ and $\hat{L}_-$ are step-up and step-down operators, respectively. We let $Y_{\alpha,\beta}(\theta, \varphi)$ represent the yet unknown simultaneous eigenfunctions of $\hat{L}^2$ and $\hat{L}_z$ which satisfy the eigenvalue equations

$$\hat{L}^2Y_{\alpha,\beta} = \alpha Y_{\alpha,\beta} \qquad \text{and} \qquad \hat{L}_zY_{\alpha,\beta} = \beta Y_{\alpha,\beta} \tag{4-34}$$

where α and β are the eigenvalues (in units of $\hbar^2$ and $\hbar$, respectively) of $\hat{L}^2$ and $\hat{L}_z$, respectively. Operating on $\hat{L}_+Y_{\alpha,\beta}$ with $\hat{L}_z$ produces

$$\hat{L}_z(\hat{L}_+Y_{\alpha,\beta}) = (\hat{L}_z\hat{L}_+)Y_{\alpha,\beta} = \hat{L}_+(\hat{L}_z + 1)Y_{\alpha,\beta} = (\beta + 1)\hat{L}_+Y_{\alpha,\beta} \tag{4-35}$$

Similarly, operating on $\hat{L}_-Y_{\alpha,\beta}$ with $\hat{L}_z$ leads to

$$\hat{L}_z(\hat{L}_-Y_{\alpha,\beta}) = (\beta - 1)Y_{\alpha,\beta} \tag{4-36}$$

Thus, $\hat{L}_+$ and $\hat{L}_-$ are step-up and step-down operators, respectively, with respect to the eigenvalues of $\hat{L}_z$.

As the reader can readily verify, both $\hat{L}_+$ and $\hat{L}_-$ commute with $\hat{L}^2$. Thus we obtain

$$\begin{aligned}\hat{L}^2(\hat{L}_+Y_{\alpha,\beta}) &= \hat{L}_+\hat{L}^2Y_{\alpha,\beta} = \alpha(\hat{L}_+Y_{\alpha,\beta})\\ \hat{L}^2(\hat{L}_-Y_{\alpha,\beta}) &= \alpha(\hat{L}_-Y_{\alpha,\beta})\end{aligned} \tag{4-37}$$

which shows that the step-up and step-down operators have no effect on the eigenvalues of $\hat{L}^2$.

The results of Eqs. (4-35) through (4-37) imply the relationships

$$\hat{L}_+ Y_{\alpha,\,\beta} = C_+ Y_{\alpha,\,\beta+1}$$
$$\hat{L}_- Y_{\alpha,\,\beta} = C_- Y_{\alpha,\,\beta-1} \tag{4-38}$$

where C_+ and C_- are numerical constants.

EXERCISES

4-9. Express $\hat{L}_+$ and $\hat{L}_-$ in spherical polar coordinates. These representations will be used in Sec. 4-6.

4-10. Find the commutators of
(*a*) $\hat{L}_+$ and $\hat{L}_-$ (*b*) $\hat{L}^2$ and $\hat{L}_+$ (*c*) $\hat{L}^2$ and $\hat{L}_-$

4-11. Show that the operators $\hat{L}_+\hat{L}_-$ and $\hat{L}_-\hat{L}_+$ have positive (or zero) expectation values only.

4-12. If $\hat{\pi}$ is the parity operator (see Exercise 3-30), show that $\hat{\pi}r = r$, $\hat{\pi}\theta = \pi - \theta$, and $\hat{\pi}\varphi = \varphi + \pi$ (r, θ, and φ are spherical coordinates; *warning*: Do not confuse $\hat{\pi}$ and π).

4-5 THE EIGENVALUES OF $\hat{L}_z$ AND $\hat{L}^2$

Since the operators $\hat{L}_+$ and $\hat{L}_-$ are adjoints of each other, it follows (see Exercise 4-11) that

$$\langle Y_{\alpha,\,\beta} | \hat{L}_-\hat{L}_+ | Y_{\alpha,\,\beta} \rangle \geq 0 \tag{4-39}$$

Substituting for $\hat{L}_-\hat{L}_+$ from Eq. (4-31), we obtain

$$\langle Y_{\alpha,\,\beta} | \hat{L}^2 - \hat{L}_z(\hat{L}_z + 1) | Y_{\alpha,\,\beta} \rangle = \alpha - \beta(\beta+1) \geq 0 \tag{4-40}$$

where we have assumed, without any loss in generality, that $Y_{\alpha,\,\beta}$ is normalized. Similarly, we obtain

$$\langle Y_{\alpha,\,\beta} | \hat{L}_+\hat{L}_- | Y_{\alpha,\,\beta} \rangle = \langle Y_{\alpha,\,\beta} | \hat{L}^2 - \hat{L}_z(\hat{L}_z - 1) | Y_{\alpha,\,\beta} \rangle = \alpha - \beta(\beta-1) \geq 0 \tag{4-41}$$

Adding Eqs. (4-40) and (4-41), we get (after simplification)

$$\alpha \geq \beta^2 \tag{4-42}$$

This implies that for a given value of α there exists a minimum and a maximum value of β; these we designate as $\underline{\beta}$ and $\bar{\beta}$, respectively.

From the ladder properties of $\hat{L}_+$ and $\hat{L}_-$ and the implications of Eq. (4-42), we conclude that

$$\hat{L}_+ Y_{\alpha,\,\bar{\beta}} = 0 \qquad \hat{L}_- Y_{\alpha,\,\underline{\beta}} = 0 \tag{4-43}$$

This means that $\hat{L}_+$ and $\hat{L}_-$ annihilate the eigenfunctions having the maximum and minimum eigenvalues of $\hat{L}_z$, respectively, for a given value of the $\hat{L}^2$

eigenvalue. Thus, from Eqs. (4-40) and (4-41) we obtain

$$\alpha - \bar{\beta}(\bar{\beta}+1) = 0 \qquad \alpha - \underline{\beta}(\underline{\beta}-1) = 0 \tag{4-44}$$

or, alternatively,

$$\alpha = \bar{\beta}(\bar{\beta}+1) = \underline{\beta}(\underline{\beta}-1) \tag{4-45}$$

This is a quadratic equation in $\bar{\beta}$ (or $\underline{\beta}$):

$$\bar{\beta}^2 + \bar{\beta} - \underline{\beta}(\underline{\beta}-1) = 0 \tag{4-46}$$

The solutions are $\bar{\beta} = \underline{\beta} - 1$ and $-\underline{\beta}$. The first root is rejected as extraneous (since it leads to a maximum value *smaller* than the minimum value), so we are left with

$$\bar{\beta} = -\underline{\beta} \tag{4-47}$$

This means that the eigenvalues of $\hat{L}_z$ are symmetric about zero. Since successive application of $\hat{L}_+$ to $Y_{\alpha,\underline{\beta}}$ will generate eigenfunctions of $\hat{L}_z$ with eigenvalues $\underline{\beta}+1, \underline{\beta}+2, \ldots, \bar{\beta}-1, \bar{\beta}$, we see that

$$\bar{\beta} - \underline{\beta} = 2\bar{\beta} = 0, 1, 2, \ldots, \text{ an integer} \tag{4-48}$$

Two different possibilities now exist: if $2\bar{\beta}$ is even, the β's are integral; otherwise, they are half-integral. These relationships are illustrated in Fig. 4-3 for the specific cases of $2\bar{\beta} = 3$ and $2\bar{\beta} = 4$.

To conform to customary symbolic usage, we replace $\bar{\beta}$ with l and β with m_l. To summarize:

$$\begin{aligned} \bar{\beta} = -\underline{\beta} = l &= 0, 1, 2, \ldots \qquad \text{or} \qquad \tfrac{1}{2}, \tfrac{3}{2}, \tfrac{5}{2}, \ldots \\ \beta = m_l &= 0, \pm 1, \pm 2, \ldots \qquad \text{or} \qquad \pm\tfrac{1}{2}, \pm\tfrac{3}{2}, \pm\tfrac{5}{2}, \ldots \end{aligned} \tag{4-49}$$

Thus, the eigenvalues of $\hat{L}^2$ are

$$\alpha = l(l+1) \tag{4-50}$$

and those of $\hat{L}_z$ are m_l, with the restriction

$$-l \le m_l \le l \tag{4-51}$$

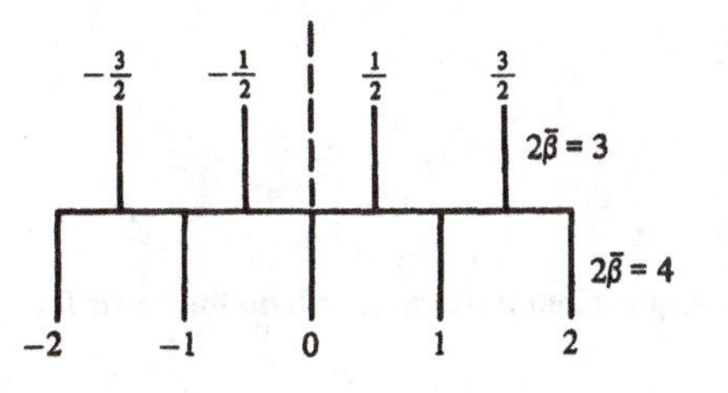

FIGURE 4-3
Eigenvalues of $\hat{L}_z$ for the two cases: $2\bar{\beta} = 3$ and $2\bar{\beta} = 4$.

Since each $\hat{L}^2$ eigenvalue is associated with $2l+1$ different values of m_l, we note that the $\hat{L}^2$ eigenfunctions are $(2l+1)$-fold degenerate. The numbers l and m_l are quantum numbers, since they denote the multiples of $\hbar^2$ and $\hbar$, respectively, which are the allowed values of L^2 and its z component, respectively.

The eigenvalue $m_l\hbar$ may be interpreted physically as the projection of the total angular momentum $|\mathbf{L}| = [\sqrt{l(l+1)}]\hbar$ onto the z axis. Looking at Fig. 4-4, we see that the angle θ between $\mathbf{L}$ and the z axis is given in general by

$$\cos\theta = \frac{m_l}{\sqrt{l(l+1)}} \tag{4-52}$$

Note that only for the trivial case of $l=0$ will the angle θ be zero. If this angle were generally zero, m_l would always equal $\sqrt{l(l+1)}$ and we would be able to determine not only L^2 but all three components of $\mathbf{L}$ as well. This would imply that the axis of rotation is known precisely. If we call this axis z, then the *linear* momentum parallel to this axis (p_z) is exactly zero and the uncertainty Δp_z is zero. Since the uncertainty of the particle's position (Δz) along the z axis is finite, $\Delta p_z \Delta z$ would be zero—a violation of the uncertainty principle.

If we use the turnover rule and Eq. (4-38), Eq. (4-39) may be written

$$\langle Y_{l,\,m_l} | \hat{L}_-\hat{L}_+ | Y_{l,\,m_l}\rangle = \langle \hat{L}_+ Y_{l,\,m_l} | \hat{L}_+ Y_{l,\,m_l}\rangle = |C_+|^2 \tag{4-53}$$

where it is assumed that $Y_{l,\,m_l}$ is normalized. Then, using Eqs. (4-50) and (4-40) we obtain

$$|C_+|^2 = l(l+1) - m_l(m_l+1) \tag{4-54}$$

Similarly, beginning with Eq. (4-41) we can get

$$|C_-|^2 = l(l+1) - m_l(m_l-1) \tag{4-55}$$

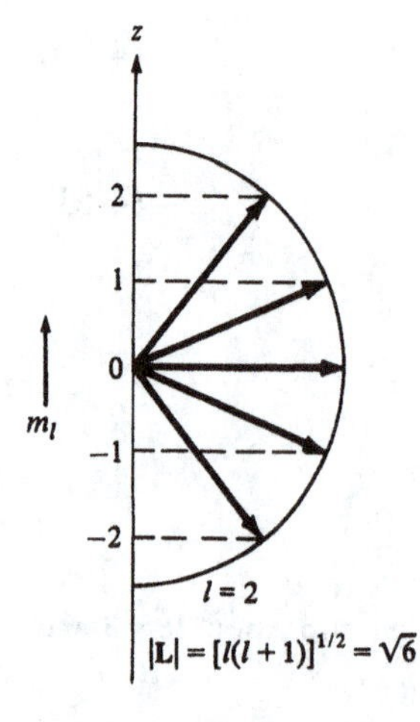

FIGURE 4-4
The eigenvalue m_l as a quantized projection of $\mathbf{L}$ on the z axis for $l=2$.

Letting $C_+ = |C_+|$ and $C_- = |C_-|$, we may rewrite the expressions in Eq. (4-38) as

$$\hat{L}_+ Y_{l,\,m_l} = [l(l+1) - m_l(m_l+1)]^{1/2} Y_{l,\,m_l+1}$$
$$\hat{L}_- Y_{l,\,m_l} = [l(l+1) - m_l(m_l-1)]^{1/2} Y_{l,\,m_l-1} \tag{4-56}$$

In general, C_+ and C_- have complex phase factors of unit modulus which may be ignored. The above relationships are useful for generating eigenfunctions of $\hat{L}^2$ and $\hat{L}_z$ from a given starting eigenfunction.

Note that the relationships leading to the eigenvalues of $\hat{L}^2$ and $\hat{L}_z$ were derived solely on the basis of how $\hat{L}^2$ and the components of $\hat{\mathbf{L}}$ commuted. Thus, the results are quite general for any set of operators satisfying an analogous set of commutation relations.

EXERCISES

4-13. Calculate the possible values of the angle θ between $\mathbf{L}$ and the z axis when $l = 2$. What do these angles represent?

4-14. What are the eigenvalues, if any, of the operator $\hat{L}_x^2$? of $\hat{L}_y^2$? of $\hat{L}_x^2 + \hat{L}_y^2$?

4-6 THE SIMULTANEOUS EIGENFUNCTIONS OF $\hat{L}_z$ AND $\hat{L}^2$

The simultaneous eigenfunctions of $\hat{L}_z$ and $\hat{L}^2$ are obtained by methods similar to those employed in the harmonic oscillator case (Sec. 3-6), but the algebra is somewhat more involved and tedious. Hence, only a sketch of the general approach is given here.

Consider the special case $l = m_l$. From Eq. (4-43) we can write

$$\hat{L}_+ Y_{l,\,l}(\theta, \varphi) = 0 \tag{4-57}$$

If $\hat{L}_+$ is written in terms of spherical polar coordinates (see Exercise 4-9), this equation becomes

$$e^{i\varphi}\left(\frac{\partial}{\partial\theta} + i\cot\theta\,\frac{\partial}{\partial\varphi}\right) Y_{l,\,l} = 0 \tag{4-58}$$

Since $e^{i\varphi} \neq 0$, we can simplify the above to

$$\left(\frac{\partial}{\partial\theta} + i\cot\theta\,\frac{\partial}{\partial\varphi}\right) Y_{l,\,l} = 0 \tag{4-59}$$

We now solve this partial differential equation by the separation of variables technique. The separation for the general case is

$$Y_{l,\,m_l}(\theta, \varphi) = \Theta_{l,\,m_l}(\theta)\Phi_{m_l}(\varphi) \tag{4-60}$$

The distribution of subscripts is not obvious at this point; only hindsight will indicate how this must be done. Substituting $Y_{l,\,l} = \Theta_{l,\,l}\Phi_l$ into Eq. (4-59) and separating the variables leads to

$$(\cot\theta\Theta_{l,\,l})^{-1}\frac{\partial\Theta_{l,\,l}}{\partial\theta} = -\frac{i}{\Phi_l}\frac{\partial\Phi_l}{\partial\varphi} = l \tag{4-61}$$

The separation constant is identified as l, since it satisfies the eigenvalue equation

$$\frac{1}{i}\frac{\partial\Phi_l}{\partial\varphi} = \hat{L}_z\Phi_l = l\Phi_l \tag{4-62}$$

This leads to the two ordinary differential equations:

$$\frac{d\Phi_{m_l}}{d\varphi} - m_l\Phi_{m_l} = 0 \qquad \frac{d\Theta_{l,\,l}}{d\theta} - l\cot\theta\Theta_{l,\,l} = 0 \tag{4-63}$$

The first expression in Eqs. (4-63) has the general solution

$$\Phi_{m_l} = N_{m_l}e^{im_l\varphi} \tag{4-64}$$

where N_{m_l} is a normalization constant given by

$$|N_{m_l}| = \left[\int_0^{2\pi}(e^{im_l\varphi})^*(e^{im_l\varphi})\,d\varphi\right]^{-1/2} = \left(\int_0^{2\pi}d\varphi\right)^{-1/2} = \frac{1}{\sqrt{2\pi}} \tag{4-65}$$

The exponential function [Eq. (4-64)] is quadratically integrable (as evidenced by the normalization) and continuous, but it is not single-valued, since the value of φ is not the same at a physically equivalent position $\varphi + 2\pi$. To avoid this ambiguity we must require that

$$\Phi_{m_l}(\varphi) = \Phi_{m_l}(\varphi + 2\pi) \tag{4-66}$$

Using Euler's relationship for the complex exponential

$$e^{im_l\varphi} = \cos m_l\varphi + i\sin m_l\varphi \tag{4-67}$$

we see that Eq. (4-66) is satisfied if

$$\cos m_l\varphi = \cos m_l(\varphi + 2\pi) \tag{4-68}$$

with the analogous situation holding for the sine portion. This will be true only if the two angles $m_l\varphi$ and $m_l(\varphi + 2\pi)$ differ by an integral number of multiples of 2π, that is, if

$$m_l\varphi - m_l(\varphi + 2\pi) = \pm n(2\pi) \tag{4-69}$$

where $n = 0, 1, 2, \ldots$. This equation has the solution

$$m_l = \pm n \qquad (=0, \pm 1, \pm 2, \ldots) \tag{4-70}$$

Note that we have obtained the solutions to the special case that l and m_l are integral [see Eqs. (4-48) and (4-49)].

The second expression in [Eq. (4-63)] may be rearranged to

$$d \ln \Theta_{l,\,l} = l \cot \theta \, d\theta = l \, d \ln (\sin \theta) \tag{4-71}$$

having the solution

$$\ln \Theta_{l,\,l} = \ln (\sin^l \theta) + \text{constant} \tag{4-72}$$

or, in the alternative antilogarithmic form,

$$\Theta_{l,\,l} = N_{l,\,l} \sin^l \theta \tag{4-73}$$

The normalization constant is given by

$$|N_{l,\,l}| = \left(\int_0^\pi \sin^{2l} \theta \sin \theta \, d\theta \right)^{-1/2} = \frac{1}{2^l l!} \left[\frac{(2l+1)!}{2} \right]^{1/2} \tag{4-74}$$

(the integration involves l integrations by parts). Combining Eqs. (4-64), (4-65), (4-73), and (4-74), we get

$$Y_{l,\,l}(\theta, \varphi) = \frac{1}{2^l l!} \left[\frac{(2l+1)!}{4\pi} \right]^{1/2} \sin^l \theta e^{il\varphi} \tag{4-75}$$

By repeated use of the step-down operator $\hat{L}_-$ on $Y_{l,\,l}$, and use of Eq. (4-56), one can get $Y_{l,\,m_l}$ for any desired value of m_l.

The general solutions $Y_{l,\,m_l}(\theta, \varphi)$ are called *spherical harmonics*. These functions were known and used by mathematicians and mathematical physicists long before the advent of quantum theory. For example, these functions arise as the normal modes of vibration of an elastic sphere. Also, as we will show later, the spherical harmonics arise in the solution of Laplace's equation—an equation used in the classical physics of gravitation and electrostatics.

The spherical harmonics are usually tabulated in terms of the new variable

$$y = \cos \theta \qquad \sqrt{1 - y^2} = \sin \theta \tag{4-76}$$

The general form of $Y_{l,\,m_l}$ then becomes[2]

$$Y_{l,\,m_l}(\theta, \varphi) = \frac{(-1)^l}{2^l l!} \left[\frac{(2l+1)(l-|m_l|)!}{4\pi(l+|m_l|)!} \right]^{1/2} (1-y^2)^{|m_l|/2} \frac{d^{l+|m_l|}}{dy^{l+|m_l|}} (1-y^2)^l e^{im_l\varphi} \tag{4-77}$$

The absolute value of m_l is required in the θ part, since it turns out that

[2] The factor of $(-1)^l$, which did not appear in Eq. (4-74), is included for convenience; it simply eliminates a minus sign for even values of l. *De gustibus non est disputandem!*

$\Theta_{l,\, m_l} = \Theta_{l,\, -m_l}$. The spherical harmonics may also be written

$$Y_{l,\, m_l}(\theta, \varphi) = N_{l,\, m_l} P_l^{|m_l|}(y) e^{im_l\varphi} \tag{4-78}$$

where $P_l^{|m_l|}$ is called the *associated Legendre function of degree l and order* $|m_l|$. This function is defined by

$$P_l^{|m_l|}(y) = (1 - y^2)^{|m_l|/2} \frac{d^{|m_l|}}{dy^{|m_l|}} P_l(y) \tag{4-79}$$

where $P_l(y)$ is the *Legendre polynomial* defined by the relation

$$P_l(y) = (2^l l!)^{-1} \frac{d^l (y^2 - 1)^l}{dy^l} \tag{4-80}$$

This relation is known as *Rodrigue's formula*. The Legendre polynomials also result when one carries out a Schmidt orthogonalization (see Sec. 2-5) of the functions $1, y, y^2, \ldots$ in the interval $-1 \le y \le 1$. The first few Legendre polynomials are

$$\begin{aligned} P_0(y) &= 1 & P_3(y) &= \tfrac{3}{2}(\tfrac{5}{3}y^3 - y) \\ P_1(y) &= y & P_4(y) &= \tfrac{3}{8}(\tfrac{35}{3}y^4 - 10y^2 + 1) \\ P_2(y) &= \tfrac{1}{2}(3y^2 - 1) \end{aligned} \tag{4-81}$$

It is evident from Theorem 2-3 that the spherical harmonics are orthogonal. If these functions have been normalized, the orthonormality condition is indicated in the compact expression

$$\langle Y_{l,\, m_l} | Y_{l',\, m_{l'}} \rangle = \delta_{l,\, l'} \delta_{m_l,\, m_{l'}}$$

In addition, the functions $\Phi_{m_l}(\varphi)$, $P_l(y)$, and $P_l^{|m_l|}(y)$ are individually orthogonal. In particular

$$\int_{-1}^{1} P_l^{|m_l|}(y) P_{l'}^{|m_l|}(y)\, dy = \begin{cases} 0 & \text{for } l \ne l' \\ \dfrac{2(l + |m_l|)!}{(2l+1)(l - |m_l|)!} & \text{for } l = l' \end{cases} \tag{4-82}$$

EXERCISES

4-15. Use $\hat{L}_- Y_{l,\, l} = 0$ to derive an expression for $Y_{l,\, -l}$ analogous to Eq. (4-75).

4-16. Demonstrate explicitly that the functions $\Phi_{m_l}(\varphi)$ are orthogonal.

4-17. Prove the orthogonality of the Legendre polynomials; i.e., show that

$$\int_{-1}^{1} P_l(y) P_{l'}(y)\, dy = \begin{cases} 0 & \text{for } l \ne l' \\ \dfrac{2}{2l+1} & \text{for } l = l' \end{cases}$$

4-7 THE SPHERICAL HARMONICS IN REAL FORM

Laplace's equation is given by

$$\nabla^2 u = 0 \tag{4-83}$$

where $u = u(x, y, z)$. The laplacian operator in spherical polar coordinates may be written

$$\nabla^2 = \frac{1}{r^2}(\hat{D} - \hat{L}^2) \tag{4-84}$$

where $\hat{L}^2$ is the angular momentum operator given in Eqs. (4-27) and $\hat{D}$ is the operator

$$\hat{D} = \frac{\partial}{\partial r} r^2 \frac{\partial}{\partial r} \tag{4-85}$$

For a fixed, nonzero value of r, Laplace's equation may be written

$$(\hat{D} - \hat{L}^2)u(r, \theta, \varphi) = 0 \tag{4-86}$$

This equation has well-behaved solutions of the form

$$u_l = x^\alpha y^\beta z^\gamma = r^l g(\theta, \varphi) \tag{4-87}$$

which are homogeneous of degree $\alpha + \beta + \gamma = l$ with $l = 0, 1, 2, \ldots$. Thus, each term of u_l has a factor r^l and additional dimensionless terms in θ and φ. Substituting Eq. (4-87) into Eq. (4-86) and carrying out the indicated operations, we get

$$\hat{D}u_l = \left(\frac{\partial}{\partial r} r^2 \frac{\partial}{\partial r}\right) r^l g(\theta, \varphi) = g(\theta, \varphi)\left(\frac{\partial r^2}{\partial r}\right) l r^{l-1}$$

$$= g(\theta, \varphi)\frac{\theta}{\theta r} l r^{l+1} = g(\theta, \varphi) l(l+1) r^l$$

$$\hat{L}^2 u_l = r^l \hat{L}^2 g(\theta, \varphi)$$

Adding the above two equations and dividing by the common factor r^l leads to

$$\hat{L}^2 g(\theta, \varphi) = l(l+1) g(\theta, \varphi) \tag{4-88}$$

Thus, the functions $g(\theta, \varphi)$ are eigenfunctions of $\hat{L}_2$. Since it is not known at this point that these are also eigenfunctions of $\hat{L}_z$, we denote them as $Y_l(\theta, \varphi)$. Using Eq. (4-87), we write

$$Y_l(\theta, \varphi) = \frac{u_l(r, \theta, \varphi)}{r^l} \tag{4-89}$$

Next, we systematically consider all homogeneous functions $x^\alpha y^\beta z^\gamma$ of degree l which satisfy Laplace's equation. We will consider the procedure for the specific cases of $l = 0, 1$, and 2.

For $l = 0$ we set $u_0 = 0$. Thus $Y_0 = u_0/r^0 = 1$. The normalization constant for this function is

$$\left(\int_0^{2\pi}\int_0^{\pi} Y_0^2 \sin\theta \, d\theta \, d\varphi\right)^{-1/2} = \frac{1}{\sqrt{2\pi}} \tag{4-90}$$

Thus, $Y_0(\theta, \varphi) = 1/\sqrt{2\pi}$, which is just the spherical harmonic $Y_{0,0}(\theta, \varphi)$ obtained in the previous section. Note that in this particular case, this is also an eigenfunction of $\hat{L}_z$.

For $l = 1$ there are three different functions which are homogeneous of degree 1 and which satisfy $\nabla^2 u_1 = 0$. These are

$$u_1 = \begin{cases} x \\ y \\ z \end{cases} \tag{4-91}$$

Thus

$$Y_1 = \begin{cases} \dfrac{x}{r} = \sin\theta\cos\varphi \\[2mm] \dfrac{y}{r} = \sin\theta\sin\varphi \\[2mm] \dfrac{z}{r} = \cos\theta \end{cases} \tag{4-92}$$

It is readily verified that only the last function is also an eigenfunction of $\hat{L}_z$, and this has an eigenvalue of $m_l = 0$. Thus we can identify the spherical harmonic (in normalized form)

$$Y_{1,0} = \frac{1}{2}\left(\frac{3}{\pi}\right)^{1/2} \cos\theta \tag{4-93}$$

Let us label the remaining functions (also normalized) as follows:

$$\begin{aligned} Y_{1,\cos\varphi} &= \frac{1}{2}\left(\frac{3}{\pi}\right)^{1/2} \sin\theta\cos\varphi \\ Y_{1,\sin\varphi} &= \frac{1}{2}\left(\frac{3}{\pi}\right)^{1/2} \sin\theta\sin\varphi \end{aligned} \tag{4-94}$$

It is easy to show that these are real forms of the spherical harmonics $Y_{1,1}$ and $Y_{1,-1}$. If we take the linear combinations

$$\begin{aligned} \frac{Y_{1,\cos\varphi} + iY_{1,\sin\varphi}}{\sqrt{2}} &= \frac{1}{2}\left(\frac{3}{\pi}\right)^{1/2} \frac{\sin\theta(\cos\varphi + i\sin\varphi)}{\sqrt{2}} \\ \frac{Y_{1,\cos\varphi} - iY_{1,\sin\varphi}}{\sqrt{2}} &= \frac{1}{2}\left(3\pi\right)^{1/2} \frac{\sin\theta(\cos\varphi - i\sin\varphi)}{\sqrt{2}} \end{aligned} \tag{4-95}$$

we obtain the spherical harmonics

$$
\begin{aligned}
Y_{1,1} &= \frac{1}{2}\left(\frac{3}{2\pi}\right)^{1/2} \sin\theta\, e^{i\varphi} \\
Y_{1,-1} &= \frac{1}{2}\left(\frac{3}{2\pi}\right)^{1/2} \sin\theta\, e^{-i\varphi}
\end{aligned}
\tag{4-96}
$$

In matrix transformation notation this becomes

$$
\frac{1}{\sqrt{2}}\begin{bmatrix} 1 & i \\ 1 & -i \end{bmatrix}\begin{bmatrix} Y_{1,\cos\varphi} \\ Y_{1,\sin\varphi} \end{bmatrix} = \begin{bmatrix} Y_{1,1} \\ Y_{1,-1} \end{bmatrix}
\tag{4-97}
$$

The simple functions which are homogeneous of degree 2 are

$$
u_2 = \begin{cases} x^2 \\ y^2 \\ z^2 \\ xy \\ xz \\ yz \end{cases}
\tag{4-98}
$$

However, only the last three satisfy Laplace's equation. Since we can have only $2l+1=5$ linearly independent functions, we must use the first three to construct two more linearly independent solutions to Laplace's equation. It is easy to verify that the three functions

$$
u_2 = \begin{cases} x^2 - y^2 \\ x^2 - z^2 \\ y^2 - z^2 \end{cases}
\tag{4-99}
$$

satisfy Laplace's equation. We now find two linearly independent combinations of these. Arbitrarily choosing $x^2 - y^2$ as one of these, the sum of the remaining two becomes the second:

$$
(x^2 - z^2) + (y^2 - z^2) = x^2 + y^2 - 2z^2 = x^2 + y^2 + z^2 - 3z^2 = r^2 - 3z^2
\tag{4-100}
$$

Thus, five linearly independent solutions to Laplace's equation which are homogeneous of degree 2 are

$$
u_2 = \begin{cases} r^2 - 3z^2 \\ xz \\ yz \\ x^2 - y^2 \\ xy \end{cases}
\tag{4-101}
$$

The spherical harmonics (normalized and in the same order as above) are

$$Y_{2,\,0} = \frac{1}{4}\left(\frac{5}{\pi}\right)^{1/2}(1 - 3\cos^2\theta)$$
$$Y_{2,\,\cos\varphi} = \frac{1}{2}\left(\frac{15}{\pi}\right)^{1/2}\sin\theta\cos\theta\cos\varphi$$
$$Y_{2,\,\sin\varphi} = \frac{1}{2}\left(\frac{15}{\pi}\right)^{1/2}\sin\theta\cos\theta\sin\varphi \qquad (4\text{-}102)$$
$$Y_{2,\,\cos 2\varphi} = \frac{1}{2}\left(\frac{15}{\pi}\right)^{1/2}\sin^2\theta\cos 2\varphi$$
$$Y_{2,\,\sin 2\varphi} = \frac{1}{2}\left(\frac{15}{\pi}\right)^{1/2}\sin^2\theta\sin 2\varphi$$

Spherical harmonics for which $l = 0, 1, 2, \ldots$ are called $s, p, d, f, \ldots$ modes, respectively. We shall encounter this terminology in the next chapter with respect to the wavefunctions of the one-electron atom. In the order in which they appear in Eq. (4-102), the d modes are called d_{z^2}, d_{xz}, d_{yz}, $d_{x^2-y^2}$, and d_{xy}. Similarly, $Y_{0,\,0}$ is called an s mode, and $Y_{1,\,0}$, $Y_{1,\,\cos\varphi}$, and $Y_{1,\,\sin\varphi}$ are called p_z, p_x, and p_y modes, respectively. The subscripts derive from the cartesian coordinate forms of the spherical harmonics.

EXERCISES

4-18. Use Eq. (4-77) to demonstrate that the functions in Eqs. (4-93) and (4-95) are actually the spherical harmonics $Y_{1,\,0}$ and $Y_{1,\,\pm 1}$.

4-19. Use the functions in Eqs. (4-102) to form the complex spherical harmonics $Y_{2,\,\pm 1}$ and $Y_{2,\,\pm 2}$.

4-20. Use the Laplace equation to generate the seven f modes.

4.8 RECURRENCE RELATIONS FOR LEGENDRE POLYNOMIALS AND ASSOCIATED LEGENDRE FUNCTIONS

The evaluation of integrals containing spherical harmonics is greatly facilitated by the use of various generating functions, similar to those employed earlier for the hermite polynomials. The Legendre polynomials [Eq. (4-80)] may be expressed in terms of the generating function

$$T(y, t) \equiv \sum_{l=0}^{\infty} P_l(y)t^l \equiv (1 - 2ty + t^2)^{-1/2} \qquad (4\text{-}103)$$

Differentiating with respect to the auxiliary variable t, we obtain

$$\frac{\partial T}{\partial t} = \sum_l P_l(y)t^l = (1 - 2ty + t^2)^{-3/2}(y - t) \qquad (4\text{-}104)$$

Multiplying the second and third expressions in Eq. (4-104) by $1-2ty+t^2$ and using Eq. (4-103), we get

$$(1-2ty+t^2)\sum_l P_l(y)t^{l-1} = (y-t)\sum_l P_l(y)t^l \tag{4-105}$$

Comparing coefficients of the term t^l, we obtain the recurrence relation

$$(l+1)P_{l+1}(y) - (2l+1)yP_l(y) + lP_{l-1}(y) = 0 \tag{4-106}$$

Similarly, differentiating Eq. (4-103) with respect to y and equating coefficients of t^l, we obtain a recurrence relationship involving derivatives of the Legendre polynomials, namely,

$$P'_{l+1}(y) - 2yP'_l(y) + P'_{l-1}(y) - P_l(y) = 0 \tag{4-107}$$

A generating function for the associated Legendre function is found by differentiating Eq. (4-103) $|m_l|$ times with respect to y and multiplying by $(1-y^2)^{|m_l|/2}$. One obtains

$$T^{|m_l|}(y,t) = \sum_{l=|m_l|}^{\infty} P_l^{|m_l|}(y)t^l = \frac{C(|m_l|)(1-y^2)^{|m_l|/2}t^{|m_l|}}{(1-2yt+t^2)^{|m_l|+1/2}} \tag{4-108}$$

where $C(|m_l|)$ is defined as

$$C|m_l| = 1\times 3\times 5\times\cdots(2|m_l|-1) = \frac{(2|m_l|)!}{2^{|m_l|}(|m_l|)!}$$

By the same technique used to obtain Eq. (4-106) from Eq. (4-103), one can obtain the recurrence relations for the associated Legendre functions. These are

$$\begin{gathered}(l-|m_l|+1)P_{l+1}^{|m_l|}(y) - (2l+1)yP_l^{|m_l|}(y) + (l+|m_l|)P_{l-1}^{|m_l|}(y) = 0 \\ (2l+1)\sqrt{1-y^2}P_l^{|m_l|-1}(y) = P_{l+1}^{|m_l|}(y) - P_{l-1}^{|m_l|}(y) = 0\end{gathered} \tag{4-109}$$

Both expressions in Eqs. (4-109) are useful in the determination of selection rules.

EXERCISE

4-21. Obtain Eq. (4-80), Rodrigue's formula, by use of the generating function seen in Eq. (4-103).

4-9 THE RIGID ROTATOR

Consider two masses m_1 and m_2 which rotate about their common center of mass while maintaining a fixed intermass distance R. This rotating dumbell is called a *rigid rotator* (see Fig. 4-5). According to classical mechanics, the

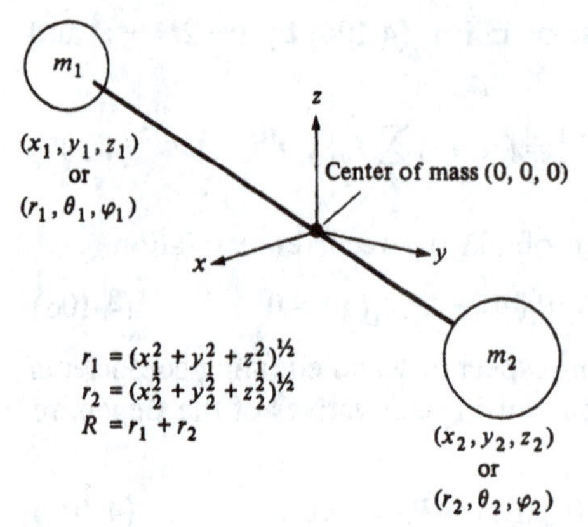

FIGURE 4-5
Coordinate system for the rigid rotator.

kinetic energy due to the rotation is

$$T = \frac{m_1 v_1^2}{2} + \frac{m_2 v_2^2}{2} \tag{4-110}$$

Letting ω be the angular velocity of the rotation and r_i the distance of the mass m_i from the center of mass of the rotator so that $v_i = \omega r_i$, this becomes

$$T = \frac{m_1 \omega^2 r_1^2 + m_2 \omega^2 r_2^2}{2} = \frac{\omega^2 (m_1 r_1^2 + m_2 r_2^2)}{2} \tag{4-111}$$

The quantity in parentheses in the right-hand expression is called the *moment of inertia*. It is defined in general for a system of n particles as

$$I = \sum_i^n m_i r_i^2 \tag{4-112}$$

where r_i is the distance of mass m_i from the center of mass of the system. Thus the kinetic energy becomes

$$T = \frac{I\omega^2}{2} \tag{4-113}$$

The angular momentum $\mathbf{L}$ may be shown to be[3]

$$\mathbf{L} = I\boldsymbol{\omega} \tag{4-114}$$

so that

$$\hat{L}^2 = \mathbf{L} \cdot \mathbf{L} = I^2 \boldsymbol{\omega} \cdot \boldsymbol{\omega} = (I\omega)^2 \tag{4-115}$$

[3] The details are sketched out by J. M. Anderson, *Introduction to Quantum Chemistry*, Benjamin, New York, 1969, pp. 91–99.

Thus, the kinetic energy can also be written

$$T = \frac{I\omega^2}{2} = \frac{I^2\omega^2}{2I} = \frac{L^2}{2I} \tag{4-116}$$

With Eq. (4-116), the hamiltonian operator for the rotational motion becomes

$$\hat{H} = \frac{\hat{L}^2}{2I} + \hat{V} \tag{4-117}$$

Since the potential energy V is constant, we omit it from the hamiltonian (just as was done in the particle in a box in Sec. 3-3). If we let $x = x_1 - x_2$, $y = y_1 - y_2$, and $z = z_1 - z_2$ and transform to spherical polar coordinates, $\hat{L}^2$ in the above equations is just the angular momentum operator given by Eq. (4-27). If we put $\hbar^2$ back into the equation, the Schrödinger equation for the rigid rotator becomes

$$-\frac{\hbar^2}{2I}\hat{L}^2\Phi(\theta, \varphi) = E\Phi(\theta, \varphi) \tag{4-118}$$

For a two-particle rotator, the moment of inertia I may also be expressed in terms of the reduced mass, i.e., as

$$I = \mu R^2 \tag{4-119}$$

where

$$\mu = \frac{m_1 m_2}{m_1 + m_2} \qquad \text{and} \qquad R = r_1 + r_2 \tag{4-120}$$

We already know that the eigenfunctions of $\hat{L}^2$ are the spherical harmonics and that the eigenvalues are $l(l+1)\hbar^2$; where $l = 0, 1, 2, \ldots$. We now replace l with the symbol J and write the Schrödinger equation as

$$-\frac{\hbar^2}{2I}\hat{L}^2 Y_{J, m_J}(\theta, \varphi) = E_J Y_{J, m_J}(\theta, \varphi) \tag{4-121}$$

where m_J (analogous to m_l) is $0, \pm 1, \pm 2, \ldots, \pm J$. The rotational energy is given by

$$E_J = \frac{J(J+1)\hbar^2}{2I} \tag{4-122}$$

It is customary to define the *rotational constant* B as

$$B = \frac{\hbar^2}{2I} \tag{4-123}$$

so that the energy may be expressed in the compact form

$$E_J = J(J+1)B \tag{4-124}$$

This shows that the rotational energy is quantized; the ground state has the energy of zero ($J = 0$), and the successive excited states have the energies:

$$E_1 = 1(1+1)B = 2B \qquad E_2 = 2(2+1)B = 6B \qquad E_3 = 3(3+1)B = 12B$$

This is also shown diagramatically in Fig. 4-6. Note that each state is $(2J+1)$-fold degenerate. This degeneracy is a consequence of the spherical symmetry of the system. For $J = 0$ there is no rotation, so the state is nondegenerate (there is only *one way* in which *not* to rotate), but for higher values of J, more and more indistinguishable modes of rotation arise, and, consequently, degeneracies abound.[4]

Selection rules for the rigid rotator are derived by examining under which conditions the following components of the transition moment do not vanish:

$$\begin{aligned} \mu^{(x)}_{J,m_J \to J',m'_J} &= \langle \psi_{J,m_J} | x | \psi_{J',m'_J} \rangle \\ \mu^{(y)}_{J,m_J \to J',m'_J} &= \langle \psi_{J,m_J} | y | \psi_{J',m'_J} \rangle \\ \mu^{(z)}_{J,m_J \to J',m'_J} &= \langle \psi_{J,m_J} | z | \psi_{J',m'_J} \rangle \end{aligned} \tag{4-125}$$

Let us examine the z component first. Using $z = R\cos\theta$, this becomes

$$R\int_0^{\pi} \Theta_{J,m_J}(\theta)\cos\theta\,\Theta_{J',m'_J}(\theta)\sin\theta\,d\theta \int_0^{2\pi} e^{im_J\varphi}e^{-im'_J\varphi}\,d\varphi \tag{4-126}$$

[4] Some excellent discussions of the physical meaning of degeneracies (in classical cases) are found in Kauzmann (see the readings at the end of this chapter).

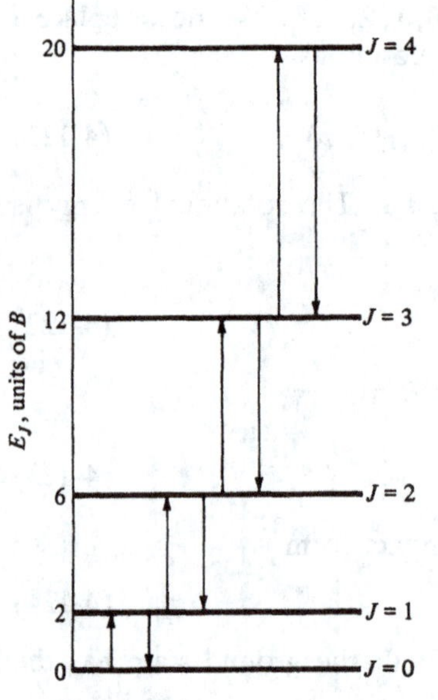

FIGURE 4-6
Energy-level diagram for the rigid rotator, showing the transitions allowed by an electric dipole mechanism.

The integral in φ vanishes unless $m_J = m_J'$ (see Exercise 4-16), so we get the selection rule

$$\Delta m_J = 0 \tag{4-127}$$

Now setting $m_J = m_J'$ in both integrals, we get

$$\int_0^{\pi} \Theta_{J,\,m_J}(\theta) \cos\theta\, \Theta_{J',\,m_J'}(\theta) \sin\theta\, d\theta = \int_{-1}^{1} P_J^{|m_J|}(y) y P_{J'}^{|m_J|}(y)\, dy \tag{4-128}$$

Using the first recursion relation in Eqs. (4-109) for $P_{J'}^{|m_J|}$, the above integral becomes

$$\int_{-1}^{1} P_J^{|m_J|} \left[\frac{J' - |m_J| + 1}{2J' + 1} P_{J'+1}^{|m_J|} + \frac{J' + |m_J|}{2J' + 1} P_{J'-1}^{|m_J|} \right] dy \tag{4-129}$$

The orthogonality of the $P_J^{|m_J|}$ implied in Eq. (4-82) shows that the above integral vanishes unless $J = J' + 1$ or $J' - 1$. Thus, the selection rule on J for the z component of the transition moment integral is

$$\Delta J = \pm 1 \tag{4-130}$$

The x and y components are most conveniently evaluated together. We look at the component

$$\langle \psi_{J,\,m_J} | x \pm iy | \psi_{J',\,m_J'} \rangle = \langle \psi_{J,\,m_J} | R \sin\theta e^{\pm i\varphi} | \psi_{J',\,m_J'} \rangle \tag{4-131}$$

The integrals in φ become

$$\int_0^{2\pi} e^{im_J\varphi} e^{\pm i\varphi} e^{-im_J'\varphi}\, d\varphi = \int_0^{2\pi} e^{im_J\varphi} e^{-i(m_J' \mp 1)\varphi}\, d\varphi \tag{4-132}$$

We see that the integrals are nonzero only if $m_J = m_J' \pm 1$. Thus we get the selection rule

$$\Delta m_J = \pm 1 \tag{4-133}$$

Combining Eqs. (4-127) and (4-133), we get as the total selection rules on m_J

$$\Delta m_J = 0,\ \pm 1 \tag{4-134}$$

As we will see later, the m_J selection rules are relevant only when the rigid rotator is subjected to an external interaction which removes some of or all the $(2J + 1)$-fold degeneracy.[5]

---

[5] The $\Delta m_J = 0$ selection rule applies to radiation which is polarized along the axis of a directional external interaction (e.g., a magnetic or electric field), whereas $\Delta m_J = \pm 1$ refers to radiation polarized in a plane perpendicular to this axis.

The integral in θ is

$$\int_{-1}^{1} P_J^{|m_J|}\sqrt{1-y^2}\, P_{J'}^{|m_J'|}\, dy \tag{4-135}$$

Using the second of the recursion relations in Eqs. (4-109), the above becomes

$$\int_{-1}^{1} P_J^{|m_J|}\left(\frac{P_{J'+1}^{|m_J'|-1} - P_{J'-1}^{|m_J'|+1}}{2J'+1}\right) dy \tag{4-136}$$

which is nonzero only if $J = J' + 1$ or $J' - 1$. Thus, the selection rule on J is the same as given by Eq. (4-130): $\Delta J = \pm 1$ (see Fig. 4-6).

We will have occasion to refer later to the selection rules for m_J; for the moment we consider $\Delta J = \pm 1$ only. The energy of a J to $J+1$ transition ($\Delta J = +1$) is given in general by

$$\Delta E = E_{J+1} - E_J = (J+1)(J+2)B - J(J+1)B = 2(J+1)B \qquad J \geq 0 \tag{4-137}$$

For a J to $J-1$ transition ($\Delta J = -1$) we get

$$\Delta E = E_J - E_{J-1} = J(J+1) - (J-1)JB = 2JB \qquad J \geq 1 \tag{4-138}$$

The simplest system of chemical interest for which the rigid rotator serves as a model is the interaction of microwave radiation with a heteronuclear diatomic molecule. As discussed in Sec. 3-1, the oscillating electric dipole of a rotating molecule has a frequency of the correct magnitude to undergo resonance coupling with the electronic component of microwave (or, occasionally, far infrared) radiation. For example, gaseous hydrogen fluoride, HF, absorbs microwave radiation of wavelength 0.0239 cm ($41.8\ \text{cm}^{-1}$). No wavelength shorter than this is absorbed until one gets into the infrared region, where vibrational oscillations of the electric dipole begin to couple. If it is assumed that the microwave interaction represents the transition of HF from its ground rotational state ($J = 0$) to its lowest excited state ($J = 1$), we can use the absorption wavelength (or frequency) to calculate the HF bond distance in this molecule. Of course, HF is also vibrating while it is rotating, so that the bond distance we calculate must be interpreted as some sort of average value. The moment of inertia of HF is calculated from

$$\Delta E = 2B = 2\frac{\hbar^2}{2I} = \frac{\hbar}{I} = \frac{hc}{\lambda}$$

which gives

$$I = \frac{\hbar^2\lambda}{hc} = \frac{h\lambda}{4\pi^2 c} = \frac{(6.626\times10^{-34}\ \text{J}\cdot\text{s})(2.39\times10^{-4}\ \text{m})}{4\pi^2(3.000\times10^{8}\ \text{m}\cdot\text{s}^{-1})} = 1.337\times10^{-47}\ \text{kg}\cdot\text{m}^2 \tag{4-139}$$

Assuming ^{1}H and ^{19}F as the nuclear species, the reduced mass of HF is

$$\mu = \frac{m_H m_F}{m_H + m_F} = \frac{(1.007825 \text{ amu})(18.99840 \text{ amu})}{1.007825 \text{ amu} + 18.99840 \text{ amu}}$$

$$= 0.957055 \text{ amu} \tag{4-140}$$

which is equivaent to 1.589×10^{-27} kg.[6] Thus the H–F bond distance is given by

$$R = \left(\frac{I}{\mu}\right)^{1/2} = \left(\frac{1.337 \times 10^{-47} \text{ kg} \cdot \text{m}^2}{1.589 \times 10^{-27} \text{ kg}}\right)^{1/2}$$

$$= 9.17 \times 10^{-11} \text{ m} \quad \text{or} \quad 91.7 \text{ pm} \quad \text{or} \quad 0.917 \text{ Å} \tag{4-141}$$

EXERCISES

4-22. Prove that, for a diatomic molecule, $I = m_1 r_1^2 + m_2 r_2^2 = \mu R^2$, where $R = r_1 + r_2$ and r_i is the distance of mass m_i from the center of mass (see Fig. 4-5).

4-23. Show that the rigid rotator hamiltonian commutes with $\hat{L}^2$ and $\hat{L}_z$. What is the physical significance of this?

4-24. (*a*) Derive the uncertainty relationship $\Delta L_z \Delta\varphi \geq \hbar/2$.
(*b*) In the ground state of the rigid rotator, $\mathbf{L} = 0$ and hence $\Delta \mathbf{L}$ (and ΔL_z) = 0. Does this violate the uncertainty principle? Explain.

4-25. Show that a zero-point energy of zero for the rigid rotator does not violate the uncertainty principle.

4-26. Show that the Schrödinger equation for a two-dimensional (planar) rigid rotator is

$$-\frac{\hbar^2}{2I}\frac{d^2}{d\varphi^2}\psi(\varphi) = E\psi(\varphi)$$

and determine the eigenfunctions and eigenvalues. Discuss the degeneracies of the eigenfunctions, and give physical reasons for the pattern of degeneracies found. What are the selection rules for this system?

4-27. The lowest microwave frequency absorbed by carbon monoxide (^{12}C^{16}O) is 115,271 s^{-1}.
(*a*) Compute the moment of inertia of CO and the average value of the C–O bond distance.
(*b*) Estimate the frequency (in s^{-1} and cm^{-1}) at which the $J = 1$ to $J = 2$ transition is expected to occur.

4-28. The average H–Cl bond distance in gaseous hydrogen chloride is about 127 pm. Estimate the frequencies (in cm^{-1}) at which the three lowest microwave interactions occur.

---

[6] Recall that 6.02252×10^{23} amu is equivalent to 1 g [the atomic mass unit (amu) is defined as exactly one-twelfth the mass of one ^{12}C atom].

4-29. Using data given in the preceding section, estimate the lowest microwave frequency absorbed by DF ($D = {}^2H$).

4-30. Construct a horizontal line marked off in units of $\tilde{B}$ (equal to B/hc), beginning with zero at the left-hand end up to $8\tilde{B}$. Mark those points which correspond to a frequency $\tilde{\nu}$ for which the selection rules allow a rigid rotator to absorb. Indicate the J values involved in each absorption. What pattern do the absorption lines fit?

4-31. Carry out a treatment of the six pi electrons of benzene by assuming the particle-on-a-ring model, i.e., that each electron follows a circular path of radius R while in a constant potential field. The Schrödinger equation for the particle is given in Exercise 4-26. Your treatment should include the following:

(*a*) An energy-level diagram showing the six pi electrons in the appropriate energy levels.

(*b*) Selection rules for electronic transitions.

(*c*) Calculation of the lowest-energy spectroscopic transition wavelength using a path radius of 0.14 nm.

(*d*) Comparison of this treatment with that of Exercise 3-27 in which the benzene pi electrons are placed into a square box. In particular, how do the two treatments differ qualitatively, e.g., nature of the energy-level diagram, electron distribution, predicted angular momentum, etc.?

SUGGESTED READINGS

Eyring, H., J. Walter, and G. E. Kimball: *Quantum Chemistry*, Wiley, New York, 1944. A tough book for most people to digest but worth the effort. Gives details on the spherical harmonics, ladder operators, etc.

Kauzmann, W.: *Quantum Chemistry*, Academic, New York, 1957. Provides much useful physical insight into the mathematics of quantum theory.

Pauling, L., and E. B. Wilson, Jr.: *Introduction to Quantum Mechanics*, McGraw-Hill, New York, 1935. Still one of the best sources for details on simple quantum systems.

CHAPTER 5

THE HYDROGEN ATOM

Hydrogen is truly a remarkable element. This lightest of all atoms serves as fuel for the nuclear furnaces of the sun and many other stars, its presence in carbohydrates accounts for much of their metabolically releasable energy, its atoms may have been one of the building blocks of all matter, and its quantum mechanical wavefunctions provide convenient building blocks for the approximate treatment of the remaining elements of the periodic table and the molecules they form. The present chapter will discuss the Schrödinger nonrelativistic treatment of the hydrogen atom and, later, show some of the consequences of including effects attributable to relativity.

5-1 SCHRÖDINGER TREATMENT OF THE ONE-ELECTRON ATOM

Let us consider the general case of a one-electron species (atom or ion) of nuclear charge Ze. If we let m_N and m_e represent the masses of the nucleus and electron, respectively, the Schrödinger equation is

$$\left(-\frac{\hbar^2}{2m_N}\nabla_N^2 - \frac{\hbar^2}{2m_e}\nabla_e^2 - \frac{Ze^2K}{r}\right)\psi_{Ne} = E\psi_{Ne} \tag{5-1}$$

where $\psi_{Ne} = \psi(x_e, y_e, z_e, x_N, y_N, z_N)$

$$r = [(x_e - x_N)^2 + (y_e - y_N)^2 + (z_e - z_N)^2]^{1/2} \tag{5-2}$$

and the coordinate system is defined as in Fig. 5-1. As illustrated before, we can separate out the translational motion of the total mass $M = m_N + m_e$ by the coordinate transformation

$$\begin{aligned} x &= x_e - x_N \qquad x_{cm} = \frac{m_e x_e + m_N x_N}{M} \\ y &= y_e - y_N \qquad y_{cm} = \frac{m_e y_e + m_N y_N}{M} \\ z &= z_e - z_N \qquad z_{cm} = \frac{m_e z_e + m_N z_N}{M} \end{aligned} \tag{5-3}$$

This allows one to employ the separation of variables

$$\psi_{Ne}(x_N, \ldots, z_e) = \psi_N(x_N, y_N, z_N)\psi_e(x_e, y_e, z_e)$$

so that we obtain the two separated differential equations

$$-\frac{\hbar^2}{2M}\nabla^2_{cm}\psi_N = E_N\psi_N \qquad \text{and} \qquad \left(-\frac{\hbar^2}{2\mu}\nabla^2_e - \frac{Ze^2K}{r}\right)\psi_e = E_e\psi_e \tag{5-4}$$

where $\mu = m_e m_N/M$ (the reduced mass) and $E = E_N + E_e$. The first equation is, of course, just that of a free particle; the second describes the motions of the electron and the nucleus relative to their mutual center of mass. Concentrating all our attention on the latter equation, we drop subscripts to write the Schrödinger equation

$$\left(-\frac{\hbar^2}{2\mu}\nabla^2 - \frac{Ze^2K}{r}\right)\psi = E\psi \tag{5-5}$$

Further separation is not possible in cartesian coordinates, but it is possible in other coordinate system, e.g., spherical polar coordinates (defined in Chap. 4).

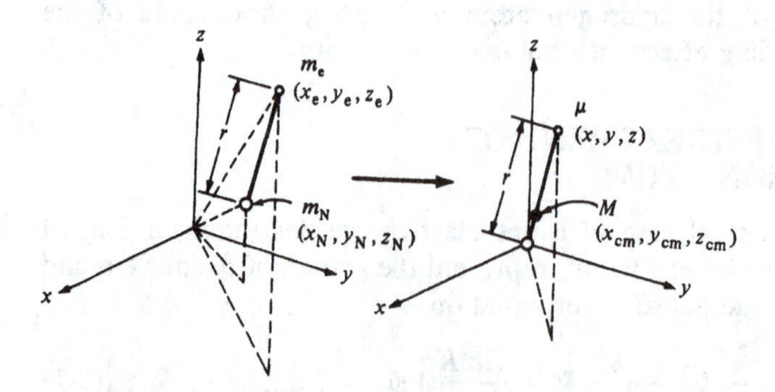

FIGURE 5-1
Transformation of the cartesian coordinates $(x_1, y_1, z_1, x_2, y_2, z_2)$ of the hydrogen atom (left) to relative and center-of-mass coordinates $(x, y, z, x_{cm}, y_{cm}, z_{cm})$ (right). The masses of the nucleus and of the electrons are represented by m_N and m_e, respectively, and μ and M represent the reduced mass and total mass, respectively.

The hamiltonian operator in Eq. (5-5) in spherical polar coordinates is

$$\hat{H} = -\frac{\hbar^2}{2\mu r^2}(\hat{D} - \hat{L}^2) - \frac{Ze^2K}{r} \tag{5-6}$$

where both $\hat{D}$ and $\hat{L}^2$ are the same operators introduced via Eqs. (4-85) and (4-27), respectively. Also, recall that the latter is just the angular momentum operator. The wavefunction ψ can now be written $\psi(r, \theta, \varphi)$ and assumed to satisfy the partial separation

$$\psi(r, \theta, \varphi) = R(r)Y(\theta, \varphi) \tag{5-7}$$

The separation leads to the two differential equations

$$\begin{aligned} \hat{L}^2 Y(\theta, \varphi) &= \gamma Y(\theta, \varphi) \\ \hat{D}R(r) + \frac{2\mu r^2}{\hbar^2}\left(E + \frac{Ze^2K}{r}\right) R(r) &= \gamma R(r) \end{aligned} \tag{5-8}$$

where γ is a separation constant. The solution to the first equation was discussed in Chap. 4; the eigenvalues are

$$\gamma = l(l+1)\hbar^2 \qquad l = 0, 1, 2, \ldots \tag{5-9}$$

and the eigenfunctions are the spherical harmonics

$$Y_{l,m_l}(\theta, \varphi) = \Theta_{l,|m_l|}(\theta)\Phi_{m_l}(\varphi) \tag{5-10}$$

The remaining equation, called the *radial equation*, now becomes

$$\left[\hat{D} + \frac{2\mu r^2}{\hbar^2}\,\frac{E + Ze^2K}{r} - l(l+1)\right]R = 0 \tag{5-11}$$

As shown by Schrödinger in one of his later papers,[1] the radial equation may be solved by ladder operators just as was done for the harmonic oscillator and angular momentum. However, the procedure is considerably messier in this present case, and we will examine instead the standard polynomial method usually employed in many areas of differential equations and which was used by Schrödinger himself in his earlier papers. First we introduce the constants

$$\alpha^2 = -\frac{2\mu E}{\hbar^2} \qquad \beta = \frac{\mu Ze^2K}{\alpha\hbar^2} \tag{5-12}$$

and the new independent variable

$$\rho = 2\alpha r \tag{5-13}$$

In defining α^2 we assume that $E < 0$ so that our solution will be restricted to so-called *bound states*, i.e., states for which no ionization occurs.[2] When we

---

[1] E. Schrödinger, *Proc. Roy. Irish Acad.*, **46A**:9 (1940).

[2] When $E > 0$, the energy spectrum becomes continuous; the interaction of the electron and the nucleus is called *scattering*.

replace $R(r)$ with $T(\rho)$, the radial equation (5-11) now assumes the dimensionless form

$$\frac{d^2T}{d\rho^2}+\frac{2}{\rho}\frac{dT}{d\rho}-\left[\frac{1}{4}-\frac{\beta}{\rho}+\frac{l(l+1)}{\rho^2}\right]T=0 \tag{5-14}$$

Let us begin by considering an asymptotic solution valid as $\rho\rightarrow\infty$. The radial equation for this case is

$$\frac{d^2T}{d\rho^2}-\frac{T}{4}=0 \tag{5-15}$$

Standard differential equation techniques show that one general solution to the asymptotic equation is

$$T(\rho)=e^{-\rho/2}+e^{\rho/2} \tag{5-16}$$

One now assumes general solutions to the nonasymptotic Eq. (5-15) of the general form

$$T(\rho)=G(\rho)e^{\pm\rho/2} \tag{5-17}$$

where $G(\rho)$ is a power series (polynomial) in ρ. It makes no difference which we choose, $e^{-\rho/2}$ or $e^{\rho/2}$, but we arbitrarily choose the former.[3] Thus $G(\rho)$ must be chosen such that

$$\lim_{\rho\rightarrow\infty}\frac{G(\rho)}{e^{-\rho/2}}=0 \tag{5-18}$$

Substituting $T(\rho)=G(\rho)e^{-\rho/2}$ into Eq. (5-14), we get

$$\rho^2\frac{d^2G}{d\rho^2}+\rho(2-\rho)\frac{dG}{d\rho}+[(\beta-1)\rho-l(l+1)]G=0 \tag{5-19}$$

Since this equation must be valid for *all* values of ρ, it must certainly be valid for $\rho=0$. This means that

$$l(l+1)G(0)=0 \tag{5-20}$$

where, if $l\neq0$, it is necessary that $G(0)=0$. This behavior suggests that $G(\rho)$ shoud have the power series form

$$G(\rho)=\rho^s\sum_{j=0}^{\infty}a_j\rho^j \tag{5-21}$$

where $s>0$ for $l\neq0$ and $a_0\neq0$. It is convenient to introduce the definition

$$L(\rho)=\sum_{j=0}^{\infty}a_j\rho^j \tag{5-22}$$

[3] The frequently encountered assertion that the $e^{\rho/2}$ choice necessarily leads to a divergent solution is incorrect. See B. F. Gray, G. Hunter, and H. O. Pritchard, *J. Chem. Phys.*, **38**:2790 (1963).

so that $G(\rho)$ may be written

$$G(\rho) = \rho^s L(\rho) \tag{5-23}$$

The first and second derivatives of $G(\rho)$ then are

$$\begin{aligned} G'(\rho) &= s\rho^{s-1}L + \rho^s \\ G''(\rho) &= \rho^s L'' + 2s\rho^{s-1}L' + s(s-1)\rho^{s-2}L \end{aligned} \tag{5-24}$$

Substituting the expressions in Eqs. (5-24) into Eq. (5-19) and combining terms of the same power of ρ, one obtains

$$\begin{aligned} (L'' - L')\rho^{s+2} &+ [2sL' + 2L' - sL + (\beta - 1)L]\rho^{s+1} \\ &+ [s(s-1) + 2s - l(l+1)]L\rho^s = 0 \end{aligned} \tag{5-25}$$

This relationship can be valid in the general interval $0 \le \rho \le \infty$ only if each coefficient of each power of ρ vanishes individually. Thus, for ρ^s

$$\begin{aligned} [s(s-1) + 2s - l(l+1)]L = [s(s-1) + 2s - l(l+1)] \\ \times (a_0 + a_1\rho + a_2\rho^2 + \cdots) = 0 \end{aligned} \tag{5-26}$$

But since we have chosen $a_0 \neq 0$, it is necessary that

$$s(s-1) + 2s - l(l+1) = 0 \tag{5-27}$$

This is a quadratic expression in s:

$$s^2 + s - l(l+1) = 0 \tag{5-28}$$

having the two roots

$$s = \begin{cases} l \\ -(l+1) \end{cases} \tag{5-29}$$

Only the first root satisfies $s > 0$ for $l \neq 0$ and thus leads to $G(0) = 0$ for $l \neq 0$. Thus, the power series $G(\rho)$ is

$$G(\rho) = \rho^l \sum_{j=0}^{\infty} a_j \rho^j \tag{5-30}$$

Substituting Eq. (5-30) into Eq. (5-25), setting $l = s$, and dividing through by ρ^{l+1}, we obtain

$$\sum_{j=0}^{\infty} [j(j-1)a_j\rho^{j-1} + 2(l+1)ja_j\rho^{j-1} - ja_j\rho^j + (\beta - l - 1)a_j\rho^j] = 0 \tag{5-31}$$

The above equation will be satisfied only if each coefficient of each power of ρ vanishes. The coefficient of ρ^j is

$$(j+1)(j+2l+2)a_{j+1} + (\beta - l - 1 - j)a_j = 0 \tag{5-32}$$

The ratio of successive coefficients is given by the recursion formula

$$\frac{a_{j+1}}{a_j} = \frac{l+j+1-\beta}{(j+1)(j+2l+2)} \tag{5-33}$$

The limiting value as $j \rightarrow \infty$ is

$$\lim_{\rho\rightarrow\infty} \frac{a_{j+1}}{a_j} = \frac{1}{j} \tag{5-34}$$

The Taylor series expansion of e^{ρ} is

$$e^{\rho} = \sum_{j=1}^{\infty} \frac{\rho^j}{j!} = \sum_{j=1}^{\infty} b_j \rho^j$$

where

$$\frac{b_j}{b_{j+1}} = j+1 \tag{5-35}$$

This shows that $G(\rho)$ increases with ρ much as $\rho^l e^{\rho}$ does. Thus

$$T(\rho) = G(\rho)e^{-\rho/2} \tag{5-36}$$

which is *approximately* equal to

$$\rho^l e^{\rho} e^{-\rho/2} = \rho^l e^{\rho/2} \tag{5-37}$$

But this is not well-behaved, since it does not vanish as $\rho \rightarrow \infty$. This deficiency can be remedied by requiring that the series $G(\rho)$ terminate at some finite index j (which we will call k) so that the numerator in Eq. (5-33) vanishes. Thus

$$l+k+1-\beta = 0 \tag{5-38}$$

Since $l+k+1$ must be an integer having the possible values $1, 2, 3, \ldots$, we set $l+k+1 = n$, where n is an integer. Then the recursion formula [Eq. (5-33)] becomes

$$n - \beta = 0 \qquad \text{or} \qquad \beta = n = 1, 2, 3, \ldots \tag{5-39}$$

Now going back to the definition of β in Eqs. (5-12), we get

$$\beta^2 = \frac{\mu^2 Z^2 e^4 K^2}{\alpha^2 \hbar^4} = -\frac{\hbar^2}{2\mu E}\frac{\mu^2 Z^2 e^4 K}{\hbar^4} = -\frac{\mu^2 Z^2 e^4 K}{2\hbar^2 E} = n^2 \tag{5-40}$$

Letting $\hbar^2/e^2\mu = a_{\mathrm{N}}$ [which is a_{H} when $Z = 1$; see Eq. (1-25)] and solving for the energy, we get

$$E_n = -\frac{Z^2 e^2 K}{2n^2 a_{\mathrm{N}}} \tag{5-41}$$

which is just the energy expression predicted by the earlier Bohr theory [Eq. (1-26)]. The integer n is called the *principal quantum number*. From the fact

$n = l + k + 1 = 1, 2, 3, \ldots,$ we see that $l = n - k - 1 = 0, 1, 2, \ldots, (n-1)$. Thus the *azimuthal* or orbital angular momentum quantum number satisfies

$$0 \leq l \leq n - 1 \tag{5-42}$$

The polynomials $G(\rho)$ introduced in Eq. (5-17) are closely related to the *associated Laguerre polynomials of degree $r - s$ and order s* defined by

$$L_r^s(\rho) = \frac{d^s}{d\rho^s} L_r(\rho) \tag{5-43}$$

where $L_r(\rho)$ is the *Laguerre polynomial of degree r* given by

$$L_r(\rho) = e^{\rho} \frac{d^r}{d\rho^r} \rho^r e^{-\rho} \tag{5-44}$$

After some manipulation, it can be shown that

$$G(\rho) = \rho^l L_{n+1}^{2l+1}(\rho) \tag{5-45}$$

The solution to the radial equation can now be written in the form

$$R_{nl}(r) = N_{nl} G(\rho) e^{-\rho/2} = N_{nl} \rho^l L_{n+1}^{2l+1}(\rho) e^{-\rho/2} \tag{5-46}$$

where N_{nl} is a normalization constant. By the use of suitable generating functions (which we will not consider here), the general form of N_{nl} is shown to be[4]

$$N_{nl} = -\left\{ \left(\frac{2Z}{na_{\mathrm{N}}} \right)^3 \frac{(n - l - 1)!}{2n[(n + l)!]^3} \right\}^{1/2} \tag{5-47}$$

The radial functions also satisfy the orthonormalization condition

$$\int_0^{\infty} R_{nl}(r) R_{n'l}(r) r^2 \, dr = \delta_{nn'} \tag{5-48}$$

We will now construct the radial function $R_{10}(r)$. This requires the evaluation of the associated Laguerre polynomial of degree 0 and order 1, that is, $L_1^1(\rho)$. The Laguerre polynomial of degree 1 is

$$L_1(\rho) = \frac{e^{\rho} d(\rho e^{-\rho})}{d\rho} = e^{\rho}(e^{-\rho} - \rho e^{-\rho}) = 1 - \rho \tag{5-49}$$

A single differentiation of $1 - \rho$ with respect to ρ leads to

$$L_1^1(\rho) = \frac{d(1 - \rho)}{d\rho} = -1 \tag{5-50}$$

---

[4] Details are given in Pauling and Wilson, *Introduction to Quantum Mechanics* (see the readings at the end of this chapter).

Thus

$$G(\rho) = \rho^0(-1) = -1$$

$$N_{10} = -\frac{1}{\sqrt{2}}\left(\frac{2Z}{a_N}\right)^{3/2} = -2\left(\frac{Z}{a_N}\right)^{3/2} \tag{5-51}$$

$$R_{10}(r) = -N_{10}e^{-\rho/2} = 2\left(\frac{Z}{a_N}\right)^{3/2} e^{-Zr/a_N}$$

In summary, wavefunctions of the one-electron atom may be written in the general form

$$\psi_{n,\,l,\,m_l}(r,\,\theta,\,\varphi) = R_{nl}(r)\Theta_{l,\,m_l}(\theta)\Phi_{m_l}(\varphi) \tag{5-52}$$

where the θ and φ functions are the spherical harmonics introduced in Chap. 4 with respect to angular momentum. Since each of the component functions making up ψ is separately orthonormal, it follows that

$$\langle \psi_{n,\,l,\,m_l}(r,\,\theta,\,\varphi) \,|\, \psi_{n',\,l',\,m_l'}(r,\,\theta,\,\varphi)\rangle = \delta_{nn'}\delta_{ll'}\delta_{m_lm_l'} \tag{5-53}$$

The ground-state wavefunction is readily shown to be

$$\psi_{100} = R_{10}\Theta_{00}\Phi_0 = 2\left(\frac{Z}{a_N}\right)^{3/2} e^{-Zr/a_N}\,\frac{\sqrt{2}}{2}\,\frac{1}{\sqrt{2\pi}}$$

$$= \frac{1}{\sqrt{\pi}}\left(\frac{Z}{a_N}\right)^{3/2} e^{-Zr/a_N} \tag{5-54}$$

As introduced in Sec. 4-7, the $l = 0, 1, 2, 3, 4, \ldots$ values are often indicated by the letters $s, p, d, f, g, \ldots$, respectively. Thus ψ_{100} is often called the $1s$ wavefunction, and the three ψ_{21m_l} are called $2p$ wavefunctions.

In the following section we will discuss these wavefunctions in some detail—in particular, how to represent them graphically and what they represent physically.

EXERCISES

5-1. Show that the radial functions for $n = 2$ are

$$R_{20}(r) = R_{2s} = \frac{1}{2\sqrt{2}}\left(\frac{Z}{a_N}\right)^{3/2}\left(2 - \frac{Zr}{a_N}\right)e^{-Zr/2a_N}$$

$$R_{21}(r) = R_{2p} = \frac{1}{2\sqrt{6}}\left(\frac{Z}{a_N}\right)^{3/2} e^{-Zr/2a_N}$$

Show also that $\Theta_{1,\,\pm1} = (\sqrt{3}/2)\sin\theta$ and that $\Phi_{\pm1} = (1/\sqrt{2\pi})e^{\pm i\varphi}$ and obtain ψ_{200} and $\psi_{21\pm1}$.

5-2. Given the solution of the Schrödinger equation for the one-electron atom, what are the ground-state wavefunction and energy of positronium (e^+e^-)?

5-3. Show that if r and θ are constant ($\theta = 0$), then the one-electron Schrödinger equation reduces to the particle-on-a-ring Schrödinger equation (see Exercise 4-31). Compare the solutions to this system with those of the two-dimensional square box (see Sec. 3-3). Note the degeneracies especially. Which quantum number of the one-electron atom ends up quantizing the energy of the particle on a ring?

5-4. Calculate the minimum energy of the process $Be^{3+} \rightarrow Be^{4+} + e^-$.

5-5. Without actually doing an integration, show why s and p wavefunctions must be orthogonal and why the three np wavefunctions are also orthogonal.

5-6. The wavelength 486.27 nm appears in the Balmer series of hydrogen. What state emits to produce this line? The line has a weak satellite line at 486.14 nm. Show that this is due to a small amount of the isotope ^2H. (This is how deuterium was actually discovered.)

5-2 WAVEFUNCTIONS OF THE ONE-ELECTRON ATOM

For many purposes it is convenient to represent the wavefunctions of the one-electron atom with the spherical harmonic components in their real (rather than complex) forms. Furthermore, notational simplicity is achieved if we represent all distances in units of the first Bohr radius a_H and if we replace Zr with ρ. Thus, where r/a_H appeared previously, we now write simply r. Note that this means that a_H itself now becomes equal to unity. Table 5-1 lists some

TABLE 5-1
Real forms of one-electron wavefunctions

Wavefunction symbol	Orbital symbol	Wavefunction equation*
ψ_{100}	$1s$	$N_1 e^{-\rho}$
ψ_{200}	$2s$	$N_2(2-\rho)e^{-\rho/2}$
$\psi_{21,\cos\varphi}$	$2p_x$	$N_2\rho \sin\theta \cos\varphi e^{-\rho/2}$
$\psi_{21,\sin\varphi}$	$2p_y$	$N_2\rho \sin\theta \sin\varphi e^{-\rho/2}$
ψ_{210}	$2p_z$	$N_2\rho \cos\theta e^{-\rho/2}$
ψ_{300}	$3s$	$N_3(27-18\rho+2\rho^2)e^{-\rho/3}$
$\psi_{31,\cos\varphi}$	$3p_x$	$N_3(6\rho-\rho^2)\sin\theta\cos\varphi\, e^{-\rho/3}$
$\psi_{31,\sin\varphi}$	$3p_y$	$N_3(6\rho-\rho^2)\sin\theta\sin\varphi\, e^{-\rho/3}$
ψ_{310}	$3p_z$	$N_3(6\rho-\rho^2)\cos\theta e^{-\rho/3}$
ψ_{320}	$3d_{z^2}$	$N_4\rho^2(3\cos^2\theta-1)e^{-\rho/3}$
$\psi_{32,\cos\varphi}$	$3d_{xz}$	$N_5\rho^2\sin\theta\cos\theta\cos\varphi e^{-\rho/3}$
$\psi_{32,\sin\varphi}$	$3d_{yz}$	$N_5\rho^2\sin\theta\cos\theta\sin\varphi e^{-\rho/3}$
$\psi_{32,\cos 2\varphi}$	$3d_{x^2-y^2}$	$N_6\rho^2\sin^2\theta\cos 2\varphi\, e^{-\rho/3}$
$\psi_{32,\sin 2\varphi}$	$3d_{xy}$	$N_6\rho^2\sin^2\theta\sin 2\varphi\, e^{-\rho/3}$

* The explicit forms of the normalization constants for one-electron wavefunctions are

$N_1 = Z^{3/2}/\sqrt{\pi}$ $\qquad N_4 = N_3/2$
$N_2 = N_1/4\sqrt{2}$ $\qquad N_5 = \sqrt{6}\,N_4$
$N_3 = 2N_1/81\sqrt{3}$ $\qquad N_6 = N_5/2$

of the lower-energy wavefunctions ($n = 1$, 2, and 3) in their real forms (note that all the wavefunctions are automatically real when $m_l = 0$). Recall also, that the real forms with $m_l \neq 0$ are not eigenfunctions of $\hat{L}_z$; in general, the real forms are linear combinations of the complex forms. Specifically, the complex functions with $m_l = a$ and $m_l = -a$ can be combined to produce two real functions, neither of which is an eigenfunction of $\hat{L}_z$.

The wavefunctions of the one-electron atom are usually referred to as *orbitals*, a term introduced by Mulliken.[5] Unfortunately, the terms "wavefunction" and "orbital" have been used to mean many different things, so we will adopt somewhat arbitrary definitions here and attempt to use them consistently thereafter. The term "wavefunction" will be used to designate an eigenfunction (or an approximation thereof) of the hamiltonian of an isolated system, while the term "orbital" will mean any mathematical function ϕ whose square $|\phi|^2$ is used to represent (exactly or approximately) the probability density (usually three-dimensional) of a single electron. Thus, the terms "wavefunction" and "orbital" are synonymous only for a one-electron system such as the hydrogen atom; for many-electron systems orbitals are not wavefunctions themselves but may be used as "building blocks" of approximate wavefunctions. Note that if the particle in a box is an electron, then the functions $\psi_n(x) = \sqrt{2/L} \sin(n\pi x/L)$ are both orbitals and wavefunctions.

It is useful to note that the $1s$, $2s$, and $2p$ orbitals have general forms suggestive of the relationships [see Eq. (4-21)] between (x, y, z) and (r, θ, φ) coordinate systems:

$$\begin{aligned} &\psi(1s) = f(r) \qquad && \psi(2p_x) = h(r) \sin\theta \cos\varphi = xh(r) \\ &\psi(2s) = g(r) \qquad && \psi(2p_y) = h(r) \sin\theta \sin\varphi = yh(r) \\ & && \psi(2p_z) = h(r) \cos\theta = zh(r) \end{aligned} \tag{5-55}$$

Since the orbitals represent the distribution of positive and negative magnitudes in a three-dimensional space, it would require a *four-dimensional* plot to give a *total* graphical representation of them—a physical impossibility when one is limited to the two dimensions of a sheet of paper. Consequently, various subterfuges must be used in order to convey just what this three-dimensional distribution looks like. One way to do this is to assign fixed values to one or two of the x, y, and z coordinates and depict how the distribution varies with respect to the remaining coordinate or coordinates. This may be very simply illustrated for the ns orbitals by setting $z = 0$ and then showing magnitudes of the orbital at various points on the xy plane. Note that this is tantamount to representing the numerical values of the orbital in a plane containing the nucleus; if one were to fix z to a value of 1 (instead of 0), the plane used to

[5] The term "orbital" may be regarded as a quantum counterpart of the classical term "orbit."

depict the orbital magnitudes would be parallel to the nuclear plane but 1 a.u. away from it.

The orbital magnitudes produced by the above technique represent a *surface* relative to the xy plane; this surface is generally represented by two sets of intersecting lines which follow the surface and impart a three-dimensional effect to it. Each intersecting line is itself a *profile* of the surface for a fixed value of one of the two coordinates (x or y). As a first example, consider the $1s$ and $2s$ orbitals as depicted in the xy plane containing the nucleus (the nucleus is at $0, 0, 0$). In the case of the $1s$ orbital, we wish to show what the exponential function e^{-r} looks like in the xy plane when $z = 0$. Ignoring the normalization constant (which is irrelevant for this purpose), we must construct a surface plot of the function

$$f(x, y) = e^{-(x^2+y^2)^{1/2}} \tag{5-56}$$

This is done in the following way:

1. A grid is chosen, e.g., both x and y values will range from -5 to $+5$ in increments of 0.1.
2. Setting $y = -5$, $f(x, y)$ values are computed for $x = -5$ to $+5$ in increments of 0.1.
3. All these points are connected by a line; this constitutes a profile of $f(x, y)$ for all x values for fixed $y = -5$.
4. Steps 1, 2, and 3 are now repeated for $y = -4.9$ to produce a profile parallel to the first. This is continued in increments of 0.1 until $y = +5$ is reached. At this point one has a set of parallel profiles which define a surface. However, the surface can be made to appear more distinct by constructing another set of profiles at right angles to the first. The first such profile would be obtained by setting $x = -5$ and computing $f(x, y)$ for all values of y from -5 to $+5$; the process would then be continued for all fixed values of x in increments of 0.1 until $x = +5$ was reached.

The intersection of profiles produced by the above constitutes a representation of a surface; the value of $f(x, y)$ at any point (x, y) represents the magnitude of the orbital at this point in the nuclear plane.

Figures 5-2 and 5-3 show the results of the procedure for the $1s$ and $2s$ orbitals, respectively. In both cases the surfaces are symmetrical about the origin (independent of direction) and both show maximum values at the origin, i.e., at the nucleus. [*Note*: The arrows in these and the following surface plots indicate only the general orientation of the axes; they are not the axes themselves (which in the case of Figs. 5-2 and 5-3, for example, intersect beneath the maxima of the "peaks").] The $2s$ orbital surface plot also shows a depression surrounding the main "peak"; this represents a region in which the orbital magnitude is negative.

The main features of Figs. 5-2 and 5-3 are customarily shown by the use of simple line graphs. Since the functions are symmetric about the origin

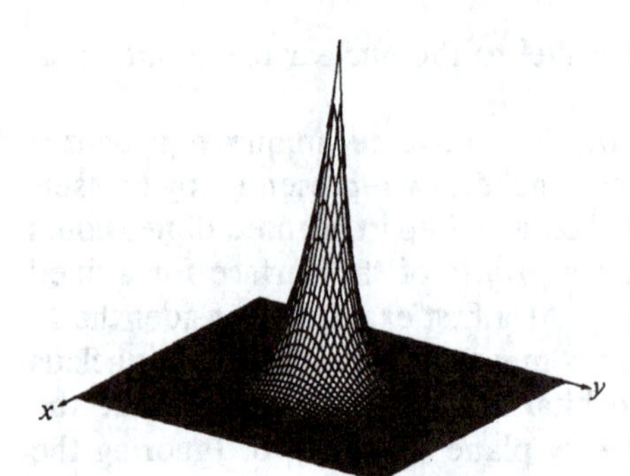

FIGURE 5-2
Surface plot of the $1s$ wavefunction (orbital) of the hydrogen atom. The height of any point on the surface above the xy plane (the nuclear plane) represents the magnitude of the $1s$ function at the point (x, y) in the nuclear plane. The nucleus is located in the xy plane immediately below the "peak."

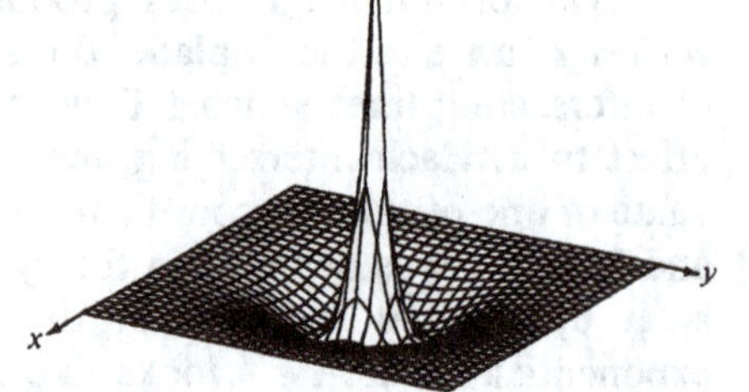

FIGURE 5-3
Surface plot of the $2s$ wavefunction (orbital) of the hydrogen atom. The height of any point on the surface above the xy plane (the nuclear plane) represents the magnitude of the $2s$ function at the point (x, y) in the nuclear plane. Note that there is a negative region (depression) about the nucleus; the negative region begins at $r = 2$ and goes asymptotically to zero at $r = \infty$.

(independent of the angles θ and φ), one can use the forms

$$\begin{aligned} 1s(r) &= e^{-r} \\ 2s(r) &= (2 - r)e^{-r/2} \end{aligned} \tag{5-57}$$

to represent the basic features of the surface plots (again, the normalization constants are immaterial in this context). Such plots are shown in Fig. 5-4*a* and *b*; these are simply profiles of the $f(x, y)$ plots which pass through the nucleus.

The same techniques can be used to construct surface plots of $(1s)^2$ and $(2s)^2$, the probability densities associated with each orbital. These are shown in Figs. 5-5 and 5-6. The $1s$ and $(1s)^2$ surface plots do not look very different from each other, but note that what was a negative "depression" in $2s$ shows up as a positive "hump" in $(2s)^2$. Figure 5-7 shows the $(1s)^2$ and $(2s)^2$ profiles which pass through the nucleus.

Functions for which $l \neq 0$ do not have the symmetry about the origin characteristic of the *ns* orbitals, since angular functions impart a directional effect. This effect is most simply illustrated for the case of $2p$ orbitals. First, all the $2p$ orbitals have a radial part which has the general form

$$R_{2p} = re^{-r} \tag{5-58}$$

If we choose either the xz or yz plane for the plot and let $r = (x^2 + z^2)^{1/2}$ or $(y^2 + z^2)^{1/2}$, respectively, a surface plot of this function would have the appearance of Fig. 5-8. In order to show clearly that this surface has a "hollow" interior, the plot was constructed such that the grid of intersecting lines is "transparent." A profile through the nucleus is shown in Fig. 5-9.

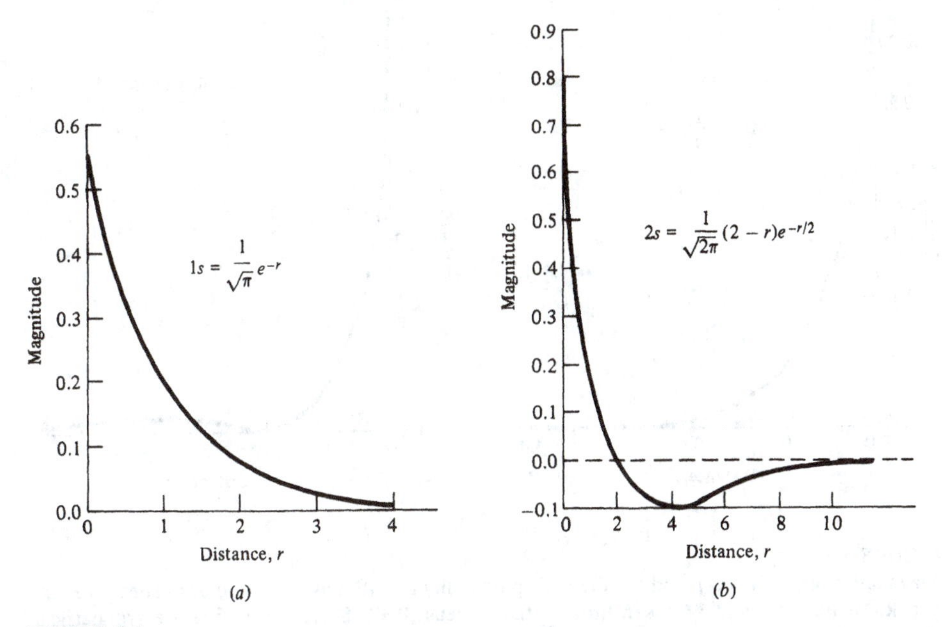

FIGURE 5-4
Magnitudes of (*a*) the 1*s* and (*b*) the 2*s* orbitals as a function of r. These are just profiles of Figs. 5-2 and 5-3 taken through the nucleus (0, 0). Since the profiles are symmetrical about $r = 0$, it is necessary to show only one of their equivalent halves.

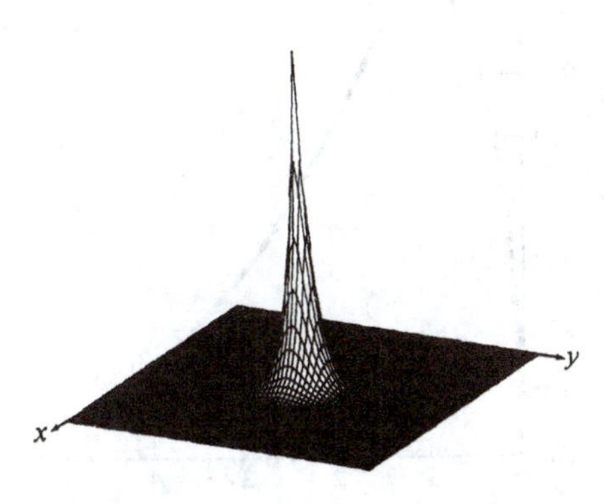

FIGURE 5-5
Surface plot of $(1s)^2$; the probability density associated with the 1*s* wavefunction of the hydrogen atom.

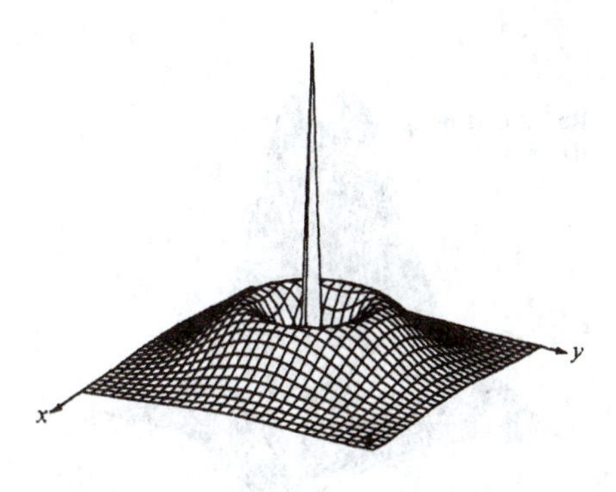

FIGURE 5-6
Surface plot of $(2s)^2$; the probability density associated with the 2*s* wavefunction of the hydrogen atom. Note that the negative region which appeared in the 2*s* plot of Fig. 5-3 now appears as a positive region.

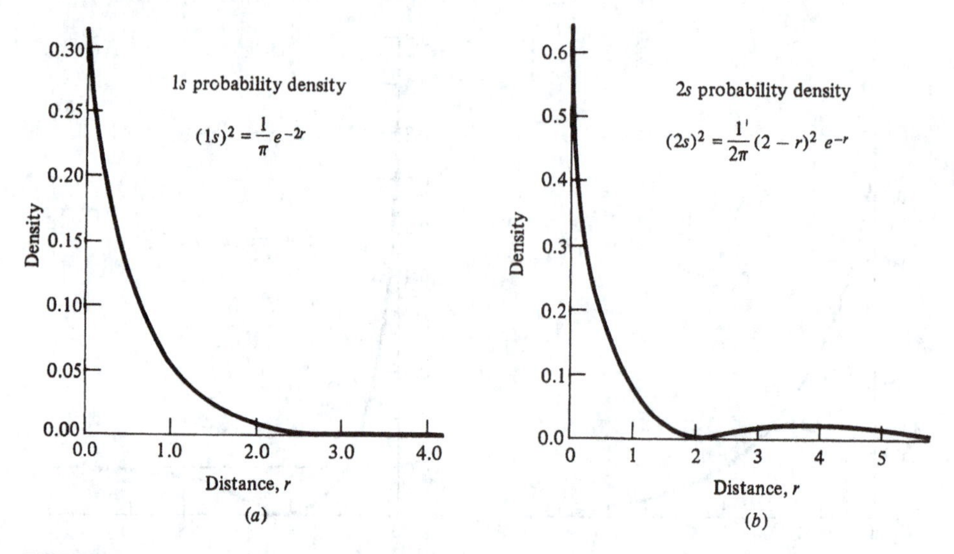

FIGURE 5-7
Magnitudes of (*a*) the $(1s)^2$ and (*b*) the $(2s)^2$ probability densities as a function of r. These are just profiles of Figs. 5-5 and 5-6 taken through the nucleus (0, 0). Since the profiles are symmetrical about $r = 0$, it is necessary to show only one of their equivalent halves.

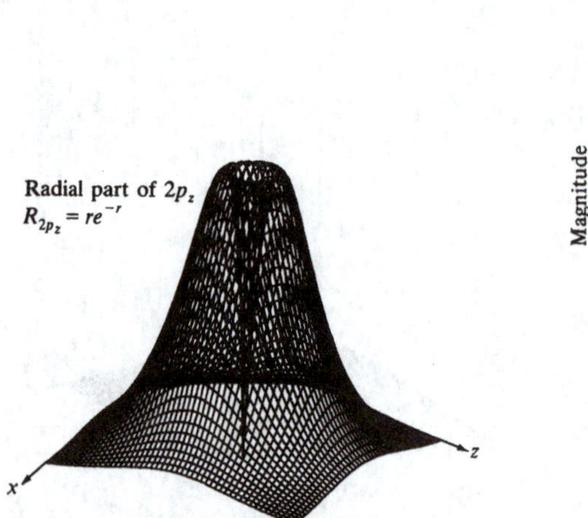

FIGURE 5-8
Surface plot of the radial portion of a $2p$ wavefunction of the hydrogen atom. The grid lines have been left transparent so that the inner, "hollow" portion is visible.

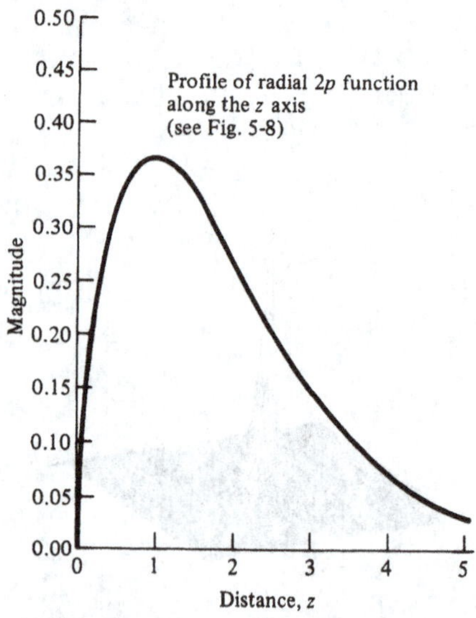

FIGURE 5-9
Profile of the radial portion of a $2p$ wavefunction of the hydrogen atom.

However, if we multiply R_{2p} by $\cos\theta$, we obtain the $2p_z$ orbital. The effect of multiplying by $\cos\theta$ is to multiply the radial part by a different numerical factor depending on the angle θ. Thus, for $\theta = 0$, the radial function is multiplied by $+1$; at $\theta = \pi$, the radial function is multiplied by -1. This latter multiplication inverts the surface through the xy plane. For $\theta = \pi/2$ and $3\pi/2$, the radial function is multiplied by zero. The overall result is shown in Fig. 5-10, which represents a surface plot of the $2p_z$ orbital in the yz or xz plane. Note that the $2p_z$ orbital shows a positive lobe (the "hill") and a negative lobe (the "pit"). A profile of this orbital will look different depending on where it is taken. For example, a profile along the z axis and including the nucleus will appear like the profile in Fig. 5-9 except that one of the halves becomes negative. This new profile is shown in Fig. 5-11.

Figure 5-12 shows the appearance of $(2p_z)^2$; note that the major effect is to turn the pit into a hill. This figure shows that the electron probability density due to a $2p_z$ orbital is not spherically symmetrical about the nucleus (as in ns orbitals) but is concentrated along a particular axis. Furthermore, the probability density has its maximum away from the nucleus and is zero along any axis perpendicular to the z axis and passing through the nucleus.

Figure 5-13 shows the polar plots of the $3d$ orbitals which are customarily shown in most chemistry textbooks. The surface plot counterparts are not familiar to most students and, thus, will be illustrated here. A surface plot of the $3d_{z^2}$ orbital is shown in Fig. 5-14. Note that the + lobes of Fig. 5-13 become large peaks while the − lobes become small pits. Figure 5-15 shows the square of the $3d_{z^2}$ orbital in which all the lobes are positive (the figure has been rotated in the yz plane by 90° in order to make the small peak easier to see).

The $3d_{xz}$, $3d_{xy}$, and $3d_{yz}$ orbitals produce exactly the same surface plot (except for labeling of axes). Figures 5-16 and 5-17 show the plots for $3d_{xy}$ and $(3d_{xy})^2$, respectively. Figures 5-18 and 5-19 show the surface plots for $3d_{x^2-y^2}$

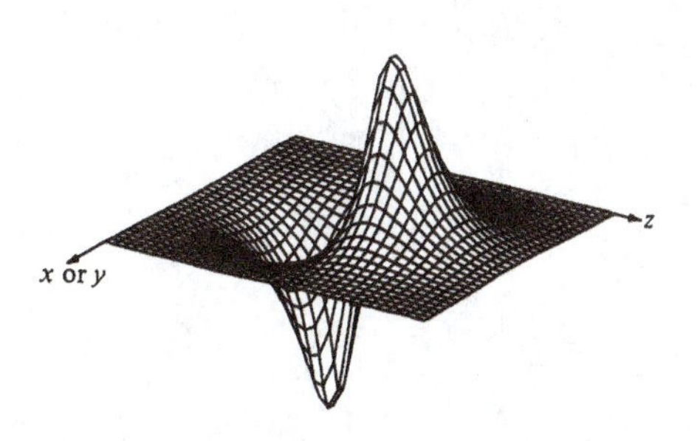

FIGURE 5-10
Surface plot of the $2p_z$ wavefunction (orbital) in the xz (or yz) plane for the hydrogen atom. The "pit" represents the negative lobe and the "hill" the positive lobe of a $2p$ orbital.

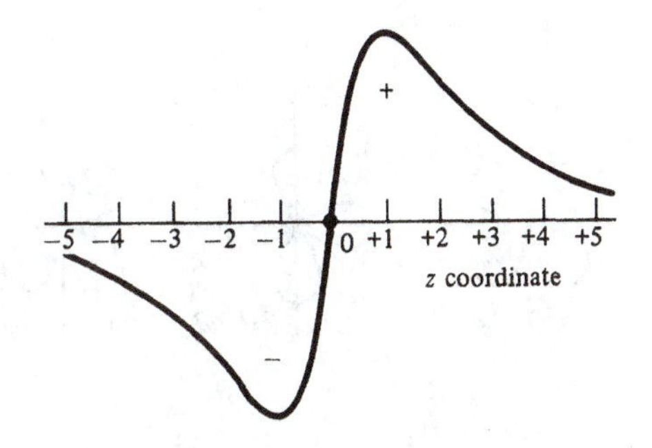

FIGURE 5-11
Profile of the $2p_z$ orbital along the z axis.

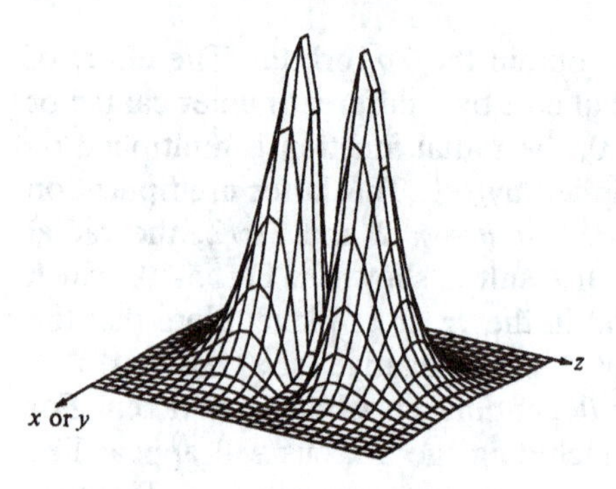

FIGURE 5-12
Surface plot of $(2p_z)^2$; the probability density represented by the $2p_z$ wavefunction of the hydrogen atom. Each of the hills represents an area in the xz (or yz) plane where the probability density is the highest. The probability density along the x (or y) axis passing through the nucleus (0, 0) is everywhere zero.

and $(3d_{x^2-y^2})^2$, respectively. Comparing Figs. 5-16 and 5-18, note that the $3d_{x^2-y^2}$ plot is the same as for the $3d_{xz}$ (and $3d_{xy}$ and $3d_{yz}$) orbitals but rotated by 45°.

When there is no unique axis, the complex forms of the orbitals imply the same probability densities as do the real form. However, when there is a unique axis, the complex forms provide some additional information. This is illustrated for the $2p$ orbitals in Fig. 5-20. The effect of a unique axis, such as from a magnetic field, is to allow for three different responses of the electron to the field. For $m_l = 0$, the electron distribution is concentrated along the axis of the unique field and has an angular momentum component of zero about this axis. For $m_l = \pm 1$, the electron can have an angular momentum component of either $-\hbar$ (counterclockwise motion about the unique axis) or $+\hbar$ (clockwise motion about the unique axis). As we will discuss later, each of these three situations represents a different energy state of the hydrogen atom. We might also note at this point that one effect of the magnetic field (or of any unique axis) is to lessen the symmetry of the system; thus, some degeneracy disappears.

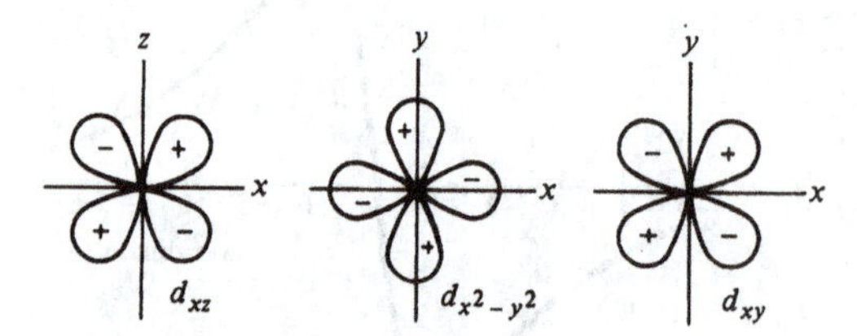

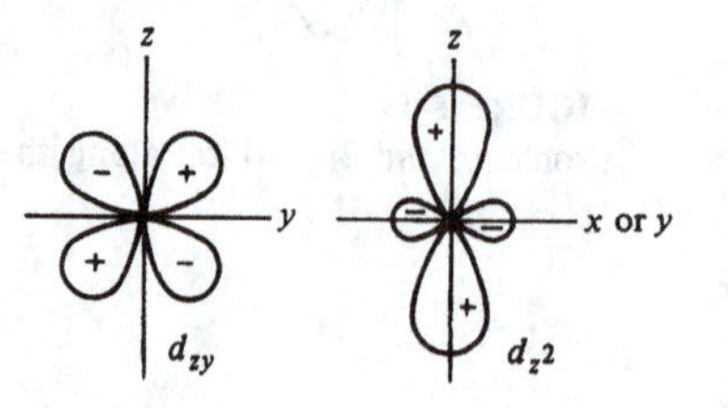

FIGURE 5-13
Conventional polar plots of the five $3d$ wavefunctions (orbitals) of the hydrogen atom.

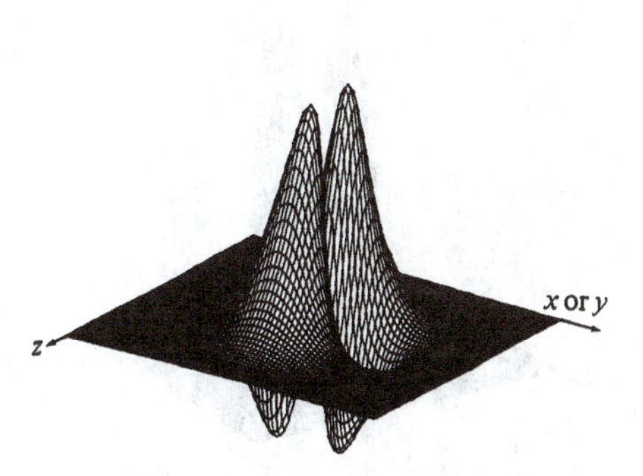

FIGURE 5-14
Surface plot of the $3d_{z^2}$ wavefunction (orbital) in the yz plane for the hydrogen atom. The large hills correspond to the positive lobes of Fig. 5-13 and the small pits correspond to the negative lobes of Fig. 5-13.

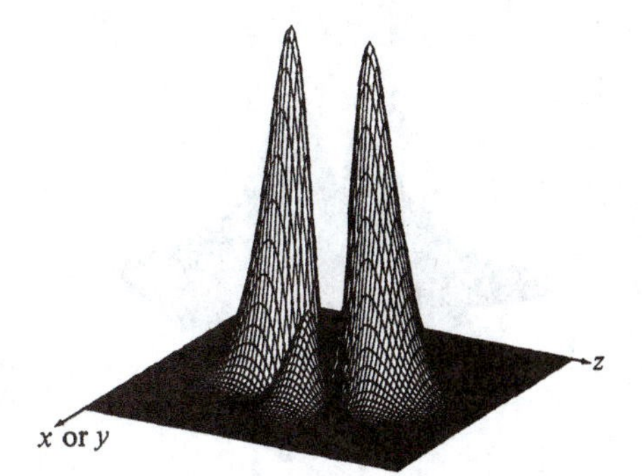

FIGURE 5-15
Surface plot of $(3d_{z^2})^2$, the probability density associated with the $3d_{z^2}$ orbital of the hydrogen atom. The figure has been rotated 90° from that of Fig. 5-14 in order to show the small hill more clearly; the second hill is hidden behind the right-hand large hill.

There is another form in which the probability densities associated with the hydrogen atom wavefunctions are often depicted. To illustrate these we define the *radial distribution function*

$$D_{nl}(r) = N_{nl} r^2 R_{nl}^2 \tag{5-59}$$

where

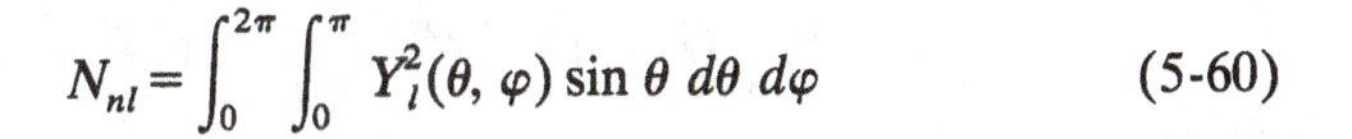

$$N_{nl} = \int_0^{2\pi} \int_0^{\pi} Y_l^2(\theta, \varphi) \sin\theta \, d\theta \, d\varphi \tag{5-60}$$

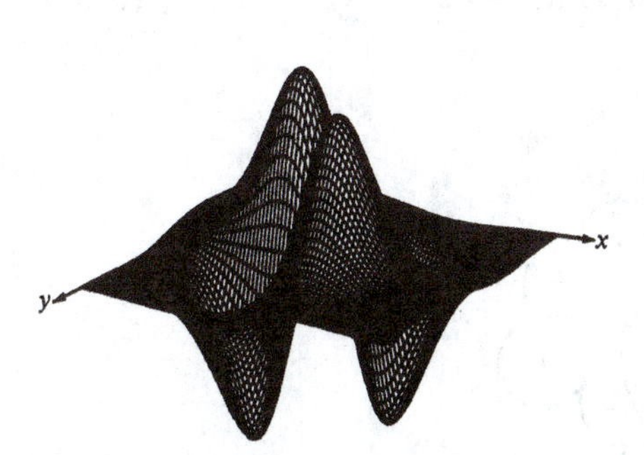

FIGURE 5-16
Surface plot of the $3d_{xy}$ wavefunction (orbital) in the xy plane for the hydrogen atom. Compare the hills and pits of this figure with the positive and negative lobes shown in Fig. 5-13.

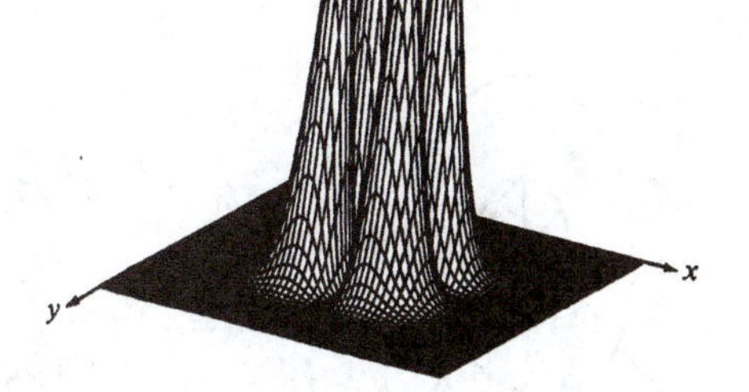

FIGURE 5-17
Surface plot of $(3d_{xy})^2$, the probability density associated with the $3d_{xy}$ wavefunction of the hydrogen atom. Note that the pits of Fig. 5-16 now appear as hills.

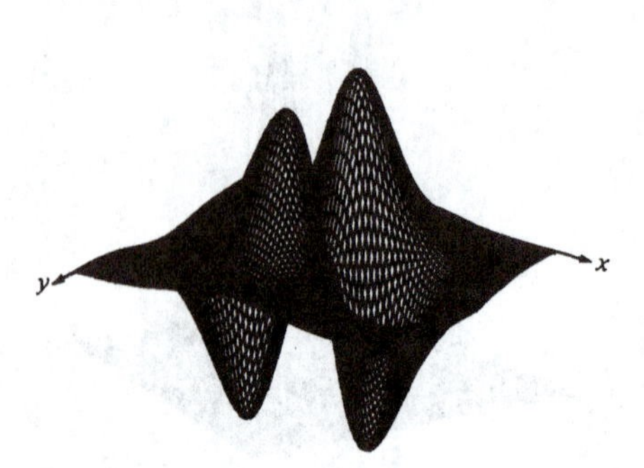

FIGURE 5-18
Surface plot of the $3d_{x^2-y^2}$ wavefunction (orbital) in the xy plane for the hydrogen atom. This surface is exactly like that of the $3d_{xy}$ orbital of Fig. 5-16 if the diagram is rotated by 45° about the vertical axis containing the nucleus.

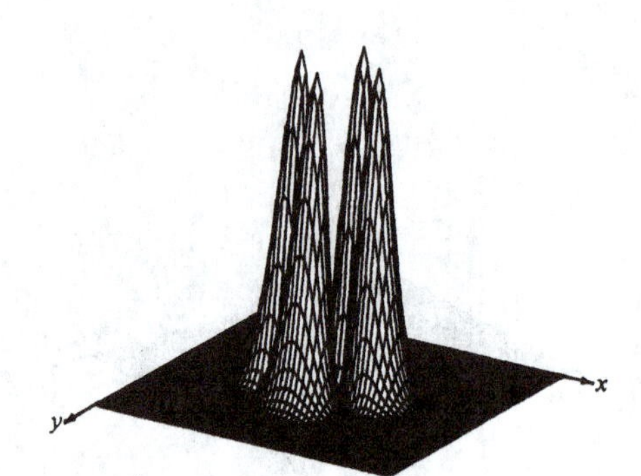

FIGURE 5-19
Surface plot of $(3d_{x^2-y^2})^2$, the probability density associated with the $3d_{x^2-y^2}$ wavefunction of the hydrogen atom. Note that the pits of Fig. 5-18 now appear as hills.

Whereas R_{nl}^2 is a probability density (per unit volume), $D_{nl}(r)$ represents the *population* of a spherically symmetrical surface located a distance r from the nucleus. More accurately, $D_{nl}(r)\,dr$ is the probability of finding the electron within an annular volume whose inner radius is r and whose outer radius is $r + dr$. The function $D_{nl}(r)$ goes through a maximum; this represents the *distance* from the nucleus at which the electron is most likely to be found *independent of direction*. For the $1s$ function

$$D_{1s}(r) = 4\pi r^2 R_{1s}^2 \tag{5-61}$$

where

$$N_{1s} = \int_0^{2\pi} \int_0^{\pi} \sin\theta\, d\theta\, d\varphi = 4\pi \tag{5-62}$$

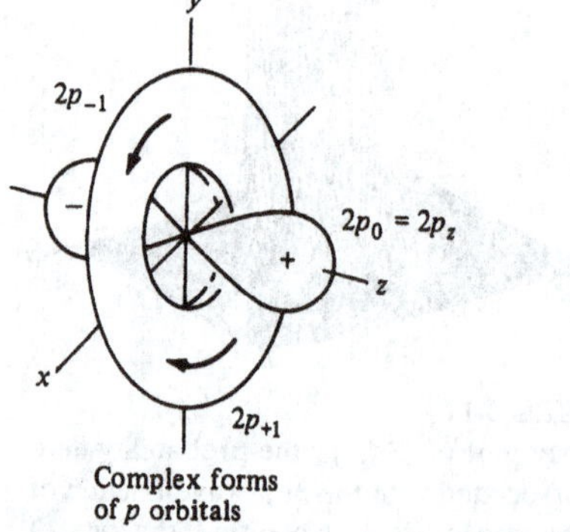

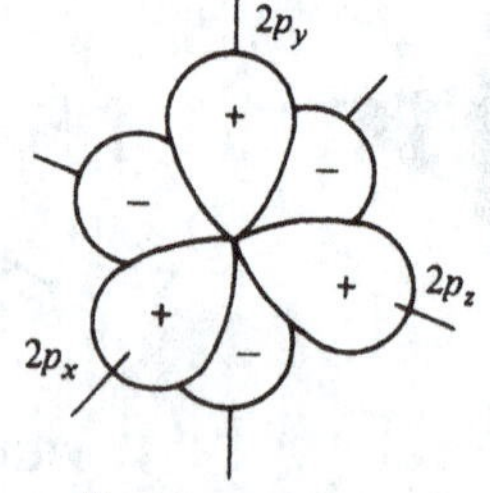

FIGURE 5-20
Complex and real forms of the $2p$ orbitals. The $2p_{+1}$ and $2p_{-1}$ complex forms may be thought of as the $2p_x$ and $2p_y$ real forms spinning in opposite directions.

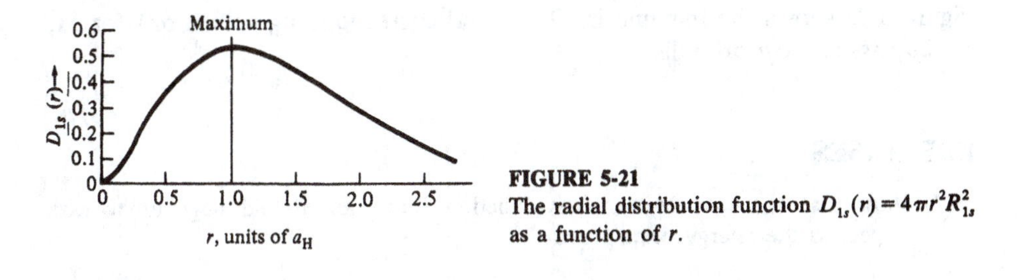

FIGURE 5-21
The radial distribution function $D_{1s}(r) = 4\pi r^2 R_{1s}^2$ as a function of r.

A graph of this function is shown in Fig. 5-21. Note that the maximum is found as follows:

$$\frac{\partial D_{1s}}{\partial r} = (2r - 2r^2)e^{-2r} = 0 \qquad \text{maximum at } r = 1 \tag{5-63}$$

(Two other extremum points are $r = 0$ and $r = \infty$; these are minima.) This means that the most likely distance to find the electron is at $r = 1$ (i.e., at $r = a_H$). Note that this is the first Bohr radius. In the simple Bohr model, the electron was in a planar orbit with precisely this radius; in the Schrödinger model, this distance becomes the most probable distance in a three-dimensional orbit.

The *average* distance of the electron from the nucleus is given by the expectation value $\langle r \rangle$. For the 1s state this quantity becomes

$$\langle r \rangle = \frac{1}{\pi} \int_0^\infty \int_0^{2\pi} \int_0^\pi r^3 e^{-2r} \sin\theta \, d\theta \, d\varphi \, dr$$

$$= 4 \int_0^\infty r^3 e^{-2r} \, dr = 4\left(\frac{3!}{2^4}\right) = \frac{3}{2} \tag{5-64}$$

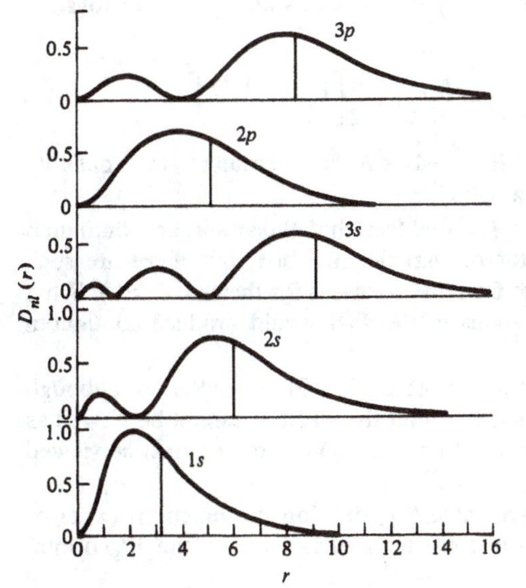

FIGURE 5-22
Radial distribution functions for 1s, 2s, 2p, 3s, and 3p orbitals. The solid vertical lines indicate values of $\langle r \rangle$.

Figure 5-22 shows the maxima in $D_{nl}(r)$ and corresponding values $\langle r \rangle$ for 1s, 2s, 2p, 3s, and 3p orbitals.

EXERCISES

5-7. Show that the one-electron wavefunctions have an n^2-fold degeneracy with respect to the energy; i.e., prove

$$\sum_{l=0}^{n-1} (2l+1) = n^2$$

5-8. Let a, b, and c be the direction cosines of a point (r, θ, φ). Show that the function $\psi_{2p} = a\psi_{2p_x} + b\psi_{2p_y} + c\psi_{2p_z}$ has its maximum values in the direction defined by θ and φ.

5-9. Show that the function $\psi^2(2p_x) + \psi^2(2p_y) + \psi^2(2p_z)$ implies spherical symmetry and that $\psi^2(2p_x) + \psi^2(2p_y) = |\psi(2p_1)|^2 + |\psi(2p_{-1})|^2$ implies cylindrical symmetry (about the z axis). How does this relate to the directional properties of the 2p wavefunctions?

5-10. Show analytically that r^2e^{-2r} has the extremum points $r = 0$, 1, and ∞.

5-11. Compute the extremum points for the D_{2s} and D_{2p} radial distribution functions. Also compute $\langle r \rangle$ for these states and compare.

5-12. The function $f(r) = (27 - 18r + 2r^2)e^{-r}$ is essentially like the 3s wavefunction of hydrogen, except for a normalization factor and a factor of $\frac{1}{3}$ in the exponent. Thus, this function will have the same general graphical appearance as the 3s wavefunction.

(*a*) Construct a plot of $f(r)$ versus r.

(*b*) Determine analytically the location of the following for $f(r)$:

(i) The extremum points.

(ii) The points at which $f(r)$ crosses the r axis (these are the nodes of the function).

(*c*) Sketch the appearance of the surface plots of the 3s and $(3s)^2$ functions.

5-13. Show that in general

$$\langle r \rangle = \langle \psi_{n,l,m_l} | r | \psi_{n,l,m_l} \rangle = \frac{n^2}{Z}\left\{1 + \frac{1}{2}\left[1 - \frac{l(l+1)}{n^2}\right]\right\}$$

5-14. Given that $\langle r^{-1} \rangle = Z^2/n^2$ and $\langle \hat{H} \rangle = -Z^2e^2K/2n^2$, evaluate the quantity $-(\hbar^2/2\mu)\langle \partial^2/\partial x^2 \rangle$ and state its meaning.

5-15. Give a simple physical explanation for the fact that there are no continuum functions for the harmonic oscillator or particle in a box, but there are such functions for the H atom (these arise from the solutions for the case $E > 0$). What modification of the boundary conditions in the PIB would produce continuum functions?

5-16. Use the fact that for an electron in an ns orbital ($l = 0$), $\mathbf{L} = \mathbf{r} \times \mathbf{p} = 0$ although neither **r** nor **p** is equal to zero. This means that the electron cannot be viewed as traveling in a circular orbit (as in the old Bohr theory) but rather must be viewed as undergoing an in-out motion.

5-17. Using the appropriate equation from Table 5-1, develop the function $f(x, z)$ or $f(y, z)$, which must be used in order to construct a surface plot of the $3d_{z^2}$ orbital in the xz or yz plane.

5-3 SELECTION RULES FOR ONE-ELECTRON ATOMS

The selection rules governing electronic transitions in the hydrogen atom and other one-electron species are obtained by examination of the integrals

$$\langle nlm_l|x|n'l'm'_l\rangle \qquad \langle nlm_l|y|n'l'm'_l\rangle \qquad \langle nlm_l|z|n'l'm'_l\rangle \tag{5-65}$$

where we let nlm_l represent the wavefunction associated with the quantum numbers n, l, and m_l.

Insofar as the l and m_l quantum numbers are concerned, we have already derived the relevant selection rules in a previous treatment of the rigid rotator (Sec. 4-9). If we replace J with l and m_J with m_l, these rules [previously stated in Eqs. (4-130) and (4-133), respectively] are

$$\Delta l = \pm 1 \qquad \text{and} \qquad \Delta m_l = 0, \pm 1 \tag{5-66}$$

The selection rule involving the principal quantum number n is not as straightforward to derive. This was first carried out by Pauli and reported in a paper by Schrödinger.[6] A more abstract but mathematically simpler method has been developed by Dirac[7] and has recently been promoted by Sannigrahi and Das.[8] This general method for the determination of selection rules avoids the need to have detailed knowledge of the relevant wavefunctions and the various recursion relations among them. The method requires two commuting operators, $\hat{A}$ and $\hat{B}$, corresponding to two constants of the motion of the system. One then finds an algebraic relationship which connects $\hat{A}$, $\hat{B}$, and the transition operator $\hat{\mu}$. Such a general equation is

$$\sum_i \sum_j f_i(\hat{A})g_i(\hat{B})\hat{\mu}f_j(\hat{A})g_j(\hat{B}) = 0 \tag{5-67}$$

One then takes the matrix element of Eq. (5-67) for two simultaneous eigenfunctions of $\hat{A}$ and $\hat{B}$. The selection rules arise upon expansion and examination of this matrix element. Thus, the selection rules for n arise by using $\hat{A} = \hat{H}$ and $\hat{B} = \hat{L}_z$. A specific relationship of the form of Eq. (5-67) is

$$[\hat{H}, [\hat{L}_z, Z]] = 0 \tag{5-68}$$

Expanding and taking the matrix element between two arbitrary states ψ_1 and ψ_2, leads to

$$\begin{aligned}\langle \psi_1|\hat{H}\hat{L}_z z - \hat{H}z\hat{L}_z - \hat{L}_z\hat{H} + z\hat{L}_z\hat{H}|\psi_2\rangle \\ = \hbar\langle \psi_1|z|\psi_2\rangle(E_1 m_{l_1} - E_1 m_{l_2} - E_2 m_{l_1} + E_2 m_{l_2}) \\ = \hbar\langle \psi_1|z|\psi_2\rangle(E_1 - E_2)(m_{l_1} - m_{l_2}) = 0\end{aligned} \tag{5-69}$$

---

[6] E. Schrödinger, *Ann. Physik*, **79**:361 (1926).

[7] See pp. 159 ff. in P. A. M. Dirac, *Quantum Mechanics*, cited in the suggested readings at the end of this chapter.

[8] A. B. Sannigrahi and R. Das, *J. Chem. Ed.*, **57**:786 (1980).

Since $\langle \psi_1 | z | \psi_2 \rangle \neq 0$ for an *allowed* transition and $m_{l_1} = m_{l_2}$ (since $\Delta m_l = 0$ for radiation polarized in the z direction), it follows that $E_1 - E_2$ can have any arbitrary value (excluding the trivial case of zero). Thus we can write

$$\Delta n = \text{anything} \tag{5-70}$$

This selection rule is not, of course, surprising. The older Bohr theory placed no restrictions on what energy levels could be involved in the spectral lines of hydrogen, and the experimental observation of series such as the Lyman, Balmer, Paschen, and Brackett series supported this. However, if the hydrogen atom spectrum is examined at increasingly better resolution, details begin to appear which neither the Bohr nor the Schrödinger treatments can resolve. For example, the Lyman series line attributed to the $n = 1$ to $n = 2$ transition is predicted to occur at a wavelength of 121.566 nm (82,259.6 cm^{-1}). What one actually finds under sufficient resolution are *two* lines spaced 0.3652 cm^{-1} apart. Apparently, there is some effect operative that the nonrelativistic Schrödinger equation does not account for. Furthermore, if the resolving power of the spectroscopic instrumentation is increased, additional details emerge—none of which is explicable on the basis of the Schrödinger equation. As we will see in the following section, one needs to incorporate relativity theory into the treatment in order to resolve some of these discrepancies. That this is the source of the fine structure came as no surprise to Schrödinger and his contemporaries. Even at the time of the Bohr theory, it was known that the model did not account for anything beyond the gross features of the spectrum, and attempts were made—particularly by Sommerfeld[9]—to account for some of these using relativity theory. Although some success was obtained by this approach, it is now known that the Bohr model itself forms an inadequate basis for a relativistically corrected treatment. That this is not true for the Schrödinger model will be illustrated in the following section.

5-4 SPIN ANGULAR MOMENTUM IN ONE-ELECTRON ATOMS: A RELATIVISTIC EFFECT

Although the Schrödinger equation provides a very good value for the ionization energy of the hydrogen atom (and other one-electron species such as He^+ and Li^{2+}) and also accounts for the line spectrum obtained by low-resolution instruments, it fails to account for several details revealed by higher-resolution studies. In particular, spectroscopic lines which appear to be *single* lines at low resolution turn out to be a *pair* of slightly separated lines (called a *doublet*) when viewed under higher resolution. As mentioned in the previous section,

[9] This approach is discussed by G. Herzberg, *Atomic Spectra and Atomic Structure*, Dover, New York, 1944, p. 19.

the doublet separation for the longest wavelength in the Lyman series is $0.3652\ \text{cm}^{-1}$. The separation decreases considerably as one goes to progressively shorter wavelength lines. Also, the separation decreases as one goes to the longest wavelength of the next series, the Balmer series. Use of much higher resolution instruments reveals a further fine structure; additional very small splittings appear in some of the lines.

In 1925 Goudsmit and Uhlenbeck[10] pointed out that the existence of doublets is inconsistent with an angular momentum of *zero* for an s state ($l = 0$) and of *one* for a p state ($l = 1$). Furthermore, they argued, if one ascribed to the electron an *additional* angular momentum of magnitude $\hbar/2$, then the doublet anomaly could be rationalized.[11] This additional angular momentum was thought to be an *intrinsic* property of the electron and could be physically related to the *spinning* of the electron about some axis; hence the property became known as *spin*. A modernized account of how this postulate accounts for doublet lines is as follows: Whereas the Schrödinger model considers the angular momentum of the electron to be due solely to an *orbital* component $\mathbf{L}$ whose magnitude is given by $\sqrt{l(l+1)}\,\hbar$, the existence of an additional angular momentum (the *spin* angular momentum, which we label $\mathbf{S}$) means that the electron has a *total* angular momentum (which we label $\mathbf{J}$) given by the vector sum

$$\mathbf{J} = \mathbf{L} + \mathbf{S} \tag{5-71}$$

Since $\mathbf{J} \geq 0$, the quantum number associated with the total angular momentum would have to be given by

$$j = l \pm s \tag{5-72}$$

where s is a spin angular momentum quantum number assumed to define the magnitude of $\mathbf{S}$ as

$$|\mathbf{S}| = \sqrt{s(s+1)}\,\hbar \tag{5-73}$$

(analogous to the corresponding relationship for $\mathbf{L}$). Continuing the analogy with $\mathbf{L}$ further, if $s = \frac{1}{2}$, then the spin angular momentum has the components $\pm\hbar/2$ associated with the quantum numbers $m_s = \pm\frac{1}{2}$ (or $-s \leq m_s \leq s$ just as orbital angular momentum l has the components $-l \leq m_l \leq l$). The spinning motion of the electron would contribute an angular momentum of

$$|\mathbf{S}| = \sqrt{\tfrac{1}{2}(\tfrac{1}{2}+1)}\,\hbar = \frac{\sqrt{3}}{2}\,\hbar \tag{5-74}$$

The z component of this is

$$S_z = |\mathbf{S}| \cos\theta \tag{5-75}$$

---

[10] S. Goudsmit and G. E. Uhlenbeck, *Naturwiss.*, **13**:953 (1925); *Nature*, **117**:264 (1962).

[11] The unsuccessful Sommerfeld model (described by Herzberg) ascribed an angular momentum of $\hbar$ to the s state.

where [following Eq. (4-52)], $\cos\theta$ is given by

$$\cos\theta = \frac{\pm m_s\hbar}{\sqrt{s(s+1)}} = \frac{\pm 1}{\sqrt{3}} \tag{5-76}$$

Then, substituting Eq. (5-76) into Eq. (5-75) gives

$$S_z = \frac{\pm\hbar}{2} \tag{5-77}$$

Now going back to Eq. (5-72), the s state ($l = 0$) would have a total angular momentum quantum number of

$$j = 0 + \tfrac{1}{2} = \tfrac{1}{2} \tag{5-78}$$

and the p state ($l = 1$) would have

$$j = \begin{cases} 1 + \frac{1}{2} = \frac{3}{2} \\ 1 - \frac{1}{2} = \frac{1}{2} \end{cases} \tag{5-79}$$

Assuming that the two p states ($j = \frac{1}{2}$ and $\frac{3}{2}$) have different (but only slightly separated) energies, the doublet lines are explained on the basis of the two transitions

$$\begin{aligned} s(l=0, s=\tfrac{1}{2}, j=\tfrac{1}{2}) &\rightarrow p(l=1, s=\tfrac{1}{2}, j=\tfrac{1}{2}) \\ s(l=0, s=\tfrac{1}{2}, j=\tfrac{1}{2}) &\rightarrow p(l=1, s=\tfrac{1}{2}, j=\tfrac{3}{2}) \end{aligned} \tag{5-80}$$

Note that these transitions satisfy the selection rule $\Delta l = \pm 1$ but also require a new selection rule $\Delta j = 0, \pm 1$.

Forced and specious as the above reasoning may appear, it nevertheless allows the rationalization of another puzzling phenomenon: the Stern-Gerlach effect.[12] When a beam of vaporized silver atoms ($5s$ valence electrons) is sent through an inhomogeneous magnetic field (see Fig. 5-23), the beam is split into two beams of equal intensity: one beam is diverted toward the N pole and the other toward the S pole of the magnetic field. According to classical theory, one would expect only a single "smeared-out" beam. Experiment further shows that the z component of the magnetic moment (the z axis connects the N and S poles) is 1 μ_B (μ_B is a *Bohr magneton*, given by $e\hbar/2m$ and numerically equal to 9.27408×10^{-24} J · T^{-1}). The magnetic moment is related to angular momentum by the classical expression

$$\boldsymbol{\mu} = -\mu_B \mathbf{L} \tag{5-81}$$

However, if $\mathbf{L} = 0$, then $\boldsymbol{\mu}$ (and hence μ_z) should be zero, contrary to what is

---

[12] O. Stern, *Zeit. f. Physik*, **7**:249, (1921); W. Gerlach and O. Stern, *Zeit. f. Physik*, **8**:110 and **9**:349 (1922); *Ann. Phys.*, **74**:673 (1924). An excellent discussion of this experiment is found in R. S. Berry, S. A. Rice, and J. Ross, *Physical Chemistry*, Wiley, New York, 1980, app. 5A, pp. 200–202.

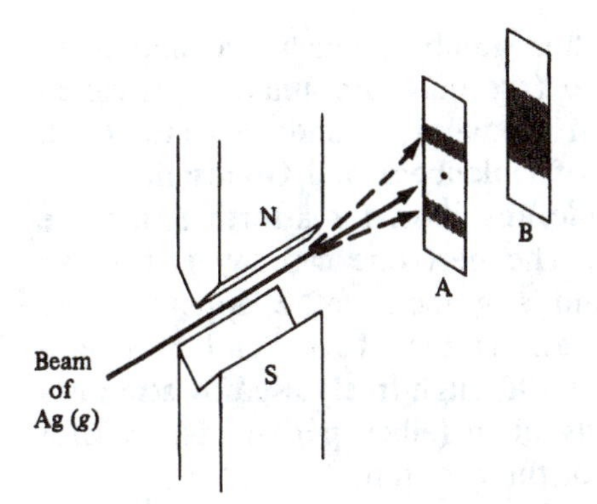

FIGURE 5-23
The Stern-Gerlach effect. A beam of silver atoms (5*s* valence electrons) is split into two beams when passed through an inhomogeneous magnetic field. Pattern A illustrates the observed splitting of the beam; pattern B illustrates the classical prediction.

actually found. If we now assume that the magnetic moment is due to *spin* angular momentum, we can write

$$\boldsymbol{\mu} = -\mu_B \mathbf{S} \tag{5-82}$$

where $|\mathbf{S}| = (\sqrt{3}/2)\hbar$, as assumed for the hydrogen doublet case. Thus

$$\boldsymbol{\mu} = \frac{\sqrt{3}}{2}\mu_B \quad \text{and} \quad \mu_z = |\boldsymbol{\mu}|\cos\theta = \frac{\sqrt{3}}{2}\frac{1}{\sqrt{3}} = \frac{1}{2}\mu_B \tag{5-83}$$

Although this accounts *qualitatively* for the results of the experiment, it is in error *quantitatively* by a factor of 2. Uhlenbeck and Goudsmit then proposed to replace Eq. (5-82) with

$$\boldsymbol{\mu} = -g\mu_B S \tag{5-84}$$

where g (now called the *Landé g factor*) equals 2 for an electron. This, of course, leads to the correct value of μ_z—but at the price of adding a second arbitrary assumption to the one previously introduced to explain doublets.

In 1928 Dirac[13] reformulated the Schrödinger treatment of the one-electron problem to make the equations consistent with the requirements of special relativity: specifically, invariance under a Lorentz-Fitzgerald transformation.[14] The solutions to the equation (now called the *Dirac equation*) produce a fourth quantum number s, with a value of $\frac{1}{2}$, which represents an angular momentum **S** which is in addition to that represented by the quantum number l. The projection of **S** onto an arbitrary axis is described by a quantum number m_s whose values are restricted by

$$-s \le m_s \le s \tag{5-85}$$

---

[13] P. A. M. Dirac, *Proc. Roy. Soc.* (*London*), **A117**:610 and **A118**:315 (1928). See also C. Darwin, *Proc. Roy. Soc.* (*London*), **A118**:654 (1928).

[14] Schrödinger considered this in his original approach but was not successful in unraveling the necessary mathematics.

Since $s = \frac{1}{2}$, $m_s = -\frac{1}{2}$ or $+\frac{1}{2}$. The m_s quantum number, which is analogous to m_l, is now called the *spin quantum number* (the quantum number s is called the *spin*). In addition, the Dirac equation produces a Landé g factor of 2, thereby vindicating the ad hoc arguments of Uhlenbeck and Goudsmit.

The Dirac treatment does not provide any basis for asserting that the electron actually spins in a literal sense. The best one can say is that an additional angular momentum of $\hbar/2$ and a g factor of 2 appear to be fundamental properties of the electron just as the charge and mass are fundamental properties. Thus the word *spin*, although freely used by scientists today, should be interpreted only as a convenient (albeit perhaps inaccurate) name for a relativistically based property of the electron.[15]

The modern method for describing the quantum states of the hydrogen atom is to define the *spectroscopic term symbol*

$$n^2 l_j \tag{5-86}$$

where n and l have their previous meanings, j is the *total* angular momentum quantum number (equal to $l \pm \frac{1}{2}$), and the left superscript "2" is the *multiplicity* defined in general as $2s + 1$, where $s = \frac{1}{2}$ in the specific case of hydrogen (the full significance of this latter quantity will be discussed in a later chapter when we extend the spectroscopic notation to atoms in general). It is also customary to replace numerical values of l with the following letter equivalents:

$$l = \begin{matrix} 0, & 1, & 2, & 3, & 4, & \ldots \\ S, & P, & D, & F, & G, & \ldots \end{matrix}$$

Lowercase letters may also be used for one-electron species, but for atoms in general, the uppercase is mandatory; thus we will begin using the uppercase.

The ground state of hydrogen ($n = 1$) is symbolized by $1^2S_{1/2}$. This symbol tells us that $n = 1$, $l = 0$, $s = \frac{1}{2}$, and $j = \frac{1}{2}$. For $n = 2$ the quantum states are $2^2S_{1/2}$, $2^2P_{1/2}$, and $2^2P_{3/2}$. For $n = 3$ the quantum states are $3^2S_{1/2}$, $3^2P_{1/2}$, $3^2P_{3/2}$, $3^2D_{3/2}$, and $3^2D_{5/2}$. States of given n and j and different l are assumed to have the same energy.[16] Two states of given n and l but different j values are separated in energy by

$$\Delta E_{j+1,\, j} = A_{nl}(j + 1) \tag{5-87}$$

(called the *Landé interval rule*) where A_{nl} is the *spin-orbit coupling constant* defined by

$$A_{nl} = \hbar^2 \int_0^\infty [R_{nl}(r)]^2 \xi(r) r^2 \, dr \tag{5-88}$$

---

[15] It is not true, however, that spin is a purely relativistic phenomenon. See p. 153 in Jammer, cited in the readings at the end of this chapter

[16] In Chap. 8 we will examine the effect of l on the energy.

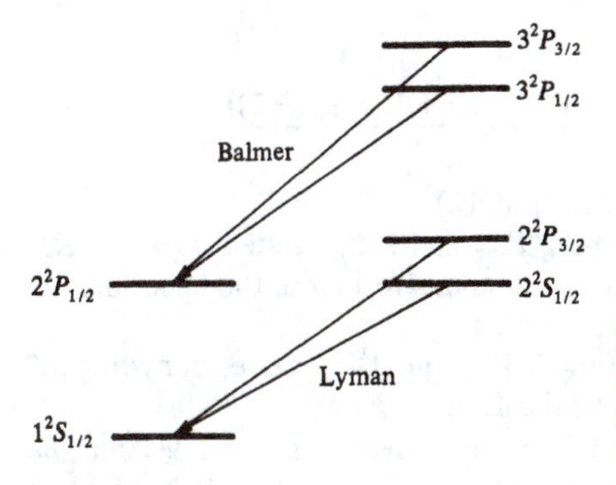

FIGURE 5-24
Energy-level diagram of the hydrogen atom showing the origins of the doublets in the first lines of the Lyman and Balmer series.

The quantity $\xi(r)$ is defined by

$$\xi(r) = -\frac{e}{2m_e^2c^2}\,\frac{1}{r}\,\frac{\partial\phi}{\partial r} \tag{5-89}$$

where ϕ (equal to $-Ve$) is the *Coulomb potential.* In addition to the selection rule $\Delta l = \pm 1$, we have

$$\Delta j = 0,\ \pm 1 \tag{5-90}$$

All of this allows one to construct the energy-level diagram shown in Fig. 5-24. The doublet in the Lyman $n = 2$ to $n = 1$ line is seen to be due to the transitions

$$2^2P_{1/2} \rightarrow 1^2S_{1/2} \qquad \text{and} \qquad 2^2P_{3/2} \rightarrow 1^2S_{1/2} \tag{5-91}$$

Similarly, there will be a doublet in the Balmer $n = 3$ to $n = 2$ line, namely,

$$3^2P_{1/2} \rightarrow 2^2S_{1/2} \qquad \text{and} \qquad 3^2P_{3/2} \rightarrow 2^2S_{1/2} \tag{5-92}$$

EXERCISES

5-18. Show that the spin-orbit coupling constant [Eq. (5-87)] is given by

$$A_{nl} = \left(\frac{Z^4e^2K}{2m_e^2c^2a_0^3}\right)\frac{1}{D_{nl}} = \alpha^2 R_\infty Z^4\,\frac{hc}{D_{nl}}$$

where $a_0 = \hbar^2/m_e e^4 K$
R_∞ = Rydberg constant [see Eq. (1-12)] for infinite nuclear mass ($\mu = m_e$)
$D_{nl} = n^3 l(l + \frac{1}{2})(l + 1)$
$\alpha = e^2K/\hbar c$

The constant α is a dimensionless constant called the *fine-structure constant*; it has a numerical value of 1/137.03604 (very close to $\frac{1}{137}$).

5-19. Use the results of Exercise 5-18 to explain why the doublet separations are most noticeable for the longest wavelength in a given series and how these change from series to series.

5-20. The energy of a state n^2l_j is given by

$$E_{n,l,j,s} = E_n + \frac{\alpha^2 hcR_\infty[j(j+1) - l(l+1) - s(s+1)]}{2D_{nl}}$$

(See Exercise 5-18 for a definition of the symbols.)

(*a*) Show that the separation between the $2^2P_{1/2}$ and $2^2P_{3/2}$ states is $(\alpha^2/16)R_\infty$. Convert this to its equivalent in cm^{-1}, and compare it with the experimental value quoted at the beginning of this section.

(*b*) What is the separation (in cm^{-1}) of the $3^2P_{1/2}$ and $3^2P_{3/2}$ states of hydrogen? These are involved in the first line of the Balmer series.

5-21. Calculate the wavelengths of the first Lyman and Balmer series lines for the spectrum of ionized helium, He^+, and estimate the magnitudes of the doublet separations.

5-22. Use the equation in Exercise 5-20 to derive the Landé interval rule [Eq. (5-87)].

5-23. The prominent yellow line in the emission spectrum of sodium vapor is also a doublet with wavelengths 589.0 and 589.6 nm. The other alkali metals have similar prominent lines. Suggest a qualitative explanation for the origin of these doublets.

5-5 THE ZEEMAN EFFECT

Since an electron has a magnetic moment (even if spin is not taken into account), there should be an interaction between the electron of the hydrogen atom and an external magnetic field. The effect of this interaction is to remove some of the energy-level degeneracies; this is because the magnetic field imposes a unique axis and thereby lowers the symmetry from spherical to cylindrical. Since the removal of degeneracy creates new, distinct energy levels, the spectrum of the hydrogen atom should change accordingly. In fact, one finds that if a moderate-strength magnetic field (less than about 10,000 G) is used, the Lyman doublet is replaced by 10 lines. This effect is known as the *Zeeman effect*.[17]

The interaction between the magnetic moment of the electron and the magnetic field may be expressed in terms of the *interaction energy*

$$E' = g_j \mu_B H m_j \tag{5-93}$$

where $H = |\mathbf{H}|$ (magnetic field strength)

m_j = quantum number for the components of the total angular momentum **J**

g_j = generalized Landé g factor

The quantum number m_j satisfies

$$-m_j \leq j \leq m_j \tag{5-94}$$

[17] P. Zeeman, *Phil. Mag.*, **43**:226 (1897).

and the generalized Landé g factor is given by

$$g_j = 1 + \frac{j(j+1) + s(s+1) - l(l+1)}{2j(j+1)} \tag{5-95}$$

Note that $g = 2$ only for the $^2S_{1/2}$ state ($l = 0$, $s = j = \frac{1}{2}$). For $^2P_{1/2}$ and $^2P_{3/2}$ states, $g_j = \frac{2}{3}$ and $\frac{4}{3}$, respectively.

Equation (5-93) shows that the effect of the magnetic field is to remove the $(2j+1)$-fold degeneracy of the state n^2l_j. Thus, a $^2S_{1/2}$ state now becomes *two* states, one with $m_j = \frac{1}{2}$ and the other with $m_j = -\frac{1}{2}$. The former will have an energy $g_j\mu_B H m_j = 2\mu_B H(\frac{1}{2}) = \mu_B H$ *above* the zero-field value of the $^2S_{1/2}$ energy, and the other will have a value of $\mu_B H$ *below* this. Since g_j for a $^2P_{1/2}$ state is $\frac{2}{3}$ (not 2 as in $^2S_{1/2}$), the $^2P_{1/2}$ state is split by a different magnitude; the $m_j = \frac{1}{2}$ state will occur $(\frac{2}{3})\mu_B H(\frac{1}{2}) = \mu_B H/3$ *above* the zero-field value of $^2P_{1/2}$, and the $m_j = -\frac{1}{2}$ state will occur at $-\mu_B H/3$ *below* the zero-field value. Similarly, the separations of the $^2P_{3/2}$ state (relative to the zero-field state) will be $\pm\frac{2}{3}\mu_B H$ (for $|m_j| = \frac{1}{2}$) and $\pm 2\mu_B H$ (for $|m_j| = \frac{3}{2}$).

The selection rules for the transitions between the new levels are

$$\Delta m_j = 0, \ \pm 1 \tag{5-96}$$

Using these selection rules, one can construct an energy-level diagram such as shown in Fig. 5-25 and predict the number of lines found. This leads to the 10 lines found by experiment.

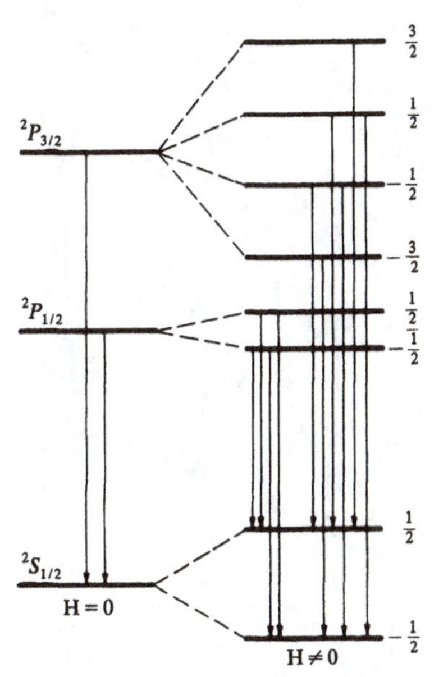

FIGURE 5-25
Zeeman effect on the doublet lines of the hydrogen atom.

EXERCISES

5-24. When the magnetic field is very strong, the orbital and spin angular momenta couple separately with the magnetic field rather than with each other (Paschen-Back effect). The result is that the interaction energy [Eq. (5-93)] now becomes $E' = \mu_B H m_j$. Redraw the energy-level diagram of Fig. 5-25, and determine how many distinct lines will be found in the spectrum.

5-25. Draw an energy-level diagram for the Zeeman transitions between a $^2P_{3/2}$ state and a $^2D_{3/2}$ state of the hydrogen atom.

SUGGESTED READINGS

Dirac, P. A. M.: *Quantum Mechanics*, Oxford University Press, Fairlawn, N.J., 1958.

Eyring, H., J. Walter, and G. E. Kimball: *Quantum Chemistry*, Wiley, New York, 1944.

Jammer, M.: *The Conceptual Development of Quantum Mechanics*, McGraw-Hill, New York, 1966.

Kauzmann, W.: *Quantum Mechanics*, Academic, New York, 1957.

Pauling, L., and E. B. Wilson, Jr.: *Introduction to Quantum Mechanics*, McGraw-Hill, New York, 1935.

Streitwieser, A., Jr., and P. H. Owens: *Orbital and Electron Density Diagrams*, Macmillan, New York, 1973. A very complete discussion of surface plots of orbitals, orbital densities, and constructs related to these. Many, many excellent illustrations of electron distributions in atoms and molecules.

CHAPTER 6

APPROXIMATE SOLUTIONS TO THE SCHRÖDINGER EQUATION

The Schrödinger equation can be solved exactly for only a few simple systems: the particle in a box, the harmonic oscillator, the rigid rotator, the hydrogen atom, and a few others.[1] Nevertheless, approximate solutions can be obtained for a very large variety of systems, and these solutions have contributed inestimably to the practical application of quantum theory to the development of chemical principles. This chapter will describe a systematic general approach to the construction and interpretation of approximate solutions to the Schrödinger equation; later chapters will exploit these principles in a variety of explicit systems.

There are some who believe that the exact formal solution to the wave equation, even if it were obtainable, might be of such a mathematically complicated nature as to preclude facile physical interpretation of the system it described. If this is so, one can assert that there are *physical* as well as *mathematical* grounds for constructing approximate solutions to the wave

[1] A survey of the exactly solvable systems is given in D. R. Bates (ed.), *Quantum Theory*, vol. 1, Academic, New York, 1961.

equation. Specifically—as we will see later—there are techniques for designing approximate solutions in such a way that physical interpretations are expedited.

6-1 THE VARIATION METHOD

The variation method may be regarded as an alternative formulation of Schrödinger's equation, a formulation which has the advantage of suggesting a route to approximate solutions of any desired degree of accuracy. Let us consider the time-independent Schrödinger equation in the compact symbolic form

$$\hat{H}\psi = E\psi \tag{6-1}$$

where ψ and E are an exact eigenfunction and eigenenergy, respectively, of the hamiltonian operator $\hat{H}$. In general, the mathematical form of the hamiltonian will be such that Eq. (6-1) will not be separable in any known coordinate system. Thus one cannot hope to obtain an analytical closed-form expression for the eigenfunction ψ. Let us now examine the properties of the functional[2] $\varepsilon[\xi]$ defined by

$$\varepsilon[\xi] = \frac{\langle \xi | \hat{H} | \xi \rangle}{\langle \xi | \xi \rangle} \tag{6-2}$$

where $\hat{H}$ is the exact hamiltonian of the system and ξ is an arbitrary function of the system coordinates, subject only to the restriction that it is well-behaved over the configuration space of the system whose exact wavefunction is ψ. It is apparent that if ξ is identical with ψ, the functional $\varepsilon[\xi]$ is the energy E (a constant). However, suppose ξ is not identical with ψ but is a trial approximation to it. Furthermore, assume that ξ differs from ψ by no more than a first-order variation,[3] namely,

$$\xi = \psi + \delta\psi \tag{6-3}$$

Now consider the expectation value of the operator $\hat{H} - E$ with respect to the trial function ξ. First, note that the application of the operator $\hat{H} - E$ to the trial function ξ leads to the result

$$(\hat{H} - E)\xi = (\hat{H} - E)(\psi + \delta\psi) = (\hat{H} - E)\delta\psi \tag{6-4}$$

[2] Whereas the domain of a *function*, say, $f(x)$, is a region of coordinate space, the domain of a *functional*, say, $F[f(x)]$, is a space of admissible functions. Thus one may think of a functional as a function of a function, i.e., a *superfunction*. See R. Courant and D. Hilbert, *Methods of Mathematical Physics*, vol. 1, Interscience, New York, 1953, chap. 4, and also L. M. Graves, *Bull. Am. Math. Soc.*, **55**:467 (1948).

[3] Two functions differ to the first order if their slopes differ; they differ to the second order if their curvatures differ.

i.e., the operator $\hat{H} - E$ annihilates the exact eigenfunction ψ. The desired expectation value then is

$$\langle \hat{H} - E \rangle = \frac{\langle \xi | \hat{H} - E | \xi \rangle}{\langle \xi | \xi \rangle} = \frac{\langle \xi | \hat{H} - E | \delta\psi \rangle}{\langle \xi | \xi \rangle} = \frac{\langle (\hat{H} - E)\xi | \delta\psi \rangle}{\langle \xi | \xi \rangle}$$
$$= \frac{\langle (\hat{H} - E)\delta\psi | \delta\psi \rangle}{\langle \xi | \xi \rangle} = \frac{\langle \delta\psi | \hat{H} - E | \delta\psi \rangle}{\langle \xi | \xi \rangle} \tag{6-5}$$

where the turnover rule has been used twice. By simple rearrangement of the above equation

$$\varepsilon[\xi] = \langle \hat{H} \rangle = E + \frac{\langle \delta\psi | \hat{H} - E | \delta\psi \rangle}{\langle \xi | \xi \rangle} \tag{6-6}$$

This result shows that the functional $\varepsilon[\xi]$ is an approximation of the exact energy E and differs from E by only a second-order term in $\delta\psi$. In other words, although the trial function ξ is in error to the first order, the total energy is in error to only the second order. Thus we can write

$$\delta\varepsilon[\xi] = \delta\langle \hat{H} \rangle = 0 \tag{6-7}$$

This is just a formal, mathematical way of saying that there is no first-order error in the energy when the corresponding eigenfunction is approximated by a function differing from it to the first order. Equation (6-7) may be regarded as a statement of the *variation principle*. Furthermore, this statement may be shown to lead to Schrödinger's equation; i.e., the condition presented in Eq. (6-7) establishes that the trial function ξ is an eigenfunction of $\hat{H}$. The proof of this fact is best established by use of the *calculus of variations*,[4] an area of mathematics concerned with a generalization of maxima and minima problems. The calculus of variations states the general problem as follows: Find the functions ξ such that the functional

$$J[\xi] = \langle \xi | \hat{H} | \xi \rangle \tag{6-8}$$

is stationary, i.e., has an extremal value (maximum or minimum) with respect to arbitrary, small variations in ξ. To ensure consistency with probabilistic interpretations, it is also necessary to impose on the functions ξ the normalization condition

$$\langle \xi | \xi \rangle = 1 \tag{6-9}$$

The above variational problem can be cast in a more convenient form by defining the additional functional

$$K[\xi] = \langle \xi | \xi \rangle - 1 \tag{6-10}$$

[4] See R. Courant and D. Hilbert, op. cit., and also F. B. Hildebrand, *Methods of Applied Mathematics*, 2d ed., Prentice-Hall, Englewood Cliffs, N.J., 1965.

and finding, instead, the functions ξ which make the new functional

$$L[\xi] = J[\xi] - \lambda K[\xi] \tag{6-11}$$

stationary *without restrictions*. The quantity λ is a real parameter known as a *lagrangian multiplier*. The stationary values of $L[\xi]$ will be those for which $\delta L[\xi]$ vanishes, and provided ξ is always normalized, this means that $\delta\langle \hat{H} \rangle = 0$ is also satisfied.

The first-order variation of $L[\xi]$ is

$$\delta L[\xi] = \langle \delta\xi | \hat{H} | \xi \rangle + \langle \xi | \hat{H} | \delta\xi \rangle - \lambda \langle \delta\xi \mid \xi \rangle - \lambda \langle \xi \mid \delta\xi \rangle - \delta\lambda K[\xi] \tag{6-12}$$

It is convenient to introduce the notational simplification

$$\langle \delta\xi | \hat{H} | \xi \rangle - \lambda \langle \delta\xi \mid \xi \rangle = Q \tag{6-13}$$

Then, since

$$(\langle \delta\xi | \hat{H} | \xi \rangle - \lambda \langle \delta\xi \mid \xi \rangle)^{\dagger} = \langle \xi | \hat{H} | \delta\xi \rangle - \lambda \langle \xi \mid \delta\xi \rangle \tag{6-14}$$

we can write Eq. (6-12) as

$$\delta L[\xi] = Q + Q^{*} - \delta\lambda K[\xi] \tag{6-15}$$

If $\langle \xi \mid \xi \rangle = 1$, the coefficient of $\delta\lambda$ vanishes and the functional $L[\xi]$ will be stationary only if Q (and thus also its complex conjugate Q^*) vanishes, that is, if

$$\langle \delta\xi | \hat{H} | \xi \rangle - \lambda \langle \delta\xi \mid \xi \rangle = 0 \tag{6-16}$$

The only way this equation can be satisfied for *any arbitrary variation* $\delta\xi$ is if the following relationship holds:

$$\hat{H}\xi - \lambda\xi = 0 \tag{6-17}$$

This is *Schrödinger's equation*. Thus, the condition that $\delta\langle \hat{H} \rangle$ vanish is *sufficient* to make ξ an eigenfunction of the exact hamiltonian $\hat{H}$. It is apparent from Eq. (6-2) that the lagrangian multiplier λ is to be identified with the functional $\varepsilon[\xi] = \langle \hat{H} \rangle$.

Note, however, that the condition $\delta\langle \hat{H} \rangle = 0$ is necessary but not sufficient to establish that λ is a minimum. The problem of establishing sufficiency of the conditions leading to true extrema of functionals is a difficult one and has not been solved in general. In practice, however, λ is usually found to be a true minimum, and one accepts the fundamental theorem on faith in most physical applications.

At this point the variation principle does not appear to have any practical usefulness. Nevertheless, it is easy to show that it does lead to some specific procedures for constructing increasingly better approximations to the exact solutions and, in the limit, to the exact solutions themselves. To demonstrate this, let $\{\psi_i\}$ represent a complete orthonormal set of eigenfunctions of the hamiltonian $\hat{H}$, and let $\{E_i\}$ be the corresponding eigenvalues. In general, we will not know these eigenfunctions, eigenvalues, or both (if we did, we probably would not be looking for *approximations* to them), but we can employ them in a purely symbolic sense. Since these eigenfunctions are complete, the trial function ξ can be expressed in terms of them by the expansion

$$\xi = \sum_{i=1}^{\infty} \psi_i c_i + \int \psi_\lambda C_\lambda \, d\lambda \tag{6-18}$$

where the integral term allows for the possibility of a continuum, i.e., a set of continuous eigenvalues[5] (we will not encounter this situation in this text). If we ignore the continuum situation, Eq. (6-18) becomes

$$\varepsilon = \frac{\langle \xi | \hat{H} | \xi \rangle}{\langle \xi | \xi \rangle} = \frac{\sum_i \sum_j c_i^* c_j H_{ij}}{\sum_i \sum_j c_i^* c_j \Delta_{ij}} \tag{6-19}$$

where ε has been used to replace $\varepsilon[\xi]$ and where H_{ij} and Δ_{ij} are the matrix elements of $\hat{H}$ and the unit operator, respectively. Using Theorem 2-5 and the fact that the $\{\psi_i\}$ are orthonormal, we can reduce Eq. (6-19) to

$$\varepsilon = \frac{\sum_i |c_i|^2 E_i}{\sum_i |c_1|^2} \tag{6-20}$$

If the trial function ξ is normalized, it follows at once that

$$\langle \xi | \xi \rangle = \sum_i |c_i|^2 = 1 \tag{6-21}$$

so that Eq. (6-20) becomes

$$\varepsilon = \sum_i |c_1|^2 E_i \tag{6-22}$$

Substracting E_0, the exact energy of the ground state, from both sides of Eq. (6-22) and using Eq. (6-21), we obtain

$$\varepsilon - E_0 = \sum_i |c_i|^2 (E_i - E_0) \tag{6-23}$$

[5] Refer back to Chap. 5 and note that the solutions to the hydrogen atom Schrödinger equation were restricted to the *bound state*, i.e., to $E < 0$. However, to describe the case of a free electron scattered by the nucleus ($E > 0$) requires a set of continuum functions, since the energy of this electron, like that of a free particle, is not quantized.

Since, in general, $E_i \geq E_0$ and $|c_i|^2 \geq 0$ for all i, it follows that

$$\varepsilon \geq E_0 \tag{6-24}$$

This very important result, first proved by Eckart,[6] is an *upper-limit theorem* for the energy. The theorem shows that any trial wavefunction ξ (which is normalizable) leads to a value of the energy which is never lower than the true ground-state energy of the system. In this result resides the power (and also the weakness) of the variation method of approximating solutions to the Schrödinger wave equation. The power lies in the fact that one can choose the "best" wavefunction from several alternatives on the basis of the criterion of lowest energy; the weakness is that the energy is an insensitive criterion with respect to the "best" wavefunction for other physical properties. Unfortunately, Eq. (6-6), which shows that a first-order error in the wavefunction leads to no first-order error in the expectation value of the hamiltonian, does not hold for the expectation values of other operators in general. The first-order error in ψ may be averaged over the coordinates in such a way that even though a very good energy is obtained, the expectation values of other operators may be exceedingly poor.

Explicit forms for trial wavefunctions are obtained in a variety of ways. One convenient way is to express ξ in terms of the members of a complete set of functions $\{\phi_i\}$. In theory, this could lead to ξ's which are the exact eigenfunctions $\{\psi_i\}$, provided that enough members of the set (generally infinite in number) are used in the expansion. In practice, the basis set $\{\phi_i\}$ is truncated to a finite number of members (n); thus, ξ will not be represented exactly. The trial wavefunction may be written

$$\xi_k = \sum_{i=1}^{n} \phi_i c_{ik} \tag{6-25}$$

We assume that the functions $\{\phi_i\}$ are linearly independent but not necessarily orthonormal. If we define the matrices (see App. 5)

$$\begin{aligned} \mathbf{\Phi} &= [|\phi_i\rangle \quad |\phi_2\rangle \quad \cdots \quad |\phi_n\rangle] \\ \mathbf{\Delta} &= \mathbf{\Phi}^\dagger \mathbf{\Phi} \\ \mathbf{H} &= \mathbf{\Phi}^\dagger \hat{H} \mathbf{\Phi} \end{aligned} \qquad \mathbf{C}_k = \begin{bmatrix} c_{1k} \\ c_{2k} \\ \vdots \\ c_{nk} \end{bmatrix} \tag{6-26}$$

then the energy ε may be written

$$\varepsilon = \frac{\mathbf{C}_k^\dagger \mathbf{H} \mathbf{C}_k}{\mathbf{C}_k^\dagger \mathbf{\Delta} \mathbf{C}_k} \tag{6-27}$$

[6] C. E. Eckart, *Phys. Rev.*, **36**:878 (1939).

Applying the variation theorem (6-7), we obtain

$$\delta\varepsilon = 0 = \delta \frac{\mathbf{C}_k^\dagger \mathbf{H} \mathbf{C}_k}{\mathbf{C}_k^\dagger \boldsymbol{\Delta} \mathbf{C}_k} = \frac{1}{\mathbf{C}_k^\dagger \boldsymbol{\Delta} \mathbf{C}_k}\left(\delta \mathbf{C}_k^\dagger \mathbf{H} \mathbf{C}_k - \varepsilon \delta \mathbf{C}_k^\dagger \boldsymbol{\Delta} \mathbf{C}_k\right) + \text{complex conjugate} \tag{6-28}$$

Thus the variation theorem is satisfied if

$$\delta \mathbf{C}_k^\dagger (\mathbf{H}\mathbf{C}_k - \varepsilon \boldsymbol{\Delta} \mathbf{C}_k) = 0 \tag{6-29}$$

In order for Eq. (6-29) to be satisfied for any arbitrary variation $\delta \mathbf{C}_k$ of the coefficients, it is necessary that the term in parentheses vanish, namely,

$$\mathbf{H}\mathbf{C}_k - \varepsilon \boldsymbol{\Delta} \mathbf{C}_k = (\mathbf{H} - \varepsilon \boldsymbol{\Delta})\mathbf{C}_k = 0 \tag{6-30}$$

The conditions leading to Eq. (6-30) are equivalent to the n conditions

$$\frac{\partial \varepsilon}{\partial c_{ik}} = 0 \qquad \text{for all } i = 1, 2, \ldots, n \tag{6-31}$$

i.e., the expansion coefficients $\{c_{ik}\}$ are variation parameters to be chosen so as to minimize ε. The nontrivial solutions to Eq. (6-30) are given by the n roots of the secular determinant

$$\det(\mathbf{H} - \varepsilon \boldsymbol{\Delta}) = 0 \tag{6-32}$$

The n roots $\epsilon_1, \epsilon_2, \ldots, \epsilon_n$ are associated with the eigenvectors $\mathbf{C}_1, \mathbf{C}_2, \ldots, \mathbf{C}_n$, respectively. We now define the matrices

$$\mathbf{C} = [\mathbf{C}_1 \quad \mathbf{C}_2 \quad \cdots \quad \mathbf{C}_n] \qquad \text{and} \qquad (\boldsymbol{\varepsilon})_{ij} = \epsilon_i \delta_{ij} \tag{6-33}$$

and write the Schrödinger equation in the matrix form

$$\mathbf{H}\mathbf{C} = \boldsymbol{\Delta}\mathbf{C}\boldsymbol{\varepsilon} \tag{6-34}$$

In the event that the basis $\{\phi_i\}$ is orthonormal, the above equation becomes

$$\mathbf{H}\mathbf{C} = \mathbf{C}\boldsymbol{\varepsilon} \tag{6-35}$$

Solution of the matrix equation produces the n different trial functions

$$\boldsymbol{\xi} = \boldsymbol{\Phi}\mathbf{C} \tag{6-36}$$

where

$$\boldsymbol{\xi} = [\xi_1 \quad \xi_2 \quad \cdots \quad \xi_n] \tag{6-37}$$

Equation (6-36) is just the matrix form of the n equations given by Eq. (6-25) when $i = 1, 2, \ldots, n$.

If ϵ_1 is the lowest root, then by Eq. (6-24) we have

$$\epsilon_1 \geq E_1 \tag{6-38}$$

where E_1 is the exact energy of the lowest state of the system the wavefunction of which is approximated by ξ_1.

It has been shown[7] that if the truncated basis set of n functions is extended by one more function (to $n+1$ functions), the roots of the $n \times n$ secular equation (6-32) will separate the roots of the $(n+1) \times (n+1)$ secular equation resulting from the extended basis. Thus if λ_1, λ_2, and λ_3 are roots of the extended secular equation and ϵ_1, ϵ_2, and ϵ_3 are corresponding roots of the $n \times n$ secular equation, it is always true that

$$\lambda_1 < \epsilon_1 < \lambda_2 < \epsilon_2 < \lambda_3 < \epsilon_3 \cdots \tag{6-39}$$

Extension of this result to a complete basis set of an infinite number of members indicates that all the roots will represent exact eigenvalues of the system under investigation. Thus, in principle, the variation method provides a route to exact solution of the Schrödinger equation.

In practice, it is necessary to use a finite basis set to solve the secular equations. In the usual situation, the basis is chosen so that the lowest root of the secular equation is an acceptable approximation of the ground state of the system. Unfortunately, a basis set which leads to an excellent ground-state energy may produce poor energies for the excited states. In general, the success of this particular approach of the variation method depends strongly on a judicious choice of basis set and the number of basis functions employed.

An alternative way of utilizing the variation principle is to construct the trial wavefunction in terms of a set of variation parameters $\{\alpha_i\}$ which enter into the function in a *nonlinear* manner. This approach has the very important advantage of allowing variational parameters to be introduced in such a way that they facilitate certain physical interpretations of the wavefunction in which they appear. The general form of such a function may be written

$$\xi = \xi(\alpha_1, \alpha_2, \ldots, \alpha_n\} \tag{6-40}$$

The "best" approximation is then chosen by minimizing the total energy ε with respect to the $\{\alpha_i\}$; this is done by solving the simultaneous equations

$$\frac{\partial \varepsilon}{\partial \alpha_i} = 0 \qquad \text{for all } i = 1, 2, \ldots, n \tag{6-41}$$

Unlike the linear variation method described previously, this does not lead to a systematic procedure for finding the best parameter values and, thus, leads to considerable computational complexity.

It is also possible to use an approach which combines linear and nonlinear parameters, but this quickly leads to prohibitive computational complexity if the number of nonlinear parameters is large. Nevertheless, a number of methods are in use in which the spirit of this approach (if not the actual letter) is employed. For example, instead of formally optimizing the energy with respect to a number of nonlinear parameters, one chooses values for these

---

[7] E. A. Hylleraas and B. Undheim, *Z. Physik*, **65**:759 (1930); J. K. L. MacDonald, *Phys. Rev.*, **43**:830 (1933).

parameters on the basis of simple "rules of thumb" which have been determined by experience; once these values are fixed, the linear variational calculation can be carried out.

EXERCISES

6-1. Two different approximate wavefunctions for a given system are given by the linear combinations

$$\xi_i = c_i\phi_i + c_2\phi_2 \qquad \text{and} \qquad \xi_2 = d_1\phi_1 + d_2\phi_2 + d_3\phi_3$$

where the $\{c_i\}$ and $\{d_i\}$ are variational parameters and the $\{\phi_i\}$ are the basis functions. Letting ε_1 and ε_2 represent the corresponding approximate energies arising from the trial functions ξ_1 and ξ_2, respectively, prove that the energy arising from the three-term wavefunction is lower than (or, at worst, the same as) the energy arising from the two-term wavefunction. *Note*: This result follows from Eq. (6-39), but it can be proved rigorously for the present specific case; simply assume that the converse of the assertion is true, and show that this leads to an impossible situation.

6-2. The approximate energy of a system is given by $\varepsilon = a^2 + 3a - 6$ (in arbitrary energy units) where a is a variational parameter. What value of a leads to the lowest energy, and what is the value of the minimum energy? Interpret your results graphically using a plot of ε versus a.

6-3. The approximate energy of a system is given by $\langle \hat{H} \rangle = 3a^4 - 4a^3 - 36a^2 + 10$ where a is a variational parameter. What is the highest energy this system can possibly have in its ground state?

6-2 ATOMIC UNITS IN ELECTRONIC STRUCTURE CALCULATIONS

The calculation of various physical quantities associated with many-electron systems is made notationally simpler by introducing a system of dimensionless units called *atomic units* (a.u.). Furthermore, the use of such units obviates the difficulties associated with comparing calculations performed by different persons using different values for physical constants—and a different number of significant figures. Use of atomic units also ensures that calculated quantities will not change with future revision of the numerical values of the fundamental constants.

The atomic unit of *mass* is the rest mass of the electron, namely, $m_e = 9.109534 \times 10^{-31}$ kg. A mass which is equivalent to 4.01 times the rest mass of the electron is then written as 4.01 a.u. (of mass) or, preferably, as $4.01m_e$. The atomic unit of *length* is the radius of the first Bohr orbit in a one-electron atom with unit atomic number Z and a nucleus of infinite mass. This is defined formally by

$$a_0 = \frac{\hbar^2}{m_e e^2 K} = 5.2917706 \times 10^{-11}\ \text{m} \tag{6-42a}$$

The atomic unit of length is also called the *bohr*. A length equivalent to 2.45 times the first Bohr radius is written as 2.45 bohr, 2.45 a.u. (of length), or, best of all, $2.45a_0$.

The atomic unit of *energy* is defined as twice the energy of an electron for the $n = 1$ state of a hydrogen atom in which the nucleus has an infinite mass (i.e., the reduced mass of the electron is equal to its isolated mass). The defining relationship is

$$2\frac{Ke^2}{2a_0} = \frac{Ke^2}{a_0} = 4.3598144 \times 10^{-18}\ \text{J} \tag{6-42b}$$

The atomic unit of energy is often called the *hartree*.[8] A quantity of energy equivalent to 3.67 times the first ionization energy of a hydrogen atom with an infinitely massive nucleus is written as 3.67 a.u. (of energy), 3.67 hartrees, or, preferably, $3.67E_\text{h}$.

The atomic unit of *electrostatic charge* is the charge on an electron, namely, $1.6021892 \times 10^{-19}$ C. The atomic unit of *probability density* is $(1/a_0)^3$; this is equal to $6.7483 \times 10^{30}\ \text{m}^{-3}$. The atomic unit of *angular momentum* is given by $\hbar = 1.0546 \times 10^{-24}\ \text{J}\cdot\text{s}$.

The Schrödinger equation (and similar equations) can be expressed in terms of atomic units simply by setting $e = \hbar = m_e = K = 1$. The hamiltonian operator appearing in the Schrödinger equation of the hydrogen atom then becomes

$$\hat{H} = -\frac{\nabla^2}{2} - \frac{1}{r} \tag{6-43}$$

The eigenvalues are then expressed in terms of atomic units (units of E_h). For example:

$$E_n = -\tfrac{1}{2}n^2\ \text{a.u.} \qquad (\text{or } -\tfrac{1}{2}n^2\ E_\text{h}) \tag{6-44}$$

EXERCISES

6-4. Express the following in the appropriate atomic units: 5 nm, Planck's constant, 50 kJ, $5 \times 10^{-5}\ \text{cm}\cdot\text{s}^{-1}$, and the mass of a benzene molecule. *Note*: The atomic unit of *velocity* is the velocity of an electron in the first Bohr orbit.

6-5. Using the previous exercise as a guide, express the velocity of light ($2.997 \times 10^8\ \text{m}\cdot\text{s}^{-1}$) in atomic units.

6-6. Convert kT (k is Boltzmann's constant and T is the temperature on the absolute scale) to atomic units when the temperature is 300 K.

---

[8] Unfortunately, the quantity $Ke^2/2a_0$ has also been called the *hartree* by some workers, and some use *double hartree* for Ke^2/a_0.

6-3 THE SCHRÖDINGER EQUATION FOR MANY-ELECTRON ATOMS: THE MASS-POLARIZATION EFFECT

The total nonrelativistic Schrödinger equation for an atom with two electrons (the general case of N electrons will be introduced later) is given by

$$\left[-\frac{1}{2m_{\rm N}}\nabla_{\rm N}^2 - \frac{1}{2m_1}\nabla_1^2 - \frac{1}{2m_2}\nabla_2^2 - Z\left(\frac{1}{r_1} + \frac{1}{r_2}\right) + \frac{1}{r_{12}}\right]\psi = E\psi \tag{6-45}$$

The coordinate system is as shown in Fig. 6-1; the nucleus N and each of the two electrons, e_1 and e_2, are assigned coordinates relative to some arbitrary origin $(0, 0, 0)$. The wavefunction ψ may be written

$$\psi = \psi(v_1, v_2, v_{\rm N}) \tag{6-46}$$

where $\quad v_1 = x_1, y_1, z_1 \qquad v_2 = x_2, y_2, z_2 \qquad v_{\rm N} = x_{\rm N}, y_{\rm N}, z_{\rm N} \qquad$ (6-47)

Note that in terms of the atomic units discussed in the preceding section, $m_1 = m_2 = m_e = 1$, but we will find it convenient to retain individual electron identities for the moment. Also, the nuclear mass, $m_{\rm N}$, is in terms of atomic units.

The atom described by the Schrödinger equation above is undergoing translational motion in free space, and at the same time the two electrons are repelling each other and being attracted to the nucleus. These two types of motions (translation of the atom as a whole and internal interactions) may be separated by transforming to a new coordinate system: *relative* and *center-of-mass* coordinates. These will be symbolized by v_a and v_b for the former and $v_{\rm cm}$ for the latter. The components of the relative coordinates, $v_a = x_a, y_a, z_a$ and

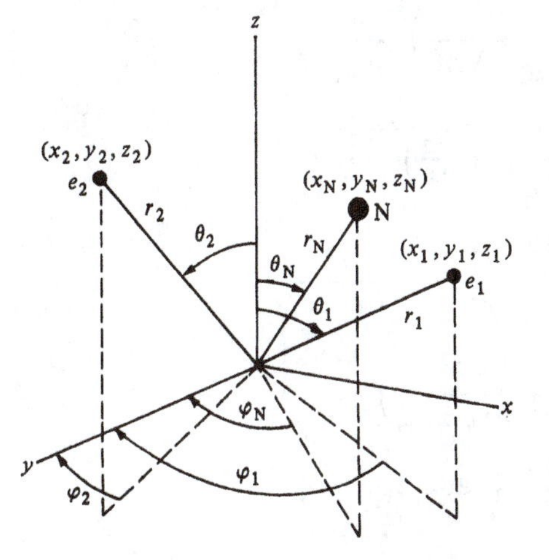

FIGURE 6-1
Coordinate system for a two-electron atom; N is the nucleus and e_1 and e_2 are the electrons. All particles are given coordinates relative to some arbitrary origin $(0, 0, 0)$.

$v_b = x_b, y_b, z_b$ are given by

$$\begin{aligned} x_a &= x_1 - x_N \qquad y_a = y_1 - y_N \qquad z_a = z_1 - z_N \\ x_b &= x_2 - x_N \qquad y_b = y_2 - y_N \qquad z_b = z_2 - z_N \end{aligned} \tag{6-48}$$

The center-of-mass coordinates are defined by

$$\begin{aligned} x_{\text{cm}} &= \frac{x_1 m_1 + x_2 m_2 + x_N m_N}{M} \\ y_{\text{cm}} &= \frac{y_1 m_1 + y_2 m_2 + y_N m_N}{M} \\ z_{\text{cm}} &= \frac{z_1 m_1 + z_2 m_2 + z_N m_N}{M} \end{aligned} \tag{6-49}$$

where $M = m_1 + m_2 + m_N$. It should be noted that

$$\begin{aligned} r_1 &= \sqrt{x_a^2 + y_a^2 + z_a^2} \\ r_2 &= \sqrt{x_b^2 + y_b^2 + z_b^2} \\ r_{12} &= \sqrt{(x_a - x_b)^2 + (y_a - y_b)^2 + (z_a - z_b)^2} \end{aligned} \tag{6-50}$$

The various partial derivatives appearing in the laplacian portions of the hamiltonian are transformed to the following:

$$\frac{\partial}{\partial x_1} = \frac{\partial}{\partial x_a} + \frac{m_1}{M}\frac{\partial}{\partial x_{\text{cm}}} \qquad \frac{\partial}{\partial x_2} = \frac{\partial}{\partial x_b} + \frac{m_2}{M}\frac{\partial}{\partial x_{\text{cm}}}$$

$$\frac{\partial}{\partial x_N} = -\frac{\partial}{\partial x_a} - \frac{\partial}{\partial x_b} + \frac{m_N}{M}\frac{\partial}{\partial x_{\text{cm}}}$$

$$\begin{aligned} \frac{\partial}{\partial x_1^2} &= \frac{\partial^2}{\partial x_a^2} + \frac{2m_1}{M}\frac{\partial^2}{\partial x_a\,\partial x_{\text{cm}}} + \left(\frac{m_1}{M}\right)^2 \frac{\partial^2}{\partial x_{\text{cm}}^2} \\ \frac{\partial^2}{\partial x_2^2} &= \frac{\partial^2}{\partial x_b^2} + \frac{2m_2}{M}\frac{\partial^2}{\partial x_b\,\partial x_{\text{cm}}} + \left(\frac{m_2}{M}\right)^2 \frac{\partial^2}{\partial x_{\text{cm}}^2} \\ \frac{\partial^2}{\partial x_N^2} &= \frac{\partial^2}{\partial x_a^2} + \frac{\partial^2}{\partial x_b^2} + \left(\frac{m_N}{M}\right)^2 \frac{\partial^2}{\partial x_{\text{cm}}^2} + 2\frac{\partial^2}{\partial x_a\,\partial x_b} \\ &\quad - 2\frac{m_N}{M}\left(\frac{\partial^2}{\partial x_a\,\partial x_{\text{cm}}} + \frac{\partial^2}{\partial x_b\,\partial x_{\text{cm}}}\right) \end{aligned} \tag{6-51}$$

with similar equations for y_1, y_2, y_N, z_1, z_2, and z_N.

The transformed Schrödinger equation now becomes

$$\left[-\frac{1}{2}\left(\frac{m_1+m_N}{m_1 m_N}\nabla_a^2+\frac{m_2+m_N}{m_2 m_N}\nabla_b^2\right)\right.$$
$$\left.-\frac{1}{m_N}\nabla_a\cdot\nabla_b-Z\left(\frac{1}{r_1}+\frac{1}{r_2}\right)+\frac{1}{r_{12}}-\frac{1}{2M}\nabla_{cm}^2\right]\psi=E\psi \tag{6-52}$$

$$\psi=\psi(v_a,v_b,v_{cm})$$

We now effect the separation of variables

$$\psi(v_a,v_b,v_{cm})\rightarrow A(v_a,v_b)B(v_{cm}) \tag{6-53}$$

where the function $A(v_a, v_b)$ describes the internal motions (interacting electrons and nucleus) and $B(v_{cm})$ describes the translational motion of the atom as a whole. If we let the total energy be given by $E = E_{cm} + E'$ (translational energy plus internal energy), the separation leads to the two equations

$$-\frac{1}{2M}\nabla_{cm}^2 B(v_{cm})=E_{cm}B(v_{cm}) \tag{6-54}$$

which describes the translational motion of the atom, and

$$\left[-\frac{1}{2\mu}(\nabla_a^2+\nabla_b^2)-Z\left(\frac{1}{r_1}+\frac{1}{r_2}\right)+\frac{1}{r_{12}}-\frac{1}{m_N}\nabla_a\cdot\nabla_b\right]A(v_a,v_b)=E'A(v_a,v_b) \tag{6-55}$$

which describes the internal motions of the nucleus and two electrons. Note that we have used the substitutions $m_1 = m_2 = m_e$ and $\mu = m_e m_N/(m_e + m_N)$ in the above. Equation (6-54) represents the free-particle situation, the solution of which is described in Sec. 3-2; we will not consider it further here. Equation (6-55) is readily generalized for an N-electron system; this has the following form:

$$\left(\sum_{\mu=1}^{N}h_\mu+\sum_{\mu<\nu}^{N}\frac{1}{r_{\mu\nu}}-\frac{1}{m_N}\sum_{\mu<\nu}^{N}\nabla_\mu\cdot\nabla_\nu\right)A(v_1,v_2,\ldots,v_N)$$
$$=E'A(v_1,v_2,\ldots,v_N) \tag{6-56}$$

where

$$h_\mu=-\frac{1}{2\mu}\nabla_\mu^2-\frac{Z}{r_\mu} \tag{6-57}$$

The last term in the hamiltonian in Eq. (6-56) is a *mass-polarization* term. In the limit of $m_N \rightarrow \infty$, the reduced mass $\mu \rightarrow m_e$, and Eq. (6-56) becomes

$$\left(\sum_{\mu=1}^{N}h_\mu+\sum_{\mu<\nu}^{N}\frac{1}{r_{\mu\nu}}\right)A=E'A \tag{6-58}$$

where the monoelectronic operator h_μ now becomes

$$h_\mu = -\frac{1}{2m_e}\nabla_\mu^2 - \frac{Z}{r_\mu} \tag{6-59}$$

From now on, unless stated otherwise, the mass-polarization effect will be ignored, and the atomic hamiltonian will be assumed to be that given by Eq. (6-58), with h_μ given by Eq. (6-59).

6-4 EXAMPLES OF VARIATIONAL CALCULATIONS

THE PARTICLE IN A BOX. Since the particle-in-a-box problem can be solved exactly (see Sec. 3-3), approximate solutions can be used as a test of the efficiency of the variational method. Using an electron as the particle, the Schrödinger equation (in atomic units) for the particle in a box may be written

$$-\frac{1}{2}\frac{d^2\psi}{dx^2} = E\psi \tag{6-60}$$

We could form a trial wavefunction ξ as a linear combination of the trigonometric functions sin kx, cos kx, or both, since these form a complete basis set, but this would quickly produce the exact solution and would not be very instructive. Instead, let us fabricate some simple algebraic function in x which satisfies the boundary conditions

$$\xi(0) = 0 \qquad \xi(L) = 0 \tag{6-61}$$

and which is otherwise well-behaved. One such simple function is the polynomial

$$\xi(x) = x^\alpha(L^2 - x^2) \tag{6-62}$$

where α is a variational parameter to be chosen such that the quantity

$$\varepsilon = \frac{\left\langle \xi \left| -\frac{1}{2}\frac{d^2}{dx^2} \right| \xi \right\rangle}{\langle \xi | \xi \rangle} \tag{6-63}$$

is a minimum. The procedure is to substitute the function [Eq. (6-62)] into Eq. (6-63) and to solve the equation

$$\frac{\partial \varepsilon}{\partial \alpha} = 0 \tag{6-64}$$

for the optimum value of the parameter α. For simplicity, the length L of the box is set equal to unity (that is, $L = 1\ a_0$). Since the algebra is straightforward (but tedious), we will not consider it here. The result is that $\alpha = 0.862$, so that the trial wavefunction becomes

$$\xi(x) = N(x^{0.862} - x^{2.862}) \tag{6-65}$$

where N is a normalization factor given by

$$N = \left[\frac{(2\alpha+1)(2\alpha+3)(2\alpha+5)}{8}\right]^{1/2} = \sqrt{10.816} \tag{6-66}$$

Substituting Eq. (6-65) into Eq. (6-63), one obtains the approximate energy

$$\varepsilon = \frac{10.3}{2} E_\mathrm{h} = 5.15\ E_\mathrm{h} \tag{6-67}$$

The exact energy of the ground state is

$$E_1 = \frac{\pi^2}{2} E_\mathrm{h} = 4.94 E_\mathrm{h} \tag{6-68}$$

The approximate energy is too high by 4.3 percent

$$\frac{\varepsilon - E_1}{E_1} = \frac{5.15 - 4.94}{4.94} = 0.043 \tag{6-69}$$

The average value of the x coordinate (which is 0.5 a_0 for the exact solution) in terms of the trial wavefunction is

$$\langle x \rangle = \langle \xi | x | \xi \rangle = 10.816 \int_0^1 x^{2\alpha+1}(1-x^2)^2\, dx = 0.520 a_0 \tag{6-70}$$

This result reflects the fact that the trial wavefunction [Eq. (6-62)] is not symmetrical (as is the exact solution, $\sin \pi x$) but is skewed to the right. It is also interesting to note that the error in this expectation value is 4 percent—lower than the error in the energy.

THE HELIUM ATOM. The hamiltonian operator for the Schrödinger equation of the helium atom is

$$\hat{H} = h_1 + h_2 + \frac{1}{r_{12}} \tag{6-71}$$

where h_1 and h_2 are monoelectronic operators of the general form

$$h_\mu = -\frac{\nabla_\mu^2}{2} - \frac{Z}{r_\mu} \tag{6-72}$$

and where μ (which is equal to 1 or 2) is an index which designates electrons. Because of the electronic repulsion term $1/r_{12}$ in the helium atom hamiltonian, the Schrödinger equation cannot be solved analytically in the usual form, i.e., one cannot separate the variables in any known coordinate system. In classical mechanics, a system of three or more interacting bodies cannot be solved exactly—not because the mathematics is too difficult, but because (as shown by Jules-Henri Poincaré) the orbital motions of the bodies may exhibit *chaotic* behavior.[9] The question now is: What should we use as a trial wavefunction?

[9] T.A. Heppenheimer, *Mosaic*, vol. 20, no. 2, summer 1989, pp. 2–11.

One simple approach is to use the wavefunction resulting from the removal of the troublesome electronic-repulsion term from the hamiltonian. In this case the Schrödinger equation is separable into two equations, each describing an independent electron. Each of these equations has the form

$$h_\mu \xi(\mu) = \epsilon(\mu)\xi(\mu) \tag{6-73}$$

where $\epsilon(\mu)$ is of the same form as the hydrogen atom energy, namely,

$$\epsilon(\mu) = -\frac{Z^2}{2} E_h = -2E_h \tag{6-74}$$

and the eigenfunctions have the hydrogen atom form

$$\xi(\mu) = \left(\frac{Z^3}{\pi}\right)^{1/2} e^{-Zr_\mu} \tag{6-75}$$

where $Z = 2$ for He. Thus, the trial wavefunction for the helium atom becomes the product of the two independent-electron wavefunctions, $\xi(1)\xi(2)$. The approximate energy arising from this trial wavefunction is

$$\varepsilon = \epsilon(1) + \epsilon(2) + \left\langle \xi(1)\xi(2 \left| \frac{1}{r_{12}} \right| \xi(1)\xi(2) \right\rangle \tag{6-76}$$

The evaluation of the electron-repulsion integral (called a *coulombic integral*) is illustrated in App. 3; this integral has the general value $\frac{5}{8}Z$ (in units of E_h). Using this value and Eq. (6-74) for the $\epsilon(\mu)$ leads to

$$\varepsilon = -2\frac{Z^2}{2} + \frac{5}{8}Z = -Z^2 + \frac{5}{8}Z = -4.00 + 1.25 = -2.75E_h \qquad \text{for } Z = 2 \tag{6-77}$$

The experimental value is $-2.90E_h$. The approximate value is thus about 5.2 percent too high.

However, we can make a simple modification in the trial wavefunction which will improve the energy and, furthermore, provide us with a crude description of how the electrons interact—a feature which is totally missing in the above approach. Since the two electrons have the same electrostatic charge, each will act as a "shield" toward the other; that is, when one of the electrons is far away from the nucleus and the other is closer, the latter will "block" part of the charge on the nucleus so that the former "sees" not a charge Z ($Z = 2$ in this specific case) but a lesser charge $Z - S$, where S may be called a *screening constant*. Of course, the effect is symmetric with respect to the two electrons; each shields the other in the same average way. But how does one assign a numerical value to this hypothetical screening constant S? Obviously, it must be given some definite numerical value if we are to be able to calculate an approximate energy. The only rational approach is to let the variation principle assign a numerical value to S, i.e., determine the expression for the approximate energy ε as a function of the parameter S and choose the best value of S by minimizing the energy with respect to it. We can make the algebra somewhat simpler by using the simple substitution

$$\eta = Z - S \tag{6-78}$$

and minimizing ε with respect to η instead; the quantity η may be interpreted as an *effective nuclear charge*. Using integrals whose evaluations are illustrated in App. 3, we can express the helium atom energy as

$$\varepsilon = \eta^2 - 2\eta Z + \tfrac{5}{8}\eta \tag{6-79}$$

The first derivative of ε with respect to the parameter η is

$$\frac{\partial \varepsilon}{\partial \eta} = 2\eta - 2Z + \tfrac{5}{8} \tag{6-80}$$

thus $\eta = \frac{27}{16} = 1.69 \quad \text{when } Z = 2$

Substituting this value of η into the energy expression (6-79) leads to an energy of $-2.85E_h$, an improvement of $0.10E_h$ over the previous calculation with $\eta = 2$.

The above calculation also allows one to make a theoretical estimate of the first ionization energy of helium; this refers to the process

$$He \rightarrow He^+ + e^-$$

Thus the first ionization energy of helium is $E_{He^+} - E_{He}$. The first quantity can be computed exactly, since it involves a one-electron species; this quantity is $-2.00E_h$ $(-Z^2/2\ E_h)$. This gives

$$E(He^+) - E(He) = -2.00 - (-2.85) = 0.85E_h$$

The experimental value is 24.5 eV $(0.90E_h)$.

The helium atom energy can be improved rather markedly by using a two-parameter function of the form

$$\xi = N[1s(1)1s'(2) + 1s'(1)1s(2)] \tag{6-81}$$

where $1s$ and $1s'$ are both $1s$ orbitals as before, but the two have *different* parameters, i.e., $1s$ has the effective nuclear charge η_1 and $1s'$ has the effective nuclear charge η_2. The reason for having a two-term wavefunction such as the above is to avoid implying that the two electrons are distinguishable. Thus $1s$ appears in the first part for electron 1 and in the second part for electron 2. Similarly, $1s'$ appears first for electron 2 and then for electron 1; this imparts a symmetry to the electron identity and wipes out any implied distinguishability. Later we will discuss the fuller implications of such symmetry for electronic systems in general. The factor N in the wavefunction above is a normalization factor to be chosen later.

A simple physical interpretation of the wavefunction [Eq. (6-81)] is as follows: the total energy can be minimized if the electrons are given additional flexibility to avoid each other. By giving the two electrons separate screening constants (implied in the η_1 and η_2 values), one electron is always assumed to be farther away from the nucleus than is the other. Such a wavefunction was first proposed by Eckart[10] and is often called a *split-shell wavefunction*. The values of the screening parameters which minimize the total energy are obtained by solving the two simultaneous equations

$$\frac{\partial \varepsilon}{\partial \eta_1} = 0 \qquad \frac{\partial \varepsilon}{\partial \eta_2} = 0 \tag{6-82}$$

The final energy expression turns out to be

$$\varepsilon = N^2(2\epsilon + 2\epsilon' + 4\Delta\epsilon'' + 2J + 2K) \tag{6-83}$$

where $N = \sqrt{2(1+\Delta^2)}$

$$\begin{aligned} \epsilon &= \langle 1s(\mu)|h_\mu|1s(\mu)\rangle \\ \epsilon' &= \langle 1s'(\mu)|h_\mu|1s'(\mu)\rangle \\ \Delta &= \langle 1s(\mu)|1s'(\mu)\rangle \\ \epsilon'' &= \langle 1s(\mu)|h_\mu|1s'(\mu)\rangle = \langle 1s'(\mu)|h_\mu|1s(\mu)\rangle \\ J &= \langle 1s(1)1s'(2)|1/r_{12}|1s(1)1s'(2)\rangle \\ K &= \langle 1s(1)1s'(2)|1/r_{12}|1s'(1)1s(2)\rangle \end{aligned} \tag{6-84}$$

Note that when $\eta_1 = \eta_2$, Eq. (6-83) reduces to the energy given in Eq. (6-79) with $J = K = \frac{5}{8}\eta$. However, when the screening parameters are evaluated separately, one obtains $\eta_1 = 1.19$ and $\eta_2 = 2.18$ (average = 1.69 as in the one-parameter case). This leads to an energy of $-2.8757E_h$—a significant improvement over the one-parameter value of $-2.85E_h$.

In the case of the helium atom (or any two-electron ion), it is possible in practice to extend the number of variation parameters until one obtains an energy value which is a more reliable estimate of the nonrelativistic energy of the ground state that can be obtained experimentally. Much of the pioneering work with such calculations was carried out by Hylleraas,[11] who investigated trial functions of the form

$$\xi = e^{-\eta s}[1 + f(s, t, u)] \tag{6-85}$$

where $f(s, t, u)$ is a power series in the variables s, t, and u, defined by

$$s = r_1 + r_2 \qquad t = r_1 - r_2 \qquad u = r_{12} \tag{6-86}$$

These variables are usually called *Hylleraas variables*. A specific example of a function investigated by Hylleraas is

$$\xi = e^{-\eta s}(1 + c_1 u + c_2 t^2 + c_3 s + c_4 s^2 + c_5 u^2) \tag{6-87}$$

---

[10] C. E. Eckart, *Phys. Rev.*, **36**:878 (1930). A general discussion of Eckart-type wavefunctions for two-electron systems is given by Wai-Kee Li, *J. Chem. Ed.*, **64**:128 (1987).

[11] E. A. Hylleraas, *Z. Physik.*, **65**:209 (1930). His personal account of the invention of the variables *s*, *t*, and *u* is in *Rev. Mod. Phys.*, **35**:421 (1963). See also *Advan. Quantum Chem.*, **1**:1 (1964) for an interesting discussion of the two-electron problem.

where η, c_1, c_2, c_3, c_4, and c_5 are variational parameters. The above function leads to an energy of $-2.90324E_h$. The experimental energy (to the same number of significant figures) is $-2.90372E_h$.

The effect of the variable u (an interelectronic coordinate) can be readily appreciated by considering a simplified Hylleraas function

$$\xi = e^{-\eta s}(1 + cu) \tag{6-88}$$

Since the two electrons repel each other strongly when u is small, the effect is to make ξ^2 small; i.e., the probability of finding the two electrons close together is diminished. On the other hand, if u is large (electrons far apart), the repulsion is small and the probability of finding the two electrons at such a distance increases. This particular wavefunction leads to an energy of $-2.8911E_h$ for $c = 0.3658$ and $\eta = 1.8497$.

Hylleraas-type calculations have been extended by many others, notably by Kinoshita[12] and Pekeris.[13] The latter used a function of 1078 terms expressed in perimetric coordinates and obtained an energy of $-2.903724375E_h$ for the ground state of the helium atom (with an uncertainty of 1 in the last figure).

THE HYDROGEN ATOM. Since the Schrödinger equation for the hydrogen atom can be solved exactly, this system affords an excellent opportunity to demonstrate the use of the linear variational method [Eq. (6-25)] to improve estimates of the ground-state energy. First, let us carry out an approximate solution using a trial wavefunction of the normalized *gaussian* form

$$\xi = \left(\frac{2\alpha}{\pi}\right)^{3/4} e^{-\alpha r^2} \tag{6-89}$$

where α is a parameter, the value of which will be determined variationally. The evaluation of the energy expression and the resulting integrals is straightforward and leads to

$$\varepsilon = \left\langle \xi \left| -\frac{\nabla^2}{2} - \frac{1}{r} \right| \xi \right\rangle = \frac{3\alpha}{2} - 2\left(\frac{2\alpha}{\pi}\right)^{1/2} \tag{6-90}$$

Solving the minimization problem $\partial\varepsilon/\partial\alpha = 0$ leads to a parameter value of $\alpha = 0.2827$. This produces the energy

$$\varepsilon = -0.424E_h \tag{6-91}$$

Since the exact value is $-0.500E_h$, this result is in error by $100(0.500 - 0.424)/0.500 = 15.2$ percent. Note also that in this case, as in all the variational

[12] T. Kinoshita, *Phys. Rev.*, **115**:366 (1959). See also P. Pluvinage, *J. Phys. Radium*, **16**:675 (1955) and **18**:474 (1957).

[13] C. L. Pekeris, *Phys. Rev.*, **112**:1649 (1958) and **115**:1216 (1959). See also E. A. Hylleraas, *Advan. Quantum Chem.*, **1**:1 (1964) for a discussion of Pekeris's method.

cases illustrated so far, the approximate energy is *higher* than the exact energy; this is, of course, the result of the upper-limit theorem.

Examination of Figs. 6-2 and 6-3 reveals why the gaussian function gives such a poor approximation to the energy. The exact solution peaks very sharply in the region of the nucleus, whereas the gaussian function is very rounded-off in this area. Also, the gaussian function does not behave properly at large distances from the nucleus. Thus, the approximate function is limited in its ability to represent the exact solution even when the variational parameter is optimized.

Let us now consider a simple way of attempting to improve the above energy. One way is to use the two-term trial wavefunction

$$\xi = c_1\phi_1 + c_2\phi_2 \tag{6-92}$$

where the functions ϕ_1 and ϕ_2 are the gaussian functions

$$\phi_1 = N_1 e^{-\alpha r^2} \qquad \phi_2 = N_2 r e^{-\alpha r^2} \tag{6-93}$$

where N_1 and N_2 are normalization constants given explicitly by

$$N_1 = \left(\frac{2\alpha}{\pi}\right)^{3/4} \qquad N_2 = \frac{2}{\sqrt{3}}\left(\frac{2\alpha}{\pi}\right)^{3/4} \tag{6-94}$$

As discussed generally in Sec. 6-1, the energy of the system is obtained by solving the determinantal equation (6-32). The explicit form of the relevant equation for our two-term wavefunction is

$$\det(\mathbf{H} - \varepsilon\mathbf{\Delta}) = \begin{vmatrix} H_{11} - \epsilon & H_{12} - \epsilon\Delta \\ H_{21} - \epsilon\Delta & H_{22} - \epsilon \end{vmatrix} = 0 \tag{6-95}$$

where the matrix elements are defined as follows:

$$\begin{aligned} H_{11} &= \left\langle \phi_1 \left| -\frac{\nabla^2}{2} - \frac{1}{r} \right| \phi_1 \right\rangle & H_{12} = H_{21} &= \left\langle \phi_1 \left| -\frac{\nabla^2}{2} - \frac{1}{r} \right| \phi_2 \right\rangle \\ H_{22} &= \left\langle \phi_2 \left| -\frac{\nabla^2}{2} - \frac{1}{r} \right| \phi_2 \right\rangle & \Delta = \langle \phi_1 | \phi_2 \rangle &= \langle \phi_2 | \phi_1 \rangle \end{aligned} \tag{6-96}$$

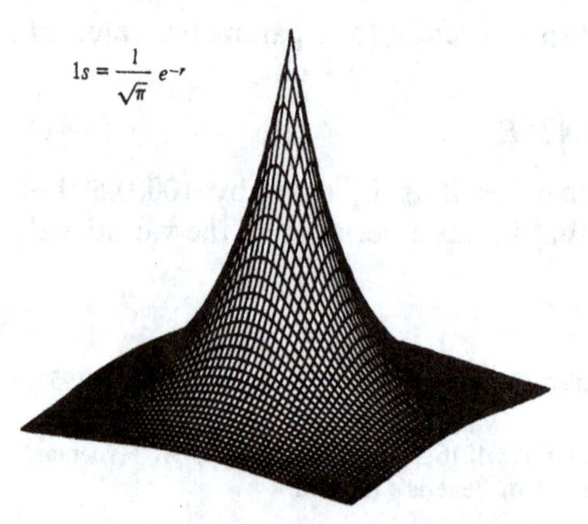

FIGURE 6-2
Surface plot of the exact 1*s* wavefunction of the hydrogen atom. Note especially that the wavefunction peaks sharply at the nucleus. This plot differs from Fig. 5-5 in the horizontal scale used.

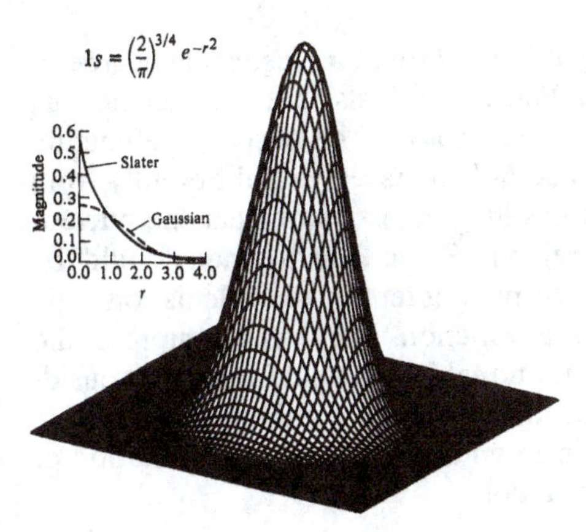

FIGURE 6-3
Surface plot of a gaussian approximation to the ground-state wavefunction of the hydrogen atom. Note especially how the wavefunction differs from that of Fig. 6-2 in the neighborhood of the nucleus. The inset shows numerically accurate profiles of the hydrogen $1s$ and energy-optimized gaussian $1s$ (both normalized to unity); note that the gaussian function is too small at the nucleus but eventually begins to drop off very much as the exact wavefunction does.

The determinantal equation is equivalent to the quadratic equation

$$A\epsilon^2 + B\epsilon + C = 0 \tag{6-97}$$

where the coefficients are identified as

$$A = 1 - \Delta^2 \qquad B = 2H_{12}\Delta - H_{11} - H_{22} \qquad C = H_{11}H_{22} - H_{12}^2 \tag{6-98}$$

The two roots of this quadratic equation will represent approximations of the ground state of hydrogen and one of its excited states.

At this point we must decide what to do about the parameter α, which now enters in *nonlinearly*. For every different value of α there will be a different solution to the determinant; what we wish to find is that value of α which minimizes the value of the lower of the two roots. Since strict adherence to this procedure requires some very tedious algebra, we will take a shortcut and assume a reasonable value for α, namely, $\alpha = 0.2827$, which we obtained from the one-term calculation. This approach is actually very similar to that routinely used for modern-day calculations on much larger electronic systems. This assumed value of α produces the following numerical values for the matrix elements:

$$\begin{aligned} H_{11} &= -0.424E_\text{h} \qquad H_{22} = -0.236E_\text{h} \\ H_{12} &= H_{21} = -0.353E_\text{h} \qquad \Delta = 0.921a_0^{-3} \end{aligned} \tag{6-99}$$

Use of these values in the quadratic equation produces the two roots

$$\epsilon_1 = -0.436E_\text{h} \qquad \epsilon_2 = 0.371E_\text{h} \tag{6-100}$$

The lower energy offers a slight improvement over the one-term result ($-0.424E_\text{h}$). The percentage error in the energy is now 12.8 percent—compared with the 15.2 percent error obtained in the one-term calculation. The other root may be regarded as a *very bad* approximation for the $2s$ state of the hydrogen atom. The reason for such a poor energy value is that the function we have chosen, which is not spectacularly good for the ground state, is simply not capable of representing any other s state to a reasonable degree of approximation. In fact, examination of Fig. 5-8 reveals that the function ϕ_2 strongly resembles the radial portion of a $2p$ orbital. The exact wavefunction for the $2s$ state (see Figs. 5-3, 5-4, 5-6, and 5-7) behaves very differently from this; in particular, the exact $2s$ function has its greatest value near the nucleus whereas the gaussian is zero in this region. Some improvement could be obtained by using two *different* variational parameters in the calculation: one for the ϕ_1 function and another for the ϕ_2 function. This would improve the energy slightly, but the amount of computational labor would increase considerably. A much more fruitful approach is to use a linear combination of simple gaussian functions (all of "$1s$" type) to form an approximate representation of a true $1s$ function, i.e., the linear combination

$$\xi = \sum_{i}^{N} c_i e^{-\eta_i r^2}$$

By use of a large number of terms, one can approximate the shape of the exact $1s$ wavefunction to any desired accuracy. This can be readily understood by recognizing that the normalization factor of a gaussian determines the value of the function at the nucleus ($r = 0$); i.e., the larger this factor (which depends directly on the scale parameter) the larger the value of the function at the nucleus. At the same time, the larger the scale parameter (which also appears in the exponential) the faster the value of the function drops off with distance. Thus, in the simple case of a two-term function, one would adjust the two scale parameters so that behavior at the nucleus is optimized at the same time that the drop-off distance is optimized. Also, the linear variational parameters c_1 and c_2 would be optimized at the same time to make the fit to a single $1s$ orbital as good as possible. It is not difficult to see that if enough terms are used, the fit between the linear combination of gaussians and the exact wavefunction can be made very, very good. Calculations of this type have been reported by Rendell and Arents.[14] For a two-term wavefunction, an energy of $-0.485812E_h$ was obtained. A 10-term wavefunction gave very close to the exact energy, namely, $-0.4999990E_h$.

It is left as an exercise for the reader to determine numerical values for the two linear coefficients of the trial wavefunction [see Eq. (6-92) and Exercise 6-11]. You will find that c_2 is much smaller than c_1, indicating that the second function contributes very little to the ground state.

EXERCISES

6-7. Consider the following trial wavefunction for the particle in a box: $\xi = Nx$ (N is a normalization constant).
(*a*) Show that $N = \sqrt{3}$ (for a box of unit length).
(*b*) Show that the approximate energy due to this wavefunction is zero.

[14] M. Rendell and J. Arents, *J. Chem. Phys.*, **49**:5366 (1968).

(*c*) The result in (*b*) appears to violate the upper-limit theorem. Explain the apparent discrepancy.

6-8. Considering the functions $\xi_1 = x(1-x)$ and $\xi_2 = \sin kx$ as approximate wavefunctions for the particle in a box:
(*a*) Which of the above functions would lead to the lower energy?
(*b*) What is the value of $\langle \xi_2|\hat{H}|\xi_2\rangle / \langle \xi_2|\xi_2\rangle$?
(*c*) Can you use $\partial\varepsilon/\partial k = 0$ to choose k in (*b*)? Explain.
(*d*) How would you recognize ξ_2 as the exact solution?

6-9. Calculate the expectation value of the linear momentum p_x for the particle in a box using the trial wavefunction given in Eq. (6-65). Compare this with the exact result and explain what you find.

6-10. Carry out a variational calculation on the hydrogen atom using a trial wavefunction of the form $\xi = Ne^{-\alpha r}$, where α is a variational parameter. How would you tell that this trial wavefunction is the exact solution?

6-11. Show that the hydrogen atom trial wavefunction [Eq. (6-92)] can be put into the form

$$\xi = \frac{1}{\sqrt{1 + 2\lambda\Delta + \lambda^2}} (\phi_1 + \lambda\phi_2)$$

where λ is a linear variational parameter given by $\lambda = c_2/c_1$.
Then show that the two coefficients have the form

$$c_1 = \frac{1}{\sqrt{1 + 2\lambda\Delta + \lambda^2}} \qquad \text{and} \qquad c_2 = \lambda c_1$$

Also, suggest a systematic procedure for minimizing the energy simultaneously with respect to three parameters: λ, α (in ϕ_1), and β (in ϕ_2).

6-12. Calculate the approximate ground-state energy of the lithium monopositive ion, Li^+, using a trial wavefunction of the form $\xi = 1s(1)1s(2)$, where $1s(\mu) = \eta\sqrt{\eta/\pi}\, e^{-r_\mu}$. The variational parameter η is an effective nuclear charge. Also calculate the second ionization energy of Li and compare it with the experimental value of 75.619 eV.

6-13. When the $1/r_{12}$ term is ignored in the hamiltonian of the helium atom, the calculated approximate energy is $-4.00E_h$, which is *lower* than the exact energy. Is this a violation of the upper-limit theorem?

6.14. Calculate the energy of the helium atom using a gaussian function $1s = Ne^{-\alpha r^2}$ in place of the hydrogenlike function used in Eq. (6-75). Why does this give such a poor energy value?

6-15. A trial wavefunction of a system is given by $\xi = c_1\phi_1 + c_2\phi_2 + c_3\phi_3$, where $\langle \phi_i|\phi_j\rangle = \delta_{ij}$ ($i, j = 1, 2$, or 3). The matrix elements of the system in the basis $\{\phi_i\}$ are $H_{22} = H_{11} + H_{12}$, $H_{33} = H_{11} + 2H_{12}$, $H_{12} = H_{22} = H_{23} = H_{21}$, and $H_{12} = H_{31} = 1.5H_{12}$. Calculate the energies and normalized wavefunctions for the ground state and the first two excited states of the system. *Hint*: It is convenient to use the substitution $x = (H_{11} - \epsilon)/H_{12}$. Also, assume that H_{11} and H_{12} are negative. The matrix eigenvalues (in relation to H_{11} and in units of H_{12}) are 3.54, 0.38, and −0.92.

6-16. Using the examples illustrated in Sec. 6-4, choose the best way to calculate the ground-state energy of the hydride ion, H^-. The experimental energy is not known, but the first ionization energy is believed to be around 0.75 eV.

6-17. Verify Eq. (6-83) for the energy of the split-shell calculation on the helium atom.

6-5 BASIS FUNCTIONS FOR ELECTRONIC CALCULATIONS

Most of the approximate calculations on electronic systems make use of one-electron functions called *orbitals* (see Sec. 5-2). Later we will discuss various ways in which such orbitals are combined to form approximations to wavefunctions. The three most commonly used orbitals are described in this section.

HYDROGENLIKE ORBITALS. These orbitals arise as exact nonrelativistic solutions of the one-electron atom, e.g., hydrogen. They have the general form

$$\psi_{n,l,m}(r, \theta, \varphi) = N_{n,\,l}\rho^l L_{n+1}^{2l+1}(\rho)e^{-2\rho}Y_{l,\,m}(\theta, \varphi) \qquad (6\text{-}101)$$

where $\rho = 2Zr/n$

$N_{n,\,l}$ = normalizing factor

$L_{n+1}^{2l+1}(\rho)$ = associated Laguerre polynomials

$Y_{l,\,m}(\theta, \varphi)$ = spherical harmonics

n, l, m = one-electron quantum numbers

These functions suffer from two serious deficiencies: although they are orthogonal, they are not complete unless continuum functions are included, and they lead to awkward calculations because of their complicated form.

SLATER-TYPE ORBITALS. The Slater-type orbitals (henceforth called *STO*s) are similar in form to the hydrogenlike orbitals but are considerably simpler to use. The STOs have the general form

$$u_{\alpha,\,n,\,l,\,m}(r, \theta, \varphi) = [(2n)!]^{-1/2}(2\alpha)^{n+1/2}r^{n-1}e^{-\alpha r}Y_{l,\,m}(\theta, \varphi) \qquad (6\text{-}102)$$

where n and α are variational parameters and l and m are quantum numbers. The parameter n may be regarded as an "effective principal quantum number." Note that if we set $\alpha = Z/n$ and $n = 1$, then the resulting STO is just the hydrogenlike 1*s* orbital. In general, the STOs differ from the hydrogenlike orbitals in two ways: they are *complete*, and they are *not mutually orthogonal*. Also, the STOs and hydrogenlike orbitals have a different number of nodes. Whereas the STOs have nodeless radial portions, the hydrogenlike orbitals have $(n - l - 1)$ nodes. Note that there is some orthogonality in the STOs; for example, functions with different values of the quantum number l are orthogonal because of the spherical harmonics portions.

Although the best general procedure is to treat the parameters n and α as variational parameters, a simpler approach was originally proposed by Slater[15] and this method is sometimes used in certain applications.

---

[15] J. C. Slater, *Phys. Rev.*, 36:57 (1930).

GAUSSIAN-TYPE ORBITALS. Gaussian-type orbitals (henceforth called *GTOs*) may be expressed in the form

$$g(\alpha, n, l, m) = Nr^{n-1}e^{-\alpha r^2}Y_{l,m}(\theta, \varphi) \tag{6-103}$$

where the normalization constant is

$$N = \left[\frac{2^{2n+3/2}}{(2n-1)!\sqrt{\pi}}\right]^{1/2} \alpha^{(2n+1)/4}$$

where n and α are variational parameters. The GTOs form a complete set, but they are not mutually orthogonal. In general, the GTOs have the basic deficiency which was illustrated in their use in the hydrogen atom calculation: they do a very poor job of representing the electron probability both near the nucleus and far away from it. However, this deficiency may be overcome by using a large number of GTOs. For example, a linear combination of several GTOs will essentially replace a single STO.[16] The gaussian calculations on the hydrogen atom (Sec. 6-4) illustrate this property.

A number of unconventional basis sets have been used by various investigators over the course of time, but none has enjoyed the overall utility of STOs. Some basis sets may appear to surpass STOs when employed for certain types of simple calculations, but these quickly lose their advantages when used in more complicated wavefunctions. Some comparisons along these lines have been documented by Bishop and Leclerc.[17]

6-6 SCALING AND THE VIRIAL THEOREM

According to the discussion of Sec. 2-12, a system of charged particles obeying Coulomb's law must obey the virial theorem. This means that the expectation values of the kinetic and potential energies must satisfy the relationship

$$\langle \hat{T} \rangle = -\frac{\langle \hat{V} \rangle}{2} \tag{6-104}$$

Although this relationship is satisfied by the exact solutions to the Schrödinger wave equation, it is not necessarily satisfied by approximate solutions. In this section we will examine a simple way of treating approximate wavefunctions so that the above virial theorem is always satisfied.

The kinetic- and potential-energy operators for an N-electron atom are given by

$$T = \sum_{\mu}^{N}\left(\frac{-\nabla_\mu^2}{2}\right) \qquad V = -\sum_{\mu}^{N}\frac{Z}{r_\mu} + \sum_{\mu<\nu}^{N}\frac{1}{r_{\mu\nu}} \tag{6-105}$$

---

[16] S. Huzinaga, *J. Chem. Phys.*, **42**:1293 (1965).

[17] D. M. Bishop and J.-C. Leclerc, *Mol. Phys.*, **24**:979 (1972).

An approximate wavefunction for the N-electron atom may be designated by

$$\xi(v_1, \ldots, v_N) = \xi(v) \tag{6-106}$$

where v_μ represents the three coordinates of the μth electron, for example, x_μ, y_μ and z_μ or, equivalently, r_μ, θ_μ, and φ_μ. Now let us replace $\xi(v)$ with a new function in which each electron's coordinates v_μ are replaced with a set of coordinates ηv_μ, where η is a *scale factor* that stretches (or shrinks) each electron coordinate by a factor η; that is, x_μ now becomes ηx_μ, and η acts similarly for all the other coordinates. This new function is called a *scaled wavefunction* and may be written

$$\xi_\eta = \eta^{3N/2}\xi(\eta v) \tag{6-107}$$

such that for $\eta = 1$ the equation reduces to $\xi(v)$. The factor $\eta^{3N/2}$ ensures that ξ_η is normalized whenever $\xi(\eta v)$ is normalized.

Now let $\langle \hat{T}_\eta \rangle$ and $\langle \hat{V}_\eta \rangle$ represent the expectation values of the kinetic-energy and potential-energy operators, respectively, with respect to the scaled wavefunction given in Eq. (6-107). For the kinetic energy we get

$$\langle \hat{T}_\eta \rangle = \langle \xi_\eta | \hat{T} | \xi_\eta \rangle \tag{6-108}$$

where ξ_η is assumed to be normalized. It is now convenient to introduce a change in notation

$$\eta v = v' \tag{6-109}$$

This means that $\eta x_\mu = x'_\mu$, and similar relationships exist for the other coordinates.[18] Equation (6-108) then becomes

$$\langle \hat{T}_\eta \rangle = \langle \xi(v') | -(\tfrac{1}{2})\eta^2 \sum_\mu (\nabla'_\mu)^2 | \xi(v') \rangle = \langle \xi(v') | \hat{T}'_\eta | \xi(v') \rangle = \eta^2 \langle \hat{T}_1 \rangle \tag{6-110}$$

This result follows from the fact that v' is a dummy variable (it gets integrated over in a *definite* integral, and, thus, whatever symbol is used for it is irrelevant), so that all operators and functions in v and v' are equivalent. The quantity $\langle \hat{T}_1 \rangle$ is just the expectation value of $\hat{T}$ over the unscaled function ($\eta = 1$). In similar fashion it follows that

$$\langle \hat{V}_\eta \rangle = \eta \langle \hat{V}_1 \rangle \tag{6-111}$$

The total energy resulting from the scaled wavefunction is

$$\langle \hat{H} \rangle = \varepsilon = \eta^2 \langle \hat{T}_1 \rangle + \eta \langle \hat{V}_1 \rangle \tag{6-112}$$

Let us now choose the scale factor η so that the energy ε is minimized.

---

[18] Thus, $\partial/\partial x = (\partial/\partial x')(\partial x'/\partial x) = \eta\partial/\partial x'$, $\partial^2/\partial x^2 = \eta^2(\partial^2/\partial(x')^2$, etc.

The variation method requires that

$$\frac{\partial \varepsilon}{\partial \eta} = 2\eta\langle \hat{T}_1 \rangle + \langle \hat{V}_1 \rangle = 0 \tag{6-113}$$

Solving for the scale factor gives

$$\eta = -\frac{\langle \hat{V}_1 \rangle}{2\langle \hat{T}_1 \rangle} \tag{6-114}$$

Substituting the above result into Eq. (6-112) leads to a simple expression for the approximate energy:[19]

$$\varepsilon = -\frac{\langle \hat{V}_1 \rangle^2}{4\langle \hat{T}_1 \rangle} \tag{6-115}$$

This means that one can set $\eta = 1$ at the outset of a calculation and then minimize E with respect to all other variational parameters (if there are any). Then, Eq. (6-114) can be used to determine the value of η which minimizes the energy. In the event that $\xi(v)$ happens to be the exact wavefunction, then $\eta = 1$ and Eq. (6-114) is just a statement of the virial theorem. But if $\xi(v)$ is not the exact energy, use of the value of η given by Eq. (6-114) in Eqs. (6-110) and (6-111) leads to the result

$$\langle \hat{T}_\eta \rangle = -\frac{\langle \hat{V}_\eta \rangle}{2} \tag{6-116}$$

which is also a statement of the virial theorem.[20] Thus, the optimized value of the factor η adjusts the coordinates in such a way that the average value of the kinetic energy is equal to half the negative value of the average potential energy.

The helium atom calculation discussed in Sec. 6-4 already contains an optimized scale factor: the effective nuclear charge. Here one finds

$$\langle \hat{T}_\eta \rangle = \eta^2 \qquad \langle \hat{V}_\eta \rangle = -2\eta Z + \tfrac{5}{8} Z \tag{6-117}$$

Thus

$$\eta = -\frac{\langle \hat{V}_\eta \rangle}{2\langle \hat{T}_\eta \rangle} = \frac{4 - \frac{5}{8}}{2} = 1.69 \tag{6-118}$$

Use of this value in Eq. (6-117) leads to $-2.85E_h$, as shown before.

Since the experimental ground-state energy of the helium atom is $-2.90E_h$, the error in the total energy is

$$\Delta E = -2.90 - (-2.85) = -0.05E_h \tag{6-119}$$

---

[19] This result was first derived by E. A. Hylleraas, *Z. Physik*, **54**:347 (1929).

[20] This is a form first derived by V. A. Fock, *Z. Physik*, **63**:855 (1930).

Using the virial theorem, we see that the total energy error can be attributed to a kinetic-energy error and a potential-energy error of

$$\Delta T = 0.05E_{\text{h}} \qquad \Delta V = -0.10E_{\text{h}} \tag{6-120}$$

This shows that the approximation used to represent the ground state of the helium atom *underestimates* the kinetic energy and *overestimates* the potential energy. The underestimation of the kinetic energy indicates that the approximate wavefunction attributes simpler motions to the electrons than they undergo in reality. Or, stated in another way, the electrons resort to more complicated motions in order to avoid each other than the wavefunction is capable of describing. Similarly, the overestimation of the potential energy indicates that the electrons do not repel each other as strongly as the wavefunction implies, i.e., the electrons are more "artful dodgers" than the wavefunction implies.

EXERCISES

6-18. Noting that the one-term gaussian approximation to the ground state of the hydrogen atom (Sec. 6-4) contained a scale factor α (this should be equated with η^2), show how this calculation satisfies the virial theorem.

6-19. Prove that if $\langle \xi(v) | \xi(v) \rangle = 1$, then ξ_η given by Eq. (6-107) is also normalized. *Hint*: Use a method similar to that used to obtain Eq. (6-110).

6-7 PERTURBATION THEORY FOR NONDEGENERATE STATES

Suppose we have a system A, for which the Schrödinger wave equation can be solved exactly. Now consider a second system B, which is physically similar in many respects to system A but whose wave equation does not admit an exact solution. If system B can be imagined to have been formed from system A by the application of a small, continuous deformation, or *perturbation*, it appears reasonable that one could approximate the wavefunctions of system B by "applying" a small, continuous mathematical perturbation to the known wavefunctions of system A. The mathematical development of this simple, physically appealing idea is known as *perturbation theory*. At the start we will assume that system A (called the *unperturbed system*) has no degeneracies; this means that there is an unambiguous one-to-one correspondence between the wavefunctions of systems A and B and that a similar correspondence exists for the eigenvalues of the two systems. In the next section we will remove this restriction and show how ambiguities arising from degeneracies may be resolved.

Let $\hat{H}^{(0)}$ be the hamiltonian operator of system A, and let $\{\psi_k^{(0)}\}$ be a complete set of orthonormal eigenfunctions of $\hat{H}^{(0)}$. The corresponding set of eigenvalues is designated by $\{E_k^{(0)}\}$. Now let the hamiltonian of system B (the *perturbed system*) be given by

$$\hat{H} = \hat{H}^{(0)} + V \tag{6-121}$$

where V is called the *perturbation* and is generally thought of as a small additional term which distorts, or *perturbs*, the system from A into B. Both $\hat{H}^{(0)}$ and V are assumed to be time-independent hermitian operators. The eigenfunction and eigenvalues of the perturbed system B will be designated as $\{\psi_k\}$ and $\{E_k\}$, respectively. It is mathematically convenient to consider the perturbation V as expressible in terms of a power series in a real parameter λ, namely,

$$V = \lambda\hat{H}^{(1)} + \lambda^2\hat{H}^{(2)} + \cdots + \sum_j \lambda^j\hat{H}^{(j)} \tag{6-122}$$

The perturbation parameter λ is chosen such that

$$\lim_{\lambda\to 0} V = 0 \qquad \text{or, equivalently} \qquad \lim_{\lambda\to 0} \hat{H} = \hat{H}^{(0)} \tag{6-123}$$

i.e., removal of the perturbation from system B produces the unperturbed system A. In practice it is notationally convenient to retain only the first-order term of the perturbation and to let $\hat{H}^{(1)}$ be replaced by the symbol V. Thus the perturbation is λV. The inclusion of the perturbation parameter λ in the perturbation itself allows us to think of it as a tunable dial: when the dial is set on zero, there is no perturbation; by turning the dial up, we can make the perturbation any size we wish.

We assume further that the eigenfunctions and eigenvalues of the perturbed system can be expressed as power series in the parameter λ. Thus, we can write for the kth state of the perturbed system

$$\psi_k = \psi_k^{(0)} + \sum_{j=0}^{\infty} \lambda^j\psi_k^{(j)} \qquad E_k = E_k^{(0)} + \sum_{j=0}^{\infty} \lambda^j E_k^{(j)} \tag{6-124}$$

Note that in each case, removal of the perturbation (dialing λ to zero) leads to the unperturbed system A in its kth state. The terms $\psi_k^{(j)}$ and $E_k^{(j)}$ are called the jth-order perturbations of the wavefunction and energy, respectively, of the kth state. Some of the relationships between the unperturbed and perturbed states are depicted graphically in Fig. 6-4 for the first five states ($k = 0$, 1, 2, 3, and 4).

The Schrödinger equation for the kth state of the perturbed system is written formally as

$$(\hat{H}^{(0)} + \lambda V)\psi_k = E_k\psi_k \tag{6-125}$$

or, alternatively,

$$(\hat{H}^{(0)} + \lambda V - E_k)\psi_k = 0 \tag{6-126}$$

This is the equation we wish to solve. Replacing E_k and ψ_k by their perturba-

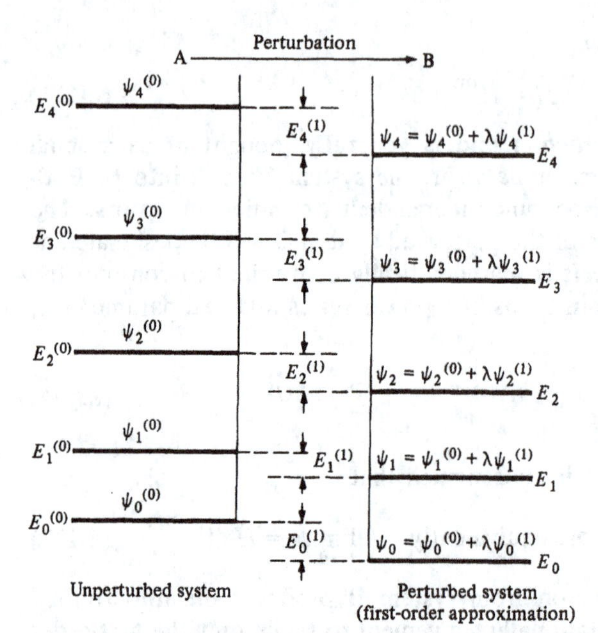

FIGURE 6-4
One-to-one correspondence between an unperturbed state A and a perturbed state B for nondegenerate states. Terms higher than the first order have been neglected.

tion expansions [Eq. (6-124)] in λ, we obtain

$$\left(\hat{H}^{(0)} + \lambda V - E_k^{(0)} - \sum_{j=1}^{\infty} \lambda^j E_k^{(j)}\right)\left(\psi_k^{(0)} + \sum_{j=1}^{\infty} \lambda^j \psi_k^{(j)}\right) = 0 \qquad (6\text{-}127)$$

Expanding the above equation and collecting terms according to ascending powers of the perturbation parameter λ, we get

$$(\hat{H}^{(0)} - E_k^{(0)})\psi_k^{(0)} + \lambda[(H^{(0)} - E_k^{(0)})\psi_k^{(1)} + (V - E_k^{(1)})\psi_k^{(0)}]$$
$$+ \lambda^2[(\hat{H}^{(0)} - E_k^{(0)})\psi_k^{(2)} + (V - E_k^{(1)}\psi_k^{(1)} - E_k^{(2)}\psi_k^{(0)}] + \cdots = 0 \qquad (6\text{-}128)$$

The first term (the coefficient of $\lambda^0 = 1$) is just the Schrödinger equation for the unperturbed state and is called the *zeroth-order equation*. Equation (6-128) is satisfied for $\lambda \neq 0$ only if each coefficient of each power of λ vanishes separately. For $n \neq 0$, the general form of the coefficient of λ_n is

$$(\hat{H}^{(0)} - E_k^{(0)})\psi_k^{(n)} = -V\psi_k^{(n-1)} + \sum_{j=0}^{n-1} E_k^{(n-j)}\psi_k^{(j)} \qquad (6\text{-}129)$$

which is called the *nth-order perturbation equation*. These equations are solved consecutively, beginning with $n = 1$ (where we assume that the zeroth-order equation has already been solved). The nth-order energy is obtained by the following procedure: first, multiply Eq. (6-129) on the left by $(\psi_k^{(0)})^*$ and integrate over all configuration space to obtain

$$\langle \psi_k^{(0)} | \hat{H}^{(0)} - E_k^{(0)} | \psi_k^{(n)} \rangle = -\langle \psi_k^{(0)} | V | \psi_k^{(n-1)} \rangle + \sum_{j=0}^{n-1} E_k^{(n-j)} \langle \psi_k^{(0)} | \psi_k^{(j)} \rangle$$
$$(6\text{-}130)$$

By use of the turnover rule and the fact that the operator $\hat{H}^{(0)} - E_k^{(0)}$ annihilates $\psi_k^{(0)}$, we can see that the left-hand side of Eq. (6-130) vanishes. It is now convenient to choose the wavefunctions of the perturbed state such that they satisfy the relationship

$$\langle \psi_k^{(0)} | \psi_k \rangle = 1 \tag{6-131}$$

which means that all the perturbation functions of the kth state are orthogonal to the wavefunction of the corresponding zeroth-order state, namely,

$$\langle \psi_k^{(0)} | \psi_k^{(n)} \rangle = \delta_{0n} \tag{6-132}$$

Thus the only term which survives in the summation part of Eq. (6-130) is the one for which $j = 0$. This produces the general expression for the nth-order perturbation-energy correction:

$$E_k^{(n)} = \langle \psi_k^{(0)} | V | \psi_k^{(n-1)} \rangle \tag{6-133}$$

In particular, the first-order energy correction is given by

$$E_k^{(1)} = \langle \psi_k^{(0)} | V | \psi_k^{(0)} \rangle \tag{6-134}$$

which is simply the perturbation averaged over the unperturbed state of the system. It is customary and convenient to incorporate the perturbation parameter λ into the definition of $E_k^{(n)}$, so that the total energy of the kth state of the perturbed system is (to a first-order approximation)

$$E_k = E_k^{(0)} + E_k^{(1)} \tag{6-135}$$

This relationship is illustrated in Fig. 6-4.

Equation (6-133) appears to indicate that one needs to know the $(n-1)$-order perturbation correction to the wavefunction in order to compute the nth-order correction to the energy. Actually, the situation is much more favorable; knowing up to and including the nth-order corrections to the wavefunction allows one to calculate up to and including the $(2n+1)$-order corrections to the energy. This is a very fortunate situation, for it turns out that corrections to the wavefunction are generally difficult to obtain. As an example of the above, let us consider the third-order correction to the energy given by

$$E_k^{(3)} = -\langle \psi_k^{(0)} | V | \psi_k^{(2)} \rangle \tag{6-136}$$

Since $E_k^{(3)}$ is a real quantity, we can interchange the bra and ket portions of the above integral. Now using Eq. (6-129) to substitute for $V\psi_k^{(0)}$ when $n = 1$, we obtain Eq. (6-136) in the form

$$E_k^{(3)} = -\langle \psi_k^{(1)} | \hat{H}^{(0)} - E_k^{(0)} | \psi_k^{(2)} \rangle \tag{6-137}$$

Substituting for $(\hat{H}^{(0)} - E_k^{(0)})\psi_k^{(2)}$ from Eq. (6-129) when $n = 2$, Eq. (6-137) becomes

$$E_k^{(3)} = \langle \psi_k^{(1)} | V | \psi_k^{(1)} \rangle - E_k^{(1)} \langle \psi_k^{(1)} | \psi_k^{(1)} \rangle \tag{6-138}$$

which requires a knowledge of only $\psi_k^{(0)}$ and $\psi_k^{(1)}$. In general, one can show that the perturbation wavefunction corrections $\psi_k^{(1)}, \psi_k^{(2)}, \ldots, \psi_k^{(n)}$ are sufficient to determine the perturbation energy corrections up to and including order $2n+1$.

The general formula for the perturbation energy correction of order $2n+1$ is given by[21]

$$E_k^{(2n+1)} = \langle \psi_k^{(n)} | V | \psi_k^{(n)} \rangle - \sum_{l,m}^{n} E_k^{(2n+1-l-m)} \langle \psi_k^{(l)} | \psi_k^{(m)} \rangle \qquad (6\text{-}139)$$

The wavefunctions of the perturbed state may be obtained in a number of different ways. We will illustrate two different ways of doing this for the first-order correction.

THE RAYLEIGH-SCHRÖDINGER METHOD. The *n*th-order perturbation correction is expanded in the complete set of eigenfunctions of the unperturbed state, namely,

$$\psi_k^{(n)} = \sum_i \psi_i^{(0)} c_{ik}^{(n)} \qquad (6\text{-}140)$$

We now wish to solve the first-order perturbation equation

$$(\hat{H}^{(0)} - E_k^{(0)})\psi_k^{(1)} = -V\psi_k^{(0)} + E_k^{(1)}\psi_k^{(0)} \qquad (6\text{-}141)$$

Expressing $\psi_k^{(1)}$ as in Eq. (6-140), we obtain

$$\sum_i c_{ik}^{(1)}(E_i^{(0)} - E_k^{(0)})\psi_i^{(0)} = -V\psi_k^{(0)} + E_k^{(1)}\psi_k^{(0)} \qquad (6\text{-}142)$$

Multiplying on the left by $\psi_j^{(0)}$ $(j \neq k)$ and integrating over all configuration space, we get (since only $i = j \neq k$ terms survive)

$$(E_i^{(0)} - E_k^{(0)})c_{ik}^{(1)} = -\langle \psi_i^{(0)} | V | \psi_k^{(0)} \rangle \qquad (6\text{-}143)$$

The coefficients $\{c_{ik}^{(1)}\}$ are then given by

$$c_{ik}^{(1)} = \frac{\langle \psi_i^{(0)} | V | \psi_k^{(0)} \rangle}{E_k^{(0)} - E_i^{(0)}} = \frac{V_{ik}}{E_k^{(0)} - E_i^{(0)}} \qquad i \neq k \qquad (6\text{-}144)$$

Using Eqs. (6-144) and (6-124), we find that the wavefunction of the *k*th state of the perturbed system is (to a first-order approximation)

$$\psi_k = \psi_k^{(0)} + \sum_{i \neq k} \frac{V_{ik}}{E_k^{(0)} - E_i^{(0)}} \psi_i^{(0)} \qquad (6\text{-}145)$$

where it is customary to incorporate the parameter λ with the matrix element V_{ik} (which equals $H_{ik}^{(1)}$).

---

[21] P. O. Löwdin, *J. Math. Phys.*, **6**:1341 (1965); A. Dalgarno and A. L. Stewart, *Proc. Roy. Soc. (London)*, **A238**:269 (1956).

The main advantage of the Rayleigh-Schrödinger method is that the perturbed-system wavefunctions are automatically expressed in terms of those wavefunctions which are most suitable for a series expansion of the correct solution. The difficulty arises, nevertheless, that one must know *all* the eigenfunctions of the unperturbed system, including a possible continuum. In many actual situations, none of the *exact* eigenfunctions of the zeroth-order state is known, but a ground-state wavefunction ($\psi_0^{(0)}$) is known to an acceptable level of approximation. But even if all the zeroth-order wavefunctions are known, it still is no simple matter to apply the method. Of course, if one is content to calculate only the first-order correction to the ground-state energy, a knowledge of $\psi_0^{(0)}$ is sufficient.

THE VARIATIONAL METHOD.[22] Assume that $\psi_k^{(0)}$ and $E_k^{(1)}$ are known, and consider the functional

$$J_k = \langle \psi_k^{(1)} | \hat{H}^{(0)} - E_k^{(0)} | \psi_k^{(1)} \rangle + 2\langle \psi_k^{(1)} | V | \psi_k^{(0)} \rangle - 2E_k^{(1)} \langle \psi_k^{(1)} | \psi_k^{(0)} \rangle \tag{6-146}$$

The first-order variation of this functional is

$$\delta J_k = 2(\langle \delta\psi_k^{(1)} | \hat{H}^{(0)} - E_k^{(0)} | \psi_k^{(1)} \rangle + \langle \delta\psi_k^{(1)} | V | \psi_k^{(0)} \rangle - E_k^{(1)} \langle \delta\psi_k^{(1)} | \psi_k^{(0)} \rangle) \tag{6-147}$$

This functional is stationary, that is, $\delta J_k = 0$, for all possible variations $\delta\psi_k^{(1)}$ if

$$(\hat{H}^{(0)} - E_k^{(0)})\psi_k^{(1)} + V\psi_k^{(0)} - E_k^{(1)}\psi_k^{(0)} = 0 \tag{6-148}$$

which is just the first-order perturbation equation. This is solved by construction of trial correction functions $\psi_k^{(1)}$ in terms of variational parameters $\alpha_1, \alpha_2, \ldots, \alpha_n$ such that

$$\frac{\partial J_k}{\partial \alpha_i} = 0 \qquad i = 1, 2, \ldots, n \tag{6-149}$$

6-8 PERTURBATION THEORY FOR DEGENERATE STATES

Next consider the situation in which the zeroth-order hamiltonian $\hat{H}^{(0)}$ has an n-fold-degenerate set of eigenfunctions $\psi_{k_1}^{(0)}, \psi_{k_2}^{(0)}, \ldots, \psi_{km}^{(0)}$ with the common energy $E_k^{(0)}$. These n wavefunctions will not generally be orthogonal, but since they are linearly independent, they can be transformed to an orthonormal set; we assume, without any loss of generality, that this has been done. Unlike the previous case of nondegenerate systems, one cannot at the outset specify just

[22] H. A. Bethe and E. E. Salpeter, *Quantum Mechanics of One- and Two-Electron Atoms*, Academic, New York, 1957.

which state of the perturbed system corresponds to any given degenerate state of the unperturbed system. Furthermore, the nondegenerate system formulation contains mathematical expressions such as Eqs. (6-144) and (6-145) which "blow up" when degeneracies occur, i.e., whenever the denominator terms $E_i^{(0)} - E_k^{(0)}$ vanish. Consequently, the mathematical approach must be modified to circumvent such difficulties.

First, let us form n different linear combinations of the n-fold-degenerate wavefunctions, namely,

$$\phi_{km}^{(0)} = \sum_{j=1}^{n} \psi_{kj}^{(0)} c_{jm} \qquad m = 1, 2, \ldots, n \tag{6-150}$$

The n different wavefunctions of the perturbed system which are to correspond to the above n-fold-degenerate functions are now written as

$$\psi_{km} = \phi_{km}^{(0)} + \sum_{j=1}^{\infty} \lambda^j \psi_{km}^{(j)} \tag{6-151}$$

and the corresponding energies as

$$E_{km} = E_k^{(0)} + \sum_{j=1}^{\infty} \lambda^j E_{km}^{(j)} \tag{6-152}$$

Note that as the perturbation is removed, the perturbed-state wavefunction ψ_{km} reduces to a particular linear combination $\psi_{km}^{(0)}$ of the unperturbed system, and all energies E_{km} reduce to the common value $E_k^{(0)}$. Of course, it is possible that the perturbed system still has some degeneracies; in fact, it may have *all* the degeneracies of the unperturbed system. In general, whether all the degeneracies, some of them, or none of them is removed by a perturbation depends on the exact nature of the perturbation. As we will see later, degeneracies are altered whenever the perturbed system has a lower state of symmetry than the unperturbed system. In Fig. 6-5 there is diagramed the specific situation where a three-fold-degenerate unperturbed system corresponds to three nondegenerate states of a perturbed system; in other words, the degeneracy is removed completely. Only the first-order effects are depicted in this illustration.

The first-order perturbation equation [see Eq. (6-128)] for the degenerate system is written in the same general form as for the nondegenerate case:

$$(\hat{H}^{(0)} - E_k^{(0)})\phi_{km}^{(1)} = E_k^{(1)}\phi_{km}^{(0)} - V\phi_{km}^{(0)} \tag{6-153}$$

Multiplying by the complex conjugate of $\psi_{k1}^{(0)}$ on the left and integrating over all configuration space, one obtains

$$\langle \psi_{k1}^{(0)} | \hat{H}^{(0)} - E_k^{(0)} | \phi_{km}^{(1)} \rangle = E_k^{(1)} \langle \psi_{k1}^{(0)} | \phi_{km}^{(0)} \rangle - \langle \psi_{k1}^{(0)} | V | \phi_{km}^{(0)} \rangle \tag{6-154}$$

Application of the turnover rule shows that the left-hand side of Eq. (6-154) vanishes. Equation (6-154) then becomes

$$\langle \psi_{k1}^{(0)} | V | \phi_{km}^{(0)} \rangle - E_k^{(1)} \langle \psi_{k1}^{(0)} | \phi_{km}^{(0)} \rangle = 0 \tag{6-155}$$

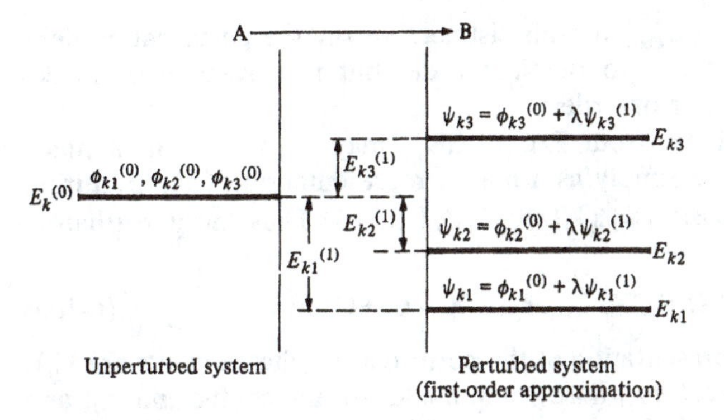

FIGURE 6-5
Correspondence between a threefold-degenerate unperturbed state of system A and the states of a perturbed system B in the special case that the perturbation completely removes the degeneracy. Note that a single subscript k identifies the unperturbed states but a double subscript (in general terms, km) identifies the three perturbed states arising from the kth unperturbed state.

If we use Eq. (6-150) for $\phi_{km}^{(0)}$ and suppress the m subscript, Eq. (6-155) becomes

$$\langle \psi_{k1}^{(0)}|V|c_1\psi_{k1}^{(0)} + c_2\psi_{k2}^{(0)} + \cdots\rangle - E_k^{(1)}\langle \psi_{k1}^{(0)}|c_1\psi_{k1}^{(0)} + c_2\psi_{k2}^{(0)} + \cdots\rangle = 0 \tag{6-156}$$

Recalling that the $\{\psi_{kj}^{(0)}\}$ are orthonormal, we can simplify Eq. (6-156) to

$$\sum_j (V_{ij} - x\delta_{ij})c_j = 0 \tag{6-157}$$

where we use the convenient notation

$$V_{ij} = \langle \psi_{k1}^{(0)}|V|\psi_{kj}^{(0)}\rangle \tag{6-158}$$

and where x refers to a value of the first-order perturbation-energy correction. Repeating the procedure for $j = 2, 3, \ldots, n$, one gets $n - 1$ additional equations of the same form as Eq. (6-157). The generalized form of such equations (n in number) is

$$\sum_j^n (V_{ij} - x\delta_{ij})c_j = 0 \qquad i, j = 1, 2, \ldots, n \tag{6-159}$$

The above n equations are *secular equations* which represent the first-order perturbation equation (6-153). The n roots of the equation ($x_1, x_2, \ldots, x_n$), which are not necessarily distinct, represent the first-order energy corrections to the perturbed system. In fact, these roots may be better labeled "$E_{k1}^{(1)}, E_{k2}^{(1)}, \ldots, E_{kn}^{(1)}$." One can now determine the coefficients $\{c_{im}\}$, corresponding to the root $E_{km}^{(1)}$ and the first-order correction to the wavefunction $\psi_{km}^{(1)}$. The perturbed system now has n energy levels $\{E_{km}\}$, some or all of which may be degenerate, corresponding to the n-fold-degenerate level of the unperturbed state. If the $\{E_{km}\}$ are all distinct, we say the perturbation does not remove any degeneracies to the first order but may remove some degeneracies at a higher order or orders.

The secular equations [from Eq. (6-159)] may be viewed in a much simpler manner; they arise simply as a matrix representation of the perturbation operator $\hat{V}$ in the basis $\{\psi_{km}^{(0)}\}$ $(m = 1, 2, \ldots, n)$. Thus the perturbation equations in matrix form are

$$\mathbf{VC} = \mathbf{CX} \qquad \text{or} \qquad \mathbf{X} = \mathbf{C}^{\dagger}\mathbf{VC} \tag{6-160}$$

where $\mathbf{V}$ is the matrix representation of the perturbation (the elements are V_{ij}), $\mathbf{C}$ is the matrix of eigenvectors whose components are the coefficients c_{ij}, and $\mathbf{X}$ is the diagonal matrix of first-order perturbation-energy corrections. The energies of the perturbed state (correct to the first order) are now given by the energy of the degenerate unperturbed system plus one of the diagonal elements of the matrix $\mathbf{X}$.

A simple example of how a perturbation removes degeneracies is illustrated in Fig. 6-6. Each stationary state of the hydrogen atom is $(2j + 1)$-fold degenerate when the atom is isolated. However, when the atom is subjected to a magnetic field (which has a unique axis of orientation), each state splits into $2j + 1$ separate states (see Sec. 5-5 for a discussion of the Zeeman effect for hydrogen). Figure 6-6 also shows what happens to the $^2S_{1/2}$, $^2P_{1/2}$, and $^2P_{3/2}$ states of the hydrogen atom when the atom is subjected to a magnetic field. The first-order perturbation energy correction in this instance, as given by Eq. (5-93), is

$$E^{(1)} = g_j \mu_{\mathrm{B}} H m_j \tag{6-161}$$

The effect of the magnetic field H is to change the symmetry of the atom from spherical to cylindrical, and it is this symmetry change which results in a loss of degeneracy.

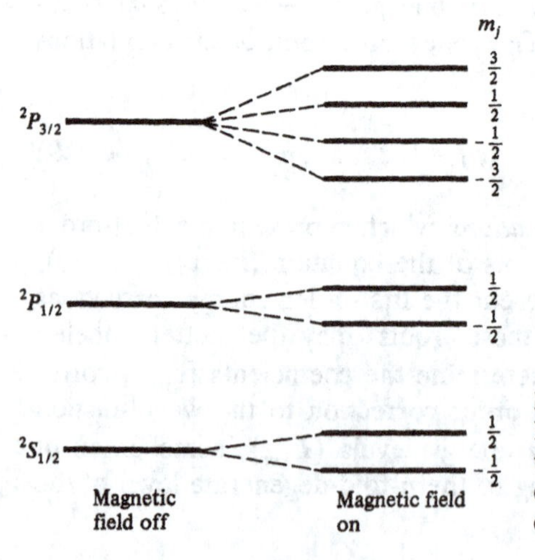

FIGURE 6-6
Effect of a magnetic field on some of the quantum states of the hydrogen atom; this is an example of the Zeeman effect. Note that all the degeneracies are removed.

The main difference between the nondegenerate and degenerate system procedures resides in the solution of the perturbation equation (6-128). Whereas the nondegenerate case involves solutions of individual equations, the degenerate case involves solutions of $n \times n$ matrix equations, each of which is a representation of the perturbation in the basis of the n-fold-degenerate unperturbed state. In a symbolic sense, the nondegenerate and degenerate systems are treated in the same way.

6-9 EXAMPLES OF PERTURBATION CALCULATIONS

In this section we discuss some simple examples of perturbation calculations: the ground-state energies of two-electron atoms, centrifugal distortion of the rigid rotator, and the effect of an electric field on the hydrogen atom (Stark effect).

THE GROUND-STATE ENERGIES OF TWO-ELECTRON ATOMS. The hamiltonian operator describing the stationary states of a two-electron atom of atomic number Z is given by

$$\hat{H} = -\frac{1}{2}\nabla_1^2 - \frac{Z}{r_1} - \frac{1}{2}\nabla_2^2 - \frac{Z}{r_2} + \frac{1}{r_{12}} = h_1 + h_2 + \frac{1}{r_{12}} \tag{6-162}$$

It is convenient to choose the unperturbed system as a two-electron atom in which the electrons do not interact. The zeroth-order hamiltonian then is

$$\hat{H}^{(0)} = h_1 + h_2 \tag{6-163}$$

As indicated previously (Sec. 6-4), the eigenfunctions of this zeroth-order hamiltonian are known exactly; the eigenfunction for the ground-state is simply the orbital product $1s(1)1s(2)$, or more explicitly,

$$\psi^{(0)} = 1s(1)1s(2) = \frac{Z^3}{\pi} e^{-Z(r_1+r_2)} \tag{6-164}$$

and the zeroth-order energy is simply

$$E^{(0)} = -Z^2 = -4.00E_h \text{ for He} \tag{6-165}$$

Note that we have suppressed the subscripts indicating the state under discussion; this is assumed to be the ground state in all cases. The perturbation is then given by

$$V = \frac{1}{r_{12}} \tag{6-166}$$

Using the results of App. 3, we derive the first-order perturbation-energy correction

$$E^{(1)} = \left\langle 1s(1)1s(2) \left| \frac{1}{r_{12}} \right| 1s(1)1s(2) \right\rangle = \tfrac{5}{8}Z \tag{6-167}$$

Thus, the total energy of the two-electron atom (to the first order) is

$$E = E^{(0)} + E^{(1)} = -Z^2 + \tfrac{5}{8}Z = 2.75E_h \text{ for He } (Z=2) \tag{6-168}$$

This is, of course, the same result obtained in the unscaled variational calculation illustrated in Sec. 6-4. Thus, the approximation does not satisfy the virial theorem (Sec. 6-6). Of course, all one need do to satisfy the virial theorem is to replace Z with a variational parameter η in the appropriate places. This also lowers the energy.

Empirical formulas proposed by Moseley[23] and Edlen[24] indicate that the atomic number Z enters the energy expressions of atoms in such a way as to suggest its use as a perturbation parameter. Such use of Z was pioneered by Hylleraas,[25] but ony relatively recently has it been extensively exploited. If one redefines the position vectors of the electrons by $Z\mathbf{v}_\mu = \bar{\mathbf{v}}_\mu$ and the total energy by $E/Z^2 = \bar{E}$, then the hamiltonian operator for the two-electron atom becomes

$$\bar{H} = \bar{h}_1 + \bar{h}_2 + \frac{1}{Z\bar{r}_{12}} \tag{6-169}$$

This means that the unit of length is now Z^{-1} and the unit of energy is $Z^2 E_h$. The reciprocal of the atomic number Z then enters the hamiltonian $\bar{H}$ as a linear perturbation parameter; i.e., the perturbation now becomes

$$\bar{V} = \frac{1}{Z\bar{r}_{12}} \tag{6-170}$$

The wavefunction of the two-electron atom now has the perturbation-expansion form

$$\psi = \psi^{(0)} + Z^{-1}\psi^{(1)} + Z^{-2}\psi^{(2)} + \cdots \tag{6-171}$$

Similarly, the total energy (in units of $Z^2 E_h$) becomes

$$\bar{E} = \bar{E}^{(0)} + Z^{-1}\bar{E}^{(1)} + Z^{-2}\bar{E}^{(2)} + \cdots \tag{6-172}$$

Multiplication of the above equation by Z^2 produces an expression for the energy in conventional atomic units (E_h):

$$E = Z^2\bar{E}^{(0)} + Z\bar{E}^{(1)} + \bar{E}^{(2)} + Z^{-1}\bar{E}^{(3)} + \cdots \tag{6-173}$$

The above formulation was used by Scherr and Knight[26] in a series of very accurate calculations on two-electron atoms. The perturbation equations were solved by the variational method. One calculation, using a 24-term trial

---

[23] H. G. J. Moseley, *Phil. Mag.*, **26**:1024 (1913) and **27**:703 (1914).

[24] B. Edlén, *J. Chem. Phys.*, **33**:98 (1960).

[25] E. A. Hylleraas, *Z. Physik.*, **65**:209 (1930).

[26] C. W. Scherr and R. E. Knight, *Rev. Mod. Phys.*, **35**:431, 436 (1963); *Phys. Rev.*, **128**:2675 (1962); *J. Chem. Phys.*, **40**:3034 (1964).

wavefunction, led to the energy (correct to the sixth order) of

$$E = -Z^2 + \tfrac{5}{8}Z - 0.157666405 + 0.008698991Z^{-1} - 0.000888587Z^{-2} - 0.001036372Z^{-3} - 0.000612917Z^{-4} \qquad (6\text{-}174)$$

The first two terms are just those obtained by the first-order perturbation calculation described previously. A 100-term trial wavefunction led to a helium atom energy of $-2.90372433E_h$ correct to the thirteenth order! This is to be compared with the value of $-2.903724375E_h$ obtained by Pekeris (Sec. 6-4) using a 1078-term variational function.

Perturbation theory may be used to estimate the mass-polarization effect discussed in Sec. 6-3. If ψ is the helium-atom wavefunction without mass polarization, then the mass-polarization correction (to the first order) is given by

$$E_{mp} = -\frac{1}{m_N}\langle\psi|\nabla_1 \cdot \nabla_2|\psi\rangle = \frac{1}{m_N}\langle\nabla_1\psi|\nabla_2\psi\rangle \qquad (6\text{-}175)$$

This has been estimated to be about $4.8\,\text{cm}^{-1}$ (about $2 \times 10^{-5}E_h$). This is a relatively small correction, but nevertheless it is generally incorporated in certain types of very precise calculations. Note that this correction is positive and thus raises the energy. Also, when the mass-polarization correction is made, the total energy should be multiplied by the factor μ/m_e (since $1\text{ a.u.} = Ke^2/a_0 = Km_e e^4/\hbar^2$). Since $\mu < m_e$, this also raises the energy slightly.

CENTRIFUGAL DISTORTION OF THE RIGID ROTATOR. The solutions of the Schrödinger equation for the rigid rotator were discussed in Sec. 4-9. The hamiltonian for this system (which will serve as the zeroth-order system) is given by

$$\hat{H}^{(0)} = \frac{\hat{L}^2}{2I} \qquad (6\text{-}176)$$

[see Eq. (4-121)]. The eigenvalues of this hamiltonian are given by Eq. (4-122) as

$$E_J = J(J+1)B \qquad (6\text{-}177)$$

where $J = 0, 1, 2, \ldots$ is the rotational quantum number. Recall that each of the quantum states is $(2J+1)$-fold degenerate. Thus the ground state ($J = 0$) is nondegenerate ($2J + 1 = 1$), the next state ($J = 1$) is triply degenerate ($2J + 1 = 3$), and so on for the higher states.

Let us now consider a rotating diatomic molecule in which the distance between atoms is not fixed (as in the rigid rotator) but increases as the energy of rotation (and thus the centrifugal distortion) increases. The operator which describes such a distortion to a fair approximation is given by

$$V = -K\hat{L}^4 \qquad (6\text{-}178)$$

where K is a positive constant and $\hat{L}^4$ is the square of the orbital angular momentum operator $\hat{L}^2$. Let us now use perturbation theory to predict what happens to the threefold-degenerate $J = 1$ state under the influence of centrifugal distortion.

The wavefunctions of the zeroth-order threefold-degenerate state may be written symbolically as ψ_1, ψ_2, and ψ_3. These satisfy the eigenvalue equation

$$\frac{\hat{L}^2}{2I}\psi_i = J(J+1)B\psi_i = 2B\psi_i \qquad i = 1, 2, \text{ or } 3 \tag{6-179}$$

where B is the rotational constant defined in Eq. (4-123). The eigenvalues of the perturbed state are now obtained by considering a matrix representation of the perturbation given in Eq. (6-178), using the above three eigenfunctions as a basis. The matrix $\mathbf{V}$ is given by

$$\mathbf{V} = \begin{bmatrix} V_{11} & V_{12} & V_{13} \\ V_{21} & V_{22} & V_{23} \\ V_{31} & V_{32} & V_{33} \end{bmatrix} \tag{6-180}$$

The first-order corrections to the energy are given by the three roots of the secular determinant

$$\det \mathbf{V} = \begin{vmatrix} V_{11} - x & V_{12} & V_{13} \\ V_{21} & V_{22} - x & V_{23} \\ V_{31} & V_{32} & V_{33} - x \end{vmatrix} = 0 \tag{6-181}$$

Evaluation of the matrix elements is simplified by use of the following relationships:

$$V\psi_1 = -K\hat{L}^4\psi_i = -K(\hat{L}^2)^2\psi_i = -4KI^2B^2J^2(J+1)^2 = -DJ^2(J+1)^2 \tag{6-182}$$

$$[V, H^{(0)}] = 0 \tag{6-183}$$

The latter relation shows that *all* the nondiagonal elements (V_{ij}, $i \neq j$) vanish. The first relation shows that the three diagonal elements V_{11}, V_{22}, and V_{33} are equal and are given by

$$V_{ii} = -DJ^2(J+1)^2 \qquad i = 1, 2, \text{ or } 3 \tag{6-184}$$

The perturbation corrections to the energy are given by the three roots of the secular determinant

$$\det \mathbf{V} = \begin{vmatrix} -DJ^2(J+1)^2 - x & 0 & 0 \\ 0 & -DJ^2(J+1)^2 - x & 0 \\ 0 & 0 & -DJ^2(J+1)^2 - x \end{vmatrix} = 0 \tag{6-185}$$

Since all the off-diagonal elements of the determinant are zero, the three roots are the diagonal elements themselves. Thus the determinant reduces to $[-DJ^2(J+1)^2 - x]^3 = 0$ and has solutions

$$x = -DJ^2(J+1)^2 \tag{6-186}$$

Note that this is an example of a case where the perturbation removes *none* of the degeneracy. This occurs because the perturbation has not altered the symmetry of the unperturbed system.

The above system is illustrated in Fig. 6-7. Note that the only effect of the perturbation is to shift each energy level downward by an amount $DJ^2(J+1)^2$. This energy shift becomes more noticeable at higher values of the rotational quantum number J.

ELECTRIC-FIELD EFFECT ON THE HYDROGEN ATOM (STARK EFFECT). The *Stark effect* is similar in some respects to the Zeeman effect (Sec. 5-5) except it involves the interaction of matter with an *electric*, rather than a *magnetic*, field. In the case of the hydrogen atom the perturbation operator representing interaction with an electric field of strength **E** is given by

$$V = -\mathbf{E} \cdot \mathbf{r} \tag{6-187}$$

If we denote the magnitude of the electric field **E** by ε, then the perturbation may also be written

$$V = -\varepsilon r \cos\theta = -\varepsilon z \tag{6-188}$$

where the electric field is assumed to be directed along the z axis. In this illustration we examine the Stark effect on the hydrogen atom in its $n = 2$ quantum state. This is a fourfold-degenerate state with zeroth-order wavefunctions of

$$\psi_1 = 2s \qquad \psi_2 = 2p_0 \qquad \psi_3 = 2p_{-1} \qquad \psi_4 = 2p_1 \tag{6-189}$$

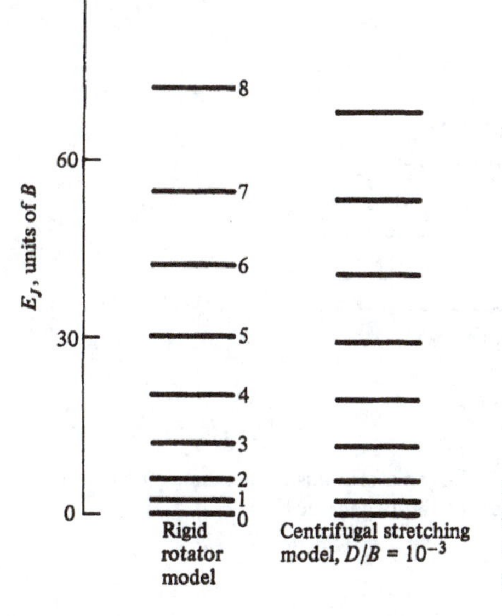

FIGURE 6-7
Centrifugal stretching of the rigid rotator: illustration of a case where the perturbation removes none of the degeneracy.

Note that the magnetic quantum numbers associated with the above states are $m_l = 0, 0, -1$, and 1, respectively. The perturbation energy corrections are obtained by solving the 4×4 secular determinant

$$\det \mathbf{V} = \begin{vmatrix} V_{11} - x & V_{12} & V_{13} & V_{14} \\ V_{21} & V_{22} - x & V_{23} & V_{24} \\ V_{31} & V_{32} & V_{33} - x & V_{34} \\ V_{41} & V_{42} & V_{43} & V_{44} - x \end{vmatrix} = 0 \qquad (6\text{-}190)$$

The matrix elements V_{ij} are readily evaluated. First we note that all matrix elements $V_{ij} = -\varepsilon \langle \psi_i | z | \psi_j \rangle$ vanish unless $\Delta l = \pm 1$ [see Eq. (5-66)]. Thus all the diagonal elements V_{11}, V_{22}, V_{33}, and V_{44} vanish. Also, the commutation relation

$$[\hat{L}_z, V] = [\hat{L}_z, -\varepsilon z] = 0 \qquad (6\text{-}191)$$

means that all the V_{ij} for which $\Delta m_l \neq 0$ will also vanish. Thus the only surviving matrix elements are V_{12} and V_{21}. It is left to the reader to demonstrate that these matrix elements have the following value:

$$V_{12} = V_{21} = 3\varepsilon \qquad (6\text{-}192)$$

The roots of the secular determinant are then readily seen to be

$$x = \begin{cases} -3\varepsilon \\ +3\varepsilon \\ 0 \\ 0 \end{cases} \qquad (6\text{-}193)$$

This shows that the electric field perturbation removes *part* of the degeneracy; one level goes up by 3ε, one goes down by the same amount, and the other two remain the same as in the zeroth-order system. This situation is shown in Fig. 6-8.

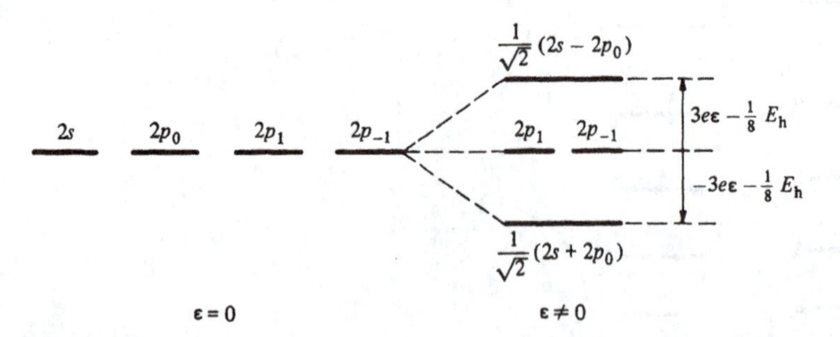

FIGURE 6-8
Effect of an electric field on the $n = 2$ state of the hydrogen atom; this is an example of the Stark effect. Note that the perturbation removes only part of the degeneracy.

EXERCISES

6-20. A trial wavefunction to a given system is given by $\xi = N(\psi_1 + \lambda\psi_2)$, where N is a normalization factor and λ is a coefficient to be determined.

(*a*) If ψ_1 represents the ground state of some unperturbed system similar to our given system and ψ_2 is a first-order perturbation correction, show how the quantity λ is determined.

(*b*) If ψ_1 and ψ_2 are arbitrary basis functions, show how to determine λ by the variation method.

(*c*) Express the normalization factor N in terms of the parameter λ. Are the results necessarily the same in cases (*a*) and (*b*)? Explain. *Hint*: Think about the orthogonality (or lack thereof) of ψ_1 and ψ_2.

6-21. Use the exact eigenfunctions and eigenvalues of the matrix

$$\mathbf{H}^{(0)} = \begin{bmatrix} 2 & 1 \\ 1 & 2 \end{bmatrix}$$

to obtain eigenvalues and eigenfunctions (correct to the first order) of the matrix

$$\mathbf{H} = \begin{bmatrix} 2+\lambda & 1-\lambda \\ 1-\lambda & 2+2\lambda \end{bmatrix} \qquad \lambda \ll 1$$

As a check, solve the problem exactly.

Answer:

$$\epsilon_1^{(1)} = \frac{\lambda}{2} \qquad \mathbf{C}_1 = \frac{1}{\sqrt{2}}\begin{bmatrix} 1-\lambda/4 \\ 1+\lambda/4 \end{bmatrix}$$

$$\epsilon_2^{(1)} = \frac{5\lambda}{2} \qquad \mathbf{C}_2 = \frac{1}{\sqrt{2}}\begin{bmatrix} 1+\lambda/4 \\ -1+\lambda/4 \end{bmatrix}$$

6-22. A particle is confined to the following box:

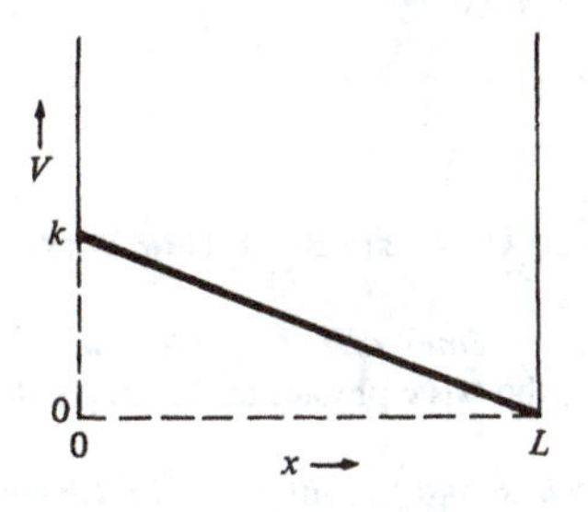

The boundary conditions are $V = 0$ outside the box and $V = k(1 - x/L)$ inside the box.

(*a*) Calculate the ground-state energy of a particle in this box, using first-order perturbation theory. Also, obtain the ground-state wavefunction for the first order and sketch its appearance. Show in what way this wavefunction differs from the corresponding wavefunction of the unperturbed system, and explain what this difference signifies.

(*b*) Repeat for the $n = 5$ state.

6-23. A particle is confined to a box which has a parabolic bottom:

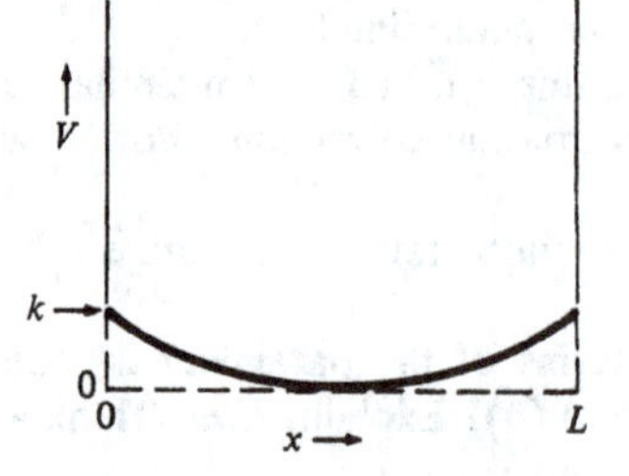

The boundary conditions are $V(0) = V(L) = k$ and $V(L/2) = 0$. The general form of the potential energy is $V = k(1 - 4x/L + 4x^2/L)$.

(*a*) Calculate the ground-state energy to the first order.

(*b*) Calculate the ground-state wavefunction to the first order and sketch its appearance. Compare it with the wavefunction of the unperturbed system.

6-24. Refer to the text treatment of the Stark effect on the hydrogen atom in performing the following exercises:

(*a*) Show that the first-order perturbation correction to the energy of the $n = 1$ state is zero.

(*b*) Show that the second-order perturbation correction to the energy of the $n = 1$ state is $\alpha\varepsilon/2$, where α is given by

$$\alpha = 2 \sum_{i \neq 1} \frac{z_{1i}^2}{E_1^{(0)} - E_i^{(0)}} \qquad z_{1i} = \langle \psi_1^{(0)} | z | \psi_i^{(0)} \rangle$$

6-25. The first seven ionization energies of fluorine are (in units of E_h) 0.64013, 1.2856, 2.3023, 3.20250, 4.19750, 5.77424, and 6.80408. Use Eq. (6-174), along with any other necessary calculations to estimate the total electronic energy of fluorine in its ground state. The accepted value is −2713.449 eV.

SUGGESTED READINGS

Atkins, P. W.: *Molecular Quantum Mechanics*, 2d ed., Oxford University Press, New York, 1983. One of the best modern texts in this area.

Bethe, H. A., and E. E. Salpeter: *Quantum Mechanics of One- and Two-Electron Atoms*, Academic, New York, 1957. A thorough treatment of the basic physics and mathematics of hydrogenlike and heliumlike systems.

Čársky, P., and M. Urban: Ab Initio *Calculations*: *Methods and Applications in Chemistry*, Springer-Verlag, New York, 1980. A survey of computer-age quantum chemistry.

Eyring, H., J. Walter, and G. E. Kimball: *Quantum Chemistry*, Wiley, New York, 1944. Probably the first thorough treatment of the fundamentals of modern quantum chemistry. A classic in the field.

Pauling, L., and E. B. Wilson, Jr.: *Introduction to Quantum Mechanics*, McGraw-Hill, New York, 1935. A classic over half-a-century old, but its treatment of the fundamentals of simple systems has never been surpassed.

Szabo, A., and N. S. Ostlund: *Modern Quantum Chemistry*: *Introduction to Advanced Electronic Structure Theory*, Macmillan, New York, 1982. Deals in depth with the modern systematic methodology of higher-order perturbation theory, especially the diagrammatic techniques used to cope with the complex algebra involved.

CHAPTER 7

ELECTRON SPIN AND MANY-ELECTRON SYSTEMS

Although consideration of electron spin is easily appended to treatments of one-electron systems, in order to be able to discuss systems containing two or more electrons, we must make a more careful examination of some fundamental relationships. In the present chapter we will see how electron spin enters into the wavefunctions used to represent various quantum states of many-electron systems, and we will see why such considerations were not explicitly necessary in discussing the ground states of one- and two-electron systems.

Readers who need a review of determinants and matrices should refer to App. 5.

7-1 THE ANTISYMMETRY PRINCIPLE

Consider two identical particles which may be part of a many-body system and which, like the electron, possess a spin angular momentum, i.e., a fourth degree of freedom. We label the coordinates of the two particles (relative to some arbitrary origin) by the position vectors $\tau_1 = (v_2, \sigma_2)$ and $\tau_2 = (v_2, \sigma_2)$, where, in general, v_i represents the cartesian or polar coordinate point (x_i, y_i, z_i or r_i, θ_i, φ_i, respectively) and σ_i is a spin coordinate. Thus, each particle has three *spatial* coordinates and one *spin* coordinate. The latter will be employed in a symbolic sense only and need not be specified further; however, it is useful to regard the σ coordinate of a single electron as having only two possible values: one value for *spin up* and another value for *spin down*. These two values arise from the fact that the total spin angular momentum of a single electron has two possible projections on an arbitrary axis (see Sec. 4-4). The probability of finding particle 1 at the point τ_1 and particle 2 at the point τ_2 at a given time is given by

$$|\psi[\tau_1(1), \tau_2(2)]|^2\, d\tau_1\, d\tau_2 \tag{7-1}$$

where ψ is a wavefunction describing the system. Since the two particles are identical, they must be physically indistinguishable. Therefore, the above probability could just as well be written

$$|\psi[\tau_2(1), \tau_1(2)]|^2\, d\tau_1\, d\tau_2 \tag{7-2}$$

which is the probability of finding particle 1 at the point τ_2 and particle 2 at the point τ_1 at the given time. Note that in the above expressions we interpret $\tau_i(j)$ as meaning a "particle labeled j is located at a point labeled i." It is apparent that expressions (7-1) and (7-2) differ only in the *exchange* of the coordinates of two identical, indistinguishable particles. In the absence of vector fields (such as a magnetic field) the wavefunction ψ can always be chosen as real, so that we can write

$$\psi^2[\tau_1(1), \tau_2(2)] = \psi^2[\tau_2(1), \tau_1(2)] \tag{7-3}$$

This very important relationship expresses a basic symmetry law of quantum mechanics; namely, if a physically measurable property depends on the coordinates (including the spin) of identical particles, the outcome of any measurement of that property must be independent of any attempt to label the particles of the system. Note that we made use of this symmetry law in writing the split-shell wavefunction (6-81) of the helium atom. The symmetry law means that the measurable property (and hence any operator used to represent it) must be a symmetric function of the coordinates.

Equation (7-3) is a quadratic function and thus can be satisfied in two different ways with respect to the symmetry of the wavefunction ψ relative to an exchange of the coordinates of a pair of indistinguishable particles. One possibility is

$$\psi[\tau_1(1), \tau_2(2)] = \psi[\tau_2(1), \tau_1(2)] \tag{7-4}$$

This expression states the wavefunction ψ is *symmetric* with respect to an interchange of the full coordinates (space and spin) of a pair of identical particles. The remaining possibility is

$$\psi[\tau_1(1), \tau_2(2)] = -\psi[\tau_2(1), \tau_1(2)] \tag{7-5}$$

This expression states that the wavefunction ψ is *antisymmetric* with respect to an interchange of the full coordinates (space and spin) of a pair of identical particles. We now consider four important postulates:[1]

---

[1] W. Pauli, *Phys. Rev.*, **58**:716 (1940).

1. All fundamental particles are described by wavefunctions which are either *symmetric* or *antisymmetric* with respect to the interchange of the full coordinates (space and spin) of a pair of identical particles.
2. Particles never go from one symmetry type to another.
3. All particles with half-integral spin are described by antisymmetric wavefunctions.
4. All particles with zero or integral spin are described by symmetric wavefunctions.

Examples of particles described by antisymmetric wavefunctions are the electron, the proton, the neutron, the α particle from helium 3, and, in general, all particles with odd mass number. Such particles obey *Fermi-Dirac statistics* and are collectively known as *fermions*. As we will see later, fermions tend to repel each other more than one would expect on the basis of otherwise similar particles. For example, a gas of fermions at a given temperature has a higher internal energy and pressure than an otherwise similar gas (e.g., a classical ideal gas) of identical particles.[2]

Particles with zero or integral spin obey *Bose-Einstein statistics* and are called *bosons*. Such particles attract each other more than one would expect on the basis of otherwise similar particles (e.g., an ideal gas), and therefore a gas of bosons at a given temperature has a lower internal energy and pressure than an otherwise similar gas of similar particles.[3] Examples of bosons are the photon (spin = 1), the deuteron (spin = 1), the α particle from helium 4 (spin = 0), and, in general, particles of even mass number. The spin is always zero if the atomic number is also even.

The postulate that electrons must be described by wavefunctions which are antisymmetric with respect to an interchange of the full coordinates (space and spin) of a pair of electrons is known as the *Pauli principle*, or the *antisymmetry principle*. The justification of the principle lies in its agreement with various experiments, some of which we will discuss later.

7-2 SPIN ANGULAR MOMENTA AND THEIR OPERATORS

Any attempt to formulate quantum mechanical operators to represent spin angular momentum is handicapped by the lack of a classical analog to use as a basis (as was done in Chap. 4 for orbital angular momentum). According to the Bohr correspondence principle, quantum mechanical orbital angular momenta reduce to classical orbital angular momenta in the limiting case of large quantum numbers. Spin angular momenta, on the other hand, simply disappear entirely in such a case, since there is no classical counterpart to them.

---

[2] A detailed discussion of such effects is given by R. C. Tolman, *The Principles of Statistical Mechanics*, Oxford University Press, New York, 1938, chap. 10.

[3] Ibid.

Nevertheless, there are good reasons to believe that the basic properties of spin angular momentum bear some similarities to those of classical orbital angular momentum; e.g., one can use spin and orbital angular momenta in analogous contexts in the Bohr model of the atom and obtain reasonable interpretations. Consequently, it is postulated that the spin angular momentum operators obey commutation relations of the same general form as do the orbital angular momentum operators. Thus, using Eqs. (4-12), (4-13), and (4-20) as analogies, and denoting spin operators by $\hat{S}$, we may write the basic spin-commutation relations as

$$[\hat{S}_x, \hat{S}_y] = i\hat{S}_z \qquad [\hat{S}_y, \hat{S}_z] = i\hat{S}_x \qquad [\hat{S}_z, \hat{S}_x] = i\hat{S}_y \qquad [\hat{S}^2, \hat{S}_z] = 0 \quad (7\text{-}6)$$

In Sec. 4-5 we saw that any operators satisfying commutation relations of the above general form would have eigenvalues belonging to one of the following two sets: either *integral* $(0, \pm 1, \pm 2, \pm 3, \ldots)$ or *half-integral* $(\pm\frac{1}{2}, \pm\frac{3}{2}, \pm\frac{5}{2}, \ldots)$. This is certainly consistent with the $\pm\frac{1}{2}$ eigenvalues associated with the spin components of the electron. Since there are only two components of the projection of electron spin, we know that the spin eigenfunctions must be representable as two-component column matrices. If these are required to be orthonormal, one simple possibility is

$$\boldsymbol{\omega}(\tfrac{1}{2}) = \begin{bmatrix} 1 \\ 0 \end{bmatrix} \qquad \boldsymbol{\omega}(-\tfrac{1}{2}) = \begin{bmatrix} 0 \\ 1 \end{bmatrix} \quad (7\text{-}7)$$

Any arbitrary 2×2 matrix, written in general form as

$$\begin{bmatrix} a & b \\ c & d \end{bmatrix} \quad (7\text{-}8)$$

may be written as a linear combination of four linearly independent basis matrices in the linear vector space in which the spin angular momenta are defined. One possible candidate for one of these basis matrices is the unit matrix

$$\mathbf{1} = \begin{bmatrix} 1 & 0 \\ 0 & 1 \end{bmatrix} \quad (7\text{-}9)$$

This matrix is hermitian and unitary. A second candidate is a matrix whose eigenvectors are $\omega(\frac{1}{2})$ and $\omega(-\frac{1}{2})$. This is the matrix

$$\begin{bmatrix} 1 & 0 \\ 0 & -1 \end{bmatrix} \quad (7\text{-}10)$$

which is also hermitian and unitary. The two remaining matrices—in order to be linearly independent of the above two—must have the general form

$$\begin{bmatrix} 0 & a \\ b & 0 \end{bmatrix} \quad (7\text{-}11)$$

Both matrices of this form must have the same eigenvalues (± 1) as Eq. (7-10). Thus the off-diagonal elements must satisfy the relationship $1 - ab = 0$. Two simple possibilities (both having $|a| = |b| = 1$ so that the matrices are unitary) are $a = b = 1$ and $a = -b = i$. These two possibilities are written in explicit form as

$$\begin{bmatrix} 0 & 1 \\ 1 & 0 \end{bmatrix} \quad \text{and} \quad \begin{bmatrix} 0 & -i \\ i & 0 \end{bmatrix} \tag{7-12}$$

The three matrices in Eqs. (7-10) and (7-12) are called the *Pauli spin matrices*; they are usually represented by the relations

$$\boldsymbol{\sigma}_x = \begin{bmatrix} 0 & 1 \\ 1 & 0 \end{bmatrix} \quad \boldsymbol{\sigma}_y = \begin{bmatrix} 0 & -i \\ i & 0 \end{bmatrix} \quad \boldsymbol{\sigma}_z = \begin{bmatrix} 1 & 0 \\ 0 & -1 \end{bmatrix} \tag{7-13}$$

The Pauli spin matrices satisfy the relationship

$$\boldsymbol{\sigma}_x^2 = \boldsymbol{\sigma}_y^2 = \boldsymbol{\sigma}_z^2 = \mathbf{1} \tag{7-14}$$

The matrices given in Eq. (7-13), as well as the unit matrix [Eq. (7-9)], satisfy the commutation relations given for $\hat{S}_x$, $\hat{S}_y$, $\hat{S}_z$, and $\hat{S}^2$ in Eq. (7-6). Since $\boldsymbol{\sigma}_x$, $\boldsymbol{\sigma}_y$, and $\boldsymbol{\sigma}_z$ have eigenvalues of ± 1 and $\hat{S}_x$, $\hat{S}_y$, and $\hat{S}_z$ have eigenvalues of $\pm \frac{1}{2}$, we may establish a formal relationship between the two as follows:

$$\begin{aligned} \mathbf{S}_x &= \frac{1}{2}\boldsymbol{\sigma}_x = \frac{1}{2}\begin{bmatrix} 0 & 1 \\ 1 & 0 \end{bmatrix} \quad & \mathbf{S}_y &= \frac{1}{2}\boldsymbol{\sigma}_y = \frac{1}{2}\begin{bmatrix} 0 & -i \\ i & 0 \end{bmatrix} \\ \mathbf{S}_z &= \frac{1}{2}\boldsymbol{\sigma}_z = \frac{1}{2}\begin{bmatrix} 1 & 0 \\ 0 & -1 \end{bmatrix} \quad & \mathbf{S}^2 &= \frac{3}{4}\begin{bmatrix} 1 & 0 \\ 0 & 1 \end{bmatrix} \end{aligned} \tag{7-15}$$

The matrix representation of $\hat{S}^2$ is most conveniently verified by use of the step-up and step-down operators whose matrix representations are

$$\mathbf{S}_+ = \mathbf{S}_x + i\mathbf{S}_y = \begin{bmatrix} 0 & 1 \\ 0 & 0 \end{bmatrix} \quad \mathbf{S}_- = \mathbf{S}_x - i\mathbf{S}_y = \begin{bmatrix} 0 & 0 \\ 1 & 0 \end{bmatrix} \tag{7-16}$$

By analogy to Eq. (4-31) for $\hat{L}^2$, we write

$$\mathbf{S}^2 = \mathbf{S}_-\mathbf{S}_+ + \mathbf{S}_z(\mathbf{S}_z + 1) = \mathbf{S}_+\mathbf{S}_- + \mathbf{S}_z(\mathbf{S}_z - 1) = \frac{3}{4}\begin{bmatrix} 1 & 0 \\ 0 & 1 \end{bmatrix} \tag{7-17}$$

If $\hat{H}$ is a hamiltonian which contains no spin coordinates, e.g., the nonrelativistic or spin-free hamiltonian generally used for the electronic structure of atoms and molecules, we have the commutation relations

$$[\hat{S}_z, \hat{H}] = 0 \qquad [\hat{S}^2, \hat{H}] = 0 \tag{7-18}$$

The fact that $\hat{S}_z$ and $\hat{S}^2$ commute with the hamiltonian means we can take advantage of Theorem 2-5 in the simplification of the matrix elements of $\hat{H}$.

The spin eigenfunctions whose matrix representations are given by Eq. (7-7) are conveniently defined as follows:

$$\omega_1, \omega_2 = \begin{cases} \alpha = \begin{cases} 1 & \text{spin } \frac{1}{2} \\ 0 & \text{spin } -\frac{1}{2} \end{cases} \\ \beta = \begin{cases} 0 & \text{spin } \frac{1}{2} \\ 1 & \text{spin } -\frac{1}{2} \end{cases} \end{cases} \tag{7-19}$$

For example, $\omega_i = \alpha$ ($i = 1$ or 2) means that the probability of finding an electron with spin $\frac{1}{2}$ is unity if the electron has that spin and is zero otherwise. This means that α is the eigenfunction of $\hat{S}_z$ with the eigenvalue $+\frac{1}{2}$ and that β is the eigenfunction of $\hat{S}_z$ with the eigenvalue $-\frac{1}{2}$. These eigenfunctions satisfy the orthonormality condition

$$\sum \omega_1^* \omega_2 = \delta_{12} \tag{7-20}$$

where the summation is over the two possible values of the m_s quantum number [see Eq. (5-85)]. More explicitly, the product $\omega_1^* \omega_2$ is evaluated for both $m_s = \frac{1}{2}$ and $m_s = -\frac{1}{2}$ and the two terms added. Thus, if $\omega_1 = \omega_2 = \alpha$ or β, the above leads to $1 + 0 = 1$ or $0 + 1 = 1$, respectively. For $\omega_1 \neq \omega_2$, one obtains $0 + 0 = 0$. An alternative notation (which we will find it convenient to use from now on) is

$$\langle \omega_1 | \omega_2 \rangle = \delta_{12} \tag{7-21}$$

where integration is interpreted as summation over m_s.

The spin function notation of Eq. (7-19) may be used to summarize the following important relationships:

$$\begin{aligned} &\hat{S}_x(\alpha, \beta) = \tfrac{1}{2}(\beta, \alpha) && \hat{S}_+(\alpha, \beta) = (0, \alpha) \\ &\hat{S}_y(\alpha, \beta) = \tfrac{i}{2}(\beta, -\alpha) && \hat{S}_-(\alpha, \beta) = (\beta, 0) \\ &\hat{S}_z(\alpha, \beta) = \tfrac{1}{2}(\alpha, -\beta) && \hat{S}^2(\alpha, \beta) = \tfrac{3}{4}(\alpha, \beta) \end{aligned} \tag{7-22}$$

Note that the operator $\hat{S}_+$ annihilates the α eigenfunction and converts the β eigenfunction into the α eigenfunction, whereas the operator $\hat{S}_-$ does the converse. These two operators will become very useful to us later, in the construction of spin eigenfunctions for many-electron systems.

For purposes of direct comparison, the following list shows in side-by-side fashion several analogous relations for spin angular momenta and orbital angular momenta for the one-electron case:

$$\begin{aligned} &\hat{S}_z \omega = m_s \omega && \hat{L}_z Y = m_l Y \\ &\hat{S}^2 \omega = s(s+1)\omega && \hat{L}^2 Y = l(l+1) Y \\ &-s \leq m_s \leq s && -l \leq m_l \leq 1 \\ &m_s = \pm \tfrac{1}{2} && m_l = 0, \pm 1, \pm 2, \ldots, \pm \bar{m}_l \\ &s = \tfrac{1}{2} && l = \bar{m}_l \end{aligned} \tag{7-23}$$

Next, we make the same comparisons for the case in which ω is a many-electron spin function

$$
\begin{aligned}
&\hat{S}_z = \sum_\mu \hat{S}_z(\mu) && \hat{L}_z = \sum_\mu \hat{L}_z(\mu) \\
&\hat{S}_z \omega = M_S \omega && \hat{L}_z Y = M_L Y \\
&\hat{S}^2 \omega = S(S+1)\omega && \hat{L}^2 Y = L(L+1)Y \\
&-S \le M_S \le S && -L \le M_L \le L \\
&M_S = \sum_\mu m_{s\mu} && M_L = \sum_\mu m_{l\mu} \\
&S = \sum_\mu s_\mu, \sum_\mu s_\mu - 1, \ldots \ge 0 && L = \sum_\mu l_\mu, \sum_\mu l_\mu - 1, \ldots \ge 0
\end{aligned}
\tag{7-24}
$$

where the summation is over the electrons.

EXERCISES

7-1. Find the normalized eigenvectors of the matrix representations of the operators $\hat{S}_x$ and $\hat{S}_y$. Show that the operators $\hat{S}_x$, $\hat{S}_y$, and $\hat{S}_z$ are related by a similarity transformation.

7-2. Verify all the relationships given in Eq. (7-22).

7-3. Show that the matrix representations of the step-down and step-up operators $\hat{S}_-$ and $\hat{S}_+$ are nilpotent. Why is this so?

7-4. Find the eigenvalues and normalized eigenvectors of the matrix representation of the $\hat{S}_x$ spin operator [see Eq. (7-15)]. If we let $\mathbf{C} = (\mathbf{C}_1, \mathbf{C}_2)$, where $\mathbf{C}_1$ and $\mathbf{C}_2$ are the eigenvectors, carry out the operators $\mathbf{C}^\dagger \mathbf{S}_x \mathbf{C}$ and explain what this operation does.

7-3 THE ORBITAL APPROXIMATION

The nonrelativistic, time-independent Schrödinger equation for an atom of N electrons may be written symbolically in the standard form

$$\hat{H}\psi = E\psi \tag{7-25}$$

where $\hat{H}$ is called the *spin-free hamiltonian*. This hamiltonian was introduced earlier in Eq. (6-56) and employed several times in Chap. 6; its general form is

$$\hat{H} = \sum_{\mu=1}^{N} h_\mu + \sum_{\mu<\nu}^{N} \frac{e^2 K}{r_{\mu\nu}} \tag{7-26}$$

The summations are over electron coordinates; each electron is designated by an index μ and has three spatial coordinates (x_μ, y_μ, z_μ or r_μ, θ_μ, φ_μ) and one spin coordinate σ_μ. The hamiltonian is called *spin-free*, since it does not explicitly include the spin coordinates of the electrons. The operator h_μ is called the *monoelectronic operator*; its general form [as defined earlier in Eq. (6-57)] is

$$h_\mu = -\frac{\hbar^2 \nabla_\mu^2}{2m} - \frac{Ze^2 K}{r_\mu} \tag{7-27}$$

or, if atomic units are used,

$$h_\mu = -\frac{\nabla^2}{2} - \frac{Z}{r_\mu} \tag{7-28}$$

Note also that the $e^2K/r_{\mu\nu}$ terms in Eq. (7-26) become simply $1/r_{\mu\nu}$ when atomic units are used. It should be noted that the monoelectronic operator contains the *actual* mass (not the *reduced* mass) of the electron. The Schrödinger equation using this operator is derived by separating the nuclear coordinates from electron coordinates and then neglecting certain terms called *mass-polarization terms* in the electronic equation.[4] This is tantamount to assuming that the nuclear mass is infinite so that the reduced mass of the electron may be replaced by its rest mass. Use of this approximation also avoids the awkwardness arising if one had to employ different reduced masses for each different electronic system studied.

The first term in h_μ is just the kinetic energy of electron μ, and the second is the coulombic potential energy of attraction between electron μ and the atomic nucleus. The second term in Eq. (7-26) represents repulsions between a pair of electrons, μ and ν ($r_{\mu\nu}$ is the distance between this pair of electrons). The summation index notation is designed to avoid two things: counting repulsions twice (since a $\mu\nu$ repulsion is the same as a $\nu\mu$ repulsion) and counting self-repulsion (that is, $\mu\mu$). Thus, if we use $N = 4$ (a four-electron system) as an example, the electron-repulsion operator expands as follows:

$$\sum_{\mu<\nu}^{4} \frac{1}{r_{\mu\nu}} = \frac{1}{r_{12}} + \frac{1}{r_{13}} + \frac{1}{r_{14}} + \frac{1}{r_{23}} + \frac{1}{r_{24}} + \frac{1}{r_{34}} \tag{7-29}$$

Note that we set $\mu = 1$ and let $\nu = 2$, 3, and 4; then set $\mu = 2$ and let $\nu = 3$ and 4, etc. An alternative method which some authors use is to do the summation $\mu \neq \nu$ (thereby avoiding self-repulsion but counting all other repulsions twice) and then dividing by 2 to eliminate the redundant repulsions.

As alluded to previously, whenever N is equal to or greater than 2 the Schrödinger equation cannot be solved exactly. This is because no one has yet invented a coordinate system in which the variables can be separated whenever the interparticle term $r_{\mu\nu}$ is present. This problem is not unique to quantum mechanics; classical mechanics is subject to the same restriction in dealing with the many-body problem.

---

[4] Details of this are given in H. A. Bethe and E. Salpeter, *Quantum Mechanics of One- and Two-Electron Atoms*, Academic, New York, 1957.

One of the simplest models used to describe many-electron systems is known as the *orbital approximation*. This model incorporates some aspects of the relativistic nature of the electron without introducing excessive complexity to the algebra. In this approximation the hamiltonian is assumed to be of the *spin-free* form, and the many-electron wavefunction is assumed to have the general form of a determinantal product of orbitals:

$$\psi(1,2,\ldots,N)=\frac{1}{\sqrt{N!}}\begin{vmatrix} S_1(1) & S_2(1) & \ldots & S_N(1) \\ S_1(2) & S_2(2) & \ldots & S_N(2) \\ \cdots & \cdots & \cdots & \cdots \\ S_1(N) & S_2(N) & \ldots & S_N(N) \end{vmatrix} \tag{7-30}$$

A wavefunction of this form is called a *Slater determinant*.[5] The quantity $S_i(\mu)$ is called a *spin orbital*; it is a one-electron distribution function involving the full coordinates (three spatial and one spin) of an electron. Explicit forms for such spin orbitals will be introduced later; for the present we will assume only that they form an orthonormal set. The factor $(N!)^{-1/2}$ in the Slater determinant is a normalization constant; its origin is discussed following the introduction of Eq. (7-33).

The Slater determinant is perhaps the simplest possible form for an antisymmetric wavefunction. Since a determinant will change sign when any two of its rows are interchanged, such a wavefunction will incorporate the antisymmetry required whenever two fermions have their full coordinates (space and spin) interchanged. Furthermore, since each electron labeled with a particular index μ becomes permuted among *all* the N spin orbitals, the determinantal form also avoids implying the distinguishability of individual electrons. In the special case that each spin orbital is associated with four quantum numbers (say, n, l, m_l, and m_s), the determinant will vanish if any two electrons in it have *all four* quantum numbers the same. This result follows from the fact that having two electrons with the same full set of quantum numbers means that two spin orbitals in the determinant are the same, i.e., two columns of the determinant are identical. As is readily verified, any determinant with two identical columns (or rows) is equal to zero. This restriction—that no two electrons have the same full set of quantum numbers—is known as the *Pauli exclusion principle*.

Formal manipulation of determinantal wavefunctions is facilitated by defining the simple orbital product function (often called a *Hartree product*)

$$\Phi=\prod_{i,\mu}^{N} S_i(\mu)=S_1(1)S_2(2)\cdots S_N(N) \tag{7-31}$$

This is simply the product of all the *diagonal* terms of the Slater determinant for an N-electron atom. The determinantal wavefunction Eq. (7-30) is ob-

[5] J. C. Slater, *Phys. Rev.*, **34**:1293 (1929).

tained from (7-31) by use of the *antisymmetrization operator* $\hat{A}$ defined by

$$\hat{A} = (N!)^{-1/2} \sum_P (-1)^p \hat{P} \tag{7-32}$$

where $\hat{P}$ is a *permutation* operator and p is the parity of the permutation, both to be described shortly. The determinantal function (7-30) can be written in terms of the antisymmetrization operator and the Hartree product as

$$\psi(1, 2, \ldots, N) = \hat{A}\Phi \tag{7-33}$$

One of the functions of the antisymmetrization operator (or, as it is often called, the *antisymmetrizer*) is to permute the coordinates of all the electrons among all the various spin orbitals and, thereby, "wipe out" any implied distinguishability of identical particles. There are $N!$ permutations necessary to achieve such an indistinguishability of N electrons—hence the presence of the $N!$ factor in the normalization constant of the Slater determinant for N electrons.

The permutation operator $\hat{P}$ appearing in the antisymmetrizer carries out all possible one-electron, two-electron, three-electron, . . . , N-electron permutations. Thus the summation part of the antisymmetrizer may be written as follows:

$$\sum_P (-1)^p \hat{P} = 1 - \sum_{ij} \hat{P}_{ij} + \sum_{ijk} \hat{P}_{ijk} - \cdots \tag{7-34}$$

where, for example, $\hat{P}_{ijk}$ carries out all possible three-electron permutations. The quantity $(-1)^p$ appearing with the operator $\hat{P}$ is called the *parity* of the permutation; the quantity p is simply the least number of transpositions needed to put the permuted indices back into natural order. For example, if the Hartree product is given explicitly by

$$\Phi = S_1(1)S_2(2)S_3(3) \tag{7-35}$$

then the antisymmetrizer produces

$$\begin{aligned}\hat{A}\Phi = (3!)^{-1/2}[&S_1(1)S_2(2)S_3(3) - S_2(1)S_1(2)S_3(3) - S_3(1)S_2(2)S_1(3) \\ &- S_1(1)S_3(2)S_2(3) + S_2(1)S_3(2)S_1(3) + S_3(1)S_1(2)S_2(3)]\end{aligned} \tag{7-36}$$

Note that there are a total of $3! = 6$ terms, each of which is a permuted Hartree product (the first permutation is, of course, trivial). The parity of the first term is $+1$, since it requires *no* transpositions to put the indices back into natural order for the trivial permutation, that is, $(-1)^0 = +1$. For the second term, *one* transposition (exchanging indices 1 and 2) suffices, and the parity is $(-1)^1 = -1$. Note also that the permutations are carried out only on the subscripts identifying the spin orbitals; this means we are reassigning different distribution functions among fixed electron labels. When this convention is used, the electron coordinates always appear in each permuted Hartree product in the natural order $1, 2, 3, \ldots, N$. We could, of course, use just the opposite approach: relocate labeled electrons among the various distribution functions.

Optimum notational use of the antisymmetrizer is facilitated by becoming familiar with some of its basic mathematical properties. Toward this goal we will now prove that the permutation operator $\hat{P}$ is unitary and use this fact to show that the antisymmetrizer is self-adjoint. Consider a generalized wavefunction of N electrons given by

$$\psi(\tau_1, \tau_2, \ldots, \tau_N) \tag{7-37}$$

where, as before, each τ represents the full space and spin coordinates of a specified electron. The operation of the permutation operator $\hat{P}$ is equivalent to permuting the electron coordinates to some new order $\tau_1', \tau_2', \ldots, \tau_N'$, where each τ' was originally one of the τ. This action may be symbolized by the simple notation

$$\hat{P}\psi(\tau) = \psi(\tau') \tag{7-38}$$

In terms of the inverse permutation, this relationship becomes

$$\psi(\tau) = \hat{P}^{-1}\psi(\tau') \tag{7-39}$$

Now consider the definite integral

$$\langle \hat{P}\psi_1(\tau) | \psi_2(\tau) \rangle = \langle \psi_1(\tau) | \hat{P}^{\dagger} | \psi_2(\tau) \rangle \tag{7-40}$$

where the turnover rule has been used to obtain the right-hand form of the integral. We now note that there is an alternative way to transform the left-hand integral in Eq. (7-40). Using Eq. (7-38) for $\hat{P}\psi_1(\tau)$ and Eq. (7-39) for $\psi_2(\tau)$ leads to

$$\langle P\psi_1(\tau) | \psi_2(\tau) \rangle = \langle \psi_1(\tau') | \hat{P}^{-1} | \psi_2(\tau') \rangle \tag{7-41}$$

Comparing the right-hand integrals in Eqs. (7-40) and (7-41) and realizing that τ and τ' represent the same collection of dummy variables of integration, we obtain

$$\hat{P}^{\dagger} = \hat{P}^{-1} \tag{7-42}$$

that is, $\hat{P}$ is a unitary operator. This property of $\hat{P}$ may now be employed to prove that the antisymmetrizer $\hat{A}$ is self-adjoint (hermitian). Taking the adjoint of Eq. (7-32) and using Eq. (7-42) we obtain

$$\hat{A}^{+} = (N!)^{-1/2} \sum_{P^{\dagger}} (-1)^{p} \hat{P}^{\dagger} = (N!)^{-1/2} \sum_{P^{-1}} (-1)^{p} \hat{P}^{-1} \tag{7-43}$$

Since $\hat{P}$ and $\hat{P}^{\dagger}$ carry out the same operations (but in inverse order), the two operators have the same parity. Thus, since the symbol $\hat{P}^{-1}$ is purely arbitrary, it can be replaced with any other symbol, say, $\hat{P}$, which means that the antisymmetrizer and its adjoint do the same thing. Thus we may write

$$\hat{A}^{\dagger} = \hat{A} \tag{7-44}$$

An important property of the antisymmetrizer is that it commutes with any operator which is symmetric in the system coordinates. Letting $\hat{G}$ represent such an operator, we may write

$$[\hat{G}, \hat{A}] = 0 \tag{7-45}$$

The reason for this commutation is that $\hat{G}$ will commute with each individual permutation and, thus, also with a sum of such permutations. The spin-free hamiltonian of an N-electron system is an example of such an operator.

There is another useful way of viewing the function of the antisymmetrizer. The hartree product Φ may be regarded as a function of mixed symmetry, with an antisymmetric product of spin orbitals as one of its components. The function of the antisymmetrizer is simply to project out this antisymmetric component; thus, the antisymmetrizer also has the properties of a *projection operator*. To show that this is indeed the case, let us examine the "square" of the antisymmetrizer

$$\begin{aligned}\hat{A}^2 &= (N!)^{-1}\Big[\sum_{P_i}(-1)^p\hat{P}_i\Big]\Big[\sum_{P_j}(-1)^p\hat{P}_j\Big] \\ &= (N!)^{-1}\Big[(-1)^p\hat{P}_1\sum_{P_j}(-1)^p\hat{P}_j + (-1)^p\hat{P}_2\sum_{P_j}(-1)^p\hat{P}_j \\ &\quad + \cdots + (-1)^p\hat{P}_{N!}\sum_{P_j}(-1)^p\hat{P}_j\Big]\end{aligned} \tag{7-46}$$

where $\hat{P}_1, \hat{P}_2, \ldots, \hat{P}_{N!}$ are fixed permutations. The kth term of this expansion can be written

$$(-1)^p\hat{P}_k\sum_{P_j}(-1)^{p'}\hat{P}_j = \sum_{P_j}(-1)^{p+p'}\hat{P}_k\hat{P}_j = \sum_{P_m}(-1)^p\hat{P}_m = (N!)^{1/2}\hat{A} \tag{7-47}$$

Each of the remaining $N! - 1$ terms has this same form, so that we can write

$$\hat{A}^2 = (N!)^{-1}[N!(N!)^{1/2}\hat{A}] = (N!)^{1/2}\hat{A} \tag{7-48}$$

This shows that $\hat{A}$ is related to a projection operator $\hat{O}_A$ defined by

$$\hat{O}_A = (N!)^{-1/2}\hat{A} \qquad \hat{O}_A^2 = \hat{O}_A \qquad \hat{O}_A^\dagger = \hat{O}_A \tag{7-49}$$

It is occasionally useful to write the left-hand relationship as

$$\hat{A} = (N!)^{1/2}\hat{O}_A \tag{7-50}$$

Thus, the antisymmetrizer operates on a function of mixed symmetry (the Hartree product) to project out its antisymmetric component and to multiply this by $(N!)^{1/2}$. When the Hartree product consists of orthonormal spin orbitals, the projected antisymmetric component is normalized.

Now consider any quantum mechanical operator $\hat{G}$ which commutes with the projection operator $\hat{O}_A$ (and thus also with $\hat{A}$). The expectation value of this operator for the normalized function $\hat{A}\Phi$ is given by

$$\langle\hat{G}\rangle = \langle\hat{A}\Phi|\hat{G}|\hat{A}\Phi\rangle = N!\langle\hat{O}_A\Phi|\hat{G}|\hat{O}_A\rangle \tag{7-51}$$

Using the turnover rule on the $\hat{O}_A$ appearing in the bra portion of the rightmost integral, the commutation of $\hat{O}_A$ and $\hat{G}$, and the projection operator properties of $\hat{O}_A$ (in that order), we obtain

$$\langle \hat{G} \rangle = N!\langle \Phi | \hat{G} | \hat{O}_A \Phi \rangle = \sum_P (-1)^p \langle \Phi | \hat{G} | \hat{P} \Phi \rangle \tag{7-52}$$

This relationship is very useful in the evaluation of expectation values of operators with respect to determinantal wavefunctions.

EXERCISE

7-5. Demonstrate that the determinant

$$D = \begin{vmatrix} a & b \\ c & d \end{vmatrix}$$

is antisymmetric with respect to:
(*a*) Interchange of columns
(*b*) Interchange of rows

7-4 TWO-ELECTRON WAVEFUNCTIONS

The wavefunction of a two-electron atom, in the *orbital approximation*, may be written

$$\psi(1,2) = \hat{A}S_1(1)S_2(2) = \frac{1}{\sqrt{2}} \begin{vmatrix} S_1(1) & S_2(1) \\ S_1(2) & S_2(2) \end{vmatrix} \tag{7-53}$$

The simplest and most commonly used explicit form for a spin orbital is a product of a spatial orbital $\phi_i(x_\mu, y_\mu, z_\mu)$ and a spin function $\omega(\sigma_\mu)$.

Thus

$$S_i(x_\mu, y_\mu, z_\mu, \sigma_\mu) = \phi_i(x_\mu, y_\mu, z_\mu)\omega(\sigma_\mu) \tag{7-54a}$$

or, more compactly,

$$S_i(\mu) = \phi_i(\mu)\omega(\mu) \tag{7-54b}$$

where $\omega(\mu)$ is either $\alpha(\mu)$ or $\beta(\mu)$. It is possible to construct four different spin orbitals from the two spatial orbitals ϕ_1 and ϕ_2 and the two spin functions α and β. These are

$$\phi_1(\mu)\alpha(\mu) \qquad \phi_1(\mu)\beta(\mu) \qquad \phi_2(\mu)\alpha(\mu) \qquad \text{and} \qquad \phi_2(\mu)\beta(\mu) \tag{7-55}$$

To determine how many different two-electron determinants we can form from these, we use the binomial coefficient

$$\binom{N}{m} = \frac{N!}{m!(N-m)!} \tag{7-56}$$

This gives the number of different ways N things can be combined m at a time. In this particular case we wish to use four spin orbitals two at a time; this leads to six possibilities. These are (introducing a new convenient notation for determinantal wavefunctions)

$$\hat{A}\phi_1(1)\alpha(1)\phi_2(2)\alpha(2) = |\phi_1\phi_2| \qquad \hat{A}\phi_1(1)\beta(1)\phi_2(2)\beta(2) = |\bar{\phi}_1\bar{\phi}_2|$$

$$\hat{A}\phi_1(1)\alpha(1)\phi_2(2)\beta(2) = |\phi_1\bar{\phi}_2| \qquad \hat{A}\phi_1(1)\beta(1)\phi_2(2)\alpha(2) = |\bar{\phi}_1\phi_2|$$

$$\hat{A}\phi_1(1)\alpha(1)\phi_1(2)\beta(2) = |\phi_1\bar{\phi}_1| \qquad \hat{A}\phi_2(1)\alpha(1)\phi_2(2)\beta(2) = |\phi_2\bar{\phi}_2|$$

(7-57)

This new notation for determinants is especially convenient when carrying out algebraic manipulations containing them. Each spatial orbital symbol without an overhead bar, say, ϕ_1 or ϕ_2, stands for a spin orbital with an α spin function; symbols with an overhead bar, say $\bar{\phi}_1$ or $\bar{\phi}_2$, stand for a spin orbital with a β spin function. Note also that the determinantal normalization factor, $(N!)^{-1/2}$, is omitted; it must be resupplied if needed in a calculation.

Approximate wavefunctions—if they are to be of maximum usefulness—must be constructed in such a way that they are eigenfunctions of all the operators which commute with the hamiltonian of the system. Since both the spin operators $\hat{S}_z$ and $\hat{S}^2$ commute with the spin-free hamiltonian [see Eq. (7-18)], we will wish to ensure that all our determinantal functions are eigenfunctions of these operators. This is not automatically assured for all determinants but can always be obtained by proper combinations of determinants.

According to Eq. (7-24) the $\hat{S}_z$ operator for a two-electron system is given explicitly by

$$\hat{S}_z = \hat{S}_z(1) + \hat{S}_z(2) \tag{7-58}$$

where $\hat{S}_z(1)$ operates only on the coordinates of electron 1 and $\hat{S}_z(2)$ operates only on the coordinates of electron 2. When an operator of this form is applied to a determinantal wavefunction such as Eq. (7-30), each of the $N!$ permutations of the Hartree product Φ leads to the same result; thus, it is sufficient to investigate the operation of $\hat{S}_z$ on Φ in order to deduce how $\hat{S}_z\hat{A}\Phi$ behaves. Using $|\phi_1\phi_2|$ as an example, we obtain

$$\begin{aligned}
&[\hat{S}_z(1) + S_z(2)][\phi_1(1)\alpha(1)\phi_2(2)\alpha(2)] \\
&\quad = \phi_1(1)[\hat{S}_z(1)\alpha(1)]\phi_2(2)\alpha(2) + \phi_1(1)\alpha(1)\phi_2(2)]\hat{S}_z(2)\alpha(2)] \\
&\quad = \tfrac{1}{2}\phi_1(1)\alpha(1)\phi_2(2)\alpha(2) + \tfrac{1}{2}\phi_1(1)\alpha(1)\phi_2(2)\alpha(2) \\
&\quad = \phi_1(1)\alpha(1)\phi_2(2)\alpha(2)
\end{aligned} \tag{7-59}$$

Thus $|\phi_1\phi_2|$ is an eigenfunction of $\hat{S}_z$ with an M_S eigenvalue of 1. In like fashion one obtains

$$
\begin{aligned}
&\hat{S}_z|\phi_1\phi_1| = |\phi_1\phi_1| && M_S = 1 \\
&\hat{S}_z|\bar{\phi}_1\bar{\phi}_2| = -|\bar{\phi}_1\bar{\phi}_2| && M_S = -1 \\
&\hat{S}_z|\phi_1\bar{\phi}_2| = 0 && M_S = 0 \\
&\hat{S}_Z|\bar{\phi}_1\phi_2| = 0 && M_S = 0 \\
&\hat{S}_Z|\phi_1\bar{\phi}_1| = 0 && M_S = 0 \\
&\hat{S}_Z|\phi_2\bar{\phi}_2| = 0 && M_S = 0
\end{aligned}
\tag{7-60}
$$

In general, if $\psi(1, 2, \ldots, N)$ is a determinantal wavefunction, then

$$
\begin{aligned}
\hat{S}_z\psi(1, 2, \ldots, N) &= M_S\psi(1, 2, \ldots, N) \\
M_S &= \tfrac{1}{2}(n_\alpha - n_\beta)
\end{aligned}
\tag{7-61}
$$

where n_α is the number of determinantal columns with α spins (no bars) and n_β is the number of determinantal columns with β spins (with bars).

The function of the operator $\hat{S}^2$ is most easily deduced by using the form

$$\hat{S}^2 = \hat{S}_-\hat{S}_+ + \hat{S}_z(\hat{S}_z + 1) \tag{7-62}$$

If we incorporate Eq. (7-61) (and an arbitrary determinantal wavefunction D) the last portion of Eq. (7-62) becomes

$$
\begin{aligned}
\hat{S}_z(\hat{S}_z + 1)D = \hat{S}_z^2 D + \hat{S}_z D &= \tfrac{1}{4}(n_\alpha - n_\beta)^2 D + \tfrac{1}{2}(n_\alpha - n_\beta)D \\
&= \tfrac{1}{4}[(n_\alpha - n_\beta)^2 + 2(n_\alpha - n_\beta)]D
\end{aligned}
\tag{7-63}
$$

The result of the operation $\hat{S}_-\hat{S}_+D$ is a bit harder to deduce. For the many-electron case the operator $\hat{S}_-\hat{S}_+$ can be written

$$\hat{S}_-\hat{S}_+ = \sum_{\mu,\nu}^{N} \hat{S}_-(\mu)\hat{S}_+(\nu) = \sum_{\mu}^{N} \hat{S}_-(\mu)\hat{S}_+(\mu) + \sum_{\mu\neq\nu}^{N} \hat{S}_-(\mu)\hat{S}_+(\nu) \tag{7-64}$$

The first summation in the right-hand expression behaves like a sum of single monoelectronic operators, and the second summation like a sum of single two-electron operators. We first consider the operational behavior of the monoelectronic terms. Antisymmetry is trivial for one-electron operators, so we can write at once

$$\hat{S}_-(\mu)\hat{S}_+(\mu)D = \hat{S}_-(\mu)\hat{S}_+(\mu)\prod_i S_i(\mu) \tag{7-65}$$

Only if the spin orbital $S_i(\mu)$ contains a β spin function does one obtain nonzero terms, since

$$\hat{S}_-(\mu)\hat{S}_+(\mu)\beta(\mu) = \hat{S}_-(\mu)\alpha(\mu) = \beta(\mu) \tag{7-66}$$

Thus

$$\sum_{\mu} \hat{S}_-(\mu)\hat{S}_+(\mu)D = n_\beta D \tag{7-67}$$

Next we consider the two-electron parts of $\hat{S}_-\hat{S}_+$. We note that nonzero terms arise only in the case of operations such as

$$\hat{S}_-(\mu)\hat{S}_+(\nu)\alpha(\mu)\beta(\nu) = \beta(\mu)\alpha(\nu) \tag{7-68}$$

This operation simply interchanges a column of α spins with a column of β spins within the determinant. Thus the general two-electron operation may be written

$$\sum_{\mu \neq \nu}^{N} \hat{S}_-(\mu)\hat{S}_+(\nu)D = \sum_P \hat{P}_{\alpha\beta} \tag{7-69}$$

where $\hat{P}_{\alpha\beta}$ is an operator which interchanges α and β spins in the original determinant. For example:

$$\sum_P \hat{P}_{\alpha\beta}|\phi_1\bar{\phi}_2\phi_3\bar{\phi}_4| = |\bar{\phi}_1\phi_2\phi_3\bar{\phi}_4| + |\phi_1\phi_2\bar{\phi}_3\bar{\phi}_4| + |\bar{\phi}_1\bar{\phi}_2\phi_3\phi_4| + |\phi_1\bar{\phi}_2\bar{\phi}_3\phi_4| \tag{7-70}$$

The overall effect of the operator $\hat{S}_-\hat{S}_+$ is then given by

$$\hat{S}_-\hat{S}_+D = \left(\sum_P \hat{P}_{\alpha\beta} + n_\beta\right) D \tag{7-71}$$

By combining Eqs. (7-63) and (7-71) and rearranging, we see that the overall effect of the operator $\hat{S}^2$ is

$$\hat{S}^2 D = \left\{\sum_P \hat{P}_{\alpha\beta} + \tfrac{1}{4}[(n_\alpha - n_\beta)^2 + 2n_\alpha + 2n_\beta]\right\} D \tag{7-72}$$

Going back to the two-electron functions [Eqs. (7-57)], we see that all except $|\phi_1\bar{\phi}_2|$ and $|\bar{\phi}_1\phi_2|$ are eigenfunctions of $\hat{S}^2$. Applying $\hat{S}^2$ to these two functions, we obtain

$$\begin{aligned} \hat{S}^2|\phi_1\bar{\phi}_2| &= |\bar{\phi}_1\phi_2| + |\phi_1\bar{\phi}_2| \\ \hat{S}^2|\bar{\phi}_1\phi_2| &= |\phi_1\bar{\phi}_2| + |\bar{\phi}_1\phi_2| \end{aligned} \tag{7-73}$$

This means that although neither of the individual determinants is an eigenfunction of $\hat{S}^2$, the two linear combinations $|\phi_1\bar{\phi}_2| \pm |\bar{\phi}\phi_2|$ are eigenfunctions of this operator. These results are summarized in Table 7-1. Note that the eigenvalues of $\hat{S}^2$ are of the general form $S(S+1)$ (in units of $\hbar^2$).

In carrying out algebraic manipulations with determinantal wavefunctions, it is convenient to adopt some additional simplifying notational conventions. Thus, a product such as $\phi_1(1)\alpha(1)\phi_1(2)\beta(2)$ is written "$\phi_1^2\alpha\beta$," and a product such as $\phi_1(1)\alpha(1)\phi_2(2)\alpha(2)$ is written "$\phi_1\phi_2\alpha\alpha$" or "$\phi_1\phi_2\alpha^2$." Note that the terms ϕ_1^2 and α^2 do not represent "squares" in the usual arithmetical or algebraic sense but mean $\phi_1(1)\phi_1(2)$ and $\alpha(1)\alpha(2)$, respectively, instead. The basic convention is that the electron coordinates are associated with the functions in the natural order from left to right. Expanded forms of each function in Table 7-1 are illustrated in Table 7-2 in both the explicit and shorthand versions.

TABLE 7-1
Eigenfunctions of $\hat{S}_z$ and $\hat{S}^2$*

Eigenfunction	Eigenvalue†		
	M_S	$S(S+1)$	
$\lvert\phi_1\phi_2\rvert$	1	2	Components of triplet state
$\lvert\bar{\phi}_1\bar{\phi}_2\rvert$	−1	2	
$\lvert\phi_1\bar{\phi}_2\rvert + \lvert\bar{\phi}_1\phi_2\rvert$‡	0	2	
$\lvert\phi_1\bar{\phi}_2\rvert - \lvert\bar{\phi}_1\phi_2\rvert$‡	0	0	Three different singlet states
$\lvert\phi_1\bar{\phi}_1\rvert$	0	0	
$\lvert\phi_2\bar{\phi}_2\rvert$	0	0	

* Constructed from the determinantal wavefunctions of Eq. (7-57).

† The M_S eigenvalues are in units of $\hbar$; those of $\hat{S}^2$ are in units of $\hbar^2$.

‡ Each function, assuming that ϕ_1 and ϕ_2 are orthonormal, should be multiplied by the normalization constant $1/\sqrt{2}$. This is in addition to the $1/\sqrt{2}$ already understood in each determinant.

In the simple two-electron case considered up to this point, the final wavefunctions can always be factored into a spatial wavefunction and a spin wavefunction, and this spin wavefunction always integrates to unity. In the case of the ground state, this means that no explicit account of spin need be incorporated into the wavefunction. However, even though the spin functions also integrate to unity in the excited states, they generally must be taken into account from the beginning in order to give the correct form to the spatial wavefunction. We will examine this circumstance explicitly in the following discussion. Also, as we will discover (see Exercise 7-10), whenever the system contains more than two electrons, separation of space and spin is no longer possible; expansion of the determinant to its most elemental form always leaves space and spin functions inextricably mixed.

The first three functions in Table 7-2 have the same spatial function, i.e., $\phi_1\phi_2 - \phi_2\phi_1$, but each has a different spin function. Since each of these has the same value of $S(S+1)$, we say these functions form the components of a *multiplet*; in this specific case there are *three* components, which we call a *triplet*. This set of triplets constitutes a threefold *spin degeneracy*. This degeneracy follows from the fact that the energy associated with a wavefunction depends only on the spatial functions. The remaining functions in Table 7-2 all have $S(S+1)$ values of zero, and since they have different spatial portions, they represent three *different* nondegenerate states, called *singlets* (or *singlet states*). The terms "singlet" and "triplet" refer to a general quantity called the *multiplicity*; the multiplicity is given formally by the value $2S+1$. Thus, for a singlet state, $S=0$ and $2S+1=1$; for a triplet state, $S=1$ and $2S+1=3$. A state with $S=\frac{1}{2}$ has a value of $2S+1=2$ and is called a *doublet* (recall from Chap. 5 that all the quantum states of the hydrogen atom are doublets).

It is important to note that the triplet-state wavefunctions all have spatial portions which are antisymmetric and spin functions which are symmetric. By contrast, singlet-state wavefunctions all have spatial portions which are sym-

TABLE 7-2
Expanded forms of the determinantal wavefunctions of Table 7-1

Determinantal form	Expanded form: Explicit	Expanded form: Short version
$\vert\phi_1\phi_2\vert$	$\frac{1}{\sqrt{2}}[\phi_1(1)\phi_2(2)-\phi_2(1)\phi_1(2)]\alpha(1)\alpha(2)$	$\frac{1}{\sqrt{2}}(\phi_1\phi_2-\phi_2\phi_1)\alpha\alpha$
$\vert\bar{\phi}_1\bar{\phi}_2\vert$	$\frac{1}{\sqrt{2}}[\phi_1(1)\phi_2(2)-\phi_2(1)\phi_1(2)]\beta(1)\beta(2)$	$\frac{1}{\sqrt{2}}(\phi_1\phi_2-\phi_2\phi_1)\beta\beta$
$\vert\phi_1\bar{\phi}_2\vert+\vert\bar{\phi}_1\phi_2\vert$	$\frac{1}{\sqrt{2}}[\phi_1(1)\phi_2(2)-\phi_2(1)\phi_1(2)][\alpha(1)\beta(2)+\beta(1)\alpha(2)]$	$\frac{1}{\sqrt{2}}(\phi_1\phi_2-\phi_2\phi_1)(\alpha\beta+\beta\alpha)$
$\vert\phi_1\bar{\phi}_2\vert-\vert\bar{\phi}_1\phi_2\vert$	$\frac{1}{\sqrt{2}}[\phi_1(1)\phi_2(2)+\phi_2(1)\phi_1(2)][\alpha(1)\beta(2)+\beta(1)\alpha(1)]$	$\frac{1}{\sqrt{2}}(\phi_1\phi_2-\phi_2\phi_1)(\alpha\beta-\beta\alpha)$
$\vert\phi_1\bar{\phi}_1\vert$	$\frac{1}{\sqrt{2}}\phi_1(1)\phi_1(2)[\alpha(1)\beta(2)-\beta(1)\alpha(2)]$	$\frac{1}{\sqrt{2}}\phi_1^2(\alpha\beta-\beta\alpha)$
$\vert\phi_2\bar{\phi}_2\vert$	$\frac{1}{\sqrt{2}}\phi_2(1)\phi_2(2)[\alpha(1)\beta(2)-\beta(1)\alpha(2)]$	$\frac{1}{\sqrt{2}}\phi_2^2(\alpha\beta-\beta\alpha)$

metric and spin functions which are antisymmetric. Thus one of the roles played by explicit incorporation of spin is to impart the correct symmetry to the spatial portion of the wavefunction; for ground states this occurs naturally anyway, but this is not true for the triplet states.

Either of the last two functions in Table 7-2 can be used to describe the ground state of a two-electron atom. Thus if ϕ_1 (or ϕ_2) is a $1s$ orbital, we obtain the ground-state wavefunction of the helium atom, namely,

$$\psi(1,2) = |1s\,\overline{1s}| = \frac{1}{\sqrt{2}}\,1s(1)1s(2)[\alpha(1)\beta(2) - \beta(1)\alpha(2)] = \frac{1s^2(\alpha\beta - \beta\alpha)}{\sqrt{2}} \tag{7-74}$$

If $\hat{G}$ is a spin-free operator, its expectation value for a system having the above wavefunction is given by

$$\langle \hat{G} \rangle = \langle \psi | \hat{G} | \psi \rangle = \langle 1s\,1s | \hat{G} | 1s\,1s \rangle \tfrac{1}{2} \langle \alpha\beta - \beta\alpha \,|\, \alpha\beta - \beta\alpha \rangle \tag{7-75}$$

Note that the spin functions end up separated from the spatial functions and thus are integrated over separately. This integration over the spin portion is as follows

$$\begin{aligned}\tfrac{1}{2}\langle \alpha\beta - \beta\alpha \,|\, \alpha\beta - \beta\alpha \rangle &= \tfrac{1}{2}(\langle \alpha | \alpha \rangle \langle \beta | \beta \rangle - 2\langle \alpha | \beta \rangle \langle \beta | \alpha \rangle + \langle \beta | \beta \rangle \langle \alpha | \alpha \rangle) \\ &= \tfrac{1}{2}(1 - 0 + 1) = 1 \end{aligned} \tag{7-76}$$

Thus, the spin portion integrates to unity and the expectation value of $\hat{G}$ is simply

$$\langle \hat{G} \rangle = \langle 1s(1)1s(2) | \hat{G} | 1s(1)1s(2) \rangle \tag{7-77}$$

Note that this is just the form we used in Chap. 6 for variational and perturbation calculations on the helium atom.

EXERCISES

7-6. Show that if the spatial orbitals ϕ_1 and ϕ_2 are orthogonal, the six determinantal functions in Table 7-1 are also mutually orthogonal.

7-7. Verify the $\hat{S}^2$ eigenvalues of $|\phi_1\phi_2|$ and $|\phi_1\bar{\phi}_1|$.

7-8. Show that the three different spin functions of the triplet state ($\alpha\alpha$, $\beta\beta$, and $\alpha\beta + \beta\alpha$) are linearly independent. Are the three components of the triplet state *necessarily* orthogonal? Explain.

7-9. Use Eq. (7-72) to verify that $|\phi_1\bar{\phi}_2|$ and $|\bar{\phi}_1\phi_2|$ are not eigenfunctions of $\hat{S}^2$ but that the sum and the difference of these two functions are.

7-10. An approximate wavefunction for the ground state of the lithium atom is $|1s\,\overline{1s}\,2s|$.

(*a*) Show that one cannot separate the spatial and spin functions in this wavefunction as was possible for helium.

(*b*) Use Eq. (7-72) to show this is a doublet state $[S(S+1) = \tfrac{3}{4}]$.

(*c*) What is the other wavefunction describing this doublet state?

(*d*) What are the M_S values of each doublet component?

7-11. Show that the spatial portion of the fourth wavefunction in Table 7-2 has the same general form as the split-shell wavefunction used for helium in Chap. 6 [see Eq. (6-81)].

7-12. Obtain the M_S eigenvalues of each of the following determinantal functions:

(a) $|\phi_1\bar{\phi}_1\phi_2\bar{\phi}_2\phi_3\bar{\phi}_4|$ (d) $|\phi_1\phi_2\phi_3\phi_4|$

(b) $|\phi_1\phi_2|$ (e) $|\phi_1\bar{\phi}_2\phi_3\bar{\phi}_4|$

(c) $|\bar{\phi}_1\bar{\phi}_2|$

Which of the above are eigenfunctions of $\hat{S}^2$, and what is the multiplicity of each of these?

7-13. Are the wavefunctions $|1s\,\overline{1s}|$ and $|1s\,\overline{2s}| - |\overline{1s}\,2s|$ orthogonal when $\langle 1s|2s\rangle \neq 0$? Repeat for $|1s\,\overline{1s}|$ and $|1s\,\overline{2s}| + |\overline{1s}\,2s|$. Why are the two cases different?

7-14. Given $D_1 = |\phi_1\bar{\phi}_2|$ and $D_2 = |\bar{\phi}_1\phi_2|$:

(a) Evaluate the integral $\langle D_1|D_2\rangle$ if the basis orbitals are normalized but not orthogonal (let $\langle\phi_1|\phi_2\rangle = \Delta$).

(b) Normalize $D_1 - D_2$ and $D_1 + D_2$.

(c) Show that $\langle D_1 - D_2|D_1 + D_2\rangle = 0$.

7-15. Normalize $D = |\phi_1\bar{\phi}_1\phi_2|$ assuming that the basis functions are normalized but not orthogonal. Let Δ represent the overlap integral of the two basis functions.

7-5 THE HELIUM ATOM REVISITED

It is instructive to present a brief reformulation of the helium atom calculations of Secs. 6-3 and 6-6 in terms of our recently introduced determinantal notation. Then, we can illustrate the application of this approach to some of the excited states of helium and show why the determinantal form is necessary in order to describe these states correctly.

We will designate the spatial part of the spin orbitals used to describe the ground state of helium by 1s but with the understanding that this orbital is not necessarily of hydrogenlike form. In a subsequent chapter we will consider the general problem of finding the *best* form to use for this orbital, and we will discover that it is *not* the hydrogenlike form. In any event, the orbital approximation wavefunction for the ground state of helium is

$$\psi(1,2) = \frac{1}{\sqrt{2}}\begin{vmatrix} 1s(1)\alpha(1) & 1s(1)\beta(1) \\ 1s(2)\alpha(2) & 1s(2)\beta(2) \end{vmatrix} = |1s\,\overline{1s}| \tag{7-78}$$

This describes the so-called $1s^2$ configuration of helium; thus the electron configurations first learned in high school or the freshman year of college serve as mnemonic indicators for the antisymmetric wavefunctions in determinantal form. As illustrated in Eqs. (7-74) through (7-77), the spin portions integrate out separately to unity when we are evaluating expectation values of spin-free operators, so that the total energy becomes simply

$$\langle\hat{H}\rangle = \left\langle 1s(1)1s(2)\middle| h_1 + h_2 + \frac{1}{r_{12}} \middle| 1s(1)1s(2)\right\rangle \tag{7-79}$$

As shown in Eq. (6-76), the energy may be put into the form

$$\langle \hat{H} \rangle = \varepsilon = 2\epsilon_{1s} + J_{1s,\,1s} \tag{7-80}$$

where we made use of the fact that $\epsilon(1) = \epsilon(2) = \epsilon_{1s}$ and employed the new notation

$$J_{ij} = \left\langle \phi_i(1)\phi_j(2) \left| \frac{1}{r_{12}} \right| \phi_i(1)\phi_j(2) \right\rangle \tag{7-81}$$

The integral J_{ij} is called a *coulombic integral*; it represents the electrostatic repulsion between two charge distributions $|\phi_i(1)|^2$ and $|\phi_j(2)|^2$.

In summary, nothing new has been introduced (except notation) in this treatment of the ground state of helium. Next, we look at some of the low-lying excited states of this atom.

The wavefunctions which describe the lowest-lying excited states of helium (or any other two-electron species) have their general forms illustrated in Table 7-1. These functions are all those formed from the configuration $1s2s$, or, in terms of the general orbitals used in the table, $\phi_1\phi_2$. Using the $1s$ and $2s$ designations, there are *four* wavefunctions describing these states:

$$\frac{1}{\sqrt{2}}(|1s\,\overline{2s}| - |\overline{1s}\,2s|) \qquad \frac{1}{\sqrt{2}}(1|1s\,\overline{2s}| + |\overline{1s}\,2s|)$$
$$|1s\,2s| \qquad |\overline{1s}\,\overline{2s}| \tag{7-82}$$

Note that the factor $1/\sqrt{2}$ in two of these functions is in addition to a like factor implied in the determinants themselves. The first is a singlet state [$S = 0$, $S(S+1) = 0$, $2S + 1 = 1$], and the other three are the $M_S = 0$, -1, and 1 components of a triplet state [$S = 1$, $S(S+1) = 2$, $2S + 1 = 3$]. As illustrated in Table 7-2, the spatial part of the excited singlet state is

$$\frac{1}{\sqrt{2}}[1s(1)2s(2) + 2s(1)1s(2)] \tag{7-83}$$

Similarly, the three triplet components all have the spatial portion

$$\frac{1}{\sqrt{2}}[1s(1)2s(2) - 2s(1)1s(2)] \tag{7-84}$$

The algebra of determining the final energy expression can be made considerably simpler by using some notational aids. Let us introduce the definitions:

$$u = 1s(1)2s(2) \qquad v = 2s(1)1s(2) \tag{7-85}$$

Also, let us introduce the symbol "g" to replace the electron-repulsion operator $1/r_{12}$. The energy of the singlet state is then evaluated as follows:

$$\begin{aligned}\varepsilon\ (\text{singlet}) &= \tfrac{1}{2}\langle u + v | h_1 + h_2 + g | u + v \rangle \\ &= \langle u|h_1|u\rangle + \langle v|h_1|v\rangle + \langle u|h_1|v\rangle + \langle v|h_1|u\rangle \\ &\quad + \langle u|h_2|u\rangle + \langle v|h_2|v\rangle + \langle u|h_2|v\rangle + \langle v|h_2|u\rangle \\ &\quad + \langle u|g|u\rangle + \langle v|g|v\rangle + \langle u|g|v\rangle + \langle v|g|u\rangle\end{aligned} \tag{7-86}$$

The presence of dummy variables of integration leads to considerable symmetry; i.e., integrals such as $\langle u|h_1|u\rangle$ and $\langle v|h_2|v\rangle$ are equal. Summarizing these identities:

$$\langle u|h_1|u\rangle = \langle v|h_2|v\rangle = \epsilon(1s) \qquad \langle u|h_2|u\rangle = \langle v|h_1|v\rangle = \epsilon(2s)$$
$$\langle u|g|u\rangle = \langle v|g|v\rangle = J_{1s,\,2s} \qquad \langle u|g|v\rangle = \langle v|g|u\rangle = K_{1s,\,2s} \tag{7-87}$$

The latter integral is called an *exchange integral*. Its general form is

$$K_{ij} = \left\langle \phi_i(1)\phi_j(2)\left|\frac{1}{r_{12}}\right|\phi_j(1)\phi_i(2)\right\rangle \tag{7-88}$$

This integral, like the coulombic integral, represents an electrostatic repulsion between a pair of electrons; one electron has the probability density $\phi_i^*(1)\phi_j(1)$, and the other $\phi_j^*(2)\phi_i(2)$. This integral has no simple physical interpretation; it is a mathematical consequence of antisymmetry and the orbital product form of the wavefunction.

The total energy may now be written

$$\begin{aligned}\varepsilon\ (\text{singlet}) &= \tfrac{1}{2}[2\epsilon(1s) + 2\epsilon(2s) + 2J_{1s,\,2s} + 2K_{1s,\,2s}] \\ &= \varepsilon(1s) + \epsilon(2s) + J_{1s,\,2s} + K_{1s,\,2s}\end{aligned} \tag{7-89}$$

The calculation of the triplet energy is almost identical; all one need do is change a few arithmetical signs to obtain

$$\varepsilon\ (\text{triplet}) = \epsilon(1s) + \epsilon(2s) + J_{1s,\,2s} - K_{1s,\,2s} \tag{7-90}$$

The actual numerical values for the energies will depend on just what specific forms we choose for the 1s and 2s orbitals. If hydrogenlike orbitals are used (in unscaled form), one obtains for the various energy terms

$$\begin{aligned}\epsilon(1s) &= -\frac{Z^2}{2} = -2.00E_\text{h} \\ \epsilon(2s) &= -\frac{Z^2}{8} = -0.50E_\text{h} \\ J_{1s,\,2s} &= \frac{17Z}{81} = 0.420E_\text{h} \\ K_{1s,\,2s} &= \frac{16Z}{729} = 0.0439E_\text{h}\end{aligned} \tag{7-91}$$

One can also show the general result that the triplet state always lies below the singlet state—a prediction which is in accord with experiment. Using the fact that

$$J_{ij} \geq K_{ij} \geq 0 \tag{7-92}$$

[see Eq. (A5-136) and following material (App. 5)] shows the general result

$$\varepsilon\ (\text{singlet}) - \varepsilon\ (\text{triplet}) = 2K_{1s,\,2s} \geq 0 \tag{7-93}$$

Numerically, this calculation produces $0.088E_h$ as the singlet-triplet energy separation; the experimental value is $0.029E_h$.

The foregoing discussion illustrates (for one specific example) that the inclusion of antisymmetry in electronic wavefunctions leads to a qualitatively correct ordering of singlet and triplet states in the helium atom. The reader should note that had antisymmetry not been taken into account, the singlet and triplet states would have been predicted to have the same energies, or, more correctly, it would not have been possible to identify separately these two states, since the wavefunctions would not be eigenfunctions of $\hat{S}^2$. Similar agreements are obtained for various spectroscopic states of other atoms.

EXERCISES

7-16. Repeat the calculation of the energy of the excited singlet state of helium assuming that the $1s$ and $2s$ orbitals are not orthogonal, and show that the total energy now has the form.

$$\varepsilon\ (\text{singlet}) = (1+\Delta)^{-1}[\epsilon(1s) + \epsilon(2s) + 2\Delta\langle 1s|h|2s\rangle + J_{1s,\,2s} + K_{1s,\,2s}]$$

[Compare this with Eq. (7-89).] Also, determine the corresponding triplet state energy expression, and compare with Eq. (7-90).

7-17. Show that if $\hat{G}$ is any spin-free operator which is expressible as a sum of one-electron operators, then the expectation values of this operator are given by $\langle\hat{G}\rangle = \Sigma\,(-1)^p\langle\Phi|\hat{G}|\hat{P}\Phi\rangle = \langle\Phi|\hat{G}|\Phi\rangle$ (summation over permutations) which indicates that antisymmetrization of the wavefunctions is immaterial in this case.

7-18. Use the wavefunction $|1s\,\overline{1s}\,2s|$ for the ground state of the lithium atom, and show that the energy has the general form $\varepsilon = 2\epsilon(1s) + \epsilon(2s) + J_{1s,\,1s} + 2J_{1s,\,2s} - K_{1s,\,2s}$. Assume orthogonality of the basis functions.

7-19. Calculate the singlet-triplet energy difference for the two lowest excited states of helium using *scaled* wavefunctions, and show that the results satisfy the quantum mechanical virial theorem. Although it would be better to use separate scale factors for the $1s$ and $2s$ orbitals, this leads to considerable algebraic complexity; thus assume that the two orbitals have the same scale factors. The general expressions for the coulombic and exchange integrals are just those given in Eq. (7-91) but with the scale factor η replacing Z. The general expression for the orbital energies of scaled hydrogenlike orbitals is

$$\epsilon(ns) = +\frac{\eta^2/2 - Z\eta}{n^2}$$

Note that the two states have different scaling parameters. The fact that scaling improves the energy separation so little indicates that the wavefunctions used are far too simple in form to provide adequate descriptions of these two excited states. The energies could be improved somewhat by using different scale factors for $1s$ and $2s$ and also by using orbitals other than the hydrogenlike orbitals.

7-20. The Eckart wavefunction for helium [see Eq. (6-81)] can be generalized to the following form:

$${}^{1,3}\psi = [2(1 \pm \Delta^2)]^{-1/2}(D_1 \mp D_2)$$

where the upper signs of operation refer to the singlet state (${}^1\psi$) and the lower signs of operation indicate the triplet state (${}^3\psi$); $D_1 = |1s\,\overline{1s'}|$ and $D_2 = |\overline{1s}\,1s'|$.

(*a*) Show that the singlet and triplet energies have the form

$$E_{\pm} = (1 \pm \Delta^2)^{-1}(\epsilon + \epsilon' \pm 2\Delta\beta_1 + J \pm K)$$

where $\Delta = \langle 1s|1s' \rangle = \dfrac{8(\alpha\beta)^{3/2}}{(\alpha+\beta)^3}$

$$\epsilon = \langle 1s|h|1s \rangle = \frac{\alpha^2}{2} - 2\alpha$$

$$\epsilon' = \langle 1s'|h|1s' \rangle = \frac{\beta^2}{2} - 2\beta$$

$$\beta_1 = \langle 1s|h|1s' \rangle = \langle 1s'|h|1s \rangle = \frac{4(\alpha\beta)^{3/2}}{(\alpha+\beta)^2}\left(\frac{\alpha\beta}{\alpha+\beta} - 2\right)$$

$$J = \langle 1s\,1s'|g|1s\,1s' \rangle = \alpha\beta^3\left[\frac{1}{\alpha\beta^2} - \frac{1}{(\alpha+\beta)^3} - \frac{1}{\alpha(\alpha+\beta)^2}\right]$$

$$K = \langle 1s\,1s'|g|1s'\,1s \rangle = \frac{20(\alpha\beta)^3}{(\alpha+\beta)^5}$$

The α and β terms are the scale parameters for the orbitals 1*s* and 1*s*′, respectively. These orbitals are given explicitly as

$$1s = \frac{\alpha^{3/2}}{\sqrt{\pi}}e^{-\alpha r} \qquad 1s' = \frac{\beta^{3/2}}{\sqrt{\pi}}e^{-\beta r}$$

(*b*) The scale parameters which minimize the singlet and triplet energies are as follows: (singlet) $\alpha = 1.19$, $\beta = 2.18$; (triplet) $\alpha = 0.32$, $\beta = 1.97$. Calculate the energies of each of these states, and compare your results with the experimental values of $-2.901E_h$ and $-2.176E_h$ for the singlet and triplet, respectively.

SUGGESTED READINGS

Daudel, R., R. Lefebvre, and C. Moser: *Quantum Chemistry: Methods and Applications*, Interscience, New York, 1959. Chapter 19 is especially useful.

Eyring, H., J. Walter, and G. E. Kimball: *Quantum Chemistry*, Wiley, New York, 1944. See especially chapter 9.

Kauzmann, W.: *Quantum Chemistry*, Academic, New York, 1957. See especially chapter 9, section D.

Slater, J. C.: *Quantum Theory of Atomic Structure*, vols. 1 and 2, McGraw-Hill, New York, 1960.

CHAPTER 8

THE QUANTUM STATES OF ATOMS

This chapter is an extension, to many-electron systems, of Sec. 5-4 in which the quantum states of hydrogen are described in terms of spin and orbital angular momentum quantities. Owing largely to the effects of electron repulsion (absent in hydrogen), some modifications of the earlier treatment must be made when dealing with atoms of two or more electrons. In the next chapter we will discuss the general principles and techniques needed if one is to organize and carry out explicit numerical calculations on the electronic states of atoms.

8-1 ORBITAL ANGULAR MOMENTA IN MANY-ELECTRON ATOMS

According to classical mechanics, the total orbital angular momentum of a system of N particles about a point is given by the vector sum

$$\mathbf{L} = \sum_{j=1}^{N} \mathbf{L}_j = L_x\mathbf{i} + L_y\mathbf{j} + L_z\mathbf{k} \tag{8-1}$$

where $\mathbf{L}_j$ is the orbital angular momentum of the jth particle about the given point. The components of the total orbital angular momentum are defined by

$$L_x = \sum_{j=1}^{N} L_{xj} \qquad L_y = \sum_{j=1}^{N} L_{yj} \qquad L_z = \sum_{j=1}^{N} L_{zj} \tag{8-2}$$

and the scalar square of **L** is given by

$$L^2 = \mathbf{L}\cdot\mathbf{L} = L_x^2 + L_y^2 + L_z^2 \tag{8-3}$$

Equations (8-1) and (8-3) are seen to be identical in form with the corresponding equations for the one-particle case (see Sec. 4-1); the quantum mechanical operators satisfy analogous equations. The z-component of the orbital angular momentum of an N-particle system is described by the operator

$$\hat{L}_z = \sum_{\mu=1}^{N} \hat{L}_{z\mu} = \frac{1}{i}\sum_{\mu=1}^{N} \frac{\partial}{\partial \varphi_\mu} \tag{8-4}$$

where φ_μ is the polar angle of the μth particle relative to a fixed coordinate system.

Referring to Eq. (8-3) and writing $\hat{L}_x$ and $\hat{L}_y$ in a form similar to Eq. (8-4), we find that the square of the total orbital angular momentum of an N-electron atom is represented by the operator

$$\begin{aligned}\hat{L}^2 &= \sum_{\mu,\nu}^{N} (\hat{L}_{x\mu}\hat{L}_{x\nu} + \hat{L}_{y\mu}\hat{L}_{y\nu} + \hat{L}_{z\mu}\hat{L}_{z\nu}) \\ &= \sum_{\mu}^{N} (\hat{L}_{x\mu}^2 + \hat{L}_{y\mu}^2 + \hat{L}_{z\mu}^2) + \sum_{\mu\neq\nu}^{N} (\hat{L}_{x\mu}\hat{L}_{x\nu} + \hat{L}_{y\mu}\hat{L}_{y\nu} + \hat{L}_{z\mu}\hat{L}_{z\nu})\end{aligned} \tag{8-5}$$

Note that Eq. (7-64) for $\hat{S}_-\hat{S}_+$ is of this form.

The nonrelativistic hamiltonian operator for an N-electron atom will now be written

$$\hat{H} = \hat{H}_1 + \hat{H}_{12} \tag{8-6}$$

where $\hat{H}_1$ is a sum of the monoelectronic operators h_μ and $\hat{H}_{12}$ is the sum of the two-electron repulsion operators $1/r_{\mu\nu}$. We begin by investigating the commutation relations among the operators $\hat{L}_x$, $\hat{L}_y$, $\hat{L}_z$, $\hat{L}^2$, and $\hat{H}$ for the N-electron system. From Eqs. (8-1) and (8-3) it is readily seen that the orbital angular momentum operators commute among themselves just as in the one-electron case. We also note that $\hat{L}_z$ and $\hat{L}^2$ commute with any hamiltonian which is of the *central-field* type,[1] e.g., the hydrogen atom hamiltonian. However, the presence of the interelectronic coordinates in the many-electron atoms means that, in general, the hamiltonian is *not* of the central-field type; hence, we must examine the commutation of $\hat{L}_z$ and $\hat{L}^2$ with $\hat{H}$ in the general case. Referring to Eq. (8-6) we see that $\hat{H}_1$ is of the central-field type and thus each term of $\hat{L}_z$ commutes with it; consequently $\hat{L}_z$ itself commutes with $\hat{H}_1$. Next we examine the commutation of $\hat{L}_z$ with the operator $\hat{H}_{12}$. To do this we consider the

[1] A hamiltonian is said to be of the central-field type if the potential-energy term depends only on the electron coordinates r_μ and not on the interelectronic coordinates $|\mathbf{r}_\mu - \mathbf{r}_\nu| = r_{\mu\nu}$.

operation of $\hat{L}_z$ on a two-electron repulsion operator

$$\hat{L}_z(r_{\mu\nu})^{-1} = \frac{1}{i}\sum_{\rho}\frac{\partial}{\partial\varphi_\rho}(r_{\mu\nu})^{-1} = (r_{\mu\nu})^{-1}\hat{L}_z + \frac{1}{i}\left[\frac{\partial(r_{\mu\nu})^{-1}}{\partial\varphi_\mu} + \frac{\partial(r_{\mu\nu})^{-1}}{\partial\varphi_\nu}\right] \tag{8-7}$$

From Fig. 8-1 we see that the distance $r_{\mu\nu}$ depends only on the angle $\chi = |\varphi_\mu - \varphi_\nu|$ so that

$$\frac{\partial\chi}{\partial\varphi_\mu} = -\frac{\partial\chi}{\partial\varphi_\nu} \tag{8-8}$$

and, therefore,

$$\frac{\partial(1/r_{\mu\nu})}{\partial\varphi_\mu} = -\frac{\partial(1/r_{\mu\nu})}{\partial\varphi_\nu} \tag{8-9}$$

Equation (8-7) then reduces to

$$\left[\hat{L}_z, \frac{1}{r_{\mu\nu}}\right] = 0 \tag{8-10}$$

which shows that $\hat{L}_z$ commutes with $\hat{H}$. In a similar fashion it is shown that $\hat{L}_z^2$ also commutes with $\hat{H}$. Repeating for the remaining x and y components and their squares shows that these also commute with $\hat{H}$ and, consequently, so does the operator $\hat{L}^2$. In summary, the total orbital angular momentum and one of its components are constants of the motion for the many-electron atom.

Although $\hat{L}_z$ does commute with the electron-repulsion operator $1/r_{\mu\nu}$, it is readily shown that the z component of an *individual* electron's orbital angular momentum is not a constant of the motion in the approximation we are considering above. This result follows from the relationship

$$\hat{L}_{z\mu}\frac{1}{r_{\mu\nu}} = \frac{1}{r_{\mu\nu}}\hat{L}_{z\mu} + \frac{1}{i}\frac{\partial(1/r_{\mu\nu})}{\partial\varphi_\mu} \tag{8-11}$$

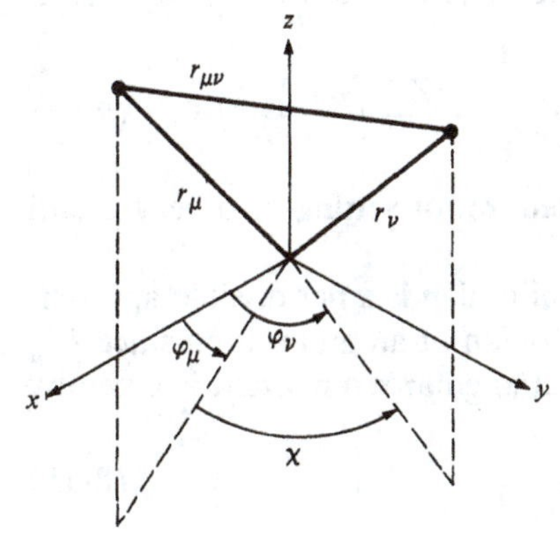

FIGURE 8-1
The interelectronic distance as a function of the difference of the polar angles of the two electrons.

and from the fact that $r_{\mu\nu}$ depends on the angle χ shown in Fig. 8-1. For this reason it is not physically meaningful to assign definite values of the magnetic quantum number m_l to individual electrons whenever electron-repulsion terms are explicitly included in the atomic hamiltonian. It then follows that azimuthal quantum numbers for individual electrons also have no sharply defined physical significance. The presence of the electron repulsions may be viewed as a *torque* on individual electrons, so that their orbital angular momenta vary with time. This torque is, of course, due to the non-central-field nature of the electron repulsions.

Below are various eigenfunction and eigenvalue relationships for the orbital angular momentum operators for both the one-electron and the many-electron cases.

$$\begin{array}{ll} \hat{L}_z Y_{l,\,m_l} = m_l Y_{l,\,m_l} & \hat{L}_z Y_{L,\,M_L} = M_L Y_{L,\,M_L} \\ \hat{L}^2 Y_{l,\,m_l} = l(l+1) Y_{l,\,m_l} & \hat{L}^2 Y_{L,\,M_L} = L(L+1) Y_{L,\,M_L} \\ -l \le m_l \le l & -L \le M_L \le L \\ l = 0, 1, 2, 3, \ldots & L = 0, 1, 2, 3, \ldots \\ \quad = s, p, d, f, \ldots & \quad = S, P, D, F, \ldots \\ |\mathbf{L}| = \sqrt{l(l+1)} & |\mathbf{L}| = \sqrt{L(L+1)} \end{array} \tag{8-12}$$

The functions $Y_{l,\,m_l}$ and $Y_{L,\,M_L}$ are the spherical harmonics (depending on the angles θ and φ). The many-electron quantum numbers L and M_L are the counterparts of the one-electron quantum numbers l and m_l, respectively.

It is sometimes convenient to replace the correct nonrelativistic hamiltonian of a many-electron atom with an approximate hamiltonian known as the *central-field hamiltonian*. This is defined by

$$\hat{H}^{\text{cf}} = \sum_{\mu=1}^{N} \left[-\frac{\nabla_\mu^2}{2} + V(r_\mu) \right] \tag{8-13}$$

where $V(r_\mu)$ is an "average," or *effective*, many-electron field in which the μth electron moves as a *quasi-independent* particle. The potential $V(r_\mu)$ is chosen such that

$$\lim_{r_\mu \to 0} V(r_\mu) = -\frac{Z}{r_\mu} \qquad \lim_{r_\mu \to \infty} V(r_\mu) = -\frac{Z - N + 1}{r_\mu} \tag{8-14}$$

In Chap. 10 we will discuss a systematic procedure for setting up a very useful model for such effective fields.

If it is assumed that the central-field hamiltonian is a permissible approximation of the nonrelativistic electronic hamiltonian of an atom, then since $\hat{L}_{z\mu}$ commutes with $\hat{H}^{\text{cf}}$, the total electronic orbital angular momentum is given by

$$\mathbf{L} = \sum_{\mu=1}^{N} \mathbf{L}_\mu \tag{8-15}$$

where $\mathbf{L}_\mu$ is the orbital angular momentum of the μth electron, that is,

$$|\mathbf{L}_\mu| = \sqrt{l_\mu(1_\mu + 1)} \tag{8-16}$$

This implies that we may regard the orbital angular momentum of an individual electron as approximating a constant of the motion (assuming the central-field approximation is valid in the first place). This approximation is justified so long as the torque on each electron due to mutual repulsions can be treated in an average, collective way. The many-electron quantum number L is now given by all the *positive* numbers

$$L = \sum_\mu^N l_\mu, \sum_\mu^N l_\mu - 1, \sum_\mu^N l_\mu - 2, \ldots \tag{8-17}$$

where l_μ is the orbital angular momentum quantum number of the μth electron. For a two-electron atom having the quantum numbers l_1 and l_2, Eq. (8-17) leads to

$$L = l_1 + l_2, l_1 + l_2 - 1, \ldots, |l_1 - l_2| \geq 0 \tag{8-18}$$

For example, if $l_1 = 2$ and $l_2 = 1$, then L can have the values 3, 2, or 1. The central-field approximation also gives the quantum number M_L by

$$M_L = \sum_\mu^N m_{l_\mu} \tag{8-19}$$

where m_{l_μ} is the magnetic quantum number of the μth electron. Obviously, M_L is the projection of $\mathbf{L}$ on an arbitrary axis, just as m_{l_μ} is the projection of $\mathbf{L}_\mu$ on an arbitrary axis. If we let $\mathbf{e}$ represent a unit vector in the direction of $\mathbf{L}$, we can write

$$\mathbf{L} = \sqrt{L(L+1)}\mathbf{e} \tag{8-20}$$

The projections of $\mathbf{L}$ on an arbitrary axis are thus given by $-L, -L + 1, \ldots, 0, 1, \ldots, L - 1$, and L (see Fig. 8-2). Note that this diagram illustrates an uncertainty relationship for $\mathbf{L}$ and its direction, that is, $\Delta L \, \Delta\varphi \geq \frac{1}{2}\hbar$.

If there is a torque acting on the individual electrons, the l_μ are no longer meaningful quantum numbers, since the vectors $\mathbf{L}_\mu$ change their directions with time. This may be viewed (see Fig. 8-3) as a precessional motion of the individual $\mathbf{L}_\mu$ about a resultant $\mathbf{L}$, much like the precession of a gyroscope about its vertical axis due to gravitational torque. The velocity of this precession varies directly as the torque, i.e., as the strength of the interactions among the electrons, and may approach the individual electron velocities. In such an extreme case the individual $\mathbf{L}_\mu$ lose their meaning entirely. It then happens that atomic states with different values of $\mathbf{L}$ have different energies, and this energy difference is a direct measure of the electronic interactions.

In the next section we will show that there is a torque acting on $\mathbf{L}$ so that not even the total orbital angular momentum is a constant of the motion. This torque—the resut of a coupling with the spin angular momentum—provides additional complexity to the quantum states of atoms.

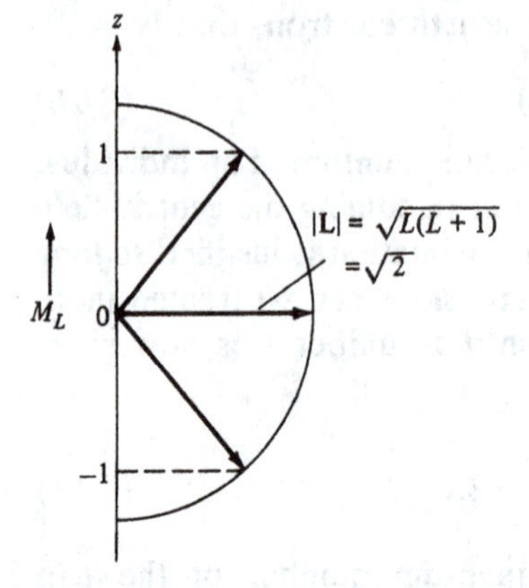

FIGURE 8-2
Projection of the total orbital angular momentum upon an arbitrary axis.

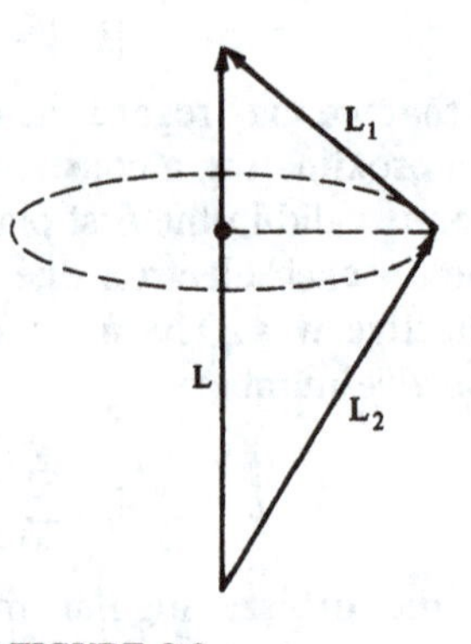

FIGURE 8-3
The precession of individual electron orbital angular momenta $\mathbf{L}_1$ and $\mathbf{L}_2$ about their resultant $\mathbf{L}$ due to interaction between the two electrons.

8-2 SPIN-ORBIT COUPLING IN MANY-ELECTRON ATOMS

In the simple case of the one-electron atom, there is a coupling between the orbital angular momentum **L** and the spin angular momentum **S** to produce a total angular momentum **J**. The coupling of **L** and **S** means that neither **L** nor **S** is a constant of the motion, since both must precess about their resultant **J** (see Fig. 8-4).

The torque which causes the precession of **L** and **S** arises from the fact that the energy of an electron depends on its orbital angular momentum and its magnetic moment. The apparent motion of the charged nucleus about the electron (if one considers a particular electron to be at rest) is responsible for the existence of a magnetic field acting upon the electron. According to

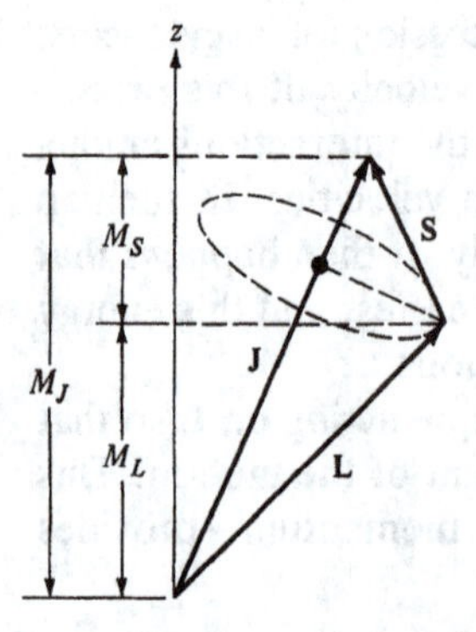

FIGURE 8-4
The precession of **L** and **S** of a light atom about the total angular momentum **J**.

classical electrodynamics, the energy of an electron with magnetic moment $\boldsymbol{\mu}$ interacting with a magnetic field $\mathbf{H}$ is given by

$$E^{\xi} = -\mathbf{H} \cdot \boldsymbol{\mu} \tag{8-21}$$

According to the Dirac relativistic theory of the electron, the magnetic moment of the electron is related to the spin angular momentum by the expression

$$\boldsymbol{\mu} = \frac{e}{mc} \mathbf{S} \tag{8-22}$$

The classical expression for the magnetic field at the position of the electron due to a charged nucleus moving about the electron is

$$\mathbf{H} = \frac{1}{c} \mathbf{E} \times \boldsymbol{v} = \frac{1}{mc} \mathbf{E} \times \mathbf{p} \tag{8-23}$$

where $\mathbf{E}$ = electric field in which the charged nucleus moves, i.e., the field arising from the nucleus
$\boldsymbol{v}$ = velocity of the charged nucleus
$\mathbf{p}$ = linear momentum of nucleus

In the special case in which the force acting on the electron is derivable from a central-field potential $\phi = \phi(r)$, the electric field is given by

$$\mathbf{E} = -\frac{1}{r} \mathbf{r} \frac{d\phi}{dr} \tag{8-24}$$

so that the magnetic field associated with the electron is

$$\mathbf{H} = -\frac{1}{mcr} \frac{d\phi}{dr} \mathbf{r} \times \mathbf{p} = -\frac{1}{mcr} \frac{d\phi}{dr} \mathbf{L} \tag{8-25}$$

The potential of an electron in the field of the nucleus is given by

$$\phi = -\frac{V}{e} = \frac{-Ze^2K}{-er} = \frac{ZeK}{r} \tag{8-26}$$

so that the magnetic field becomes

$$\mathbf{H} = \frac{ZeK}{mcr^3} \mathbf{L} \tag{8-27}$$

We now need to compute the field produced by an electron not at rest but moving about a stationary nucleus; this requires a transformation from a frame moving with the electron to a frame at rest and requires that Eq. (8-27) be multiplied by a relativistic factor of $\frac{1}{2}$.[2] Substituting Eqs. (8-22) and (8-27) into Eq. (8-21), one obtains the interaction energy

$$E^{\xi} = \frac{Ze^2K}{2m^2c^2r^3} \mathbf{L} \cdot \mathbf{S} \tag{8-28}$$

[2] L. H. Thomas, *Nature*, 117:514 (1926) and J. Frenkel, *Z. Physik.*, 37:243 (1926).

This may be regarded as arising from interaction of the spin angular momentum and the orbital angular momentum of the electron, since the former arises from the electron's magnetic moment and the latter arises from the relative orbital motion of the electron about the nucleus. The quantum mechanical operator describing the spin-orbit interaction then becomes

$$\hat{H}^{\xi} = \frac{Ze^2K}{2m^2c^2r^3}\,\hat{\mathbf{L}}\cdot\hat{\mathbf{S}} = \xi(r)\hat{\mathbf{L}}\cdot\hat{\mathbf{S}} \tag{8-29}$$

This means that the operators $\hat{L}^2$ and $\hat{S}^2$ do not commute with the hamiltonian containing the spin-orbit interaction term given by Eq. (8-29). Consequently **L** and **S** are not constants of the motion although their resultant $\mathbf{J} = \mathbf{L} + \mathbf{S}$ is a constant of the motion.

It is convenient to define a new hamiltonian which includes not only electron repulsion but also spin-orbit interactions. This hamiltonian is

$$\hat{H}^{so} = \hat{H}_1 + \hat{H}_{12} + \hat{H}^{\xi} \tag{8-30}$$

We will call this the *spin-orbit hamiltonian*. Let us now investigate the commutation of the operators $\hat{L}_z$, $\hat{L}^2$, $\hat{S}_z$, and $\hat{S}^2$ with the spin-orbit hamiltonian. Since the commutation of these operators with $\hat{H}_1$ and $\hat{H}_{12}$ has already been determined, it is necessary to consider only the commutation with the spin-orbit interaction operator $\hat{H}^{\xi}$. Since the $\xi(r)$ portion of this operator does not affect commutation properties, we need concern ourselves only with the operator $\hat{\mathbf{L}}\cdot\hat{\mathbf{S}}$. The commutation relation for $\hat{L}_z$ and $\hat{\mathbf{L}}\cdot\hat{\mathbf{S}}$ is obtained from the following relations:

$$\begin{aligned} \hat{L}_z(\hat{\mathbf{L}}\cdot\hat{\mathbf{S}}) &= \hat{L}_z\hat{L}_x\hat{S}_x + \hat{L}_z\hat{L}_y\hat{S}_y + \hat{L}_z^2\hat{S}_z \\ (\hat{\mathbf{L}}\cdot\hat{\mathbf{S}})\hat{L}_z &= \hat{L}_x\hat{S}_x\hat{L}_z + \hat{L}_y\hat{S}_y\hat{L}_z + \hat{L}_z^2\hat{S}_z \end{aligned} \tag{8-31}$$

In writing down the above relationships, use was made of the fact that $\hat{\mathbf{L}}$ and $\hat{\mathbf{S}}$ and all their components commute. Subtracting the second relationship in Eq. (8-31) from the first, we obtain

$$[\hat{L}_z, \hat{\mathbf{L}}\cdot\hat{\mathbf{S}}] = i(\hat{L}_y\hat{S}_x - \hat{L}_x\hat{S}_y) \tag{8-32}$$

In a similar fashion we get

$$[\hat{S}_z, \hat{\mathbf{L}}\cdot\hat{\mathbf{S}}] = i(\hat{S}_y\hat{L}_x - \hat{S}_x\hat{L}_y) \tag{8-33}$$

Thus neither $\hat{L}_z$ nor $\hat{S}_z$ commutes with $\hat{H}^{\xi}$. However, if we add Eqs. (8-32) and (8-33), we obtain

$$[\hat{L}_z + \hat{S}_z, \hat{\mathbf{L}}\cdot\hat{\mathbf{S}}] = 0 \tag{8-34}$$

i.e., the sum of the z components of $\hat{\mathbf{L}}$ and $\hat{\mathbf{S}}$ do commute with $\hat{H}^{\xi}$. We then define the operator

$$\hat{J}_z = \hat{L}_z + \hat{S}_z \tag{8-35}$$

such that

$$[\hat{J}_z, \hat{H}^\xi] = 0 \tag{8-36}$$

The *total* angular momentum operator and its scalar square are defined by

$$\hat{\mathbf{J}} = \hat{J}_x\mathbf{i} + \hat{J}_y\mathbf{j} + \hat{J}_z\mathbf{k} \tag{8-37}$$

One can readily show that each component of $\hat{J}^2$ commutes with $\hat{H}^\xi$, so that

$$[\hat{J}^2, \hat{H}^\xi] = 0 \tag{8-38}$$

Since $\hat{J}^2$ must also commute with the total (spin-orbit) hamiltonian, the total angular momentum **J** is a constant of the motion even though its components **L** and **S** are not.

One also finds that $\hat{L}^2$ and $\hat{S}^2$ *almost* commute with $\hat{H}^\xi$, so that it is sometimes acceptable to regard these as approximate constants of the motion.

The operators $\hat{J}_z$ and $\hat{J}^2$ are assumed to satisfy the eigenvalue equations

$$\hat{J}_z\Phi = M_j\Phi \qquad \hat{J}^2\Phi = J(J+1)\Phi \qquad -J \leq M_J \leq J$$
$$J = 0, 1, 2, \ldots \qquad \text{or} \qquad J = \tfrac{1}{2}, \tfrac{3}{2}, \tfrac{5}{2}, \ldots \tag{8-39}$$
$$|\mathbf{J}| = \sqrt{J(J+1)}$$

where Φ is a function of the θ and φ coordinates and the spin coordinates of the electrons. We will now show how the total angular momentum quantum number J can be related (in an approximate fashion) to the quantum numbers associated with the orbital and spin angular momenta.

First, we consider the special case of light atoms, i.e., atoms with a small atomic number Z. From Eq. (8-29) and the form of $\hat{H}_{12}$ we see that for sufficiently small Z

$$\hat{H}_{12} \gg \hat{H}^\xi \tag{8-40}$$

so that the total atomic hamiltonian can be approximated by

$$\hat{H}^{(0)} = \hat{H}_1 + \hat{H}_{12} \tag{8-41}$$

In the language of perturbation theory, $\hat{H}^{(0)}$ is the zeroth-order approximation to $\hat{H}$, and $\hat{H}^\xi$ is regarded as a small perturbation. Since $\hat{L}^2$ and $\hat{S}^2$ commute with $\hat{H}^{(0)}$, they also approximately commute with $\hat{H}$ whenever H^ξ is suitably small. We then regard **L** and **S** as *approximate* constants of the motion. The J quantum is then approximated by

$$J = L + S,\ L + S - 1, \ldots, |L - S| \geq 0 \tag{8-42}$$

and the total angular momentum **J** is given by the vector sum **L** + **S**. This situation represents a type of spin-orbit coupling known as *LS coupling* or *Russell-Saunders coupling.*[3]

---

[3] H. N. Russell and F. A. Saunders, *Astrophys. J.*, **61**:38 (1925).

If one makes the further approximation that $\hat{H}^{(0)}$ in Eq. (8.41) can be replaced by a central-field hamiltonian, the L quantum numbers are approximated by Eq. (8-17), that is, in terms of the azimuthal quantum numbers of individual electrons.

In the LS coupling approximation, the operator $\hat{J}^2$ becomes

$$\hat{J}^2 = (\hat{\mathbf{L}} + \hat{\mathbf{S}}) \cdot (\hat{\mathbf{L}} + \hat{\mathbf{S}}) = \hat{L}^2 + \hat{S}^2 + 2\hat{\mathbf{L}} \cdot \hat{\mathbf{S}} \tag{8-43}$$

which yields the useful expression

$$\hat{\mathbf{L}} \cdot \hat{\mathbf{S}} = \tfrac{1}{2}(\hat{J}^2 - \hat{L}^2 - \hat{S}^2) \tag{8-44}$$

This expression will be very useful in the calculation of the spin-orbit interaction energy by means of first-order perturbation theory.

For heavy atoms (ones with a large atomic number Z) the term $\hat{H}^{\xi}$ becomes much larger than $\hat{H}_{12}$, and the total atomic hamiltonian is approximated to the zeroth order by

$$\hat{H}^{(0)} = \hat{H}_1 + \hat{H}^{\xi} \tag{8-45}$$

Now it is the term $\hat{H}_{12}$ which is regarded as a small perturbation, i.e., the electron-repulsion energy is small compared with the spin-orbit interaction energy. Since $\hat{L}^2$ and $\hat{S}^2$ now do not commute with $\hat{H}^{(0)}$, it is apparent that the L and S quantum numbers are not even approximately valid. However, the operator $\hat{J}_z = \hat{L}_z + \hat{S}_z$ does commute with $\hat{H}^{(0)}$, and so does the single-electron operator

$$\hat{J}_{z\mu} = \hat{L}_{z\mu} + \hat{S}_{z\mu} \tag{8-46}$$

where

$$\hat{J}_z = \sum_{\mu}^{N} \hat{J}_{z\mu} \tag{8-47}$$

We then define the total angular momentum quantum number of the μth electron by

$$j_\mu = l_\mu + s_\mu, l_\mu + s_\mu - 1, \ldots, |l_\mu - s_\mu| \geq 0 \tag{8-48}$$

such that the total angular momentum is given by the vector sum

$$\mathbf{J} = \sum_{\mu}^{N} \mathbf{J}_\mu \tag{8-49}$$

This means that the individual electron orbital angular momentum and spin angular momentum couple rather strongly to form a resultant total angular momentum for a given electron, and that these individual total angular momenta couple further (rather weakly) to form the resultant **J** (see Fig. 8-5). This type of interaction is known as *jj coupling*. Note that in the case of *LS* coupling, the individual orbital angular momenta couple strongly to form **L**, the individual spin angular momenta couple strongly to form **S** and then **L** and **S** couple weakly to form the resultant **J**.

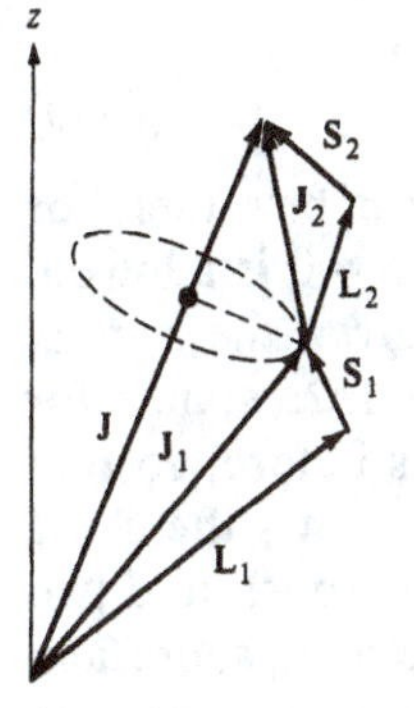

FIGURE 8-5
Spin-orbit interactions for a pair of electrons in a heavy atom (*jj* coupling).

In actuality, both *LS* and *jj* types of spin-orbit interaction represent two extreme cases of coupling; the behavior of actual many-electron systems is better described by interpolation between the two forms of coupling. As we will see later, the two coupling schemes are related by a correlation diagram which allows one to do this interpolation (at least in principle).

EXERCISES

8-1. Justify the statement that $\hat{L}^2$ and $\hat{S}^2$ *almost* commute with $\hat{H}^{\xi}$.
8-2. Justify the use of the quantum numbers l_μ and s_μ in Eq. (8-48).

8-3 ATOMIC TERM SYMBOLS

The spectroscopic state of an atom is conveniently summarized by the use of *spectroscopic term symbols*. For light atoms in which *LS* coupling is an acceptable approximation, the complete term symbol has the general form

$$^{2S+1}L_J$$

where L, S, and J are the angular momentum quantum numbers discussed in the previous sections. The quantity $2S+1$ is called the *multiplicity* and represents the number of independent spin functions which one can associate with a given spatial wavefunction. For $2S+1=1, 2, 3, 4, 5, 6$, etc., the associated states are called *singlet*, *doublet*, *triplet*, *quartet*, *quintet*, *sextet*, etc. The multiplicity then refers to the spin degeneracy, discussed in Chap. 7. It is customary to represent the quantum number L by the letters S, P, D, F, G, etc., when $L = 0, 1, 2, 3, 4$, etc., respectively. Note that one must be careful to avoid confusion of the symbol S for the spin quantum number with use of the same symbol for $L = 0$.

In general, the electron configuration of an atom is a listing of those orbitals to be used in the construction of approximate wavefunctions of that atom. In the case of the helium atom, the ground-state electron configuration is

$1s^2$; this implies the wavefunction

$$\xi(1,2) = |1s\,\overline{1s}| \tag{8-50}$$

Since $l_1 = l_2 = 0$, we see that $L = 0$. Also, since $s_1 = s_2 = \frac{1}{2}$, S can be either 1 or 0, but since $S = 1$ violates the Pauli exclusion principle, only $S = 0$ is allowed. Thus J can be only zero, and the ground state of helium is given the symbol $^1S_0$. This is called a *singlet-ess-zero* state. The configuration $1s2s$ stands for more than one state. In this case $l_1 = l_2 = 0$ and $s_1 = s_2 = \frac{1}{2}$ as before, so that $L = 0$ and $S = 1$ or 0. However, the exclusion principle now allows the $S = 1$ value (since the n quantum number is different for the two electrons), and two different quantum states arise: when $L = 0$ and $S = 0$ we have a $^1S_0$ state, and when $L = 0$ and $S = 1$ we have a $^3S_1$ state (triplet-ess-one). Both of these are excited states with higher energies than the ground state $^1S_0$. The wavefunctions of these states have already been illustrated [see, for example, Table 7-1 and Eqs. (7-83) and (7-84)], but they are repeated below for convenience of reference:

$$\xi(^1S_0) = \frac{1}{\sqrt{2}}\left(|1s\,\overline{2s}| - |\overline{1s}\,2s|\right)$$

$$\xi(^3S_1) = \begin{cases} \dfrac{1}{\sqrt{2}}\left(|1s\,\overline{2s}| + |\overline{1s}\,2s|\right) & M_S = 0 \\ |1s\,2s| & M_S = 1 \\ |\overline{1s}\,\overline{2s}| & M_S = -1 \end{cases} \tag{8-51}$$

The electron configuration of an atom generally consists of two parts: a *closed-shell* part and an *open-shell* part. The closed-shell part always consists of pairs of spatial orbitals in which each orbital is used $4l + 2$ times: $2l + 1$ times with α spin and $2l + 1$ times with β spin. For example, the helium $1s^2$ configuration is entirely closed-shell, and the lithium $1s^2 2s$ configuration has $1s^2$ as the closed-shell part and $2s$ as the open-shell part. The helium atom configuration $1s2s$ is entirely open-shell. The carbon atom configuration $1s^2 2s^2 2p^2$ has $1s^2 2s^2$ as the closed-shell part and $2p^2$ as the open-shell part. In the most commonly employed scheme, the fundamental characteristic of the closed-shell portion of the determinantal wavefunction is that it always consists of doubly occupied spatial orbitals; adjacent pairs of spin orbitals have the general forms

$$S_i(\mu) = \phi_i(\mu)\alpha(\mu) \qquad S_{i+1}(\nu) = \phi_i(\nu)\beta(\nu) \tag{8-52}$$

A given spatial orbital then appears *twice* in the determinant, namely, as ϕ_i and as $\bar{\phi}_i$. Whenever a determinantal wavefunction constructed from nl orbitals is entirely closed-shell with the form

$$|\phi_1\bar{\phi}_1\phi_2\bar{\phi}_2\cdots\phi_N\bar{\phi}_N| \tag{8-53}$$

(for an atom of $2N$ electrons), it invariably refers to a $^1S_0$ state ($L = S = J = 0$). Such a wavefunction is always an eigenfunction of $\hat{S}^2$ with an eigenvalue of zero. Consequently, given a specific electron configuration, it is necessary to examine only the open-shell part (if any) to deduce what spectroscopic state or states are represented. Thus for lithium $1s^2 2s$ we need look only at the $2s$ part. For this $L = 0$, $S = \frac{1}{2}$, and $J = \frac{1}{2}$ so that the spectroscopic term symbol is ${}^2S_{1/2}$. The determinantal wavefunctions are

$$\xi({}^2S_{1/2}) = \begin{cases} |1s\,\overline{1s}\,\overline{2s}| & M_S = \frac{1}{2} \\ |1s\,\overline{1s}\,2s| & M_S = -\frac{1}{2} \end{cases} \tag{8-54}$$

Both of the determinants are eigenfunctions of $\hat{S}^2$ with eigenvalues of $\frac{1}{2}(\frac{1}{2} + 1) = \frac{3}{4}$. Since $2S + 1 = 2$, these are the components of a *doublet* state.

The ground state of boron is described by the electron configuration $1s^2 2s^2 2p$. Since $l = 1$ for $2p$, this leads to $L = 1$ (a P state). For a single electron in an open shell, $S = \frac{1}{2}$, so this is a doublet state. From $L = 1$ and $S = \frac{1}{2}$ we can form two different J values: $J = 1 + \frac{1}{2} = \frac{3}{2}$ and $J = 1 - \frac{1}{2} = \frac{1}{2}$. Thus the boron ground-state configuration represents two different quantum states labeled ${}^2P_{1/2}$ and ${}^2P_{3/2}$. In the absence of spin-orbit interaction, these two states would be degenerate; in actuality the state with the lower J value is slightly lower in energy.

When there are two or more electrons in the open-shell portion of the electron configuration, it is convenient to use a more systematic formal procedure to deduce the possible spectroscopic states represented. We will examine the procedure using carbon $1s^2 2s^2 2p^2$ as an example. Looking at the $2p^2$ part only, we see that $l_1 = l_2 = 1$, so that L can have all the positive values beginning with $1 + 1 = 2$ down to $1 - 1 = 0$, that is, $L = 2$, 1, or 0. Similarly, $S = 1$, or 0. Up to this point the states ${}^1S$, ${}^3S$, ${}^1P$, ${}^3P$, ${}^1D$, and ${}^3D$ might be expected, but we will see that some of these violate the exclusion principle. An easy way to do this is to construct a table such as Table 8-1. In this table we

TABLE 8-1
Allowed spatial orbital-spin function combinations for a wavefunction represented by a $2p^2$ open-shell configuration*

	m_l			
	−1	**0**	**+1**	$M_L = \sum_{\mu=1}^{3} M_{l_\mu}$
1.	↑↓	—	—	−2
2.	—	↑↓	—	0
3.	—	—	↑↓	+2
4.	↑	↓	—	−1
5.	↑	—	↓	0
6.	—	↑	↓	+1
7.	↑	↑	—	−1
8.	↑	—	↑	0
9.	—	↑	↑	+1

* Each $2p$ can be $2p_{-1}$, $2p_0$, or $2p_1$.

simply write down all the different ways we can place two electrons into three $2p$ orbitals without violating the exclusion principle. This means that any time a spatial orbital of given n, l, and m_l value is used twice, it must be used once with an α spin and once with a β spin. Thus, arrangement 2 implies a ^1S state. Similarly, arrangements 1, 4, 5, 6, and 3 are the $M_L = -2, -1, 0, +1$, and $+2$ components of a ^1D state, while 7, 8, and 9 are the $M_L = -1, 0$, and $+1$ components of a ^3P state. Now that spurious states have been eliminated (note there are no ^3S, ^1P, or ^3D states), we can deduce the possible J values for each state. For the ^1S case, $L = 0$ and $S = 0$, so that $J = 0$. For the ^1D state, $L = 2$ and $S = 0$, so that $J = 2$. For the ^3P state, $L = 1$ and $S = 1$, so J can be 2, 1, or 0. The possible spectroscopic states then are $^1S_0$, $^1D_2$, $^3P_2$, $^3P_1$, and $^3P_0$.

Next we determine the relative orders of the energies of the above five states. This can be accomplished by the use of *Hund's rules*: a distillation of various empirical observations and theoretical calculations on atomic spectra. These are given below:

Hund's rule 1. Of the states arising from a given electron configuration, the lowest in energy is the one having the highest multiplicity. This is known as *Hund's rule of maximum multiplicity*.

In the case where there are two or more states with the same multiplicity, the following rule is used:

Hund's rule 2. Of two or more states of the same multiplicity, the lower in energy is the one with the higher L value.

The energy separation of two states with the same multiplicity but different L values is due to removal of degeneracy by the electron-repulsion term in the hamiltonian. When $\hat{H}$ contains only one-electron terms, the potential energy is of the central-field type and a $(2L + 1)$-fold degeneracy exists (as in the one-electron atom). However, the $1/r_{\mu\nu}$ terms destroy the central field, because of inclusion of non-central-field coordinates, and remove this degeneracy. Thus, a simple rationalization of Hund's rule 2 follows: The total energy of an atom is lower if electron repulsions (which are *positive* contributions to the total energy) are made as small as possible. Now, two electrons will repel each other less if they follow paths which allow them to stay out of each others way as much as possible. This will happen when the two electrons travel in the same direction about the nucleus (clockwise or counter-clockwise) rather than in opposite directions. In the former case the orbital angular momentum is maximized, since the individual momenta are additive, and in the latter case the orbital angular momentum is minimized, since the individual momenta cancel each other.

Hund's rule 1 indicates that one of the ^3P states represents the ground state of carbon. Since there are no other triplet states in this particular example of carbon, Hund's rule 2 is of no help in further designating the ground state.

Consequently, we must examine what role the *total* angular momentum plays in determining relative energies of states.

Levels such as $^3P_2$, $^3P_1$, and $^3P_0$ which differ only in the total angular momentum are called *multiplets*. The multiplets will have slightly different energy values because of precession of **L** and **S** about their resultant **J** (see Fig. 8-4). The more that **L** and **S** interact, the greater the precession and the greater the multiplet separation. In the *LS* coupling approximation, the spin-orbit interaction operator is given by

$$\hat{H}^{\xi} = A\hat{\mathbf{L}}\cdot\hat{\mathbf{S}} \tag{8-55}$$

[see Eq. (8-29)], where A is a constant. Using the results of Eq. (8-44), one can obtain the first-order perturbation estimation of the spin-orbit coupling energy as

$$\begin{aligned} E^{\xi} &= \tfrac{1}{2}A\langle\psi|\hat{J}^2 - \hat{L}^2 - \hat{S}^2|\psi\rangle \\ &= \tfrac{1}{2}A[J(J+1) - L(L+1) - S(S+1)] \end{aligned} \tag{8-56}$$

Since all members of a multiplet have the same L and S quantum numbers, the energy separation between the J and $J+1$ levels is given by

$$\Delta E^{\xi} = E_{J+1} - E_J = A(J+1) \tag{8-57}$$

The constant A is positive and increases with Z. The above equation is known as the *Landé interval rule*; it states that the energy separation due to spin-orbit coupling is proportional to the larger of the two J values involved. Thus, the multiplet of lowest energy is $^3P_0$. Using the Landé interval rule, we can predict the ^3P multiplet separations in carbon to be as follows:

$$\frac{E(^3P_2) - E(^3P_1)}{E(^3P_1) - E(^3P_0)} = \frac{A(2)}{A(1)} = 2 \tag{8-58}$$

The results are depicted in Fig. 8-6.

Unfortunately, first-order perturbation theory is not adequate to handle the multiplet separation problem in general; when the open-shell part of the electron configuration is more than half-filled, the energy order of the multiplets inverts; i.e., the lowest state is the one with the *highest* J value. Taking all this into account leads to the multiplet rule:

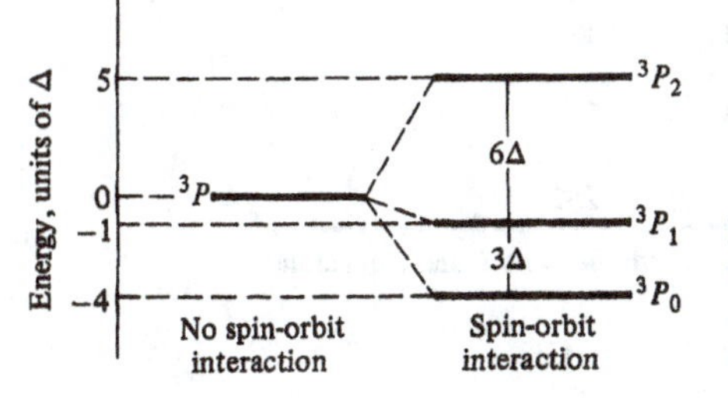

FIGURE 8-6
Splitting of the ^3P term by spin-orbit interaction according to the Landé interval rule. The quantity Δ is an arbitrary unit of energy.

Hund's rule 3. For configurations containing the open-shell portion $(nl)^x$, if x is less than $2l+1$, the lowest state is the one with the lowest J value (normal Landé order); if x is greater than $2l+1$, the lowest state is the one with the highest J value (inverted Landé order).

It is easy to deduce by use of a table such as 8-1 that a configuration with an open-shell portion $(nl)^x$ will lead to the same states as will $(nl)^y$ whenever $x+y=4l+2$. This fact follows at once from recognition that there is a symmetry between the number of different ways of assigning electrons to orbitals and the number of ways of assigning the same number of vacancies to the same orbitals. Thus, both carbon $(1s^2 2s^2 2p^2)$ and oxygen $(1s^2 2s^2 2p^4)$ configurations lead to ${}^1S_0$, ${}^1D_2$, ${}^3P_2$, ${}^3P_1$, and ${}^3P_0$ states. However, due to Hund's rule 3, the ground state of carbon is ${}^3P_0$ and that of oxygen is ${}^3P_2$. This fact makes it simple to deduce the states implied by a configuration such as $1s^2 2s^2 2p^5$ of fluorine without making up a table such as 8-1. Since boron $(1s^2 2s^2 2p)$ is immediately seen to lead to ${}^2P_{1/2}$ and ${}^2P_{3/2}$ states, Hund's rule 3 allows one to say at once that the ground state of boron is ${}^2P_{1/2}$ and that of fluorine is ${}^2P_{3/2}$.

For exactly half-filled shells, there will be only one J value for the state of highest multiplicity, and thus no ambiguity arises.

When the open-shell part of an electron configuration is of the form $(nl)^x$, the x electrons in the nl orbital are said to be *equivalent*. It is in such configurations that care must be taken to exclude states violating the exclusion principle. Table 8-2 lists the electron configurations arising from various equivalent electronic configurations of atoms.

TABLE 8-2
Spectroscopic terms arising from equivalent electronic configurations of atoms

Configuration	*LS* terms	Number of independent states
ns^2	${}^1S$	1
np, np^5	${}^2P$	6
np^2, np^4	${}^1S$, ${}^1D$, ${}^3P$	15
np^3	${}^2P$, ${}^2D$, ${}^4S$	20
nd, nd^9	${}^2D$	10
nd^2, nd^8	${}^1S$, ${}^1D$, ${}^1G$, ${}^3P$, ${}^3F$	45
nd^3, nd^7	${}^2D(2)$,* ${}^2P$, ${}^2F$, ${}^2G$, ${}^2H$, ${}^4P$, ${}^4F$	120
nd^4, nd^6	${}^1S(2)$, ${}^1D(2)$, ${}^1F$, ${}^1G(2)$, ${}^1I$, ${}^3P(2)$, ${}^3D$, ${}^3F(2)$, ${}^3G$, ${}^3H$, ${}^5D$	210
nd^5	${}^2S$, ${}^2P$, ${}^2D(3)$, ${}^2F(2)$, ${}^2G(2)$, ${}^2H$, ${}^2I$, ${}^4P$, ${}^4D$, ${}^4F$, ${}^4G$, ${}^6S$	252

* The number in parentheses is the number of distinct terms with the same L and S quantum numbers.

In dealing with nonequivalent electrons, it is convenient to have a systematic means of analyzing the types of states which can arise. For example, when there are two nonequivalent electrons, one can have the two possibilities $S = \frac{1}{2} + \frac{1}{2} = 1$ and $S = \frac{1}{2} - \frac{1}{2} = 0$, that is, either a triplet or a singlet. For three electrons there are a total of *eight* different couplings of spins. These are

1. $\frac{1}{2} + \frac{1}{2} + \frac{1}{2} = \frac{3}{2}$
2. $\frac{1}{2} + \frac{1}{2} - \frac{1}{2} = \frac{1}{2}$
3. $\frac{1}{2} - \frac{1}{2} - \frac{1}{2} = -\frac{1}{2}$
4. $-\frac{1}{2} - \frac{1}{2} - \frac{1}{2} = -\frac{3}{2}$
5. $-\frac{1}{2} - \frac{1}{2} + \frac{1}{2} = -\frac{1}{2}$
6. $-\frac{1}{2} + \frac{1}{2} + \frac{1}{2} = \frac{1}{2}$
7. $-\frac{1}{2} + \frac{1}{2} - \frac{1}{2} = -\frac{1}{2}$
8. $\frac{1}{2} - \frac{1}{2} + \frac{1}{2} = \frac{1}{2}$

For N electrons there would be 2^N different spin couplings. In the case of the $2^3 = 8$ couplings shown above, these consist of (1) the group $-\frac{3}{2} \le M_S \le \frac{3}{2}$ which corresponds to $S = \frac{3}{2}$ and thus represents a state of multiplicity $2S + 1 = 4$ and (2) two groups $-\frac{1}{2} \le M_S \le \frac{1}{2}$ which correspond to $S = \frac{1}{2}$ and thus represent states of multiplicity $2S + 1 = 2$. Thus an atom (or molecule) with 3 electrons outside a closed shell may exist in a quartet state or in one of two different doublet states. For $N > 3$ this procedure is rather tedious to carry out, but use of the branching diagram (Fig. 8-7) makes it easier. This branching diagram shows the number of states of different multiplicities obtainable for a given number of independent electrons.[4] The diagram is very easy to construct beginning with a single electron and successively coupling other electron spins to it in all possible algebraic ways. In the diagram the number of states of a given multiplicity is indicated within a circle whose abscissa is the number of

[4] For N electrons and N different spatial orbitals (that is, $2N$ different spin orbitals) one can write $(2N)!/(N!)^2$ different determinants. Of these, 2^N are constructed from N different spatial orbitals. The branching diagram refers to the latter.

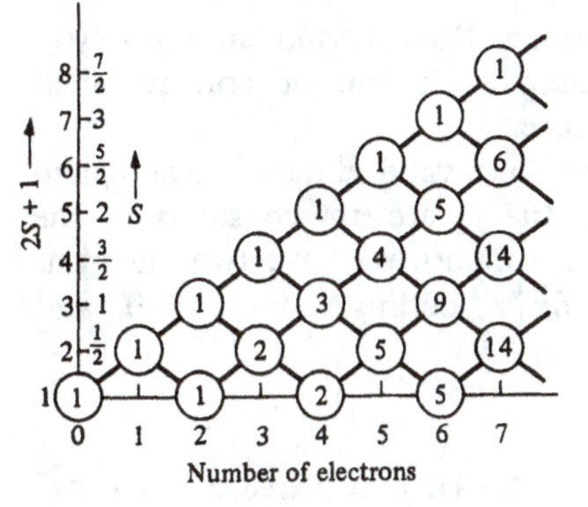

FIGURE 8-7
Branching diagram showing the number of states of a given multiplicity obtainable from N independent electrons.

electrons and whose ordinate is the multiplicity. The diagram has a rather simple structure; each encircled number is the sum of the two adjacent encircled numbers to its left. We thus see very quickly that for four electrons one would have two single states ($M_S = 0$), three triplet states ($M_S = -1, 0, 1$), and one quintet state ($M_S = -2, -1, 0, 1, 2$). Thus, for four electrons not in closed shells one could construct 16 different wavefunctions leading to six different energies.

Let us now consider the deduction of the states arising from a configuration with three nonequivalent electrons, for example, $2p3p3d$. From the branching diagram we see this produces one quartet and two doublets. First, combining $2p$ and $3p$, we get $L = 1+1, 1+1-1, 1-1 = 2, 1, 0 = D, P, S$, respectively. Next we couple in the $3d$ electron to get

$$S + d = 0 + 2, \ldots, |0 - 2| = 2 = D$$

$$P + d = 1 + 2, \ldots, |1 - 2| = 3, 2, 1 = F, D, P$$

$$D + d = 2 + 2, \ldots, |2 - 2| = 4, 3, 2, 1, 0 = G, F, D, P, S$$

The two sets of doublets will have 17 members each, and the quartet will have 31 members, a total of 65 distinct levels. These are as follows:

Doublets: ${}^2S_{1/2}$, ${}^2P_{1/2,\,3/2}(2)$, ${}^2D_{3/2,\,5/2}(3)$, ${}^2F_{5/2,\,7/2}(2)$, ${}^2G_{7/2,\,9/2}$

Quartets: ${}^4S_{3/2}$, ${}^4P_{1/2,\,3/2,\,5/2}(2)$, ${}^4D_{1/2,\,3/2,\,5/2,\,7/2}(3)$, ${}^4F_{3/2,\,5/2,\,7/2,\,9/2}(2)$, ${}^4G_{5/2,\,7/2,\,9/2,\,11/2}$

The number in parentheses is the number of times a given set of multiplets appears.

We note from the above that whenever L is less than S, the number of multiplets is given by $2L + 1$ and is less than the multiplicity; an example would be ${}^4S_{3/2}$, in which $2L + 1 = 1$, but $2S + 1 = 4$. Whenever L is equal to or greater than S, we find that the number of multiplets equals the multiplicity. It is customary to describe states in terms of their $2S + 1$ values, whether or not this equals the number of multiplets. Thus one still refers to the ${}^4S_{3/2}$ state as a quartet.

Rapid, systematic procedures for obtaining Russell-Saunders term symbols from electron configurations (containing equivalent or nonequivalent electrons) have been discussed in the literature.[5]

For a heavy atom in which LS coupling is not valid, the individual l_μ and s_μ couple to form resultants j_μ. Although the j_μ are not constants of the motion, their resultant $\mathbf{J}$ is. As an example, suppose we have two electrons described by the open-shell configuration $6p7s$. Letting $l_1 = 1$, $l_2 = 0$, and

[5] See, for example, K. E. Hyde, *J. Chem. Ed.*, **52**:87 (1975) and E. A. Castro, *J. Chem. Ed.*, **54**:367 (1977).

$s_1 = s_2 = \frac{1}{2}$, we get

$$j_1 = 1 + \tfrac{1}{2}, \ldots, |1 - \tfrac{1}{2}| = \tfrac{3}{2}, \tfrac{1}{2}$$
$$j_2 = 0 + \tfrac{1}{2} = \tfrac{1}{2} \tag{8-59}$$

Thus, the J values are given by

$$J = \tfrac{3}{2} + \tfrac{1}{2}, \ldots, |\tfrac{3}{2} - \tfrac{1}{2}| = 2, 1 \quad \text{and} \quad J = \tfrac{1}{2} + \tfrac{1}{2}, \ldots, |\tfrac{1}{2} - \tfrac{1}{2}| = 1, 0$$

Note that the $J = 1$ value occurs twice. Using the notation $(j_1, j_2)_J$ to designate the jj coupling states, we obtain the four states $(\frac{1}{2}, \frac{3}{2})_1$, $(\frac{1}{2}, \frac{3}{2})_2$, $(\frac{1}{2}, \frac{1}{2})_1$, and $(\frac{1}{2}, \frac{1}{2})_0$.

EXERCISES

8-3. Verify the spectroscopic term symbols arising from np^3 and nd^2 configurations (see Table 8-2). What do you predict to be the ground states of nitrogen, titanium, and nickel?

8-4. Show that for an sp^2 configuration, the allowed spectroscopic states are ^2S, ^2D, ^2P, and ^4P, and determine the J values for each.

8-5. Show that the carbon configuration $1s^2 2s^2 2p 3s$ leads to the LS states shown in Fig. 8-8, and that the configuration for lead (Pb) $1s^2 2s^2 2p^6 3s^2 3p^6 4s^2 3d^{10} 4p^6 5s^2 4d^{10} 5p^6 6s^2 4f^{14} 5d^{10} 6p 7s$ leads to the jj states shown in the same figure. This figure is a correlation diagram which is sometimes used to show how states can be described in terms of LS coupling even though jj coupling occurs.

8-6. What are the possible J values for a ^4G state?

8-7. What are the ground-state spectroscopic term symbols for
(*a*) Sb: (Kr)$5s^2 4d^{10} 5p^3$
(*b*) S:(Ne)$3s^2 3p^4$

8-8. What spectroscopic states arise from the (Ne)$3s^2 3d 4s$ configuration of silicon?

8-9. What is the ground-state term symbol for the Mn(II) cation, (Ar)$3d^5$?

8-10. If vanadium, (Ar)$3d^3 4s^2$, has a $^4F_{3/2}$ ground state, what is the ground state of cobalt, (Ar)$3d^7 4s^2$?

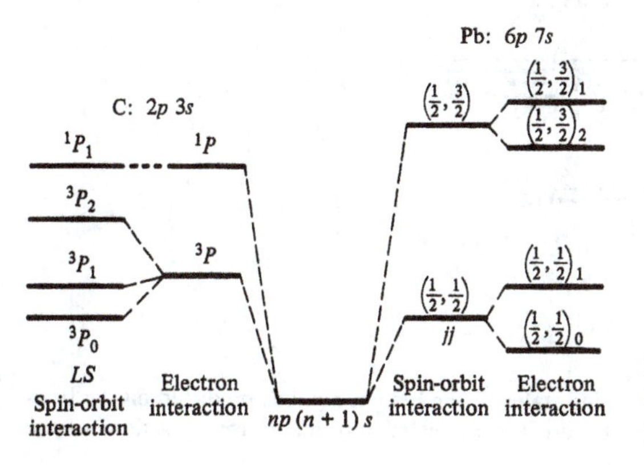

FIGURE 8-8
Correlation diagram for LS and jj coupling schemes.

8-4 ENERGY-LEVEL DIAGRAMS FOR ATOMS

By the use of a few simple rules, it is possible to set up a qualitatively correct energy-level diagram for the low-lying quantum states of most light atoms. We will examine this procedure for the helium atom. We begin with the ground-state electron configuration $1s^2$ which leads to the $^1S_0$ state. Then we generate new configurations from $1s^2$ by changing one of the spin orbitals to a new, higher-energy orbital; this is generally referred to as "promoting" an electron to a higher orbital. This leads to excited-state configurations such as $1s2s$, $1s2p$, $1s3s$, $1s3p$, and $1s3d$. As a rough rule—valid for lower-lying orbitals—all the states represented by a configuration $1snl$ are lower in energy than those generated from $1s(n+1)l$, and those generated from $1snl$ are lower in energy than those generated from $1sn(l+1)$. Thus the configurations given above lead to the following states:

$1s2s$: $^1S_0$, $^3S_1$
$1s2p$: $^1P_1$, $^3P_{2,1,0}$
$1s3s$: $^1S_0$, $^3S_1$
$1s3p$: $^1P_1$, $^3P_{2,1,0}$
$1s3d$: $^1D_2$, $^3D_{3,2,1}$

Using the rules stated in the previous paragraph and two previous rules (Hund's rule 1 and the Landé interval rule) leads to the energy-level diagram of Fig. 8-9.

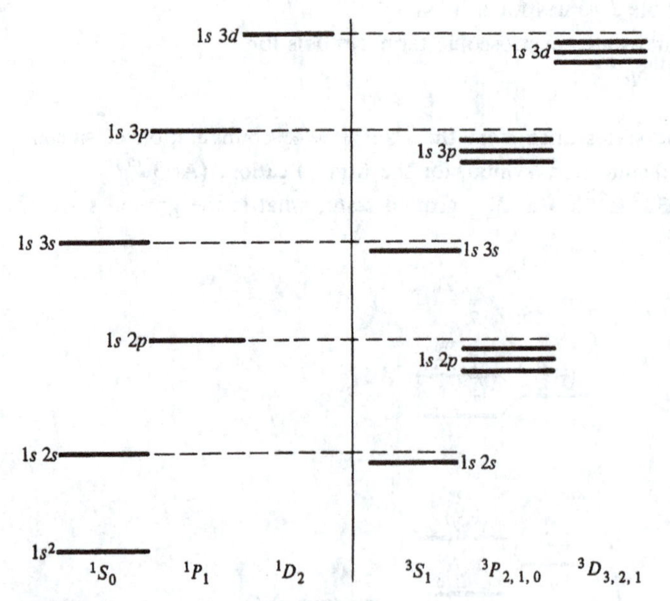

FIGURE 8-9
Energy-level diagram of some of the low-lying states of the helium atom as predicted on the basis of one-electron "promotions." The spacings are not to scale; only the relative positions of the states are represented.

When dealing with atoms in period 4 (K, Ca, Sc, etc.), one must remember that a configuration such as (Ar)4s leads to a lower energy than a configuration such as (Ar)3d.

EXERCISES

8-11. Beginning with $1s^22s$, predict the qualitative appearance of the energy-level diagram of lithium for all states arising by promoting the 2s electron to all orbitals up to and including 4f. What kind of promotion would have to be included in order to predict quartet states?

8-12. Repeat the above for calcium, beginning with (Ar)$4s^2$ and carrying out promotions of 4s up to 5f.

8-5 SELECTION RULES IN MANY-ELECTRON ATOMS

The probability of an electric dipole–induced transition from a stationary state k to another stationary state l is given by the Einstein coefficient [see Eqs. (3-22) and (3-23)] as

$$B_{k\to l} = \frac{2\pi}{3\hbar^2}|\boldsymbol{\mu}_{kl}|^2 \tag{8-60}$$

where $\boldsymbol{\mu}_{kl}$ is the transition moment given by

$$\boldsymbol{\mu}_{kl} = e\langle\psi_l|\mathbf{r}|\psi_k\rangle \tag{8-61}$$

It is convenient to define a quantity called the *dipole strength* of the transition by

$$D_{kl} = \frac{|\boldsymbol{\mu}_{kl}|}{e} = |\langle\psi_l|\mathbf{r}|\psi_k\rangle| \tag{8-62}$$

The dipole strength has the three cartesian components

$$\begin{aligned} D_{kl}^{(x)} &= \langle\psi_l|x|\psi_k\rangle \\ D_{kl}^{(y)} &= \langle\psi_l|y|\psi_k\rangle \\ D_{kl}^{(z)} &= \langle\psi_l|z|\psi_k\rangle \end{aligned} \tag{8-63}$$

In order for a transition $k\to l$ to be allowed, it is necessary for at least one of the components of D_{kl} to be nonzero. If only one of the components of D_{kl} vanishes, the transition is allowed and the emitted or absorbed radiation is *plane-polarized* in the plane defined by the nonvanishing components. If two of the components vanish, the transition is allowed and is polarized along the axis of the nonzero component. The former is usually called *σ polarization*, and the latter *π polarization*. Radiation which is σ-polarized appears to be circularly polarized when viewed along an axis perpendicular to the plane of polarization, whereas π-polarization radiation is absent when viewed along the axis of polarization but appears plane-polarized when viewed transversely. The state of polarization is useful in the analysis of atomic spectra.

Obtaining selection rules for many-electron atoms involves procedures similar to those used in obtaining the hydrogen atom selection rules in Sec. 5-3 except that one must now take both spin-orbit coupling and electron repulsions into account. When *LS* coupling is a valid approximation, the following selection rules obtain for the many-electron atom:

$$\Delta S = 0$$

$$\Delta L = 0, \pm 1$$

$$\Delta J = 0, \pm 1 \qquad 0 \not\leftrightarrow 0$$

$$\Delta M_J = 0, \pm 1 \qquad 0 \not\leftrightarrow 0 \text{ for } \Delta J = 0$$

Note that the selection rule on L is different from that in the one-electron atom where $\Delta l = \pm 1$. The rule $\Delta L = 0$ is not allowed in the one-electron atom, since **L** must be in a single plane, a restriction not present in the many-electron case.

The selection rule $\Delta S = 0$ arises from the orthogonality of spin wavefunctions and the fact that $\hat{S}^2$ commutes with **r** in the transition-moment integral (see Theorem 2-5). This selection rule is strictly valid only in the limited case of vanishingly small spin-orbit interaction and, consequently, is often violated. Transitions for which $\Delta S \neq 0$ are called *spin-forbidden*. Such spin-forbidden transitions occur more and more commonly as the spin-orbit interaction increases. In a heavy atom such as mercury they become very intense. The selection rule $\Delta L = 0, \pm 1$ is also violated as a result of increased spin-orbit coupling. For example, there is a ${}^2S \rightarrow {}^2D$ transition in the sodium atom at 342.71 nm for which $\Delta L = 2$. Apparent violations also occur as a result of interaction of radiation and charge by means of higher-multipole mechanisms. For example, electric quadrupole and magnetic dipole transitions satisfy the selection rules $g \rightarrow g$ and $u \rightarrow u$. Magnetic dipole transitions are observed in the anomalous Zeeman effect, since atoms for which $S \neq 0$ have an oscillating magnetic dipole in the presence of a magnetic field.

Figure 8-10 shows a portion of the energy-level diagram of the neutral sodium atom. The diagram is constructed from the experimental spectrum with the aid of the selection rules.

For *jj* coupling we have the following selection rules:

$$\Delta J = 0, \pm 1 \qquad 0 \not\leftrightarrow 0$$

$$\Delta M_J = 0, \pm 1 \qquad 0 \not\leftrightarrow 0 \text{ for } \Delta J = 0$$

We note that the selection rules on S, L, and l are absent, since these quantum numbers have no validity whenever the spin-orbit coupling is large.

The selection rule for M_J is common to both *LS* and *jj* coupling schemes. This rule will be of relevance when discussing the Zeeman effect in the following section.

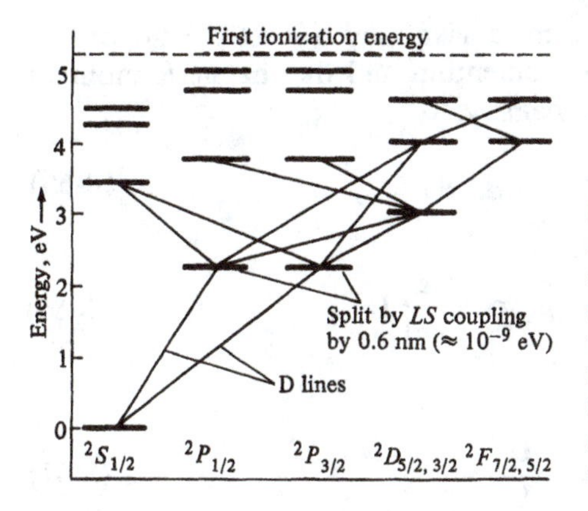

FIGURE 8-10
A portion of the energy-level diagram of the neutral sodium atom. The diagram was constructed from the experimental spectrum, and the transitions were deduced from the selection rules. Only a few of the allowed transitions are illustrated. The sodium D lines (which occur around 589.3 nm) are split by about 0.6 nm by spin-orbit coupling.

EXERCISES

8-13. Using the energy-level diagram of helium shown in Fig. 8-9, sketch in all allowed spectroscopic transitions. Repeat for the lithium atom energy-level diagram constructed in Exercise 8-11.

8-14. Construct that portion of the energy-level diagram of beryllium represented by the configuration $1s^2 2s2p$, and indicate all the allowed transitions (if any).

8-15. Referring to Fig. 8-8, show that there are two allowed transitions in the *LS* scheme and five in the *jj* scheme. This is one way of telling what kind of coupling predominates in a given atom.

8-6 THE ZEEMAN EFFECT IN MANY-ELECTRON ATOMS

When a radiation source originating from an excited atom is subjected to a magnetic field, the $(2J+1)$-fold degeneracy of each multiplet is removed. Each multiplet (in the absence of a magnetic field) is associated with $2J+1$ independent wavefunctions leading to the same energy. The multiplet splitting is due to the interaction of the atom's magnetic moment $\boldsymbol{\mu}_J$ with the external magnetic field and is known as the *Zeeman effect* (see Sec. 5-5 for the Zeeman effect in the hydrogen atom). The total angular momentum of magnitude $\sqrt{J(J+1)}$ now precesses about the direction of the applied magnetic field. Although $\mathbf{J}$ is no longer conserved, its components in the field direction are quantized; that is, J_z (using z as the field direction) is a constant of the motion.

The classical expression for the energy of interaction of the magnetic moment of an atom and a magnetic field is given by an equation of the same form as Eq. (8-21), but now $\boldsymbol{\mu}$ is the total magnetic moment of the atom and $\mathbf{H}$ is the external magnetic field. The interaction energy due to the component of the magnetic moment in the z direction is

$$E' = -|\mathbf{H}|\,\mu_z \tag{8-64}$$

The total magnetic moment of an atom is given by a vector sum of the magnetic moment due to spin angular momentum and the magnetic moment due to the orbital angular momentum, namely,

$$\boldsymbol{\mu} = \boldsymbol{\mu}_S + \boldsymbol{\mu}_L \tag{8-65}$$

where

$$\boldsymbol{\mu}_S = -\frac{e}{m}\mathbf{S} \qquad \boldsymbol{\mu}_L = -\frac{e}{2m}\mathbf{L} \tag{8-66}$$

It is noted that

$$\frac{\boldsymbol{\mu}_S}{\mathbf{S}} = 2\,\frac{\boldsymbol{\mu}_L}{\mathbf{L}} \tag{8-67}$$

The factor of 2 in the above relationship is the Landé g factor for an electron (see Sec. 5-4). From Fig. 8-11 we see that, in general, the magnetic moment $\boldsymbol{\mu}$ does not have exactly the opposite direction of the vector $\mathbf{J}$ as would be true if the Landé g factor were unity. This means that the magnetic moment $\boldsymbol{\mu}$ precesses about the direction of $\mathbf{J}$ and is not a constant of the motion; i.e., the atom has an oscillating magnetic dipole. Now if the magnetic field is not too large, the precession of $\boldsymbol{\mu}$ about $\mathbf{J}$ will be faster than the precession of $\mathbf{J}$ about the field direction. In this case we will assume that the magnetic interaction energy can be calculated from the component of $\boldsymbol{\mu}$ in the direction of $\mathbf{J}$. From Fig. 8-11 we also see that this component, which we call $\boldsymbol{\mu}_J$, has the magnitude

$$\boldsymbol{\mu}_J = \boldsymbol{\mu}_S \cos\alpha + \boldsymbol{\mu}_L \cos\beta \tag{8-68}$$

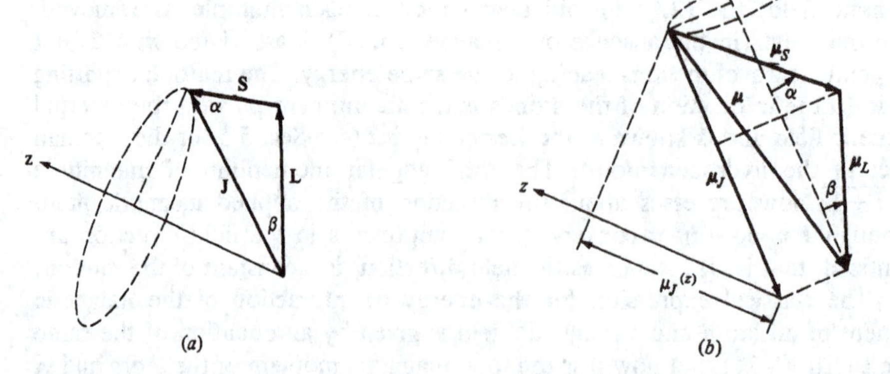

FIGURE 8-11
The magnetic moment of an atom. Note that the magnetic moment vectors (*b*) are oppositely directed to the corresponding momenta (*a*). The magnetic moment μ precesses about **J**, and **J** itself precesses about the direction of the magnetic field.

The projection of $\boldsymbol{\mu}_J$ onto the axis of the magnetic field is labeled $\mu_J^{(z)}$. Equation (8-64) now becomes

$$E' = -|\mathbf{H}|\mu_J^{(z)} \tag{8-69}$$

It is now necessary to deduce an expression for $\mu_J^{(z)}$ in terms of the L, S, J, and M_J quantum numbers. When $S = 0$ and $J = 1$, we see that $\cos\beta = 1$ and $\boldsymbol{\mu}_J = \boldsymbol{\mu}_L$. Thus

$$\mu_J^{(z)} = -\frac{e}{2m}J_z \tag{8-70}$$

If $L = 0$ and $J = S$, then $\cos\alpha = 1$ and $\boldsymbol{\mu}_J = 2\boldsymbol{\mu}_S$. Thus

$$\mu_J^{(z)} = -\frac{e}{m}J_z \tag{8-71}$$

The latter two relationships suggest that the magnitude of $\boldsymbol{\mu}_J$ is defined as

$$\mu_J = g\sqrt{J(J+1)} \tag{8-72}$$

where g is a *generalized Landé g factor*. Then $\mu_J^{(z)}$ becomes, in general,

$$\mu_J^{(z)} = -\frac{e}{2m}gJ_z \tag{8-73}$$

The quantum mechanical operator for the interaction energy can be written

$$\hat{H}' = \frac{e}{2m}g|\mathbf{H}|\hat{J}_z \tag{8-74}$$

having the expectation value

$$E' = \frac{e}{2m}g|\mathbf{H}|M_J \tag{8-75}$$

It is left as an exercise (Exercise 8-18) for the student to show that the above treatment produces the following general expression for the Landé g factor:

$$g = 1 + \frac{J(J+1) - L(L+1) + S(S+1)}{2J(J+1)} \tag{8-76}$$

Classical mechanics predicts that the frequency of precession of $\mathbf{J}$ about the magnetic field direction is given by

$$\nu_L = \frac{1}{2\pi}\frac{e|\mathbf{H}|}{2m} \tag{8-77}$$

which is known as the *Larmor frequency*. Since M_J is given in units of $\hbar$, we can write the interaction energy as

$$E' = 2\pi\nu_L g M_J \tag{8-78}$$

A multiplet with quantum number J having an energy $E_J^{(0)}$ in the absence of a magnetic field will be split by a magnetic field into $2J+1$ levels having the energies

$$E_{M_J} = E_J^{(0)} + 2\pi\nu_L g M_J \qquad -J \leq M_J \leq J \tag{8-79}$$

The energy splitting is thus proportional to the field strength (via the Larmor frequency). Actually, it is found that if the second-order perturbation is taken into account, the splitting actually depends on $|\mathbf{H}|^2$. The latter effect will normally be negligible at low magnetic field strengths.

For singlet states ($S=0$) one finds that $g=1$, so that the energy splitting is independent of J and L. This is called the *normal* Zeeman effect. For other multiplicities, g depends on J, L, and S and leads to what is called the *anomalous* Zeeman effect.

The selection rules on J and M_J are common to both LS and jj coupling. We will now derive the selection rules for M_J and use them to describe the Zeeman effect. For simplicity we consider only the one-electron case; the selection rules then arise in the same way as the selection rules for m_l in the hydrogen atom (see Sec. 5-3).

The Zeeman effect is observed experimentally by placing the atom (excited so as to emit radiation) in a magnetic field; this imposes a unique symmetry axis upon the system, an axis of *cylindrical* symmetry defined by the direction of the field. It is then convenient to write the atomic wavefunction in terms of cylindrical coordinates z, r, and φ defined by

$$\begin{aligned} x &= r\cos\varphi \qquad && 0 \leq r \leq \infty \\ y &= r\sin\varphi \qquad && 0 \leq \varphi \leq 2\pi \\ z &= z \qquad && -\infty \leq z \leq \infty \end{aligned} \tag{8-80}$$

where φ is the angle of rotation about the axis of the magnetic field (the z axis) and r is the distance from the z axis (see Fig. 8-12). We consider two states ψ_l and ψ_k which are eigenfunctions of $\hat{J}_z$ and which have the form

$$\psi_l = f(z,r)e^{iM_{Jl}\varphi} \qquad \psi_k = g(z,r)e^{iM_{Jk}\varphi} \tag{8-81}$$

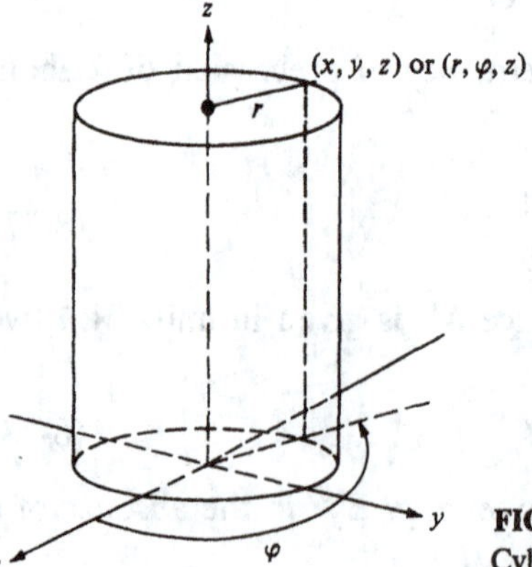

FIGURE 8-12
Cylindrical coordinates used to describe the Zeeman effect.

The x component of the dipole strength is

$$D_{kl}^{(x)} = \langle \psi_l | x | \psi_k \rangle = \int_0^\infty \int_{-\infty}^\infty \int_0^{2\pi} f^* g e^{i(M_{Jk}-M_{Jl})\varphi} \cos\varphi \, r^2 \, d\varphi \, dz \, dr$$
$$= \int_0^\infty \int_{-\infty}^\infty f^* g r^2 \, dz \, dr \int_0^{2\pi} e^{i(M_{Jk}-M_{Jl})\varphi} \cos\varphi \, d\varphi \tag{8-82}$$

If we substitute $\cos\varphi = \frac{1}{2}(e^{i\varphi} + e^{-i\varphi})$, the integral over φ becomes

$$\frac{1}{2}\left[\int_0^{2\pi} e^{i(M_{Jk}-M_{Jl}+1)\varphi} \, d\varphi + \int_0^{2\pi} e^{i(M_{Jk}-M_{Jl}-1)\varphi} \, d\varphi\right] \tag{8-83}$$

Since

$$\int_0^{2\pi} e^{ia\varphi} \, d\varphi = 0 \qquad \text{if } a \neq 0 \tag{8-84}$$

it follows that $D_{kl}^{(x)}$ will not vanish only if

$$\Delta M_J = \pm 1 \tag{8-85}$$

It is readily verified that the y component of D_{kl} leads to the same selection rule. Thus this selection rule [Eq. (8-85)] refers to radiation polarized in the xy plane, that is, σ polarization.

The z component of D_{kl} is given by

$$D_{kl}^{(z)} = \langle \psi_l | z | \psi_k \rangle = \int_0^\infty \int_{-\infty}^\infty f^* g z r^2 \, dz \, dr \int_0^{2\pi} e^{i(M_{Jk}-M_{Jl})\varphi} \, d\varphi \tag{8-86}$$

In order for this integral in φ not to vanish, it is necessary that

$$\Delta M_J = 0 \tag{8-87}$$

This selection rule applies to π-polarized radiation.

One can also show that if $\Delta J = 0$, the transition $M_{Jk} = 0$ to $M_{Jl} = 0$ is not allowed.

Figure 8-13 is an energy-level diagram for Zeeman transitions between two singlet states, ${}^1D_2$ and ${}^1F_3$. Since $g = 2$ for each of these levels, the splitting of the multiplets is the same for both states. One can see from the diagram that all transitions with the same value of ΔM_J have the same frequency, and thus one obtains three spectral lines from this particular transition.

When the two states involved in a Zeeman transition are not singlets, the g factor is different for the two levels and one finds that a given ΔM_J does not lead to transitions of the same energies; consequently, more than three lines are observed. This is shown in Fig. 8-14 for transitions between ${}^2S_{1/2}$ and ${}^2P_{1/2}$ levels. Since $g = \frac{2}{3}$ for the upper level and $g = 2$ for the lower level, the two transitions with $\Delta M_J = 0$ have different energies and lead to two different π-polarized spectral lines. For transitions between two states k and l, the transition energy is given in general by

$$\Delta E = E_{M_{Jl}} - E_{M_{Jk}} = E_l^{(0)} - E_k^{(0)} + 2\pi\nu_L(g_l M_{Jl} - g_k M_{Jk}) \tag{8-88}$$

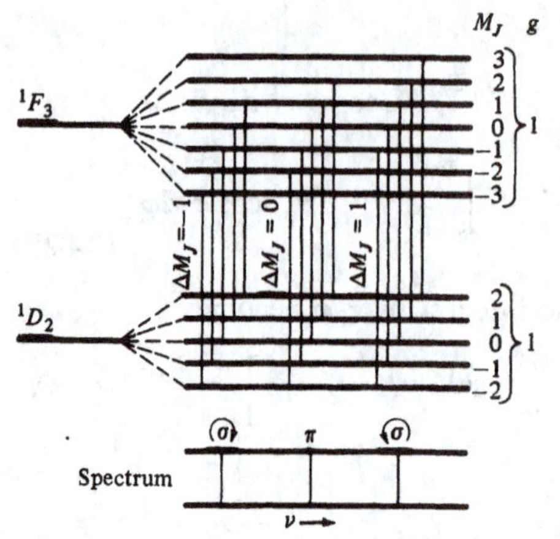

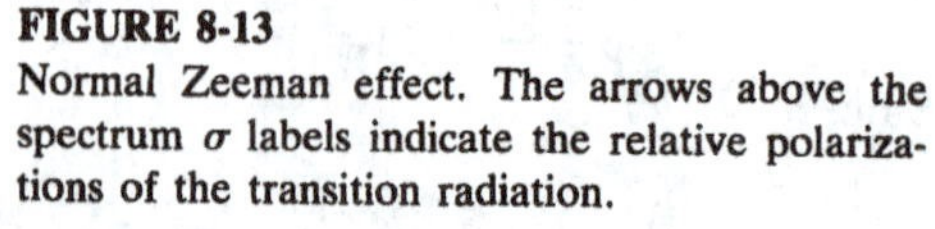
FIGURE 8-13
Normal Zeeman effect. The arrows above the spectrum σ labels indicate the relative polarizations of the transition radiation.

Therefore, the spacing between the two π-polarized lines in Fig. 8-14 is $2\pi\nu_L(\frac{4}{3})$.

The Zeeman effect is very useful in the analysis of atomic spectra, enabling one to obtain experimental information concerning the electronic structure of atoms. In particular, the Zeeman effect is of value in the identification of terms in the experimental spectrum.

At higher magnetic field strengths, the anomalous Zeeman effect begins to resemble the normal Zeeman effect. Specifically, one now obtains three sets of lines, each set consisting of very closely spaced lines. This is known as the *Paschen-Back effect*[6] and is due to the fact that at higher values of **H**, the rate of precession of **J** about the field direction becomes higher than the rate of

[6] The Paschen-Back and Stark effects are described by G. Herzberg, *Atomic Spectra and Atomic Structure*, Dover, New York, 1944, and by H. G. Kuhn, *Atomic Spectra*, Longmans, Green, New York, 1962.

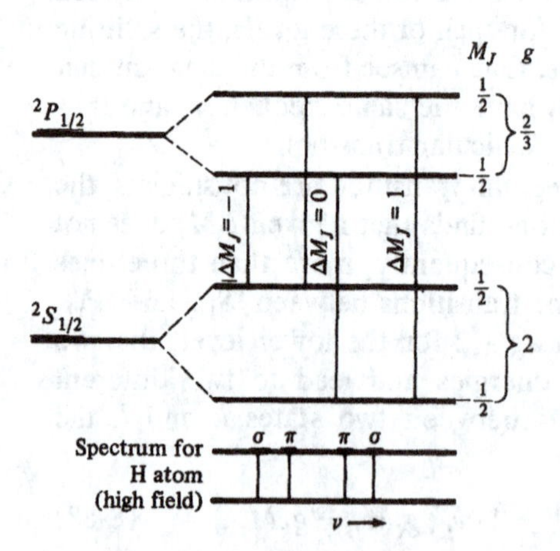

FIGURE 8-14
Anomalous Zeeman effect.

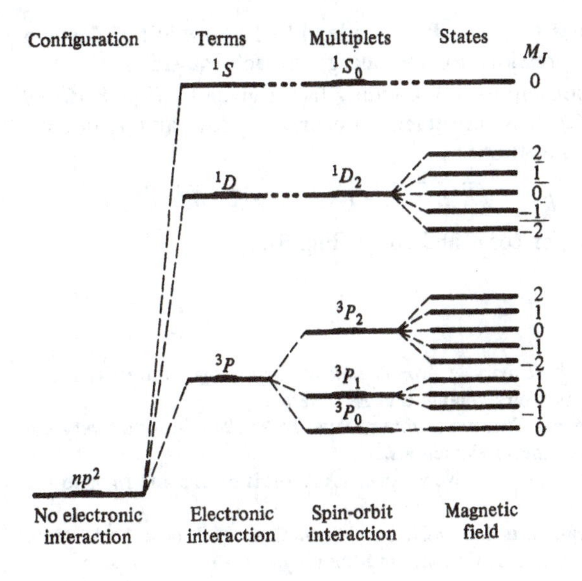

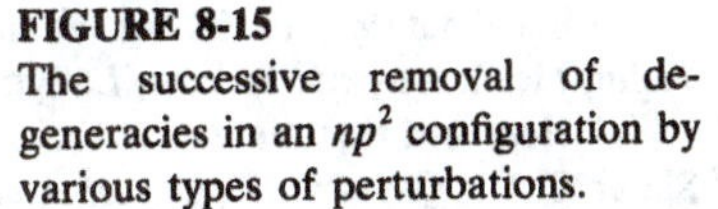
FIGURE 8-15
The successive removal of degeneracies in an np^2 configuration by various types of perturbations.

precession of **μ** about **J**, so that uncoupling of **L** and **S** occurs. If **L** and **S** were uncoupled completely, the resulting spectrum would lead to exactly three lines; however, the presence of a small amount of *LS* coupling splits these three lines slightly.

Spectral lines are also split by electric fields, a phenomenon known as the *Stark effect* (see Sec. 6-9 on the Stark effect in the hydrogen atom). Here the energy splitting depends not on M_J but on $|M_J|$. This occurs because in an electric field, a reversal in the sense of rotation leaves the energy unchanged. Since the Stark effect is of less value in the analysis of spectra than the Zeeman effect is, we will not consider it here. However, the Stark effect is very useful in the elucidation of molecular structure.

Figure 8-15 summarizes some of the main features of the present chapter. The diagram shows that if an electronic system described by the np^2 configuration is treated by a model in which electron repulsion is ignored, then a single quantum state is predicted. The addition of electron-repulsion terms to the hamiltonian immediately leads to three terms; these arise because the presence of $r_{\mu\nu}$ terms destroys the central-field nature of the potential energy, and thus the $(2l+1)$-fold degeneracy characteristic of the one-electron atom is removed. Next, interaction of **L** and **S** remove additional degeneracy. Finally, interaction of the atom with a magnetic field removes the $(2J+1)$-fold degeneracy associated with each multiplet.

EXERCISES

8-16. Show that the energies of the normal Zeeman transitions between two states with energies $E_k^{(0)}$ and $E_l^{(0)}$ are given by $\Delta E = E_k^{(0)} - E_l^{(0)} + 2\pi\nu_L M_J$. Show that this provides an experimental method for the measurement of $E_k^{(0)} - E_l^{(0)}$.

8-17. Show that the sodium D lines ($^2S_{1/2} \rightarrow {}^2P_{3/2,\,1/2}$) lead to 10 lines in the Zeeman effect. Obtain theoretical expressions for the energy of each transition.

8-18. Derive the theoretical relationship for the Landé g factor given by Eq. (8-76) by use of Eqs. (8-68) and (8-72). It is convenient to express μ_J, μ_L, and μ_S in units of $e/2m$ and to employ the substitutions

$$\mu_J = g\sqrt{J(J+1)} = ga \qquad \mu_L = \sqrt{L(L+1)} = b \qquad \mu_S = 2\sqrt{S(S+1)} = 2c$$

Also, use the law of cosines for $\cos\alpha$ and $\cos\beta$ (Fig. 8-11).

SUGGESTED READINGS

Condon, E. U., and G. H. Shortley: *The Theory of Atomic Spectra*, Cambridge University Press, New York, 1957. An old book but very complete in its coverage.

Herzberg, G.: *Atomic Spectra and Atomic Structure*, Dover, New York, 1944. Another very old book but very useful for learning the fundamentals.

Kuhn, H. G.: *Atomic Spectra*, Longmans, Green, New York, 1962. Adds details and rigor to the Herzberg treatment.

Slater, J. C.: *Quantum Theory of Atomic Structure*, vols. 1 and 2, McGraw-Hill, New York, 1960. May be regarded as a modern update of Condon and Shortley.

CHAPTER 9

THE ALGEBRA OF MANY-ELECTRON CALCULATIONS

The ready availability of powerful, versatile computer programs makes it possible for chemists of all specializations to carry out state-of-the-art quantum mechanical calculations on electronic structures of atoms and molecules. This chapter begins an examination of some of the mathematical and organizational structure of modern computer programs of electronic structure so that users of these programs are freed from a total "black box" approach. The general procedures required to do a many-electron calculation are not completely obvious extensions of the helium atom calculations previously discussed; the fact that the wavefunction of an atom with three or more electrons cannot be factored into a product of a spatial part and a spin part leads to considerable algebraic complexity when working out expectation values over determinantal wavefunctions. Furthermore, even though chemists are generally more interested in molecules than in atoms, a reasonably thorough understanding of atoms is a prerequisite for the understanding of molecules.

9-1 CONSTRUCTION OF DETERMINANTAL EIGENFUNCTIONS OF $\hat{S}^2$

In Chapter 7 we learned that determinantal wavefunctions describing the electronic states of atoms are chosen such that they are eigenfunctions of the spin operator $\hat{S}^2$. Such wavefunctions are referred to as *pure spin states*. Chapter 7 also showed how pure spin states could be constructed for the simple case of a two-electron system, namely, the helium atom. Now we will see how to extend this to the general case of more than two electrons.

First we consider the case of a three-electron system; from this, it is rather easy to induce enough generality to treat any number of electrons. Letting ω represent an N-electron function containing spin (and spatial) coordinates, we have the eigenvalue equation

$$\hat{S}_z \omega = M_S \omega \tag{9-1}$$

When $N = 3$, we obtain the following possibilities for the M_S values: $\frac{3}{2}, \frac{1}{2}, -\frac{1}{2}$, and $-\frac{3}{2}$ (see Sec. 8-3); i.e., there are four different spin alignments. As shown earlier in Sec. 8-3, there are $2^3 = 8$ different ways in which these four different M_S values can arise. These eight possibilities can be grouped as follows: the group $-\frac{3}{2} \leq M_S \leq \frac{3}{2}$, which corresponds to $S = \frac{3}{2}$ and thus represents a state with multiplicity $2S + 1 = 4$, and two groups $-\frac{1}{2} \leq M_S \leq \frac{1}{2}$ which correspond to $S = \frac{1}{2}$ and thus represent states of multiplicity $2S + 1 = 2$. This means that an atom or molecule having three electrons outside a closed shell may exist in a quartet state or in one of two different doublet states. For $N > 3$ the procedure is rather tedious to carry out, since there will be 2^N different spin couplings to identify. However, as shown earlier, use of the branching diagram (Fig. 8-7), greatly simplifies the identifications. Thus one can deduce rather quickly that for $N = 4$ one would have two singlet states ($M_S = 0$), three triplet states ($M_S = -1, 0, 1$), and one quintet state ($M_S = -2, -1, 0, 1, 2$). Thus for four electrons one can write down 16 linearly independent wavefunctions leading to six different energies.

Equation (7-72) is the basic equation for determining whether or not a given determinantal wavefunction represents a pure spin state or is a mixture of spin states. If $n_\alpha > n_\beta$, then $\Sigma \hat{P}_{\alpha\beta}$ will be a constant only if the number of doubly occupied orbitals (once with α spin and once with β spin) is equal to n_β. Thus for $|\phi_1 \bar{\phi}_1|$, $n_\alpha = n_\beta = 1$, and there is one doubly occupied orbital; consequently, this determinant represents a pure spin state (a singlet). Similarly, $|\phi_1 \bar{\phi}_1 \phi_2|$ is a pure spin state (a doublet), whereas $|\phi_1 \phi_2 \bar{\phi}_3|$ is not.

Suppose we have N different spatial orbitals $\phi_1, \phi_2, \ldots, \phi_N$. From these we can form $2N$ different spin orbitals (N with α spin and N with β spin). These, in turn, can be used to produce

$$\binom{2N}{N} = \frac{(2N)!}{N!(2N-N)!}$$

different determinants that describe a system of N electrons. Of these, 2^N will contain N *different* spatial orbitals [recall that it is these to which the branching diagram (Fig. 8-7) refers]. For $N = 3$ one obtains $6!/[3!(6-3)!] = 20$ different determinants, and $2^3 = 8$ of these will contain each of the three spatial functions ϕ_1, ϕ_2 and ϕ_3.[1] These eight (in decreasing order of their $\hat{S}_z$ eigenvalues) are

$$\begin{aligned}
D_1 &= |\phi_1\ \phi_2\ \phi_3| \qquad M_S = \tfrac{3}{2}\\
D_2 &= |\phi_1\ \phi_2\ \bar{\phi}_3| \qquad M_S = \tfrac{1}{2}\\
D_3 &= |\phi_1\ \bar{\phi}_2\ \phi_3| \qquad M_S = \tfrac{1}{2}\\
D_4 &= |\bar{\phi}_1\ \phi_2\ \phi_3| \qquad M_S = \tfrac{1}{2}\\
D_5 &= |\bar{\phi}_1\ \bar{\phi}_2\ \phi_3| \qquad M_S = -\tfrac{1}{2}\\
D_6 &= |\bar{\phi}_1\ \phi_2\ \bar{\phi}_3| \qquad M_S = -\tfrac{1}{2}\\
D_7 &= |\phi_1\ \bar{\phi}_2\ \bar{\phi}_3| \qquad M_S = -\tfrac{1}{2}\\
D_8 &= |\bar{\phi}_1\ \bar{\phi}_2\ \bar{\phi}_3| \qquad M_S = -\tfrac{3}{2}
\end{aligned} \tag{9-2}$$

The other twelve determinants are all eigenfunctions of $\hat{S}^2$ with $S(S+1) = \frac{3}{4}$ (doublet states), since each has a number of doubly occupied orbitals equal to the number of columns of α spin or β spin (whichever is smaller). Note that D_1 and D_8 in Eqs. (9-2) are also eigenfunctions of $\hat{S}^2$ by the same criterion.

From the branching diagram we see that these eight determinants must represent a quartet and two doublets. Applying $\hat{S}^2$ to D_1, we obtain

$$\begin{aligned}
&\sum P_{\alpha\beta} D_1 = 0\\
&\tfrac{1}{4}[(n_\alpha - n_\beta)^2 + 2n_\alpha + 2n_\beta]D_1 = \tfrac{1}{4}(3^2 + 6)D_1 = \tfrac{15}{4}D_1
\end{aligned} \tag{9-3}$$

Therefore

$$\hat{S}^2 D_1 = \tfrac{15}{4} D_1 \tag{9-4}$$

Similarly, we get

$$\hat{S}^2 D_8 = \tfrac{15}{4} D_8 \tag{9-5}$$

Therefore, D_1 and D_8 are two components of the quartet state [since $2S+1=4$ when $S(S+1) = \frac{15}{4}$].

For D_2 we obtain

$$\hat{S}^2 D_2 = D_3 + D_4 + \tfrac{7}{4} D_2 \tag{9-6}$$

which means that D_2 is not an eigenfunction of $\hat{S}^2$. Similarly, we find that none of the remaining determinants is an eigenfunction of $\hat{S}^2$. However, we can always form linear combinations of determinants of the same M_S value to

[1] We assume these spatial orbitals are orthonormal so that $\langle D_i | D_j \rangle = \delta_{ij}$ for the functions in Eq. (9-2).

obtain eigenfunctions of $\hat{S}^2$.[2] For example, from D_2, D_3, and D_4 one can form three different linear combinations which are such eigenfunctions. In matrix form such a linear combination may be written

$$D_i = \mathbf{D}\mathbf{C}_i \qquad i = 2, 3, 4 \tag{9-7}$$

where

$$\mathbf{D} = [D_2 \quad D_3 \quad D_4] \qquad \mathbf{C}_i = \begin{bmatrix} c_{1i} \\ c_{2i} \\ c_{3i} \end{bmatrix} \tag{9-8}$$

and where D_i is to satisfy

$$\hat{S}^2 D_i = \lambda_i D_i \tag{9-9}$$

Letting the matrix representation of $\hat{S}^2$ in the basis D_2, D_3, and D_4 be given by

$$\mathbf{S}^2 = \mathbf{D}^\dagger \hat{S}^2 \mathbf{D} \tag{9-10}$$

we obtain the matrix equation

$$(\mathbf{S}^2 - \lambda_i \mathbf{1})\mathbf{C}_i = \mathbf{0} \tag{9-11}$$

However, the three roots $\{\lambda_i\}$ are already known from the branching diagram. From D_2, D_3, and D_4 we will obtain one more quartet component and single components of two different doublet states. The roots of Eq. (9-11) then are $\frac{15}{4}$ and $\frac{3}{4}$ (twice). The matrix elements of $\mathbf{S}^2$ are readily evaluated by the aid of Eq. (7-72), and the linear combinations which diagonalize $\mathbf{S}^2$ are then constructed from the eigenvectors in a straightforward fashion.

A simpler procedure, which we now examine in some detail, involves the use of *projection operators* (see App. 5, Sec. 8). The appropriate projection operator for this purpose is

$$\hat{O}_k = \prod_{i \neq k} \frac{\hat{S}^2 - \lambda_i}{\lambda_k - \lambda_i} \tag{9-12}$$

Any arbitrary three-electron determinant constructed from the orbitals ϕ_1, ϕ_2, and ϕ_3 can be regarded as a linear combination of determinants, each of which is an eigenfunction of $\hat{S}^2$. Thus, application of the projection operator $\hat{O}_k$ to such an arbitrary determinant will project that determinantal component which

[2] The restriction to the same M_S values in a particular linear combination is a result of Theorem 2-5; i.e., since $\hat{S}_z$ and $\hat{S}^2$ commute $\langle D_i|\hat{S}^2|D_j\rangle = 0$ unless D_i and D_j have the same S_z eigenvalues. It is not too difficult to see by inspection that the 8×8 matrix representation of $\hat{S}^2$ in the basis of the eight determinantal functions given in Eq. (9-2) does indeed factor to four different matrix representations of $\hat{S}^2$, that is, to two 1×1 representations and two 3×3 representations. Theorem 2-5 enables one to recognize this result ahead of time: thus the restriction to the same M_S values.

has the $\hat{S}^2$ eigenvalue λ_k. The operator which will project out the quartet state is

$$\hat{O}_1 = \frac{\hat{S}^2 - \frac{3}{4}}{\frac{15}{4} - \frac{3}{4}} = \frac{1}{3}(\hat{S}^2 - \tfrac{3}{4}) \tag{9-13}$$

(The factor of $\frac{1}{3}$ could just as well be omitted at this time, but we will retain it to avoid possible confusion.) Applying this operator to D_2 (or to D_3 or D_4) and using Eq. (7-72), one obtains

$$\hat{O}_1 D_2 = \tfrac{1}{3}(D_3 + D_4 + \tfrac{7}{4}D_2 - \tfrac{3}{4}D_2) = \tfrac{1}{3}(D_2 + D_3 + D_4) \tag{9-14}$$

This is an eigenfunction (not normalized) of $\hat{S}^2$ with the eigenvalue $\frac{15}{4}$.

The doublet-state projection operator is

$$\hat{O}_2 = \frac{\hat{S}^2 - \frac{15}{4}}{\frac{3}{4} - \frac{15}{4}} = -\tfrac{1}{3}(\hat{S}^2 - \tfrac{15}{4}) \tag{9-15}$$

Applying this operator to D_2, D_3, and D_4, we get

$$O_2 \begin{cases} D_2 = \frac{1}{3}(2D_2 - D_3 - D_4) \\ D_3 = \frac{1}{3}(2D_3 - D_4 - D_2) \\ D_4 = \frac{1}{3}(2D_4 - D_2 - D_3) \end{cases} \tag{9-16}$$

Since D_2, D_3, and D_4 contain two single components of two different doublets, the above functions must represent *two* doubly spin-degenerate functions. From the cyclic symmetry of the indices we see that these three functions can be visualized as three vectors 120° apart in the same plane. Thus the vector formed by the difference of any two functions will be orthogonal to the remaining function. One possible set of orthogonal functions representing the doublet components is

$$\begin{aligned} \hat{O}_2(D_2 - D_3) &= D_2 - D_3 \\ \hat{O}_2 D_4 &= \tfrac{1}{3}(2D_4 - D_2 - D_3) \end{aligned} \tag{9-17}$$

Similarly, we find that the functions D_5, D_6, and D_7 lead to the remaining quartet component and to the two remaining components of the two doublets.

EXERCISES

9-1. Supply the 12 determinants which complement those given in Eq. (9-2) and show explicitly why these are eigenfunctions of $\hat{S}^2$.

9-2. Construct the eigenfunctions of $\hat{S}^2$ arising from D_5, D_6, and D_7 of Eq. (9-2).

9-3. Prove the orthogonality of the functions given in Eq. (9-17). Also, normalize each of these functions.

9-4. Show that $|\phi_1\,\bar{\phi}_2\,\phi_3\,\bar{\phi}_4|$, $|\bar{\phi}_1\,\phi_2\,\bar{\phi}_3\,\phi_4|$, $|\phi_1\,\phi_2\,\bar{\phi}_3\,\bar{\phi}_4|$, $|\bar{\phi}_1\,\bar{\phi}_2\,\phi_3\,\phi_4|$, $|\bar{\phi}_1\,\phi_2\,\phi_3\,\bar{\phi}_4|$, and $|\phi_1\,\bar{\phi}_2\,\bar{\phi}_3\,\phi_4|$ are not eigenfunctions of $\hat{S}^2$. Use spin-projection operators to construct six orthogonal eigenfunctions of $\hat{S}^2$ from these six functions.

9-2 MANIPULATION OF DETERMINANTS IN MANY-ELECTRON CALCULATIONS

The most general form of an N-electron atomic wavefunction in the orbital approximation is a linear combination of Slater determinants:

$$\xi_i(1, 2, \ldots, N) = \sum_k^m D_k \gamma_{ki} \tag{9-18}$$

where the $\{\gamma_{ki}\}$ are variationally determined numerical coefficients. If the spin orbitals used to make up the determinants belong to a complete basis set, then the above can—at least in principle—lead to an exact solution of the Schrödinger equation. In practice, however, only a limited number of Slater determinants are used in the expansion, so that the expectation value $\langle \xi_i | \hat{H} | \xi_i \rangle$ is but an upper limit to the exact energy.

The variational problem can be put into matrix form by defining the determinantal basis

$$\mathbf{D} = [D_1 \quad D_2 \quad \cdots \quad D_m] \tag{9-19}$$

The matrix representation of the hamiltonian in this basis is

$$\mathbf{H} = \mathbf{D}^\dagger \hat{H} \mathbf{D} \tag{9-20}$$

The linear variational coefficients which diagonalize the above matrix are of the form

$$\boldsymbol{\gamma} = [\boldsymbol{\gamma}_1 \quad \boldsymbol{\gamma}_2 \quad \cdots \quad \boldsymbol{\gamma}_m] \qquad \boldsymbol{\gamma}_i = \begin{bmatrix} \gamma_{1i} \\ \gamma_{2i} \\ \vdots \\ \gamma_{mi} \end{bmatrix} \tag{9-21}$$

If the spin orbital basis $\{S_i\}$ is orthonormal, then $\mathbf{D}^\dagger \mathbf{D} = \mathbf{1}$ and the matrix equation to be solved is

$$\mathbf{H}\boldsymbol{\gamma} = \boldsymbol{\gamma}\mathbf{E} \tag{9-22}$$

where $\mathbf{E}$ is a diagonal matrix of m eigenvalues of $\mathbf{H}$. These m eigenvalues of $\mathbf{H}$ are approximations of the exact eigenvalues of the system hamiltonian $\hat{H}$. The m eigenvalues of $\mathbf{H}$ are determined as the m roots of the secular determinant

$$\det |\mathbf{H} - x\mathbf{1}| = 0 \tag{9-23}$$

The heart of the calculation resides in obtaining explicit expressions for the matrix elements of $\mathbf{H}$ and determining their numerical values. It is the former we consider in this section. Once this is done, actual diagonalization of the matrix representation of $\hat{H}$ to obtain numerical values of the roots is generally carried out with a computer; a large variety of programs to do this are now readily available.

In obtaining explicit forms for the matrix elements of $\mathbf{H}$, it is convenient to define the matrices

$$\mathbf{h} = \mathbf{D}^\dagger \left(\sum_\mu^N h_\mu \right) \mathbf{D} \qquad \mathbf{G} = \mathbf{D}^\dagger \left(\sum_{\mu<\nu}^N \frac{1}{r_{\mu\nu}} \right) \mathbf{D} \tag{9-24}$$

so that the matrix representation of the hamiltonian is written as the matrix sum

$$\mathbf{H} = \mathbf{h} + \mathbf{G} \tag{9-25}$$

The general matrix element of **H** is then written as

$$H_{pq} = h_{pq} + G_{pq} \tag{9-26}$$

where

$$h_{pq} = \left\langle D_p \left| \sum_{\mu}^{N} h_\mu \right| D_q \right\rangle \qquad G_{pq} = \left\langle D_p \left| \sum_{\mu<\nu}^{N} \frac{1}{r_{\mu\nu}} \right| D_q \right\rangle \tag{9-27}$$

Recall that h_μ is the monoelectronic operator $-\nabla_\mu^2/2 - Z/r_\mu$.

Evaluation of the matrix elements is based on recognition of the fact that every pair of determinants D_p and D_q occurring in the basis $\{D_i\}$ can be placed into one of the following classifications:

Type A. $D_p = D_q$; that is, the two determinants have the same spin orbitals in their corresponding columns.

Type B. D_p and D_q differ in *one* spin orbital; i.e., where D_p has S_I, D_q has S_L.

Type C. D_p and D_q differ in *two* spin orbitals; i.e., where D_p has S_I and S_J, D_q has S_L and S_M, respectively.

Type D. D_p and D_q differ in *three or more* spin orbitals.

In the above and in later uses we will find it notationally convenient to use uppercase subscripts $I, J, K, \ldots$ to indicate spin orbitals and lowercase subscripts $i, j, k, \ldots$ to indicate the spatial components of spin orbitals. Electron coordinates will be designated (when necessary) by lowercase Greek subscripts $\mu, \nu, \sigma, \ldots$ as before. This notation enables one to tell at a glance that a quantity $\langle IJ|g|KL\rangle$ refers to an integral over spin orbitals and that $\langle ij|g|kl\rangle$ refers to an integral over spatial orbitals; i.e., there are no spin variables to integrate over in the latter. In classifying pairs of determinants according to the above types, it is necessary to exclude trivial differences between a pair of determinants which arise owing to constant multiplicative factors. For example, the determinants $D_1 = |\phi_1\, \bar{\phi}_1\, \phi_2|$ and $D_2 = |\phi_1\, \phi_2\, \bar{\phi}_1|$ appear to differ in *two* columns and thus to belong to type C. However, by exchanging columns 2 and 3 in D_2, one obtains $-|\phi_1\, \bar{\phi}_1\, \phi_2| = -D_1$. Thus the two determinants actually belong to type A. Such relationships become easier to spot if one first places the entire set of determinants into maximum coincidence of columns by an appropriate permutation of columns. Any changes of sign arising from such permutations must be kept track of; i.e., the sign must accompany the determinant.

From Eq. (7-52) we recall that if $\hat{F}$ is any operator which satisfies the commutation relation $[\hat{F}, \hat{A}] = 0$ (where $\hat{A}$ is the antisymmetrization operator), its matrix element in the basis $\{D_i\}$ will be given by

$$F_{pq} = \langle D_p|\hat{F}|D_q\rangle = \langle \hat{A}\Phi_p|\hat{F}|\hat{A}\Phi_q\rangle = \sum_P (-1)^m \langle \Phi_p|\hat{F}|\hat{P}\Phi_q\rangle \tag{9-28}$$

where we assume $\langle S_I|S_J\rangle = \delta_{IJ}$, so that $\langle D_p|D_q\rangle = \delta_{pq}$ and where m is the same as p in Eq. (7-34). The matrix element h_{pq} involves a sum of monoelectronic operators, and thus only the identity permutations in Eq. (9-28) lead to nonzero integrals. For type A, h_{pq} becomes

$$\begin{aligned} h_{pq} &= \sum_\mu^N \langle S_1(1)S_2(2)\cdots S_I(\mu)\cdots S_N(N)|h_\mu|S_1(1)S_2(2)\cdots S_I(\mu)\cdots S_N(N)\rangle \\ &= \sum_\mu^N \langle S_I(\mu)|h_\mu|S_I(\mu)\rangle = \sum_I^N \alpha_I \end{aligned} \tag{9-29}$$

where we introduce the convenient notation

$$\langle S_I(\mu)|h_\mu|S_I(\mu)\rangle = \langle I|h|I\rangle = \alpha_I \tag{9-30}$$

Note that the summation in Eq. (9-29) is over the N spin orbitals in D_p (equal to D_q). By integrating over the spin functions in the usual way, we can reduce each of these N integrals to integrals over spatial functions. In fact, we see that if $S_I(\mu) = \phi_i(\mu)\omega(\mu)$, integration over spin leads to

$$\alpha_I = \langle i|h|i\rangle = \epsilon_i^{(0)} \tag{9-31}$$

The notation $\epsilon_i^{(0)}$, which we will employ from now on, is a convenient one; if the orbital ϕ_i is the exact eigenfunction of the monoelectronic operator h (equal to $-\nabla^2/2 - Z/r$), then $\epsilon_i^{(0)}$ is an exact eigenvalue of that operator and, more importantly, represents a contribution to the *zeroth-order approximation* to the energy of any system in which the monoelectronic operator appears in $\hat{H}^{(0)}$. For example, the zeroth-order hamiltonian of an N-electron atom is usually chosen to be of the form

$$\hat{H}^{(0)} = \sum_\mu^N h_\mu \tag{9-32}$$

In such a case the zeroth-order energy of the system is given by a sum of $\epsilon_i^{(0)}$ terms, namely,

$$E^{(0)} = \sum_i^N \epsilon_i^{(0)} \tag{9-33}$$

We will use this notation even when ϕ_i is not an exact eigenfunction of h.

If D_p and D_q belong to type B, there will be just one term in h_{pq}, and this will involve the particular spin orbitals in which the two determinants differ, namely,

$$\begin{aligned} &\langle S_1(1)S_2(2)\cdots S_I(\mu)\cdots S_N(N)|h_\mu|S_1(1)S_2(2)\cdots S_L(\mu)\cdots S_N(N)\rangle \\ &\quad = \langle S_I(\mu)|h_\mu|S_L(\mu)\rangle = \langle I|h|L\rangle = \beta_{IL} \end{aligned} \tag{9-34}$$

If $S_I(\mu) = \phi_i(\mu)\omega_1(\mu)$ and $S_L(\mu) = \phi_j(\mu)\omega_2(\mu)$, then

$$\beta_{IL} = \begin{cases} \langle i|h|j\rangle \delta_{12} = \beta_{ij}\delta_{12} & \text{if } i \neq j \\ 0 & \text{if } i = j \end{cases} \tag{9-35}$$

The latter relationship means that if the spatial orbitals of S_I and S_L are the same ($i = j$), then the spin functions must be different; hence, orthogonality of spin functions causes the entire integral to vanish. Also, one can see that β_{ij} will vanish whenever ϕ_i and ϕ_j are two different nondegenerate eigenfunctions of h.

For pairs of determinants belonging to type C or to type D, matrix elements of h vanish; the proof is left to the reader (see Exercise 9-5).

Next, we consider the two-electron matrix elements G_{pq}. Here, nonzero integrals arise not only for identity permutations but also for two-electron permutations as well. For pairs of determinants of type A, identity permutations leads to integrals such as

$$\left\langle S_1(1)S_2(2)\cdots S_I(\mu)S_J(\nu)\cdots S_N(N)\left|\frac{1}{r_{\mu\nu}}\right|S_1(1)S_2(2)\cdots S_I(\mu)S_J(\nu)\cdots S_N(N)\right\rangle$$

$$= \left\langle S_I(\mu)S_J(\nu)\left|\frac{1}{r_{\mu\nu}}\right|S_I(\mu)S_J(\nu)\right\rangle = \langle IJ|g|IJ\rangle \tag{9-36}$$

Each of the above integrals is a coulombic integral in terms of spin orbitals. For the two-electron permutations, we obtain

$$-\left\langle S_I(\mu)S_J(\nu)\left|\frac{1}{r_{\mu\nu}}\right|S_J(\mu)S_I(\nu)\right\rangle = -\langle IJ|g|JI\rangle \tag{9-37}$$

which can be regarded as a sort of exchange integral in terms of spin orbitals. The total contribution to G_{pq} from type A determinants is then given by

$$G_{pq} = \sum_{I<J}^{N} (\langle IJ|g|IJ\rangle - \langle IJ|g|JI\rangle) \tag{9-38}$$

From type B pairs of determinants we obtain

$$\sum_{J\neq I,\, L} \left\langle S_1(1)S_2(2)\cdots S_I(\mu)S_J(\nu)\cdots S_N(N)\left|\frac{1}{r_{\mu\nu}}(1-\hat{P}_{LJ})\right| \right.$$

$$\left. \times \left|S_1(1)S_2(2)\cdots S_L(\mu)S_J(\nu)\cdots S_N(N)\right\rangle\right.$$

$$= \sum_{J\neq I,\, L} \left\langle S_I(\mu)S_J(\nu)\left|\frac{1}{r_{\mu\nu}}(1-\hat{P}_{LJ})\right|S_L(\mu)S_J(\nu)\right\rangle$$

$$= \sum_{J\neq I,\, L} (\langle IJ|g|LJ\rangle - \langle IJ|g|JL\rangle) \tag{9-39}$$

The summation is over only the S_J which D_p and D_q have in common.

For type C pairs of determinants it is readily seen that one obtains only the two integrals

$$\langle IJ|g|LM\rangle - \langle IJ|g|ML\rangle \tag{9-40}$$

All terms in G_{pq} which involve type D pairs of determinants vanish.

Summarizing the results, the matrix elements H_{pq} satisfy the following relationships:[3]

Type A: $$H_{pq} = \sum_{I=1}^{N} \alpha_I + \sum_{I<J}^{N} (\langle IJ|g|IJ\rangle - \langle IJ|g|JI\rangle) \tag{9-41}$$

Type B: $$H_{pq} = \beta_{IL} + \sum_{J\neq I,\,L}^{N} (\langle IJ|g|LJ\rangle - \langle IJ|g|JL\rangle) \tag{9-42}$$

Type C: $$H_{pq} = \langle IJ|g|LM\rangle - \langle IJ|g|ML\rangle \tag{9-43}$$

Type D: $$H_{pq} = 0 \tag{9-44}$$

It is important to note that any sign changes resulting from putting the determinants into maximum coincidence must be carried along. Further reduction to integrals over spatial orbitals is carried out by performing the spin integrations in the usual manner. In the special case when the wavefunction is expressed as a single determinant of N doubly occupied spatial orbitals, integration of Eq. (9-41) over the spin coordinates leads to the result

$$E = 2\sum_{i=1}^{N} \epsilon_i^{(0)} + \sum_{i,\,j}^{N} (2J_{ij} - K_{ij}) \tag{9-45}$$

The first summation term may be regarded as equivalent to a zeroth-order energy—it is the energy the atom would have in the absence of electron repulsion, provided this is calculated from the same basis orbitals. The second summation term is then the energy due to electron repulsion; i.e., it represents the perturbation energy.

For the simple case of the helium atom using the wavefunction $|1s\,\overline{1s}|$, the total energy may be written down at once as

$$E = 2\langle 1s|h|1s\rangle + \langle 1s(1)1s(2)|g|1s(1)1s(2)\rangle = 2\epsilon_{1s}^{(0)} + J_{1s,\,1s} \tag{9-46}$$

For beryllium, with the wavefunction $|1s\,\overline{1s}\,2s\,\overline{2s}|$, the energy is

$$E = 2\epsilon_{1s}^{(0)} + 2\epsilon_{2s}^{(0)} + J_{1s,\,1s} + J_{2s,\,2s} + 4J_{1s,\,2s} + 2K_{1s,\,2s} \tag{9-47}$$

Here we have used the readily deducible relationships:

$$J_{ii} = K_{ii} \qquad J_{ij} = J_{ji} \qquad K_{ij} = K_{ji} \tag{9-48}$$

[3] These rules were originally developed by J. C. Slater, *Phys. Rev.*, **34**:1293 (1929) and **38**:1109 (1931) and by E. U. Condon, *Phys. Rev.*, **36**:1121 (1930).

Integration of Eq. (9-41) over spin coordinates also produces the total energy in terms of spatial orbitals for any one-determinant wavefunction: open- or closed-shell. For example, consider the lithium atom in the $1s^2 2s$ configuration. The determinantal wavefunction (a doublet eigenfunction of $\hat{S}^2$) is $|1s\,\overline{1s}\,2s|$ (for the $M_S = \frac{1}{2}$ component), where we will assume that the spatial functions are orthonormal, hydrogenlike orbitals. Letting $S_1(1) = 1s(1)\alpha(1)$, $S_2(2) = 1s(2)\beta(2)$, and $S_3(3) = 2s(3)\alpha(3)$ (or, more simply, $S_1 = 1s\alpha$, $S_2 = 1s\beta$, and $S_3 = 2s\alpha$), we obtain from Eq. (9-41)

$$
\begin{aligned}
E = \alpha_1 + \alpha_2 + \alpha_3 &+ \langle 12|g|12\rangle + \langle 13|g|13\rangle \\
&+ \langle 23|g|23\rangle - \langle 12|g|21\rangle - \langle 13|g|31\rangle - \langle 23|g|32\rangle
\end{aligned} \tag{9-49}
$$

where all integrals are over spin orbitals. Integrating the above over spin coordinates and adopting some simplified notation ($1s$ subscript becomes 1; $2s$ subscript becomes 2), we obtain for the various integrals:

$$
\begin{aligned}
\alpha_1 = \alpha_2 &= \langle 1s|h|1s\rangle = \epsilon_1^{(0)} \\
\alpha_3 &= \langle 2s|h|2s\rangle = \epsilon_2^{(0)} \\
\langle 12|g|12\rangle &= \langle 1s(1)1s(2)|g|1s(1)2s(2)\rangle = J_{11} \\
\langle 13|g|13\rangle &= \langle 23|g|23\rangle = \langle 1s(1)2s(2)|g|1s(1)2s(2)\rangle = J_{12} \\
\langle 12|g|12\rangle &= \langle 23|g|32\rangle = 0 \\
\langle 13|g|31\rangle &= \langle 1s(1)2s(2)|g|2s(1)1s(2)\rangle = K_{12}
\end{aligned} \tag{9-50}
$$

The total energy [Eq. (9-49)] then reduces to

$$
E = 2\epsilon_1^{(0)} + \epsilon_2^{(0)} + J_{11} + 2J_{12} - K_{12} \tag{9-51}
$$

EXERCISES

9-5. Prove that the matrix elements of the monoelectronic part of the hamiltonian vanish in the cases of type C and type D.

9-6. Equations (9-41) to (9-44) are valid only if one uses a set of orthonormal spatial orbitals as a basis (so that $\langle D_p | D_q \rangle = \delta_{pq}$). Show how the results would change if one had to deal with nonorthogonal (but normalized) basis functions.

9-7. Use Eq. (9-41) to obtain a general expression for the energy of boron in the $1s^2 2s^2 2p$ configuration.

9-8. Consider the wavefunctions $\psi_1 = |1s\,\overline{1s}|$ and $\psi_2 = N(|1s\,\overline{2s}| - |\overline{1s}\,2s|)$, where $\langle 1s | 1s \rangle = \langle 2s | 2s \rangle = 1$ and $\langle 1s | 2s \rangle = \Delta \neq 0$.

(*a*) Show that $N = [2(1 + \Delta^2)]^{-1/2}$ and $\langle \psi_1 | \psi_2 \rangle = 2\Delta N$.

(*b*) What is the *physical* meaning of the fact that ψ_1 and ψ_2 are not orthogonal?

9-3 THE GROUND-STATE ENERGY OF THE LITHIUM ATOM

Once a general energy expression such as Eq. (9-46), (9-47), or (9-51) is determined, it remains to obtain explicit forms for all the integrals appearing in the expression and—most difficult of all—to optimize the energy with respect to various possible parameters occurring in the energy expression. The latter task, which is very simple for a one-parameter expression, becomes increasingly difficult as the number of parameters increases. Without the use of a high-speed electronic computer, many such calculations would be too time-consuming to be done.[4]

The first step is to decide just what basis orbitals to employ. Since the rules given in Sec. 9-2 are based on the assumption of orthonormal orbitals, the energy expressions obtained from these are valid only for orthonormal orbitals. For the case of lithium we will choose the hydrogenlike orbitals

$$1s = ae^{-\alpha r} \qquad 2s = b(1 - \lambda\beta r)e^{-\beta r} \tag{9-52}$$

where α and β are scaling parameters, λ is chosen to make $1s$ and $2s$ orthogonal, and a and b are normalization constants. In the one-electron atom we have $\alpha = 2\beta$, $\lambda = 1$, $a = Z^{3/2}/\sqrt{\pi}$, and $b = (Z/2)^{3/2}/\sqrt{\pi}$. It should be noted that if $\lambda = 1$, then $1s$ and $2s$ are orthogonal only if $\alpha = 2\beta$, that is, if only *one* scale parameter is employed. It is readily established that $1s$ and $2s$ are orthonormal if we choose

$$a = \frac{\alpha^{3/2}}{\sqrt{\pi}} \qquad b = \frac{\beta^{3/2}}{\sqrt{\pi(3\lambda^2 - 3\lambda + 1)}} \qquad \lambda = \frac{\alpha + \beta}{3\beta} \tag{9-53}$$

Alternatively, one could have chosen the two scaled Slater orbitals

$$1s = \frac{1}{\sqrt{\pi}}\,\alpha^{3/2}e^{-\alpha r} \qquad 2s = \frac{\beta^{5/2}}{\sqrt{3\pi}}\,re^{-\beta r/2} \tag{9-54}$$

Since these are not orthogonal, they may be orthogonalized (for example, by the Schmidt procedure). This produces the two new orbitals

$$1s' = 1s \qquad 2s' = N(2s - \Delta 1s) \tag{9-55}$$

where $\Delta = \langle 1s|2s\rangle$ and N is a normalization constant. As we will see later, these two types of orthogonalization do not lead to the same final energy.

The one-electron integrals in Eq. (9-51) are evaluated in a straightforward manner (although the algebra is tedious). One obtains (for the case of general atomic number Z)

$$\begin{aligned} \epsilon_1^{(0)} &= \frac{\alpha^2}{2} - Z\alpha \\ \epsilon_2^{(0)} &= \frac{\beta^2}{6} + \frac{\beta^4}{\alpha^2 + \beta^2 - \alpha\beta} - \frac{Z\beta}{2} + \frac{Z\beta^2(\alpha - 2\beta)/2}{\alpha^2 + \beta^2 - \alpha\beta} \end{aligned} \tag{9-56}$$

[4] Before the advent of the modern digital computer, such calculations were carried out by armies of secretaries using mechanical calculators—and laboring perhaps a year or more on a single problem which one can do today in a few seconds! Most computer programs in use today can be obtained from the Quantum Chemistry Program Exchange (QCPE), Chemistry Department, Indiana University, Bloomington, IN 47401.

The two-electron integrals can be evaluated by a variety of procedures (see App. 3). Alternatively, one can save a great deal of labor using general formulas found in several different compilations frequently used by quantum chemists.[5] Since these compilations list formulas for integrals over Slater orbitals, one must rewrite the functions given in Eq. (9-52) in terms of such orbitals. The 1*s* function leads to no difficulty, since it has the same form as a Slater orbital, but to rewrite 2*s* one first defines the 1*s* and 2s Slater orbitals

$$\phi_1' = b_1 e^{-\beta r} \qquad \phi_2' = b_2 r e^{-\beta r} \tag{9-57}$$

where b_1 and b_2 are normalization constants given by

$$b_1 = \frac{\beta^{3/2}}{\sqrt{\pi}} \qquad b_2 = \frac{\beta^{5/2}}{\sqrt{3\pi}} \tag{9-58}$$

The hydrogenlike 2*s* function may then be written in terms of the Slater ϕ_1' and ϕ_2' orbitals as

$$2s = \frac{b}{b_1}\phi_1' - \frac{b\lambda\beta}{b_2}\phi_2' \tag{9-59}$$

This enables one to express all integrals involving the hydrogenlike function 2*s* in terms of the Slater orbitals ϕ_1' and ϕ_2'. In any event one obtains the results

$$\begin{aligned} J_{11} &= \frac{5\alpha}{8} \\ J_{12} &= \alpha - \frac{\alpha^3}{(\alpha+\beta)^2} - \frac{1}{2}\frac{\alpha^4\beta(3\alpha+\beta)}{(\alpha+\beta)^3(\alpha^2+\beta^2-\alpha\beta)} \\ K_{12} &= \frac{4\alpha^3\beta^5}{(\alpha+\beta)^5(\alpha^2+\beta^2-\alpha\beta)} \end{aligned} \tag{9-60}$$

The total energy can now be written in terms of the scale parameters α and β. However, instead of minimizing such an expression directly with respect to α and β, it is simpler to effect the substitution

$$\alpha = k\beta \tag{9-61}$$

and to minimize the resulting energy expression with respect to β and k. The substitution indicated in Eq. (9-61) is suggested once one notes that terms appearing in the total energy have "dimensions" either of a scale parameter or of the square of a scale parameter. The above substitution enables one to write

[5] The most well known are J. Miller, J. M. Gerhausen, and F. A. Matsen, *Quantum Chemistry Integrals and Tables*, University of Texas Press, Austin, 1959; M. Kotani, A. Amemiya, E. Ishiguro, and T. Kimura, *Tables of Molecular Integrals*, Maruzen, Tokyo, 1963; and H. Preuss, *Integraltafeln zur Quantenchemie*, 4 vols., Springer-Verlag, Berlin, 1956.

the total energy of the lithium atom in the form (setting $Z = 3$)

$$E = \beta^2 f_1(k) + \beta f_2(k) \tag{9-62}$$

where

$$f_1(k) = \frac{6k^4 - 6k^3 + 7k^2 - k + 7}{6(k^2 - k + 1)}$$

$$f_2(k) = (k^2 - k + 1)^{-1}\left[\frac{3}{2}(k - 2) - k^4 \frac{3k + 1}{(k + 1)^3} - \frac{4k^3}{(k + 1)^5}\right]$$

$$- \frac{2k^3}{(k + 1)^2} - \frac{3}{2} - \frac{27k}{8} \tag{9-63}$$

If one uses $\alpha = 2\beta = 3$ (no scaling), the energy turns out to be $-7.056E_h$ compared with the experimental nonrelativistic energy of $-7.478E_h$. This is an error of 5.6 percent. Not only is this a very poor approximation of the energy (as is typical when unscaled hydrogenlike orbitals are used in atomic calculations), but also the virial theorem is not satisfied (see Sec. 6-6). The scaled calculation (which leads to satisfaction of the virial theorem) may be carried out as follows:

$$\frac{\partial E}{\partial \beta} = 2\beta f_1(k) + f_2(k) \qquad \text{and thus} \qquad \beta = -\frac{f_2(k)}{2f_1(k)} \tag{9-64}$$

and the energy becomes

$$E = -\frac{f_2^2}{4f_1} \tag{9-65}$$

Minimizing E with respect to k (which can be done graphically) leads to $k = 3.512$. This, in turn, produces $\beta = 0.767$ and $\alpha = 2.694$. This leads to an energy of $-7.414E_h$; this is $0.064E_h$ higher than the experimental energy—an error of 0.86 percent.[6]

If one uses a single scale factor $\alpha = 2\beta$, the energy obtained is $-7.289E_h$ (an error of 2.5 percent) when $\alpha = 2.795$.

If one uses the Schmidt-orthogonalized Slater functions [Eq. (9-55)], the energy turns out to be $-7.418E_h$; an error of 0.80 percent. The parameter values leading to this minimum energy are $\alpha = 2.686$ and $\beta = 0.637$. Note that

[6] The reader should not be misled into thinking that this is a very accurate calculation. As we will discover later, when calculations of this type are used to provide theoretical values for molecular binding energies, an error of this magnitude is greater than the quantity one is trying to calculate! This is because the binding energy of a molecule is a small difference between two very large numbers. The late C. A. Coulson liked to compare the procedure to trying to determine the weight of the captain of the *Queen Mary* by doing separate weighings of the *Queen Mary* with the captain aboard and with him ashore—and then taking the difference of the two weights.

this is a slightly *lower* energy than obtained from orthogonalized hydrogenlike orbitals. Use of nonorthogonalized Slater orbitals leads (in this particular case) to the same energy as for orthogonalized Slater orbitals.[7]

Calculations involving more than two nonlinear variational parameters become very time-consuming to carry out. The difficulty resides in the fact that there is no general solution to the problem of minimizing a function of two or more variables. In the case of linear parameters, one has a general solution in terms of the secular equations, but for nonlinear parameters no comparable procedure is known.

EXERCISES

9-9. Consider the lithium atom wavefunction $\xi(1,2,3) = N(|1s\,\overline{1s}'\,2s| - |1s\,1s'\,\overline{2s}|)$, where N is a normalization constant and $1s$ and $1s'$ have different scale factors.

(*a*) Show that $\hat{S}^2\xi = \frac{3}{4}\xi$; this describes a doublet state.

(*b*) Find the explicit form of the normalization constant N; note that a factor of $1/\sqrt{3!}$ is understood in each determinantal function.

Answer: $N = (2\Delta_{13}^2 - \Delta_{12}^2 + 2\Delta_{23}^2 - 2\Delta_{12}\Delta_{23}\Delta_{13})^{-1/2}$. *Note*: Subscripts refer to orbitals as follows: $1 = 1s$, $2 = 1s'$, $3 = 2s$.

(*c*) Determine the energy associated with the above wavefunction.

Answer:

$$\begin{aligned}
E = \langle \xi|\hat{H}|\xi\rangle = N^2[&2\epsilon_1^{(0)}(1+\Delta_{23}^2) + \epsilon_2^{(0)}(2-\Delta_{13}^2) \\
&+ \epsilon_3^{(0)}(2-\Delta_{12}^2) - \beta_{12}(2\Delta_{12} + \Delta_{23}\Delta_{13}) + 2\beta_{23}(\Delta_{23} - \Delta_{12}\Delta_{13}) \\
&- \beta_{13}(2\Delta_{13} + 2\Delta_{12}\Delta_{23}) + 2(J_{12} + J_{13} + J_{23} + K_{23}) - K_{12} \\
&- K_{13} - 2(\langle 12|g|32\rangle\Delta_{13} + \langle 13|g|23\rangle\Delta_{12} - 2\langle 12|g|13\rangle\Delta_{23} \\
&+ \langle 12|g|23\rangle\Delta_{13} + \langle 13|g|21\rangle\Delta_{23} + \langle 23|g|31\rangle\Delta_{12})]
\end{aligned}$$

where $\beta_{ij} = \langle i|h|j\rangle$ for $i \neq j$.

9-10. The experimental electronic energy of the lithium atom (nonrelativistic) is $-7.478E_h$, and the first ionization energy is 5.39 eV. Calculate the second and third ionization energies of lithium.

9-4 THE METHOD OF CONFIGURATION INTERACTION

Equation (9-18) shows that the wavefunction of an atom in the orbital approximation may be written in general as a linear combination of Slater determinants. In this section we will specify in greater detail the restrictions which the determinants appearing in such a wavefunction must satisfy. To do

[7] The effect of orthogonalization on the energy is discussed by W. Joy, L. J. Schaad, and G. S. Handler, *J. Chem. Phys.*, **41**:2026 (1964).

this, it is convenient to express Eq. (9-18) in slighly different form:

$$\xi_i(1, 2, \ldots, N) = \sum_{k}^{m} \psi_k c_{ki} \tag{9-66}$$

where ψ_k is in itself a linear combination of determinants and, furthermore, each ψ_k must represent the same term symbol as the wavefunction ξ. For example, if the wavefunction ξ is to be an approximation to the $^2S_{1/2}$ ground state of lithium, then each ψ_k must also represent such a state. This means that the linear combination of determinants representing ψ_k must be chosen to be an eigenfunction of $\hat{S}^2$ with $S = \frac{1}{2}$; furthermore, each ψ_k must lead to $L = 0$, so that the state is $^2S_{1/2}$. Once these conditions are met, the expectation value of the energy is determined in the same way as detailed in Eqs. (9-19) thru (9-23) except that the basis is no longer regarded as the individual determinants D_i but rather their linear combinations ψ_k. The roots of the $m \times m$ determinant will now represent m *different* states, all having the term symbol $^2S_{1/2}$. The lowest of these is an approximation to the ground state, and the remaining $m - 1$ roots will represent excited states having the same L, S, and J quantum numbers as the ground state. The technique employing a wavefunction of the above form is called the method of *configuration interaction* (CI). The name comes from the fact that each linear combination of determinants represented by ψ_k generally arises from a particular electronic configuration of the atom and that a lowering of energy results from allowing these configurations to mix, i.e., interact. For instance (using lithium as an example), one ψ_k would arise from the ground-state configuration $1s^2 2s$, and others would arise from excited configurations such as $1s^2 3s$, $1s^2 4s$, etc.; in general, any configurations leading to a $^2S_{1/2}$ state could be included. Note that if one were to include a configuration such as $1s^2 2p$, this would produce a ψ_k representing a ^2P state. This would show up in the final calculation as zero values for matrix elements such as $\langle \psi_k | \hat{H} | \psi_l \rangle$, in which ψ_k belongs to a ^2S state and ψ_l belongs to a ^2P state. In addition one would obtain only $m - 1$ roots representing the $^2S_{1/2}$ states plus an additional root representing the ^2P state. Since the matrix elements between the two states are zero, there would be no interaction between the corresponding configurations; in effect, one would have two separate sets of calculations: a CI calculation on the $^2S_{1/2}$ states and a single-wavefunction calculation on the ^2P state.

Note that although the matrix elements now become of the form $\langle \psi_k | \hat{H} | \psi_l \rangle$, each matrix element can be expressed in terms of integrals of the form $\langle D_p | \hat{H} | D_q \rangle$. Consequently, all the procedures developed in Sec. 9-2 are still useful.

The basis of the noninteraction of wavefunctions belonging to different states resides in Theorem 2-5 which states that if the ψ_k are chosen to be nondegenerate eigenfunctions of some normal operator Λ which commutes with an arbitrary operator $\hat{F}$, then matrix elements of $\hat{F}$ between two different ψ_k having different eigenvalues of Λ will vanish. In the present context we will identify $\hat{F}$ with the hamiltonian $\hat{H}$ and Λ with various spin and orbital angular momentum operators. In the lithium example, the wavefunctions for ^2S states and ^2P states are both eigenfunctions of $\hat{L}^2$ (which commutes with $\hat{H}$ in the absence of spin-orbit interaction), but both have different L values ($L = 0$ for ^2S and $L = 1$ for ^2P). A similar situation arises when two wavefunctions ψ_k and ψ_l have different multiplicities; in this case the relevant operator commuting with $\hat{H}$ is the spin operator $\hat{S}^2$. Since ψ_k and ψ_l will have different values of S, their matrix elements vanish. Later (in Sec. 9-5) we will study a more powerful application of this theorem. There we will see how its use turns a problem of apparently frightening complexity into one which is actually rather easy. For the present, however, we will consider a much simpler example: a simple CI calculation on the helium atom.

Let us begin with a two-configuration wavefunction for the helium atom having the form

$$\xi(1, 2) = c_1\psi_1 + c_2\psi_2 \tag{9-67}$$

where ψ_1 arises from the configuration $1s^2$ and ψ_2 arises from $1s2s$. Specifically, the two wavefunctions are

$$\psi_1 = |1s\,\overline{1s}| \tag{9-68}$$

$$\psi_2 = \frac{1}{\sqrt{2}}(|1s\,\overline{2s}| - |\overline{1s}\,2s|) \tag{9-69}$$

Since both ψ_1 and ψ_2 describe $^1S_0$ states ($S = L = 0$), there will be no vanishing matrix elements of $\hat{H}$. If we represent these matrix elements by $H_{ij} = \langle\psi_i|\hat{H}|\psi_j\rangle$ ($i, j = 1$ or 2), the secular determinant to be solved is

$$\begin{vmatrix} H_{11} - E & H_{12} \\ H_{12} & H_{22} - E \end{vmatrix} = 0 \tag{9-70}$$

The diagonal matrix elements H_{11} and H_{22} are just the energies of single-configurational calculations for the $^1S_0$ ground state and the $^1S_0$ excited state. Their values can be readily deduced from Eqs. (6-76) and (6-83), respectively, as

$$H_{11} = 2\epsilon_1^{(0)} + J_{11} \tag{9-71}$$

$$H_{22} = \epsilon_1^{(0)} + \epsilon_2^{(0)} + J_{12} + K_{12} \tag{9-72}$$

where we represent the 1s orbital by the subscript "1" and the 2s orbital by the subscript "2." Although the matrix element H_{22} is easily evaluated directly, it is instructional to illustrate how the rules given in Sec. 9-2 are used for this purpose. Let us make the convenient notational substitutions

$$D_1 = |1s\,\overline{2s}| \qquad \text{and} \qquad D_2 = |\overline{1s}\,2s| \tag{9-73}$$

The matrix element then becomes

$$H_{22} = \tfrac{1}{2}\langle D_1 - D_2|\hat{H}|D_1 - D_2\rangle \tag{9-74}$$

This is readily expanded to the form

$$H_{22} = \tfrac{1}{2}(\langle D_1|\hat{H}|D_1\rangle - 2\langle D_1|\hat{H}|D_2\rangle + \langle D_2|\hat{H}|D_2\rangle) \tag{9-75}$$

Putting in the actual orbitals (as we did in the earlier calculation) readily leads to the result in Eq. (9-72). If we let D_0 represent $|1s\,\overline{1s}|$, the off-diagonal matrix element becomes

$$H_{12} = \left\langle D_0|\hat{H}|\ \frac{D_1 - D_2}{\sqrt{2}} \right\rangle = \frac{1}{\sqrt{2}}(\langle D_0|\hat{H}|D_1\rangle - \langle D_0|\hat{H}|D_2\rangle) \tag{9-76}$$

Note that D_0 and D_1 differ by one spin orbital (type B) and that D_0 and D_2 differ by two spin orbitals (type C). Thus one obtains

$$H_{12} = \sqrt{2}\left\langle 1s(1)1s(2)\left|\frac{1}{r_{12}}\right||1s(1)2s(2)\right\rangle = \sqrt{2}\,\langle 11|g|12\rangle \tag{9-77}$$

This calculation is made much more difficult if we decide to put scale factors into the orbitals, since these are nonlinear parameters. Consequently, we will set $Z = 2$ and use orthogonal hydrogenlike orbitals for $1s$ and $2s$. Of course, the virial theorem will not be satisfied in this calculation. The results for the unscaled calculation are

$$\begin{array}{lll} \epsilon_1^{(0)} = -2.00E_h & J_{11} = 1.25E_h & K_{12} = 0.044E_h \\ \epsilon_2^{(0)} = -0.50E_h & J_{12} = 0.419E_h & \langle 11|g|12\rangle = 0.179E_h \end{array} \tag{9-78}$$

Combining the appropriate integrals, the matrix elements become

$$H_{11} = -2.75E_h \qquad H_{22} = -2.037E_h \qquad H_{12} = 0.253E_h \tag{9-79}$$

and the secular determinant is

$$\begin{vmatrix} -2.75 - E & 0.253 \\ 0.253 & -2.037 - E \end{vmatrix} = 0 \tag{9-80}$$

Conversion to quadratic form and use of the quadratic formula produce the two roots: $E = -2.831E_h$ and $-1.956E_h$. The lower root represents the ${}^1S_0$ ground state whose experimental energy is $-2.903E_h$. Note that this improves the result of the single-configuration, unscaled calculation ($H_{11} = -2.75E_h$). However, this does not do so well as the single-configuration, *scaled* calculation ($-2.85E_h$). The higher root represents the lowest excited state of the ${}^1S_0$ type (experimental energy $= -2.15E_h$).

Note that although the upper-limit theorem is satisfied for both states, the CI calculation improves the ground-state energy (relative to the single-configuration calculation), but the energy estimate of the excited state is made worse (the $-1.956E_h$ root is *higher* than the value of the single-configuration result: $H_{22} = -2.037E_h$). This poorer performance is a simple mathematical consequence of the fact that the trace of the matrix representation of $\hat{H}$ is invariant under a similarity transformation. Thus the sum $H_{11} + H_{22}$ will equal $-4.787E_h$ both before and after **H** is diagonalized. Of course, if scale factors are included and optimized, the sum of the diagonal elements will change from that of the unscaled value, since the operations of optimizing nonlinear parameters are not part of the similarity transformation.

The actual CI wavefunctions for the two states may be obtained as follows. The secular equations are

$$c_1(-2.75 - E) + c_2(0.253) = 0 \qquad c_1(0.253) + c_2(-2.037 - E) = 0 \tag{9-81}$$

Considering the ground state first, set $E = -2.831$ to obtain, for the first secular equation,

$$0.085c_1 + 0.253c_2 = 0$$

Solving for c_2 in terms of c_1 produces

$$c_2 = -0.318c_1 \tag{9-82}$$

The normalization condition is

$$c_1^2 + c_2^2 = 1$$

Thus, substituting for c_2 from Eq. (9-82), we get

$$c_1^2 + (0.318c_1)^2 = 1$$

Thus

$$c_1^2 = 0.908 \qquad \text{and} \qquad c_1 = 0.953$$

Therefore $c_2 = (-0.318)(0.953) = -0.303$, and the ground-state wavefunction is

$$\xi_1(^1S_0) = 0.953\psi_1 - 0.303\psi_2 \tag{9-83}$$

This means that a wavefunction which is 90.8 percent $1s^2$ and 9.2 percent $1s2s$ is better than a wavefunction which is 100 percent $1s^2$ (all other factors such as scaling being the same). Similarly, using $E = -1.956$ in the same procedure leads to the excited-state wavefunction

$$\xi_2(^1S_0) = 0.303\psi_1 + 0.953\psi_2 \tag{9-84}$$

EXERCISES

9-11. Prove that the wavefunctions $\xi_1(^1S_0)$ and $\xi_2(^1S_0)$ given by Eqs. (9-83) and (9-84), respectively, are indeed orthogonal. What would it mean physically if this were not true?

9-12. Use the wavefunctions $\psi_1 = |1s\,\overline{1s}|$, $\psi_2 = |2s\,\overline{2s}|$, and $\psi_3 = (1/\sqrt{2})(|1s\,\overline{2s}| - |\overline{1s}\,2s|)$ as the basis for a three-configuration CI calculation on helium under the following conditions, and develop explicit expressions for each matrix element involved:

(*a*) The basis is orthonormal, and $1s$ and $2s$ are eigenfunctions of the monoelectronic operator h.

(*b*) The basis is normalized but not orthogonal ($\langle 1s|2s\rangle = \Delta \neq 0$), and 1*s* and 2*s* are not eigenfunctions of *h*.
(*c*) Show that (*b*) reduces to (*a*) when $\Delta = 0$ and 1*s* and 2*s* are eigenfunctions of *h*.

Answer [part (*b*)]:

$$H_{11} = 2\epsilon_1^{(0)} + J_{11}$$

$$H_{22} = 2\epsilon_2^{(0)} + J_{22}$$

$$H_{33} = (1 - \Delta^2)^{-1}[\epsilon_1^{(0)} + \epsilon_2^{(0)} + J_{12} + 2\Delta\beta_{12} + K_{12}]$$

$$H_{12} = 2\Delta\beta_{12} + \langle 11|g|22\rangle$$

$$H_{13} = [2(1 + \Delta^2)]^{-1/2}[2\Delta\epsilon_1^{(0)} + 2\beta_{12} + 2\langle 11|g|12\rangle]$$

$$H_{23} = [2(1 + \Delta^2)]^{-1/2}[2\Delta\epsilon_2^{(0)} + 2\beta_{12} + \langle 22|g|12\rangle]$$

9-5 THE WAVEFUNCTIONS OF THE $1s^2 2s^2 2p^2$ CONFIGURATION OF THE CARBON ATOM

In Sec. 8-3 we saw that the $1s^2 2s^2 2p^2$ electron configuration of carbon led to the states ${}^3P_{2,1,0}$, ${}^1D_2$, and ${}^1S_0$. If spin-orbit coupling is neglected, this leads to three different energy states: one for each different L value. We will now look at how to set up wavefunctions for each of these states and how to set up the calculations needed to determine their energies.

The wavefunctions which one can obtain from the above configuration arise from the various combinations of $2p_0$, $2p_1$, and $2p_{-1}$ orbitals used to represent the $2p^2$ portion. Thus we have six spin orbitals (each of the 2*p* orbitals with either α or β spin) to be used two at a time. Thus the number of different combinations is

$$\binom{6}{2} = \frac{6!}{2!(6-2)!} = 15$$

Table 9-1 lists the 15 different determinantal functions arising from all these combinations. At first glance, it appears as if one should form a 15-determinant-term wavefunction [as in Eq. (9-18)] and solve a 15×15 determinant in order to extract all the energies. This is indeed one possible approach, but we will see that Theorem 2-5 enables one to do all the calculations without solving a determinant at all.

As we saw earlier, the M_S value of each determinant is derived very simply by application of the equation

$$\hat{S}_z D_i = \sum_{\mu=1}^{N} m_{s\mu} D_i = M_S D_i \tag{9-85}$$

TABLE 9-1.
Determinantal functions representing the $1s^2 2s^2 2p^2$ electron configuration of the carbon atom

Determinant	M_S	M_L	Terms
$D_1 = \lvert 1s\,\overline{1s}\,2s\,\overline{2s}\,2p_0\overline{2p_0}\rvert$	0	0	
$D_2 = \lvert 1s\,\overline{1s}\,2s\,\overline{2s}\,2p_1\,\overline{2p_{-1}}\rvert$	0	0	$^1S, {}^1D, {}^3P$
$D_3 = \lvert 1s\,\overline{1s}\,2s\,\overline{2s}\,\overline{2p_1}\,2p_{-1}\rvert$	0	0	
$D_4 = \lvert 1s\,\overline{1s}\,2s\,\overline{2s}\,2p_0\,\overline{2p_1}\rvert$	0	1	$^1D, {}^3P$
$D_5 = \lvert 1s\,\overline{1s}\,2s\,\overline{2s}\,\overline{2p_0}\,2p_1\rvert$	0	1	
$D_6 = \lvert 1s\,\overline{1s}\,2s\,\overline{2s}\,2p_0\,\overline{2p_{-1}}\rvert$	0	−1	$^1D, {}^3P$
$D_7 = \lvert 1s\,\overline{1s}\,2s\,\overline{2s}\,\overline{2p_0}\,2p_{-1}\rvert$	0	−1	
$D_8 = \lvert 1s\,\overline{1s}\,2s\,\overline{2s}\,2p_1\,\overline{2p_1}\rvert$	0	2	^1D
$D_9 = \lvert 1s\,\overline{1s}\,2s\,\overline{2s}\,2p_{-1}\,\overline{2p_{-1}}\rvert$	0	−2	^1D
$D_{10} = \lvert 1s\,\overline{1s}\,2s\,\overline{2s}\,2p_1\,2p_{-1}\rvert$	1	0	^3P
$D_{11} = \lvert 1s\,\overline{1s}\,2s\,\overline{2s}\,2p_0\,2p_1\rvert$	1	1	^3P
$D_{12} = \lvert 1s\,\overline{1s}\,2s\,\overline{2s}\,2p_0\,2p_{-1}\rvert$	1	−1	^3P
$D_{13} = \lvert 1s\,\overline{1s}\,2s\,\overline{2s}\,\overline{2p_1}\,\overline{2p_{-1}}\rvert$	−1	0	^3P
$D_{14} = \lvert 1s\,\overline{1s}\,2s\,\overline{2s}\,\overline{2p_0}\,\overline{2p_1}\rvert$	−1	1	^3P
$D_{15} = \lvert 1s\,\overline{1s}\,2s\,\overline{2s}\,\overline{2p_0}\,\overline{2p_{-1}}\rvert$	−1	−1	^3P

Similarly, the M_L values are determined by an exactly analogous equation

$$\hat{L}_z D_i = \sum_{\mu=1}^{N} m_{l\mu} D_i = M_L D_i \tag{9-86}$$

Note that since the closed-shell portion $1s^2 2s^2$ contributes zero to both M_S and M_L, we need to consider only the $2p^2$ portion in determining these quantum numbers.

Earlier (Sec. 9-1) we learned how to construct determinantal eigenfunctions of $\hat{S}^2$; now we will see that determinantal eigenfunctions of $\hat{L}^2$ are constructed in a completely analogous manner. To do this we need the relationships

$$\begin{gathered}
\hat{L}^2 = \hat{L}_-\hat{L}_+ + \hat{L}_z(\hat{L}_z + 1) \\
\hat{L}_+ = \sum_{\mu}^{N} \hat{L}_+(\mu) \qquad \hat{L}_- = \sum_{\mu}^{N} \hat{L}_-(\mu) \\
\hat{L}_+(\mu) Y_{l\mu,\, m_{l\mu}} = [l_\mu(l_\mu + 1) - m_{l\mu}(m_{l\mu} + 1)]^{1/2} Y_{l\mu,\, m_{l\mu}+1} \\
\hat{L}_-(\mu) Y_{l\mu,\, m_{l\mu}} = [l_\mu(l_\mu + 1) - m_{l\mu}(m_{l\mu} - 1)]^{1/2} Y_{l\mu,\, m_{l\mu}-1}
\end{gathered} \tag{9-87}$$

Since only the $2p$ orbitals need be considered, $\hat{L}^2$ may be written as a two-electron operator, namely,

$$\hat{L}^2 = [\hat{L}_-(1) + \hat{L}_-(2)][\hat{L}_+(1) + L_+(2)] + \hat{L}_z^2 + \hat{L}_z \tag{9-88}$$

From Eq. (9-86) we see that

$$(\hat{L}_z^2 + \hat{L}_z)D_i = (M_L^2 + M_L)D_i \tag{9-89}$$

The projection operators which are needed for $L = 0$, 1, and 2 (S, P, and D) are entirely analogous to those we used previously for the spin operator $\hat{S}^2$ [see Eqs. (9-12) to (9-15)]. Specific expressions for these are

$$\hat{O}_S = (\hat{L}^2 - 6)(\hat{L}^2 - 2) \qquad \hat{O}_P = \hat{L}^2(\hat{L}^2 - 6) \qquad \hat{O}_D = \hat{L}^2(\hat{L}^2 - 2) \tag{9-90}$$

The subscripts on the projection operators denote the $\hat{L}^2$ eigenvalue characteristic of the eigenfunctions they will project. For example, $\hat{O}_S$ will project out the $S = 0$ [that is, the $S(S+1) = 0$] eigenfunction. Note that certain constant terms which multiply the projection operators have been omitted, since they have no role in the projection and, besides, eventually become changed when the projected function is normalized.

It is convenient to look at the $\hat{S}^2$ eigenfunctions first. Examining the first three determinants, we obtain

$$\hat{S}^2 D_1 = 0 \qquad \hat{S}^2(D_2 - D_3) = 0 \qquad \hat{S}^2(D_2 + D_3) = 2(D_2 + D_3) \tag{9-91}$$

Since the only triplet state is the ^3P term, we see that the combination which has the eigenvalue of 2 [that is, $S(S+1) = 2$, $S = 1$] represents this particular state. Thus the $M_S = M_L = 0$ component of this state has the eigenfunction

$$\psi_{0,0}(^3P) = \frac{1}{\sqrt{2}}(D_2 + D_3) \tag{9-92}$$

where the subscripts on ψ indicate M_S and M_L values, respectively. One can easily check this result by carrying out the operation $\hat{L}^2\psi_{0,0}(^3P)$. Since $M_S = M_L = 0$ for the example chosen, we need consider only the $\hat{L}_-\hat{L}_+$ portion of $\hat{L}^2$. One obtains

$$\hat{L}^2\psi_{0,0}(^3P) = 2\psi_{0,0}(^3P) \tag{9-93}$$

that is, $L(L+1) = 2$ and, thus, $L = 1$. From Eq. (9-91) we see that the functions $\psi_{0,0}(^1S)$ and $\psi_{0,0}(^1D)$ must be linear combinations of D_1 and $(D_2 - D_3)$. The linear combinations must be chosen such that

$$\hat{L}^2\psi_{0,0}(^1S) = 0 \qquad \text{and} \qquad \hat{L}^2\psi_{0,0}(^1D) = 6\psi_{0,0}(^1D) \tag{9-94}$$

that is, $L = 0$ and 2, respectively. If we use projection operators on D_1, we can leave out the $(\hat{L}^2 - 2)$ portions of $\hat{O}_S$ and $\hat{O}_D$, since this annihilates the P component, which we know is not there anyway [we have already found it; see Eq. (9-92)]. Thus, we can set $\hat{O}_S = \hat{L}^2 - 6$ (this will annihilate the D component of D_1) and $\hat{O}_D = \hat{L}^2$ (this will annihilate the S component of D_1).

Applying $\hat{O}_D$ first, we obtain

$$\begin{aligned}
\hat{O}_D D_1 &= \hat{L}^2 D_1 = \hat{L}_- \hat{L}_+ D_1 \\
&= \hat{L}_-[\hat{L}_+(1) + \hat{L}_+(2)]|2p_0 \overline{2p}_0| \\
&= \hat{L}_-[\hat{L}_+(1)|2p_0 \overline{2p}_0| + \hat{L}_+(2)|2p_0 \overline{2p}_0|] \\
&= \hat{L}_-(\sqrt{2}|2p_1 \overline{2p}_0| + \sqrt{2}|2p_0 \overline{2p}_1|) \\
&= \sqrt{2}[\hat{L}_+(1) + \hat{L}_+(2)](|2p_1 \overline{2p}_0| + |2p_0 \overline{2p}_1|) \\
&= \sqrt{2}(\sqrt{2}|2p_0 \overline{2p}_0| + \sqrt{2}|2p_{-1} \overline{2p}_1| + \sqrt{2}|2p_1 \overline{2p}_{-1}| + \sqrt{2}|2p_0 \overline{2p}_0|) \\
&= 2D_1 + 2(-D_3) + 2D_2 + 2D_1 = 4D_1 + 2D_2 - 2D_3 \qquad (9\text{-}95)
\end{aligned}$$

Thus, we obtain (after normalization)

$$\psi_{0,0}(^1D) = \frac{1}{\sqrt{6}}(2D_1 + D_2 - D_3) \qquad (9\text{-}96)$$

Similarly,

$$\hat{O}_S D_1 = (\hat{L}^2 - 6)D_1 = -2D_1 + 2D_2 - 2D_3 \qquad (9\text{-}97)$$

so that

$$\psi_{0,0}(^1S) = \frac{1}{\sqrt{3}}(D_1 - D_2 + D_3) \qquad (9\text{-}98)$$

Note that we have changed the sign of the projected linear combination of determinants in Eq. (9-97); this of course has no effect on the wavefunction.

The determinants D_4 through D_7 could be treated similarly to construct additional ^1D and ^3P wavefunctions. From Table 9-1 it is seen that

$$\begin{aligned}
E(^3P) &= \langle D_i|\hat{H}|D_i\rangle \qquad i = 10 \text{ through } 15 \\
E(^1D) &= \langle D_i|\hat{H}|D_i\rangle \qquad i = 8 \text{ or } 9
\end{aligned} \qquad (9\text{-}99)$$

To find the energy of the ^1S state, let us form a matrix representation of $\hat{H}$ in the basis D_1, D_2, and D_3. Since the trace of a matrix is invariant under a similarity transformation (unitary in this case), we can write

$$E(^1S) + E(^1D) + E(^3P) = H_{11} + H_{22} + H_{33} \qquad (9\text{-}100)$$

where

$$H_{ii} = \langle D_i|\hat{H}|D_i\rangle \qquad i = 1, 2, \text{ or } 3 \qquad (9\text{-}101)$$

Thus, once $E(^3P)$ and $E(^1D)$ are calculated by Eq. (9-99), $E(^1S)$ may be calculated from Eq. (9-100).

An alternative approach (or one which may be combined with the above) is to make use of Theorem 2-5 in factoring the original 15×15 determinant into smaller determinants. Since six of the basis determinants describe singlet states (^1S and ^1D) and nine describe a triplet state (^3P), there will be no nonzero matrix elements in which ψ_k is a singlet and ψ_l is a triplet. This has the effect of reducing the original 15×15 matrix representation of $\hat{H}$ to one 6×6 representation and one 9×9 representation. From a practical standpoint, this means we now have to solve two smaller determinants instead of one larger one. However, we can reduce the sizes of the matrix representations even further. The 6×6 matrix of the singlet states contains one ψ_k with $L = 0$ (^1S) and five with $L = 2$ (^1D). Thus there will be no matrix elements between these different $\hat{L}^2$ eigenfunctions, resulting in a 1×1 matrix for ^1S and a 5×5 matrix for ^1D. Furthermore, the 5×5 matrix for ^1D contains five functions having the different M_L values -2, -1, 0, 1, and 2. Thus, there will be no matrix elements between these different M_L values, and the 5×5 determinant now becomes five 1×1 determinants. Consequently, the original 6×6 determinant for singlet states becomes six 1×1 "determinants"; i.e., the energies can be obtained by solving a single matrix element for each. Of course, in the absence of spin-orbit coupling, all five ^1D states have the same energy, so one need solve only one of these.

The 9×9 matrix representation of the ^3P state is handled in the same way. First, this matrix can be factored into three 3×3 matrices on the basis of the three different M_S values -1, 0, and 1. Furthermore, each of these can be factored into three 1×1 "matrices" on the basis of the three different M_L values -1, 0, and 1. Consequently, this particular problem—which appeared hopelessly complex at the outset—becomes almost trivial when simplified by Theorem 2-5.[8]

EXERCISES

9-13. Obtain normalized forms for the carbon atom wavefunctions $\psi_{0,1}(^1D)$, $\psi_{0,1}(^3P)$, $\psi_{0,-1}(^1D)$, and $\psi_{0,-1}(^3P)$.

9-14. Construct a table like Table 9-1 for the nitrogen atom configuration $1s^2 2s^2 2p^3$.

9-15. Construct a table like Table 9-1 for the oxygen atom configuration $1s^2 2s^2 2p^4$ and compare the results with that for carbon $1s^2 2s^2 2p^2$.

9-16. Referring to Table 9-1, construct wavefunctions for the ^1D and ^3P states of carbon from the determinants D_4, D_5, D_6, and D_7.

SUGGESTED READINGS

Daudel, R., R. Lefebvre, and C. Moser: *Quantum Chemistry: Methods and Applications*, Interscience, New York, 1959.

Slater, J. C.: *Quantum Theory of Atomic Structure*, vols. 1 and 2, McGraw-Hill, New York, 1960.

---

[8] Numerical calculations on the low-lying states of carbon have been reported by P. S. Bagus and C. M. Moser, *Phys. Rev.*, **167**:13 (1968).

CHAPTER 10

THE HARTREE-FOCK SELF-CONSISTENT FIELD METHOD

Considerable use has been made in previous chapters of the *orbital approximation*: the expression of approximate wavefunctions as antisymmetrized products of one-electron distribution functions called *orbitals*. We have observed that not all orbitals are equally good (all other factors equal) in approximating a wavefunction. For example, gaussian orbitals appear to be much poorer than either hydrogenlike or Slater orbitals. Furthermore, regardless of what orbitals are used to construct a wavefunction, scaling parameters must be incorporated in order to improve calculation of the energy and, of course, to satisfy the virial theorem. In theory one could obtain increasingly better approximations of the energies of many-electron systems by simply increasing the number of variational parameters imbedded in the wavefunction. For example, the split-shell calculation on helium achieves an improvement in the energy estimate by using two different scale factors for $1s$-type orbitals. Conceivably, one could imbed additional nonlinear parameters in the wavefunction and improve the energy estimates still further. Also, one could form a wavefunction which is a linear combination of several configurations (the CI method of Chap. 9) and then minimize the energy with respect to the additional linear parameters thus introduced. And, of course, one could employ a combination of the two methods.

In the present chapter we will consider a simple way of carrying out what at first sight appears to be a hopelessly impossible task: optimizing a wavefunction of given arbitrary form with respect to *all possible variational parameters*. Furthermore, we will see that the model which accomplishes this task establishes the theoretical structure of much of modern chemistry. Many of the theorems and basic concepts arising from this model are used by chemists on a day-to-day basis, albeit often unwittingly.

10-1 THE GENERATION OF OPTIMIZED ORBITALS

Instead of being concerned about imbedding specific parameters into the orbitals which we will use to form a wavefunction, let us consider instead the problem of finding the *best* orbitals to use in the wavefunction. "Best" means those orbitals which minimize the energy.[1] In order to make the treatment more tractable, we will limit the form of the wavefunction to that of a single determinant, namely,

$$\xi(1,2,\ldots,N) = \frac{1}{\sqrt{N!}} \begin{vmatrix} S_1(1) & S_2(1) & \cdots & S_N(1) \\ S_1(2) & S_2(2) & \cdots & S_N(2) \\ \cdots & \cdots & \cdots & \cdots \\ S_1(N) & S_2(N) & \cdots & S_N(N) \end{vmatrix} \tag{10-1}$$

where each $S_i(\mu)$ is a spin orbital. The general problem is to seek those spin orbitals which make the total energy $\langle \xi|\hat{H}|\xi\rangle$ a minimum. This means that there are no restrictions whatsoever on the spin orbitals other than that they lead to a well-behaved wavefunction; for example, the spin orbitals need not be eigenfunctions of any particular operator. Nor need the determinant itself be an eigenfunction of $\hat{S}^2$, $\hat{S}_z$, $\hat{L}^2$, $\hat{L}_z$, or any other operator. However, in order to make the mathematics of the treatment more tractable, it is customary to back off somewhat from the original strategy and impose certain restrictions. These restrictions can be incorporated in one statement: The determinant is to be of closed-shell form, with each spin orbital expressed as a product of a spatial orbital ϕ and a spin function (α or β); furthermore each spatial orbital is to be used twice, once with α spin and once with β spin. This means that we will be restricted to systems with an even number of electrons ($2N$). The double-occupancy restriction can be symbolized by

$$S_{2i-1}(\mu) = \phi_i(\mu)\alpha(\mu) \qquad S_{2i}(\nu) = \phi_i(\nu)\beta(\nu) \tag{10-2}$$

Thus if $S_1(\mu) = \phi_1(\mu)\alpha(\mu)$, then $S_2 = (\nu) = \phi_1(\nu)\beta(\nu)$, etc. The determinan-

[1] Recall that a wavefunction which gives the best energy estimate does not necessarily do the same for other properties of the system.

tal wavefunction will then have the form

$$\xi(1,2,\ldots,2N) = |\phi_1\,\bar{\phi}_1\,\phi_2\,\bar{\phi}_2\cdots\phi_N\,\bar{\phi}_N| \tag{10-3}$$

For closed-shell atoms, this wavefunction satisfies $\hat{S}^2\xi = 0$ and $\hat{L}^2\xi = 0$ and thus describes a $^1S_0$ state. As an additional restriction, introduced to facilitate numerical computations, the spatial orbitals are chosen to be orthonormal, namely,

$$\langle \phi_i | \phi_j \rangle = \delta_{ij}$$

This restriction means that no overlap integrals need be considered, and hence, considerable algebraic simplification results.

According to Eq. (9-45) the total energy of a system having a wavefunction of the form of Eq. (10-1) and consisting of N doubly occupied spatial orbitals is given by

$$E = 2\sum_{i=1}^{N} \epsilon_i^{(0)} + \sum_{i,j}^{N} (2J_{ij} - K_{ij}) \tag{10-4}$$

We now wish to find the best possible orbitals (subject to the restrictions mentioned) to use in a wavefunction of single-determinantal form; "best possible" means "leading to the *lowest energy*." In the language of the variation method (see Sec. 6-1), we wish to find those orbitals which lead to a minimum value of the energy given by Eq. (10-4), subject to the restriction that these orbitals are orthonormal. Thus we construct the functional

$$F[E] = 2\sum_{i=1}^{N} \epsilon_i^{(0)} + \sum_{i,j}^{N} (2J_{ij} - K_{ij}) - 2\sum_{i,j}^{N} \lambda_{ij}(\langle \phi_i | \phi_j \rangle - \delta_{ij}) \tag{10-5}$$

where the $\{\lambda_{ij}\}$ are the lagrangian multipliers.[2] We now wish to find the conditions that the orbitals $\{\phi_i\}$ must satisfy in order for the functional $F[E]$ to have a minimum value; i.e., we require that $\delta F[E] = 0$ for arbitrarily small variations $\delta\phi_i$ of these optimized orbitals.

The treatment is greatly simplified in a notational sense if we define coulomb and exchange operators as follows:

$$\begin{aligned} J_i(\mu)\phi_j(\mu) &= \left\langle \phi_i(\nu) \left| \frac{1}{r_{\mu\nu}} \right| \phi_i(\nu) \right\rangle \phi_j(\mu) \\ K_i(\mu)\phi_j(\mu) &= \left\langle \phi_i(\nu) \left| \frac{1}{r_{\mu\nu}} \right| \phi_j(\nu) \right\rangle \phi_i(\mu) \end{aligned} \tag{10-6}$$

[2] The factor "2" as a multiplier of the summation over lagrangian multipliers is put in to make the final equations simpler in form. Had this factor been omitted at this point, the final equations would have contained terms such as $\frac{1}{2}\lambda_{ij}$. This factor "$\frac{1}{2}$" would then have been incorporated into the lagrangian multipliers, thereby accomplishing the same notational simplification as using the "2" factor ahead of time.

Examine these two operators closely and you will note that the exchange operator $K_i(\mu)$ exchanges electrons μ and ν between the two spatial orbitals ϕ_i and ϕ_j. With the above definitions of the coulomb and exchange operators, the coulomb and exchange integrals (both representing electron repulsions) occurring in the energy expression given by Eq. (10-4) can be written

$$\begin{aligned} J_{ij} &= \langle \phi_i(\mu)|J_j(\mu)|\phi_i(\mu)\rangle = \langle \phi_j(\nu)|J_i(\nu)|\phi_j(\nu)\rangle \\ K_{ij} &= \langle \phi_i(\mu)|K_j(\mu)|\phi_i(\mu)\rangle = \langle \phi_j(\nu)|K_i(\nu)|\phi_j(\nu)\rangle \end{aligned} \tag{10-7}$$

It is seen that the coulomb operator is just the operator for the potential energy which would arise from an electron distribution $|\phi_i|^2$. Such operators represent effective potentials for an electron moving in the repulsive field of other electrons. The exchange operator, on the other hand, has no classical analog, since it arises from the nonclassical antisymmetry principle.

The first-order variation in the functional $F[E]$ is

$$\begin{aligned} \delta F[E] = 2\sum_{i}^{N} &(\langle \delta\phi_i|h_\mu|\phi_i\rangle + \langle \phi_i|h_\mu|\delta\phi_i\rangle) \\ &+ \sum_{i,j}^{N} (\langle \delta\phi_i|2J_j - K_j|\phi_i\rangle + \langle \phi_i|2J_j - K_j|\delta\phi_i\rangle) \\ &+ \sum_{i,j}^{N} (\langle \delta\phi_j|2J_i - K_i|\phi_j\rangle + \langle \phi_j|2J_i - K_i|\delta\phi_j\rangle) \\ &- 2\sum_{i,j}^{N} (\lambda_{ij}\langle \delta\phi_i|\phi_j\rangle + \lambda_{ij}\langle \phi_i|\delta\phi_j\rangle) \end{aligned} \tag{10-8}$$

where h_μ is the monoelectronic part of the atomic hamiltonian operator. The first and second double summations are symmetric in their indices and lead to the same final sums. Thus Eq. (10-8) can be written

$$\begin{aligned} \delta F[E] = 2\sum_{i}^{N} &\left[\left\langle \delta\phi_i \middle| h_\mu + \sum_{j}^{N}(2J_j - K_j) \middle| \phi_i \right\rangle\right] \\ &+ 2\sum_{i}^{N} \left[\left\langle \phi_i \middle| h_\mu + \sum_{j}^{N}(2J_j - K_j) \middle| \delta\phi_i \right\rangle\right] \\ &- 2\sum_{i,j}^{N} (\lambda_{ij}\langle \delta\phi_i|\phi_j\rangle + \lambda_{ij}\langle \phi_i|\delta\phi_j\rangle) \end{aligned} \tag{10-9}$$

Since h_μ, J_j, and K_j are hermitian, we see that the first and second summations are just the adjoints of each other. Furthermore, we can interchange summation indices in the last term of the double summation. Then using the fact that $\langle \phi_j|\delta\phi_i\rangle$ and $\langle \delta\phi_i|\phi_j\rangle$ are adjoints of each other, we can write

$$\sum_{i,j}^{N} \lambda_{ij}\langle \phi_i|\delta\phi_j\rangle = \sum_{i,j}^{N} \lambda_{ji}\langle \delta\phi_i|\phi_j\rangle^* \tag{10-10}$$

Equation (10-9) then becomes

$$\delta F[E] = 2\sum_i^N \left[\left\langle \delta\phi_i \middle| h_\mu + \sum_j^N (2J_j - K_j) \middle| \phi_i \right\rangle - \sum_j^N \lambda_{ij} \langle \delta\phi_i | \phi_j \rangle \right]$$

$$+ 2\sum_i^N \left[\left\langle \delta\phi_i \middle| h_\mu + \sum_j^N (2J_j - K_j) \middle| \phi_i \right\rangle^* - \sum_j^N \lambda_{ji} \langle \delta\phi_i | \phi_j \rangle^* \right] \tag{10-11}$$

The vanishing of $\delta F[E]$ for an arbitrary variation $\delta\phi_i$ is now satisfied by the conditions

$$\left[h_\mu + \sum_j^N (2J_j - K_j) \right] \phi_i = \sum_j^N \phi_j \lambda_{ij}$$

$$\left[h_\mu + \sum_j^N (2J_j - K_j) \right] \phi_i^* = \sum_j^N \phi_j^* \lambda_{ji} \tag{10-12}$$

Taking the complex conjugate of the second equation and subtracting it from the first equation leads to

$$\sum_j^N \phi_j (\lambda_{ij} - \lambda_{ji}^*) = 0 \tag{10-13}$$

Since the orbitals $\{\phi_i\}$ are linearly independent, it follows that

$$\lambda_{ij} = \lambda_{ji}^* \tag{10-14}$$

i.e., the lagrangian multipliers are the elements of a hermitian matrix. Consequently, the two expressions in Eqs. (10-12) are complex conjugates of each other and are equivalent. These equations are known as the *Hartree-Fock equations*. They were proposed simultaneously and independently by Fock[3] and by Slater.[4] Earlier, a similar set of equations (but lacking the requirement of an antisymmetrized wavefunction) was proposed by Hartree.[5] The derivation given here follows a version developed by Roothaan.[6]

The Hartree-Fock equations can be written in the matrix form

$$\hat{F}\mathbf{\Phi} = \mathbf{\Phi}\boldsymbol{\lambda} \tag{10-15}$$

---

[3] V. Fock, *Z. Physik*, **61**:126 (1930).

[4] J. C. Slater, *Phys. Rev.*, **35**:210 (1930).

[5] D. R. Hartree, *Proc. Cambridge Phil. Soc.*, **24**:89 (1928).

[6] C. C. J. Roothaan, *Rev. Mod. Phys.*, **23**:69 (1951). See also G. G. Hall, *Proc. Roy. Soc. (London)*, **A205**:541 (1951).

where $\hat{F}$ is the Hartree-Fock operator defined by

$$\hat{F} = h_\mu + \sum_j^N (2J_j - K_j) \tag{10-16}$$

and $\mathbf{\Phi}$ and $\boldsymbol{\lambda}$ are given by

$$\mathbf{\Phi} = [\phi_1 \quad \phi_2 \quad \cdots \quad \phi_N] \qquad \boldsymbol{\lambda} = \begin{bmatrix} \lambda_{11} & \lambda_{12} & \cdots & \lambda_{1N} \\ \lambda_{21} & \lambda_{22} & \cdots & \lambda_{2N} \\ \cdots & \cdots & \cdots & \cdots \\ \lambda_{N1} & \lambda_{N2} & \cdots & \lambda_{NN} \end{bmatrix} \tag{10-17}$$

It is evident from Eq. (10-16) that the Hartree-Fock operator is a monoelectronic operator. The summation term over the coulomb and exchange operators represents a one-electron approximation to the behavior of one electron in the field of the others. Thus, $\Sigma_j (2J_j - K_j)$ represents an *average potential* experienced by a single electron while moving in the field provided by the other $2N - 1$ electrons. The coulombic part of this potential is analogous to a classical electrostatic potential, but the exchange part is not interpretable in a simple way, since it arises from the nonclassical antisymmetry principle.

It is important to note that the sole function of the Hartree-Fock operator is to generate the orbitals to be used in the wavefunction given by Eq. (10-3). Once these orbitals are obtained, they are used in Eq. (10-4) to calculate the energy of the system.

The orbitals generated by the Hartree-Fock equations are not unique; given a particular set of such orbitals, there are many other sets obtainable from them by various unitary transformations which leave the total wavefunction invariant. One particular unitary transformation is of especial interest; this is the unitary transformation which diagonalizes the matrix of lagrangian multipliers. This transformation is written as

$$\mathbf{U}^\dagger \boldsymbol{\lambda} \mathbf{U} = \boldsymbol{\epsilon} \tag{10-18}$$

where $\boldsymbol{\epsilon}$ is a diagonal matrix, namely,

$$(\boldsymbol{\epsilon})_{ij} = \epsilon_i \delta_{ij} \tag{10-19}$$

Multiplying Eq. (10-15) on the right by $\mathbf{U}$ and using Eq. (10-18), we obtain

$$\hat{F}\mathbf{\Phi}' = \mathbf{\Phi}'\boldsymbol{\epsilon} \tag{10-20}$$

where

$$\mathbf{\Phi}' = \mathbf{\Phi}\mathbf{U} \tag{10-21}$$

However, the operator $\hat{F}$ is defined in terms of the original Hartree-Fock orbitals $\{\phi_i\}$ and not in terms of the transformed orbitals $\{\phi_i'\}$, so that Eqs. (10-15) and (10-20) are not of the same form. We now show that the operator [Eq. (10-16)] is invariant under a unitary transformation. Since h_μ does not depend on the Hartree-Fock orbitals, we need consider only the coulomb and exchange operators. Considering the coulomb operator first, we write

$$J_j'(\mu)\phi_i(\mu) = \left\langle \phi_j'(\nu) \left| \frac{1}{r_{\mu\nu}} \right| \phi_j'(\nu) \right\rangle \phi_i(\mu) \tag{10-22}$$

Letting u_{ij} represent an element of the unitary transformation matrix **U**, we obtain

$$\begin{aligned} \sum_J J_j'(\mu)\phi_i(\mu) &= \sum_{k,l} \left\langle \phi_k(\nu) \left| \frac{1}{r_{\mu\nu}} \right| \phi_l(\nu) \right\rangle \sum_j u_{kj}^* u_{lj} \phi_i(\mu) \\ &= \sum_{k,l} \left\langle \phi_k(\nu) \left| \frac{1}{r_{\mu\nu}} \right| \phi_l(\nu) \right\rangle \phi_i(\mu) \delta_{kl} \\ &= \sum_k J_k(\mu)\phi_i(\mu) \end{aligned} \tag{10-23}$$

A completely analogous procedure for the exchange operator produces

$$\sum_j K_j'(\mu)\phi_i(\mu) = \sum_k K_k(\mu)\phi_i(\mu) \tag{10-24}$$

Combining the results of Eqs. (10-23) and (10-24) gives

$$\sum_j (2J_j' - K_j') = \sum_j (2J_j - K_j) \tag{10-25}$$

where use is made of the fact that summation indices are merely "tags" and can be denoted by any symbol one pleases.

The advantage of the particular unitary transformation [Eq. (10-18)] is that it permits the Hartree-Fock equation [Eqs. (10-12)] to be written in pseudo-eigenvalue form

$$\hat{F}\phi_i = \epsilon_i \phi_i \tag{10-26}$$

where we have dropped the primes on the transformed orbitals for notational convenience. The $\{\epsilon_i\}$ which arise from diagonalization of the matrix of lagrangian multipliers are called the *Hartree-Fock eigenvalues* and are the energies associated in a specific way with the Hartree-Fock orbitals. The physical significance of these energies will be discussed in Sec. 10-2. However, let us state at the outset that the energies $\epsilon_i^{(0)}$ occurring in the energy expression given by Eq. (10-4) and the Hartree-Fock energies ϵ_i are *not* equivalent.

Since the operators $J_j(\mu)$ and $K_j(\mu)$ appearing in the Hartree-Fock operator depend on the orbitals generated by this operator, it is not possible to solve the Hartree-Fock equations in the same manner as true eigenvalue equations are solved, hence the reference to these as *pseudo-eigenvalue* equations. One of the oldest procedures used to solve these equations is the *self-consistent field* (*SCF*) *method*. The basic procedure is to choose some beginning set of orbitals $\{\phi_i^{(0)}\}$, which are used to construct an initial approximation of the coulombic and exchange operators found in $\hat{F}$. A first approximation to the optimized orbitals is then obtained by solving

$$\hat{F}^{(0)}\phi_i^{(1)} = \epsilon_i^{(1)}\phi_i^{(1)} \tag{10-27}$$

Such solutions generally involve numerical integration techniques, and the orbitals themselves are not expressible in the usual analytical form but rather in terms of tabulated numerical values over a grid of spatial positions. The new orbitals $\{\phi_i^{(1)}\}$ are now used to redefine the Hartree-Fock operator, and a second, further improved set of orbitals is obtained by solution of

$$\hat{F}^{(1)}\phi_i^{(2)} = \epsilon_i^{(2)}\phi_i^{(2)} \tag{10-28}$$

Since a given set of orbitals $\{\phi_i^{(n)}\}$, obtained from the nth iteration, represents a *field* in which the electrons move, there is an internal inconsistency in that the field assumed for the operator $\hat{F}^{(n)}$ always generates a new field $\{\phi_i^{(n+1)}\}$. However, at some point the functions obtained in an iteration become virtually the same as those of the subsequent iteration (to within some specified tolerance limit), and we say a *self-consistent field* has been reached. It is this final set of SCF orbitals which constitute the optimized orbitals used to calculate the energy via Eq. (10-4).[7]

It should be noted that since the Hartree-Fock SCF orbitals are *fully* optimized, the energy obtained from them must satisfy the virial theorem. In fact, during execution of Hartree-Fock calculations, satisfaction of the virial theorem is often used as a necessary (but not sufficient) criterion of self-consistency.[8]

There are several alternative expressions for the total Hartree-Fock energy of closed-shell systems which are frequently useful. Solving Eq. (10-26) for the Hartree-Fock eigenvalue, one obtains

$$\epsilon_i = \langle \phi_i | \hat{F} | \phi_i \rangle = \epsilon_i^{(0)} + \sum_j (2J_{ij} - K_{ij}) \tag{10-29}$$

Adding $\epsilon_i^{(0)}$ to both sides and summing over all spatial orbitals, one gets

$$\sum_{i=1}^{N} (\epsilon_i + \epsilon_i^{(0)}) = 2\sum_{i=1}^{N} \epsilon_i^{(0)} + \sum_{i,j}^{N} (2J_{ij} - K_{ij}) = E \tag{10-30}$$

This latter equation shows that the Hartree-Fock energy is a sum of the

[7] It should be noted that there are equivalent but superior ways of reaching self-consistency other than by a literal adherence to the method described above. However, the above method is better-suited for obtaining a conceptual understanding of self-consistency than are some of the more practical computational methods.

[8] Nor is satisfaction of the virial theorem a sufficient condition for energy minimization. See W. K. Li, *J. Chem. Ed.*, **65**:963 (1988) for an interesting example.

Hartree-Fock eigenvalues and the energies of interaction of the electrons with the bare nucleus (assuming each electron is itself described by an SCF orbital).[9]

It is useful to write the Hartree-Fock operator in the form

$$\hat{F} = h + \hat{G} \tag{10-31}$$

where

$$\hat{G} = \sum_j (2J_j - K_j) \tag{10-32}$$

Functional dependence of the operators on electron coordinates (μ) is suppressed for notational convenience. Using this simplified notation, one can readily show that various alternative expressions for the total Hartree-Fock energy for a closed-shell system are of the form

$$E = \sum_i \langle \phi_i | \hat{B} | \phi_i \rangle \tag{10-33}$$

where the operator $\hat{B}$ has the alternative forms

$$\hat{B} = \hat{F} + h = 2\hat{F} - \hat{G} = 2h + \hat{G} \tag{10-34}$$

Elementary texts occasionally state, or at least imply, that the total energy of an atom can be written as a sum of orbital energies. Obviously, this possibility is inconsistent with any of the energy expressions given in the foregoing discussion. In actuality, as the energy based on Eq. (10-34) using $\hat{B} = 2\hat{F} - \hat{G}$ shows, the total energy is the sum of the orbital energies *minus* the electron-repulsion energy. This implies that simply adding the orbital energies leads to counting the electron repulsions *twice*. Thus, the correct expression subtracts the electron repulsion *once* from the orbital energy sum in order to delete the extraneous inclusion.

It is of some interest to note, however, that the total energy of an atom can be written as a sum of *modified* orbital energies if one uses the form

$$E = K \sum_i n_i \epsilon_i \tag{10-35}$$

where K is a constant and n_i is the occupation number of the orbital ϕ_i; n_i is thus always 0 (for an unoccupied orbital) and 1 or 2 otherwise. Various values have been proposed for K; generally these are around $\frac{3}{2}$.[10]

---

[9] In actuality, the SCF orbital does not provide an accurate description of how a single, isolated electron interacts with the bare nucleus, since such orbitals are not eigenfunctions of the monoelectronic operator h_μ.

[10] P. Politzer, *J. Chem. Phys.*, **64**:4239 (1976); K. Ruedenberg, *J. Chem. Phys.*, **66**:375 (1977); E. A. Castro, L. Villata, and F. M. Fernandez, *J. Chem. Phys.*, **74**:592 (1981).

EXERCISES

10-1. Show that for a single-determinantal wavefunction for a $2N$-electron system (each spatial orbital doubly occupied)

$$\left\langle \sum_{\mu<\nu}^{2N} \frac{1}{r_{\mu\nu}} \right\rangle = \sum_{i,j}^{N} (2J_{ij} - K_{ij})$$

Hint: An easy approach is to do the specific case of $N = 1$ and then induce the general result. Note that $J_{ii} = K_{ii}$.

10-2. Hartree-Fock calculations on the $1s^2 2s^2 p^4$ (^3P) ground state of the oxygen atom show that the average kinetic energy is $74.809E_h$ and the average potential energy is $-149.618E_h$.
(*a*) Show that the above satisfies the virial theorem (see Sec. 6-6).
(*b*) What is the total electronic energy of the ^3P ground state of oxygen?

10-2 KOOPMANS' THEOREM: THE PHYSICAL SIGNIFICANCE OF ORBITAL ENERGIES

The physical significance of the eigenvalues of the Hartree-Fock operator $\hat{F}$ is made clear by a theorem first proved by the Dutch physicist T. C. Koopmans.[11] Koopmans showed that the optimized orbitals used to construct single-determinantal wavefunctions of an atom X and its two ions X^+ and X^- were the same to within first order. Thus, the energies of the ions are related to that of the neutral atoms by the simple relationships

$$\begin{aligned} E(X^+) &= E(X) - \epsilon_k \qquad k = 1, 2, \ldots, N \\ E(X^-) &= E(X) + \epsilon_m \qquad m = N+1, N+2, \ldots \end{aligned} \tag{10-36}$$

where k refers to an orbital used to describe the ground state of the neutral atom and m refers to what is popularly called an *unoccupied*, or *excited-state*, orbital; the latter is generally called a *virtual* orbital by quantum chemists. The expressions given in Eqs. (10-36) thus imply that the energies of the occupied orbitals should be approximations to the negatives of various ionization energies of the neutral atom. Thus, if ϕ_k is the highest (in terms of energy) occupied orbital of the ground state, $-\epsilon_k$ is an approximation of the first ionization energy of the atom, i.e., of the minimum energy required to carry out the process

$$X_{(g)} \rightarrow X^+_{(g)} + e^- \tag{10-37}$$

[11] T. C. Koopmans, *Physica*, **1**:104 (1934). Koopmans won the Nobel prize in economics in 1975. A humorous essay on misspellings of Koopmans' name and uncertainty about his middle initial and the year he published his theorem has been published by W. A. St. Cyr, L. N. Domelsmith, and K. N. Hauk in *J. Irr. Results*, **26**(1):23 (1980).

Similarly, the energies of the virtual orbitals are approximations of the electron affinities of the atom. In particular, if ϕ_m is the lowest energy virtual orbital, ϵ_m should be an approximation of the energy of the process

$$X_{(g)} + e^- \rightarrow X^-_{(g)} \tag{10-38}$$

The Hartree-Fock energy of the helium atom is $-2.862E_h$ (only about $0.01E_h$ lower than that due to a scaled 1s hydrogenlike orbital). The energy of the 1s Hartree-Fock orbital, ϵ_{1s}, is $-0.9180E_h$ (-24.98 eV). The experimental value of the first ionization energy is 24.5 eV.

The lithium atom has a Hartree-Fock energy of $-7.433E_h$ with $\epsilon_{1s} = -2.478E_h$ (-67.43 eV) and $\epsilon_{2s} = -0.1963E_h$ (-5.34 eV). The experimental first ionization energy of lithium is 5.39 eV. The energy of the 1s orbital should correlate with an x-ray term value for lithium. When lithium is used as a target cathode in an x-ray tube, radiation of wavelength 22.6 nm is produced; this radiation is related to the energy lost when an electron is removed from the K shell and an electron from the L shell "drops down" to replace it. The x-ray wavelength emitted by the dropping electron must satisfy

$$\frac{hc}{\lambda} = \epsilon_{2s} - \epsilon_{1s} \tag{10-39}$$

so that the 1s orbital energy is given experimentally by

$$\epsilon_{1s} = \epsilon_{2s} - \frac{hc}{\lambda} \tag{10-40}$$

Calculation of the last term on the right gives

$$\frac{(6.626 \times 10^{-34}\ \mathrm{J \cdot s})(3.000 \times 10^{8}\ \mathrm{m \cdot s^{-1}})}{22.6 \times 10^{-9}\ \mathrm{m}} = 8.80 \times 10^{-18}\ \mathrm{J}\ (54.9\ \mathrm{eV})$$

Using the experimental value of ϵ_{2s} in Eq. (10-40), one obtains

$$\epsilon_{1s} = -5.39 - 54.9 = -60.3\ \mathrm{eV}$$

This correlates reasonably well with the Hartree-Fock value of -67.43 eV.[12]

Table 10-1 shows a typical comparison between Hartree-Fock orbital energies and experimental values of orbital energies calculated from visible and ultraviolet spectroscopic term values.[13]

The Hartree-Fock SCF wavefunction predicts electron densities which are often in fair agreement with experimental methods such as electron diffraction.

[12] An extensive tabulation of Hartree-Fock calculations on atoms and some of their ions is given by E. Clementi in a supplement to a paper, "Tables of Atomic Functions," *IBM Journal of Research and Development*, **9**:2 (1965).

[13] Orbital energies computed from x-ray versus spectroscopic term values will not be identical, since different reference points are used for the energy. The former uses the zero point as the Fermi energy in the crystal; the latter uses infinite separation of electrons and nucleus.

TABLE 10-1
Koopmans' theorem for the Cu^+ ion and Na atom

Orbital	Energy, a.u. Calculated	Experimental*	Percent difference
	$Cu^{+\dagger}$		
1*s*	658.4	662.0	0.54
2*s*	82.30	81.3	1.2
2*p*	71.83	69.6	3.3
3*s*	10.651	9.6	11
3*p*	7.279	6.1	19
3*d*	1.613	0.79	104
	$Na^{\ddagger}$		
1*s*	79.4	81.2	2.2
2*s*	5.2	6.0	13
2*p*	2.80	3.66	23
3*s*	0.378	0.372	1.6

* J. C. Slater, *Phys. Rev.*, **98**:1039 (1955).

† D. R. Hartree and W. Hartree, *Proc. Roy. Soc. (London)*, **A157**:490 (1936).

‡ V. Fock and M. Petrashen, *Physik. Z. Sowjetunion*, **6**:632 (1934) and **8**:359 (1935).

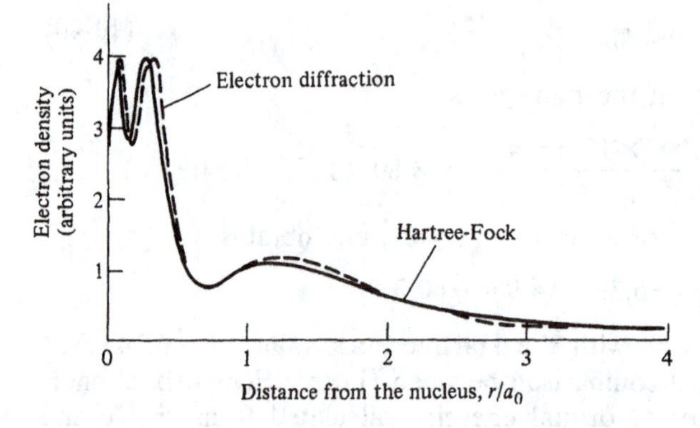

FIGURE 10-1
Comparison between experimental and theoretical electron densities in the argon atom. Both the Hartree-Fock and the experimental values were determined by L. S. Bartell and L. O. Brockway, *Phys. Rev.*, 90:833 (1953).

Figure 10-1 illustrates such a comparison for the argon atom as described by the ground-state configuration $1s^2 2s^2 2p^6 3s^2 3p^6$ ($^1S_0$ state).

EXERCISES

10-3. Given that the corresponding orbitals used to describe an atom X and its ions X^+ and X^- are the same to the first order, prove the relationships given in Eqs. (10-36).

10-4. Demonstrate that Koopmans' theorem is valid for helium by determining the energy expressions of He^+ (1*s*), He ($1s^2$), and He^- ($1s^2 2s$) and taking the appropriate differences. What assumption is needed in order for this demonstration to be valid?

10-5. Show that if the proper assumptions are made (those leading to Koopmans' theorem), one can show that

$$E(\mathrm{Li}) - E(\mathrm{Li}^+) = \epsilon_{2s}^{(0)} + J_{1s,\,2s} - K_{1s,\,2s} = \epsilon_{2s}$$

10-6. The Hartree-Fock orbital energies of the ^3P ground state of carbon are $1s$ ($-11.33E_h$), $2s$ ($-0.7056E_h$), and $2p$ ($-0.4333E_h$). The total electronic energy of this state in the Hartree-Fock approximation is $-37.689E_h$.

(*a*) Estimate the first ionization energy of carbon in its ground state. The experimental value is 11.3 eV.

(*b*) What is the total electron-repulsion energy of carbon in its ground state according to the Hartree-Fock approximation?

(*c*) An emission line of 4.47 nm is observed when carbon is employed as an x-ray cathode. What process (as described by the orbital model) corresponds to this emission?

(*d*) The Hartree-Fock energies for the two other states arising from the $1s^22s^22p^2$ configuration are $-37.631E_h$ (^1D) and $-37.550E_h$ (^1S). Show that these values (and that for the ^3P state) are in accord with Hund's rules.

10-3 THEORETICAL BASIS OF THE AUFBAU PRINCIPLE

Beginning as early as high school, chemistry students learn the order in which orbitals are to be used to predict the ground-state electron configurations of the elements: $1s$, $2s$, $2p$, $3s$, $3p$, $4s$, $3d$, etc. It is instructive to see how Hartree-Fock calculations of atomic energies for various electron configurations correlate with the order used in the Aufbau principle. Of particular interest is the apparent reversal of the $4s$ and $3d$ orbitals; just what does this represent, and can Hartree-Fock calculations shed any light on this?

First, Hartree-Fock calculations do indeed show that potassium and calcium exhibit their lowest energies when calculated from determinantal wavefunctions based on the electron configurations $1s^22s^22p^63s^24s$ and $1s^22s^22p^63s^24s^2$, respectively. If it were possible to say that the total electronic energy of an atom is the sum of the energies of the orbitals constituting the electron configuration, then one would be forced to conclude that the energy of a $4s$ orbital is less than that of a $3d$ orbital; hence, the $4s$ orbital is used before the $3d$ in order to minimize the total energy. Although this is essentially what elementary (and advanced) textbooks state, this interpretation is at variance with the results of Hartree-Fock calculations. Furthermore, this interpretation makes it necessary to "invent" some rather fanciful explanations for why the first ionization energies correlate with the supposedly lower-energy $4s$ orbitals rather than with the higher-energy $3d$ orbitals.

Table 10-2 contains a listing of the energies of the $3d$ and $4s$ orbitals of all the elements (except potassium and calcium) belonging to that period containing the first transition elements, i.e., all the elements from atomic number 21 through 36, inclusive. In all cases the energy of the $3d$ orbital is significantly lower than is the energy of the $4s$ orbital. Figure 10-2 illustrates this same data

TABLE 10-2
Energies of 4s and 3d orbitals for elements 21 to 36*

Element	Atomic no.	State	$-\epsilon_{4s}/E_h$	$-\epsilon_{3d}/E_h$	$\Delta = (\epsilon_{4s} - \epsilon_{3d})/E_h$
Sc	21	^2D	0.210	0.344	0.134
Ti	22	^3F	0.221	0.440	0.219
V	23	^4F	0.231	0.509	0.278
Cr	24	^5D	0.240	0.569	0.329
Mn	25	^6S	0.248	0.639	0.391
Fe	26	^5D	0.258	0.647	0.389
Co	27	^4F	0.267	0.676	0.409
Ni	28	^3F	0.276	0.707	0.431
Cu	29	^2S	0.237	0.493	0.256
Zn	30	^1S	0.292	0.783	0.491
Ga	31	^2P	0.425	1.193	0.768
Ge	32	^3P	0.553	1.635	1.082
As	33	^4S	0.686	2.113	1.427
Se	34	^2P	0.837	2.649	1.812
Br	35	^2P	0.993	3.220	2.227
Kr	36	^1S	1.153	3.825	2.672

* Data taken from E. Clementi, "Tables of Atomic Functions," *IBM Journal of Research and Development*, 9:2 (1965). Each of the SCF orbitals was expressed as a linear combination of Slater-type orbitals. A similar calculation by A. J. H. Wachters, *J. Chem. Phys.*, 52:1033 (1970) expressed each SCF orbital as a linear combination of gaussians; the two sets of calculations are in very close agreement.

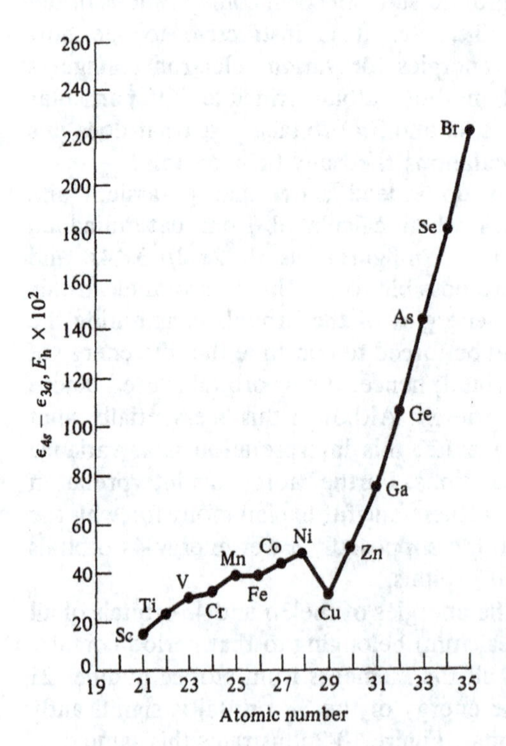

FIGURE 10-2
The 4*s* and 3*d* orbital energy differences for elements 21 through 36 based on Hartree-Fock calculations reported by E. Clementi, "Tables of Atomic Functions," *IBM Journal of Research and Development*, 9:2 (1965).

in graphic form; especially noteworthy are the following features:

1. The energy separation $\epsilon_{4s} - \epsilon_{3d}$ increases slowly with atomic number up until element 30 (zinc), whereupon it increases very rapidly. For the elements up to zinc, the 4*s* energy decreases very slowly; thus the increase in the 4*s*-3*d* energy difference is due mostly to a more rapid decrease in the 3*d* energies. This general trend continues beyond zinc but becomes exaggerated as 4*p* electrons are added.
2. There is a sharp dip in the energy difference in the case of element 29 (copper). Recall that copper is one of those atoms in which the simple rules would predict a ground-state electron configuration $1s^2 2s^2 2p^6 3s^2 3p^6 4s^2 3d^9$ instead of the $1s^2 2s^2 2p^6 3s^2 3p^6 4s^1 3d^{10}$ one must use. This is generally explained on the basis of an extra stability imparted by a fully filled set of *d* orbitals. Chromium (element 24) should exhibit an analogous anomaly owing to the extra stability of a half-filled set of *d* orbitals, but there is nothing in Fig. 10-2 to reflect this feature.[14]

The data exhibited in Table 10-2 (or in the equivalent Fig. 10-2) do not provide any information about the 3*d* and 4*s* orbital energies of potassium and calcium. If one assumes that the 4*s*-3*d* energy difference follows a regular pattern from about element 28 (nickel) down to element 19 (potassium), it appears that the 4*s*-3*d* energy differences for potassium and calcium must be almost zero—or perhaps even slightly negative. In any event, even if this difference is negative (so that 4*s* is indeed below 3*d*), this does not justify a *general* rule which implies this is true throughout the period. Worse still, a serious contradiction arises: If 4*s* is indeed below 3*d*, why do all the monopositive ions from atomic numbers 21 through 36 exhibit their lowest energies when a 4*s* electron is lost by ionization? Clearly, the Hartree-Fock results are consistent: 4s is always above 3*d*, and thus the 4*s* electron is the easiest to remove by ionization as predicted by Koopmans' theorem.

As for potassium and calcium: Is 4*s* also above 3*d*? Unfortunately, it is difficult to give a completely definitive answer to this question. Since these two atoms do not use both 3*d* and 4*s* orbitals, one cannot examine the energies directly. If one calculated Hartree-Fock energies for these atoms using a 4*s* orbital one time and a 3*d* orbital another time, one could compare the 4*s* and 3*d* energies directly— provided all other orbital energies did not change as a result of exchanging a 4*s* orbital with a 3*d* orbital. Virtually all attempts to

---

[14] However, Hartree-Fock calculations by Herman and Skillman, *Atomic Structure Calculations*, Prentice-Hall, Englewood Cliffs, N.J., 1963, shows similar "dips" for *both* chromium and copper. It is not clear why there should be these differences between different calculations, but they may be due to differences in some of the numerical methods employed to do the calculations. In all other respects, however, the Clementi calculations and the Herman and Skillman calculations are in reasonable agreement.

determine the $4s$ and $3d$ energies from separate calculations contain the same flaw and thus cannot prove the point.

Another question arises: If $4s$ is indeed above $3d$ in potassium and calcium, how can it be that the total energy is minimized by using the *higher-energy* orbital? Fortunately, this question is easily answered. The total energy of an atom can be written in a variety of forms (see Sec. 10-1), but one particularly convenient form for this purpose is

$$E = 2\sum_{i}^{N} \epsilon_i - \sum_{i,j}^{N} (2J_{ij} - K_{ij})$$

One can see at once that using a $4s$ orbital instead of a $3d$ orbital will *raise* that part of the total energy arising from the orbital energy sum. Thus, if $4s$ is actually to minimize the energy, it is necessary that it lead to an energy-repulsion energy which more than compensates for the increased orbital energy sum. (Note that the repulsion-energy term is *positive* and that it is *subtracted* from the sum of orbital energies.) It appears that this is what actually happens: if a $4s$ orbital is used, the repulsion of a $4s$ electron distribution with the rest of the electron distribution (an argon-core distribution) is greater than the analogous repulsion between a $3d$ electron distribution and the argon core. Presumably, this is because the electron distribution due to $4s$ penetrates the argon core more than does a $3d$ electron distribution; the former has a considerable value near the nucleus, whereas the latter has a node at the nucleus. Of course, this is not a rigorous proof, merely a demonstration that there is nothing intrinsically contradictory about the notion that $4s$ is *always* above $3d$.

Attempts to rationalize the periodic table beyond the first full period of elements begins to be fraught with ever-increasing difficulties as the atomic number becomes larger. Probably the most important reason for this is that relativistic effects become very important for atoms with large atomic numbers, and thus the simple nonrelativistic Schrödinger equation is inadequate to describe their electronic structures. However, when relativistic effects are taken into account, the heavier elements appear to lend themselves much better to a quantum mechanical description.

EXERCISE

10-7. The experimental first ionization energy of scandium is 6.5 eV. Does this correlate with the $4s$ or the $3d$ orbital energy of this atom? Explain.

10-4 THE ELECTRON CORRELATION ENERGY

The Hartree-Fock energy is usually within about 1 percent of the experimental value. Although this may strike one as very good agreement, it must be pointed out that total atomic energies per se are not of much significance to the most important problems of physics and chemistry. Rather, one is usually interested in *differences* between total energies, e.g., the energy difference between two spectroscopic states. Unfortunately, these energy differences themselves amount to no more than a small percent of the total energies of either state involved in the calculation. Thus small absolute errors in the total energies can lead to very large relative errors in their differences. For this reason there is considerable interest in quantum mechanical calculations which give better energies than does the Hartree-Fock method.

Since the Hartree-Fock method is perhaps the best-defined approximation method employed by quantum chemists, it is convenient to use the energies calculated by this method as a reference point for judging the quality of methods which attempt to bridge the gap between it and the "exact" solutions of the Schrödinger equation toward which we strive. Thus it is customary to define a quantity called the *electron correlation energy*, which is a measure of the ability (or inability) of the Hartree-Fock wavefunction to provide an accurate description of the electronic structure of an atom (or molecule). This quantity can then be used as a reference point against which improved calculations can be compared. After Löwdin, we define the electron correlation energy as follows:

> The correlation energy for a certain state with respect to a specified Hamiltonian is the difference between the exact eigenvalue of the Hamiltonian and its expectation value in the Hartree-Fock approximation for the state under consideration.[15]

The correlation energy defined in this way depends on the hamiltonian and may be expressed mathematically as

$$E_{\text{corr}} = \langle \hat{H} \rangle \text{ (exact)} - \langle \hat{H} \rangle \text{ (Hartree-Fock)} \tag{10-41}$$

From now on, unless otherwise stated, we will assume that $\hat{H}$ is the nonrelativistic hamiltonian, and we will use the term *correlation energy* to mean *nonrelativistic* correlation energy. It must be emphasized that the correlation energy defined in this way is *not* the difference between the exact experimental energy and the Hartree-Fock energy, since the former contains contributions from relativistic effects. Thus the correlation energy defined in this way must be regarded as a strictly mathematical quantity which serves as a useful criterion for the acceptability of wavefunctions over Hartree-Fock wavefunctions *insofar as the total energy is concerned*. It should be recalled that wavefunctions leading to good energies do not necessarily predict other properties as well. However, we will refer to the "goodness" of a wavefunction in direct proportion to the amount of the correlation energy it can account for. Thus if the correlation energy of a given atom is $-0.04E_{\text{h}}$ and a calculation

[15] P. O. Löwdin, *Advan. Chem. Phys.*, **2**:207 (1959).

produces an energy which is just $0.02E_h$ higher than the exact nonrelativistic energy, we say that the wavefunction accounts for 50 percent of the correlation energy. If another wavefunction comes within $0.01E_h$ of the exact nonrelativistic energy, we say it accounts for 75 percent of the correlation energy. Furthermore, we say that the latter wavefunction is "better" than the former.

The correlation error in the Hartree-Fock method reflects the fact that coulombic interaction between pairs of electrons is not properly accounted for. This inadequacy in the description of coulombic interaction is especially noticeable in the case of electron pairs with antiparallel spins. Electron pairs with parallel spins are kept apart by the antisymmetry principle, an effect which apparently overrides the coulombic repulsion, and thus are described somewhat better than electron pairs of antiparallel spin. Wavefunctions which lead to lower energies than the Hartree-Fock energy must somehow account for details of electronic motions in a more sophisticated manner than by simply allowing each electron to move in a "smeared out" or "average" field of the others. It is this assumption of an average potential field which leads to the prediction that electrons with antiparallel spins avoid each other less assiduously than is actually the case.

According to the variation principle, it is evident that the correlation energy defined by Eq. (10-41) is a *negative* quantity, since the Hartree-Fock energy lies *above* the exact energy. From the virial theorem (Sec. 6-6), which the Hartree-Fock approximation satisfies, we see that the kinetic and potential components of the correlation energy must satisfy

$$T_{\text{corr}} = -\frac{V_{\text{corr}}}{2} \tag{10-42}$$

Since $E_{\text{corr}} = T_{\text{corr}} + V_{\text{corr}} < 0$, it follows that $T_{\text{corr}} > 0$ and $V_{\text{corr}} < 0$. The positive nature of T_{corr} means that the Hartree-Fock method always underestimates the kinetic energies of the electrons. A simple physical interpretation of this fact is that the electrons actually undergo more complicated motions to avoid each other than the Hartree-Fock method is able to account for. The negative nature of the potential-energy error results primarily from allowing two electrons with different spins to occupy the same spatial region, leading to a higher repulsion energy than is actually the case.

In order to make an accurate estimate of the correlation energy as we have chosen to define it, one needs accurate values of the exact nonrelativistic energies of atoms. Unfortunately, such values are not generally known and must be estimated. It is apparent that if one took the experimental electronic energy of an atom and subtracted from this the relativistic contribution, the result would be the exact nonrelativistic energy. However, generally, the relativistic contributions can be estimated only. Only in the special case of the helium atom (where very accurate calculations of the nonrelativistic energy are available)[16] can the correlation energy be calculated with high accuracy.

---

[16] For example, the Pekeris or Scherr and Knight calculations discussed in Secs. 6-3 and 6-9.

One of the most-used methods of estimating the nonrelativistic energy of an atom is to determine the experimental energy by summation of successive ionization energies[17] and then estimating the relativistic contribution by perturbation theory.[18]

The Hartree-Fock energy of the ground state of the helium atom is $-2.862E_h$. The Pekeris "exact" nonrelativistic value (rounded off to the same number of significant figures) is $-2.904E_h$. Thus the correlation energy is $[-2.862-(-2.904)]\ E_h = -0.042E_h$ (this is often quoted as -1.14 eV). The virial theorem shows that this represents an error of $0.042E_h$ (1.14 eV) in the kinetic energy (an underestimation) and an error of $-0.084E_h$ (-2.28 eV) in the potential energy (an overestimation). In Table 10-3 are listed the theoretical and semiempirical correlation energies obtained by Fröman for several two-electron species.

Table 10-3 shows that the correlation energy is remarkably constant for two-electron systems. The slight discrepancy between the theoretical and semiempirical values is within the limits of experimental error but may possibly indicate the existence of some small, not yet understood effect. For atoms and ions with more than two electrons, the correlation energy is on the order of

[17] Such ionization energies are tabulated in C. E. Moore, *Atomic Energy Levels*, Natl. Bur. Std. (U.S.) Circular 467, 1949.

[18] See A. Fröman, *Phys. Rev.*, **112**:870 (1958) and *Rev. Mod. Phys.*, **32**:317 (1960).

TABLE 10-3
Correlation energies in electronvolts of some heliumlike systems*

	Theoretical		Semiempirical		
System	**Z**	**E_{corr}**	**E_{corr}**	**$E_{exp} - E_{HF}$**	**E_R**
H^-	1	-1.08†			
He	2	-1.14†	-1.142	-1.145	-0.003‡
Li^+	3	-1.18†	-1.182	-1.197	-0.015‡
Be^{2+}	4	-1.20†	-1.194	-1.250	-0.056‡
B^{3+}	5	-1.22†	-1.196	-1.345	-0.149‡
C^{4+}	6	-1.23†	-1.197	-1.521	-0.324‡
	∞	-1.28§			

Note: Z = nuclear charge; E_{corr} = correlation energy; E_{exp} = experimentally determined energy; E_{HF} = Hartree-Fock energy; E_R = relativistic energy.

* P. O. Löwdin, *Advan. Chem. Phys.*, **2**:207 (1959). (*Adapted by permission of the author and Interscience Publishers.*)

† Exact nonrelativistic energies from E. A. Hylleraas and J. Midtdal, *Phys. Rev.*, **103**:829 (1956), except for the He energy, which is from C. L. Pekeris, *Phys. Rev.*, **115**:1216 (1959). Hartree-Fock energies from L. C. Green et al., *Phys. Rev.*, **93**:757 (1954)

‡ Calculated by the perturbation method discussed by A. Fröman, *Rev. Mod Phys.*, **32**:317 (1960).

§ Theoretical limit estimated by P. O. Löwdin, *J. Mol. Spectry.*, **3**:46 (1959).

$-0.07E_h$ (-2 eV) per doubly filled orbital.[19] This is a rather large error, since this corresponds to about 183 kJ · mol^{-1}—a quantity which is of the same order of magnitude as chemical binding energies.

For heavier ions, for example, Al^{3+} the correlation energy is about $-0.4E_h$ (-11 eV, or -1062 kJ · mol^{-1}). In this case, this is also just about the magnitude of the relativistic contribution to the total energy. As reflected in Table 10-3, the relativistic energy contribution in two-electron systems increases roughly as Z^4. For $Z = 13$ (Al^{11+}), the two-electron relativistic energy is estimated to be about $0.33E_h$ (9 eV). For $Z = 92$ (U^{90+}) the relativistic energy contribution is estimated to be about 15 percent of the total energy.[20]

10-5 INTRODUCTION TO DENSITY MATRICES

The quantum mechanical state of a time-independent system of N electrons is represented by a wavefunction which may be written

$$\psi(\tau) = \psi(\tau_1, \tau_2, \ldots, \tau_N) \tag{10-43}$$

where, as in Sec. 7-1, τ_μ represents the space and spin coordinates $(x_\mu, y_\mu, z_\mu, \sigma_\mu)$ of the μth electron. Such a wavefunction is a function of $4N$ variables: $3N$ spatial variables and N spin variables. However, there are certain properties of the system which do not depend on all these $4N$ degrees of freedom. For example, in the nonrelativistic treatments we have used so far the total energy depends only on averages involving no more than two electrons at a time—regardless of the number of electrons in the system. This suggests the possibility that there exists some function of two electrons which we could use instead of the N-electron wavefunction in order to compute the energy. Since wavefunctions describe isolated systems (i.e., systems which do not interact with their surroundings), this new function cannot itself be a wavefunction; two electrons in an N-electron system ($N > 2$) do not constitute an isolated system, since they interact with all the other electrons. Consequently, we need to introduce a new mathematical construct which is more general than a wavefunction and of which a wavefunction is a special case. Such a construct is the *density matrix*. In the following we examine an abbreviated account of this construct and indicate its usefulness in discussing electronic structures. More complete discussions can be found elsewhere.[21]

---

[19] E. Clementi, *J. Chem. Phys.*, **38**:2248 and **39**:175 (1963).

[20] A list of experimental and Hartree-Fock energies (including estimates of mass corrections and relativistic energies) is given by E. Clementi for first- and second-row atoms in *Chem. Revs.*, **68**:341 (1968).

[21] See, for example, P. O. Löwdin, *Phys. Rev.*, **97**:1474 (1955); R. McWeeny, *Rev. Mod. Phys.*, **32**:335 (1960); D. ter Haar, *Rept. Prog. Phys.*, **24**:304 (1961); U. Fano, *Rev. Mod Phys.*, **29**:74 (1957).

We begin by defining the *density operator* for an N-electron system described by the wavefunction $\psi_k(\tau)$; this wavefunction is one of the members of a complete set of eigenfunctions $\psi_1, \psi_2, \ldots$ of the system hamiltonian. This operator may be written in Dirac bra-ket notation as

$$\rho_k = |\psi_k\rangle\langle\psi_k| \tag{10-44}$$

The most important properties of this operator are that it is hermitian and idempotent. These properties are illustrated in the following:

$$\begin{aligned} \rho_k^\dagger &= (|\psi_k\rangle\langle\psi_k|)^\dagger \\ \rho_k^2 &= |\psi_k\rangle\langle\psi_k|\psi_k\rangle\langle\psi_k| = |\psi_k\rangle\langle\psi_k| = \rho_k \end{aligned} \tag{10-45}$$

(assuming ψ_k is normalized). Thus ρ_k functions as a projection operator. If $|\phi\rangle$ represents an arbitrary vector in the space of the system, we see that the operation of ρ_k on this vector projects out the component $|\psi_k\rangle$:

$$\rho_k|\phi\rangle = |\psi_k\rangle\langle\psi_k|\phi\rangle = |\psi_k\rangle\left\langle\left|\sum_i \psi_i c_i\right\rangle = \left|\psi_k\right\rangle c_k \tag{10-46}$$

In the Schrödinger representation, the density operator is expressed as an integral operator, namely,

$$\rho_k\phi(\tau) = \int \rho_k(\tau'|\tau)\phi(\tau)\, d\tau \tag{10-47}$$

where

$$\rho_k(\tau'|\tau) = \psi_k(\tau')\psi_k^*(\tau) \tag{10-48}$$

The function $\rho_k(\tau'|\tau)$, where τ (and τ') represent the entire set of spatial and spin coordinates, is called the *density matrix* or, more completely, the *full*, or N*th-order*, *density matrix* of the system. We use the term "matrix," since the τ' and τ are analogous to the double indices of matrix elements.[22] The diagonal elements of this matrix ($\tau' = \tau$) produce the usual probability distribution function

$$\rho_k(\tau|\tau) = \psi_k(\tau)\psi_k^*(\tau) = |\psi_k(\tau)|^2 \tag{10-49}$$

The expectation value in the state k of a general operator $\hat{A}$ is given by

$$\langle\hat{A}\rangle = \langle\psi_k|\hat{A}|\psi_k\rangle = \int \hat{A}\rho_k(\tau'|\tau)\, d\tau \tag{10-50}$$

[22] The correct general mathematical term for a quantity such as $p_k(\tau'|\tau)$ which occurs in an integral equation is a *kernel*. Thus, one should refer to this as the *density kernel*; however, the name "matrix" has been sanctified by long usage and is undoubtedly here to stay.

In evaluating the above integral, the following conventions must be used:

1. The operator $\hat{A}$ operates only on the unprimed variables.
2. After operation with $\hat{A}$ and before integration, set $\tau' = \tau$.

This procedure is analogous to computing the trace of a matrix product, and often one replaces Eq. (10-50) with the symbolic form

$$\langle \hat{A} \rangle = \mathrm{Tr}[\hat{A}\rho_k(\tau' \mid \tau)] \tag{10-51}$$

It is convenient at this point to drop the subscript k—keeping in mind, however, that $\psi(\tau)$ and $\rho(\tau' \mid \tau)$ refer to some particular quantum state. If the wavefunction $\psi(\tau)$ is represented by some basis $\{\phi_i\}$, namely

$$\psi(\tau) = \sum_i \phi_i(\tau)c_i = \mathbf{\Phi C} \tag{10-52}$$

then the density matrix becomes

$$\rho(\tau' \mid \tau) = \mathbf{\Phi C C^\dagger \Phi^\dagger} = \mathbf{\Phi R \Phi^\dagger} \tag{10-53}$$

where $\mathbf{R} = \mathbf{CC^\dagger}$ is the matrix representation of the density matrix in the basis $\{\phi_i\}$. The general matrix element of $\mathbf{R}$ is given by

$$R_{ij} = c_i c_j^* \tag{10-54}$$

The expectation value of the operator $\hat{A}$ can now be written

$$\langle \hat{A} \rangle = \sum_i \sum_j c_i c_j^* \langle \phi_i | \hat{A} | \phi_j \rangle = \sum_i \sum_j R_{ij} A_{ij} = \mathrm{Tr}\,(\mathbf{AR}) = \mathrm{Tr}\,(\mathbf{RA}) \tag{10-55}$$

where

$$\mathbf{A} = \mathbf{\Phi}^\dagger \hat{A} \mathbf{\Phi} \tag{10-56}$$

If $\psi(\tau)$ is normalized and the basis $\{\phi_i\}$ is orthonormal, it follows that $\mathbf{C^\dagger C} = \mathbf{1}$ and thus

$$\mathbf{R}^2 = \mathbf{CC^\dagger CC^\dagger} = \mathbf{CC^\dagger} = \mathbf{R} \tag{10-57}$$

that is, $\mathbf{R}$ is idempotent. From Eq. (10-55) for the special case that $\hat{A} = 1$ (unit operator) we see that

$$\mathrm{Tr}\,\mathbf{R} = 1 \tag{10-58}$$

The diagonal element ($\tau'_\mu = \tau_\mu$ for $\mu = 1, 2, \ldots, N$) of the density matrix [Eq. (10-48)] expresses the probability that electron 1 lies in the volume element $d\tau_1$ at the same time that electron 2 lies in the volume element $d\tau_2$, and so on for the remaining $N-2$ electrons. Since the electrons are indistinguishable, any one of them is equally likely to be in *any* volume element, so that the probability of N electrons' occupying N selected volume elements $d\tau_1, d\tau_2, \ldots, d\tau_N$ *in any order* is given by $N!\rho(\tau' \mid \tau)$. Thus it is convenient to renormalize $\rho(\tau'|\tau)$ to $N!$ by defining a new density matrix

$$\Gamma'_N = N!\rho(\tau'|\tau) \tag{10-59}$$

In this simplified notation, the diagonal element of the renormalized density matrix becomes Γ_N; that is, we simply drop the prime. Note that whereas $\mathrm{Tr}\,\rho(\tau'|\tau) = 1$, we now have $\mathrm{Tr}\,\Gamma'_N = N!$

Next we introduce the *reduced density matrix* of order p $(1 \le p \le N)$, defined in such a way that the diagonal elements express the probability of finding p electrons *in any order* in the volume elements $d\tau_1, d\tau_2, \ldots, d\tau_p$. This definition is

$$\Gamma'_p = p!\binom{N}{p}\int \rho(\tau'|\tau)\, d\tau_{p+1}\, d\tau_{p+2}\cdots d\tau_N \tag{10-60}$$

where the binomial coefficient is the number of ways of taking N things p at a time and $p!$ is the number of indistinguishable arrangements.[23] Note that when $N = p$, Γ'_p reduces to Γ'_N.

The utility of the pth-order reduced density matrix (from here on we will drop the adjective "reduced" for convenience) is in the evaluation of expectation values of operators which do not depend on all the $4N$ degrees of freedom of the system. Thus if the operator $\hat{A}$ depends on only p electrons at a time, its expectation value is given in terms of the pth-order density matrix as

$$\langle \hat{A} \rangle = \frac{1}{p!}\int \hat{A}(1, 2, \ldots, p)\Gamma'_p\, d\tau_1\, d\tau_2\cdots d\tau_p \tag{10-61}$$

Thus, the pth-order density matrix is the mathematical construct we sought at the outset; it provides a complete description of all system properties which depend on no more than p electrons at a time. Once one has the pth-order density matrix, all other orders less than p can be obtained by successive integration; for example, the $(p-1)$-order density matrix is given in terms of the pth-order density matrix by

$$\Gamma'_{p-1} = (N - p + 1)^{-1}\int \Gamma'_p\, d\tau_p \tag{10-62}$$

The preintegral factor is just the ratio of the Γ'_p and Γ'_{p-1} normalization factors.

In principle, one should be able to dispense with the wavefunction itself and deal exclusively with the appropriate reduced density matrices. Unfortunately, it is not known what conditions reduced density matrices must satisfy, and, consequently, no way of obtaining them is known except by integration of the wavefunction.[24] Nevertheless, as we will see, reduced density matrices

---

[23] The normalization factor follows the convention of McWeeny; Löwdin employs only the binomial coefficient part of this, and others normalize to unity for all orders.

[24] A. J. Coleman, *Rev. Mod. Phys.*, **35**:665 (1963).

provide a useful means of analyzing many important aspects of electronic structure.

Of particular interest for the study of electronic systems are the first- and second-order density matrices defined by

$$\begin{aligned}\Gamma_1' &= \gamma' = N\int \rho(\tau'|\tau)\, d\tau_2\, d\tau_3 \cdots d\tau_N \\ \Gamma_2' &= \Gamma' = N(N-1)\int \rho(\tau'|\tau)\, d\tau_3\, d\tau_4 \cdots d\tau_N\end{aligned} \tag{10-63}$$

Note that we have once again streamlined the notation by introducing new symbols for density matrices; to denote the diagonal elements of γ' and Γ' we simply omit primes as before

According to Eq. (10-62) the first- and second-order density matrices are related as follows:

$$\gamma' = (N-1)^{-1}\int \Gamma'\, d\tau_2 \tag{10-64}$$

We now examine the obtaining of expectation values of one- and two-electron operators in terms of γ' and Γ', respectively. First consider a one-electron operator of the general form

$$\hat{A} = \sum_{\mu=1}^{N} \hat{A}_\mu \tag{10-65}$$

The expectation value of this operator is

$$\langle \hat{A} \rangle = \left\langle \psi \middle| \sum \hat{A}_\mu \middle| \psi \right\rangle = N\langle \psi|\hat{A}_1|\psi\rangle = N\int \hat{A}_1 \rho(\tau'|\tau)\, d\tau_1\, d\tau_2 \cdots d\tau_N \tag{10-66}$$

Integrating over $N-1$ electron coordinates in all possible positions leads to

$$\langle \hat{A} \rangle = N\,\frac{(N-1)!}{N!}\int \hat{A}_1 \gamma'\, d\tau_1 = \int \hat{A}_1 \gamma'\, d\tau_1 \tag{10-67}$$

Note that if the operator $\hat{A}$ does not involve differentiation (for example, $1/r_\mu$) the prime is not needed; i.e., only the diagonal elements of the first-order density matrix are relevant. Thus

$$\left\langle \sum_{\mu}^{N} \frac{1}{r_\mu} \right\rangle = \int \frac{1}{r_1}\, \gamma\, d\tau_1 \tag{10-68}$$

However, if the operator $\hat{A}$ involves differentiation (for example, ∇_μ^2), the prime is needed:

$$\left\langle \sum_{\mu}^{N} \nabla_\mu^2 \right\rangle = \int \nabla_1^2 \gamma'\, d\tau_1 \tag{10-69}$$

Next we consider two-electron operators of the form

$$\hat{G} = \sum_{\mu<\nu}^{N} \hat{G}_{\mu\nu} \tag{10-70}$$

which do not involve differentiation (for example, $1/r_{\mu\nu}$). The expectation value of $\hat{G}$ is

$$\langle \hat{G} \rangle = \langle \psi | \sum_{\mu<\nu} \hat{G}_{\mu\nu} | \psi \rangle = \binom{N}{2} \int \hat{G}_{12} \rho(\tau' | \tau)\, d\tau_1\, d\tau_2 \cdots d\tau_N \tag{10-71}$$

(The binomial coefficient represents the number of electron pairs.) Integrating over the coordinates of $N-2$ electrons in all possible position leads to

$$\langle \hat{G} \rangle = \frac{(N-2)!}{N!} \binom{N}{2} \iint \hat{G}_{12} \Gamma\, d\tau_1\, d\tau_2 = \frac{1}{2} \iint \hat{G}_{12} \Gamma\, d\tau_1\, d\tau_2 \tag{10-72}$$

Note that only the diagonal elements of Γ' are needed.[25]

From the foregoing we can see that the energy of an atom in the nonrelativistic approximation is given by

$$\langle \hat{H} \rangle = -\frac{1}{2} \int \nabla_1^2 \gamma'\, d\tau_1 - Z \int \frac{\gamma\, d\tau_1}{r_1} + \frac{1}{2} \iint \frac{\Gamma\, d\tau_1\, d\tau_2}{r_{12}} \tag{10-73}$$

Note that the kinetic energy of the electrons depends on the nondiagonal elements of the first-order density matrix, whereas the potential energy depends on the diagonal elements of both the first- and the second-order density matrices.

Since γ' can be determined from Γ' [see Eq. (10-62) or (10-64)], it is apparent that knowing only the second-order density matrix provides sufficient information for computing the total energy of an n-electron atom. However, how to obtain this second-order density matrix on the basis of first principles remains an unsolved problem. Since the wavefunction of the helium atom (a two-electron system) is the second-order density matrix for that system, studies on two-electron systems are of great interest to those who seek a general solution to the second-order density matrix problem.

In the following we will examine in some detail how to construct the first-order density matrix of the helium atom for two different approximate wavefunctions: the simple single-determinantal wavefunction and the split-shell wavefunction.

[25] The reader should note how the primes in density matrices are used; essentialy they serve as "flags" to identify which terms in the integrand are to be operated on before integration. Once the operator has acted on these terms, the flags are no longer needed and may be removed so that integration proceeds over all terms in the integrand.

The simple single-determinantal wavefunction describing the $^1S_0$ ground state of helium is given by

$$\psi(1,2) = |1s\,\overline{1s}| = \frac{1}{\sqrt{2}}\begin{vmatrix} 1s(1)\alpha(1) & 1s(1)\beta(1) \\ 1s(2)\alpha(2) & 1s(2)\beta(2) \end{vmatrix}$$

$$= \frac{1}{\sqrt{2}}[1s(1)1s(2)\alpha(1)\beta(2) - 1s(1)1s(2)\beta(1)\alpha(2)] \qquad (10\text{-}74)$$

This is, of course, simply related to the second-order density matrix

$$\rho(1',2'\,|\,1,2) = \tfrac{1}{2}[1s(1')1s(2')\alpha(1')\beta(2') - 1s(1')1s(2')\beta(1')\alpha(2')] \times [1s(1)1s(2)\alpha(1)\beta(2) - 1s(1)1s(2)\beta(1)\alpha(2)] \qquad (10\text{-}75)$$

Algebraic manipulation is made easier if we write the above in the form

$$\rho(1',2'\,|\,1,2) = \tfrac{1}{2}(a'-b')(a-b) = \tfrac{1}{2}(a'a - b'a - a'b + b'b) \qquad (10\text{-}76)$$

where a' is the first term in the first set of brackets in Eq. (10-75), b' is the second term in the first set of brackets, a is the first term in the second set of brackets, and b is the second term in the second set of brackets. The first-order density matrix is given by integrating the above over $d\tau_2$:

$$\gamma' = \frac{1}{2}\int (a'a - b'a - a'b + b'b)\,d\tau_2$$

$$= \tfrac{1}{2}[1s(1')1s(1)\alpha(1')\alpha(1) + 1s(1')1s(1)\beta(1')\beta(1) \qquad (10\text{-}77)$$

where terms such as $b'a$ and $a'b$ vanish because of spin orthogonality. The diagonal elements of γ' are given by

$$\gamma = \tfrac{1}{2}[1s^2(1)\alpha^2(1) + 1s^2(1)\beta^2(1)] \qquad (10\text{-}78)$$

This means that 50 percent of the electron density comes from a $1s$ spatial function and α spin and 50 percent comes from the same spatial function and β spin. If the spin is integrated over, we get simply $1s^2(1)$; this is sometimes called the *diagonal element of the spinless first-order density matrix*.

The split-shell wavefunction is

$$\psi(1,2) = \frac{1}{\sqrt{2}}(|1s\,\overline{1s'}| - |\overline{1s}\,1s'|)$$

$$= \tfrac{1}{2}[1s(1)1s'(2) + 1s'(1)1s(2)][\alpha(1)\beta(2) - \beta(1)\alpha(2)]$$

$$= \tfrac{1}{2}[1s(1)1s'(2)\alpha(1)\beta(2) + 1s'(1)1s(2)\alpha(1)\beta(2) - 1s'(1)1s(2)\beta(1)\alpha(2) - 1s(1)1s'(2)\beta(1)\alpha(2)] \qquad (10\text{-}79)$$

The second-order density matrix is

$$\begin{aligned}\rho(1',2'|1,2) &= \tfrac{1}{4}(a'+b'-c'-d')(a+b-c-d)\\ &= \tfrac{1}{4}(a'a+a'b-a'c-a'd+b'a+b'b\\ &\quad -b'c-b'c-c'a-c'b+c'c\\ &\quad +c'd-d'a-d'b+d'c+d'c)\end{aligned} \tag{10-80}$$

The terms $a'c$, $a'd$, $b'c$, $b'd$, $c'a$, $d'a$, $c'b$, and $d'b$ vanish upon integration owing to spin orthogonality when the integration over $d\tau_2$ is carried out to obtain γ'. The first-order density matrix becomes

$$\begin{aligned}\gamma' = \tfrac{1}{4}\{&[1s(1')1s(1)+1s(1')1s'(1)\Delta+1s'(1')1s(1)\Delta\\ &+1s'(1')1s'(1)]\alpha(1')\alpha(1)+[1s(1')1s(1)+1s(1')1s'(1)\Delta\\ &+1s'(1')1s(1)\Delta+1s'(1')1s'(1)]\beta(1')\beta(1)\}\end{aligned} \tag{10-81}$$

The diagonal element is (after suitable rearrangement)

$$\gamma = \tfrac{1}{4}[1s^2(1)+(1s')^2(1)+2\Delta 1s(1)1s'(1)][\alpha^2(1)+\beta^2(1)] \tag{10-82}$$

Integrating over the spin we obtain the spinless first-order density matrix

$$\gamma = \tfrac{1}{2}[1s^2(1)+(1s')^2(1)+2\Delta 1s(1)1s'(1)] \tag{10-83}$$

This shows that the electron density is distributed about the nucleus as follows: a fraction $1/(2+2\Delta)$ due to $1s$, a fraction $1/(2+2\Delta)$ due to $1s'$, and a fraction $2\Delta/(2+2\Delta)$ due to $1s(1)1s'(1)$.

EXERCISES

10-8. Work out the first-order density matrices for the singlet and triplet states of helium arising from the $1s2s$ configuration (assume $\langle 1s|2s\rangle = 0$). Explain why the results turn out as they do.

10-9. Show that the diagonal elements of the spinless first-order density matrix of the lithium atom in the configuration $1s^2 2s$ is given by

$$\gamma = \tfrac{2}{3}\{\tfrac{1}{2}[1s^2(1)][\alpha^2(1)+\beta^2(1)]\} + \tfrac{1}{3}[2s^2(1)\alpha^2(1)]$$

Let the wavefunction be $|1s\,\overline{1s}\,2s|$ and assume $\langle 1s\,|\,2s\rangle = 0$. Explain the significance of the density matrix in terms of the total electron density of the atom.

10-10. Use the density matrices arising from the helium wavefunction (10-74) and the energy expression (10-73) to obtain the ground-state energy expression for this atom. Compare it with the result quoted in Eq. (7-80).

10-6 DENSITY MATRIX ANALYSIS OF THE HARTREE-FOCK APPROXIMATION

The quantity $\gamma\, d\tau_1$ is seen to be the probability (normalized to N) of finding any one of the electrons within the volume element $d\tau_1$, that is, around the point (x_1, y_1, z_1) with spin σ_1, while all the other electrons have arbitrary positions and spin. Similarly, $\Gamma\, d\tau_1\, d\tau_2$ represents the probability [normalized to $N(N-1)$] of finding any one of the electrons within the volume element $d\tau_1$ and any other electron within $d\tau_2$, while all the other electrons have arbitrary positions and spins. The diagonal element of the second-order density matrix must become relatively small whenever the two electrons have the same spatial coordinates but opposite spins, i.e., whenever $v_1 = v_2$ but $\sigma_1 \neq \sigma_2$. This effect is as if there were an excluded region about each electron with respect to all other electrons; this apparent excluded region is called a *coulomb hole*. The coulomb hole owes its existence to the $1/r_{\mu\nu}$ coulombic-repulsion terms in the hamiltonian.

Also, as a consequence of the antisymmetry principle, the second- and higher-order density matrices must be antisymmetric with respect to *each set* of their indices. Thus

$$\Gamma(\tau_1', \tau_2' \mid \tau_1, \tau_2) = -\Gamma(\tau_2', \tau_1' \mid \tau_1, \tau_2) = -\Gamma(\tau_1', \tau_2' \mid \tau_2, \tau_1) \tag{10-84}$$

This implies that

$$\Gamma(\tau_1', \tau_2' \mid \tau_1, \tau_2) = 0 \tag{10-85}$$

whenever $\tau_1' = \tau_2'$, $\tau_1 = \tau_2$, or both. The latter condition also implies an excluded region around each electron insofar as all other electrons of the *same* spin are concerned; this excluded region is called a *fermi hole*. The diagonal element $\Gamma(\tau_1, \tau_2 \mid \tau_1, \tau_2)$ may be said to represent both the coulomb hole for all electrons and the fermi hole for electrons of a given spin. For this reason the diagonal element is often called the *correlation density* function. The appearance of this correlation density for electrons of the same spin (fermi hole) is shown in Fig. 10-3.

A major failing of the Hartree-Fock approximation is that it overestimates electron repulsion by giving the fermi hole too much weight. At the same time the method underestimates the coulomb hole—in fact the coulomb

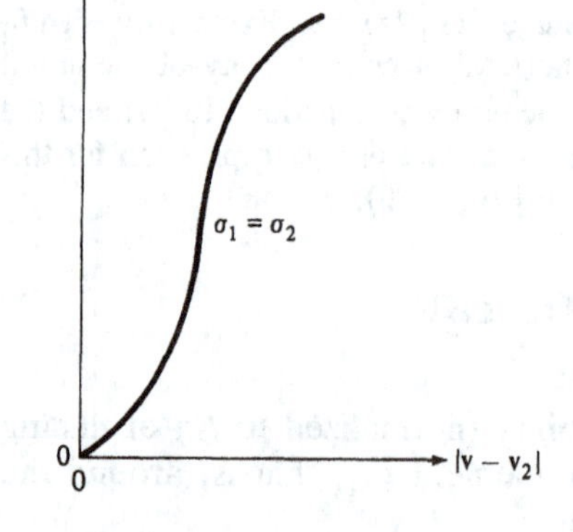

FIGURE 10-3
Schematic representation of the fermi hole for electrons with parallel spin. The Hartree-Fock method overestimates electron repulsion by giving the fermi hole too much weight. This more than compensates for an almost complete neglect of the coulomb hole. (*Adapted from P. O. Löwdin,* Advan. Chem. Phys., *2:218 (1959) and used by permission of the author and Interscience Publishers, Inc.*)

hole is neglected almost completely. In the final analysis the overestimation of the fermi hole more than compensates for the neglect of the coulomb hole; the net result is an overestimation of electron repulsion (recall that $V_{\text{corr}} < 0$). The search for wavefunctions which are better than the Hartree-Fock wavefunctions may be said to be an attempt to find improved descriptions of the coulomb hole.

The full density matrix defined in terms of the single-determinantal Hartree-Fock wavefunction is

$$\Gamma'_N = \begin{vmatrix} \rho(\tau'_1, \tau_1) & \rho(\tau'_1, \tau_2) & \cdots & \rho(\tau'_1, \tau_N) \\ \rho(\tau'_2, \tau_1) & \rho(\tau'_2, \tau_2) & \cdots & \rho(\tau'_2, \tau_N) \\ \cdots & \cdots & \cdots & \cdots \\ \rho(\tau'_N, \tau_1) & \rho(\tau'_N, \tau_2) & \cdots & \rho(\tau'_N, \tau_N) \end{vmatrix} \tag{10-86}$$

where the general determinantal element is

$$\rho(\tau'_i, \tau_j) = \sum_{k=1}^{N} S_k(\tau'_i) S_k^*(\tau_j) \tag{10-87}$$

This latter element is known as the *Fock-Dirac density matrix*; this quantity is invariant under a unitary transformation and thus constitutes the fundamental invariant of the self-consistent field method. Making use of the fact that a determinant can be expanded in terms of minors and cofactors, one can obtain reduced density matrices of any order from the full density matrix. In general, the pth-order (reduced) density matrix is

$$\Gamma'_p = \begin{vmatrix} \rho(\tau'_1, \tau_1) & \rho(\tau'_1, \tau_2) & \cdots & \rho(\tau'_1, \tau_p) \\ \rho(\tau'_2, \tau_1) & \rho(\tau'_2, \tau_2) & \cdots & \rho(\tau'_2, \tau_p) \\ \cdots & \cdots & \cdots & \cdots \\ \rho(\tau'_p, \tau_1) & \rho(\tau'_p, \tau_2) & \cdots & \rho(\tau'_p, \tau_p) \end{vmatrix} \tag{10-88}$$

In particular, the second- and first-order density matrices are

$$\Gamma' = \begin{vmatrix} \rho(\tau'_1, \tau_2) & \rho(\tau'_1, \tau_2) \\ \rho(\tau'_2, \tau_1) & \rho(\tau'_2, \tau_2) \end{vmatrix} \tag{10-89}$$

The latter result shows that the Fock-Dirac density matrix given in Eq. (10-87) is simply the first-order density matrix.

Examination of the steps between Eqs. (10-86) and (10-89) reveals a significant relationship: all higher orders of density matrices ($p = 2$ to $p = N$) are completely determined by the first-order density matrix. This fact may be regarded as the basic theorem of the Hartree-Fock approximation: all properties of the electronic system are determined by the first-order (one-electron) density matrix.

The Fock-Dirac density matrix defined by Eq. (10-89) is readily shown to be idempotent. Using the expression for the product of two matrices, we can write

$$(\gamma')^2 = \int \gamma(\tau_1' \mid \xi)\gamma(\xi \mid \tau_1)\, d\xi = \sum_k \sum_l S_k(\tau_1')S_l^*(\tau_1) \int S_k(\xi)S_l^*(\xi)\, d\xi$$

$$= \sum_k \sum_l S_k(\tau_1')S_l^*(\tau_1)\, \delta_{kl} = \sum_k S_k(\tau_1')S_l^*(\tau_1) = \gamma' \qquad (10\text{-}90)$$

The trace of the Fock-Dirac density matrix for an N-electron system is

$$\mathrm{Tr}[\gamma(\tau_1' \mid \tau_1)] = \int \sum_{k=1}^{N} S_k(\tau_1')S_k^*(\tau_1)\, d\tau_1 = N \qquad (10\text{-}91)$$

It is also readily verified that $\gamma(\tau_1' \mid \tau_1)$ is hermitian; thus the Fock-Dirac density matrix has the properties of a projection operator.

The first- and second-order density matrices can be resolved into components which express the probabilities of given spin assignments to electrons, namely,

$$\gamma = \gamma^\alpha + \gamma^\beta \qquad \Gamma = \Gamma^{\alpha\alpha} + \Gamma^{\beta\beta} + \Gamma^{\alpha\beta} + \Gamma^{\beta\alpha} + \Gamma^{\text{mix}} \qquad (10\text{-}92)$$

(primes have been omitted in the above, but the relationships are valid for diagonal as well as nondiagonal elements). Here γ^α is the probability distribution function for electrons of α spin, $\Gamma^{\alpha\beta}$ is the probability distribution function for pairs of electrons with unlike spin (α for 1 and β for 2), and so on for the other terms, some of which may be zero in particular cases. The last term in Γ (or Γ'), Γ^{mix}, refers to a term with mixed spin, e.g., the $b'a$ and $a'b$ terms in Eq. (10-76). For example, if $\psi(1, 2) = |S_1\ S_2|$, where S_1 has α spin and S_2 has β spin, then

$$\Gamma' = 2!\psi(1', 2')\psi^*(1, 2) = S_1(1')S_1^*(1)S_2(2')S_2^*(2) - S_1(1')S_2^*(1)S_2(2')S_1^*(2)$$
$$- S_2(1')S_1^*(1)S_1(2')S_2^*(2) + S_2(1')S_2^*(1)S_1(2')S_1^*(2) \qquad (10\text{-}93)$$
$$\gamma' = S_1(1')S_1^*(1) + S_2(1')S_2^*(1) = \gamma^\alpha + \gamma^\beta$$

We see that the first term in Γ' is $\Gamma^{\alpha\beta}$, the fourth term is $\Gamma^{\beta\alpha}$, and the second and third terms are Γ^{mix}. Terms such as $\Gamma^{\alpha\alpha}$ and $\Gamma^{\beta\beta}$ are absent, and $\int \Gamma^{\text{mix}}\, d\tau_2 = 0$. Note also that the following relations are true:

$$\gamma^\alpha(\tau_1' \mid \tau_1)\gamma^\beta(\tau_2' \mid \tau_2) = \Gamma^{\alpha\beta}$$

but

$$\gamma^\alpha(\tau_1' \mid \tau_1)\gamma^\alpha(\tau_2' \mid \tau_2) \neq \Gamma^{\alpha\alpha}$$

The latter relationships clearly indicate the fact that the Hartree-Fock approximation does not adequately correlate antiparallel spins. The fact that for antiparallel spins the second-order density matrix component $\Gamma^{\alpha\beta}$ is simply a product of the two first-order density matrix components γ^α and γ^β implies that the product law for combination of independent events is obeyed. In actuality, the two electrons are not precisely *independent*; they are actually *quasi-independent* by virtue of the fact that γ' of a one-electron system is not precisely the same as γ' of a many-electron system.

The reduced density matrices we have used so far include spin coordinates. For some purposes it is more convenient to integrate over the spin coordinates to produce *spinless* density matrices.[26] These are defined simply as

$$\Gamma(v_1', v_2' | v_1, v_2) = \int\int \Gamma(\tau_1', \tau_2' | \tau_1, \tau_2)\, d\sigma_1\, d\sigma_2 \tag{10-94}$$

and analogously for other orders. Expectation values of spin-free operators remain unchanged, but the volume elements of integration now become $dv_1\, dv_2 \ldots$ instead of $d\tau_1\, d\tau_2 \ldots$.

Using Eqs. (10-73) and (10-89), we find that the Hartree-Fock energy in the spin-free formulation becomes

$$E = -\frac{1}{2}\int \nabla_1^2 \gamma'\, dv_1 - Z\int \frac{\gamma}{r_1}\, dv_1 + \frac{1}{2}\int\int \frac{\gamma(v_1|v_1)\gamma(v_2|v_2) - \gamma(v_2|v_1)\gamma(v_1|v_2)}{r_{12}}\, dv_1\, dv_2 \tag{10-95}$$

Thus, an alternative way of stating the Hartree-Fock problem is that one must find the first-order density matrix which minimizes the energy, subject to the restriction of Eqs. (10-90) and (10-91).

The exact nonrelativistic energy of an atom can be written in the form

$$E = -\frac{1}{2}\int \nabla_1^2 \gamma'\, dv_1 - Z\int \frac{\gamma}{r_1}\, dv_1 + \frac{1}{2}\int\int \frac{\gamma(v_1|v_1)\gamma(v_2|v_2)}{r_{12}}\, dv_1\, dv_2 + \frac{1}{2}\int\int \frac{g(1,2)}{r_{12}}\, dv_1\, dv_2 \tag{10-96}$$

where we have used the *correlation function*, defined as

$$g(1,2) = \Gamma(v_1, v_2 | v_1, v_2) - \gamma(v_1|v_1)\gamma(v_2|v_2) \tag{10-97}$$

where Γ and γ are the *exact* second- and first-order density matrices, respectively. The correlation function represents an excluded region or hole about each electron, relative to all the other electrons. The next-to-last term in Eq. (10-96) is just the zeroth-order repulsion energy in which correlation is ignored, and the last term is the correlation energy. Each term in Eq. (10-96) is well defined and is invariant under any transformation which does not affect the total wavefunction. Thus, estimates of the terms of this energy expression from a variety of widely differing approximations should be strictly comparable.

---

[26] R. McWeeny, *Rev. Mod. Phys.*, loc. cit.

EXERCISE

10-11. Referring to the results of Exercise 10-8, use Eq. (10-92) to compare the first-order density matrices of the ^1S and ^3S states of helium arising from the $1s2s$ configuration.

10-7 NATURAL ORBITALS

The Hartree-Fock SCF wavefunction is often used as the starting point of configuration interaction calculations (Sec. 9-4) to improve the energy. The approach is to form a new wavefunction in which the ground-state wavefunction is one term and then to add wavefunctions representing excited states of the same symmetry type as the ground state; such excited-state wavefunctions are made up as determinantal functions of virtual orbitals obtained from the ground-state calculation. Thus, if ground-state calculations on the helium atom produce the ground-state configuration $1s^2$ and, in addition, the virtual orbital $2s$, one can also form excited-state configurations $1s2s$ and $2s^2$. The new wavefunction then becomes a linear combination of three configurations: $1s^2$, $1s2s$, and $2s^2$. In actuality the second configuration (said to be "singly excited," since it arises by promotion, or excitation, of one electron from the ground state to a $2s$ orbital) does not mix directly with the ground state. This is an example of Brillouin's theorem,[27] which states in general that matrix elements such as $\langle \psi_0 | \hat{H} | \psi' \rangle$ vanish whenever ψ' represents a singly excited configuration (ψ_0 is the Hartree-Fock ground-state wavefunction). However, matrix elements between ψ' and doubly excited configurations (for example, $2s^2$) are not zero and, consequently, do have a slight effect on the energy.

Unfortunately, configuration-interaction calculations often turn out to be disappointingly slow in converging to really good wavefunctions; i.e., a very large number of configurations must be superposed in order to obtain a significant lowering of the energy. Occasionally, if one happens to pick the right configurations and employs the right basis functions, a fairly good energy can be obtained with relatively few configurations. Unfortunately, there is no general rule for choosing such configurations, and one must rely on previous computational experience and intuition. Löwdin[28] has suggested an approach which finds that set of basis functions which leads to the fastest convergence for a specific state. Such functions are known as *natural orbitals*. These are defined in such a way that a wavefunction constructed from N natural orbitals has a smaller total deviation from the exact solution than a wavefunction constructed from any alternative set of N orbitals. As shown by Löwdin, the natural orbitals arise as eigenfunctions of the first-order density matrix of the system.

---

[27] L. Brillouin, *Les Champs 'self-consistents' de Hartree et de Fock*, Hermann & Cie, Paris, 1934, p. 19.

[28] P. O. Löwdin, *Phys. Rev.*, **97**:1474 (1955).

Since any antisymmetric wavefunction can be expanded as a linear combination of Slater determinants constructed from an arbitrary basis set $\{\phi_i\}$, the first-order density matrix can be written in the general form

$$\gamma(\tau_1'\,|\,\tau_1) = \sum_{k,l} S_k(\tau_1')S_l^*(\tau_1)\gamma_{kl} \tag{10-98}$$

where the spin orbitals $\{S_k\}$ are assumed to be orthonormal. The coefficients $\{\gamma_{kl}\}$ form a hermitian matrix and are readily evaluated from the coefficients $\{C_K\}$ of the CI wavefunction. The diagonal elements of the matrix of the coefficients satisfy

$$\gamma_{kk} = \sum_K |C_K|^2 \tag{10-99}$$

where the summation is over all the configurations K containing the specific index k. This diagonal coefficient is interpreted as the occupation number of the orbital S_k and depends only on S_k's spatial part and the total wavefunction. Summation over all configurations leads to

$$0 \le \gamma_{kk} \le 1 \tag{10-100}$$

(provided the total wavefunction is normalized). This relationship is simply a statement of the Pauli exclusion principle. We now let $\{\chi_k\}$ represent a basic orthonormal set of functions for which the occupation numbers are a maximum. We define these new orbitals by the unitary transformation

$$\chi_k = \sum_r S_r u_{rk} \tag{10-101}$$

or, in matrix notation,

$$\boldsymbol{\chi} = \mathbf{SU} \tag{10-102}$$

where

$$\boldsymbol{\chi} = [\chi_1 \quad \chi_2 \quad \cdots] \qquad \mathbf{S} = [S_1 \quad S_2 \quad \cdots] \qquad \mathbf{U}^\dagger\mathbf{U} = \mathbf{1} \tag{10-103}$$

The unitary transformation is chosen such that it diagonalizes the matrix of coefficients $\{\gamma_{kl}\}$, namely,

$$\mathbf{U}^\dagger\boldsymbol{\gamma}\mathbf{U} = \mathbf{n} \tag{10-104}$$

where

$$(\mathbf{n})_{kl} = n_k\,\delta_{kl} \quad \text{and} \quad 0 \le n_k \le 1 \tag{10-105}$$

The functions $\{\chi_k\}$ are the natural spin orbitals. The coefficients $\{n_k\}$, the eigenvalues of the matrix $\boldsymbol{\gamma}$, can be ordered such that $n_1 \ge n_2 \ge n_3 \cdots$. Then the natural spin orbital χ_1 will have the highest possible occupation number; i.e., it will be the dominant contributor to the overall wavefunction. Similarly, χ_2 will have the highest occupation number of all spin orbitals orthogonal to χ_1, χ_3 will have the highest occupation number of all spin orbitals orthogonal to χ_1

and χ_2, etc. The total wavefunction can then be written as a linear combination of determinantal configurations constructed from the natural spin orbitals.

For two-electron systems the spin functions can always be factored out of the spin orbitals, and so we will speak of *natural orbitals* for such systems. Shull and Löwdin[29] considered the six-term wavefunction for the ground state of helium constructed from six configurations of 1*s*, 2*s*, and 3*s*, basis functions (STOs, see Sec. 6-5). Table 10-4 lists these configurations and the coefficients describing them in the optimized CI wavefunction. This wavefunction produced an energy of $-2.878116E_h$ compared with the Hartree-Fock value of $-2.8616800E_h$[30] and, therefore, accounts for about one-quarter of the correlation energy. Diagonalization of the first-order density matrix leads to the following natural orbitals, whose properties are summarized in Table 10-5.

Thus, the total wavefunction in terms of the natural orbitals is

$$\xi(1,2) = \sqrt{n_1}|\chi_1\,\overline{\chi_1}| + \sqrt{n_2}|\chi_2\,\overline{\chi_2}| + \sqrt{n_3}|\chi_3\,\overline{\chi_3}| \tag{10-106}$$

In general, the use of N basis orbitals to construct natural orbitals leads to only N diagonal configurations instead of $N(N+1)/2$ "ordinary" configurations. If the above wavefunction is truncated by dropping χ_3, one obtains an energy of $-2.877236E_h$, almost as good an energy value as that of the six-configuration wavefunction from which the natural orbitals were derived. This is because forming the natural orbital configurations regroups basis functions in such a way as to form very efficient basis sets. The two-configuration natural orbital wavefunction also gives an energy which is slightly better than the Eckart function described earlier [see Eq. (6-81); the energy in this case is $-2.8757E_h$]. If the natural orbital wavefunction is truncated to only the first configuration, the energy turns out to be $-2.861530E_h$—a value which is very close to that of the Hartree-Fock energy. The basis for this similarity has been discussed by Nazaroff and Hirschfelder.[31] Davidson[32] has analyzed some very accurate wavefunctions of helium in terms of natural orbitals.

[29] H. Shull and P. O. Löwdin, *J. Chem. Phys.*, **23**:1565 (1955) and **30**:617 (1959).

[30] E. Clementi, *J. Chem. Phys.*, **40**, 1944 (1964).

[31] G. V. Nazaroff and J. O. Hirschfelder, *J. Chem. Phys.*, **39**:715 (1963).

[32] E. R. Davidson, *J. Chem. Phys.*, **37**:577 (1962) and **39**:875 (1963).

TABLE 10-4
Six-term CI wavefunction for the helium atom

Configuration	CI coefficient
$1s^2$	0.963175
$1s2s$	−0.250360
$2s^2$	−0.032613
$1s3s$	0.092211
$2s3s$	−0.000310
$3s^2$	−0.006692

TABLE 10-5
Natural orbital expansion of the helium CI wavefunction

Natural orbital	u_1	u_2	u_3	n
χ_1	0.983545	−0.168992	0.063880	0.995660
χ_2	0.178369	0.964488	−0.194800	0.004265
χ_3	−0.028690	0.202991	0.978760	0.000075

10-8 ROOTHAAN'S EQUATIONS: THE MATRIX SOLUTION OF THE HARTREE-FOCK EQUATIONS

When the Hartree-Fock equations are solved by numerical integration methods, the procedure is unwieldy, is incapable of being extended to molecules, and leads to orbitals which have the form of extensive tables of grid points. All these undesirable features are removed by converting the equations to matrix form. The first complete systematic treatment of this method was described by Roothaan.[33]

The set of basis functions for the matrix representation of the Hartree-Fock equations is denoted by $\{\chi_p\}$. Each Hartree-Fock orbital is then expanded in terms of this basis set by

$$\phi_i = \sum_p \chi_p c_{pi} = \boldsymbol{\chi}\mathbf{C}_i \tag{10-107}$$

In the limit of a complete basis set, this can lead to an exact solution of the Hartree-Foch equations; that is, the ϕ_i so expressed are the self-consistent field orbitals. In practice, however, it is necessary to truncate the expansion to m finite members of the basis set; $\mathbf{C}_i$ is then a column matrix of m rows. In order to be able to construct at least N linearly independent solutions, it is necessary that $m \geq N$ (recall that N is the number of doubly occupied spatial orbitals in the single-determinantal wavefunction of a $2N$-electron system).

The Roothaan treatment involves the following matrices and matrix elements:

$$\begin{gathered}
\boldsymbol{\chi} = [|\chi_1\rangle \quad |\chi_2\rangle \quad \cdots \quad |\chi_m\rangle] \qquad \boldsymbol{\Delta} = \boldsymbol{\chi}^\dagger\boldsymbol{\chi} \qquad \Delta_{pq} = \langle \chi_p | \chi_q \rangle \\
\epsilon_i^{(0)} = \langle \phi_i | h | \phi_i \rangle = \mathbf{C}_i^\dagger \boldsymbol{\chi}^\dagger h \boldsymbol{\chi} \mathbf{C}_i = \mathbf{C}_i^\dagger \mathbf{h} \mathbf{C}_i \\
(\mathbf{J}_i)_{pq} = \langle \chi_p | J_i | \chi_q \rangle \qquad (\mathbf{K}_i)_{pq} = \langle \chi_p | K_i | \chi_q \rangle \\
J_{ij} = \mathbf{C}_i^\dagger \mathbf{J}_j \mathbf{C}_i = \mathbf{C}_j^\dagger \mathbf{J}_i \mathbf{C}_j \qquad K_{ij} = \mathbf{C}_i^\dagger \mathbf{K}_j \mathbf{C}_i = \mathbf{C}_j^\dagger \mathbf{K}_i \mathbf{C}_j
\end{gathered} \tag{10-108}$$

Using the above definitions, the total energy of a closed-shell atom with $2N$

[33] C. C. J. Roothaan, *Rev. Mod. Phys.*, **23**:69 (1951).

electrons in doubly occupied orbitals is

$$E = 2\sum_{i}^{N} \mathbf{C}_i^\dagger \mathbf{h} \mathbf{C}_i + \sum_{i,j}^{N} \mathbf{C}_i^\dagger (2\mathbf{J}_j - \mathbf{K}_j)\mathbf{C}_i \tag{10-109}$$

We now consider the functional

$$F[E] = E - 2\sum_{i,j}^{N} \lambda_{ij}(\mathbf{C}_i^\dagger \mathbf{\Delta} \mathbf{C}_j - \delta_{ij}) \tag{10-110}$$

and find the conditions to be satisfied by the $\{\mathbf{C}_i\}$ such that $\delta F[E] = 0$ for an arbitrary variation δC_i. The resulting algebra is very similar to that used in Sec. 10-1 to obtain the Hartree-Fock equations and will not be repeated here. One finds that the coefficient matrices $\{\mathbf{C}_i\}$ must satisfy

$$\mathbf{F}\mathbf{C}_i = \sum_{j} \mathbf{\Delta} \mathbf{C}_j \lambda_{ij} \tag{10-111}$$

where

$$\mathbf{F} = \boldsymbol{\chi}^\dagger \hat{F} \boldsymbol{\chi} \tag{10-112}$$

and where $\hat{F}$ is the Hartree-Fock operator previously defined in Eq. (10-16). The expressions in equation (10-111) are known as *Roothaan's equations*; they are the matrix representations of the Hartree-Fock equations given by Eq. (10-12). By use of a properly chosen unitary transformation on the $\{\phi_i\}$, one can rewrite Roothaan's equations in the pseudo-eigenvalue form

$$\mathbf{F}\mathbf{C}_i = \epsilon_i \mathbf{\Delta} \mathbf{C}_i \tag{10-113}$$

or, more generally,

$$\mathbf{F}\mathbf{C} = \mathbf{\Delta}\mathbf{C}\boldsymbol{\epsilon} \tag{10-114}$$

where

$$\mathbf{C} = [\mathbf{C}_1 \quad \mathbf{C}_2 \quad \cdots \quad \mathbf{C}_m] \qquad \mathbf{C}_i = \begin{bmatrix} C_{1i} \\ C_{2i} \\ \vdots \\ C_{mi} \end{bmatrix} \qquad \boldsymbol{\epsilon} = \mathbf{C}^\dagger \mathbf{F} \mathbf{C} \tag{10-115}$$

The m eigenvectors $\{\mathbf{C}_i\}$ are orthogonal in the sense that $\mathbf{C}_i^\dagger \mathbf{\Delta} \mathbf{C}_i = \delta_{ij}$. The nontrivial solutions of Eq. (10-111) are obtained by solving for the m roots of the secular determinant

$$\det|\mathbf{F} - \epsilon \mathbf{\Delta}| = 0 \tag{10-116}$$

where each matrix element is of the form

$$F_{pq} - \epsilon \Delta_{pq} = \langle \chi_p | \hat{F} | \chi_q \rangle - \epsilon \langle \chi_p | \chi_q \rangle \tag{10-117}$$

Since the matrix $\mathbf{F}$ depends on the $\{\mathbf{C}_i\}$, the Roothaan equations are nonlinear and must be solved by an iterative process; this is analogous to the situation existing in the numerical Hartree-Fock equations in which $\hat{F}$ depends on the orbitals. The usual approach to the solution is to assume some starting matrix $\mathbf{C}^{(1)}$ as a first approximation, use it in $\mathbf{F}$ to generate a new matrix $\mathbf{C}^{(2)}$, and continue in this fashion until, finally, an assumed matrix $\mathbf{C}^{(n)}$ generates a new matrix $\mathbf{C}^{(n+1)}$ which differs from it by no more than some previously chosen increments.[34] At this point, one has reached self-consistency with respect to the matrix **C**. Note that this does not lead to the true Hartree-Fock solutions unless the basis employed is very large. The N lowest roots of the secular determinant are assumed to be those occupied by the $2N$ electrons and describe the ground state of the system. The remaining solutions, called *virtual orbitals*, can be used to construct excited-state configurations for use in improving the energy value by the configuration-interaction method (see Sec. 9-4).

Both Slater orbitals and gaussian orbitals (see Sec. 6-4) are commonly used for the basis functions $\{\chi_p\}$. The Hartree-Fock calculations by Clementi given in Table 10-2 employed up to 20 STOs to represent a single Hartree-Fock orbital. To represent a Hartree-Fock orbital nl, each basis function χ_p is of the form

$$N_i r^{n-1} e^{-\alpha_i r} Y_{lm}(\theta, \varphi) \tag{10-118}$$

where the nonlinear parameter α_i is different for each basis function. Thus, a $1s$ Hartree-Fock orbital would have the form

$$1s = c_1\chi_1 + c_2\chi_2 + \cdots \tag{10-119}$$

The $2s$ Hartree-Fock orbital would use the same values of $\alpha_1, \alpha_2, \ldots$, but the variational coefficients $c_1, c_2, \ldots$ would differ.

The calculations by Wachters referred to in Table 10-2 employed gaussian basis functions for the $\{\chi_p\}$. The basis set consisted of 14 s-type, 9 p-type, and 5 d-type gaussian functions.

If the Hartree-Fock matrix elements are written analogously to Eqs. (10-31) and (10-32), one obtains

$$\mathbf{F} = \mathbf{h} + \mathbf{G} \tag{10-120}$$

The corresponding matrix elements are

$$h_{pq} = \langle \chi_p | h | \chi_q \rangle$$

$$\begin{aligned} G_{pq} = \langle \chi_p | \hat{G} | \chi_q \rangle &= \sum_{j=1}^{N} (2\langle x_p | J_j | \chi_q \rangle - \langle \chi_p | K_j | \chi_q \rangle) \\ &= \sum_{j}^{N} \sum_{r,s}^{m} c_{rj} c_{sj}^* [2\langle \chi_p(\mu)\chi_s(\nu) | g | \chi_q(\mu)\chi_r(\nu) \rangle \\ &\quad - \langle \chi_p(\mu)\chi_s(\nu) | g | \chi_r(\mu)\chi_q(\nu) \rangle] \\ &= \sum_{r,s}^{m} R_{rs}(2\langle ps | g | qr \rangle - \langle ps | g | rq \rangle) \end{aligned} \tag{10-121}$$

[34] Commonly used starting matrices are based on a very simple method discussed in Sec. 15-1 (the extended Hückel method).

The quantity R_{rs} in the last expression is defined by

$$R_{rs} = \sum_{j=1}^{N} c_{rj} c_{sj}^{*} \tag{10-122}$$

The elements R_{rs} are those of a matrix **R** defined as

$$\mathbf{R} = \mathbf{T}\mathbf{T}^{\dagger} \qquad \mathbf{T} = [C_1 \quad C_2 \quad \cdots \quad \mathbf{C}_N] \tag{10-123}$$

where **T** is an $m \times N$ matrix of the eigenvectors associated with the N lowest eigenvalues. This quantity **R** is a matrix representation of the density matrix: a quantity introduced earlier in Eq. (10-53). Using Eqs. (10-33) and (10-34) (the second alternative for $\hat{B}$), we can write the total energy of the system as

$$E = \sum_{i}^{N} \mathbf{C}_i^{\dagger} \boldsymbol{\chi}^{\dagger} (2\hat{F} - \hat{G}) \boldsymbol{\chi} \mathbf{C}_i = \sum_{i}^{N} \mathbf{C}_i^{\dagger} (2\mathbf{F} - \mathbf{G}) \mathbf{C}_i$$

$$= 2\,\mathrm{Tr}\,(\mathbf{T}^{\dagger}\mathbf{F}\mathbf{T}) - \mathrm{Tr}\,(\mathbf{T}^{\dagger}\mathbf{G}\mathbf{T}) = 2\,\mathrm{Tr}\,(\mathbf{R}\mathbf{F}) - \mathrm{Tr}\,(\mathbf{R}\mathbf{G}) \tag{10-124}$$

where use has been made of the fact that the trace of a matrix is invariant under a cyclic permutation of the matrices. Since the matrices **F** and **G** depend only on the matrix **R**, it is evident from the above equation that the total energy depends only on the density matrix **R**. This density matrix is particularly useful in certain types of molecular calculations, where it is called a *charge and bond-order matrix*.

10-9 OPEN-SHELL HARTREE-FOCK CALCULATIONS

Since not all atoms or molecules—nor all states of closed-shell atoms or molecules—can be described in terms of a single determinant of N doubly occupied orbitals, modifications must be made to the procedures discussed in the preceding sections when dealing with more general systems. The Hartree-Fock method discussed earlier is often known as the *restricted Hartree-Fock* (*RHF*) *method*. If the wavefunction is assumed to be a single determinant in which there is one or more open-shell (unpaired) electrons, there are two variants of the RHF approach often used: the *restricted open-shell Hartree-Fock* (*ROHF*) and the *unrestricted Hartree-Fock* (*UHF*) *methods*.

In the ROHF method, all electrons—except those explicitly required to occupy open-shell orbitals—are in closed-shell (doubly occupied) orbitals. Such wavefunctions have the advantage of being eigenfunctions of the spin operator $\hat{S}^2$, but the constraint of double occupancy on some of the orbitals raises the variational energy.

In the UHF method, there are no restrictions on occupancy; in general, each electron is assigned to a different orbital. However, the wavefunction is now not an eigenfunction of $\hat{S}^2$; in general, it is a linear combination of different spin states. However, the undesired spin states (referred to as "contaminants") generally make only a small numerical contribution to the overall wavefunction, and their effect is often ignored. Another way to correct for the "contamination" is to use a projection operator to project out the desired spin state from the total wavefunction and to evaluate the total energy from this projection. There is some question, however, as to which is better: project first, then optimize the energy, or optimize the energy first and then project out the desired spin state.

Various methodologies for going beyond the RHF level have been proposed by a number of workers,[35] but most of these methods suffer from lack of generality; i.e., they apply only to specific couplings. An alternative formulation of the Hartree-Fock method—one not utilizing lagrangian multipliers—has been given by Binkley, Pople, and Dobosh[36] and extended by Caldwell and Gordon.[37] The advantage of such methods is that they are more easily extended to electronic systems which cannot be described by a single determinant of doubly occupied orbitals. The method is based on a partitioning technique which divides the available one-electron space (for a given basis) into three mutually orthogonal subspaces for doubly occupied, singly occupied, and empty orbitals.

SUGGESTED READINGS

Hartree, D. R.: *The Calculation of Atomic Structures*, Wiley, New York, 1957.

Roothaan, C. C. J., and P. Bagus: *Methods in Computational Physics*, vol. 2, Academic, New York, 1963.

Slater, J. C.: *Quantum Theory of Atomic Electronic Structure*, vols. 1 and 2, McGraw-Hill, New York, 1960.

Szabo, A., and N. S. Ostlund: *Modern Quantum Chemistry: Introduction to Advanced Electronic Structure Theory*, Macmillan, New York, 1982. An excellent complete treatment of the various Hartree-Fock methods and other methods but written at an advanced level suitable for practicing theoretical chemists.

---

[35] C. C. J. Roothaan, *Rev. Mod. Phys.*, **32**:179 (1960); F. W. Birss and S. Fraga, *J. Chem. Phys.*, **38**:2252 (1963) and **40**:3203, 3207, 3212 (1964); and G. Berthier in P. O. Löwdin and B. Pullman (eds.), *Molecular Orbitals in Chemistry, Physics and Biology*, Academic, New York, 1964, p. 57.

[36] J. S. Binkley, J. A. Pople, and P. A. Dobosh, *Mol. Phys.*, **28**:1423 (1974).

[37] J. W. Caldwell and M. S. Gordon, *Chem. Phys. Lett.*, **43**:493 (1976).

CHAPTER 11

INTRODUCTION TO MOLECULAR STRUCTURE

A molecule is an aggregate of atomic nuclei surrounded by electrons. The positively charged nuclei repel each other but are attracted by the negatively charged electrons, which also repel each other. The aggregate is stable only if the total attractive forces between nuclei and electrons just balance the mutual repulsive forces among nuclei and among electrons. Furthermore, this collection of mutually interacting charged particles is not a static one; the nuclei are in constant motion, executing mutual vibrational motions and rotating in space while the entire mass undergoes translational motion. At the same time, the electrons are responding to changing nuclear positions while avoiding each other. This dynamic melee appears, at first sight, to constitute a hopelessly complex problem for quantum mechanics to describe—after all, even the much simpler helium atom is far from easy to treat in a highly accurate manner. Nevertheless, owing mainly to the large disparity in mass between nuclei and electrons (on the order of 2000 to 1 or more), it is possible to develop some very satisfactory models of molecular dynamics. This chapter will consider some of the basic foundations on which the quantum mechanical descriptions of molecular structure depend.

11-1 THE BORN-OPPENHEIMER APPROXIMATION

After the motion of the center of mass has been separated out, the nonrelativistic hamiltonian operator of a molecule may be written in the symbolic form

$$\hat{H} = \hat{T}_n + \hat{V}_n + \hat{H}_e \tag{11-1}$$

The first term ($\hat{T}_n$) represents the kinetic energy of the nuclei, the second term ($\hat{V}_n$) represents the potential energy of nuclear-nuclear repulsions, and the third term ($\hat{H}_e$) represents all terms involving electrons: electron kinetic energies, electron-nuclear attractions, electron-electron repulsions, and mass polarization (see Sec. 6-3). Restricting ourselves to the special case of a diatomic molecule, AB, with nuclear masses m_A and m_B, explicit expressions for these various operators are

$$\hat{T}_n = -\frac{1}{2\mu}\nabla_n^2 \qquad \hat{V}_n = \frac{Z_A Z_B}{R}$$
$$\hat{H}_e = \sum_{\mu=1}^{N} h_\mu + \sum_{\mu<\nu}^{N} \frac{1}{r_{\mu\nu}} - \frac{1}{2M}\left[\sum_{\mu}^{N} \nabla_\mu^2 + 2\sum_{\mu<\nu}^{N} \nabla_\mu \cdot \nabla_\nu\right] \tag{11-2}$$

where $\mu = m_A m_B / M$ (the reduced mass of the nuclei) and $M = m_A + m_B$ (the total mass of the nuclei). The internuclear distance between nuclei A and B is denoted by R. The monoelectronic operator h_μ appearing in $\hat{H}_e$ is analogous to its atomic counterpart but now contains interactions with more than one nucleus. In general, if there are S nuclei, this operator has the form

$$h_\mu = -\frac{1}{2}\nabla_\mu^2 - \sum_{k=1}^{S} \frac{Z_k}{r_{\mu k}} \tag{11-3}$$

For the specific case of two nuclei, A and B, the summation term in Eq. (11-3) becomes simply $Z_A/r_{\mu A} + Z_B/r_{\mu B}$. Since the translational motion of the molecule as a whole has been separated out, the hamiltonian defined in Eq. (11-1) is often called the *internal hamiltonian*.

Suppose we had a motion picture camera capable of photographing the actual nuclear and electronic motions of a molecule. We could then make a film showing how the nuclei vibrate and rotate and how the electrons constantly readjust their motions to those of the nuclei while at the same time avoiding each other. We would observe that since the nuclei are much more massive than the electrons, they tend to move considerably more slowly than do the electrons. Consequently, the electrons would readjust their motions very rapidly to match the nuclear motions, so that the two would always be *almost*—but not quite—at equilibrium.

Next, suppose we had no movie camera but only a single-frame camera with a very slow shutter speed at our disposal. We could still make a motion picture of the molecule as follows: At some time t we stop all nuclear motions, i.e., freeze the nuclei, and allow the electrons to come to equilibrium with the stationary nuclei. After snapping a picture, we unfreeze the nuclei momentarily and quickly refreeze them at slightly displaced positions.[1] We again wait for establishment of equilibrium, snap another picture, thaw the nuclei momentarily, refreeze, etc., until we have a sequence of stills of a progression of frozen nuclear positions. If the stills were then joined to make a film strip and the film strip run through a projector at the proper number of frames per minute, we would have a motion picture of the molecule. The question is: How would the true motion picture and the sequence of stills compare? The main difference between the two should be due to a lag in equilibrium attainment between nuclear and electron motions; this would appear in the true motion picture but not in the sequence of stills. However, if this lag is very small—as it would be if the electrons adjust very rapidly to nuclear motions—the two films would be virtually identical.

This admittedly crude and incomplete motion picture analogy can be put into mathematical form as follows: the exact solution of the Schrödinger equation containing the internal molecular hamiltonian [Eq. (11-1)] may be written

$$\hat{H}\psi(r, R) = E\psi(r, R) \tag{11-4}$$

where E is the total energy of the molecule (all the energy except translational energy) and $\psi(r, R)$ is the molecular wavefunction depending on both the electron coordinates (r) and the nuclear coordinates (R). Now let us consider a possible separation of electron and nuclear coordinates of the form

$$\psi(r, R) = \psi_R(r)\phi(R) \tag{11-5}$$

where $\psi_R(r)$ (the *electronic wavefunction*) describes the electron motions for a fixed set of nuclear coordinates R and $\phi(R)$ (the *nuclear wavefunction*) describes the nuclear motions (vibrational and rotational) in a "sea" of electrons. Thus, the nuclear coordinates R enter the electronic wavefunction only in a *parametric* sense. Using Eqs. (11-5) and (11-1) in Eq. (11-4) leads to

$$\hat{T}_{\rm n}\psi_R(r)\phi(R) + \hat{V}_{\rm n}\psi_R(r)\phi(R) + \hat{H}_{\rm e}\psi_R(r)\phi(R) = E\psi_R(r)\phi(R) \tag{11-6}$$

We note that if the nuclei are frozen, then $\hat{V}_{\rm n}$ is a constant and $\hat{V}_{\rm n}\psi_R(r)\phi(R) = \psi_R(r)\phi(R)\hat{V}_{\rm n}$. We also make the approximation

$$\hat{T}_{\rm n}\psi_R(r)\phi(R) = \psi_R(r)\phi(R)\hat{T}_{\rm n} \tag{11-7}$$

The basis of this approximation is that we are neglecting terms of the form

$$-\frac{1}{2\mu}\nabla_{\rm n}^2\psi_R(r) \qquad \text{and} \qquad -\frac{1}{\mu}\nabla_{\rm n}\psi_R(r) \tag{11-8}$$

[1] Of course, the electrons are still moving, so they would photograph as a blur or "charge cloud."

A justification of this neglect is as follows: the nuclear operator ∇_n^2 (which is $\mathbf{\nabla}_n \cdot \mathbf{\nabla}_n$) may be essentially replaced with its electron counterpart ∇_μ^2, and since the nuclear mass (m_A or m_B) is at least 1840 times the electron mass m_e, the operators in Eq. (11-8) appear in terms which are approximately equal to

$$-\frac{1}{3680}\nabla_\mu^2\psi_R \quad \text{and} \quad -\frac{1}{1840}\mathbf{\nabla}_n\psi_R(r) \tag{11-9}$$

which are very small. Thus, we treat $\hat{T}_n$ as if it does not operate on $\psi_R(r)$. Use of the above approximations allows us to rewrite Eq. (11-6) as

$$\psi_R\hat{T}_n\phi(R) + \psi_R(r)\phi(R)\hat{V}_n + \phi(r)\hat{H}_e\psi_R(r) = E'\psi_R(r)\phi(R) \tag{11-10}$$

where E' is now an approximation to the exact eigenvalue E of $\hat{H}$. Dividing both sides of Eq. (11-10) by $\psi_R(r)\phi(R)$ leads to

$$\frac{\hat{T}_n\phi(R)}{\phi(R)} + \hat{V}_n + \frac{\hat{H}_e\psi_R(r)}{\psi_R(r)} = E' \tag{11-11}$$

We now introduce the convenient definition

$$\frac{\hat{H}_e\psi_R(r)}{\psi_R(r)} = \varepsilon \tag{11-12}$$

where ε is the energy due to the electrons' interacting with the fixed nuclei. Thus the quantity

$$\varepsilon + \hat{V}_n = E(R) \tag{11-13}$$

represents the total energy of the molecule for a *fixed set of nuclear coordinates*. If we solve Eq. (11-12) for a series of fixed nuclear coordinates (which is analogous to taking a series of stills), the quantity $E(R)$ as a function of these coordinates represents the potential field in which the nuclei vibrate and rotate. Thus Eq. (11-11) becomes

$$\frac{\hat{T}_n\phi(R)}{\phi(r)} + E(R) = E' \tag{11-14}$$

or, in eigenvalue form,

$$[\hat{T}_n + E(R)]\phi(R) = E'\phi(R) \tag{11-15}$$

Solutions of this equation—once $E(R)$ is determined by a series of solutions of Eq. (11-12)—describe how the nuclei vibrate and rotate in the sea of electrons.

To summarize, the quantum mechanical treatment of molecular motions is carried out in two parts. First, one solves the electronic Schrödinger equation for a series of fixed nuclear coordinates

$$(\hat{H}_e + \hat{V}_n)\psi_R(r) = E(R)\psi_R(r) \tag{11-16}$$

and then—once $E(R)$ is known as a function of R—one solves the nuclear equation (11-15). Note that the term $\hat{V}_n$ occurring in Eq. (11-16) is a constant during each solution for fixed R and, thus, can be omitted during the solution and then added afterwards.

If mass polarization is neglected in the electronic hamiltonian $\hat{H}_e$ and the resulting energy ε is used to define $E(R)$ for the nuclear equation via Eq. (11-13), the approach is called the *Born-Oppenheimer approximation.*[2] However, if $E(R)$ in the nuclear equation (11-15) is replaced with

$$E(R) + E'(R) \tag{11-17}$$

where $E'(R)$ is defined by

$$E'(R) = \left\langle \psi_R(r) \left| \frac{-\nabla_n^2}{2\mu} \right| \psi_R(r) \right\rangle + \sum_{\mu}^{N} \left\langle \psi_R(r) \left| \frac{\nabla_\mu^2}{2M} \right| \psi_R(r) \right\rangle$$
$$- \frac{1}{M} \sum_{\mu<\nu}^{N} \langle \psi_R(r) | \nabla_\mu \cdot \nabla_\nu | \psi_R(r) \rangle \tag{11-18}$$

we have what is known as the *adiabatic approximation.*[3] The first integral represents a correction for coupling between nuclear and electronic motions, and the last two represent mass polarization.

It is instructive to illustrate the Born-Oppenheimer approximation for a specific molecule, e.g., the simplest molecular species, the hydrogen molecule ion, H_2^+. The total hamiltonian of this ion is conveniently written in terms of the coordinates r_A, r_B, and R shown in Fig. 11-1, where A and B refer to the two atomic nuclei of the molecule ion.

$$\hat{H} = -\frac{1}{2m_p}(\nabla_A^2 + \nabla_B^2) - \frac{1}{2}\nabla_e^2 - \frac{1}{r_A} - \frac{1}{r_B} + \frac{1}{R} \tag{11-19}$$

The first term is $\hat{T}_n$, the next three form $\hat{H}_e$, and the last is $\hat{V}_n$. The quantity m_p is the mass of the proton in atomic units (about 1840 m_e). If one effects the substitution of Eq. (11-5) into Eq. (11-19) for a *fixed* internuclear distance R,

[2] M. Born and J. R. Oppenheimer, *Ann. Physik*, **84**:457 (1927). A modern treatment is given by M. Born and K. Huang, *Dynamical Theory of Crystal Lattices*, Oxford University Press, New York, 1954.

[3] Details are given by J. O. Hirschfelder and W. J. Meath, *Advan. Chem. Phys.*, **12**:3 (1967).

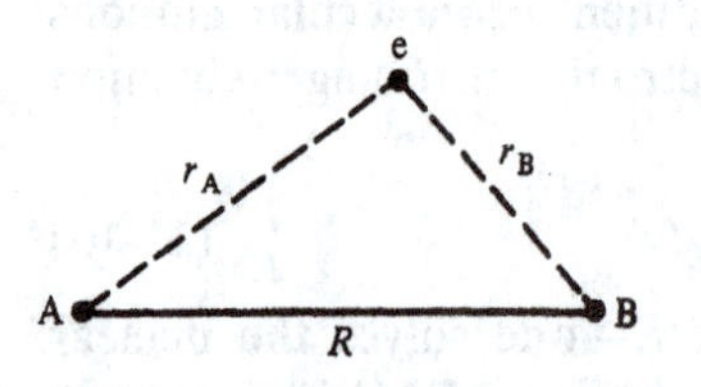

FIGURE 11-1
Relative coordinate system for the hydrogen molecule ion, H_2^+. The two nuclei are labeled A and B.

the operator ∇_A^2 (which is part of $\hat{T}_n$) operates on the wavefunction as follows:

$$\nabla_A^2 \psi_R(r)\phi(R) = \psi_R(r)\nabla_A^2\phi(R) + \phi(R)\nabla_A^2\psi_R(r) + 2\nabla_A\psi_R(r)\cdot\nabla_A\phi(R) \tag{11-20}$$

A similar term is obtained for the ∇_B^2 part of $\hat{T}_n$. The second and third terms of Eq. (11-20) are negligible, by the same arguments given previously [see Eqs. (11-7) through (11-9)]. It is of interest to note that the third term represents the rate at which nuclear motions induce electronic motions. If we now make use of this approximation concerning $\hat{T}_n$, the Schrödinger equation for the hydrogen molecule ion, analogous to Eq. (11-11), becomes

$$\frac{-\dfrac{1}{2m_p}(\nabla_A^2 + \nabla_B^2)\phi(R)}{\phi(R)} + \frac{\left[-\dfrac{1}{2}\nabla_e^2 - \left(\dfrac{1}{r_A} + \dfrac{1}{r_B} - \dfrac{1}{R}\right)\right]\psi_R(r)}{\psi_R(r)} = E' \tag{11-21}$$

We now define

$$\left[-\frac{1}{2}\nabla_e^2 - \left(\frac{1}{r_A} + \frac{1}{r_B} - \frac{1}{R}\right)\right]\psi_R(r) = E(R)\psi_R(r) \tag{11-22}$$

where $E(R) = \varepsilon + 1/R$. Thus

$$\frac{-(1/2m_p)(\nabla_A^2 + \nabla_B^2)\phi(R)}{\phi(R)} + E(R) = E' \tag{11-23}$$

or, in normal eigenvalue form,

$$\left[-\frac{1}{2m_p}(\nabla_A^2 + \nabla_B^2) + E(R)\right]\phi(R) = E'\phi(R) \tag{11-24}$$

The consecutive solutions of the two equations (11-22) and (11-24) instead of the single equation utilizing the hamiltonian (11-19) constitutes the Born-Oppenheimer approximation for the hydrogen molecule ion.

Solution of Eq. (11-22) for a diatomic molecule AB at various values of internuclear distance R produces a curve such as shown in Fig. 11-2. If E_A and E_B are the electronic energies of atoms A and B, respectively (in their ground states), then the dissociation energy, D_e, of the molecule is given by

$$D_e = E_A + E_B - E(R_e) \tag{11-25}$$

where R_e is the value of R at the minimum point of the curve. As we will discuss later, the quantity D_e must be corrected for the zero-point energy in order to be identified with the experimental dissociation energy of the molecule.

One should note that the total energy E' obtained by use of the Born-Oppenheimer approximation is not equal to the exact molecular energy,

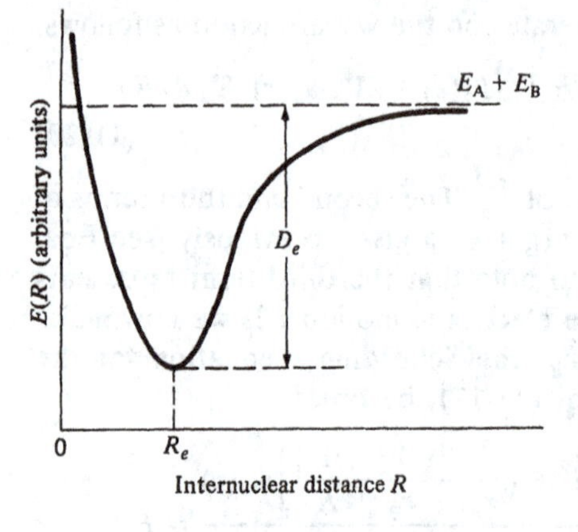

FIGURE 11-2
Schematic plot of the electronic energy $E(R)$ of a diatomic molecule AB as a function of internuclear distance R according to the Born-Oppenheimer approximation.

even if all steps in the procedure are carried out without further approximation. In fact, the total energy E' provides an *upper bound* to the exact energy; i.e., the Born-Oppenheimer energy is always equal to or *higher* than the exact energy.

11-2 SOLUTION OF THE NUCLEAR EQUATION

In this section we will use the specific example of a diatomic molecule to illustrate how the nuclear equation (11-15) can be solved to describe the vibrational and rotational motions of the molecule. The approach used will be to consider the simplest (and most drastic) approximations first and to follow this up with stepwise modifications which remove some of the more unrealistic of the approximations. The nuclear equation we consider is that given earlier by Eq. (11-24), namely,

$$\left[-\frac{1}{2m_A}\nabla_A^2 - \frac{1}{2m_B}\nabla_B^2 + V(R)\right]\phi(R) = E\phi(R) \tag{11-26}$$

where we have dropped the prime on E' to simplify the notation and rewritten the term $E(R)$ (the potential energy of the nuclei) as $V(R)$. Also, we have allowed for the possibility that the two nuclei have different masses. In practice it is not convenient to determine $V(R)$ by solutions of the electronic equation (11-12) for a series of fixed values of R; instead $V(R)$ is replaced with an analytical function which serves as an approximation of the potential field in which the nuclei move. Just what particular approximation to use depends on the particular system under study. In the following we will consider some specific cases in which $V(R)$ is assumed to have some simple functional form.

The nuclear equation (11-26) may be expressed in spherical polar coordinates and in terms of the reduced mass of the system; the latter is defined by

$$\mu = \frac{m_A m_B}{M} \qquad M = m_A + m_B \tag{11-27}$$

where m_A and m_B are the masses of the two nuclei. The polar coordinate system is defined in a manner completely analogous to that for the hydrogen atom (see Fig. 5-1). If we define the operator

$$\hat{D} = \hbar^2 \frac{\partial}{\partial R} R^2 \frac{\partial}{\partial R} \tag{11-28}$$

which is analogous to the operator used for the hydrogen atom [see Eqs. (4-85) and (5-6)], we obtain the nuclear equation in the form

$$\left[-\frac{1}{2\mu R^2} (\hat{D} - \hat{L}^2) + V(R) \right] \psi(r, \theta, \varphi) = E\psi(r, \theta, \varphi) \tag{11-29}$$

where $\hat{L}^2$ is an operator describing the angular momentum of the nuclear system. Note again the close similarity to the hydrogen atom equation (5-6).

Equation (11-29) can be solved by assuming a separation of variables of the form

$$\psi(r, \theta, \varphi) = S(R)Y(\theta, \varphi) \tag{11-30}$$

The solution in θ and φ leads to the spherical harmonics introduced in Chap. 4 for angular momentum (see Sec. 4-6), namely,

$$\hat{L}^2 Y_{J,K}(\theta, \varphi) = J(J+1)\hbar^2 Y_{J,K}(\theta, \varphi) \tag{11-31}$$

where J is the rotational quantum number ($J = 0, 1, 2, \ldots$) and K (previously written m_J) is given by $-J \le K \le J$ (see Sec. 4-9). The radial equation can then be written

$$\{\hat{D} + 2\mu R^2[E - V(R)] - J(J+1)\}S(R) = 0 \tag{11-32}$$

which can be solved for the vibrational-rotational energy once an explicit form for $V(R)$ is known. In the following we will look at solutions of the nuclear equation under various special conditions and with a number of different levels of approximation. For this purpose it is convenient to divide the discussion into three separate areas depending on the specific area of spectroscopy involved, namely, (1) the microwave (about 1 mm to 10 cm) and far infrared (between normal infrared and microwave) regions, (2) the infrared region [about 10^3 nm to 10^6 nm (1 mm)], and (3) the Raman region [ultraviolet to the visible region (about 100 nm to 500 nm)].

MICROWAVE AND FAR INFRARED REGION. Assuming that the diatomic molecule AB has a dipole moment, microwave and far infrared radiation can interact with a dipole fluctuation which has a frequency on the order of that produced by a rotating molecule (see Fig. 3-3). This molecule will also be undergoing vibrational motion, but the frequency of the accompanying oscillating dipole moment will be much higher than that due to rotation, and hence microwave or far infrared radiation will not couple with this oscillation to produce transitions between vibrational states. Consequently, we can treat the rotating molecule as if it were a rigid dumbbell whose interatomic distance R is some average of that produced by the vibrations. This model leads to the treatment described in Sec. 4-9 on the rigid rotator.

Experiments show that the rigid rotator approximation—which predicts a spectrum in which successive absorption lines are uniformly spaced at intervals of $2B$ (see Fig. 4-6)—is not strictly correct; in actuality the spacings become progressively smaller as the rotational quantum number J increases. One can show that this departure from uniform spacings can be partially accounted for by assuming a "centrifugal stretching" correction; i.e., the higher the J quantum number, the faster the molecule rotates and, hence, the more it stretches owing to centrifugal motion. Perturbation theory shows that this centrifugal stretching should be roughly proportional to $(\hat{L}^2)^2$, that is, to $\hat{L}^4$. Thus one writes the total rotational energy in the form

$$E_J = J(J+1)B - DJ^2(J+1)^2 \tag{11-33}$$

where D is a constant (the centrifugal stretching constant) which is determined empirically. This new equation provides an improved fit to the microwave spectrum, at least for the lower-energy transitions.

INFRARED REGION. Infrared radiation has the correct frequency to couple with the oscillating dipole moment due to molecular vibration (see Fig. 3-2). Since the molecule will be rotating at the same time it is vibrating, transitions which represent this coupling will represent not only changes in vibrational states but also changes in rotational states. Naturally, this leads to a more complicated spectrum than when only rotational changes are occurring. However, if the molecule under study is dissolved in an inert solvent (a solvent which is infrared-inactive in the region in which AB is active), the rotational motions will be largely damped and we can observe what is tantamount to pure vibrational transitions. If we assume further that the vibrational motions can be approximated by a harmonic oscillator, then Eq. (11-32) can be replaced by the harmonic oscillator Schrödinger equation (3-80):

$$\left(-\frac{\hbar^2}{2\mu}\frac{d^2}{dx^2} + \frac{kx^2}{2}\right)\psi_v = E_v\psi_v = \left(v + \frac{1}{2}\right)h\nu_0\psi_v \tag{11-34}$$

where ν_0 is the fundamental vibration frequency of the molecule. Given the relationships $\nu = c\tilde{\nu}$ (where $\tilde{\nu}$ is the reciprocal of the wavelength λ) and $\tilde{\nu} = E_v/hc$ along with the selection rule $\Delta v = \pm 1$, the energy change accompanying an allowed transition is given by

$$\Delta E = h\nu = hc\tilde{\nu} = hc\tilde{\nu}_0 \tag{11-35}$$

where $\tilde{\nu}$ is the "frequency" (in wave numbers) at which vibrational transitions occur. The fundamental vibrational frequency is related to the force constant k through the equation

$$\tilde{\nu}_0 = \frac{1}{2\pi c}\left(\frac{k}{\mu}\right)^{1/2} \tag{11-36}$$

Since the lowest possible energy ($v = 0$) of a harmonic oscillator is nonzero (E_0), the actual dissociation energy of a molecule is not correctly represented by the quantity D_e given in Eq. (11-25) and Fig. 11-2 but by

$$D_0 = D_e - E_0 \tag{11-37}$$

where E_0 is called the zero-point energy. It is this quantity, D_0, which correlates with the enthalpy of dissociation defined by thermodynamics.

If the harmonic oscillator model were correct, then one would obtain only a single absorption line in the infrared spectrum for a diatomic molecule. This is because the selection rule is $\Delta v = \pm 1$ (equal to $+1$ for absorption) and all energy levels are separated by the same amount, namely, $h\nu_0$ (or $hc\tilde{\nu}_0$). In practice, one obtains several additional absorptions with frequencies corresponding approximately to $2\tilde{\nu}_0$, $3\tilde{\nu}_0$, $4\tilde{\nu}_0$, etc. (the intensities decrease sharply, however, as the frequency increases). The frequency $2\tilde{\nu}_0$ is called either the second harmonic or the first overtone, with a similar nomenclature being adopted for the remaining lines. For example, in HCl, absorptions are obtained at 2886, 5668, 8347, 10,923, 13,397, etc. (all in cm^{-1}). Note that the second and following numbers are almost multiples of the first; exact multiples would be 5772, 8658, 11,544, 14,430, etc. The existence of these overtone absorptions and the fact that the spacings become progressively less than twice, three times, etc., the fundamental is accounted for reasonably well by applying an anharmonic correction to the energy expression. Thus, Eq. (11-34) is replaced by

$$E_v = (v + \tfrac{1}{2})hc\tilde{\nu}_e - x_e\tilde{\nu}_e(v + \tfrac{1}{2})^2 \tag{11-38}$$

where x_e is an anharmonicity constant and $\tilde{\nu}_e$ is related to the fundamental vibration frequency $\tilde{\nu}_0$ by

$$\tilde{\nu}_0 = \tilde{\nu}_e - 2x_e\tilde{\nu}_e \tag{11-39}$$

The selection rules for the anharmonic oscillator become $\Delta v =$ anything, but the larger the quantum number change, the weaker the transition.

The force constant of a molecule is generally calculated from the fundamental vibration frequency. Thus when hydrogen fluoride (HF) is dissolved in a solvent which is inactive in the same spectral region, the strongest absorption occurs at $3958\ \mathrm{cm}^{-1}$. This leads to a force constant value of

$$\begin{aligned} k &= 4\pi^2\mu\tilde{\nu}_0^2c^2 \\ &= (4)(3.1416)^2(1.590 \times 10^{-27}\ \mathrm{kg})(3598 \times 10^2\ \mathrm{m}^{-1})^2(3.00 \times 10^8\ \mathrm{m \cdot s^{-1}})^2 \\ &= 8.13 \times 10^2\ \mathrm{N \cdot m^{-1}} \end{aligned}$$

This is almost twice the value obtained for HCl (see Exercise 11-6), indicating that HF has a much "stiffer" bond than HCl. The vibrational transitions $0 \rightarrow 2$, $0 \rightarrow 3$, etc. (the first, second, etc., overtones or the second, third, etc., harmonics, respectively) lead to frequencies which are roughly two times, three times, etc., the fundamental frequency, respectively. Lines for other transitions such as $1 \to 2, 1 \to 3$, etc., also occur but are usually very weak, since only the $v = 0$ state (the ground vibrational state) usually has a large population at relatively low temperatures.

If the molecule is studied in the gas phase, damping of rotational motions no longer occurs and the resulting infrared spectrum becomes much more complicated. The simplest treatment of this situation involves the coupled harmonic oscillator–rigid rotator. The energy levels, energy differences, and absorption frequencies are now given by

$$E(v, J) = (v + \tfrac{1}{2})hc\tilde{\nu}_0 + J(J+1)B \tag{11-40}$$

For $\Delta v = +1$, $\Delta J = -1$ $(J \to J-1)$

$$\Delta E = hc\tilde{\nu} = hc\tilde{\nu}_0 - 2JB \qquad \tilde{\nu} = \frac{\Delta E}{hc} = \tilde{\nu}_0 - 2JB_e$$

For $\Delta v = +1$, $\Delta J = +1$ $(J \to J+1)$

$$\Delta E = hc\tilde{\nu} = hc\tilde{\nu}_0 + 2JB_e \qquad \tilde{\nu} = \frac{\Delta E}{hc} = \tilde{\nu}_0 + 2JB_e$$

where $B_e = B/hc$ ($\Delta J = 0$ is also allowed for molecules such as NO which have an odd number of electrons). This produces a relatively simple spectrum in which all lines are centered symmetrically about $\tilde{\nu}_0$. The center of the spectrum is a region $4B_e$ in width; the center of this is where $\tilde{\nu}_0$ is located. On the low-frequency side of this region (called the "null gap") is a series of lines spaced $2B_e$ apart; the first is the $J = 1$ to $J = 0$ transition, the next is the $J = 2$ to $J = 1$ transition, etc. This part of the spectrum is called the P branch. On the high-frequency side of this region is a series of lines also spaced $2B_e$ apart; the first is the $J = 0$ to $J = 1$ transition, the next is the $J = 1$ to $J = 2$ transition, etc. This part of the spectrum is called the R branch. The $\Delta J = 0$ transition (which occurs only for a molecule with an odd number of electrons) would be in the exact center of the null gap and is called the Q branch. A diagram showing the relative positions of the absorption lines is given in Fig. 11-3.

In actuality, the above spectrum is not exactly symmetrical about the null gap, and the Q branch is not in the exact center. Experiment shows that the spacings are not uniformly $2B_e$ (with $4B_e$ for the null gap) but, rather, become progressively smaller as one proceeds from the P branch toward the R branch.

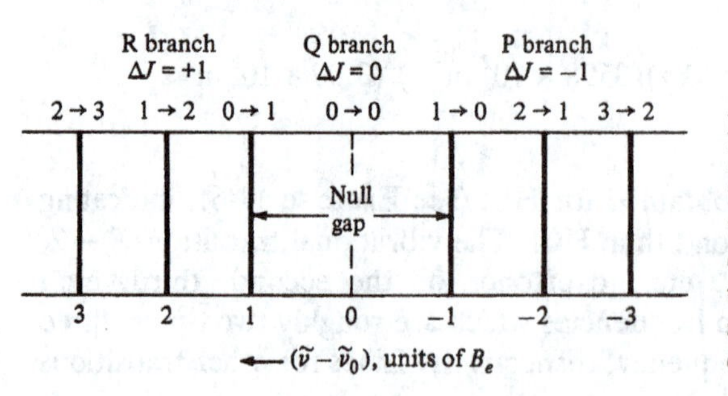

FIGURE 11-3
Relative positions of vibrational-rotational absorption lines for a hypothetical molecule AB. This type of diagram (which shows only the relative positions of the absorption lines and not their relative intensities) is often called a *picket fence spectrum*.

This effect is due to many factors: anharmonicity, centrifugal stretching, and coupling between vibrational and rotational motions. A much-improved fit to the spectrum is obtained by replacing Eq. (11-40) with

$$\frac{E(v, J)}{hc} = \tilde{\nu}_e(v + \tfrac{1}{2}) - x_e\tilde{\nu}_e(v + \tfrac{1}{2})^2 + J(J+1)B_0 - D_eJ^2(J+1)^2 - \alpha_e(v + \tfrac{1}{2})J(J+1) \qquad (11\text{-}41)$$

The terms in the energy equation can be identified as follows: the first term is of harmonic oscillator form, the second represents an anharmonic correction, the third is of rigid rotator form, the fourth represents centrifugal distortion, and the fifth represents a coupling between vibration and rotation. The last term is introduced by replacing B_e in Eq. (11-40) with

$$B_e = B_0 - \alpha_e(v + \tfrac{1}{2}) \qquad (11\text{-}42)$$

where B_0 is the rotational constant in the absence of vibrational stretching. Note the form of this correction makes the internuclear distance proportional to the quantum number of the vibrational state; thus, the more intense the vibration, the more the internuclear distance is stretched. Generally, the centrifugal stretching correction is much less important than is the vibration-rotation coupling correction.

Figure 11-4 shows an actual experimental spectrum of the HCl molecule whose fundamental vibrational frequency is 2886 cm^{-1}. Note that the spacings become slightly narrower as one progresses from the P branch to the R branch;

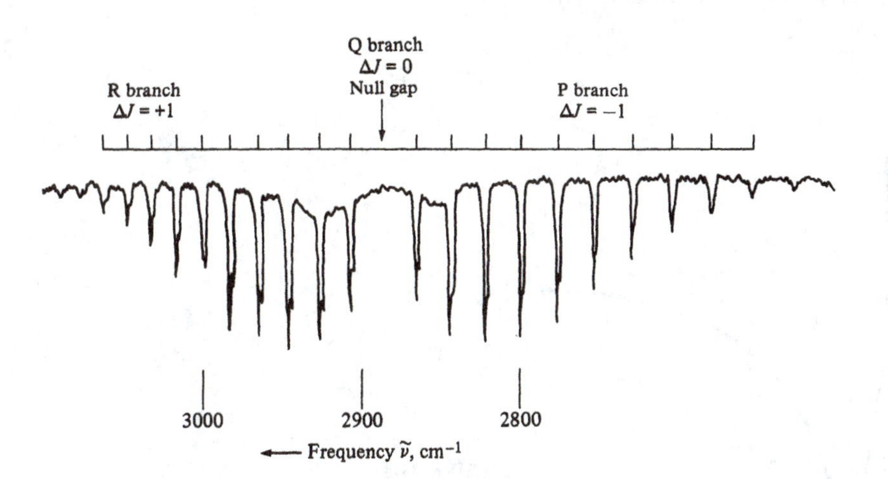

FIGURE 11-4
Experimental infrared spectrum of gaseous HCl for the fundamental vibrational transition ($v = 0$ to $v = 1$). Note that each absorption has a small shoulder appearing along its right-hand edge; this is due to the chlorine 37 isotope which is present in 24.47% abundance. The spectrum was taken by the author on a Beckmann 1112 IR spectrometer.

this can be seen more easily by examining the picket fence representation above the actual spectrum. The spectra of the overtone bands ($v = 0$ to $v = 2, 3, \ldots$) would have a similar appearance, but the null gap would appear at roughly two times, three times, etc., the frequency of the fundamental. The variation in intensities depends on the Maxwell-Boltzmann distribution $n_J = n_0(2J+1)\exp(-E_J/kT)$, where n_J is the number of states with rotational energy E_J and n_0 is the number of states with zero rotational energy. Thus, the number of states is not a maximum at $J = 0$ but at some intermediate J value. Note that if rotational energy levels were nondegenerate (as are vibrational levels), the maximum population would be at $J = 0$.

A number of explicit analytical forms for $V(R)$ occurring in Eq. (11-26) have been proposed; one of the first (and best known) is the *Morse potential*[4] given by

$$V(R) = D[1 - e^{-\beta(R-R_e)}]^2 \tag{11-43}$$

where D and β are empirical constants and R_e is the value of R for which $V(R)$ is a minimum (see Fig. 11-5). Equation (11-29) can be solved exactly for such a potential; the result can be expressed in the same general form as Eq. (11-41).[5] It should be noted that the Morse potential predicts a finite value for $V(R)$ at $R = 0$, whereas the true potential approaches infinity as R approaches zero. This discrepancy is ordinarily of little consequence, however.

[4] P. M. Morse, *Phys. Rev.*, **34**:57 (1929).

[5] C. L. Pekeris, *Phys. Rev.*, **45**:98 (1934).

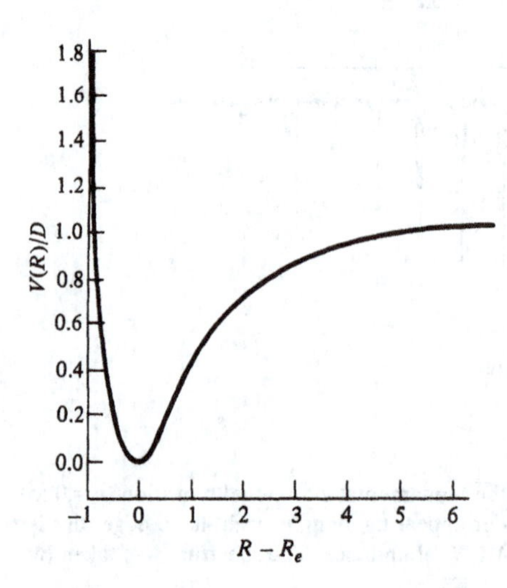

FIGURE 11-5
The Morse potential-energy function plotted as $V(R)/D$ versus $R - R_e$. The parameter β has been set equal to 1. This curve provides an analytical form for the $E(R)$ shown in Fig. 11-2.

For $^{1}H^{35}Cl$, with a fundamental vibrational frequency of 2885.9 cm^{-1}, the infrared spectrum is described quite well by the use of the following parameters:

$$x_e = 0.0174 \qquad B_e = 10.5909\ \text{cm}^{-1} \qquad \tilde{\nu}_e = 2989.74\ \text{cm}^{-1}$$

$$D_e = 0.0004\ \text{cm}^{-1} \qquad \alpha_e = 0.3019\ \text{cm}^{-1}$$

For polyatomic molecules the treatment of vibrational and rotational motions and their interactions becomes much more complex and takes various different forms depending on what particular class the molecule belongs to. Interested readers are referred to texts on molecular spectroscopy for descriptions and details of such treatments.

REGION OF RAMAN SPECTROSCOPY. It should also be mentioned that only molecules with a permanent dipole moment will exhibit absorption spectra due to rotational transitions. Also, only those (unsymmetrical) vibrations of nonpolar molecules which produce transient dipoles will exhibit vibrational-rotational spectra. For such molecules one must use Raman spectroscopy to obtain information on rotations and on some of the vibrations. In Raman spectroscopy, one observes the change in wavelength due to inelastic scattering of ultraviolet or visible radiation photons by the molecule; the gain or loss of energy due to scattering represents differences between vibrational and rotational states. Such an interaction depends on the nature of the polarizability ellipsoid of the molecule and not on the electric dipole moment. Thus a Raman study of dihydrogen (which has zero electric dipole) will provide information about the force constant of this molecule and also its equilibrium interatomic separation. In general, one must use both electric dipole and Raman spectroscopy to obtain a complete picture of the vibrational and rotational structure of a molecule. In some cases the two areas provide overlapping information. The interested reader should consult a specialized text for details of the Raman effect. Readers should also consult the excellent elementary reviews by Hollenberg[6] and Hoskins[7] for further details on electric dipole and Raman spectroscopy of diatomic molecules.

EXERCISES

11-1. Verify Eq. (11-40) as an approximate solution to Eq. (11-29) with the potential $V(R) = k(R - R_e)^2/2$. *Hint*: It is convenient to make the substitution $S(R) = T(R)/R$ and to transform to the relative coordinate $q = R - R_e$. For small vibrations, $q \ll R_e$ and may be neglected in an appropriate term.

---

[6] L. J. Hollenberg, *J. Chem. Ed.*, **47**:2 (1970).

[7] L. C. Hoskins, *J. Chem. Ed.*, **52**:568 (1975).

11-2. One of the consequences of the Born-Oppenheimer approximation is the prediction that isotopic substitution does not change the force constant of a diatomic molecule. If the fundamental vibrational frequency of HCl is 2886 cm^{-1}, what would you predict to be the fundamental vibrational frequency of DCl? The accepted value of $\tilde{\nu}_e$ (which is not the same as $\tilde{\nu}$) is 2090.78 cm^{-1}. The two quantities are related by $\tilde{\nu} = \tilde{\nu}_e - 2x_e\tilde{\nu}_e$. The value of x_e for DCl is 0.1118.

11-3. An emission line at $115.271 \times 10^9 \text{ s}^{-1}$ from Titan shows the presence of CO on that body. This emission is attributed to the $J = 0$ to $J = 1$ rotational transition of CO.[8] Calculate the following for CO:
(*a*) The moment of inertia
(*b*) The average C to O bond distance assuming ^{12}C and ^{16}O

11-4. (*a*) The $J = 0$ to $J = 1$ transition in the pure rotational spectrum of gaseous hydrogen chloride ($^1H^{35}Cl$) occurs at 20.68 cm^{-1}. Calculate the moment of inertia of HCl and the average H to Cl bond distance.
(*b*) The transition $J = 3$ to $J = 4$ in HCl is associated with radiation of 83.03 cm^{-1}. Use the rigid rotator approximation to calculate the moment of inertia and internuclear bond distance of HCl and compare these results with those in (*a*).

11-5. The infrared spectrum of CO ($^{12}C^{16}O$) leads to the parameters $x_e = 0.0062$ and $\tilde{\nu}_e = 2170 \text{ cm}^{-1}$. Calculate the fundamental vibrational frequency of this molecule.

11-6. The fundamental absorption frequency of HCl (obtained from infrared spectroscopy of a HCl/CCl_4 solution) is 2886 cm^{-1}. Calculate the force constant of HCl and compare it with the HF value quoted in the previous section. Assume $^1H^{35}Cl$, and look up nuclear masses in a publication such as the *Chemical Rubber Handbook*.

11-3 SELECTION RULES FOR MOLECULAR ELECTRONIC TRANSITIONS

Selection rules for electronic transitions *in atoms* are based on detailed considerations of spin and orbital angular momenta of the electrons and on the interactions of these two momenta. The situation *in molecules* is quite different; except for the case of linear molecules, angular momentum is of limited usefulness in the interpretation of electronic spectra.

For a molecule of S nuclei and N electrons, the z component of the total orbital angular momentum is a sum of the orbital angular momenta of individual nuclei and electrons; analogous relationships hold for the x and y components. If we are using the Born-Oppenheimer approximation, only the electronic orbital angular momenta are of importance. Thus we consider the z component of the electronic orbital angular momentum and the scalar square of the total electronic orbital angular momentum. These satisfy the relation-

[8] D. O. Muhleman, G. L. Berge, and R. T. Clancy, *Science*, **223**:393 (1984).

ships previously discussed in Chap. 4:

$$\hat{L}_z = \sum_{\mu} \hat{L}_{z_\mu} \qquad \hat{L}_x = \sum_{\mu} \hat{L}_{x_\mu}$$
$$\hat{L}_y = \sum_{\mu} \hat{L}_{y_\mu} \qquad \hat{L}^2 = \hat{L}_x^2 + \hat{L}_y^2 + \hat{L}_z^2 \tag{11-44}$$

Referring back to Sec. 8-1, we see that $\hat{L}_z$ will commute with all terms of the electronic hamiltonian in Eq. (11-2) with the possible exception of the electron-nucleus attraction terms $1/r_{\mu k}$. To investigate commutation of $\hat{L}_z$ and $1/r_{\mu k}$, we write

$$\hat{L}_z r_{\mu k}^{-1} = \frac{1}{i} \sum_{\rho}^{N} \frac{\partial}{\partial \varphi_\rho} r_{\mu k}^{-1} = r_{\mu k}^{-1} \hat{L}_z + \frac{1}{i} \frac{\partial r_{\mu k}^{-1}}{\partial \varphi_\mu} \tag{11-45}$$

It is apparent that $\hat{L}_z$ and $1/r_{\mu k}$ will commute only if the last term in Eq. (11-45) vanishes. This term can vanish only if the electron-nucleus distance $r_{\mu k}$ is independent of the polar angle of the μth electron. Such is the case only for linear molecules, since in that particular case one can define a unique axis, namely, the line formed by the nuclear centers. However, even though $\hat{L}_z$ may commute with $1/r_{\mu k}$, the other components $\hat{L}_x$ and $\hat{L}_y$ definitely do not; hence, the total electronic orbital angular momentum $\hat{L}^2$ is *not* a constant of the motion. Thus only $\hat{L}_z$ (where z is chosen along the bond axis) is a constant of the motion for linear molecules.

For a linear molecule of one electron (say, the hydrogen molecule ion, H_2^+), the operator $\hat{L}_z$ satisfies

$$\hat{L}_z \psi = \lambda \psi \tag{11-46}$$

where ψ is the electronic wavefunction of the molecule and λ is a molecular quantum number analogous to the m_l quantum number of a one-electron atom. Note, however, that the role of λ in molecular electronic structure is much like that of the l quantum number in atoms. In the case of a many-electron linear molecule in which LS coupling is assumed valid, $\hat{L}_z$ satisfies

$$\hat{L}_z \psi = \Lambda \psi \tag{11-47}$$

where Λ is given by

$$\Lambda = \sum_{\mu=1}^{N} \lambda_\mu \tag{11-48}$$

Analogous to the $s, p, d, f, \ldots$ and $S, P, D, F, \ldots$ notation for atoms, it is customary to write

$$|\lambda| = 0,\ 1,\ 2,\ 3, \ldots$$
$$\sigma,\ \pi,\ \delta,\ \phi, \ldots$$
$$|\Lambda| = 0,\ 1,\ 2,\ 3, \ldots$$
$$\Sigma,\ \Pi,\ \Delta,\ \Phi, \ldots$$

The spin operators $\hat{S}_z$ and $\hat{S}^2$ commute with the molecular electronic hamiltonian if spin-orbit interaction is neglected, so that the spectroscopic term symbol of a linear molecule can be written in the general form

$$^{2S+1}|\Lambda|^{\pm}_{u \text{ or } g} \tag{11-49}$$

The subscripts (u or g) occur only if the molecule has a center of symmetry, for example, H_2, O_2, CO_2, $HC \equiv CH$, and other molecules of $D_{\infty h}$ point-group symmetry (see App. 6 for a discussion of point-group symmetry). This subscript refers to the inversion symmetry of the electronic wavefunction. The right superscript applies only to Σ states, i.e., those for which $\Lambda = 0$, and refers to the reflection symmetry of the electronic wavefunction with respect to the σ_v plane containing the principal axis.[9] The left superscript refers to the spin multiplicity; the usage is the same as in atoms. Linear molecules without a center of symmetry, for example, CO, HCN, OCS, etc., belong to the $C_{\infty v}$ point group. Thus we see that the molecular term symbol corresponds to an irreducible representation of the symmetry group to which the molecule belongs. We will find that the ground states of linear molecules with a closed-shell electron configuration are always symbolized by the term symbols $^1\Sigma_g^+$ or $^1\Sigma^+$, respectively, depending on whether the molecule has a center of symmetry or not. For nonlinear molecules (to be discussed later) we will see that orbital angular momentum plays no role; however, the various electronic states will be labeled according to the irreducible representations of the symmetry groups to which they belong.

We recall from Sec. 3-1 that a transition from a stationary state k to a stationary state l can occur only if at least one component of the dipole strength [see Eq. (3-23)] does not vanish. Obtaining selection rules for molecular electronic transitions is most easily carried out by use of the character table for the point group to which the molecule belongs. The character tables for linear molecules with a center of symmetry ($D_{\infty h}$ point group) and linear molecules without a center of symmetry ($C_{\infty v}$ point group) are found at the end of App. 7. We will now see how to obtain electronic selection rules for $D_{\infty h}$ molecules; the procedure for $C_{\infty v}$ molecules is entirely analogous.

Let the state k be the ground state of a molecule with a closed-shell electron configuration, so that the spectroscopic state is either $^1\Sigma_g^+$ or $^1\Sigma^+$, depending on whether the molecule has or does not have a center of symmetry. Considering the $D_{\infty h}$ point group specifically, we see that the functions x and y transform like the Π_u irreducible representation and that z transforms like Σ_u^+. The allowed transitions originating at the ground state then are

$$^1\Sigma_g^+ \rightarrow \begin{cases} ^1\Pi_u & \sigma\text{-polarized} \\ ^1\Sigma_u^+ & \pi\text{-polarized} \end{cases} \tag{11-50}$$

---

[9] F. L. Pilar, *J. Chem. Ed.*, **58**:758 (1981).

Letting the initial state k be any arbitrary state, the general selection rules for $D_{\infty h}$ molecules become

$$\begin{gathered} \Delta\Lambda = 0, \pm 1 \\ g \leftrightarrow u \qquad g \not\leftrightarrow g, \qquad u \not\leftrightarrow u \\ - \rightarrow - \qquad + \rightarrow + \qquad + \not\leftrightarrow - \end{gathered} \tag{11-51}$$

In addition we have the multiplicity selection rule $\Delta S = 0$, which follows from the commutation of $\hat{S}^2$ with the dipole operators.

For the $C_{\infty v}$ point group one obtains

$$\begin{gathered} {}^1\Sigma^+ \rightarrow \begin{cases} {}^1\Pi & \sigma\text{-polarized} \\ {}^1\Sigma^+ & \pi\text{-polarized} \end{cases} \\ \Delta\Lambda = 0, \pm 1 \\ - \rightarrow - \qquad + \rightarrow + \qquad + \not\leftrightarrow - \end{gathered} \tag{11-52}$$

Spin-forbidden transitions, i.e., those for which $\Delta S \neq 0$, occur as a result of spin-orbit coupling. Symmetry-forbidden transitions, i.e., those for which $|\mu_{kl}| = 0$ when $\Delta S = 0$, occur as a result of a distortion of the molecular geometry during vibration of the nuclei. The existence of such transitions indicate that the Born-Oppenheimer approximation is not totally correct. The calculation of the intensities of forbidden transitions can be carried out by the use of perturbation theory in which the zeroth-order wavefunctions are those of the molecule in its equilibrium geometrical conformation. The general theory of the absolute intensities of symmetry-forbidden bands (and relative intensities of vibrational components) has been developed by Herzberg and Teller.[10] The treatment involves letting the nondiagonal elements of the perturbation operator represent the distortion of the molecular symmetry. Suppose ψ_0, ψ_1, and ψ_2 represent three zeroth-order states such that the transition $\psi_0 \rightarrow \psi_1$ is symmetry-forbidden but $\psi_0 \rightarrow \psi_2$ is allowed. If the perturbation has the proper symmetry, one will obtain a new state described by the function

$$\psi = \psi_1 + \lambda\psi_2 \tag{11-53}$$

such that the transition $\psi_0 \rightarrow \psi$ will be allowed during the course of a vibration. The intensity of this transition will be proportional to the square of the mixing parameter λ. The effect is as if the forbidden transition $\psi_0 \rightarrow \psi_1$ borrowed intensity from the allowed transition $\psi_0 \rightarrow \psi_2$. The spectrum will now include the forbidden transition $\psi_0 \rightarrow \psi_1$ with an intensity proportional to λ^2. Such symmetry-forbidden transitions usually have considerable vibrational fine structure.

---

[10] G. Herzberg and E. Teller, *Z. Phys. Chem.*, **B21**:410 (1933). See also an excellent review by A. Liehr, *Advan. Chem. Phys.*, **5**:241 (1963).

It is often convenient to use an additional set of symbols to distinguish between different multiplicities and among different excited states of linear molecules. The ground-state spectroscopic symbol is preceded with the symbol X, and successively higher energy states of the same multiplicity are preceded by the symbols $A, B, C, \ldots$. Higher-energy states of a different multiplicity are preceded by the lowercase letters $a, b, c, \ldots$. The ground state of H_2 is labeled $X^1\Sigma_g^+$, and the two lowest excited singlets are labeled $B^1\Sigma_u^+$ and $C^1\Pi_u$ (located 91,689.9 and 100,043.0 cm^{-1} above the ground state, respectively).[11] The lowest triplet is labeled $a^3\Sigma_g^+$ and lies 95,085 cm^{-1} above the singlet ground state. Similarly, the ground state of O_2 is written $X^3\Sigma_g^-$, and the lowest excited triplets are written $A^3\Sigma_u^+$ and $B^3\Sigma_u^-$, located at 36,096 and 49,802.1 cm^{-1} above the ground state, respectively. The lowest singlets are $a^1\Delta_g$ and $b^1\Sigma_g^+$, located 7918.1 and 13,195.2 cm^{-1} above the ground state, respectively. An exception to this notation occurs in N_2, for which long-established usage describes excited singlets by $a, b, c, \ldots$ and excited triplets by $A, B, C, \ldots$.

11-4 MOLECULAR HARTREE-FOCK CALCULATIONS

The orbital approximation may be applied to molecules in a way which is formally similar to the treatment of atoms. Molecular wavefunctions in this approximation are written in terms of Slater determinants having the form

$$D = |S_1 \quad S_2 \quad \cdots \quad S_N| \tag{11-54}$$

except that the $\{S_i(\mu)\}$ are now *molecular spin orbitals*, i.e., they are one-electron distribution functions describing an electron interacting with two or more nuclei simultaneously. Each molecular spin orbital (MSO) is itself expressed as a product of a spatial orbital $\psi_i(\mu)$ and a spin function $\omega(\mu)$, namely,

$$S_i(\mu) = \psi_i(\mu)\omega(\mu) \tag{11-55}$$

These spatial orbitals are known as *molecular orbitals*. Each molecular orbital (MO) is itself expressed in terms of some basis set of functions; one of the most convenient expansions utilizes atomic orbitals (AOs), which are generally centered on the nuclei which the MSO describes:

$$\psi_i(\mu) = \sum_{p=1}^{m} \phi_p(\mu)c_{pi} \tag{11-56}$$

where, in general, the coefficients $\{c_{pi}\}$ are determined variationally. The number of terms in the expansion (m) must always be equal to or greater than the number of different spatial orbitals (N); that is, $m \geq N$. This is necessary in

[11] An exception to the rules occurs in H_2; there is no state labeled A.

order to guarantee linear independence of the MOs. The general practice is to use as a basis those atomic orbitals which are necessary to describe the ground states of the component atoms; this is known as the *minimal basis set*. For example, to describe the molecule CO one would use as a basis the $1s$, $2s$, $2p_x$, $2p_y$ and $2p_z$ AOs of both carbon and oxygen. Each MO of carbon monoxide would then be a linear combination of these 10 basis AOs, except that, as we will see later, certain orbitals will appear in certain combinations with zero coefficients. Also, there is nothing to prevent one from using as many additional orbitals as desired; this will, in general, improve the energy calculation, but at the expense of additional computational labor. The general method utilizing the above format is known as the *linear combination of atomic orbitals approximation*, or LCAO approximation.

If the molecule has $2N$ electrons and has a closed-shell electron configuration $\psi_1^2\psi_2^2\cdots\psi_N^2$ described by N doubly occupied molecular orbitals, then the single-determinantal approximation to the wavefunction is

$$D_0 = |\psi_1\,\bar{\psi}_1\,\psi_2\,\bar{\psi}_2\cdots\psi_N\,\bar{\psi}_N| \tag{11-57}$$

This wavefunction leads to a total energy expression entirely analogous to that of an atom having $2N$ electrons and described by N doubly occupied atomic orbitals, namely,

$$E = 2\sum_i^N \epsilon_i^{(0)} + \sum_{i,j}^N (2J_{ij} - K_{ij}) \tag{11-58}$$

where $\epsilon_i^{(0)} = \langle \psi_i(\mu)|h_\mu|\psi_i(\mu)\rangle$

$$J_{ij} = \left\langle \psi_i(\mu)\psi_j(\nu)\left|\frac{1}{r_{\mu\nu}}\right|\psi_i(\mu)\psi_j(\nu)\right\rangle$$

$$K_{ij} = \left\langle \psi_i(\mu)\psi_j(\nu)\left|\frac{1}{r_{\mu\nu}}\right|\psi_j(\mu)\psi_i(\nu)\right\rangle \tag{11-59}$$

Note that the monoelectronic operator in the $\epsilon_i^{(0)}$ expression is given by Eq. (11-3); it differs from the analogous atomic operator only in that the electron-nuclear attractions are summed over several nuclei. In addition, there is another term which should be added to the energy expression (11-58); this is the nuclear-repulsion term

$$\hat{V}_n = \sum_{k<1}^{S} \frac{Z_k Z_l}{r_{kl}} \tag{11-60}$$

[see the middle expression in Eqs. (11-2)]. However, this term is constant (in the fixed-nuclei approximation) and can be added to expression (11-58) at the end of the calculation; consequently, we will not always explicitly include it in all of our discussions.

The best single-determinantal wavefunction which can be written for a molecule described by Eq. (11-57) is that composed of molecular orbitals which satisfy the Hartree-Fock pseudo-eigenvalue equations

$$\hat{F}\psi_i = \sum_j \psi_i \lambda_{ji} \tag{11-61}$$

The molecular Hartree-Fock operator is as defined in Eq. (10-16) for atoms except that the monoelectronic operator h is given by its molecular counterpart [Eq. (11-3)]. Straightforward solution of the molecular Hartree-Fock equations is too difficult to carry out; consequently, use is made of Roothaan's matrix approximation (see Sec. 10-8). If the LCAO expansion (11-56) is written in matrix form as

$$\psi_i = \mathbf{\phi C}_i \tag{11-62}$$

where $\mathbf{\Phi}$ is a $1 \times m$ row matrix of basis functions (AOs) and $\mathbf{C}_i$ is an $m \times 1$ column matrix of variational coefficients, then Roothaan's equations become

$$\mathbf{FC}_i = \epsilon_i \mathbf{\Delta C}_i \qquad \text{or} \qquad \mathbf{FC} = \mathbf{\Delta C \epsilon} \tag{11-63}$$

where ϵ_i is the energy associated with the MO ψ_i [see Eqs. (10-113), (10-114), and (10-115)]. The total molecular (electronic) energy can be written in terms of these orbital energies by an alternative expression to Eq. (11-58), namely,

$$E = 2\sum_i^N \epsilon_i - \sum_{i,j}^N (2J_{ij} - K_{ij}) \tag{11-64}$$

Whether one uses the energy expression (11-58) or (11-64) to evaluate the molecular electronic energy, one needs to evaluate a number of one-electron and two-electron matrix elements. The possible one-electron matrix elements are

$$h_{pq} = \langle \phi_p | h | \phi_q \rangle \qquad \text{or} \qquad F_{pq} = \langle \phi_p | \hat{F} | \phi_q \rangle \tag{11-65}$$

and the two-electron matrix elements (representing electron repulsions) are of the form

$$G_{pq} = \langle \phi_p | \hat{G} | \phi_q \rangle = \sum_{r,s}^m R_{rs}(2\langle ps|g|qr\rangle - \langle ps|g|rq\rangle) \tag{11-66}$$

The quantity R_{rs} is an element of the first-order density matrix of the system [see Sec. 10-8, especially Eq. (10-121)]. This matrix is defined in terms of the eigenvectors $\mathbf{C}_i$ associated with the MOs occupied in the ground state, i.e., with the N lowest-energy MOs. Thus

$$\mathbf{R} = \mathbf{TT}^\dagger \tag{11-67}$$

where the matrix $\mathbf{T}$ is defined as a $1 \times N$ array of $\mathbf{C}_1, \mathbf{C}_2, \ldots, \mathbf{C}_N$. Note that since each eigenvector $\mathbf{C}_i$ is itself an m-component column matrix, $\mathbf{T}$ is actually an $m \times N$ matrix. Thus $\mathbf{R}$ will be of order $(m \times N)(N \times m) = m \times m$. When the normal LCAO approximation is used, the matrix $\mathbf{R}$ is also called a *charge and bond-order matrix*; we will see later that the elements R_{rs} can be used to describe how the total electronic population of a molecule can be apportioned among the various atoms and bonds of the molecule. As shown in Eq.

(10-124), the energy (11-64) can now be written in terms of **R** as

$$E = 2T_r(\mathbf{RF}) - T_r(\mathbf{RG}) \tag{11-68}$$

where **F** and **G** are matrix representations of the Hartree-Fock operator and electron repulsion operator, respectively.

The one-electron integrals in Eq. (11-65) will contain either one or two atomic centers, depending on whether ϕ_p and ϕ_q are centered on the same or different atoms. Similarly, the two-electron integrals in Eq. (11-66) will have from one to four atomic centers. The one- and two-center integrals are reasonably easy to compute, but the three- and four-center integrals can be very troublesome if expressed in certain types of basis functions, e.g., Slater orbitals. However, many efficient techniques have been devised for such evaluations, and this no longer constitutes a problem for modern high-speed computational methods. One particularly simple stratagem is to employ gaussian approximations to Slater-type orbitals (STOs) for the basis functions ϕ_p. Such an approach was introduced in Sec. 6-4 in illustrating the variational principle for the hydrogen atom. The method is called the *STO-NG method*: each STO is represented by a linear combination of N gaussians. The reason for using gaussian functions resides in one of their mathematical properties: a product of two different gaussian functions, each centered on a different atom, say, A and B, is a new gaussian centered at a single intermediate point P. This mathematical property enhances numerical computations considerably. For example, for the special case of s-type gaussians, the product of two gaussians (one centered on atom A and the other on atom B) is given by

$$\exp(-\alpha r_A^2)\exp(-\beta r_B^2) = K \exp[-(\alpha + \beta) r_P^2] \tag{11-69}$$

where P is a point located along the line segment AB. The quantity K is a constant given by

$$K = \exp\left(-\frac{\alpha\beta}{\alpha + \beta} R^2\right) \tag{11-70}$$

where

$$R^2 = (A_x - B_x)^2 + (A_y - B_y)^2 + (A_z - B_z)^2 \tag{11-71}$$

A_x, A_y, A_z are the x, y, z coordinates of atom A, and B_x, B_y, B_z, are the x, y, z coordinates of atom B. The coordinates of the intermediate point P are given by

$$P_i = \frac{\alpha A_i + \beta B_i}{\alpha + \beta} \qquad i = x, y, z \tag{11-72}$$

When the gaussians are not of the s-type, extra factors appear in their product.[12]

[12] Pioneering calculations with gaussian functions were discussed by S. F. Boys, *Proc. Roy. Soc.* (*London*), **A200**:542 (1950). See also P. Čársky and M. Urban, Ab Initio *Calculations: Methods and Applications in Chemistry*, Springer-Verlag, New York, 1980.

SUGGESTED READINGS

Barrow, G.: *Introduction to Molecular Spectroscopy*, McGraw-Hill, New York, 1962. One of the most readable treatments available; covers a wide range of topics.

Herzberg, G.: Three volume series on "Molecular Spectra and Molecular Structure": Vol. I. *Spectra of Diatomic Molecules*; Vol. II. *Infrared Spectra and Raman Spectra*; Vol. III. *Electronic Spectra and Electronic Structure of Polyatomic Molecules*. Van Nostrand, Princeton, N.J., 1950, 1945 and 1966, respectively. Very complete treatments but at an advanced level. Volumes I and II are authoritative sources of spectroscopic constants for a large number of diatomic molecules.

Levine, I. N.: *Molecular Spectroscopy*, Wiley, New York, 1975.

CHAPTER 12

THE ELECTRONIC STRUCTURE OF LINEAR MOLECULES

There are two major types of linear molecule based on symmetry properties (see App. 6): those with a center of symmetry (the $D_{\infty h}$ point group) and those without a center of symmetry (the $C_{\infty v}$ point group). In both types of molecules, $\hat{L}^2$ does not commute with the $\hat{H}_1 + \hat{H}_{12}$ part of the electronic hamiltonian [see Eq. (8-6)] but $\hat{L}_z$ does. Thus, as pointed out in Sec. 11-3, the eigenvalues of $\hat{L}_z$ (symbolized by Λ) may be used to describe molecular electronic states.

This chapter deals most extensively with one particular model of molecular electronic structure: the *molecular orbital (MO) approximation* and extended methods based on this model, e.g., the configuration-interaction (CI) method. The molecular orbital approximation has its general basis in Roothaan's approach to solving the Hartree-Fock self-consistent field equations and has proved to be one of the most successful models of molecular electronic structure. Few chemists—no matter how practical and applied their interests—fail to come under the influence of the molecular orbital approximation; its terms and concepts permeate the roots of chemical science, and this influence is likely to become more pronounced in the future. It is the goal of this chapter to illustrate the application of the molecular orbital approximation to simple molecules in sufficient detail so that the student obtains a realistic idea of how reasonably good calculations on molecules are carried out. Since the field of computational chemistry is changing so rapidly, no attempt is made to reflect the state of the art; readers interested in the latest developments should consult the extensive research literature which has arisen—most of it during the last two decades. It is hoped that such coverage will be of value to that majority of chemists who do not do calculations themselves or who do their calculations via one or more of the computer programs currently available from the Indiana University Quantum Chemistry Program Exchange or from comparable sources.

This chapter is not strictly limited to the molecular orbital approximation; there is also a brief discussion of the *valence bond (VB) approximation* and its modern extension, the generalized valence bond (GVB) method. Although the average chemist no longer encounters the VB approximation very often, some familiarity with this approach leads to an increased appreciation of the general problems associated with electronic structure calculations.

12-1 ELECTRONIC TRANSITIONS IN MOLECULES

In order to provide an adequate perspective and background to molecular electronic structure, it is helpful to summarize some basic features of electronic spectroscopic transitions in molecules. A number of excellent specialized texts (see the suggested readings at the end of this chapter) may be consulted for detailed treatments.

The rapidly fluctuating electron densities of atoms and molecules produce an oscillating electric dipole which can couple with radiation in the visible and ultraviolet regions (see Sec. 3-1). This leads to excitation of the atom or molecule from one electronic quantum state to another and, in the case of molecules, is also accompanied by changes in the vibrational and rotational states as well. Because of the latter, molecular electronic spectra can be quite complex and difficult to interpret. In principle, however, such spectra are rich in information; the problem is how to extract it. In general, one can obtain more information about the vibrational-rotational structure of a molecule from the electronic spectrum (visible and ultraviolet regions) than from the much simpler vibrational-rotational spectrum (infrared region) or from the pure rotational spectrum (microwave region). In some cases one gets a pattern of vibrational levels almost to the limit of chemical dissociation. From this it is possible to obtain an estimate of D_e, the spectroscopic heat of dissociation of the molecule. Whereas infrared spectroscopy can give information only about the vibrational structure of the ground electronic state, the electronic spectrum also provides information about the vibrational structures of excited electronic states. However, extracting this information is not often an easy matter.

A clearer idea of the nature of electronic transitions in molecules can be obtained by considering the following simple example of a molecule in the gas phase. Consider a typical electronic transition between two electronic levels differing in energy by 1×10^{-18} J (about 6 eV). This corresponds to a transition frequency given by

$$\tilde{\nu} = \frac{\Delta E}{hc} = \frac{1 \times 10^{-18}\ \text{J}}{(6.6 \times 10^{-34}\ \text{J} \cdot \text{s})(3 \times 10^{8}\ \text{m} \cdot \text{s}^{-1})} = 5 \times 10^{6}\ \text{m}^{-1} \text{ or } 5 \times 10^{4}\ \text{cm}^{-1}$$

The corresponding wavelength is 200 nm and lies in the ultraviolet region of the spectrum. If this were an atom and the spectral resolution low, one would observe this transition as a single line at a wavelength of 200 nm. However, the molecular spectrum consists of a plethora of lines, many of them crowded together in a seemingly unsystematic manner. The sources of the additional lines are the vibrational and rotational states which accompany each of the electronic states involved in the transition. Each of the electronic states is associated with a number of different vibrational levels, typically spaced about 0.1 eV, or 1.6×10^{-20} J, apart. Furthermore, each of these vibrational levels is associated with a distinct set of rotational levels; typically the lower ones of these are spaced approximately 0.006 eV, or 1×10^{-21} J, apart. Consequently, the electronic spectrum will consist of a number of transitions centered roughly around 200 nm. At spacings of about 5 nm there will be groups of lines representing transitions between pairs of vibrational levels, each such level belonging to a different electronic state. The lines within this group represent rotational transitions. Each such line represents a transition between two rotational levels, and each of these two rotational levels belongs to a different vibrational level. Furthermore, each vibrational level belongs to a different electronic state. A simple description of the transition is that the bulk of the energy difference comes from a change in the electronic state with fine-tuning from changes in vibrational energy and additional finer-tuning from changes in rotational energy.

Each of the different vibrational changes yields a set, or *band*, of rotational changes. The entire collection of such vibrational bands for the electronic transition in question constitutes a *band group, or band system*. Each specific electronic change leads to its own *band group*, and all these band groups form the *electronic spectrum*.

The situation is somewhat different if the spectrum is of a molecule dissolved in an "inert" solvent; the rotational transitions become quenched, and only the vibrational transitions appear. The latter are not well resolved, however, and the spectrum acquires the appearance of a rugged mountain range.

Whereas infrared and microwave spectra are generally analyzed in terms of vibrational and rotational selection rules which involve changes in quantum numbers, electronic transitions are analyzed somewhat differently (see Sec. 11-3). Furthermore, the vibrational changes accompanying the electronic transitions are also analyzed quite differently; instead of using quantum number changes as a criterion, one employs the *Franck-Condon principle* to predict which vibrational transitions are most likely between a pair of electronic states. Simply stated, the Franck-Condon principle says that the most probable transitions are those for which there is a large overlap of the wavefunctions of initial and final vibrational states.[1] This is illustrated in Fig. 12-1. Only those vibrational states connected by vertical lines would be expected to participate in strong absorptions. Any levels not so connected by a vertical line will be associated with transitions of weak intensity. If the two electronic states have their minima at almost the same value of the internuclear separation, and if the potential-energy curves are otherwise comparable, the most likely vibrational transitions will be of the type $0 \rightarrow 0$, $1 \rightarrow 1, \ldots$, that is, those for which $\Delta v = 0$. Usually only the lower vibrational states of the ground electronic state are appreciably populated, so that the spectrum will contain a strong $0 \rightarrow 0$ transition with progressively weaker lines for $0 \rightarrow 1$, $0 \rightarrow 2$, etc.

The possibility exists that the molecule may acquire sufficient radiant energy to dissociate or to ionize. The former will occur if enough vibrational energy is acquired to separate the molecule into fragments (usually atoms or radicals in various excited states). If ionization of the molecule occurs (photoionization), one observes transitions to such ions. Such transitions are called *Rydberg transitions* and are usually observed (with some difficulty) in the vacuum ultraviolet region ($\lambda < 200$ nm). The Rydberg series is frequently

[1] A classical explanation of the Franck-Condon principle is as follows: Absorption of a photon increases the total energy by increasing the potential energy of the vibrating molecule; the kinetic energy changes very little. The vibrating molecule spends most of its time near the extreme points of the vibration (just like a swinging pendulum)—that is, where the kinetic energy is very small (and where the probability density ψ_v^2 is the largest). Thus, electronic-vibrational transitions are accompanied by very small changes in internuclear distances. A fairly rigorous discussion is given by S. E. Schwartz, *J. Chem. Ed.* **50**:608 (1973).

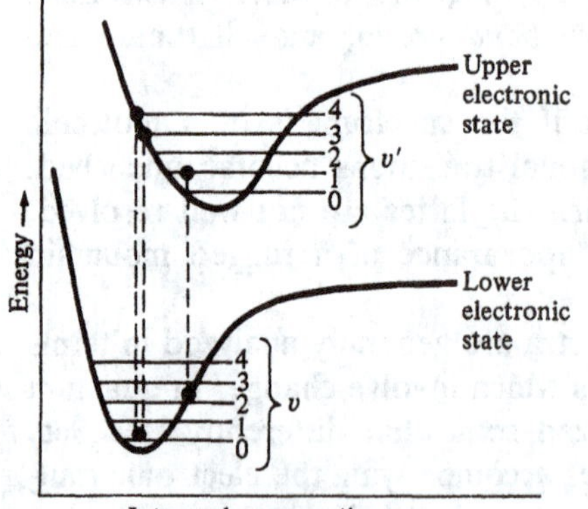

FIGURE 12-1
Schematic representation of two electronic states and their associated vibrational levels. According to the Franck-Condon principle, the most probable vibrational transitions are those whose energy levels are connected by vertical lines. Such vertical lines represent maximum overlapping of vibrational wavefunctions between the two electronic states. Note that transitions from $v = 0$ take place from the center of the level and not from the end.

masked by the continuum arising from the dissociation of the molecule in one of its low-lying electronic states. Even a simple diatomic molecule such as O_2 gives rise to more than one Rydberg series, corresponding to the excitation and ionization of electrons belonging to different energy levels. For a nonlinear molecule of n atoms, a full representation of the energy states would require $3n - 6$ potential wells in a $(3n - 6)$-dimensional space. Each potential well would have its own dissociation energy, and each Rydberg series would have its own ionization energy.

It sometimes appears as if potential-energy curves due to different electronic states interact; i.e., a molecule excited to one state may change to another, adjacent overlapping state via a radiationless process. This can lead to a phenomenon known as *predissociation*. Predissociation is evidenced in a spectrum by a lack of rotational fine structure in some of the bands. One probable explanation is that the molecule is excited from one electronic level to another with a subsequent redistribution of the energy so that dissociation can occur. If the redistribution occurs more rapidly than the period of the rotation, the rotation will not be quantized and some of the vibrational bands will be diffuse. One can also use the quantum mechanical tunneling effect to provide a rationale for predissociation.

In some cases a radiationless transition occurs between two electronic states without involving dissociation. One such process, known as *phosphorescence*, involves spontaneous emission from excited triplet states to lower-lying singlet states. As a simple example, suppose a molecule undergoes a transition from a singlet electronic level to a higher singlet level whose potential curve overlaps with that of a triplet state (see Fig. 12-2). The excited singlet may now lose vibrational energy by collisional deactivation, e.g., by transferring some of its vibrational energy via thermal collisions to other

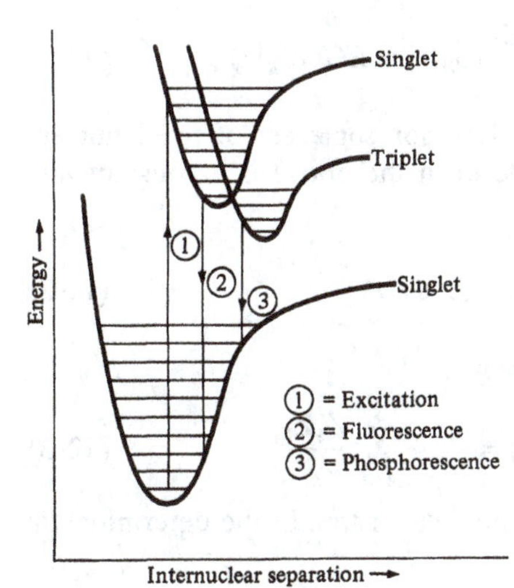

FIGURE 12-2
The mechanism of phosphorescent and fluorescent emission.

molecules. When the vibrational energy of the excited singlet reaches the point where the singlet-triplet potential curves cross, a radiationless transition to the triplet state can occur (provided the collisional deactivation is not too rapid). The triplet state will also become collisionally deactivated until the lowest vibrational level is reached. Now the triplet state can decay to the singlet state by a spontaneous process known as *phosphorescent emission*. Since this is a *spin-forbidden* transition, the emission may be very slow; such singlet-triplet transition may have half-lives in the range of 0.01 s to several seconds. The reverse of phosphorescence, i.e., a transition from the lower singlet state to the higher triplet state, can also occur but does so with much lower intensity. Such a transition occurs in benzene at about 330 nm, the same wavelength at which phosphorescent emission is observed.

A process similar to the above is called *fluorescence*. It involves a spontaneous emission between two states of the same multiplicity. For example, if the collisional deactivation in the excited singlet state in Fig. 12-2 did not lead to a triplet state, this singlet state could emit radiation from its lowest vibrational state to return to the singlet ground state. Such processes are usually very fast—on the order of 10^{-4} s or less following excitation. Generally, there exist many overlapping potential curves, so that it is even possible for the excited singlet to return to the ground state by passing through two or more overlapping states. It is only when the collisional deactivation process is very inefficient that phosphorescence and fluorescence (collectively known as *luminescence*) can occur.

12-2 SUMMARY OF THE MO-LCAO APPROXIMATION

The general problem is to obtain approximate solutions to the molecular Schrödinger equation as expressed by the Born-Oppenheimer approximation:

$$\left(\sum_{\mu}^{N} h_{\mu} + \sum_{\mu<\nu}^{N} \frac{1}{r_{\mu\nu}} + \sum_{k<l}^{S} \frac{Z_k Z_l}{r_{kl}}\right)\psi_R(r) = E(R)\psi_R(r) \qquad (12\text{-}1)$$

(see Sec. 11-1). Since the equation is solved for some set of fixed nuclear coordinates $\{r_{kl}\}$, the nuclear-repulsion term in the above is a constant and may be added after one has solved

$$\left(\sum_{\mu}^{N} h_{\mu} + \sum_{\mu<\nu}^{N} \frac{1}{r_{\mu\nu}}\right)\psi_R(r) = E_{\text{el}}\psi_R(r) \qquad (12\text{-}2)$$

The molecular energy $E(R)$ is then given by

$$E(R) = E_{\text{el}} + \sum_{k<l}^{S} \frac{Z_k Z_l}{r_{kl}} \qquad (12\text{-}3)$$

As in atoms, the electronic wavefunction may be written in the determinantal form

$$\psi(1,2,\ldots,N) = |S_1(1) \quad S_2(2) \quad \cdots \quad S_N(N)| \qquad (12\text{-}4)$$

where the $\{S_i(\mu)\}$ are *molecular spin orbitals* and where the subscript "R" on the wavefunction ψ is omitted. Considering only molecules with closed-shell structures and using double occupancy of spatial orbitals, we write the molecular determinantal wavefunction as

$$\psi(1, 2, \ldots, 2N) = |\phi_1 \quad \bar{\phi}_1 \quad \phi_2 \quad \bar{\phi}_2 \quad \cdots \quad \phi_N \quad \bar{\phi}_N| \tag{12-5}$$

where the spatial functions $\{\phi_i\}$ are called *molecular orbitals* (MOs).[2]

The MOs themselves must generally be constructed from a set of convenient basis functions. In the most commonly used procedure, each MO is represented as a linear combination of atomic orbitals:

$$\phi_i = \sum_r^m \chi_r c_{ri} \tag{12-6}$$

where $m \geq N$ (this assures linear independence of the MOs). The AOs $\{\chi_r\}$ are most simply chosen as those AOs appearing in the ground-state electron configurations of the atoms making up the molecule. This is known as the *minimal basis set*, and the entire approach is called the *molecular orbital, linear combination of atomic orbitals (MO-LCAO) approximation*. Various refinements of the MO-LCAO approximation are possible; for example, the minimal basis set may be extended by including AOs not appearing in the ground-state electron configuration of the constituent atoms. Other, more far-reaching refinements, will be introduced later.

Solution of Eq. (12-1) by means of Roothaan's equations and the LCAO approximation proceeds as follows:

1. A set of fixed nuclear coordinates is chosen.
2. A basis set (minimal or extended) $\{\chi_r\}$ is chosen, and Roothaan's matrix equations [see Eq. (11-63)] are solved iteratively to produce an energy-optimized set of expansion coefficients $\{c_{ri}\}$. The electronic energy [see Eq. (11-64)] may be written in the form

$$E_{\rm el} = 2\,\mathrm{Tr}\,(\mathbf{RF}) - \mathrm{Tr}\,(\mathbf{RG}) \tag{12-7a}$$

where $\mathbf{R}$ is the density and bond-order matrix (or first-order density matrix), $\mathbf{F}$ is the matrix representation (in the basis $\{\phi_i\}$) of the Hartree-Fock operator, and $\mathbf{G}$ is the matrix representation of the electron-repulsion energy. Adding the nuclear-repulsion term to the above produces $E(R)$.

The above procedure is repeated for several values of fixed nuclear coordinates. For a diatomic molecule, $E(R)$ may be plotted as a function of the

[2] Whereas atomic orbitals (AOs) describe an electronic distribution about a single positive center (the nucleus), MOs describe an electronic distribution about two or more positive centers (the nuclei of the atoms forming the molecule).

internuclear distance to produce the familiar potential-energy curve shown in Fig. 11-2. The minimum energy of the molecule is then obtained from the minimum of the $E(R)$ versus R curve.

If a minimal basis set is chosen, then several options are generally available. The ones we will consider explicitly are the following:

1. The basis functions are STOs, and the scale factors are chosen by one of several techniques:
   a. Slater's rules [see J. C. Slater, *Phys. Rev.*, **36**:57 (1939); also see Pauling and Wilson, cited in the suggested readings for Chap. 3].
   b. Variational optimization by calculations on the constituent atoms.
   c. Full optimization during the molecular calculation.
2. The basis functions are nominally STOs, but these are represented by a linear combination of gaussians; this is the STO-NG method introduced earlier in Sec. 11-4. We will consider primarily the $N=3$ and $N=6$ cases; the latter usually gives almost as good energies as using STOs directly, while improving computation times significantly.
3. Two functions of the same symmetry type are present in the same basis set for each occupied AO in the separated atoms; i.e., each of the minimal basis functions is replaced by two functions, each with its own scale parameter. This approach is generally known as the *double zeta* (DZ) *method* [the symbol "ζ" (zeta) is used by most workers in lieu of the "η" (eta) used in this text].

None of the above will produce energies corresponding to the true Hartree-Fock limit; the third will generally come the closest. To approach the Hartree-Fock limit more closely, one must employ *extended basis sets*, i.e., use orbitals of higher order than those forming the minimal basis set. Such additional basis functions are often called *polarization functions*. For example, the minimal basis set contributed by a nitrogen atom in a calculation is $1s$, $2s$, and $2p$ (five basis functions total). Extending this by including $3s$, $3p$, and $3d$ functions (or any one of these) constitutes the use of polarization functions. In practice, not all these polarization functions would be efficacious; for example, using a $3s$ for a nitrogen atom is virtually ineffective, whereas $3p$ and $3d$ may improve the energy and other properties significantly.

Another common practice is to use what is known as the *contracted basis function method*. The basis of this method is to group basis functions together so that each group can be manipulated as only one function. As an example, suppose we wish to describe a given atom in a molecule by the use of n s orbitals and m p orbitals; this basis may be written in the form "(ns mp)." These orbitals ($n+m$ in number) are called the *primitive functions*, and they could be used in a straightforward manner to form the wavefunction. However, a far better approach is to use these primitive functions to form *contracted functions* via the linear combinations

$$\chi_i = N_i \sum_j c_{ji} p_i \qquad (12\text{-}7b)$$

where $\{p_i\}$ represents the primitive functions. One would then represent all the s orbitals in the basis by linear combination of the appropriate primitive functions. For example, a nitrogen atom may contribute the primitive functions ($7s\,3p$) as a basis (this means seven s orbitals and three p orbitals). One could then use the seven primitive functions of the s type to form a $1s$ AO and a $2s$ AO, and could use a similar process for the $2p$ AOs. Generally, it is found to be convenient to use gaussian functions as the primitives; in fact this constitutes a very efficient way of utilizing gaussian functions in molecular calculations—far better than using gaussian functions directly as primitive bases and even better than via the STO-NG approach. The efficacy of the contracted basis functions resides in the fact that their use leads to a reduction in the number of integrals which must be manipulated. Specifically, if there are n primitive functions and m contracted functions, the number of integrals to be manipulated is reduced by the factor $(m/n)^4$. In practice, m is generally roughly equal to $n/2$ so that there is a 16-fold reduction in the number of integrals.

A combination of several of the previously described techniques is the *split-valence basis method*, which may be described as a valence double zeta method. The general representation of this approach has the form "$x - yz$G," where x, y, and z are integers defined as follows: x is the number of gaussian functions whose sum is used to represent each inner-shell basis AO, and yz implies that each valence AO ϕ is to be represented by a sum of *two* AOs (ϕ' and ϕ''), with ϕ' represented by a sum of y gaussians and ϕ'' by a sum of z gaussians (generally, $z = 1$). The number of "splits" is indicated by the presence of two integers, y and z. If more splits than two are desired, one must indicate so by using more integers; for example, 31 implies two splits, and 311 implies three splits. The ϕ' AOs are called *nondiffuse* and the ϕ'' AOs (and any additional AOs) are called *diffuse*. For example, a 4-31G basis for carbon AOs implies that the $1s$ AOs (inner shell) are represented by a linear combination of four gaussian functions and that each of the valence AOs ($2s$ and $2p$) become split into two parts: ($2s' + 2s''$) for $2s$ and ($2p' + 2p''$) for $2p$, where each part has different scaling parameters. Then $2s'$ and $2p'$ each become represented by a linear combination of three gaussians and $2s''$ and $2p''$ each become represented by a single gaussian. If polarization functions are added to nonhydrogen AOs, this is indicated by $x - yz$G*; if polarization functions are also added to hydrogen AOs, this is indicated by an additional *, that is, $x - yz$G**. The most commonly used split-valence basis sets are 3-21G, 4-31G, 6-21G, and 6-31G; the first two are perhaps the most popular, since they provide a balance between size and efficiency.

As mentioned previously, all the above methods approach the Hartree-Fock energy as a limit. A variety of methods are available to obtain energies lower than this, i.e., to account for some of the correlation energy. One of the most used is the method of *configuration interaction* (CI) (also called *superposition of configurations*). In its simplest form, this consists of expressing the wavefunction as a linear combination of functions, each one of which nominally represents a simple approximation to an electronic state of the molecule (all these functions must represent the same symmetry type). Thus, the dihydrogen molecule ($^1\Sigma_g^+$ ground state) may be represented by a linear combination of determinantal functions, each of which formally represents a state of $^1\Sigma_g^+$ symmetry. Depending on just what functions are chosen—and the number of functions—the CI method can give energies lower than the Hartree-Fock value. In fact, this method is capable (in principle) of leading to the exact nonrelativistic energy if enough terms of proper type are included in the wavefunction.

Full details of the above methods are given in a number of publications; particularly useful are the texts of Čársky and Urban[3] and of Hehre, Radom, Schleyer, and Pople.[4]

12-3 THE HYDROGEN MOLECULE ION, H_2^+

The simplest molecular species is the singly ionized hydrogen molecule. This one-electron molecule ion plays a role in the electronic structure of diatomic molecules which is analogous, in some respects, to the role of the hydrogen atom in atomic structure. Just as the nonrelativistic Schrödinger equation of the hydrogen atom can be solved exactly, so can the nonrelativistic Schrödinger equation of H_2^+, provided the Born-Oppenheimer approximation is assumed. The hamiltonian operator describing the motion of the single electron moving in the field of two positive nuclei constrained to maintain an internuclear distance R is given by

$$h = -\frac{\nabla^2}{2} - \frac{1}{r_A} - \frac{1}{r_B} \tag{12-8}$$

where the coordinate system is as shown in Fig. 12-3. Note that this system is

[3] P. Čársky and M. Urban, Ab Initio *Calculations: Methods and Applications in Chemistry*, Springer-Verlag, New York, 1980.

[4] W. Hehre, L. Radom, P. v.R. Schleyer, and J. A. Pople, Ab Initio *Molecular Orbital Theory*, Wiley-Interscience, New York, 1986.

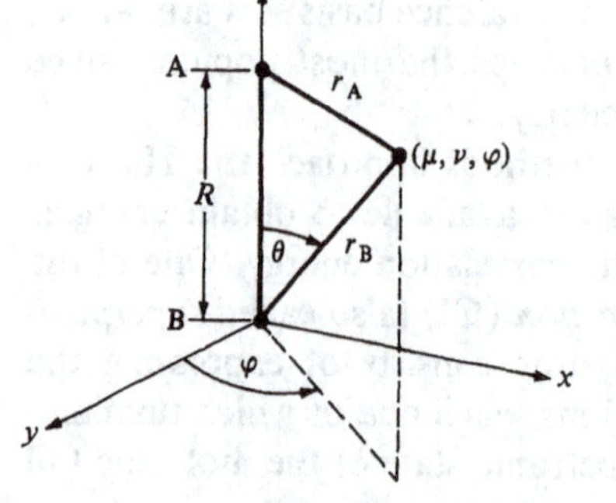

FIGURE 12-3
Coordinate system for H_2^+. The nuclei lie along the z axis at $(0, 0, 0)$ and $(0, 0, R)$. For a given value of the internuclear distance R, the system is described by the three coordinates r_A, r_B, and φ. The angle θ is related to the R, r_A, and r_B by the law of cosines and can be eliminated.

essentially that of the interaction of an electron and a "dumbell" which has two positively charged ends. If the repulsion between the two fixed nuclei is included, the total Born-Oppenheimer hamiltonian becomes

$$\hat{H} = h + \frac{1}{R} \tag{12-9}$$

In 1927, Burrau[5] showed that the H_2^+ wave equation was separable in the confocal elliptical coordinates μ and ν, defined by

$$\mu = \frac{r_A + r_B}{R} \qquad \nu = \frac{r_A - r_B}{R} \tag{12-10}$$

and the azimuthal angle φ (rotation about the internuclear axis). The variable μ is defined in the interval 1 to ∞, and ν is defined in the interval -1 to 1. The variable μ describes confocal ellipsoids of revolution with the nuclei as foci, and ν describes confocal hyperboloids. Carrying out the substitution,

$$\psi(\mu, \nu, \varphi) = M(\mu)N(\nu)F(\varphi) \tag{12-11}$$

leads to a separation of the variables to obtain the three ordinary differential equations

$$\left\{\frac{d}{d\mu}\left[(\mu^2 - 1)\frac{d}{d\mu}\right] + \epsilon\mu^2 + 2R\mu - \frac{\lambda^2}{\mu^2 - 1} + \kappa\right\}M(\mu) = 0$$

$$\left\{\frac{d}{d\nu}\left[(1 - \nu^2)\frac{d}{d\nu}\right] + \epsilon\nu^2 + \frac{\lambda^2}{1 - \nu^2} - \kappa\right\}N(\nu) = 0 \tag{12-12}$$

$$\left(\frac{d^2}{d\varphi^2} + \lambda^2\right)F(\varphi) = 0$$

The parameters λ and κ are separation constants, and the parameter ϵ is defined by

$$\epsilon = -\frac{R^2 E_{el}}{2} \tag{12-13}$$

The expressions in Eqs. (12-12) have well-behaved solutions only if the parameters λ, κ, and ϵ have certain definite values. The equation in φ is exactly of the same form as the φ portion of the spherical harmonics (see Sec. 4-6) and is well-behaved only if $\lambda = 0, \pm 1, \pm 2, \ldots$. Thus λ is a quantum number associated with the component of the electronic orbital angular momentum along the axis of nuclear centers (this provides a *unique* z axis). Recall, also, that the total electronic orbital angular momentum of a linear molecule is not quantized even though one of its components is (see Sec. 11-3). The case of $\lambda = 0$ corresponds to no rotation about the bond axis; for $\lambda \neq 0$, the two possible integral values (differing only in sign) represent the fact that clockwise and counterclockwise rotations about the bond axis are possible. Since we

[5] Φ. Burrau, *Kgl. Danske Videnskab. Selskab.*, **7**:1 (1927).

cannot distinguish between these two senses of rotation, these two values represent a double degeneracy. The solution of the equations in μ and ν must be carried out by numerical methods; these equations contain the energy parameter ϵ, which in turn involves λ^2. Consequently, the energy depends on $|\lambda|$, and this means that for all values of $|\lambda|>0$ the energy levels will be doubly degenerate.[6] The quantum number λ can be seen to reduce to the atomic magnetic quantum number m_l as the internuclear distance R approaches zero. However, $|\lambda|$ plays much the same role in H_2^+ as the azimuthal quantum number l does in a one-electron atom, for example, He^+, to which H_2^+ reduces as $R \to 0$. The various states of H_2^+ are usually symbolized by a notation analogous to the hydrogen atom spectroscopic notation $1s, 2s, 2p, \ldots$, namely,

$$nl|\lambda| = 1s\sigma_g, 2s\sigma_g, 2p\sigma_u, 2p\pi_u, \ldots \tag{12-14}$$

where nl refers to the one-electron AO to which the H_2^+ MOs reduce as $R \to 0$. The value of $|\lambda|$ is indicated by the Greek symbols σ, π, δ, $\phi, \ldots$ when $|\lambda| = 0, 1, 2, 3, \ldots$, respectively. The u and g refer to the inversion symmetry of the MO (see App. 6).

The ground state of H_2^+ ($|\lambda| = 0$) is designated by $1s\sigma_g$, which reduces to $1s$ of He^+ when $R = 0$. Referring to the $D_{\infty h}$ character table, we see that this state belongs to the $^2\Sigma_g^+$ irreducible representation (the left superscript is the multiplicity and is customarily appended to the point-group symbol). This identification comes from $|\lambda| = 0$ (which establishes the Σ) and the $1s$, which means $\chi(\sigma_v) = 1$ ("+" superscript) and $\chi(i) = 1$ ("g" subscript).[7] The next state is designated $2p\sigma_u$ and belongs to the $^2\Sigma_u^+$ irreducible representation. Similarly, the $2p\pi_u$ state belongs to the $^2\Pi_u$ irreducible representation (doubly degenerate).

A relatively recent updated solution of the H_2^+ problem is that of Wind.[8] The total energy ($E = E_{el} + 1/R$) is $-0.6026342E_h$ (-16.398 eV) at $R = R_e =$ 2.00 a_0. The spectroscopic dissociation energy of H_2^+ is then calculated as follows:

$$\begin{aligned} D_e &= E(H) + E(H^+) - E(H_2^+) = -0.50000 - (-0.60263) \\ &= 0.102653E_h \ (2.7928 \text{ eV}) \end{aligned} \tag{12-15}$$

The thermodynamic dissociation energy [see Eq. (11-37)] is given by

$$D_0 = D_e - E_0 \tag{12-16}$$

---

[6] Since H_2^+ belongs to the $D_{\infty h}$ point group, this degenerate behavior would be expected a priori, i.e., before the actual calculation of the wavefunction.

[7] The symbol $\chi(\sigma_v)$ represents the character of the σ_v symmetry operation in the $D_{\infty h}$ point group for the irreducible representation to which the $1s\sigma_g$ function belongs. Similarly, $\chi(i)$ is the character for the inversion symmetry operation.

[8] H. Wind, *J. Chem. Phys.*, **42**:2371 (1965).

where E_0 is the zero-point vibrational energy. If one takes into account the first anharmonicity correction, then E_0 (in cm^{-1}) is given by

$$E_0(\tilde{\nu}) = \frac{\tilde{\nu}}{2} - x_e\tilde{\nu} \tag{12-17}$$

According to Herzberg,[9] $\tilde{\nu} = 2297\ cm^{-1}$ and $x_e\tilde{\nu} = 62\ cm^{-1}$. Thus $E_0(\tilde{\nu}) = 1133\ cm^{-1}$, which corresponds to a zero-point energy of $0.00516E_h$. This leads to a chemical dissociation energy of $0.09748E_h$ (2.6524 eV). Herzberg quotes an indirect experimental value of $0.09732E_h$ (2.6481 eV) for D_0.[10]

Figure 12-4 shows a plot of the total energy $(E_{el} + 1/R)$ of H_2^+ as a function of R for the $X^2\Sigma_g^+(1s\sigma_g)$ and $^2\Sigma_u^+(2p\sigma_u)$ states as calculated by Teller.[11] Note that the latter state is unstable, since it exhibits no minimum. In fact, all the excited states of H_2^+ either are unstable or have very shallow minima. Consequently there are no direct spectroscopic measurements on H_2^+, and one has to rely heavily on quantum mechanical calculations for detailed information concerning this species. The chemical dissociation energy, normally an observable quantity, is obtained indirectly from the experimental D_0 of H_2 and the ionization energies of H_2 and H.

In Fig. 12-5 we see the appearance of the electronic distribution function $|\psi|^2$ along the H_2^+ internuclear axis. It is seen that the $X^2\Sigma_g^+(1s\sigma_g)$ state shows a high electron density between the two nuclei, whereas the unstable $^2\Sigma_u^+(2p\sigma_u)$ state shows an apparent decrease of electron density in this region. Using the electron density curve for the two noninteracting H atoms as a basis

[9] G. Herzberg, *Spectra of Diatomic Molecules*, Van Nostrand, Princeton, N.J., 1950.

[10] A survey of accurate experimental and theoretical work is given by D. R. Bates, K. Ledsham, and A. L. Stewart, *Phil. Trans. Roy. Soc. (London)*, **A246**:215 (1953).

[11] E. Teller, *Z. Physik*, **61**:458 (1930); see also E. A. Hylleraas, *Z. Physik*, **71**:739 (1931) and G. Jaffe, *Z. Physik*, **87**:535 (1934).

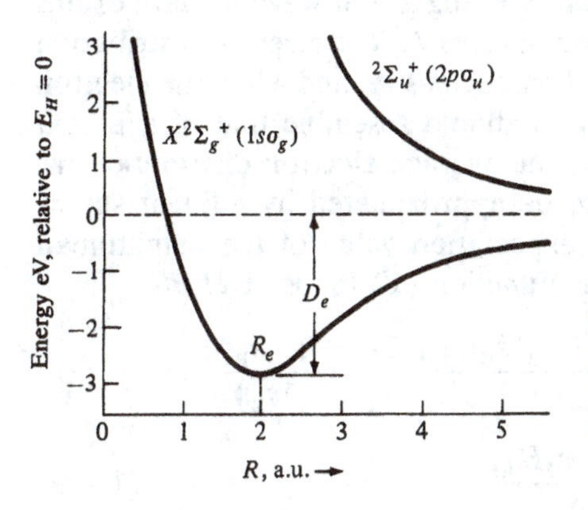

FIGURE 12-4
Potential-energy curves for the two lowest electronic states of H_2^+. [*After E. Teller*, Z. Physik, *61:474 (1930).*]

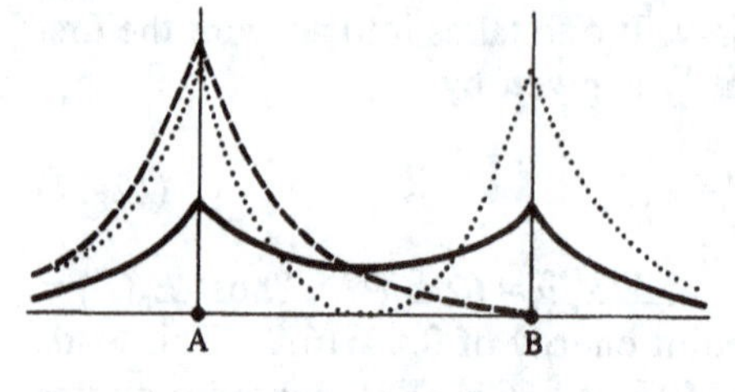

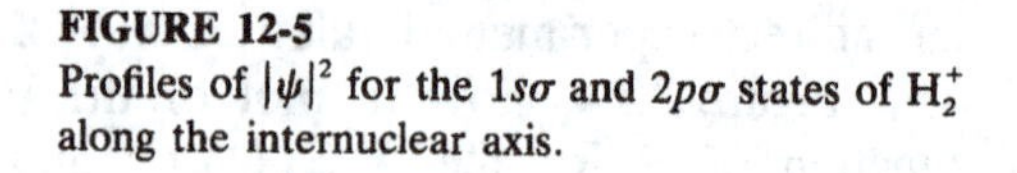
FIGURE 12-5
Profiles of $|\psi|^2$ for the $1s\sigma$ and $2p\sigma$ states of H_2^+ along the internuclear axis.

for comparison, it is evident that the electron density "flows" from the extranuclear region into the internuclear region in the case of the stable state and in the opposite direction in the unstable state. States in which there is an increased internuclear electron population (relative to the noninteracting case) are called *bonding* states, and those in which there is a decreased internuclear electron population are called *antibonding* states.

The minimal basis set for a Roothaan-type calculation on H_2^+ consists of two $1s$ AOs: one centered on nucleus A ($1s_A$) and the other centered on nucleus B ($1s_B$). In this very simple case the LCAO coefficients are determined entirely by symmetry (we know at the outset that $c_1 = \pm c_2$), and only one cycle of computations is needed. Not surprisingly, this does not produce a very good energy estimate, but since the procedure is instructive in a pedagogical sense, we will consider it in some detail.

Since this is a one-electron species, there is no real procedural difference between the Hartree-Fock treatment using Roothaan's equations with a minimal basis set and a straightforward variational approach in which the trial wavefunction is assumed to have the form

$$\psi = c_1 1s_A + c_2 1s_B \tag{12-18}$$

where c_1 and c_2 are to be formally treated as variational parameters. Also, one can give a plausible physical argument for choosing a trial wavefunction of this particular form: when the electron is near nucleus A, the electron distribution should resemble that of a $1s$ AO centered on nucleus A, and when the electron is near nucleus B, the electron distribution should resemble that of a $1s$ AO centered on nucleus B. Thus, in general, the average electron distribution (as represented by the wavefunction) should be approximated by a linear superposition of the $1s_A$ and $1s_B$ AOs. The expectation value of the hamiltonian [Eq. (12-9)] with respect to the trial wavefunction (12-18) is given by

$$\begin{aligned}\langle \hat{H} \rangle = \frac{\langle \psi|\hat{H}|\psi\rangle}{\langle \psi|\psi\rangle} &= \frac{\langle c_1 1s_A + c_2 1s_B|\hat{H}|c_1 1s_A + c_2 1s_B\rangle}{\langle c_1 1s_A + c_2 1s_B|c_1 1s_A + c_2 1s_B\rangle} \\ &= \frac{c_1^2 H_{AA} + c_2^2 H_{BB} + 2c_1c_2H_{AB}}{c_1^2 + c_2^2 + 2c_1c_2\Delta}\end{aligned} \tag{12-19}$$

$$
\begin{aligned}
\text{where } H_{AA} &= \langle 1s_A|\hat{H}|1s_A\rangle \\
H_{BB} &= \langle 1s_B|\hat{H}|1s_B\rangle \\
H_{AB} &= \langle 1s_A|\hat{H}|1s_B\rangle = \langle 1s_B|\hat{H}|1s_A\rangle = H_{BA} \\
\Delta &= \langle 1s_A|1s_B\rangle = \langle 1s_B|1s_A\rangle
\end{aligned} \tag{12-20}
$$

Minimization of the energy requires the conditions

$$\frac{\partial\langle \hat{H}\rangle}{\partial c_i} = 0 \qquad i = 1, 2 \tag{12-21}$$

which leads to the secular equations

$$c_1(H_{AA} - E) + c_2(H_{AB} - E\Delta) = 0 \qquad c_1(H_{AB} - E\Delta) + c_2(H_{BB} - E) = 0 \tag{12-22}$$

The corresponding secular determinant is

$$\begin{vmatrix} H_{AA} - E & H_{AB} - E\Delta \\ H_{AB} - E\Delta & H_{BB} - E \end{vmatrix} = 0 \tag{12-23}$$

It is convenient to employ the substitutions

$$a = \langle 1s_A|h|1s_A\rangle \qquad b = \langle 1s_A|h|1s_B\rangle \tag{12-24}$$

so that the hamiltonian matrix elements become

$$H_{AA} = H_{BB} = a + \frac{1}{R} \qquad H_{AB} = b + \frac{\Delta}{R} \tag{12-25}$$

The secular determinant (12-23) now becomes

$$\begin{vmatrix} a + \dfrac{1}{R} - E & b + \dfrac{\Delta}{R} - E\Delta \\ b + \dfrac{\Delta}{R} - E\Delta & a + \dfrac{1}{R} - E \end{vmatrix} = 0 \tag{12-26}$$

The two roots are

$$E_1 = \frac{a+b}{1+\Delta} + \frac{1}{R} \qquad \text{and} \qquad E_2 = \frac{a-b}{1-\Delta} + \frac{1}{R} \tag{12-27}$$

Since a and b are negative and Δ is positive, it is seen that E_1 is the lower of the two roots. This lower root is an approximation of the energy of the ground state, and the other is an approximation of the energy of an excited state.

The coefficients are determined in the standard manner; the first secular equation in (12-22) gives us

$$c_2 = \frac{\left[E - \left(\dfrac{a+1}{R}\right)\right]c_1}{b + \dfrac{\Delta}{R} - E\Delta} = Qc_i \tag{12-28}$$

Substituting the above relationship into the normalization condition

$$c_1^2 + c_2^2 + 2c_1c_2\Delta = 1 \tag{12-29}$$

leads to the general relationship

$$c_1^2 = \frac{1}{1 + Q^2 + 2Q\Delta} \tag{12-30}$$

Using the E_1 root gives $Q = 1$; the other root E_2 gives $Q = -1$. Thus for the ground state, $c_{11} = c_{21} = [2(1 + \Delta)]^{-1/2}$, and the wavefunction is given by

$$\psi_1 = \frac{1}{\sqrt{2(1+\Delta)}}(1s_A + 1s_B) \tag{12-31}$$

For the excited state, $c_{12} = -c_{22} = [2(1 - \Delta)]^{-1/2}$, and the wavefunction is

$$\psi_2 = \frac{1}{\sqrt{2(1-\Delta)}}(1s_A - 1s_B) \tag{12-32}$$

Note that the results of Eqs. (12-31) and (12-32) could have been deduced by symmetry arguments alone. Since the two nuclei A and B are indistinguishable, it follows at once that $c_1 = \pm c_2$. This, plus the normalization condition (12-29), produces Eqs. (12-31) and (12-32).

The integrals appearing in the secular determinant have the following general forms (in atomic units):

$$a = -\frac{1}{2} + e^{-2R} + \frac{e^{-2R} - 1}{R} \qquad b = -\frac{\Delta}{2} - e^{-R}(1 + R)$$
$$\Delta = e^{-R}\left(1 + R + \frac{R^2}{3}\right) \tag{12-33}$$

If $\langle \hat{H} \rangle$ for the lower root E_1 is plotted as a function of the internuclear distance R, a minimum value of the molecular energy is obtained at $R = R_e = 2.49\ a_0$. This produces a value of $D_e = 1.76$ eV for the $X^2\Sigma_g^+$ ground state (the experimental values are $R_e = 2.00\ a_0$ and $D_e = 2.79$ eV). An analogous plot for the higher root E_2 ($^2\Sigma_u^+$ state) exhibits no minimum. A comparison of these calculations with the exact results is shown in Fig. 12-6.

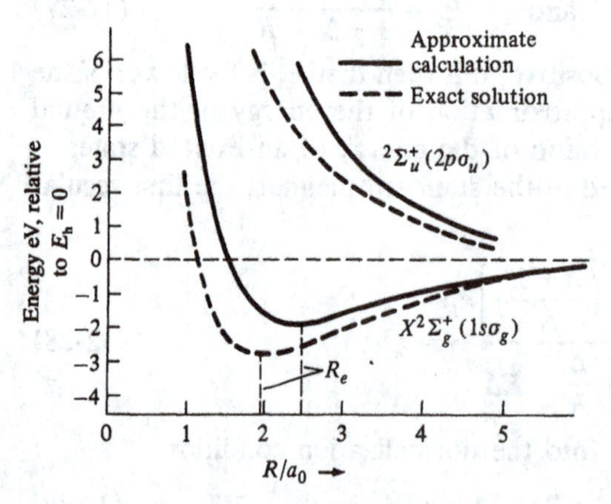

FIGURE 12-6
Comparison of the exact theoretical curves and approximate curves for the two lowest states of H_2^+.

Figure 12-7 shows surface plots of the probability densities $|\psi_1|^2$ and $|\psi_2|^2$ for the $X^2\Sigma_g^+$ and $^2\Sigma_u^+$ states of H_2^+ in the plane of the nuclei. This figure illustrates that the stable ground state is characterized by a buildup of electron density in the internuclear region at the expense of the extranuclear region, whereas the unstable excited state exhibits an opposite migration of electron density. Figure 12-8 illustrates this reallocation of electron density in a different way. The difference in electron density in the nuclear plane is given by the quantity

$$\delta_i = \psi_i^2 - 0.5(1s_A^2 + 1s_B^2) \qquad (12\text{-}34)$$

where $i = 1$ for the bonding state and $i = 2$ for the antibonding state. Thus δ_1 represents the positive buildup of electron density between the nuclei and a

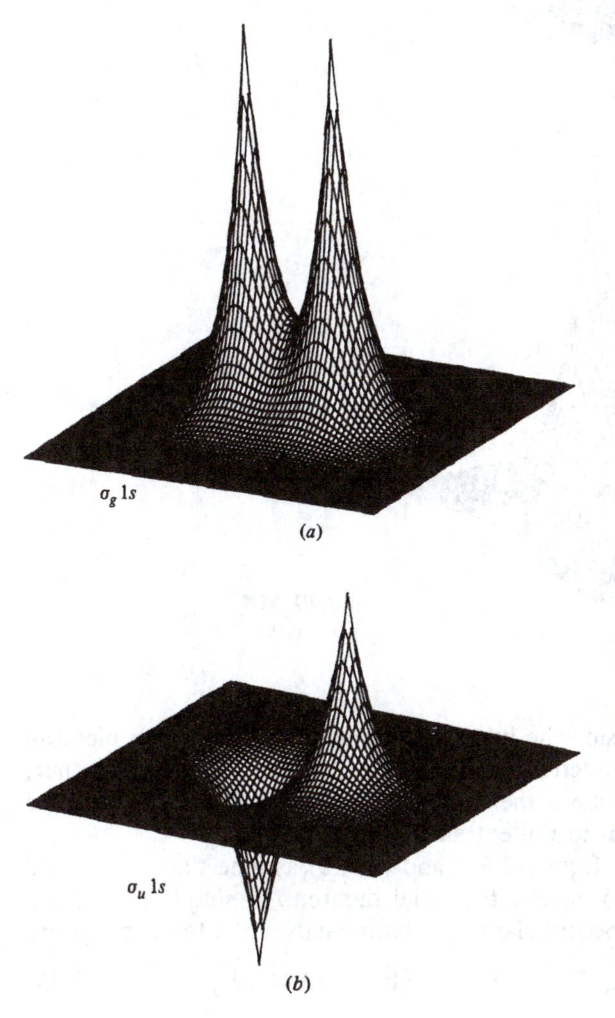

FIGURE 12-7
Surface plots of (*a*) the $\sigma_g 1s$ and (*b*) the $\sigma_u 1s$ wavefunctions of the H_2^+, together with (*c*) and (*d*), the corresponding probability densities $(\sigma_g 1s)^2$ and $(\sigma_u 1s)^2$. The $\sigma_g 1s$ and $(\sigma_g 1s)^2$ plots look very similar in that both indicate a buildup of the function between the two nuclei. In the case of $(\sigma_g 1s)^2$, this buildup may be interpreted as bonding of the two nuclei. The $\sigma_u 1s$ and $(\sigma_u 1s)^2$ plots show no buildup between the nuclei; in fact, there is a node between the two parts of the functions and a buildup *outside* the internuclear region. Thus $\sigma_u 1s$ describes an *antibonding* state.

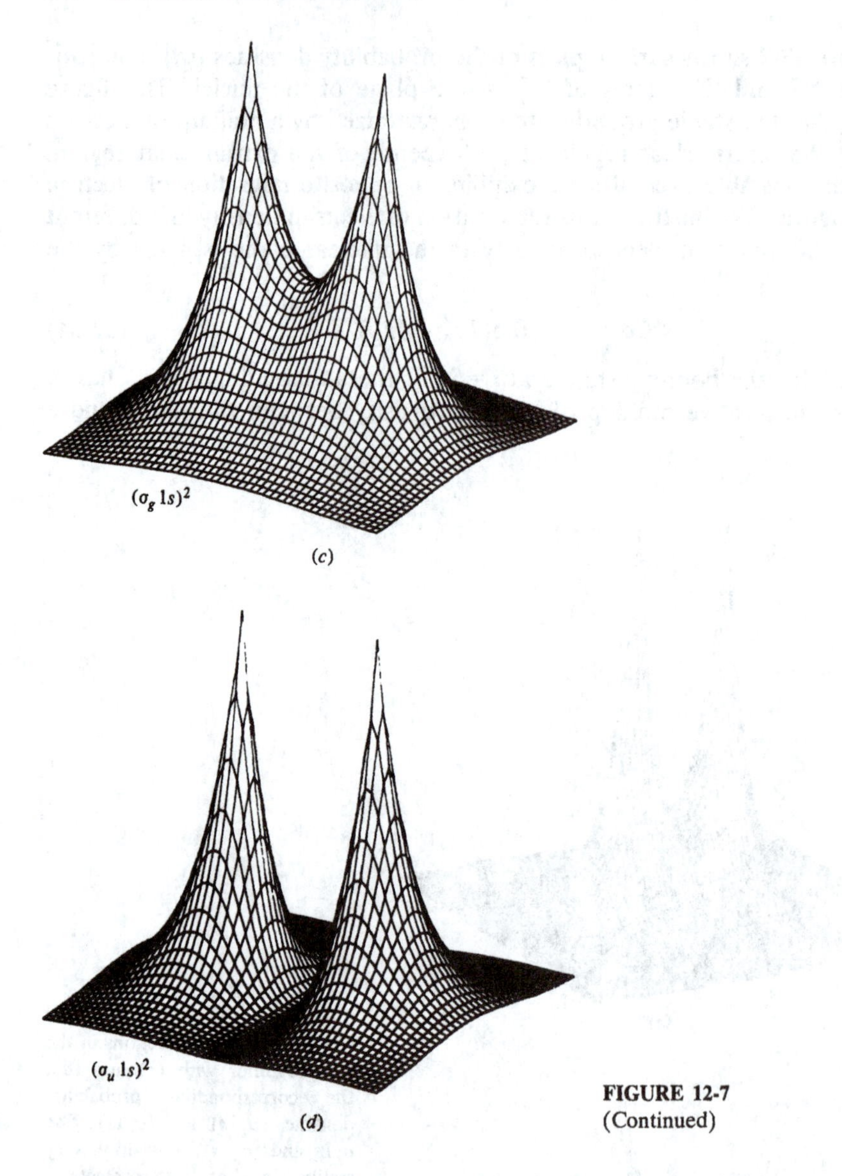

FIGURE 12-7
(Continued)

corresponding decrease outside the internuclear region relative to the electron densities found in the unbonded atoms located a distance R from each other. Similarly, the quantity δ_2 shows the opposite effect—a migration of charge from the internuclear region to the extranuclear region.

The basis functions in Eqs. (12-31) and (12-32) assume that $Z = 1$, and thus the calculation does not satisfy the virial theorem. A simple calculation shows that the kinetic- and potential-energy components of the total energy are

$$\langle \hat{T} \rangle = 0.3827 E_h \qquad \text{and} \qquad \langle \hat{V} \rangle = -0.9475 E_h \qquad (12\text{-}35)$$

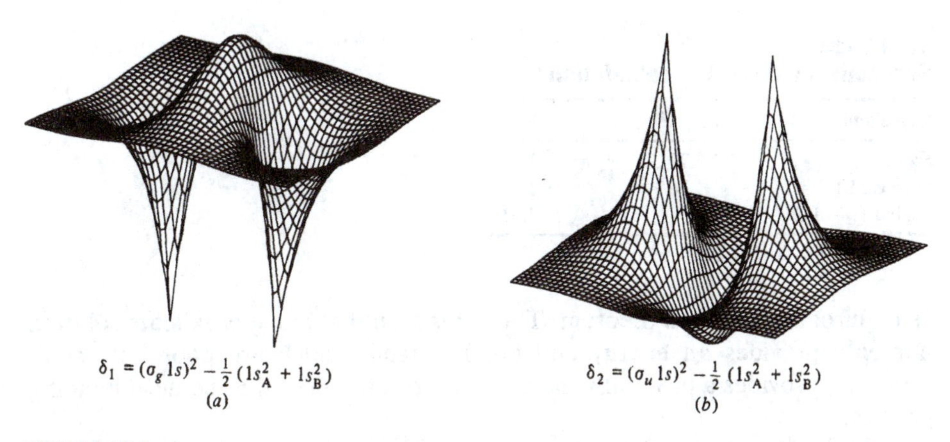

$\delta_1 = (\sigma_g 1s)^2 - \frac{1}{2}(1s_A^2 + 1s_B^2)$
(a)
$\delta_2 = (\sigma_u 1s)^2 - \frac{1}{2}(1s_A^2 + 1s_B^2)$
(b)

FIGURE 12-8
Differences in probability densities between (*a*) the bonding $\sigma_g 1s$ function and (*b*) the antibonding $\sigma_u 1s$ function of H_2^+ according to Eq. (12-34). The two upward-pointing peaks of (*b*) represent a buildup of probability density *outside* the internuclear region; the sag between these peaks represents a decrease in probability density between the nuclei. Note that this decrease and increase are relative to the density due to two noninteracting $1s^2$ atomic probability densities. The two downward-pointing peaks of (*a*) represent a decrease of probability density outside the internuclear region, and the mound between these indicates an increase of probability density between the nuclei.

Thus

$$\frac{\langle \hat{T} \rangle}{-\frac{1}{2}\langle \hat{V} \rangle} = 0.8078 \tag{12-36}$$

According to Eq. (2-87) the above quantity must equal unity if the virial theorem is satisfied. The calculated dissociation energy is given by

$$D_e = -(\Delta T + \Delta V) = 0.0648 E_h \tag{12-37}$$

where ΔT and ΔV are the changes in kinetic and potential energies, respectively, in going from a hydrogen atom and a bare proton to the hydrogen molecule ion. These two quantities may be computed individually as follows:

$$\begin{aligned} \Delta T &= \langle \hat{T} \rangle - 0.5000 = -0.1173 E_h \\ \Delta V &= \langle \hat{V} \rangle - (-1.0000) = 0.0525 E_h \end{aligned} \tag{12-38}$$

This result violates Theorem 2-7, which states that ΔT must be positive and ΔV must be negative. By contrast, the exact solution produces

$$\Delta T = 0.0126 E_h \quad \text{and} \quad \Delta V = -0.2052 E_h \tag{12-39}$$

and leads to

$$\frac{\Delta T}{-\frac{1}{2}\Delta V} = 1.0000 \tag{12-40}$$

TABLE 12-1
Summary of three H_2^+ calculations

Calculation	D_e, eV	R_e/a_0
Exact	2.79	2.00
Unscaled ($Z = 1$)	1.76	2.49
Scaled ($\eta = 1.24$)	2.35	2.01

as required by the virial theorem. Thus, it is seen that the approximate solution not only provides an energy and bond distance which are poor but, more seriously, provides a physically incorrect interpretation of the chemical bonding process.

In the following section we will discuss how scaling and the virial theorem apply to diatomic molecules in general, and we will return to an analysis of the unscaled and scaled calculations on the dihydrogen molecule ion.

Table 12-1 summarizes the results of the three H_2^+ calculations we have discussed.

EXERCISES

12-1. Using the formulas for a, b, and Δ, evaluate E_1 for $R = 1.0$, 2.0, 2.5, 3.0, 5.0, and 7.0 a_0. Plot the results. Repeat for E_2 and $R = 1.0$, 2.0, 5.0, and 7.0 a_0.

12-2. Consider the trial wavefunction for H_2^+ given by $\psi = e^{-b\mu}(1 + \lambda\nu^2)$ where b and λ are variational parameters and μ and ν are confocal elliptical coordinates.

(*a*) Normalize this wavefunction.

(*b*) Using $R = 2.00\ a_0$ and $b = 1.35$, find λ by the variation method and calculate the energy. This calculation was first carried out by H. M. James, *J. Chem. Phys.*, **3**:9 (1935).

12-4 SCALING AND THE VIRIAL THEOREM FOR DIATOMIC MOLECULES

The previous discussion of scaling and the virial theorem for calculations on atoms (Sec. 6-6) will now be extended to diatomic molecules in the Born-Oppenheimer (fixed nuclei) approximation. If the nuclei are fixed to a distance R, the external force holding the nuclei in this position is given by

$$\mathbf{F} = -\text{grad}\ E(R) = -\frac{dE(R)}{dR} \tag{12-41}$$

where $E(R)$ is the total energy at the internuclear separation R. The virial theorem then becomes

$$2\langle\hat{T}\rangle = -\langle\hat{V}\rangle - R\,\frac{dE(R)}{dR} \tag{12-42}$$

where $\langle\hat{T}\rangle$ is the average kinetic energy of the electrons (the nuclei are stationary and have no kinetic energy) and $\langle \hat{V} \rangle$ is the average electrostatic energy. Such a relationship was first discussed by Slater.[12]

When $R = R_e$ or ∞, the last term in Eq. (12-42) vanishes and we obtain the ordinary virial theorem (6-116). We can thus apply the ordinary theorem to the H_2^+ calculations of the preceding section. Considering the unscaled calculation, the expectation value of the kinetic energy in the ground state is

$$\langle \hat{T} \rangle = \left\langle \psi_1 \left| -\frac{\nabla^2}{2} \right| \psi_1 \right\rangle = [2(1+\Delta)]^{-1} \left\langle 1s_A + 1s_B \left| -\frac{\nabla^2}{2} \right| 1s_A + 1s_B \right\rangle$$
$$= (1+\Delta)^{-1} \left(\left\langle 1s_A \left| -\frac{\nabla^2}{2} \right| 1s_A \right\rangle + \left\langle 1s_A \left| -\frac{\nabla^2}{2} \right| 1s_B \right\rangle \right) \tag{12-43}$$

Using the integral values (in atomic units)

$$\left\langle 1s_A \left| -\frac{\nabla^2}{2} \right| 1s_A \right\rangle = \frac{1}{2} \qquad \left\langle 1s_A \left| -\frac{\nabla^2}{2} \right| 1s_B \right\rangle = -\frac{1}{2} [\Delta - 2(1+R)e^{-R}] \tag{12-44}$$

and Eq. (12-33) for Δ, one obtains (at $R = 2.49\ a_0$)

$$\langle \hat{T} \rangle = 0.3827 E_h \tag{12-45}$$

Since the total energy at $R = 2.49\ a_0$ is $-0.5648 E_h$, the average potential energy is

$$\langle \hat{V} \rangle = -0.5648 - 0.3827 = -0.9475 E_h \tag{12-46}$$

This leads to the ratio

$$-\frac{2\langle \hat{T} \rangle}{\langle \hat{V} \rangle} = 0.8078 \tag{12-47}$$

and thus, the virial theorem is not satisfied by the unscaled calculation. Furthermore, we can show that the unscaled calculation provides an incorrect interpretation of the chemical binding process. The calculated dissociation energy D_e can be written

$$D_e = -(\Delta T + \Delta V) = 0.0648 E_h \tag{12-48}$$

where ΔT and ΔV are the changes in kinetic and potential energies, respectively, in going from a hydrogen atom and a bare proton to H_2^+, that is,

$$\begin{aligned} \Delta T &= \langle \hat{T} \rangle - 0.5000 = -0.1173 E_h \\ \Delta V &= \langle \hat{V} \rangle - (-1.000) = 0.0525 E_h \end{aligned} \tag{12-49}$$

[12] J. C. Slater, *J. Chem. Phys.*, **1**:687 (1933).

The expressions in Eq. (12-49) imply that the process of forming the dihydrogen molecule ion involves a decrease in the kinetic energy and an increase in the potential energy. However, Wind's exact calculation led to a total energy of $-0.6026E_h$, so that $\langle \hat{T} \rangle = 0.6026E_h$ and $\langle \hat{V} \rangle = -1.2052E_h$. Thus, the correct values of ΔT and ΔV are given by

$$\begin{aligned} \Delta T &= 0.6026 - 0.5000 = 0.1026E_h \\ \Delta V &= -1.2052 - (-1.000) = -0.2052E_h \end{aligned} \tag{12-50}$$

Clearly this shows just the opposite behavior to that predicted from the unscaled calculation. Consequently, it is dangerous to use unscaled calculations to provide details of chemical bonding, since the total energy may be partitioned incorrectly.

The role of scaling in the general case ($R \neq R_e$ or ∞) has been discussed by Hirschfelder and Kincaid[13] and by Löwdin.[14] The following discussion is based on the latter's work.

If we let τ represent the set of electron coordinates $\tau_1, \tau_2, \ldots, \tau_N$ for N electrons, the general wavefunction for an N-electron diatomic molecule may be written in the symbolic form

$$\psi_\eta = \eta^{3N/2}\psi(\eta^3\tau, \rho) \tag{12-51}$$

where $\rho = \eta R$. The relationship in Eq. (12-51) means that the internuclear distance R is scaled by the same factor as are the electron coordinates. Note that for the unscaled wavefunction, $\eta = 1$ and ψ_η reduces to ψ_1 (the unscaled wavefunction). The counterparts of Eqs. (6-110) and (6-111) now become

$$\begin{aligned} \langle \hat{T} \rangle^{(\eta, R)} &= \langle \psi_\eta | \hat{T} | \psi_\eta \rangle = \eta^2 \langle \hat{T} \rangle^{(1, \rho)} \\ \langle \hat{V} \rangle^{(\eta, R)} &= \langle \psi_\eta | \hat{V} | \psi_\eta \rangle = \eta \langle \hat{V} \rangle^{(1, \rho)} \end{aligned} \tag{12-52}$$

The total energy associated with the scaled wavefunction is

$$E(\eta, R) = \eta^2 \langle \hat{T} \rangle^{(1, \rho)} + \eta \langle \hat{V} \rangle^{(1, \rho)} \tag{12-53}$$

Letting R be a fixed parameter and applying the variational principle with η as a variational parameter, we obtain

$$\frac{\partial E(\eta, R)}{\partial \eta} = 2\eta \langle \hat{T} \rangle^{(1, \rho)} + \langle \hat{V} \rangle^{(1, \rho)} + \eta^2 R \frac{\partial \langle \hat{T} \rangle^{(1, \rho)}}{\partial \rho} + \eta R \frac{\partial \langle \hat{V} \rangle^{(1, \rho)}}{\partial \rho} = 0 \tag{12-54}$$

The last two terms in the above arise because the expectation values depend on

[13] J. O. Hirschfelder and J. F. Kincaid, *Phys. Rev.*, **52**:658 (1937).

[14] P. O. Löwdin, *J. Mol. Spectry.*, **3**:46 (1959). See also A. D. McLean, *J. Chem. Phys.*, **40**:2774 (1964).

η through ρ. The form of these terms arises when we employ the substitution

$$\left(\frac{\partial \rho}{\partial \eta}\right)_R = R \tag{12-55}$$

When ψ_η is the exact solution, then $\eta = 1$ and $\rho = R$, so that Eq. (12-54) reduces to

$$2\langle \hat{T} \rangle^{(1,R)} + \langle \hat{V} \rangle^{(1,R)} + R\,\frac{dE(1,R)}{dR} = 0 \tag{12-56}$$

which is just the virial theorem (12-42). The last term in Eq. (12-56) arises from Eq. (12-54) when $\eta = 1$ and $\rho = R$, since $E(1, R)$ then is the exact energy and does not depend on η. The last two terms in Eq. (12-54) then become

$$R\,\frac{d[\langle \hat{T} \rangle^{(1,R)} + \langle \hat{V} \rangle^{(1,R)}]}{dR} = R\,\frac{dE(1,R)}{dR} \tag{12-57}$$

If one uses the relationship (12-53) for $\eta = 1$ and $\rho = R$ along with the virial theorem (12-56), one can obtain two useful relationships:

$$\begin{aligned} \langle \hat{T} \rangle^{(1,R)} &= -E(1,R) - R\,\frac{dE(1,R)}{dR} \\ \langle \hat{V} \rangle^{(1,R)} &= 2E(1,R) + R\,\frac{dE(1,R)}{dR} \end{aligned} \tag{12-58}$$

These relationships allow one to calculate potential energies and kinetic energies individually and in a unique manner if the total energy is known as a function of R. Figure 12-9 is a plot of total, kinetic, and potential energies for a diatomic molecule whose total energy is closely approximated by the Morse function [see Eq. (11-43) and Fig. 11-5]

$$E(R) = D_e[1 - e^{-\beta(R-R_e)}]^2 \tag{12-59}$$

Next, we investigate the case when $\psi(\tau, R)$ is not the exact wavefunction, so that the virial theorem is not automatically satisfied. We let $\rho = \eta R$ as before, but now we regard ρ as an auxiliary basic parameter in terms of which

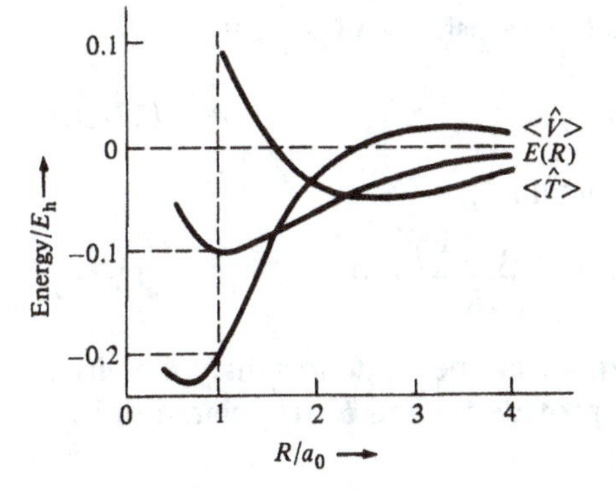

FIGURE 12-9
Total, kinetic, and potential energies of a diatomic molecule as a function of internuclear distance using Eqs. (12-58) and (12-59).

R and η may be expressed, that is,

$$R = R(\rho) \qquad \eta = \eta(\rho) \tag{12-60}$$

It is convenient to introduce the simplified notation

$$X_\rho = \frac{\partial\langle\hat{X}\rangle^{(1,\rho)}}{\partial\rho} \tag{12-61}$$

where $\hat{X}$ is either $\hat{T}$ or $\hat{V}$. Then, solving Eq. (12-54) for η one obtains

$$\eta = \eta(\rho) = -\frac{\langle\hat{V}\rangle^{(1,\rho)} + \rho V_\rho}{2\langle\hat{T}\rangle^{(1,\rho)} + \rho T_\rho} \tag{12-62}$$

The internuclear distance R is given in terms of ρ as

$$R = \rho\eta^{-1} = \rho\left(\frac{-2\langle\hat{T}\rangle^{(1,\rho)} + \rho T_\rho}{\langle\hat{V}\rangle^{(1,\rho)} + \rho V_\rho}\right) \tag{12-63}$$

From Eq. (12-63) we can obtain the inverse relationship $\rho = \rho(R)$, so that η and R can be related through Eq. (12-62) as well as through $\rho = \eta R$. This inverse relationship is most easily found by using Eq. (12-63) to plot a graph of R versus ρ.

Multiplying Eq. (12-54) by η, we obtain

$$2\eta^2\langle\hat{T}\rangle^{(1,\rho)} + \eta\langle\hat{V}\rangle^{(1,\rho)} + R(\eta^3 T_\rho + \eta^2 V_\rho) = 0 \tag{12-64}$$

Now using Eq. (12-52), we see that the first two terms in Eq. (12-64) are

$$2\langle\hat{T}\rangle^{(\eta,R)} + \langle\hat{V}\rangle^{(\eta,R)} \tag{12-65}$$

From Eq. (12-53) we see that the term in parentheses in Eq. (12-64) is

$$\eta\left[\frac{\partial E(\eta,R)}{\partial\rho}\right]_\eta = \left[\frac{\partial E(\eta,R)}{\partial R}\right]_\eta \tag{12-66}$$

Since $E(\eta, R)$ depends on η and R, we can write

$$\frac{dE(\eta,R)}{dR} = \left[\frac{\partial E(\eta,R)}{\partial R}\right]_\eta + \left[\frac{\partial E(\eta,R)}{\partial\eta}\right]_R \frac{d\eta}{dR} \tag{12-67}$$

But since $\eta R = \rho$, the second term in (12-67) vanishes and we get

$$\frac{dE(\eta,R)}{dR} = \left[\frac{\partial E(\eta,R)}{\partial R}\right]_\eta \tag{12-68}$$

Combining Eqs. (12-65) and (12-68), we obtain

$$2\langle\hat{T}\rangle^{(\eta,R)} + \langle\hat{V}\rangle^{(\eta,R)} + R\,\frac{dE(\eta,R)}{dR} = 0 \tag{12-69}$$

which shows that the trial wavefunction can always be made to satisfy the virial theorem at any internuclear separation R, provided η and R are related to Eq. (12-62) and $\rho = \eta R$.

If we are interested only in obtaining the minimum energy (for which $R = R_e$), then, from Eqs. (12-64) and (12-68)

$$\frac{dE(\eta, R)}{dR} = 0 = \eta^3 T_\rho + \eta^2 V_\rho \tag{12-70}$$

Using the above result in Eq. (12-54) (after dividing by η) and solving for η, we obtain

$$\eta = -\frac{\langle \hat{V} \rangle^{(1,\rho)}}{2\langle \hat{T} \rangle^{(1,\rho)}} \tag{12-71}$$

which is analogous to Eq. (6-114). The value of η for the minimum in the energy curve (which we will call η_e) may be obtained from Eq. (12-71) by finding the value of ρ (equal to ρ_e) which minimizes $E(\eta, R)$. Using Eqs. (12-71) and (12-53), we can write the energy as

$$E(\eta, R) = -\frac{[\langle \hat{V} \rangle^{(1,\rho)}]^2}{4\langle \hat{T} \rangle^{(1,\rho)}} \tag{12-72}$$

An alternative and equivalent procedure is to obtain ρ_e by minimizing Eq. (12-72). From this value of ρ_e, the minimum energy and η_e can be computed by using Eqs. (12-71) and (12-70), respectively. The internuclear distance at which $E(\eta, R)$ is a minimum is then given by $R_e = \rho_e \eta_e^{-1}$.

As an example of the method just discussed, we consider the simple MO treatment of the dihydrogen molecule ion obtained by scaling the wavefunction (12-31). The results for the equilibrium separation can be obtained as follows. Using Eq. (12-71) we obtain the scale factor

$$\eta_e = -\frac{-0.9475}{2(0.3827)} = 1.238 \tag{12-73}$$

which is valid for $\rho_e = 2.49\ a_0$. Thus $R_e = \rho_e/\eta_e = 2.49/1.24 = 2.01\ a_0$. The minimum energy is given by Eq. (12-72) as $-0.5864 E_\mathrm{h}$ and leads to a dissociation energy of $0.09871 E_\mathrm{h}$ (2.35 eV), over 0.5 eV higher than that found by the unscaled function. A calculation similar to this has been reported by Finkelstein and Horowitz.[15]

The virial theorem has been generalized to polyatomic molecules by Hurley.[16]

EXERCISE

12-3. Calculate the values of ΔT and ΔV predicted by the scaled wavefunction (12-31) for the dihydrogen molecule ion. Compare these qualitatively and quantitatively with the analogous results for the unscaled wavefunction and for the exact wavefunction.

---

[15] B. N. Finkelstein and G. E. Horowitz, *Z. Physik.*, **48**:118 (1928).

[16] A C. Hurley, *J. Chem. Phys.*, **37**:449 (1962).

12-5 THE HYDROGEN MOLECULE (DIHYDROGEN)

Dihydrogen, H_2, has the distinction of being the first molecule whose dissociation energy was correctly predicted by quantum mechanical calculation (1968) before this same quantity was measured reliably by experiment (1969). Over 30 years earlier (1933), laborious calculations with primitive computational facilities were the first to demonstrate that the Schrödinger equation was capable of providing quantitative information on molecules as well as on atoms.[17]

The electronic hamiltonian of dihydrogen in the Born-Oppenheimer approximation is given by

$$\hat{H} = -\frac{1}{2}\nabla_1^2 - \frac{1}{r_{A1}} - \frac{1}{r_{B1}} - \frac{1}{2}\nabla_2^2 - \frac{1}{r_{A2}} - \frac{1}{r_{B2}} + \frac{1}{r_{12}} + \frac{1}{R}$$

$$= h_1 + h_2 + \frac{1}{r_{12}} + \frac{1}{R} \tag{12-74}$$

where we use the coordinate system shown in Fig. 12-10. Note that except for the nuclear-repulsion term, the dihydrogen hamiltonian is of exactly the same general form as the helium atom hamiltonian [see Eq. (6-71)]. Thus, there will be strong parallels between the MO treatment of dihydrogen and the AO treatment of helium.

According to the orbital approximation, the simplest wavefunction for the ground state of dihydrogen is given by the single determinant

$$\psi(1,2) = |\phi\,\bar{\phi}| \tag{12-75}$$

[17] The relevant papers for these three events, respectively, are W. Kolos and L. Wolniewicz, *J. Chem. Phys.*, **49**:404 (1968); G. Herzberg, *Phys. Rev. Lett.*, **23**:1081 (1969); and H. M. James and A. S. Coolidge, *J. Chem. Phys.*, **1**:825 (1933).

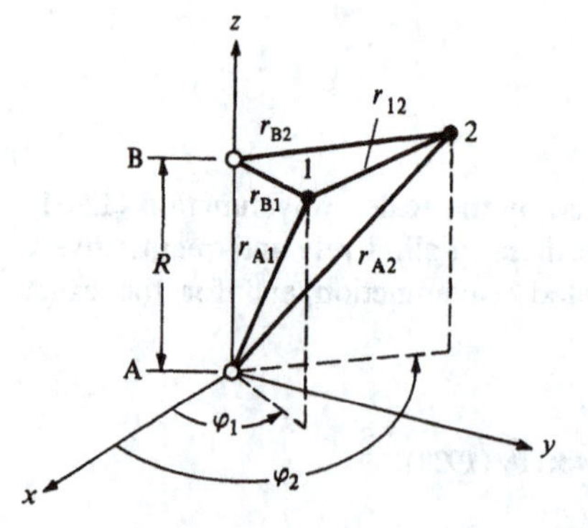

FIGURE 12-10
Coordinate system for the dihydrogen molecule.

where ϕ is a *molecular orbital*. The best MO to use in the above wavefunction is that satisfying the Hartree-Fock equations

$$\hat{F}\phi_i = \epsilon_i \phi_i \tag{12-76}$$

(or any other MOs related to those above by a unitary transformation). For notational convenience the Hartree-Fock operator may be written

$$\hat{F} = -\frac{1}{2}\nabla^2 - \sum_k \frac{Z_k}{r_k} + V^{\text{eff}} = h + V^{\text{eff}} \tag{12-77}$$

where V^{eff} is a convenient notation for the electron-repulsion portion of the Hartree-Fock operator [see Eqs. (10-16) and the following]:

$$V^{\text{eff}} = \sum_j (2J_j - K_j) \tag{12-78}$$

The term V^{eff} describes the effective or average potential experienced by a single electron while moving in the field of one or more electrons. Using the LCAO approximation with a minimal basis set, the MOs of dihydrogen may be written

$$\phi_i = c_{1i}1s_{\text{A}} + c_{2i}1s_{\text{B}} \tag{12-79}$$

where $1s_{\text{A}}$ and $1s_{\text{B}}$ are the same $1s$ AOs used previously in Eq. (12-18). Although only one MO is needed in Eq. (12-75), Roothaan's method produces extra, so-called *virtual orbitals* whenever $m > N$. Here, m (the number of basis functions) is 2, and N (the number of MOs needed) is 1. Thus $m - N = 1$ is the number of virtual orbitals obtained. Minimization of the total energy leads to the same secular equations as in H_2^+ [see Eq. (12-22)], but the secular determinant now is

$$\begin{vmatrix} F_{\text{AA}} - \epsilon & F_{\text{AB}} - \epsilon\Delta \\ F_{\text{AB}} - \epsilon\Delta & F_{\text{BB}} - \epsilon \end{vmatrix} = 0 \tag{12-80}$$

$$\begin{aligned} \text{where } F_{\text{AA}} &= F_{\text{BB}} = \langle 1s_{\text{A}}|\hat{F}|1s_{\text{A}}\rangle = \langle 1s_{\text{B}}|\hat{F}|1s_{\text{B}}\rangle \\ F_{\text{AB}} &= \langle 1s_{\text{A}}|\hat{F}|1s_{\text{B}}\rangle = \langle 1s_{\text{B}}|\hat{F}|1s_{\text{A}}\rangle = F_{\text{BA}} \\ \Delta &= \langle 1s_{\text{A}}|1s_{\text{B}}\rangle \end{aligned} \tag{12-81}$$

It is convenient to introduce the notation

$$F_{\text{AA}} = F_{\text{BB}} = a + J' \qquad F_{\text{AB}} = F_{\text{BA}} = b + J'' \tag{12-82}$$

where a and b are given by Eq. (12-25) and J' and J'' are defined by

$$\begin{aligned} J' &= \langle 1s_{\text{A}}|V^{\text{eff}}|1s_{\text{A}}\rangle = \langle 1s_{\text{B}}|V^{\text{eff}}|1s_{\text{B}}\rangle \\ J'' &= \langle 1s_{\text{A}}|V^{\text{eff}}|1s_{\text{B}}\rangle = \langle 1s_{\text{B}}|V^{\text{eff}}|1s_{\text{A}}\rangle \end{aligned} \tag{12-83}$$

The secular determinant is now written as

$$\begin{vmatrix} a + J' - \epsilon & b + J'' - \epsilon\Delta \\ b + J'' - \epsilon\Delta & a + J' - \epsilon \end{vmatrix} = 0 \tag{12-84}$$

This has the two roots

$$\epsilon_1 = \frac{a+b+J'+J''}{1+\Delta} \qquad \epsilon_2 = \frac{a-b+J'-J''}{1-\Delta} \tag{12-85}$$

These are eigenvalues of the Hartree-Fock operator, i.e., orbital energies. Substitution of ϵ_1 into the secular equations leads to

$$c_{11} = c_{21} = \frac{1}{\sqrt{2(1+\Delta)}} \tag{12-86}$$

just as for the $X^2\Sigma_g^+$ state of H_2^+. The corresponding MO is

$$\phi_1 = \frac{1}{\sqrt{2(1+\Delta)}}(1s_A + 1s_B) \tag{12-87}$$

which has the same form as the approximate H_2^+ wavefunction (12-31). The root ϵ_2 leads to

$$c_{12} = -c_{22} = \frac{1}{\sqrt{2(1-\Delta)}} \tag{12-88}$$

$$\phi_2 = \frac{1}{\sqrt{2(1-\Delta)}}(1s_A - 1s_B) \tag{12-89}$$

This has the same form as the H_2^+ excited-state wavefunction (12-32). Note that just as in H_2^+, the explicit forms of the MOs [Eqs. (12-87) and (12-89)] could have been deduced on the basis of symmetry alone [and the normalization condition given by Eq. (12-29)].

The quantities J' and J'' satisfy the relationships

$$\begin{aligned} \frac{J'+J''}{1+\Delta} &= \left\langle \phi_1(1)\phi_1(2) \left| \frac{1}{r_{12}} \right| \phi_1(1)\phi_1(2) \right\rangle = J_{11} \\ \frac{J'-J''}{1-\Delta} &= \left\langle \phi_2(1)\phi_2(2) \left| \frac{1}{r_{12}} \right| \phi_2(1)\phi_2(2) \right\rangle = J_{22} \end{aligned} \tag{12-90}$$

These are coulombic integrals analogous to $J_{1s,\,1s}$ of the helium atom but defined in terms of MOs rather than AOs. When the molecular coulombic integral J_{11} is expanded in terms of the AO basis, one obtains

$$J_{11} = \frac{1}{2(1+\Delta)^2}(\langle \mathrm{AA}|g|\mathrm{AA}\rangle + \langle \mathrm{AB}|g|\mathrm{AB}\rangle + 4\langle \mathrm{AA}|g|\mathrm{AB}\rangle + 2\langle \mathrm{AB}|g|\mathrm{BA}\rangle) \tag{12-91}$$

where $\mathrm{A} = 1s_A$
$\mathrm{B} = 1s_B$
$g = 1/r_{12}$

The integral $\langle \mathrm{AA}|g|\mathrm{AA}\rangle$ is just $J_{1s,\,1s}$ and equals $5Z/8$. The remaining quantities are two-center integrals somewhat more difficult to evaluate.

Explicit expressions for these are

$$\begin{aligned}
\langle \mathrm{AB}|g|\mathrm{AB}\rangle &= \frac{1}{R}\left[1-\left(1+\frac{11R}{8}+\frac{3R^2}{4}+\frac{R^3}{6}\right)e^{-2R}\right] \\
\langle \mathrm{AA}|g|\mathrm{AB}\rangle &= \left(R+\frac{1}{8}+\frac{5}{16R}\right)e^{-R}+\left(-\frac{1}{8}-\frac{5}{16R}\right)e^{-3R} \\
\langle \mathrm{AB}|g|\mathrm{BA}\rangle &= \frac{1}{5}\left\{-e^{-2R}\left(-\frac{25}{8}+\frac{23R}{4}+3R^2+\frac{R^3}{3}\right)\right. \\
&\quad \left.+\frac{6}{R}\left[\Delta^2(C+\ln R)+S^2E_i(-4R)-2\Delta S E_i(-2R)\right]\right\}
\end{aligned} \tag{12-92}$$

where $C=\int_0^\infty \left(\frac{1}{1+x^2}-\frac{1}{e^{-x}}\right)\frac{dx}{x}=0.57722\cdots$

$$S=e^{R}\left(1-R+\frac{R^2}{3}\right)$$

$$E_i(-nR)=\int_{-nR}^{\infty}\frac{e^{-x}}{x}\,dx \qquad n=2,4 \tag{12-93}$$

The first integral in Eqs. (12-93) is known as *Euler's constant*, and the last integral is known as an *internal logarithm* or *exponential integral*. Note that the integral S closely resembles the overlap integral Δ given by Eq. (12-33).

Since the coefficients in Eq. (12-79) depend only on symmetry, further iteration of Roothaan's equations is not necessary in this particular case. The solutions are now self-consistent with respect to the matrix $\mathbf{C}=[\mathbf{C}_1 \quad \mathbf{C}_2]$ for the particular basis chosen.

The orbital energies have the explicit forms

$$\epsilon_1=\frac{a+b}{1+\Delta}+J_{11} \qquad \epsilon_2=\frac{a-b}{1-\Delta}+J_{22} \tag{12-94}$$

Since $\epsilon_1<\epsilon_2$, the ground-state wavefunction (12-75) is

$$\psi_1=|\phi_1\,\bar{\phi}_1| \tag{12-95}$$

The energy of the ground state (labeled $X^1\Sigma_g^+$) is

$$E(X^1\Sigma_g^+)=2\epsilon_1-J_{11}+\frac{1}{R}=\frac{2(a+b)}{1+\Delta}+J_{11}+\frac{1}{R} \tag{12-96}$$

Note, in analogy to the helium case,

$$\epsilon_1^{(0)}=\langle\phi_1|h|\phi_1\rangle=\frac{a+b}{1+\Delta} \tag{12-97}$$

This illustrates that $E(H_2)$ is related to $E(H_2^+)$ in much the same way as $E(\mathrm{He})$ is related to $E(\mathrm{H})$.

When the dihydrogen calculation is carried out with unscaled AOs $(Z=1)$, D_e comes out to be 2.65 eV at $R_e=1.57\ a_0$. The corresponding experimental values are 4.75 eV and 1.4008 a_0, respectively. Introducing a scale factor (optimized to $\eta = 1.197$) leads to $D_e = 3.49$ eV at $R_e = 1.38\ a_0$. When the Hartree-Fock equations are solved using a large basis set, one obtains $D_e = 3.63$ eV at $1.40\ a_0$.[18]

The MOs ϕ_1 and ϕ_2 may be used to describe several excited states of dihydrogen. A common notation for the minimal basis set MOs is as follows:

$$\phi_1 = \sigma_g 1s \qquad \phi_2 = \sigma_u 1s \tag{12-98}$$

The σ_g and σ_u are lowercase versions of the $D_{\infty h}$ point-group symbols Σ_g^+ and Σ_u^+, respectively, and the $1s$ indicates the basis AOs used to approximate the MOs. In the method by which the MO approximation is generally carried out, the MOs belong to irreducible representations of the point group to which the molecule belongs, and the symmetry of an electronic state is the direct product of the MOs used to describe it. Thus, the ground state of dihydrogen has the electron configuration $(\sigma_g 1s)^2$, and the symmetry of this state is $\Sigma_g^+ \otimes \Sigma_g^+ = \Sigma_g^+$. With the multiplicity and the ground-state prefix added, the complete spectroscopic term symbol is $X^1\Sigma_g^+$ (see Sec. 11-3). The electron configuration $(\sigma_g 1s)\ (\sigma_u 1s)$ of dihydrogen is analogous to $1s2s$ of helium; this leads to the states $B^1\Sigma_u^+$ and $b^3\Sigma_u^+$ having the wavefunctions

$$\psi(B^1\Sigma_u^+) = \frac{1}{\sqrt{2}}\left(|\sigma_g 1s\, \overline{\sigma_u 1s}| - |\overline{\sigma_g 1s}\, \sigma_u 1s|\right) \tag{12-99}$$

$$\psi(b^3\Sigma_u^+) = \begin{cases} \dfrac{1}{\sqrt{2}}\left(|\sigma_g 1s\, \overline{\sigma_u 1s}| + |\overline{\sigma_g 1s}\, \sigma_u 1s|\right) & M_S = 0 \\ |\sigma_g 1s\, \sigma_u 1s| & M_S = 1 \\ |\overline{\sigma_g 1s}\, \overline{\sigma_u 1s}| & M_S = -1 \end{cases} \tag{12-100}$$

Another possible state is a ${}^1\Sigma_g^+$ excited state with the configuration $(\sigma_u 1s)^2$ and the wavefunction

$$\psi({}^1\Sigma_g^+) = |\sigma_u 1s\, \overline{\sigma_u 1s}| \tag{12-101}$$

This latter state is unstable (shows no minimum). This configuration is analogous to $2s^2$ of helium.

Figure 12-11 depicts some of the electronic states of dihydrogen and compares their energies with those of the corresponding molecular ion and some of its components. Figure 12-12 shows potential-energy curves for the ground state and three excited states of dihydrogen. A different depiction of the electronic structure of dihydrogen is provided by Fig. 12-13. This figure shows the $1\sigma_g$ and $1\sigma_u$ MOs of dihydrogen by means of contour lines which represent constant electron charge densities; i.e., each line (solid or dashed) represents a region in three-dimensional space along which the electron charge

[18] An SCF calculation equivalent to using a large basis set has been carried out by J. Goodisman, *J. Chem. Phys.*, **39**:2397 (1963).

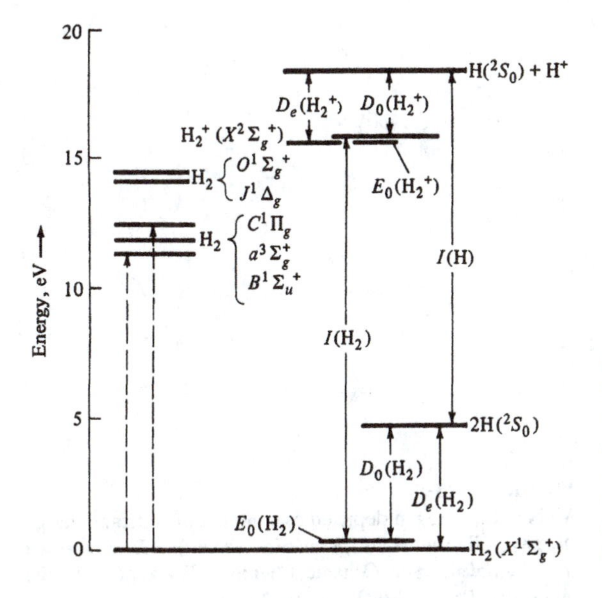

FIGURE 12-11
Some stable spectroscopic states of the dihydrogen molecule. The dashed arrows show two symmetry- and spin-allowed electronic transitions. The diagram also shows relative energies of H_2^+ and some of its fragments.

density is equal to 0.01 (in units of e/a_0^3). Such lines have been spaced to be approximately 0.4 a_0 apart and coincide with the symmetry planes of the molecule. Solid lines represent regions where the MO is positive (or negative), and dashed lines indicate regions where the MO has the opposite sign. Thus the $1\sigma_g$ MO resembles an ellipsoid surrounding the two hydrogen nuclei, and each point on the implied surface has a charge density of 0.01 e/a_0^3. We will find that such diagrams are very useful in comparing the electronic structures of isoelectronic species.

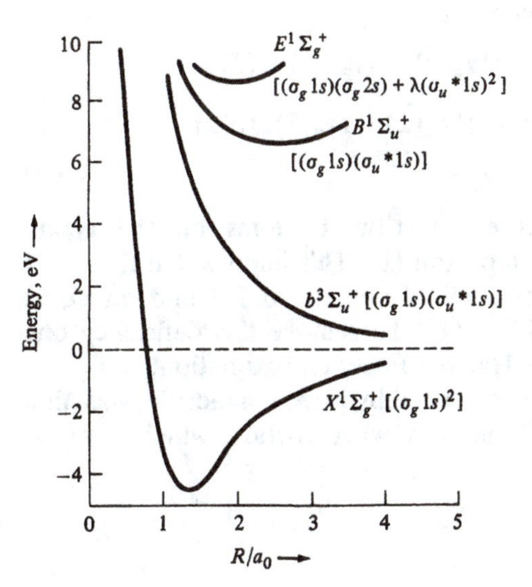

FIGURE 12-12
Potential-energy curves for the ground state and three excited states of dihydrogen. The brackets after each spectroscopic term symbol indicate the MO configuration which serves as a first approximation to these states. [*From W. Kolos and C. C. J. Roothaan*, Rev. Mod. Phys. *32:227* (*1960*). *Reproduced by permission of the authors.*]

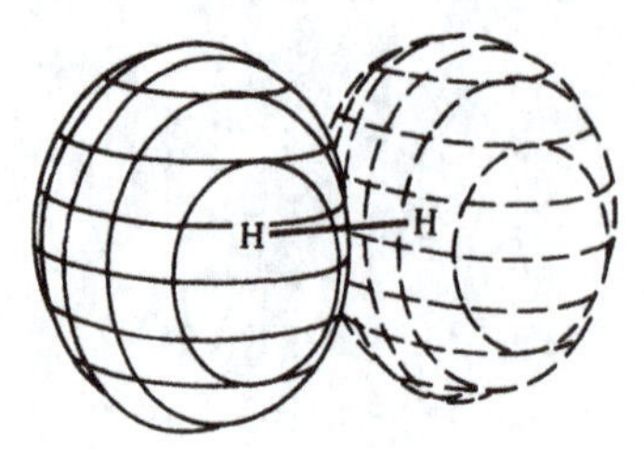

$\epsilon_{1\sigma_u} = 0.2656\ E_h$

$\epsilon_{1\sigma_g} = -0.5944\ E_h$

FIGURE 12-13
MOs of dihydrogen depicted as contours of constant charge density. (*Reprinted by permission from W. L. Jorgensen and L. Salem*, The Organic Chemist's Book of Orbitals, *Academic Press, New York, 1973.*)

The dihydrogen wavefunction $|\sigma_g 1s\ \overline{\sigma_g 1s}|$ contains a serious deficiency which is shared by Hartree-Fock wavefunctions of other molecules at all levels of accuracy. The wavefunction of a molecule AB should reduce to the wavefunctions of the separated atoms A and B as the internuclear distance R approaches infinity, that is,

$$\lim_{R\to\infty} \psi_{AB} = \psi_A + \psi_B \tag{12-102}$$

In the case of the ground-state wavefunction, as $R \to \infty$, the overlap integral goes to zero and the wavefunction becomes

$$\begin{aligned} \lim_{R\to\infty} |\sigma_g 1s\ \overline{\sigma_g 1s}| &= [1s_A(1)1s_B(2) + 1s_B(1)1s_A(2)] \\ &\quad + [1s_A(1)1s_A(2) + 1s_B(1)1s_B(2)] \\ &= \psi_{cov} + \psi_{ionic} \end{aligned} \tag{12-103}$$

The first term in square brackets represents just two H atoms, but the second term represents a hydride ion H^- and a proton H^+. This implies that there is a 50 percent probability that dihydrogen will dissociate to 2H and an equal probability that it will dissociate to $H^- + H^+$. To remove this deficiency one must go to wavefunctions beyond the Hartree-Fock approximation.

The exact nonrelativistic energy of the dihydrogen molecule was first calculated by James and Coolidge,[19] using a wavefunction which may be

---

[19] H. M. James and A. S. Coolidge, loc. cit.

regarded as an extension of the Hylleraas function for the helium atom (see Sec. 6-4). The trial wavefunction had the general form

$$\xi = \sum_i c_i \psi_i \tag{12-104}$$

where

$$\psi_i = e^{-\alpha(\mu_1+\mu_2)}(\mu_1^{p_i}\mu_2^{q_i}\nu_1^{r_i}\nu_2^{s_i}r_{12}^{t_i} + \mu_1^{q_i}\mu_2^{p_i}\nu_1^{s_i}\nu_2^{r_i}r_{12}^{t_i}) \tag{12-105}$$

with the requirement that $r_i + s_i$ be even (so as to lead to Σ_g^+ symmetry). The original calculation, in which 13 terms were used, has been extended to 50 terms by Kolos and Roothaan,[20] to obtain $D_e = 4.7467$ eV at $R_e = 1.4\ a_0$. Use of these latter calculations along with the SCF calculations of Goodisman quoted previously permit a calculation of the correlation energy of dihydrogen, namely, $-(4.75 - 3.63) = -1.12$ eV. This is just about the same value obtained for a two-electron atom (see Sec. 10-4). Kolos and Roothaan also pointed out that the correlation energy of dihydrogen increases as the internuclear distance R increases; this incorrect behavior is due to the incorrect dissociative behavior of the Hartree-Fock wavefunction.

Beginning in the early 1960s Kolos and coworkers began a series of calculations on dihydrogen within the framework of the adiabatic approximation (see Sec. 11-1). As the calculations became more accurate, it became evident that the theoretical dissociation energy of dihydrogen was higher than the experimentally accepted value, a situation which indicated that either the experimental value was incorrect or there was an error in the theoretical calculations. Intensive efforts to refine the theoretical calculations culminated in a classic paper by Kolos and Wolniewicz[21] which led to a final resolution of the uncertainty. These two workers constructed a potential-energy curve for the $X^1\Sigma_g^+$ state of dihydrogen for internuclear distances in the range $1 \leq R \leq 3.2\ a_0$ using double-precision arithmetic and a 100-term expansion for the electronic wavefunction. The vibrational equation was solved for all H isotopes, and the rotational levels $J = 0$ to 10 were included. The calculated adiabatic dissociation energy, D_0, was 36,117.3 cm^{-1} (4.4780 eV) when corrected for relativistic and radiative effects. At this time the accepted experimental value for D_0 was $36{,}113.6 \pm 0.03$ cm^{-1} (4.4776 eV). Thus the theoretical value was 3.7 cm^{-1} (0.0004 eV) *higher* than the experimental one. Accordingly, the experimental determination of the dihydrogen dissociation energy was reinvestigated by Herzberg.[22] This work revealed that the experimental value was indeed in error; the corrected experimental value was in fact about 1 cm^{-1} *larger* than the best theoretical value!

[20] W. Kolos and C. C. J. Roothaan, *Rev. Mod. Phys.*, **32**:219 (1960).

[21] W. Kolos and L. Wolniewicz, loc. cit.

[22] G. Herzberg, *Phys. Rev. Lett.*, **23**:1081 (1969).

EXERCISES

12-4. Work out energy expressions (in terms of $a, b, \Delta, R, J_{11}, J_{22}$, etc.) for each of the dihydrogen excited states arising from the $\sigma_g 1s$ and $\sigma_u 1s$ MOs.

12-5. Show that the ground-state wavefunction of dihydrogen (in the minimal basis Hartree-Fock form) may be written as $\psi(1,2) = [2(1+\Delta^2)]^{-1/2}(\psi_{\text{cov}} + \psi_{\text{ionic}})$, where ψ_{cov} and ψ_{ionic} are defined in Eq. (12-103).

12-6. Referring to Eq. (12-103), show that $[2(1+\Delta^2)]^{-1}\langle \psi_{\text{cov}} | \psi_{\text{ionic}} \rangle = 2\Delta/(1+\Delta^2)$. Using $R_e = 1.4\ a_0$ and Eq. (12-33) for Δ, evaluate the above and explain what it means physically. Is it meaningful to use the MO wavefunction of dihydrogen to talk about the "covalent" and "ionic" natures of the chemical bond? [See H. Shull, *J. Appl. Phys.*, **33**:290 (1962).]

12-7. The exact SCF value of ϵ_1, the energy of the MO ϕ_1, is $-0.59465E_{\text{h}}$. The coulombic integral J_{11} has the value $0.64797E_{\text{h}}$. Calculate:
(*a*) The estimated first ionization energy of dihydrogen by Koopmans' theorem (experimental value is 15.4 eV)
(*b*) The dissociation energy of dihydrogen at $R_e = 1.4008\ a_0$

12-8 Write in explicit form the first-order density matrix of each of the following:
(*a*) The $X^2\Sigma_g^+$ and $^2\Sigma_u^+$ states of H_2^+ (in the minimal basis set)
(*b*) The $X^1\Sigma_g^+$ state of dihydrogen (in the minimal basis set)
What is the physical significance of each term in each density matrix in part (*a*)? How does the $X^2\Sigma_g^+$ density matrix of H_2^+ differ from the $X^1\Sigma_g^+$ density matrix of dihydrogen?

12-9. Examine the simple MO wavefunction for the $B^1\Sigma_u^+$ state of dihydrogen. What does each term in the wavefunction represent?

12-6 THE AUFBAU PRINCIPLE FOR HOMONUCLEAR DIATOMIC MOLECULES

If the basis set used to expand the MOs of a homonuclear diatomic molecule AB consists of AOs of the atoms A and B, then it is convenient to combine these AOs to form symmetry-adapted functions of the general form $N(\phi_A \pm \phi_B)$. Linear combinations of all the symmetry-adapted functions of a given symmetry type will then produce MOs of that same symmetry type. For example, the symmetry-adapted functions of Σ_g^+ symmetry are as follows:

$$\begin{aligned} \sigma_g ns &= N(ns_A + ns_B) \qquad && n = 1, 2, \ldots \\ \sigma_g np &= N(np_{zA} + np_{zB}) \qquad && n = 2, 3, \ldots \ (z \text{ is the bond axis}) \\ \sigma_g nd &= N(nd_{z^2A} + nd_{z^2B}) \qquad && n = 3, 4, \ldots \end{aligned} \tag{12-106}$$

with similar relationships for *f*, *g*, . . . AOs if desired. The quantity *N* appearing in each symmetry-adapted function is often chosen as $1/\sqrt{2}$, but its actual value at this stage is immaterial, since it eventually becomes imbedded in the MO coefficients when these are normalized. The symmetry-adapted functions of Σ_u^+ symmetry are constructed from the same AOs as those of Σ_g^+ symmetry, but the coefficient of the second AO in each combination (the *B* AO) changes sign. For example,

$$\sigma_u ns = N(ns_{\text{A}} - ns_{\text{B}}) \qquad n = 1, 2, \ldots \tag{12-107}$$

The symmetry-adapted functions of Π_u symmetry are

$$\begin{aligned}\pi_u np &= \begin{cases} N(np_{x\text{A}} + np_{x\text{B}}) \\ N(np_{y\text{A}} + np_{y\text{B}}) \end{cases} \qquad n = 2, 3, \ldots \\ \pi_u nd &= \begin{cases} N(nd_{xz\text{A}} + nd_{xz\text{B}}) \\ N(nd_{yz\text{A}} + nd_{yz\text{B}}) \end{cases} \qquad n = 3, 4, \ldots \end{aligned} \tag{12-108}$$

with similar relationships for $f, g, \ldots$ functions if desired. The symmetry-adapted functions of Π_g symmetry are analogous, but each B AO enters in with a negative coefficient. Similarly, symmetry-adapted functions of Δ_g and Δ_u symmetry are formed from nd_{xy}, $nd_{x^2-y^2}$, and higher orbitals of atoms A and B.

The MOs are formed by taking linear combinations of the appropriate symmetry-adapted functions. For example, if one takes a linear combination of the Σ_g^+ symmetry-adapted functions (j in number), one obtains the j MOs

$$n\sigma_g = \sum_{n=1}^{j} \chi_n(\sigma_g) c_{nj} \tag{12-109}$$

where $\chi_n(\sigma_g)$ is a particular symmetry-adapted function of Σ_g^+ symmetry. By convention, the MO of lowest energy is labeled $1\sigma_g$, the next to the lowest is labeled $2\sigma_g$, and so on up to the highest energy MO. In analogous fashion, the Σ_u^+ MOs are given by

$$n\sigma_u = \sum_{n=1}^{j} \chi_n(\sigma_u) c_{nj} \tag{12-110}$$

These are also labeled in order of increasing energy: $1\sigma_u$ for the lowest, etc. Similarly, the Π_g, Π_u, Δ_g, and Δ_u MOs are given by

$$\begin{aligned} n\pi_g &= \sum_{n=1}^{k} \chi_n(\pi_g) c_{nk} \rightarrow 1\pi_g, 2\pi_g, \ldots, k\pi_g \\ n\pi_u &= \sum_{n=1}^{k} \chi_n(\pi_u) c_{nk} \rightarrow 1\pi_u, 2\pi_u, \ldots, k\pi_u \\ n\delta_g &= \sum_{n=1}^{l} \chi_n(\delta_g) c_{nl} \rightarrow 1\delta_g, 2\delta_g, \ldots, l\delta_g \\ n\delta_u &= \sum_{n=1}^{l} \chi_n(\delta_u) c_{nl} \rightarrow 1\delta_u, 2\delta_u, \ldots, l\delta_u \end{aligned} \tag{12-111}$$

Note that all σ MOs are nondegenerate but that the remaining MOs (π, δ, $\phi, \ldots$) are doubly degenerate. Thus for the latter MOs one will always obtain two sets with the same coefficients.

When a minimal basis set of m AOs is used in the LCAO approximation ($m > N$; $2N$ is the number of electrons, and thus N is the number of doubly occupied orbitals needed to describe the molecule), the matrix representation of the Hartree-Fock operator is an $m \times m$ matrix. Since all the symmetry operators of a molecule commute with the electronic hamiltonian, Theorem 2-5 can be used to factor the secular determinant into smaller determinants. For example, the minimal basis set for N_2 consists of the AOs $1s$, $2s$, $2p_x$, $2p_y$, and $2p_z$ (once for each atom), so that $m = 10$ and $N = 7$ (half the total number of electrons in N_2). This leads to a 10×10 matrix representation of the Hartree-Fock operator and produces 10 MOs. Of these, seven are needed to describe the ground state of the molecule. The $1s$, $2s$, and $2p_z$ AOs will form Σ_g^+ and Σ_u^+ MOs (three of each) and thus will appear in two different 3×3 matrices. Similarly, the $2p_x$ and $2p_y$ AOs occur only in Π_u and Π_g MOs and produce two more 2×2 matrices. Note also that since each 2×2 matrix contains a pair of doubly degenerate MOs, the two roots will be equal. Consequently, the original 10×10 matrix factors into two 3×3 and two 2×2 determinants. The corresponding secular determinant will then produce eight distinct orbital energy values (two roots are repeated in the Π_u and Π_g determinants).

The first SCF MO calculation on dinitrogen employing a minimal basis set of STOs was carried out by Scherr in 1955.[23] The calculation used two different scale parameters in the AOs: one for the $1s$ orbitals and a second for the $2s$ and $2p$ orbitals. The scale parameters were not optimized but were chosen by use of Slater's empirical rules (see Sec. 12-2). The calculations were carried out at the experimental value of the internuclear distance, 2.0675 a_0. The total molecular energy obtained was $-108.574E_h$. Since the scale parameters were not optimized, the virial theorem was not necessarily satisfied; nevertheless, a value of -1.9964 for $\langle \hat{V} \rangle / \langle \hat{T} \rangle$ was obtained, indicating that optimizing the parameters would not improve the energy significantly.[24]

The dissociation energy of a homonuclear diatomic molecule A_2 is given by

$$D_e = 2E_A - E(A_2) \qquad (12\text{-}112)$$

If one employs the exact atomic energies in the above calculation, the SCF MO method generally fails to predict a positive value for the molecular dissociation energy. For example, the experimental energy of a nitrogen atom is $-54.627E_h$. Thus, Scherr's calculation leads to the dissociation energy

$$D_e = -2(54.627) - (-108.574) = -0.68E_h \text{ or } -18.5 \text{ eV} \qquad (12\text{-}113)$$

[23] C. W. Scherr, *J. Chem. Phys.*, **23**:569 (1955).

[24] The scaling parameters were fully optimized by B. J. Ransil, *Rev. Mod. Phys.*, **32**:245 (1960); the total molecular energy estimate was lowered to $-108.63359E_h$, an improvement of about $0.06E_h$.

This is in sharp disagreement with the experimental value of 9.902 eV (a *positive* quantity). However, it can be argued that it is improper to do the above calculation using *experimental* values for the atomic energies but *theoretical* values for the molecular energy; more reasonable would be theoretical values for both quantities—both calculated with comparable methods. Unfortunately, it is difficult to define just what is meant by *comparable* methods in a case such as this. For example, it is apparent that the above calculation would lead to the exact dissociation energy if a sufficiently *poor* value of E_A were used.

One of the major difficulties with a calculation such as Scherr's is that STO basis functions lead to integrals which are computationally awkward and time-consuming to evaluate, and this problem becomes especially pronounced when the molecule contains more than two nuclei. For example, the electrostatic-repulsion integral

$$\langle \phi_r(1)\phi_s(2)|g|\phi_t(1)\phi_u(2)\rangle = \langle rs|g|tu\rangle \qquad (12\text{-}114)$$

may have up to four different atomic centers (r, s, t, and u). When there are only two different centers, the numerical evaluation of the integrals is relatively easy, but when there are three or four different centers, it becomes very time-consuming to compute large numbers of these. Consequently, a number of different strategies have been developed to improve computational time. Many of these are based on certain convenient properties of gaussian functions. Although using gaussian functions as direct replacements of STOs leads to poor energy estimates, there are indirect ways of using them which retain the good features of STOs but avoid the latter's computational awkwardness. The main gaussian property leading to this fortunate circumstance is the fact that a product of two gaussian functions, each on a different center, is a new gaussian function centered at a single point. This property may be illustrated mathematically as follows: Let a general gaussian function be represented by

$$g(l, \alpha, A) = f(l)e^{-\alpha r_A^2} \qquad f(0) = 1 \qquad (12\text{-}115)$$

where l = azimuthal quantum number
α = scale factor
A = atomic center of the function

The product of two such functions centered on two different nuclei, A and B, is given by

$$g(l_1, \alpha, A)g(l_2, \beta, B) = K(l_1, l_2)g(0, \gamma, P) \qquad (12\text{-}116)$$

where $\gamma = \alpha + \beta$ and P is the new single center replacing the two different centers A and B. For s-type gaussians the factor $K(l_1, l_2)$ reduces to

$$K(0, 0) = e^{-\alpha\beta P^2/\gamma} \qquad (12\text{-}117)$$

where

$$P^2 = (A_x - B_x)^2 + (A_y - B_y)^2 + (A_z - B_z)^2$$
$$P_x = \frac{\alpha A_x + \beta B_x}{\gamma} \quad \text{etc.} \tag{12-118}$$

(P_x is the x coordinate of the point P, etc.) Consequently, if one nominally employs an STO basis but represents each STO as a linear combination of gaussian functions (e.g., as in the STO-NG method discussed in Sec. 11-4), one obtains the best of two worlds (the efficiency of STOs as basis functions *and* the computational efficiency of gaussians), provided the number of gaussians per STO (the value of N) is large enough. In general, $N = 3$ leads to an energy value which is significantly higher than obtained by the straight STO calculation, whereas $N = 6$ leads to an energy value that is almost as good. For example, repetition of Scherr's calculation with an STO-6G basis[25] leads to a molecular energy of $-108.541 E_h$ as compared with the Scherr value of $-108.574 E_h$ (both calculations were made using the experimental bond distance of 2.0675 a_0).

Figure 12-14 illustrates the charge density contours of the dinitrogen MOs; these are based on the energy calculations of Ransil.[26] Note that these calculations—in contrast with the Scherr and STO-6G calculations—show that the $1\pi_u$ MO is slightly higher in energy than is the $3\sigma_g$ MO.

The true Hartree-Fock energy of dinitrogen is obtained by employing a considerably extended basis set of functions. A much-improved SCF MO energy of dinitrogen is obtained by using contracted gaussian-type functions defined by [see Eq. (12-7b)]

$$\psi_i = N_i \sum_j g_j c_{ji} \tag{12-119}$$

where the $\{g_j\}$ are (primitive) gaussian functions. In a series of calculations on dinitrogen, Dunning[27] showed that using a contracted gaussian-type function basis set [4s3p] led to an energy of $-108.8887 E_h$ (an improvement of about $0.314 E_h$ over the Scherr calculation). This calculation also showed that the $1\pi_u$ MO is higher in energy than the $3\sigma_g$ MO. Interestingly, this contracted gaussian basis led to a better energy value than did using the primitive (4s3p) basis of STOs.[28] Dunning also showed that the [4s3p] contracted gaussian basis was optimum; going to a [9s5p] contracted basis led to negligible improvement in the energy.

---

[25] Calculated by the author using GAMESS, version 1.02.

[26] Ransil, loc. cit.

[27] T. H. Dunning, *J. Chem. Phys.*, 53:2823 (1970).

[28] Dunning's notation, which we will adopt, is to use parentheses to denote a primitive basis and square brackets to denote a contracted basis.

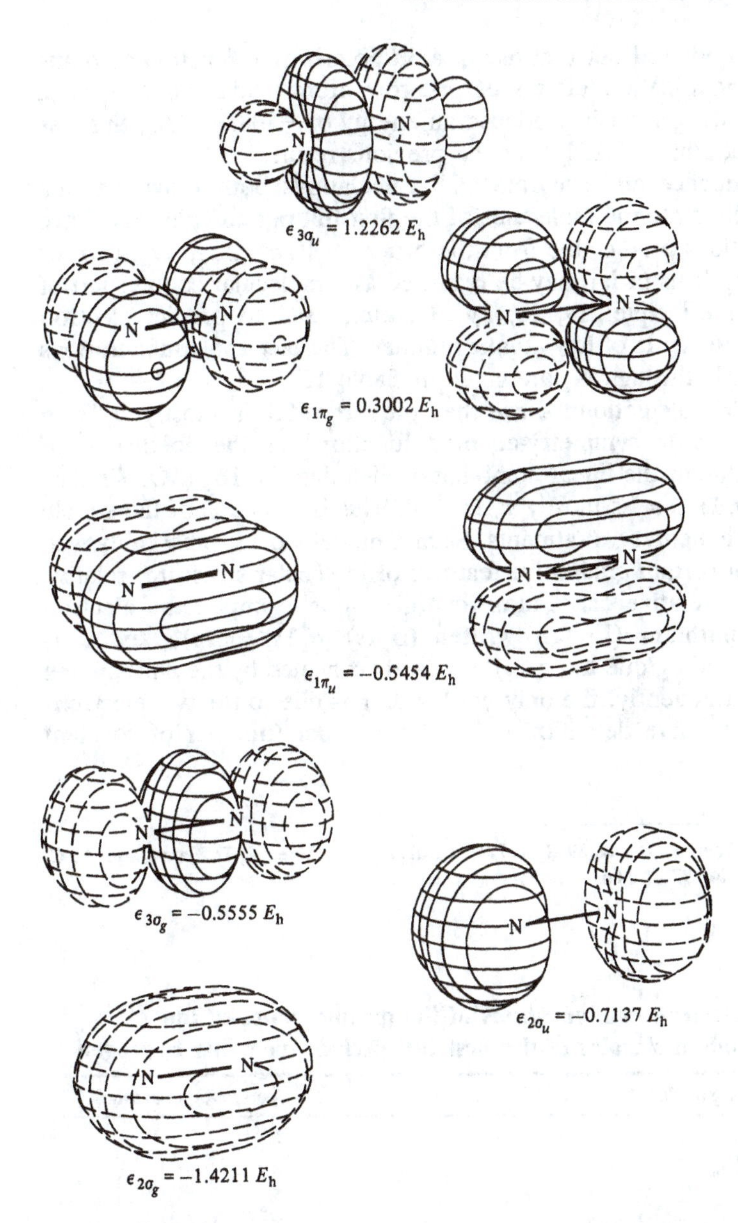

FIGURE 12-14
Contours of constant charge denisty for the MOs of dinitrogen based on calculations by B. J. Ransil. (*Reprinted, by permission, from W. L. Jorgensen and L. Salem*, The Organic Chemist's Book of Orbitals, *Academic Press, New York, 1973.*)

Dunning also pointed out that use of a $3d$ polarization function accounts for most of the polarization effects of first-row atoms; addition of such a function to the dinitrogen basis produced an energy of $-108.9732E_h$; this was an improvement of almost $0.1E_h$ over the previous result.[29]

Empirical evidence and accurate SCF calculations both show that the ground states of the diatomic molecules of the first full period (plus H_2) have electron configurations predictable from the order $1\sigma_g$, $1\sigma_u$, $2\sigma_g$, $2\sigma_u$, ($3\sigma_g$ or $1\pi_u$), $1\pi_g$, and $3\sigma_u$. This order may be regarded as a molecular counterpart of the atomic aufbau principle: $1s$, $2s$, $2p$, $3s$, etc., used to predict electron configurations for elements of low atomic number. The electron configurations for the molecules H_2 through N_2 are given in Table 12-2.

Accurate SCF calculations show that the $1\sigma_g$ MO is mostly $\sigma_g 1s$ in character, i.e., the $\sigma_g 1s$ symmetry-adapted function has the coefficient of largest absolute value in the linear combination defining the $1\sigma_g$ MO. Similarly, $1\sigma_u$ is mostly $\sigma_u 1s$, $1\pi_u$ is mostly $\pi_u 2p$, etc. This is the basis of the simple model utilized in elementary treatments. Such a model works remarkably well in the prediction of certain qualitative features of molecular structure: relative bond distances and relative dissociation energies. For example, the electron configuration of dilithium (Li_2) is written $(\sigma_g 1s)^2(\sigma_u 1s)^2(\sigma_g 2s)^2$, and it is assumed that the bonding due to $(\sigma_g 1s)^2$ is exactly canceled by the antibonding due to $(\sigma_u 1s)^2$. Consequently, the only net bonding is due to the two electrons in the $\sigma_g 2s$ MO. A naive definition of the bond order (number of covalent

[29] T. H. Dunning, *J. Chem. Phys.*, **55**:3958 (1971); see also P. E. Cade, K. D. Sales, and A. C. Wahl, *J. Chem. Phys.*, **44**:1973 (1966).

TABLE 12-2
Molecular orbital electron configurations of the ground states of the homonuclear diatomic molecules of the first full period, including hydrogen

Molecule	MO configuration	Spectroscopic symbol
H_2	$(1\sigma_g)^2$	$^1\Sigma_g^+$
He_2	$(1\sigma_g)^2(1\sigma_u)^2$	$^1\Sigma_g^+$
Li_2	$(1\sigma_g)^2(1\sigma_u)^2(2\sigma_g)^2$	$^1\Sigma_g^+$
Be_2	$(1\sigma_g)^2(1\sigma_u)^2(2\sigma_g)^2(2\sigma_u)^2$	$^1\Sigma_g^+$
B_2	$(1\sigma_g)^2(1\sigma_u)^2(2\sigma_g)^2(2\sigma_u)^2(1\pi_u)^2$	$^3\Sigma_g^-$
C_2	$(1\sigma_g)^2(1\sigma_u)^2(2\sigma_g)^2(2\sigma_u)^2(1\pi_u)^4$	$^1\Sigma_g^+$
N_2	$(1\sigma_g)^2(1\sigma_u)^2(2\sigma_g)^2(2\sigma_u)^2(3\sigma_g)^2(1\pi_u)^4$	$^1\Sigma_g^+$
O_2	$(1\sigma_g)^2(1\sigma_u)^2(2\sigma_g)^2(2\sigma_u)^2(3\sigma_g)^2(1\pi_u)^4(1\pi_g)^2$	$^3\Sigma_g^-$
F_2	$(1\sigma_g)^2(1\sigma_u)^2(2\sigma_g)^2(2\sigma_u)^2(1\pi_u)^4(3\sigma_g)^2(1\pi_g)^4$	$^1\Sigma_g^+$
Ne_2	$(1\sigma_g)^2(1\sigma_u)^2(2\sigma_g)^2(2\sigma_u)^2(1\pi_u)^4(3\sigma_g)^2(1\pi_g)^4(3\sigma_u)^2$	$^1\Sigma_g^+$

Note: The MOs are listed in order of increasing energy. Note that F_2 has $1\pi_u$ and $3\sigma_g$ in the opposite order to O_2 and N_2.

bonds) is given by

$$P = \frac{n - n^*}{2} \tag{12-120}$$

where n is the number of electrons in bonding MOs and n^* is the number of electrons in antibonding MOs. Note that all bonding MOs have both coefficients positive and that all antibonding MOs have their two coefficients opposite in sign. Thus, dilithium has the bond order

$$P = \frac{4 - 2}{2} = 1 \tag{12-121}$$

It is further assumed that unless the bond order P is greater than zero, the molecule is unstable (and thus more readily dissociates to atoms); i.e., the process

$$A_2(g) \rightarrow 2A(g)$$

will have a negative ΔG at ordinary temperatures (i.e., around 298 K). Thus, He_2, Be_2, and Ne_2 all have $P = 0$ and, in fact, are not known as stable molecules. Similarly, dioxygen (O_2) has $P = 2$ (a double bond), and dinitrogen has $P = 3$ (a triple bond). Also, dicarbon (C_2) has $P = 2$; both bonds are what are usually called *pi bonds*. Interestingly, diboron (B_2) has $P = 1$, and this single bond is a pi bond.

Figure 12-15 shows surface plots of the $\pi_u 2p$ and $\pi_g 2p$ symmetry-adapted functions. Electron densities arising from $\sigma_g 1s$ and $\sigma_u 1s$ functions are essentially those already depicted in Fig. 12-7. Insofar as general features are concerned, the densities represented by the *full* MO expansions (for example, $1\sigma_g$ as opposed to just its $\sigma_g 1s$ component) are quite similar to those exhibited by the truncated versions. Note that σ_g MOs show an increase of electron density along the internuclear axis whereas π_u MOs show the increase in an annular region about this axis.

Figure 12-16 shows surface plots for $\sigma_g 2p$ and $\sigma_u 2p$ symmetry-adapted MOs and their corresponding electron densities. The $\sigma_g 2p$ MO is involved in the triple bond of N_2 ($\sigma\pi^2$), the double bond of O_2 ($\sigma\pi$) and the single bond of F_2.

All the molecules listed in Table 12-2, except B_2 and O_2, have closed-shell configurations and, consequently, are ${}^1\Sigma_g^+$ states. The spectroscopic designations of B_2 and O_2 are determined as follows (using B_2 as a specific example): Considering only that part of the MO configuration beyond the closed shell [i.e., the $(1\pi_u)^2$ portion], we note that λ (the component of angular momentum about the bond axis) is ± 1 and thus the total angular momentum about the bond axis is either 0 or ± 2. This implies Σ and Δ states.[30]

[30] Note that the quantum number λ is analogous to m_l of atoms and that Λ is analogous to M_L.

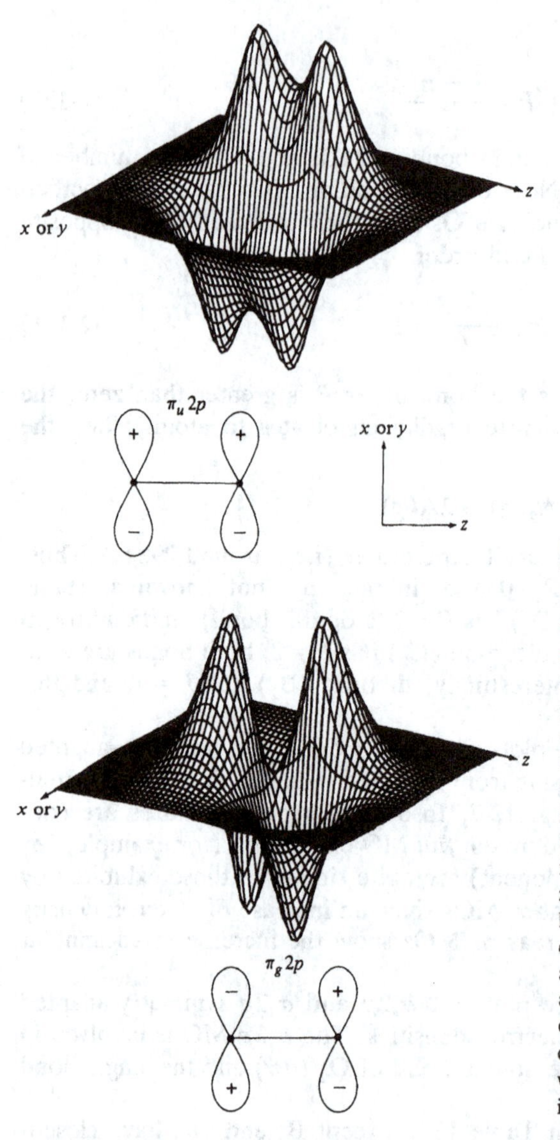

FIGURE 12-15
The $\pi_u 2p$ (bonding) and $\pi_g 2p$ (antibonding) symmetry-adapted functions. The squares (probability densities) associated with these are not shown, but both would look like four peaks above the nuclear planes indicated. However, $(\pi_u 2p)^2$ would exhibit internuclear buildup, and $(\pi_g 2p)^2$ would exhibit the opposite. The former is an example of a pi bond; such a bond has zero probability along the internuclear axis itself but has a nonzero density about this axis. Since a homonuclear diatomic molecule has cylindrical symmetry about the bond axis, the probability density $(\pi_u 2p)^2$ also has this cylindrical symmetry. The diagrams below the surface plots indicate the orientation of the $2p$ AOs which overlap to form the corresponding surface plots.

Table 12-3, which is analogous to Table 8-1 for the sp^2 part of the carbon atom electron configuration, shows the different ways two electrons can go into π_u MOs without violating the Pauli exclusion principle. The assignment of ${}^1\Sigma^+$ and ${}^3\Sigma^-$ is based on the behavior of the wavefunction under the σ_v operation.[31] This operator has the property

$$\hat{\sigma}_v f(\varphi) = f(-\varphi) \tag{12-122}$$

[31] F. L. Pilar, *J. Chem. Ed.*, **58**:758 (1981).

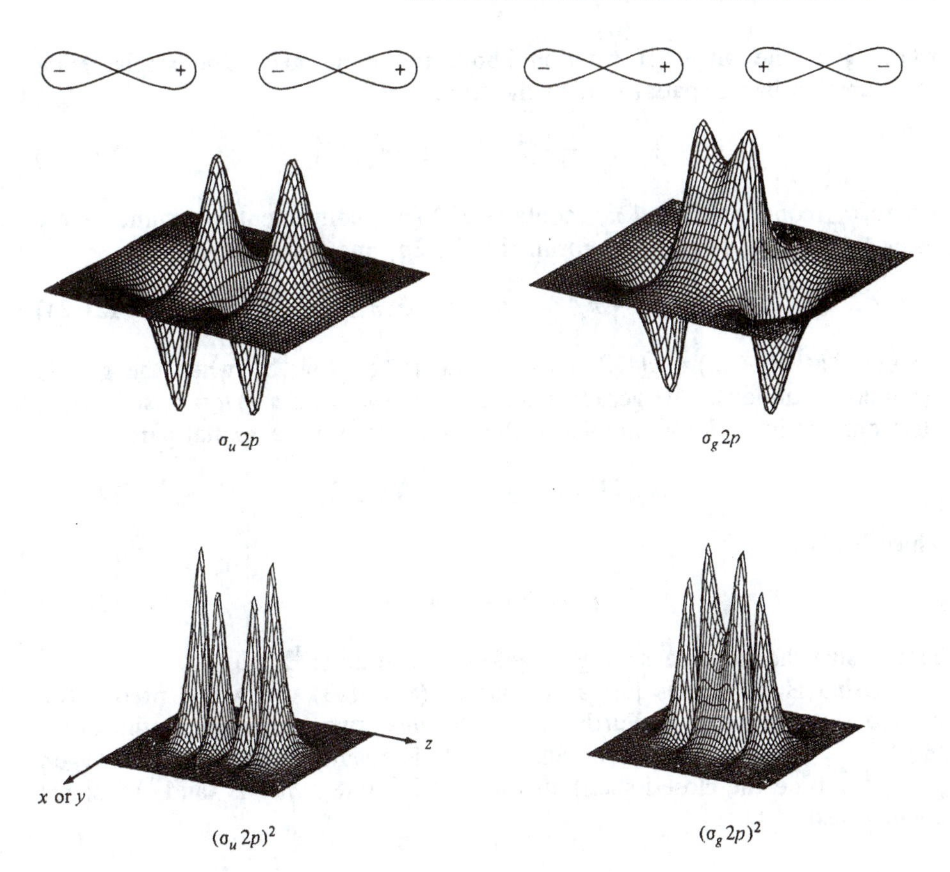

FIGURE 12-16
Surface plots of $\sigma_g 2p$ and $\sigma_u 2p$ symmetry-adapted functions and their squares for a diatomic molecule. The small diagrams at the top indicate how the $2p$ AOs making up these functions are overlapped. Note that the $\sigma_u 2p$ function shows an increase of probability density outside the internuclear region and a decrease of probability density (antibonding) inside the internuclear region whereas the $\sigma_g 2p$ function shows the opposite (bonding) effect.

TABLE 12-3
Spatial MO-spin function combinations allowed for a wavefunction containing a $(1\pi_u)^2$ open-shell configuration

	λ		$\lvert\Lambda\rvert$	State
	−1	+1		
1	↑↓	—	2	$^1\Delta_g$
2	—	↑↓	2	
3	↑	↓	0	$^1\Sigma_g^+$
4	↑	↑	0	$^3\Sigma_g^-$

where φ is the angle of rotation about the bond axis. The singlet state wavefunction has a spatial part of the form

$$\pi_{g+}(1)\pi_{g-}(2) + \pi_{g-}(1)\pi_{g+}(2) \tag{12-123}$$

where π_{g+} contains $e^{i\varphi}$ and π_{g-} contains $e^{-i\varphi}$ (it is convenient to assume we are using the orbitals in complex form, that is, $2p_1$ and $2p_{-1}$). Since

$$\hat{\sigma}_v \pi_{g+} = \pi_{g-} \qquad \text{and} \qquad \hat{\sigma}_v \pi_{g-} = \pi_{g+} \tag{12-124}$$

we see that $\hat{\sigma}_v \psi(^1\Sigma) = \psi(^1\Sigma)$, so the state is $^1\Sigma^+$ (or $^1\Sigma_g^+$ when the gerade symmetry is added). The gerade symmetry follows from $u \otimes u = g$, since both electrons are in π_u MOs. Similarly, the triplet state has a spatial part

$$\pi_{g+}(1)\pi_{g-}(2) - \pi_{g-}(1)\pi_{g+}(2) \tag{12-125}$$

which leads to

$$\hat{\sigma}_v \psi(^3\Sigma) = -\psi(^3\Sigma) \tag{12-126}$$

Thus, using the same reasoning as before, the state is $^3\Sigma_g^-$.

Using Hund's rules just as for atoms (Sec. 8-3), we would predict the ground state to be $^3\Sigma_g^-$. Furthermore, the next two states are in the order $^1\Delta_g < {}^1\Sigma_g^+$, based on rule 2 of Sec. 8-3. It is readily verified that dioxygen $[(1\pi_g)^2$ outside the closed shell] also has a $^3\Sigma_g^-$ state, with $^1\Delta_g$ and $^1\Sigma_g^+$ states coming next.

EXERCISES

12-10. Show that the double bond of dioxygen is a sigma bond and a pi bond, and that the triple bond in dinitrogen is a sigma bond and two pi bonds. What sort of bonding (in terms of sigma and pi bonds) is found in the diboron (B_2) and dicarbon (C_2) molecules?

12-11. From what you know of the bonding in ethylene and acetylene, contrast the double bonds of dioxygen with those of ethylene and contrast the triple bonds of dinitrogen with those of acetylene. What role does molecular symmetry play in this comparison?

12-12. What are the spectroscopic term symbols for the ground states of the ions N_2^+ and O_2^+?

12-13. What are the possible spectroscopic states for a configuration containing $(\delta_g 3d)^2$ in addition to a closed shell. Note that the wavefunction will contain $e^{\pm 2i\varphi}$. Arrange the states in order of increasing energies. Show that, in general, a configuration $(n\lambda_u)^2$ or $(n\lambda_g)^2$ will lead to a state $^1(2|\Lambda|)_g^1$ (where Λ is the uppercase counterpart of λ) and also to the two states $^1\Sigma_g^+$ and $^3\Sigma_g^-$.

12-14. The MO bond order defined by Eq. (12-120) is often approximately directly proportional to dissociation energy and inversely proportional to bond length.

Show that these relationships are consistent with the following data on dioxygen and two of its ions:

Molecule or ion	D_e, eV	R_e, nm
O_2^+	6.48	0.123
O_2	5.08	0.121
O_2^-	?	0.126

What would you predict as an approximate value for the missing dissociation energy?

12-15. Write the determinantal wavefunctions for the three lowest states of diboron.

12-16. Would you expect He_2^+ to be a relatively stable ion? Explain.

12-7 HETERONUCLEAR DIATOMIC MOLECULES

Heteronuclear diatomic molecules ($C_{\infty v}$ point group) do not have a center of symmetry; thus the coefficients of AOs such as $1s_A$ and $1s_B$ are no longer determined by symmetry as in homonuclear diatomic molecules ($D_{\infty h}$ point group). We will illustrate most of the general features of calculations on heteronuclear diatomic molecules by considering the simple molecule lithium hydride, LiH. This is the simplest neutral heteroatomic molecule, and calculations on it serve as models for larger heteroatomic systems. The minimal basis set for LiH comprises the 1*s* and 2*s* AOs of lithium and the 1*s* AO of hydrogen. This leads to three MOs of the general form

$$n\sigma = c_{1n}1s(\text{Li}) + c_{2n}2s(\text{Li}) + c_{3n}1s(\text{H}) \qquad n = 1, 2, \text{ or } 3 \qquad (12\text{-}127)$$

A qualitative description of the molecular bonding is represented by Fig. 12-17. Along the left-hand side of the diagram are depicted the 1*s* and 2*s* AOs of the lithium atom; the 1*s* AO of the hydrogen atom is depicted along the right-hand side of the same diagram. In the center are the MOs of lithium hydride formed

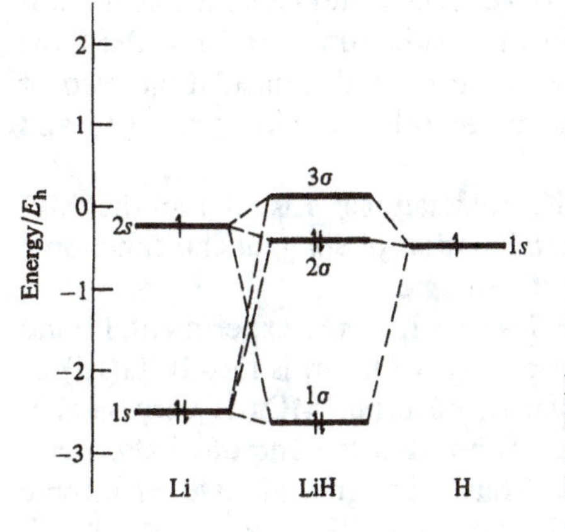

FIGURE 12-17
Schematic representation of the mixing of lithium 1*s* and 2*s* AOs with the 1*s* AO of hydrogen to form MOs of lithium hydride.

by the three possible linear combinations of the basis AOs. The lowest energy MO, 1σ, is seen to be mostly $1s$ of lithium with only a small amount of $2s$(Li) and $1s$(H) mixed in. The reason that $1s$(Li) does not combine strongly with $1s$(H) [or with $2s$(Li)] is that its energy is too different from that of either of the other two AOs; in general, two AOs combine more strongly the more similar their energies. An MO of this type is often called a *nonbonding* MO; this means that the two electrons in it do not contribute very much to binding between the Li and H atoms. Stated in another way, these two electrons in LiH behave much as if the H atom were totally absent (as in atomic lithium). The next MO, 2σ, is mostly a linear combination of $2s$(Li) and $1s$(H). This MO describes what we normally call the *single sigma bond* of lithium hydride; this is what the dash in Li–H represents when the molecule is described at an elementary level. This MO would also contain a small amount of $1s$(Li). The third MO, 3σ, is a *virtual* MO; it is not occupied in the ground state and would be used in describing excited states of lithium hydride. This MO is also a mixture of all three basis functions. Such an MO is sometimes called an *antibonding MO*, since its use destabilizes the molecule, i.e., leads to a total molecular energy higher than that of the separated atoms in their ground states. Using familiar notation, the MO electron configuration of LiH in its ground state is written as $(1\sigma)^2(2\sigma)^2$; this spectroscopic state is given the symbol $X^1\Sigma^+$.

In actuality, the minimal basis set above does not provide a good description of lithium hydride, and considerable improvement is obtained simply by adding a fourth basis function: the $2p$ AO of lithium, which is aligned along the Li-to-H bond axis [we call this the $2p_\sigma$(Li) AO]. This fact is easily rationalized on the basis that the $2s$ and $2p$ AOs of lithium have similar energies and thus should have comparable contributions to bonding (a similar situation exists in beryllium, whose minimal basis set should also include $2p$ orbitals). To illustrate more explicitly the roles of the various AOs in the LiH bonding, we will examine a numerical calculation using an STO-6G basis. Although this calculation does not represent the state of the art in LiH calculations, its simplicity permits considerable insight into general calculations of this type. Since this calculation introduces an additional AO into the basis set, there will be a total of *four* MOs resulting from the calculation; two of these (1σ and 2σ) will represent the bonding, and the remaining two (3σ and 4σ) will be virtual MOs.

Table 12-4 summarizes the STO-6G calculations. Listed are the MO coefficients of each STO (each a linear combination of six gaussian functions) for all four MOs and also the resulting MO energies.

The calculated molecular energy is $-7.95196E_h$ at the experimental bond distance of 3.015 a_0. Note that the lowest-energy MO, 1σ, is mostly $1s$(Li) as predicted (its coefficient is 0.993, and that of all other MOs is very small). Next, note that 2σ is mostly $2s$(Li) and $1s$(H) but that the $2p\sigma$ of Li also plays a significant role in this bonding MO. Thus, the ground-state electronic configuration of LiH is summarized in the notation LiH $(1\sigma)^2(2\sigma)^2$: $X^1\Sigma^+$.

TABLE 12-4
Summary of MO calculations on lithium hydride using an STO-6G basis

*n*1	*s*(Li)	2*s*(Li)	2*pσ*(Li)	1*s*(H)	Orbital energy/E_h
1	0.993	0.024	−0.006	0.004	−2.387
2	0.158	−0.446	−0.346	−0.555	−0.286
3	0.199	−0.798	0.614	0.131	0.079
4	−0.087	0.711	0.977	−1.185	0.549

This same calculation has been carried out by Ransil[32] using a straight STO basis; the results for a calculation in which the scale factors were chosen by Slater's rules are given in Table 12-5.

The molecular electronic energy at the experimental bond distance is $-7.96666E_h$; this is an improvement of about $0.015E_h$ over the STO-6G calculation. Although there are considerable quantitative differences in the values of certain orbital coefficients and orbital energies, the two calculations exhibit the same general qualitative features. The STO-6G calculation predicts an ionization energy of $0.286E_h$ (7.8 eV) compared with the Ransil value of $0.304E_h$ (8.3 eV); the experimental value is uncertain.

If we use Ransil's calculations, the dissociation energy of LiH is given by

$$D_e = E(\text{LiH}) - E(\text{Li}) - E(\text{H}) \tag{12-128}$$

The electronic part of the molecular energy $E(\text{LiH})$ is the sum of the MO energies minus the electron-repulsion energy; the former is seen from Table 12-5 to be $2(-2.447 - 0.304) = -5.502E_h$, and the latter is $3.460E_h$. The calculation was carried out at an internuclear distance of $R = 3.015\ a_0$ (the experimental value), and thus the nuclear-repulsion portion of $E(\text{LiH})$ is given by

$$\frac{Z(\text{Li})Z(\text{H})}{3.015} = \frac{3}{3.015} = 0.995E_h$$

[32] Ransil, loc. cit.

TABLE 12-5
Summary of MO calculations on lithium hydride using an STO basis

*n*1	*s*(Li)	2*s*(Li)	2*pσ*(Li)	1*s*(H)	Orbital energy/E_h
1	0.997	0.016	−0.005	0.006	−2.447
2	0.131	−0.323	−0.231	−0.685	−0.304
3	0.134	−0.805	0.599	0.148	0.017
4	−0.024	0.869	1.117	−1.286	0.349

Thus the molecular energy E(LiH) is given by

$$E(\text{LiH}) = -5.502 - (3.460) + 0.995 = -7.967E_h$$

If one uses the Slater value for the atomic energy of lithium ($-7.418E_h$) and substitutes this value into Eq. (12-128), one obtains

$$D_e = -7.967 - (-7.418) - (-0.500) = 0.049E_h \text{ (1.34 eV)}$$

The experimental value is about 2.5 eV. It is of interest to note that if the experimental value for the atomic energy of lithium is used ($-7.478E_h$), then the dissociation energy turns out to be $0.011E_h$, or 0.30 eV. The justification for using the less accurate theoretical value for the atomic energy is that the resulting calculation places both the molecule and its constituent atoms on comparable computational levels. This procedure is justified if in fact the errors in both the molecular and the atomic energies are comparable and thus effectively cancel each other out. This procedure is not without its flaws, but it is the one usually employed in the theoretical calculation of molecular dissociation energies.

Kahalas and Nesbet[33] used an extended basis set ($3d$ functions on Li and $2s$ and $2p$ on H) to obtain an energy of $-7.9859698E_h$ for LiH at the experimental bond distance. This is an improvement of almost $0.02E_h$ over the Ransil calculation.

One commonly computed physical property of heteronuclear molecules is the *electric dipole moment*. The classical definition of this quantity for a collection of N particles of charge q_i and position vectors $\mathbf{r}_i$ (equal to $\mathbf{i}x_i + \mathbf{j}y_i + \mathbf{k}z_i$) is given by

$$\boldsymbol{\mu} = \sum_i^N q_i \mathbf{r}_i \tag{12-129}$$

The quantum mechanical expression for this quantity (for a molecule of N electrons and S nuclei) is given by

$$\boldsymbol{\mu} = \left\langle \psi \left| -\sum_\mu^N \mathbf{r}_\mu \right| \psi \right\rangle + \sum_k^S Z_k \mathbf{R}_k \tag{12-130}$$

where $q_\mu = -1$ for electrons and $q_k = Z_k$ for nuclei. The quantity $\mathbf{R}_k$ is the position vector for the kth nucleus. The electronic dipole operator appearing in the first part of Eq. (12-130) is a sum of one-electron operators and thus may be expressed in terms of the charge density and bond-order matrix $\mathbf{R}$ (also called the *density matrix*) defined earlier in Secs. 10-5 and 10-6. This first integral then becomes

$$-2\sum_r \sum_s R_{rs} \langle s|\mathbf{r}|r\rangle \tag{12-131}$$

[33] S. L. Kahalas and R. K. Nesbet, *J. Chem. Phys.*, **39**:529 (1963).

where

$$R_{rs} = \sum_{\mu}^{N/2} c_{r\mu} c_{s\mu}^* \tag{12-132}$$

In the case of linear molecules (such as LiH), only the component along the bond axis (arbitrarily labeled the z axis) is nonzero. This is defined by

$$-2 \sum_r \sum_s R_{rs} \langle s|z|r \rangle \tag{12-133}$$

where the integral over the basis functions is given by

$$\langle s|z|r \rangle = \int \phi_s^*(\mathbf{r}_1) z_1 \phi_r(\mathbf{r}_1)\, d\mathbf{r}_1 \tag{12-134}$$

The STO-6G calculation leads to a dipole moment of 4.92 debyes (D) for LiH (a debye unit is 1×10^{-18} esu · cm; in SI units, this is 3.3×10^{-30} C · m); the calculation predicts a polarity Li^-H^+. By contrast, the Ransil STO calculation predicts a dipole moment of 6.41 D and the opposite polarity, Li^+H^-. The experimental value of the dipole moment is 5.882 D,[34] but the sign is unknown. In Table 12-6 are listed six different calculations on LiH, all of which lead to roughly the same dipole moment but which differ as to the molecule's polarity.

Calculations on molecules are frequently analyzed in terms of a *population analysis*. Such analyses attempt to assign fractions of the total electron

[34] L. Wharton, L. P. Gold, and W. Klemperer, *J. Chem. Phys.*, **33**:1255 (1960) and **37**:2149 (1962).

TABLE 12-6
Dipole moment and polarity calculations on LiH

Basis set	Dipole moment, D	Li charge	H charge
STO-3G*	4.86	−0.02	+0.02
STO-6G*	4.92	−0.002	+0.002
Minimal STO†	6.41	+	−
Extended STO-6G‡	5.59	+0.28	−0.28
Extended STO-6G§	5.62	+0.27	−0.27
Extended¶	5.8875	+	−

* GAMESS calculation by the author.

† B. J. Ransil, *Rev. Mod. Phys.*, **32**:245 (1960).

‡ GAMESS calculation by the author using an extended basis set (1s, 2s, 3s, 2p, and 3p on Li and 1s, 2s and 2p on H) with the STO-6G basis. The resulting energy was $-7.96746E_h$, compared with the STO-6G result of $-7.95196E_h$ and the STO-3G value of $-7.86201E_h$.

§ Same as the previous extended STO-6G calculation, but the Li basis was changed to 1s, 2s, 2p, and 3d. The total energy was $-7.96692E_h$. This shows that the 3d function is much more effective in polarization than are the 3s and 3p functions.

¶ S. L. Kahalas and R. K. Nesbet, *J. Chem. Phys.*, **39**:529 (1963). The minimal basis was extended by 3d on Li and by 2s and 2p on H. The calculated energy was $-7.98597E_h$.

density to particular atoms and interatomic regions as an aid to interpreting certain physical and chemical properties of the molecules. Although various different population analyses have been proposed (none of them unique), the most popular is that devised by Mulliken.[35] This procedure is especially useful for molecules of low symmetry. The *net atomic population* of an atom which arises from the ith MO of the normalized form $c_{Ai}\phi_A + c_{Bi}\phi_B$ is given by $n_i c_{Ai}^2$ for atom A and $n_i c_{Bi}^2$ for atom B (n_i is the occupation number of the ith MO; this is always 1 or 2 for an occupied orbital, zero otherwise). Thus $n_i c_{Ai}^2$ is the number of electrons associated with atom A in the ith MO. The *electron density* between the two nuclei arising from the ith MO is given by the *overlap population* $2n_i c_{Ai} c_{Bi} \Delta$, where $\Delta = \langle \phi_A | \phi_B \rangle$. The *gross atomic population* is found by assuming that the overlap population can be divided equally between the two atoms and then added to the net atomic population; that is, $n_i(c_{Ai}^2 + c_{Ai}c_{Bi}\Delta)$ is the gross atomic population (of electrons) on atom A which results from the ith MO. The *gross charge* on a given atom is found by summing the above over all the occupied MOs. This analysis is readily generalized to MOs which have more than one component AO with a given atomic center. If we let $c_{ri}(k)$ represent the coefficient of the rth type of AO ($1s, 2s, 2p$, etc.) of the atom k in the ith MO, then the *electron density* associated with atoms k and l due to AOs of type r and s, respectively, is given by

$$\rho_{rsi}(kl) = 2n_i c_{ri}(k) c_{si}(l) \Delta_{rs}(kl) \tag{12-135}$$

The above quantity represents that part of the overlap population due to atoms k and l and AOs r and s, respectively, in the ith MO. The *electron overlap density* due to all occupied MOs and resulting from atoms k and l described by AOs r and s, respectively, is obtained by summing over all occupied MOs, that is,

$$\rho_{rs}(kl) = \sum_i \rho_{rsi}(kl) \tag{12-136}$$

The *total overlap population* of the ith MO with respect to atoms k and l is

$$\rho_i = \sum_{r,s} \rho_{rsi}(kl) \tag{12-137}$$

When $\rho_i > 0$, the MO is said to be *bonding*; when $\rho_i < 0$, the MO is said to be *antibonding*; and when $\rho_i = 0$, the MO is called *nonbonding*. The electron density associated with an atom k due to the rth AO in the ith MO is defined by

$$\rho_{ri}(k) = n_i c_{ri}^2(k) + \frac{1}{2} n_i \sum_{l \neq k} \rho_{rsi}(kl) \tag{12-138}$$

[35] R. S. Mulliken, *J. Chem. Phys.*, **23**:1833, 1841, 2338, 2343 (1955) and **36**:3428 (1962).

The total electron density associated with an atom k due to the rth AO is given by

$$\rho_r(k) = \sum_i \rho_{ri}(k) \tag{12-139}$$

It is this quantity which was used to determine the atomic charges quoted in Table 12-6.

The Mulliken population analysis (as well as other types of population analyses) must be used with some caution, since physically incorrect predictions may be obtained, e.g., the polarity of a heteronuclear molecule may turn out to be opposite that of the calculated dipole moment.

12-8 LINEAR POLYATOMIC MOLECULES

The molecular orbital treatments of molecules such as acetylene ($D_{\infty h}$) and hydrogen cyanide ($C_{\infty v}$) are no different in principle than those of diatomic molecules, but the presence of three or more atoms leads to some practical problems which make the calculations more time-consuming. In particular, the traditional types of basis functions needed to describe a molecule with more than two atoms will involve as many different atomic centers as there are atoms in the molecule, and this will lead to electron-repulsion integrals, $\langle ij|1/r_{12}|kl\rangle$, which have as many as four different atomic centers. At one time, the numerical evaluation of such integrals (particularly those with three or four centers) was extremely time-consuming, but modern computer hardware and software have largely eliminated this bottleneck. Because of programs written by various research groups throughout the world, such integrals can now be evaluated in no more than twice the time required for a diatomic molecule described by the same number of basis functions. Furthermore, if gaussian functions are employed, computation times have dropped significantly from those of even a decade ago. Calculations in which bond lengths are optimized will necessarily take more time when there are two or more bonds to optimize simultaneously.

There are other aspects of calculations on linear polyatomic molecules which remain intrinsically harder than those on diatomic molecules; e.g., the spectra of linear polyatomic molecules are considerably more complicated than those of diatomic molecules, and, consequently, satisfactory interpretation of such features requires more elaborate calculations. Also, whereas it is relatively easy to develop simple rules for predicting MO configurations, bond orders, etc., for diatomic molecules without doing actual calculations, it is considerably more difficult to develop corresponding general methods for treating linear polyatomic molecules, except in certain special cases where a close analogy can be established with a diatomic molecule.

Since it is beyond the intended scope of this text to delve into the computational details of quantum theory calculations, we will consider only the general features of minimal basis set calculations on a pair of linear molecules (acetylene and hydrogen cyanide), each containing 14 electrons and, thus, isoelectronic with diatomic molecules such as N_2 and CO.

As the first example, we consider the acetylene (ethyne) molecule, C_2H_2. If we label the atoms from left to right (using the order HCCH) as a, b, c, and d, the minimal basis set is $1s_a$, $1s_b$, $2s_b$, $2p_{xb}$, $2p_{yb}$, and $2p_{zb}$, and so on for atoms c and d. As before, we arbitrarily let the bond axis be designated z. These 12 AOs will produce 12 MOs having the following general forms:

$$\begin{aligned} j\sigma_g &= a_{1j}(\sigma_g 1s)_H + a_{2j}(\sigma_g 1s)_C + a_{3j}(\sigma_g 2s)_C + a_{4j}(\sigma_g 2p)_C \\ j\sigma_u &= b_{1j}(\sigma_u 1s)_H + b_{2j}(\sigma_u 1s)_C + b_{3j}(\sigma_u 2s)_C + b_{4j}(\sigma_u 2p)_C \end{aligned} \qquad j = 1, 2, 3, 4$$

$$\pi_u = (\pi_u 2p)_C \qquad \pi_g = (\pi_g 2p)_C \tag{12-140}$$

where

$$\begin{aligned} (\sigma_g 1s)_H &= 1s_a + 1s_d & (\sigma_u 1s)_H &= 1s_a - 1s_d \\ (\sigma_g 1s)_C &= 1s_b + 1s_c & (\sigma_u 1s)_C &= 1s_b - 1s_c \\ (\sigma_g 2s)_C &= 2s_b + 2s_c & (\sigma_u 2s)_C &= 2s_b - 2s_c \\ (\sigma_g 2p)_C &= 2p_{zb} + 2p_{zc} & (\sigma_u 2p)_C &= 2p_{zb} - 2p_{zc} \\ (\pi_g 2p)_C &= \begin{cases} 2p_{xb} - 2p_{xc} \\ 2p_{yb} - 2p_{yc} \end{cases} & (\pi_u 2p)_C &= \begin{cases} 2p_{xb} + 2p_{xc} \\ 2p_{yb} + 2p_{yc} \end{cases} \end{aligned} \tag{12-141}$$

The results of an SCF calculation by McLean[36] are given in Table 12-7. The basis functions were STOs. The ground-state electron configuration was found to be

$$(1\sigma_g)^2(1\sigma_u)^2(2\sigma_g)^2(2\sigma_u)^2(3\sigma_g)^2(1\pi_u)^4\text{: } {}^1\Sigma_g^+ \tag{12-142}$$

From Table 12-7 it is seen that $1\sigma_g$ and $1\sigma_u$ are largely $(\sigma_g 1s)_C$ and $(\sigma_u 1s)_C$, respectively. These two MOs represent the inner-shell, nonbonding electrons of the carbon atoms. Note that the hydrogen $1s$ orbitals contribute only a small amount to these MOs. Similarly, the carbon $2s$ and $2p$ orbitals make small contributions to these essentially nonbonding MOs. The $1\pi_u$ MO represents the so-called pi bonds of acetylene. Figure 12-18 shows the appearance of a single pi MO between a pair of bonded atoms via a surface plot in a plane containing these atoms. Since acetylene is linear, there is no unique distinction between the x and y axes, and, consequently, the electron density due to pi bonds must be regarded as a region of annular charge about the bond axis (the z axis). Thus, the electron density in such a pi bond is zero along the bond axis itself, increases to a maximum value at right angles to this axis (in all directions), goes through a maximum value, then dies off exponentially to zero at larger distances. In many respects, the carbon-to-carbon bonding of

[36] A. D. McLean, *J. Chem. Phys.*, **32**:1595 (1960). See also A. D. McLean, B. J. Ransil, and R. S. Mulliken, *J. Chem. Phys.*, **32**:1873 (1960).

TABLE 12-7
Normalized molecular orbitals and orbital energies for the acetylene molecule*

MO symbol	$(\sigma_g 1s)_H$	$(\sigma_g 1s)_C$	$(\sigma_g 2s)_C$	$(\sigma_g 2p)_C$	MO energy/E_h
$1\sigma_g$	−0.0033	0.7067	0.0098	−0.0007	−11.406
$2\sigma_g$	0.1033	−0.0702	0.4557	0.1850	−1.041
$3\sigma_g$	−0.3109	−0.0202	−0.0942	0.4761	−0.683
$4\sigma_g$	1.0594	−0.1157	−0.8059	0.7016	0.484
	$(\sigma_u^* 1s)_H$	$(\sigma_u^* 1s)_C$	$(\sigma_u^* 2s)_C$	$(\sigma_u^* 2p)_C$	
$1\sigma_u$	−0.0031	0.7091	0.0188	0.0042	−11.397
$2\sigma_u$	0.2957	−0.0411	0.3159	−0.3139	−0.776
$3\sigma_u$	−0.9259	0.1220	1.1522	−0.1334	0.353
$4\sigma_u$	0.5989	0.2309	1.2970	1.6753	1.195
		$(\pi_u 2p)_C$	$(\pi_g^* 2p)_C$		
π_u		0.6071			−0.441
π_g			0.8814		0.251

* From A. D. McLean, *J. Chem. Phys.*, **32**:1595 (1960). (*By permission of the author and the American Institute of Physics.*)

acetylene resembles the nitrogen-to-nitrogen bonding of dinitrogen. Neither the nitrogen nor the acetylene pi bonds resemble the pi bond in ethylene; since unique x and y axes can be assigned to ethylene, the pi bond has no cylindrical symmetry and exhibits maximum electron densities above and below the plane defined by the four hydrogen atoms and the two carbon atoms. These points

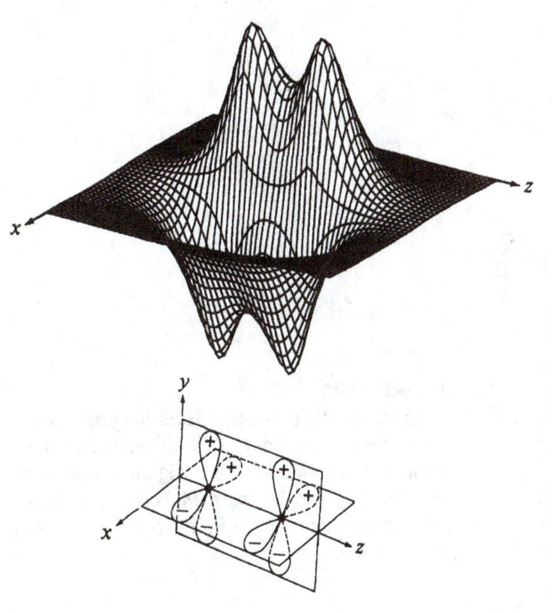

FIGURE 12-18
The pi bond of acetylene. This is just the $\pi_u 2p$ function of Fig. 12-16. The second pi bond of acetylene would look the same, except it would appear in the yz plane. Since acetylene has cylindrical symmetry about the bonding axis, the two pi bonds may be visualized by imagining the $(\pi_u 2p)^2$ function "spinning" about the C–C axis.

will be discussed again in the following chapter dealing with nonlinear polyatomic molecules.

The remaining occupied MOs ($2\sigma_g$, $2\sigma_u$, and $3\sigma_g$) must account for two C–H bonds and one C–C bond (so-called *sigma bonds*). It is not possible to relate these MOs with specific bonds, since they are not localized, but some approximate assignments can be made. For example, the $2\sigma_g$ MO has carbon $2s$ AOs as the dominant AOs, and the $3\sigma_g$ is dominated by carbon $2p_z$ AOs. Taken together, these represent the bonding often attributed to overlap of two carbon *sp* (trigonal) hybrid orbitals. However, both of these MOs also account for some carbon-to-hydrogen bonding, so the correspondence is rather crude.

According to Table 12-7, the first ionization energy of acetylene should be given by the negative of the orbital energy of the π_u MO: $0.441E_{\mathrm{h}}$, or about 12 eV. The experimental value is 11.4 eV. The good agreement is fortuitous; Koopmans' theorem should apply strictly only to truly optimized Hartree-Fock calculations, and even for these, agreement with experiment is not always satisfactory.

Figure 12-19 shows the orbital energy-level diagrams of both dinitrogen and acetylene for all but the inner-shell, nonbonding MOs. Note that there is considerable qualitative similarity between the two molecules in terms of these diagrams. In particular, both have analogous electron configurations, and in both the highest occupied MO is of π_u type. Also, there is a $3\sigma_g$ MO slightly below this π_u MO in the case of both molecules. This similarity should not be too surprising; if one considers the H–C "fragment" as electronically similar to a nitrogen atom (the two are isoelectronic, of course), then the two molecules

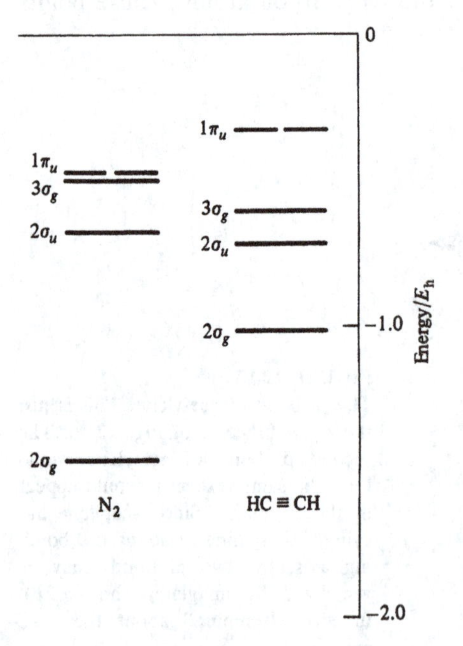

FIGURE 12-19
Comparison of MO energy-level diagrams of the isoelectronic and isosymmetric molecules dinitrogen and acetylene for the higher-energy orbitals. The $1\sigma_u$ and $1\sigma_g$ core MOs are not shown; these lie at about $-15.6\ E_{\mathrm{h}}$ for N_2 and $-11.4\ E_{\mathrm{h}}$ for acetylene.

should be similar in electronic structure. In an approximate sense the hydrogen atom attached to the carbon of acetylene behaves electronically somewhat like a lone pair of electrons (nonbonding) of a nitrogen atom in dinitrogen. An even more dramatic comparison between dinitrogen and acetylene is provided by comparing the respective electron density diagrams of the occupied MOs. The electron density diagram for acetylene is illustrated in Fig. 12-20; the comparable diagram for dinitrogen is given in Fig. 12-14.

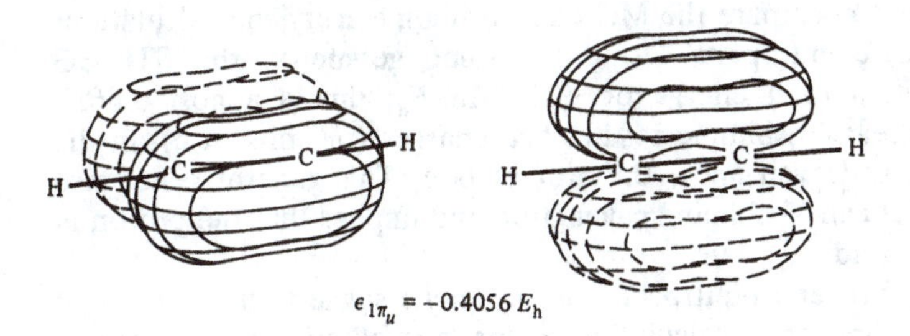

$\epsilon_{1\pi_u} = -0.4056\ E_h$

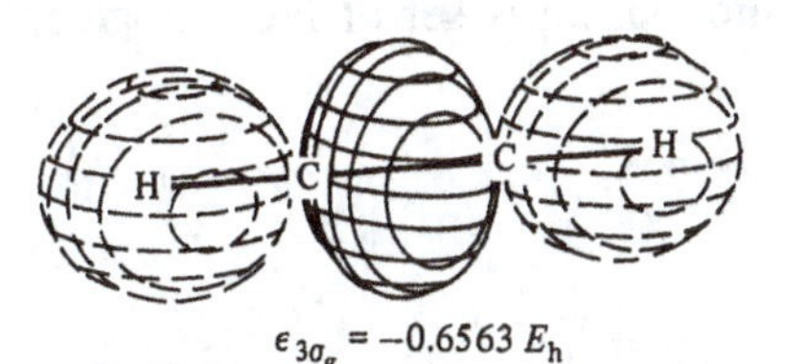

$\epsilon_{3\sigma_g} = -0.6563\ E_h$

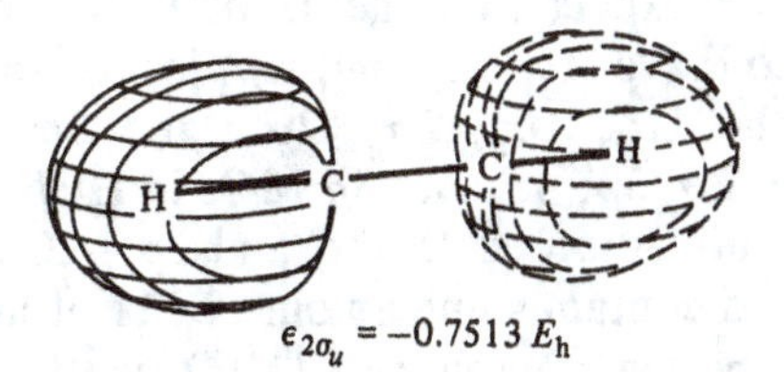

$\epsilon_{2\sigma_u} = -0.7513\ E_h$

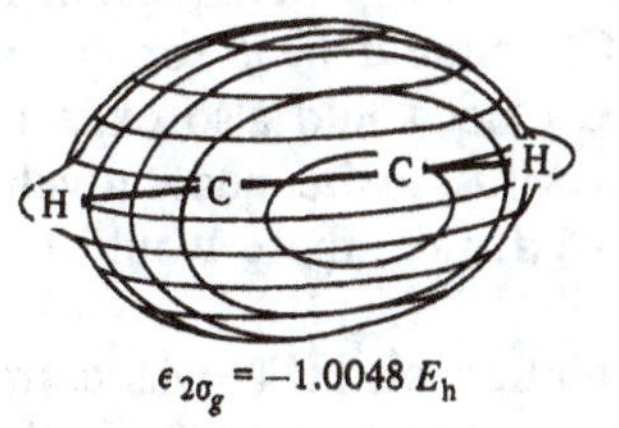

$\epsilon_{2\sigma_g} = -1.0048\ E_h$

FIGURE 12-20
Contours of electron density for some higher-energy MOs of acetylene. Compare these with Fig. 12-14 for dinitrogen and note the great similarities. (*Reprinted by permission from W. L. Jorgensen and L. Salem*, The Organic Chemist's Book of Orbitals, *Academic Press, New York, 1973.*)

A more quantitative comparison between dinitrogen and acetylene is obtained by examining the Mulliken population analyses of these two molecules. The sigma overlap population in dinitrogen (Scherr's calculation) is 0.395, whereas that in acetylene (McLean-Yoshimine calculation) is 1.42. This rather large difference in sigma overlap population is due largely to the large lone-pair repulsions in dinitrogen; in acetylene this is replaced with a C–H overlap population of 0.79. The pi overlap populations are 0.880 in dinitrogen and 1.05 in acetylene—not significantly different.

It is instructive to compare the McLean-Yoshimine acetylene calculations with their STO-6G counterpart. Using the same geometry, the STO-6G calculation produces a total energy of $-76.60189E_{\mathrm{h}}$; this is almost $0.06E_{\mathrm{h}}$ lower than the McLean-Yoshimine result. The charges (in units of e) on the atoms are 0.11 for hydrogen and -0.11 for carbon. This general result also follows from the McLean-Yoshimine calculation and implies that the proton in acetylene is slightly acidic.

The similarity between dinitrogen and acetylene suggests that we might be able to use the dinitrogen calculation to make qualitative predictions of what the orbital energy-level diagram of hydrogen cyanide, HCN, will look like in advance of a calculation. The minimal basis set of HCN is given by

H: $1s$
C: $1s, 2s, 2p_x, 2p_y, 2p_z$
N: $1s, 2s, 2p_x, 2p_y, 2p_z$

There are seven AOs which will combine to form σ MOs ($1s$ of H and the $1s$, $2s$, and $2p_z$ AOs of C and N) and four AOs which will form π MOs (the $2p_x$ and $2p_y$ AOs of C and N). We can expect that the $1s$ of H will be found predominantly in MOs which also have $1s$, $2s$, and $2p_z$ of C as well as corresponding AOs of N. Thus the $1\sigma_g$, $1\sigma_u$, $2\sigma_g$, $2\sigma_u$, and $3\sigma_g$ MOs of dinitrogen would be replaced by 1σ, 2σ, 3σ, 4σ, and 5σ MOs in HCN. Each of these σ MOs would be a linear combination of $1s$, $2s$, and $2p_z$ AOs of both carbon and nitrogen, and each would contain some amount of $1s$ of hydrogen. One would expect the 1σ to have the least amount of $1s$(H), with somewhat larger amounts in the remaining MOs. Then there would be a pair of degenerate π MOs to correspond to the π_u MOs of dinitrogen. Furthermore, the orbital energy-level diagram of HCN should bear some resemblance to those of both dinitrogen and acetylene. One would also expect the electron density contour diagrams of analogous MOs to be qualitatively similar but those of HCN to be polarized or distorted; i.e., there would no longer be a center of symmetry.

The STO-6G (minimal basis) calculation on HCN was carried out using what appears to be a reasonable compromise to uncertainties in the experimental geometry: H–C (2.0144 a_0) and C–N (2.1790 a_0). The total molecular energy turned out to be $-92.573496E_{\mathrm{h}}$; the dipole moment was 2.45 D; and atomic charges were +0.15 for H, +0.02 for C, and -0.17 for N. The

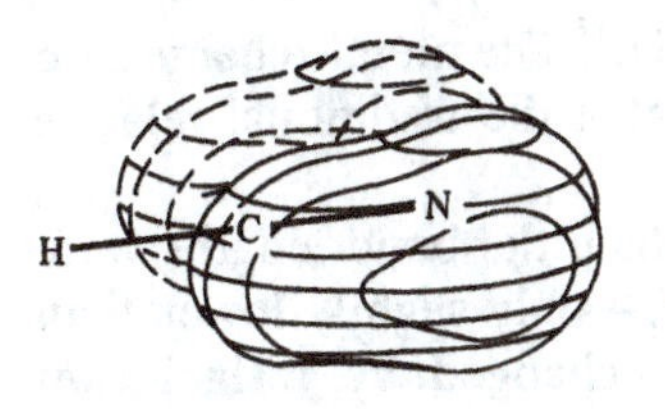

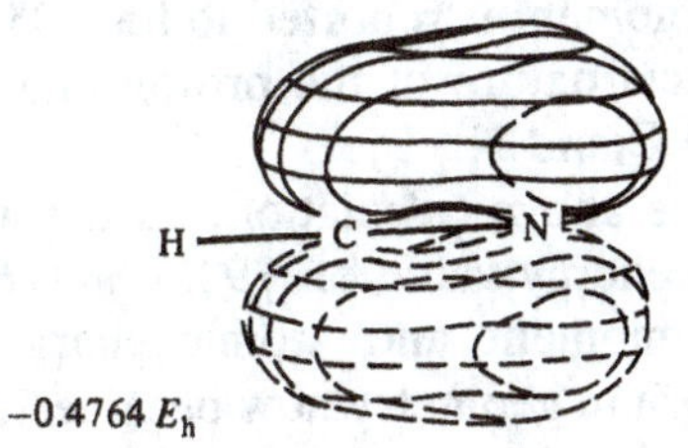

$\epsilon_{1\pi} = -0.4764\ E_h$

$(1\pi_u)$

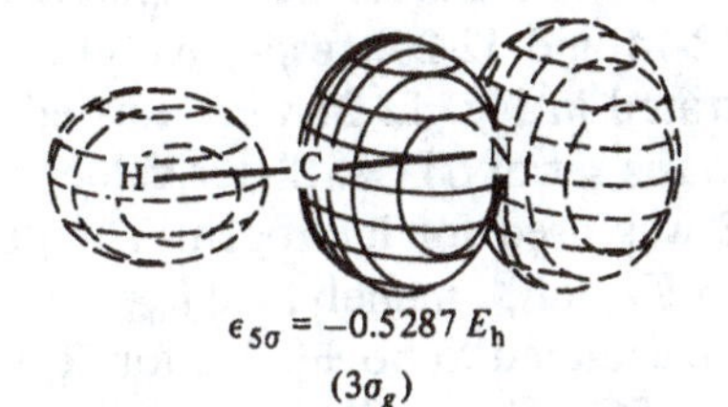

$\epsilon_{5\sigma} = -0.5287\ E_h$

$(3\sigma_g)$

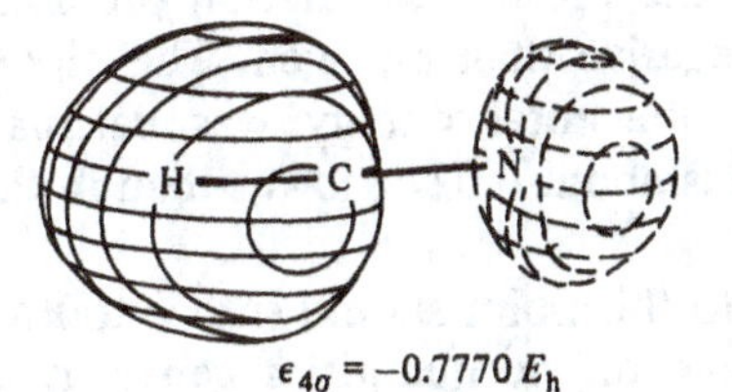

$\epsilon_{4\sigma} = -0.7770\ E_h$

$(2\sigma_u)$

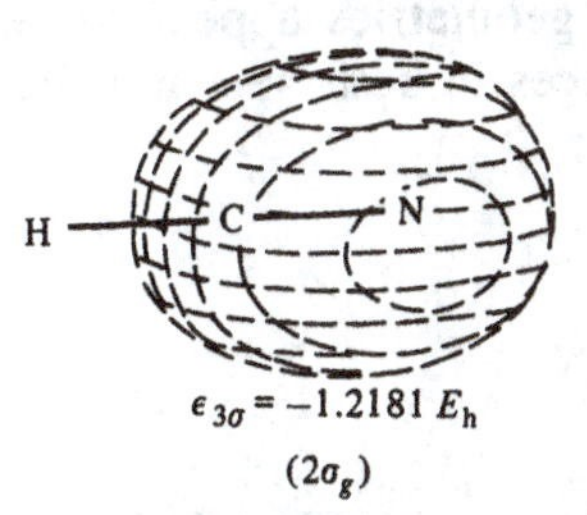

$\epsilon_{3\sigma} = -1.2181\ E_h$

$(2\sigma_g)$

FIGURE 12-21
Electron density contours of hydrogen cyanide (HCN). The numbers in parentheses below each MO symmetry label refer to the analogous MOs of the isoelectronic (but not isosymmetric) molecules dinitrogen and acetylene. Note the strong qualitative similarity of MOs of dinitrogen, acetylene, and hydrogen cyanide. Note also that the contribution of the hydrogen atoms to the overall appearance of the acetylene and hydrogen cyanide MOs is very small. (*Reprinted by permission from W. L. Jorgensen and L. Salem*, The Organic Chemist's Book of Orbitals, *Academic Press, New York, 1973.*)

experimental dipole moment is reported to be 2.95 D.[37] The atomic charges are consistent with the acid nature of the proton and with the traditional relative electronegativities of C and N.

Repetition of the above calculation, during which the bond lengths were optimized, led to an energy value of $-92.573499 E_h$—only slightly lower than before. The dipole moment and atomic charges changed by insignificant amounts. The equilibrium geometry now became 2.0172 a_0 for H–C and 2.1794 a_0 for C–N. Note that these values are insignificantly different from the assumed values in the first calculation.

Figure 12-21 shows the electron density contour diagrams of some of the higher bonding MOs of HCN; these MOs should be compared with those of dinitrogen and acetylene (Figs. 12-14 and 12-20, respectively).

The HCN calculations illustrated in Fig. 12-21 were carried out by Palke and Lipscomb[38] using a minimal basis set of STOs. Slater values were used for scale parameters, except that 1.2 was used for hydrogen. The geometry used was based on values obtained from Kaplan,[39] namely, 2.0 a_0 for C–H and 2.187 a_0 for C–N. Atomic charges were calculated to be +0.216 for H, −0.141 for C, and −0.075 for N. Note that the STO-6G calculations predicted that carbon has a net positive charge, whereas the present calculation predicts that C is negative and, furthermore, more negative than nitrogen. The dipole moment was predicted to be 2.11 D, and the total energy was calculated to be $-95.5903 E_h$. The energy value is thus about $0.02 E_h$ lower than in the STO-6G calculation.

The calculations discussed up to this point should suffice to illustrate that there are many uncertainties in some of the quantities generally derived by near-Hartree-Fock calculations on even relatively small linear molecules, particularly, values for the dipole moment, atomic charges, and ionization energies. On the other hand, equilibrium geometries appear to be more reliably predicted and less sensitive to differences in basis sets and the techniques for using them.

12-9 MOLECULAR CONFIGURATION-INTERACTION CALCULATIONS

The basic structure of configuration-interaction (CI) calculations on molecules is the same as that for atoms (see Sec. 9-4) but is complicated by the same numerical evaluation problem as found in molecular calculations in general. In this section we will consider the following types of CI calculations:

---

[37] A. L. McClellan, *Tables of Experimental Dipole Moments*, Freeman, San Francisco, 1963.

[38] W. E. Palke and W. N. Lipscomb, *J. Am. Chem. Soc.*, **88**:2384 (1966).

[39] H. Kaplan, *J. Chem. Phys.*, **26**:1704 (1957).

1. CI calculations in which the configurations are those generated by solution of the ground-state Hartree-Fock SCF equations.
2. CI calculations in which the configurations are not restricted to those generated by the above and in which the entire CI wavefunction is optimized in a single sequence of operations. This is known as the *multiconfigurational self-consistent field method* (*MCSCF*).

If one regards the Hartree-Fock SCF method as the starting point of a CI calculation, then the CI method can be viewed as a means of accounting for some of the correlation energy of the molecules. Furthermore, in the limited case of an infinitely large set of configurations *all* the correlation energy can be accounted for. Suppose the SCF calculation generates $2K$ molecular spin orbitals and that the molecule contains $2N$ electrons. The total number of $2N$-electron determinants that one can construct from the $2K$ spin orbitals is given by the binomial coefficient

$$\binom{2K}{2N} = \frac{(2K)!}{(2K-2N)!(2N)!}$$

For example, the Hartree-Fock calculation on dihydrogen using the minimal basis $1s_A$ and $1s_B$ generates the four spin orbitals: $1\sigma_g\alpha$, $1\sigma_g\beta$, $1\sigma_u\alpha$, and $1\sigma_u\beta$. Thus $2K=4$, and since $2N=2$, one can generate the following six determinants:

$$\begin{array}{lll} D_0 = |1\sigma_g\,\overline{1\sigma_g}| & D_2 = |1\sigma_g\,1\sigma_u| & D_4 = |\overline{1\sigma_g}\,1\sigma_u| \\ D_1 = |1\sigma_u\,\overline{1\sigma_u}| & D_3 = |1\sigma_g\,\overline{1\sigma_u}| & D_5 = |\overline{1\sigma_g}\,\overline{1\sigma_u}| \end{array} \tag{12-143}$$

However, if the CI calculation is intended to improve the energy of the ground state, then only those configurations having the same symmetry as the ground state will contribute. This means we can combine the following: D_0, D_1, and $D_3 - D4$. The first is the ground-state wavefunction, the second represents an excited state of the same symmetry formed by double excitation (i.e., excitation of both electrons from a $1\sigma_g$ MO to a $1\sigma_u$ MO), and the third represents another excited state formed by a single excitation (i.e., one electron is excited from $1\sigma_g$ to $1\sigma_u$). The full CI wavefunction then has the general form

$$\Psi_{CI} = C_1D_0 + C_2D_1 + C_3(D_3 - D_4) \tag{12-144}$$

Note that the remaining combinations (D_2, D_5, and $D_3 + D_4$) represent a triplet state. The CI calculation can be simplified by the use of *Brillouin's theorem*,[40] which states that matrix elements such as the following vanish:

$$\langle \psi_0|\hat{H}|\psi'\rangle = 0$$

---

[40] L. Brillouin, *Les Champs 'self-consistents' de Hartree et de Fóck*, Hermann & Cie, Paris, 1934, p. 19.

where ψ_0 is the Hartree-Fock ground-state wavefunction and ψ' is a wavefunction representing a singly excited configuration obtained from the Hartree-Fock orbitals. Thus, it is customary to omit such configurations in the CI calculation. In the case above this means that the CI wavefunction will consist of D_0 and the doubly excited configuration wavefunction D_1. Note, however, that Brillouin's theorem does not say that singly excited configurations have no effect on the energy; it merely says that there will be no *direct* effect via integrals such as the above. In actuality, singly excited configurations will have a higher-order effect on the energy through matrix elements between the singly and doubly excited configurations. In the present case, there will be nonzero matrix elements in the full CI calculation between D_1 and $(D_3 - D_4)$ which will affect the final energy somewhat. Furthermore, the singly excited configurations do have significant effects on properties other than the energy, e.g., dipole moments (not relevant for dihydrogen) and factors related to the electronic spectrum. Nevertheless, we will omit the singly excited configuration entirely and write the CI wavefunction in the form

$$\Psi_{\text{CI}} = C_1|1\sigma_g\,\overline{1\sigma_g}| + C_2|1\sigma_u\,\overline{1\sigma_u}| \tag{12-145}$$

Weinbaum[41] has carried out a calculation which is equivalent to the two-configuration CI calculation described above. Weinbaum's wavefunction has the explicit form

$$\Psi = C_1'\psi_{\text{cov}} + C_2'\psi_{\text{ionic}} \tag{12-146}$$

where ψ_{cov} describes a *covalent* bonding situation and ψ_{ionic} describes an *ionic* bonding situation; that is,

$$\begin{aligned}\psi_{\text{cov}} &= [2(1+\Delta^2)]^{-1/2}(|1s_A\,\overline{1s_B}| - |\overline{1s_A}\,1s_B|)\\ \psi_{\text{ionic}} &= [2(1+\Delta^2)]^{-1/2}(|1s_A\,\overline{1s_A}| + |1s_B\,\overline{1s_B}|)\end{aligned} \tag{12-147}$$

where $\Delta = \langle 1s_A|1s_B\rangle$. The Weinbaum calculation led to a dissociation energy of 4.026 eV at $R_e = 1.42\ a_0$ using a scale factor of $\eta = 1.193$. Since the SCF value of the dissociation energy is 3.63 eV and the experimental value is 4.746 eV, the percent of the correlation energy accounted for by the Weinbaum function is

$$100\,\frac{4.026 - 3.63}{4.746 - 3.63} = 35 \text{ percent}$$

Weinbaum also demonstrated a simple method for improving the energy by a small amount; each $1s$ AO used in the wavefunction was replaced by a new function of the form

$$1s + \lambda(2p_z) \tag{12-148}$$

[41] S. Weinbaum, *J. Chem. Phys.*, **1**:593 (1933).

where λ is determined variationally. This form has the effect of adding configurations containing $2p_z$ AOs (see Exercise 12-19). Use of these orbitals produced a dissociation energy of 4.122 eV, an improvement of 0.096 eV.

Extensive CI calculations on dihydrogen have been carried out by McLean, Weiss, and Yoshimine.[42] One of the best CI functions was the linear combination of determinants representing the following five configurations:

1. The three split-shell configurations $(\sigma_g 1s)(\sigma_g 1s')$, $(\sigma_g 2s)(\sigma_g 2p)$, and $(\sigma_u 1s)(\sigma_u 1s')$
2. The two configurations $(\pi_u 2p)^2$ and $(\pi_g 2p)^2$

(Recall that primes imply the use of different scale factors in a configuration.) The first two split-shell configurations account for *in-out correlation*; i.e., when one electron is far away from the nucleus, the other is more likely to be close to the nucleus. The third split-shell configuration accounts for *left-right correlation*; i.e., because the nodal plane passes through the center of the internuclear bond and perpendicular to its axis, the two electrons are able to remain on opposite sides of this plane. The remaining two configurations represent *angular correlation*; i.e., when one electron is above the nuclear plane, the other can be on the opposite side. With a basis of STOs, this wavefunction led to a dissociation energy of 4.54306 eV at $R_e = 1.4013\ a_0$ and thus accounted for about 82 percent of the correlation energy. The split shells accounted for about 0.08 eV of correlation energy, the in-out correlation was about 0.25 eV, the left-right correlation was about 0.5 eV, and the angular correlation was about 0.3 eV. Adding seven more configurations similar to those above improved the dissociation energy by an insignificant amount (about 0.02 eV at most). Consequently, any attempt to extend this procedure would lead to extremely slow convergence.

Note that the McLean, Weiss, and Yoshimine CI calculation did not utilize Hartree-Fock MOs but, rather, used configurations constructed from STOs. This produces a more flexible calculation in that the MOs are not constrained to be of Hartree-Fock form; in actuality MOs of a strictly Hartree-Fock type are poor basis functions to use in a CI calculation. In the following discussion, an approach is examined which capitalizes on the advantages of using MOs in CI calculations which have maximum flexibility.

Before proceeding, however, we must consider a very important requirement which all CI calculations should satisfy, particularly if they are to be used to study chemical reactions: such calculations must be *size-consistent*. Simply stated, "size consistency" is the requirement that the energy of a system consisting of interacting particles be proportional to the number of particles N

---

[42] A. D. McLean, A. Weiss, and M. Yoshimine, *Rev. Mod. Phys.*, **32**:211 (1960). This paper contains a bibliography of dihydrogen calculations up to 1960.

in the limit $N \rightarrow \infty$. Full CI, in which *all* possible excitations are included, is size-consistent, but truncated CI is not. Consequently, using truncated CI suffers from the fact that the method is not equally good for molecules with different numbers of atoms. For example, suppose we are studying the interaction of two molecules, A and B. When the molecules are very far apart (noninteracting), the total energy should be simply the sum of the individual molecular energies. But if the calculation is carried out with truncated CI (treating the molecules as one supermolecule), this will not be the case. To see why this is so, suppose we restrict the calculation to double excitations; this excludes the possibility that both molecules are *simultaneously* doubly excited—a situation which would be remedied if quadruple excitation were allowed in the *system* calculation.

The multiconfigurational SCF method can be described most simply in terms of the dihydrogen calculation based on the wavefunction (12-144). Instead of requiring that the ground-state wavefunction D_0 be obtained by the Hartree-Fock SCF method and, furthermore, that the other two configurations be constructed from Hartree-Fock MOs, let us do a full variational calculation on the CI wavefunction. This means that we must determine variationally (e.g., through the scale factors) the forms of the MOs themselves (in terms of the basis $1s_A$ and $1s_B$) in addition to varying the CI coefficients C_1, C_2, and C_3. The earliest MCSCF calculations were carried out by D. R. Hartree, W. Hartree, and B. Swirles[43] on the oxygen atom. Most of the basic methods have been discussed more recently by a variety of authors.[44] Perhaps the principal difficulty with the method at present is that the MCSCF equations are very difficult to solve and convergence is often very slow. Nevertheless, the method exhibits considerable promise, and it is very likely that many of the practical problems with the method will eventually be solved.

We shall examine here an MCSCF calculation on dihydrogen using the minimal basis set and the STO-6G method. This calculation, which was also designed to optimize the molecular geometry (i.e., determine the internuclear distance which leads to the lowest energy) produced a dissociation energy of 3.97 eV at $R_e = 1.38\ a_0$. This is not quite as good an energy value as obtained by Weinbaum with STOs (4.026 eV at $R = 1.42\ a_0$) and indicates once again that the STO-6G method is not quite as good as the straight STO method.

MCSCF calculations on LiH using the minimal basis set expanded in terms of STO-6G lead to a total energy of $-7.97159 E_h$ when the experimental bond length is used. The dipole moment is calculated to be 4.72 D (not too different from what most calculations appear to obtain), but the polarity of the molecule unexpectedly comes out as Li^-H^+. (See Table 12-6 for a comparison

---

[43] D. R. Hartree, W. Hartree, and B. Swirles, *Phil. Trans. Roy. Soc. (London)*, **A238**:229 (1939).

[44] A. C. Wahl and G. Das, *Adv. Quantum Chem.*, **5**:261 (1970); J. Hinze, *J. Chem. Phys.*, **59**:6424 (1973); A. Veillard, *Theoret. chim. Acta*, **4**:22 (1966); E. Clementi and A. Veillard, *J. Chem. Phys.*, **44**:3050 (1966); and E. Clementi, *J. Chem. Phys.*, **46**:3842 (1967).

with other calculations—most of which favor the opposite polarity as chemical sense would require.) The present MCSCF calculations utilized the following six configurations:

1. $1\sigma^2 2\sigma^2$ **4.** $1\sigma^2 2\sigma 3\sigma$
2. $1\sigma^2 3\sigma^2$ **5.** $1\sigma^2 2\sigma 4\sigma$
3. $1\sigma^2 4\sigma^2$ **6.** $1\sigma^2 3\sigma 4\sigma$

The first configuration is the ground state, the next two are doubly excited states, and the remaining three are singly excited states. Each configuration contains $1\sigma^2$; this represents inner-core, nonbonding electrons which are customarily not included in the excitations. Note that in this calculation the singly excited configurations were included; Brillouin's theorem does not apply to these, since the ground state was not constrained to consist of Hartree-Fock MOs. Note that the energy value obtained in the MCSCF (STO-6G) calculation is better than that of the Ransil Hartree-Fock (STO) calculation ($-7.96666E_h$).

One of the most extensive MCSCF calculations on LiH has been reported by Docken and Hinze.[45] These workers studied not only the $X^1\Sigma^+$ ground state but also the excited states $A^1\Sigma^+$, $B^1\Pi$, $^3\Sigma^+$, and $^3\Pi$. These workers represented the ground state in terms of 15 configurations which were chosen with respect to their anticipated ability to contribute to various aspects of electron correlation. These 15 configurations (based on an STO basis set) were

1. $1\sigma^2 2\sigma^2$ **9.** $1\sigma^2 2\sigma 3\sigma$
2. $1\sigma^2 4\sigma^2$ **10.** $1\sigma^2 4\sigma 5\sigma$
3. $1\sigma^2 5\sigma^2$ **11.** $1\sigma^2 3\sigma 4\sigma$
4. $1\sigma^2 6\sigma^2$ **12.** $1\sigma^2 3\sigma 5\sigma$
5. $1\sigma^2 1\pi^2$ **13.** $1\sigma^2 3\sigma 6\sigma$
6. $1\sigma^2 2\pi^2$ **14.** $1\sigma^2 1\pi 2\pi$
7. $1\sigma^2 1\delta^2$ **15.** $1\sigma^2 1\delta 2\delta$
8. $1\sigma^2 2\delta^2$

The interested reader should consult the original article for the rationale used to include or omit various specific configurations. The calculation produced a minimum energy of $-8.021254E_h$ at $R = 3.0\ a_0$; this leads to a dissociation energy of 2.411 eV (experimental value = 2.5154 eV). Docken and Hinze reported that the dominant configurations (along with their wavefunction coefficients) were $1\sigma^2$ (0.96), $1\sigma^2 2\sigma 3\sigma$ (−0.23), $1\sigma^2 4\sigma^2$ (−0.10), and $1\sigma^2 3\sigma 4\sigma$ (0.07). Docken and Hinze also estimated that the Hartree-Fock SCF energy of LiH is $-7.987317E_h$. Thus, the MCSCF (STO-6G) calculation ($-7.97159E_h$) does not even do as well as the Hartree-Fock calculation.

[45] K. K. Docken and J. Hinze, *J. Chem. Phys.*, **57**:4928 (1972).

EXERCISES

12-17. Let the dihydrogen wavefunctions (12-145) and (12-146) be written in the two alternative forms

$$\Psi = A_1\psi_{\text{cov}} + A_2\psi_{\text{ionic}} = B_1|\sigma_g 1s\, \overline{\sigma_g 1s}| + B_2|\sigma_u 1s\, \overline{\sigma_u 1s}|$$

where

$$\psi_{\text{cov}} = [2(1+\Delta^2)]^{-1/2}(|1s_A\, \overline{1s}_B| - |\overline{1s}_A\, 1s_B|)$$
$$\psi_{\text{ionic}} = [2(1+\Delta^2)]^{-1/2}(|1s_A\, \overline{1s}_A| + |1s_B\, \overline{1s}_B|)$$

Show that the coefficients satisfy the relationship

$$\frac{A_1}{A_2} = \frac{B_1 - B_2 - \Delta(B_1 + B_2)}{B_1 + B_2 - \Delta(B_1 - B_2)}$$

or, alternatively,

$$\frac{B_1}{B_2} = \frac{(1+\Delta)(A_1 + A_2)}{(1-\Delta)(A_2 - A_1)}$$

where $\Delta = \langle 1s_A | 1s_B \rangle$.

Also show that this wavefunction dissociates properly, i.e., approaches the wavefunction of two hydrogen atoms as $R \to \infty$. Recall that the single-determinantal wavefunction $|\sigma_g 1s\, \overline{\sigma_g 1s}|$ dissociates improperly (see Sec. 12-5).

12-18. Demonstrate that all 15 of the configurations used in the Docken and Hinze LiH calculation produce functions of ${}^1\Sigma^+$ symmetry, and show explicitly what each such function is.

12-19. Consider the MO $1\sigma_g' = 1\sigma_g + \lambda(3\sigma_g)$ which is related to the basis function defined by Eq. (12-148) and in which $3\sigma_g = 2p_A + 2p_B$ (the $2p$'s are $2p_z$ AOs). Show that this leads to

$$|1\sigma_g'\, \overline{1\sigma_g'}| = |1\sigma_g\, \overline{1\sigma_g}| + \lambda^2|3\sigma_g\, \overline{3\sigma_g}| + \lambda(|1\sigma_g\, \overline{3\sigma_g}| - |\overline{1\sigma_g}\, 3\sigma_g|)$$

Show that this is a *constrained* CI wavefunction in which single and double excitations from $1\sigma_g$ to $3\sigma_g$ are considered. Where does the constraint enter in?

12-10 THE VALENCE BOND METHOD

In addition to the molecular orbital approximation, based ultimately on the Hartree-Fock SCF method, there is another important model of molecular electronic structure: the *valence bond*, or *VB*, *method*. This model makes no use of molecular orbitals and self-consistent fields, but may be regarded as a quantitative version of the *resonance theory* of electronic structure—a theory which is still used today.

The VB method assumes that the wavefunction of a molecule may be written as a linear superposition of mathematical functions which represent *canonical* electronic structures. A canonical electronic structure may be regarded as a fictitious structure of a molecule in which electrons are assigned to specific atoms and then paired (spinwise) in specific ways. For example, in dihydrogen, one canonical structure is H–H; this canonical structure is interpreted to represent electron 1 on atom A and electron 2 on atom B, with each spin of opposite sign. For obvious reasons, this is called a *covalent (canonical) structure*. This covalent canonical structure is represented by the following wavefunction:

$$\psi_{\text{cov}} = [2(1 + \Delta^2)]^{-1/2}(|1s_A\,\overline{1s}_B| - |\overline{1s}_A\,1s_B|) \tag{12-149}$$

This wavefunction represents the fact that the two electrons have opposite spins (one AO is α and the other is β) and also represents the *indistinguishability* of the two electrons (hence, the two terms differ only in spin assignments). A second canonical structure, called an *ionic canonical structure*, is written "H^+, H^-" and assumes that both electrons are on a single atom. The wavefunction describing this situation is

$$\psi_{\text{ionic}} = [2(1 + \Delta^2)]^{-1/2}(|1s_A\,\overline{1s}_A| + |\overline{1s}_B\,1s_B|) \tag{12-150}$$

Again, indistinguishability of the two electrons and two nuclei is assured by using two terms differing only by the atoms they represent.

The covalent wavefunction (12-149) was first employed as an approximate dihydrogen wavefunction by Heitler and London[46] and is often called the *Heitler-London* wavefunction. When the orbitals used are unscaled Slater 1*s* functions, the calculation produces a dissociation energy of $D_e = 3.14$ eV at $R_e = 1.65\ a_0$. This is 0.5 eV larger than the simple unscaled MO wavefunction (12-75).

It is interesting to compare the one-electron density matrices of the simple MO calculation and the Heitler-London VB calculation. These are given, respectively, by

$$\begin{aligned}\gamma(1|1)_{\text{MO}} &= \frac{1}{1+\Delta}\,[1s_A^2(1) + 1s_B^2(1) + (2)1s_A(1)1s_B(1)] \\ \gamma(1|1)_{\text{VB}} &= \frac{1}{1+\Delta^2}\,[1s_A^2(1) + 1s_B^2(1) + 2\Delta 1s_A(1)1s_B(1)]\end{aligned} \tag{12-151}$$

The two electron density distributions are very similar, but (since $\Delta < 1$) the MO wavefunction leads to a slightly greater internuclear density (overlap population) than does the VB wavefunction.

An improved wavefunction is obtained by combining the two canonical wavefunctions (12-149) and (12-150) in the linear form

$$\psi = c_1\psi_{\text{cov}} + c_2\psi_{\text{ionic}} \tag{12-152}$$

[46] W. Heitler and F. London, *Z. Physik.*, **44**:455 (1927). See also Y. Sugiura, *Z. Physik.*, **45**:484 (1927). A scaled calculation was done by S. Wang, *Phys. Rev.*, **31**:579 (1928). This gave $D_e = 3.76$ eV and $R_e = 1.44\ a_0$ with $\eta = 1.17$.

Note that this is just the Weinbaum function described earlier; it is equivalent to the CI wavefunction which contains a doubly excited configuration in the $\sigma_g 1s$, $\sigma_u 1s$ basis. This wavefunction (when scaling is employed) leads to $D_e = 4.026$ eV at $R_e = 1.42\ a_0$ using a scale factor of $\eta = 1.193$.

The nature of the Weinbaum function is conveniently analyzed by considering the generalized Heitler-London function

$$\xi = [2(1+\Delta^2)]^{-1/2}(|\phi_A\ \bar{\phi}_B| - |\bar{\phi}_A\ \phi_B|) \tag{12-153}$$

where Δ now represents $\langle \phi_A | \phi_B \rangle$. The AOs are defined as follows:

$$\phi_A = 1s_A + k1s_B \qquad \phi_B = 1s_B + k1s_A \tag{12-154}$$

where k is a variational parameter. The AOs ϕ_A and ϕ_B are called *semilocalized* AOs; their appearance is shown in Fig. 12-22. Note that ϕ_A is primarily a $1s$ AO centered on nucleus A but contains a small admixture of a $1s$ AO centered on the other nucleus. Expansion of the generalized Heitler-London wavefunction and suitable rearrangement leads to

$$\xi = N[(1+k^2)\psi_{\text{cov}} + 2k\psi_{\text{ionic}}] \tag{12-155}$$

where N is determined by normalization. Thus, this is just the Weinbaum function in which $c_1 = N(1+k^2)$ and $c_2 = 2kN$. This illustrates the function of adding the ionic canonical wavefunction to the covalent canonical wavefunction; the resulting wavefunction is better able to represent the electronic structure of the bonding region of the molecule.

The VB method can be recast in a self-consistent framework. Using the dihydrogen molecule as a specific example, let us rewrite the Heitler-London "covalent" wavefunction in a generalized form:

$$\psi = [2(1+\Delta^2)]^{-1/2}(|u\ \bar{v}| - |\bar{u}\ v|) \tag{12-156}$$

where u and v are a pair of generalized orbitals not necessarily constrained to be orthogonal. Now instead of choosing these orbitals as definite functions (say, $1s_A$ and $1s_B$ as in the Heitler-London calculation), we allow these to be as flexible as possible by expanding them in terms of some basis set of functions, the coefficients of the expansion being determined variationally. In the simplest

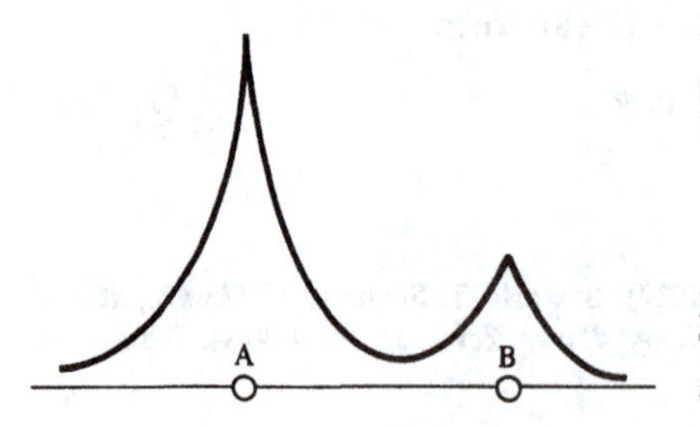

FIGURE 12-22
The semilocalized orbital $\phi_A = 1s_A + k1s_B$ $(k < 1)$ as it appears along the internuclear axis.

case, one could expand u and v in terms of the basis $1s_A$ and $1s_B$:

$$u = c_1 1s_A + c_2 1s_B \qquad v = d_1 1s_A + d_2 1s_B \tag{12-157}$$

where c_1, c_2, d_1, and d_2 are determined variationally. This particular expansion leads to a wavefunction which is identical with that obtained by the MCSCF wavefunction based on the MOs $\sigma_g 1s$ and $\sigma_u 1s$; that is, the coefficients are determined solely by symmetry, and u and v simply turn out to be $\sigma_g 1s$ and $\sigma_u 1s$ MOs.

The general version of the approach described above for the specific example of dihydrogen is known as the *generalized valence bond* (GVB) method; it has been developed principally by Goddard and coworkers.[47]

EXERCISES

12-20. Demonstrate that the normalization constant in the expanded generalized Heitler-London wavefunction has the explicit form

$$N = N'\left\{\frac{2(1+\Delta^2)}{2[1+4k^2+(k^2+1)^2\Delta^2+4k(k^2+1)\Delta]}\right\}^{1/2}$$

where $N' = [2(1+\Delta^2)]^{-1/2}$.

12-21. Verify the expression for the one-electron density matrices given in Eq. (12-151).

12-22. Show that the Heitler-London wavefunction leads to the energy expression

$$\langle \psi_{\text{cov}}|\hat{H}|\psi_{\text{cov}}\rangle = (1+\Delta^2)^{-1}(2\langle 1s_A|h|1s_A\rangle + 2\Delta\langle 1s_A|h|1s_B\rangle + \langle AB|g|AB\rangle + \langle AB|g|BA\rangle)$$

Repeat for the ionic function to show that

$$\langle \psi_{\text{ionic}}|\hat{H}|\psi_{\text{ionic}}\rangle = (1+\Delta^2)^{-1}(2\langle 1s_A|h|1s_A\rangle + 2\Delta\langle 1s_A|h|1s_B\rangle + \langle AA|g|AA\rangle + \langle AA|g|BB\rangle$$

12-23. Show that the equation for the normalization factor for the semilocalized orbital reduces to the Heitler-London normalization factor when $k = 0$.

12-11 NATURAL ORBITAL ANALYSIS OF MOLECULAR WAVEFUNCTIONS

Davidson and Jones[48] analyzed the very accurate James and Coolidge type of wavefunction [50 terms; see Eq. (12-104)] used in the Kolos and Roothaan calculation on dihydrogen. These workers found that the first four natural orbitals led to an energy of $-1.169884E_h$, which corresponds to $D_e =$

[47] W. A. Goddard, III, T. H. Dunning, Jr., W. J. Hunt, and P. J. Hay, *Acc. Chem. Res.*, **6**:368 (1973); W. A. Goddard, III, and L. B. Harding, *Ann. Rev. Phys. Chem.*, **29**:363 (1978).

[48] E. R. Davidson and L. L. Jones, *J. Chem. Phys.*, **37**:577, 2966 (1962).

4.522543 eV (experimental value = 4.746 eV). These first four natural orbitals had an overlap of 0.999667 with the complete 50-term function and accounted for about 80 percent of the correlation energy. The researchers also found that the first natural orbital configuration led to an energy within $0.00002E_h$ of the Hartree-Fock energy. The second, third, and fourth natural orbital configurations can be interpreted in terms of all three types of correlation: in-out, left-right, and angular.

Braunstein and Simpson[49] have pointed out that the covalent and ionic wavefunctions used in the VB calculation on dihydrogen [see Eq. (12-152)] have an overlap of 0.95. This means that this wavefunction does not define the terms "covalent" and "ionic" in a unique manner. Shull[50] has studied the traditional VB covalent and ionic wavefunctions by use of natural orbital expansions. Shull employed the artifice of dividing the molecular configuration space into two regions by means of a plane perpendicular to the internuclear axis and passing through its midpoint. This produces a division of a two-configuration, two-electron wavefunction into two orthogonal parts, each having optimum properties associated with the plane and intuitively corresponding to the terms "ionic" and "atomic." Physical properties which depend on the wavefunction (or, rather, its complex-conjugate "square") then involve three well-defined parts: ionic, atomic, and ionic-atomic cross terms. It is this last part which corresponds to what a chemist usually terms *covalency*. In the natural orbital expansion this division is virtually invariant to the choice of basis and thus makes possible a unique definition of the ionic and covalent characters of a bond. Obviously, what chemists normally regard as "covalent" character contains a large admixture of ionic character. This does not mean that these concepts are useless to the laboratory chemist, but it does mean that the true meanings of the concepts are not what they are ordinarily regarded to be. Klaus Ruedenberg of Iowa State University, whose research group has made many outstanding contributions toward understanding of the chemical bond, summarized the situation as follows in a 1962 review article which remains pertinent today:

> As yet, the physical nature of the chemical bond is little understood in many essential details, and the reason for this must be seen in the mathematical difficulties which are encountered in solving molecular quantum-mechanical problems. The older concepts on the subject have suffered from being based on wave functions which, by virtue of their simplicity, permitted plausible interpretations (or so it was hoped), but turned out to be inadequate approximations to the true solutions. The recent progress towards better approximations, on the other hand, is leading to increasingly complicated wave functions whose conceptual meaning is becoming less and less lucid.

---

[49] J. Braunstein and W. T. Simpson, *J. Chem. Phys.*, **23**:174, 176 (1955).

[50] H. Shull, *J. Am. Chem. Soc.*, **82**:1287 (1960) and *J. Appl. Phys.*, **33**:290 (1962).

> While there used to exist hope of arriving at satisfactory results by supplementing mathematically unjustified approximations with chemical and physical intuition in such a way as to achieve an all-around cancellation of errors, it now seems to transpire that *bona fide* solutions of the mathematical problems, based on justified approximations only, cannot be sidestepped if quantitative reliability and unambiguous predictions are to be achieved in the absence of close analogies. It has furthermore become apparent that *bona fide* wave functions must be determined according to methods which are largely influenced by considerations of mathematical practicability and computational efficiency, and that they will have complex appearances of various forms. Thus, there has arisen the need for a uniform and generally applicable procedure of interpretation leading to a meaningful analysis of the physical and chemical significance of molecular wave functions.[51]

Shull[52] considered natural orbital expansions of all dihydrogen wavefunctions whose spatial portions could be represented by the general form

$$\psi = N[u(1)v(2) + v(1)u(2)] \tag{12-158}$$

where u and v are distinct spatial orbitals and are not necessarily orthogonal. In the specific example of the Heitler-London "covalent" wavefunction, $u = 1s_A$ and $v = 1s_B$. Since these AOs are not orthogonal, it is convenient to replace them with the orthonormal set

$$u' = [2(1+\Delta)]^{-1/2}(1s_A + 1s_B) \qquad v' = [2(1-\Delta)]^{-1/2}(1s_A - 1s_B) \tag{12-159}$$

Substituting u' and v' for u and v in Eq. (12-158) leads to the wavefunction in natural orbital form

$$\psi_{NO} = [2(1+\Delta^2)]^{-1/2}[(1+\Delta)u'(1)u'(2) + (1-\Delta)v'(1)v'(2)] \tag{12-160}$$

Thus, the natural orbitals are given by

$$\chi_1 = [2(1+\Delta)]^{-1/2}(1s_A + 1s_B) \qquad \chi_2 = [2(1-\Delta)]^{-1/2}(1s_A - 1s_B) \tag{12-161}$$

and the natural orbital coefficients are

$$\sqrt{n_1} = \frac{1+\Delta}{[2(1+\Delta^2)]^{1/2}} \quad \text{and} \quad \sqrt{n_2} = \frac{1-\Delta}{[2(1+\Delta^2)]^{1/2}} \tag{12-162}$$

Using the value of $R = 1.65\ a_0$, which Heitler and London found minimized the energy, produces

$$n_1 = 0.9658 \qquad n_2 = 0.0342 \tag{12-163}$$

---

[51] K. Ruedenberg, *Rev. Mod. Phys.*, **34**:326 (1962).

[52] H. Shull, *J. Chem. Phys.*, **30**:1405 (1959).

EXERCISE

12-24. Show that the natural orbitals for the Heitler-London wavefunction can be arrived at via the following alternative route:

1. Put the first-order density matrix [Eq. (12-151)] into the matrix form:

$$\gamma = \begin{bmatrix} N^2 & N^2\Delta \\ N^2\Delta & N^2 \end{bmatrix}$$

where $N^2 = \dfrac{1}{2(1+\Delta^2)}$

$\Delta = \langle 1s_A | 1s_B \rangle$

2. Diagonalize this matrix and use the eigenvectors of the transformation matrix to form orthonormal linear combinations of $1s_A$ and $1s_B$. Verify the numerical values of n_1 and n_2 by use of Eq. (12-33) for the overlap integral Δ.

12-12 MOLECULAR PERTURBATION CALCULATIONS

The main advantage of CI (including MCSCF) is that it is variational and thus provides upper bounds to the energy. The main disadvantage of the method is that it is not size-consistent unless all possible excitations are considered. Perturbation theory, on the other hand, is not variational (except at the first order) but is size-consistent at each level, i.e., at all orders. This very important property of perturbation theory was first conjectured by K. Brueckner and later proved formally by Goldstone in his famous linked-cluster theorem.

As discussed in Chap. 6, perturbation theory (in the Rayleigh-Schrödinger formulation) expresses the hamiltonian of a system as a sum of two parts: the zeroth-order hamiltonian, $\hat{H}^{(0)}$, and the perturbation, namely,

$$\hat{H} = \hat{H}^{(0)} + V \tag{12-164}$$

One particularly useful way of partitioning the hamiltonian of a many-electron system was proposed by Møller and Plesset[53] and leads to what is now known as the *Møller-Plesset perturbation theory* (*MPPT*). The zeroth-order hamiltonian is the Hartree-Fock hamiltonian

$$\hat{H}^{(0)} = \sum_{\mu} \hat{F}(\mu) = \sum_{\mu} (h_\mu + V_\mu^{\text{eff}}) \tag{12-165}$$

[53] C. Møller and M. S. Plesset, *Phys. Rev.*, **46**:618 (1934). This paper also points out that the Hartree-Fock method gives all one-electron properties, e.g., the dipole moment, correct to the first order. See also G. G. Hall, *Phil. Mag.*, **6**:249 (1961) and M. Cohen and A. Dalgarno, *Proc. Phys. Soc.* (*London*), **77**:748 (1961).

and the perturbation is given by

$$V = \sum_{\mu<\nu} \frac{1}{r_{\mu\nu}} - \sum_{\mu} V_{\mu}^{\text{eff}} \tag{12-166}$$

It should be noted that this method gives the Hartree-Fock energy as the sum of the zeroth- and first-order corrections, namely,

$$E(\text{SCF}) = E_0^{(0)} + E_0^{(1)} \tag{12-167}$$

Thus an exact expression for the correlation energy is given by summing all the perturbation-energy corrections from the second order on:

$$E_{\text{corr}} = \sum_{n=2}^{\infty} E_0^{(n)} \tag{12-168}$$

The formal treatment of MPPT requires the use of diagrammatic techniques whose discussion is out of the scope of this text.[54] As a partial indication of the scope of this method, Table 12-8 indicates some MPPT calculations on dihydrogen using a variety of basis sets. In each of the examples listed, a Hartree-Fock calculation was carried out with the basis indicated and then used as the basis of an MPPT calculation of the second- and third-order perturbation energies. As might be expected, the ability of the MPPT method to account for the correlation energy is improved by using a larger basis set in the Hartree-Fock calculation.

[54] The interested reader will find a lucid and more detailed account in A. Szabo and N. S. Ostlund, *Modern Quantum Chemistry: Introduction to Advanced Electronic Structure Theory*, Macmillan, New York, 1982.

TABLE 12-8
MPPT calculations on dihydrogen at $R = 1.4\ a_0$

Basis	$E_0^{(2)}/E_h$	% full CI	$(E_0^{(2)} + E_0^{(3)})/E_h$	% full CI
STO-3G	−0.0132	64	−0.0180	87
4-31G	−0.0174	70	−0.0226	91
$(10s, 5p, 1d)$*	−0.0321	81	−0.0376	95
Exact†			−0.0409	

* The $(10s, 5p, 1d)$ results come from two sources: $E_0^{(2)}$ from J. M. Schulman and D. N. Kaufman, *J. Chem. Phys.*, **53**:477 (1970) and $E_0^{(3)}$ from U. Kaldor, *J. Chem. Phys.*, **62**:4634 (1975).

† The exact calculation is from W. Kolos and L. Wolniewicz, *J. Chem. Phys.*, **49**:404 (1968).

SUGGESTED READINGS

Čársky, P., and M. Urban: Ab Initio *Calculations*: *Methods and Applications in Chemistry*, Springer-Verlag, New York, 1980.

Christoffersen, R. E: *Basic Principles and Techniques of Molecular Quantum Mechanics*, Springer-Verlag, New York, 1989.

Hehre, W. J., L. Radom, P. v.R. Schleyer, and J. A. Pople: Ab Initio *Molecular Orbital Theory*, Wiley-Interscience, New York, 1986.

Herzberg, G.: *Spectra of Diatomic Molecules*, Van Nostrand, Princeton, N.J., 1950.

Schaefer, H. F., III: *The Electronic Structure of Atoms and Molecules*, Addison-Wesley, Reading, Mass., 1972.

Szabo, A., and N. S. Ostlund: *Modern Quantum Chemistry*: *Introduction to Advanced Structure Theory*, Macmillan, New York, 1982. Appendix B of this book contains an *ab initio* SCF FORTRAN program which has been translated and modified by L. H. Reed and A. R. Murphy [*J. Chem. Ed.* **64**:789 (1987)] into a highly interactive BASIC program for the Apple II + and IIe microcomputers. This program, obtainable from Project SERAPHIM, performs a Hartree-Fock-Roothaan calculation on a two-electron diatomic molecule and produces orbital energies, eigenvectors, the electronic energy, the total molecular energy (including nuclear repulsion), various matrices, and a Mulliken population analysis.

CHAPTER

13

THE ELECTRONIC STRUCTURE OF NONLINEAR MOLECULES

Fully optimized electron structure calculations on nonlinear molecules become increasingly time-consuming as the number of atoms increases. Part of this is because of the concomitant increase in the total number of electrons, but most of the increase in calculational time comes from the large number of bond distances and bond angles which must be optimized. Of course, if the molecule has sufficient symmetry, bond angles may not require optimization; e.g., the ground state of methane (CH_4) is generally assumed to be tetrahedral and seldom would one insist on verifying the tetrahedral H–C–H bond angles by a full search of all other possibilities. However, in some other molecules, e.g., hydrogen peroxide (H_2O_2), one might very well carry out detailed calculations on a rather wide range of bond angles (and bond lengths) in an attempt to predict an accurate geometry. However, considerable computational time can be saved if one can begin with a set of bond lengths and angles which are reasonable approximations to the equilibrium values. One of the best ways of coming up with a reasonable beginning geometry is to use Molecular Mechanics, a computer program which uses classical mechanics to "guess" at a most probable equilibrium geometry.[1]

---

[1] U. Burkert and N. L. Allinger, *Molecular Mechanics*, ACS monograph 177, American Chemical Society, Washington, D.C., 1982.

In this chapter we will discuss—among other topics—various *ab initio* calculations on some of the smaller nonlinear molecules. The following chapter will then discuss calculations on much larger molecules, but the calculations will not be of the *ab initio* type. Rather, the amount of computational time will be decreased rather significantly by employing various types of approximations in which some of the integrals are not computed—or are approximated in greatly simplified ways. Such calculations constitute *semiempirical* methods.

It should be noted at the outset that whereas in linear molecules one can use one of the components of the electronic angular momentum as an aid in the classification of states, this is no longer possible for nonlinear molecules, since neither $\hat{L}^2$ nor $\hat{L}_z$ commutes with the hamiltonian. Instead, we will use spin angular momentum and the point groups to which the molecules belong as the basis of classification of states.

13-1 THE AH_n MOLECULES: METHANE, AMMONIA, AND WATER

Methane (CH_4), ammonia (NH_3), and water (H_2O) are 10-electron molecules and hence isoelectronic with the neon atom ($1s^22s^22p^6$). Considering the large number of diverse chemical reactions in which these molecules participate, they are undoubtedly among the most important of all chemical species, and, furthermore, calculations involving them serve as prototypes for calculations on larger molecules. Consequently, there have been many, many calculations carried out on these molecules in an attempt to understand their electronic structures thoroughly enough to provide insight into their chemical behaviors. In this section we will consider a few of the highlights of some of the simpler calculations on these molecules.

METHANE. The minimal basis set for methane comprises the AOs $1s$, $2s$, $2p_x$, $2p_y$, and $2p_z$ of carbon and four $1s$ AOs of hydrogen, which we will label H_1, H_2, H_3, and H_4. The coordinate and labeling systems for this molecule are shown in Fig. 13-1. Since the carbon AOs contain all the symmetry elements of

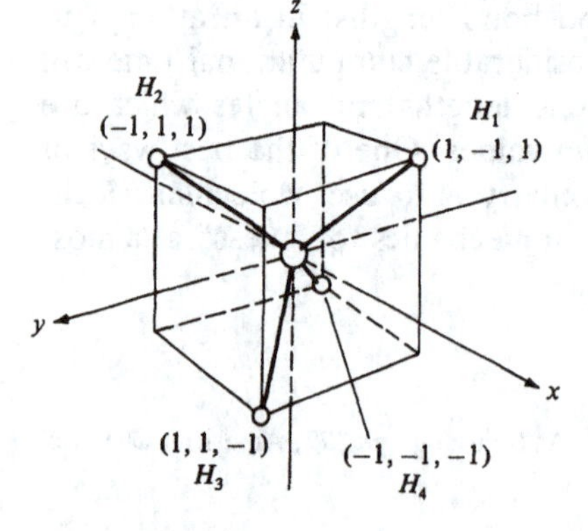

FIGURE 13-1
The tetrahedral methane molecule inscribed in a cube with sides of length 2. There are three S_4 axes coincident with the x, y, and z axes; three C_2 axes coincident with x, y, and z axes; four C_3 axes coincident with the CH_1, CH_2, CH_3, and CH_4 bonds; and six symmetry planes containing the atoms CH_1H_4, CH_1H_2, CH_2H_3, CH_2H_4, CH_3H_4, and CH_3H_1.

the T_d point group to which methane belongs, these AOs are immediately classified as to the irreducible representations to which they belong. Using the T_d character table from App. 7 leads to the classifications:

$$\begin{aligned} &A_1\text{: } 1s, 2s \\ &T_2\text{: } 2p_x, 2p_y, 2p_z \end{aligned} \tag{13-1}$$

The hydrogen AOs belong to a reducible representation, and therefore, linear combinations of them will belong to the irreducible representations contained therein. There linear combinations can be deduced in a formal fashion by the use of projection operators, but a straightforward inspection method is much simpler and faster. Using Fig. 13-1 for reference, we see that a function such as $(H_1 + H_2) - (H_3 + H_4)$ has the same symmetry as a carbon $2p_z$ AO, and, consequently, belongs to the T_2 irreducible representation. Similarly, $(H_1 + H_4) - (H_2 + H_3)$ is like $2p_y$ and $(H_1 + H_3) - (H_2 + H_4)$ is like $2p_x$. A fourth linear combination, $H_1 + H_2 + H_3 + H_4$ has the same symmetry as $1s$ and $2s$ of carbon and is of A_1 symmetry. Summarizing the above and introducing a simple notation for each linear combination:

$$\begin{aligned} G_s &= H_1 + H_2 + H_3 + H_4 \\ G_x &= (H_1 + H_3) - (H_2 + H_4) = H_1 - H_2 + H_3 - H_4 \\ G_y &= (H_1 + H_4) - (H_2 + H_3) = H_1 - H_2 - H_3 + H_4 \\ G_z &= (H_1 + H_2) - (H_3 + H_4) = H_1 + H_2 - H_3 - H_4 \end{aligned} \tag{13-2}$$

The linear combinations G_s, G_x, G_y, and G_z are often called *group orbitals*. The relationships in Eq. (13-2) can also be written in matrix form as follows:

$$[H_1 \quad H_2 \quad H_3 \quad H_4]\begin{bmatrix} 1 & 1 & 1 & 1 \\ 1 & -1 & -1 & 1 \\ 1 & 1 & -1 & -1 \\ 1 & -1 & 1 & -1 \end{bmatrix} = [G_s \quad G_x \quad G_y \quad G_z]$$

Note that the transformation matrix is unitary if multiplied by $\frac{1}{2}$; this is just the factor that would be needed to normalize the group orbitals if their components were orthonormal.

The minimal symmetry-adapted basis now is

$$\begin{aligned} &A_1\text{: } 1s, 2s, G_s \\ &T_2\text{: } 2p_x, 2p_y, 2p_z, G_x, G_y, G_z \end{aligned} \tag{13-3}$$

These basis functions can be used to form nine MOs, five of which will be used in the ground-state electron configuration. The nine MOs have the general forms

$$\begin{aligned} ja_1 &= c_{1j}(1s) + c_{2j}(2s) + c_{3j}G_s \qquad && j = 1, 2, 3 \\ jt_2 &= c_{1j}(2p_i) = jt_i && j = 1, 2 \text{ for } i = x, y, z \end{aligned} \tag{13-4}$$

Note that there are two sets of the MOs jt_x, jt_y, and jt_z and each set is triply degenerate. Note also that the method we have used to define these triply degenerate MOs produces them in orthogonal form; actual calculations with some of the currently available computer programs do not necessarily produce the forms shown here.

It is frequently possible to guess which MOs minimize the total electronic energy, without actually doing any calculations. The lowest MO is most likely $1a_1$, in which the carbon $1s$ AO plays a predominant role; i.e., the coefficient $|c_{11}|$ is much larger than either $|c_{21}|$ or $|c_{31}|$. The next MO should be $2a_1$, in which the coefficients of carbon $2s$ and the group orbital G_s predominate. One would not expect $3a_1$ to come next, since that would force the use of only two of the $1t_2$ MOs to complete the configuration and we know that the three $1t_2$ MOs ($1t_x$, $1t_y$, and $1t_z$) are degenerate and thus have the same orbital energies. Consequently, the ground-state electron configuration is predicted to be

$$(1a_1)^2(2a_1)^2(1t_x)^2(1t_y)^2(1t_z)^2 \qquad \text{or} \qquad (1a_1)^2(2a_1)^2(1t_2)^6 \tag{13-5}$$

The direct product of all the MOs shows that the resulting determinantal wavefunction is of A_1 symmetry, and the ground state is represented by the term symbol ${}^1A_1$.

Since so many calculations have been reported on the methane molecule, it would be impossible to attempt any adequate summary of them here. Consequently, we will arbitrarily choose only a few of these and discuss some of their main features. Pitzer[2] carried out Hartree-Fock calculations using a minimal basis set of Slater orbitals, and consequently, these calculations can be compared directly with the preceding discussion. As the calculation summary in Table 13-1 shows, the MOs have the basic compositions predicted on the basis of qualitative arguments. In particular, $1a_1$ is almost entirely $1s$ of

[2] R. M. Pitzer, *J. Chem. Phys.*, **46**:4871 (1967).

TABLE 13-1
Ground-state MOs and orbital energies for methane for a C–H distance of 2.05 a_0

	AO coefficients					
MO symbol	**$1s$**	**$2s$**	**G_s**	**$2p_x$**	**G_x**	**MO energy, ϵ/E_h**
$1a_1$	0.9947	0.0256	−0.0047			−11.2049
$2a_1$	−0.2158	0.6037	0.1865			−0.9253
$1t_x$				0.5539	0.3178	−0.5384

Source: R. M. Pitzer, *J. Chem. Phys.*, **46**:4871 (1967).

carbon. Consequently, the $1a_1$ MO energy is very nearly the $1s$ orbital energy of carbon ($-11.3255E_h$ for the ^3P ground state).

Pitzer obtained a molecular energy of $-40.12822E_h$ for a carbon-hydrogen bond length of $2.05\,a_0$. Parabolic interpolation led to a slightly lower energy of $-40.12827E_h$ at $R_{CH} = 2.0587\,a_0$. The spectroscopic dissociation energy of methane was calculated to be

$$\begin{aligned} D_e &= 4E_H + E_C - E_{CH_4} \\ &= 4(-0.5000) + (-37.8558) - (-40.1282) = 0.2724E_h \end{aligned} \tag{13-6}$$

where the atomic energies of hydrogen and carbon are the exact values. This value is equivalent to 7.4 eV. The experimental value for the process

$$CH_4(g) \rightarrow C(g) + 4H(g)$$

is about 17 eV. Note that use of Hartree-Fock SCF atomic energies would lead to better agreement.

The first ionization energy of a molecule may be written in the form

$$I = -\epsilon + (E_{corr} - E^+_{corr}) - E_{reorg} \tag{13-7}$$

where ϵ is the energy of the highest-occupied MO, E_{corr} is the correlation energy of the neutral molecule, E^+_{corr} is the correlation energy of the ion (with the same geometry as that of the neutral molecule), and E_{reorg} is the energy of reorganization, i.e., the energy involved when the atoms rearrange to some slightly different geometry in response to the different electronic environment. Note that if Koopmans' theorem were exact, then the last three quantities would total zero. In actuality, the correlation energy of the neutral molecule is generally greater than the correlation energy of the ion, and since E_{reorg} is generally positive, the sum of these last three terms is a relatively small nonzero quantity so that Koopmans' theorem is approximately valid. Using Pitzer's data, Koopmans' theorem identifies the first ionization energy of methane with the energy of the $1t_x$ MO, that is, $-(-0.5384)E_h$, or 14.6 eV. The experimental value is somewhat lower: 12.6 eV. One way of taking all the additional factors into account is to reject Koopmans' theorem in favor of the direct calculation

$$I = E(\text{ion}) - E(\text{neutral molecule}) \tag{13-8}$$

where both E(ion) and E(neutral molecule) are calculated by the same model and both are fully optimized. Such an approach was used by Schwartz[3] to calculate the *largest* ionization energy of CH_4, that is, the energy to remove an electron from the $1a_1$ MO. Schwartz's calculation (which employed a gaussian basis set) produced an energy of -304.9 eV for the $1a_1$ MO. The experimental value of the largest ionization energy is 290.7 eV, about 14 eV lower than

[3] M. E. Schwartz, *Chem. Phys. Lett.*, 5:50 (1967).

Koopmans' theorem predicts. The total molecular energy of CH_4 obtained by Schwartz was $-40.1812E_h$ ($0.05E_h$ lower than Pitzer's value). Schwartz then used the same basis to obtain the CH_4^+ energy [in the $(1a_1)(2a_1)^2(1t_2)^6$ configuration] of $-29.4873E_h$. This leads to a value for the largest ionization energy of

$$-29.4873-(-40.1812)=10.69E_h, \text{ or } 291 \text{ eV}$$

which is in excellent agreement with experiment.

It is of interest to compare a minimal basis set STO-6G calculation on methane with those of Pitzer and Schwartz.[4] Table 13-2 summarizes the orbital energies and MO coefficients obtained from a geometry-optimized calculation. When the experimental bond lengths were used, the total electronic energy turned out to be $-40.1090E_h$; geometry optimization lowered this to $-40.1107E_h$. The predicted C–H bond length turned out to be 2.038 a_0 compared with the experimental value of 1.997 a_0. Note that there is general agreement with the numerical values quoted in Table 13-1, which were obtained by Pitzer.

If one uses the methane geometry obtained by optimization of the minimal basis set STO-6G calculation and extends the basis by adding $2s$ and $2p$ to hydrogen and $3d$ to carbon, the total energy of methane becomes $-40.1447E_h$. The use of the extended basis also reduces the extent of the C–H charge separation; for example, the minimal basis set predicts a net charge of -0.274 on C and $+0.067$ on each H (using a Mulliken population analysis) compared with the extended basis values of -0.209 and $+0.052$, respectively. However, just what such charge distributions signify is open to question. As Turner, Saturno, Hauk, and Parr pointed out years ago,[5] the concept of a bond

[4] All STO-6G calculations referred to in this chapter were carried out by the author using the GAMESS program.

[5] A. G. Turner, A. F. Saturno, P. Hauk, and R. G. Parr, *J. Chem. Phys.*, **40**:1919 (1964).

TABLE 13-2
Ground-state MOs and orbital energies for methane for a C–H distance of 2.038 a_0

MO symbol	AO coefficients					MO energy, ϵ/E_h
	$1s$	$2s$	G_s	$2p_x$	G_x	
$1a_1$	0.9951	0.0246	−0.0046			−11.1818
$2a_1$	−0.2122	0.6235	0.1814			−0.9170
$1t_x$				0.5745	0.2923	−0.5252

Note: All calculations were made with a GAMESS program using STO-6G minimal basis with geometry optimization.

moment in a molecule such as methane is illusive, elusive, and ill defined and should be discussed only in a specific context for a particularly defined meaning.

Figure 13-2 illustrates contours of constant MO values for some of the ground-state MOs of methane.

AMMONIA. The ammonia molecule is of C_{3v} symmetry. Theoretical prediction of the geometry of this molecule is more of a challenge than is CH_4, since unlike the latter, the bond angles are not determined by symmetry alone. Using a coordinate system in which the C_3 rotation axis is the z axis, the symmetry-adapted minimal basis set is

$$\begin{aligned} &A_1\text{: } 1s, 2s, 2p_z, G_s \\ &E\text{: } 2p_x, 2p_y, G_x, G_y \end{aligned} \tag{13-9}$$

where $1s$, $2s$, $2p_x$, $2p_y$, and $2p_z$ are nitrogen AOs and the group orbitals are composed of hydrogen AOs as follows:

$$\begin{aligned} G_s &= \frac{1}{\sqrt{3}}(H_1 + H_2 + H_3) \\ G_x &= \frac{1}{\sqrt{6}}(2H_1 - H_2 - H_3) \\ G_y &= \frac{1}{\sqrt{2}}(H_2 - H_3) \end{aligned} \tag{13-10}$$

The relationship leading to the G_s group orbital is fairly obvious; the doubly degenerate G_x and G_y combinations are most easily chosen by using projection operators to form three linear combinations and then making two orthogonal combinations from these. The resulting MOs are written $1a_1$, $2a_1$, $3a_1$, $4a_1$, $1e_x$, $1e_y$, $2e_x$, and $2e_y$.

Alternatively, the group orbitals G_x and G_y may be chosen as follows:

$$\begin{aligned} G_x &= \frac{1}{\sqrt{2}}(H_1 - H_2) \\ G_y &= \frac{1}{\sqrt{2}}(H_2 - H_3) \end{aligned} \tag{13-11}$$

but these latter orbitals will not lead to orthogonal MOs in the doubly degenerate representation to which they belong. If we use arguments similar to those for methane, the ground-state electron configuration would be expected to be

$$(1a_1)^2(2a_1)^2(1e)^4(3a_1)^2$$

which is a $^1A_1$ state. Note that $(1e)^4$ refers to $(1e_x)^2(1e_y)^2$. The $1a_1$ MO should be largely nitrogen $1s$ and represents inner-shell nonbonding electrons. The $2a_1$

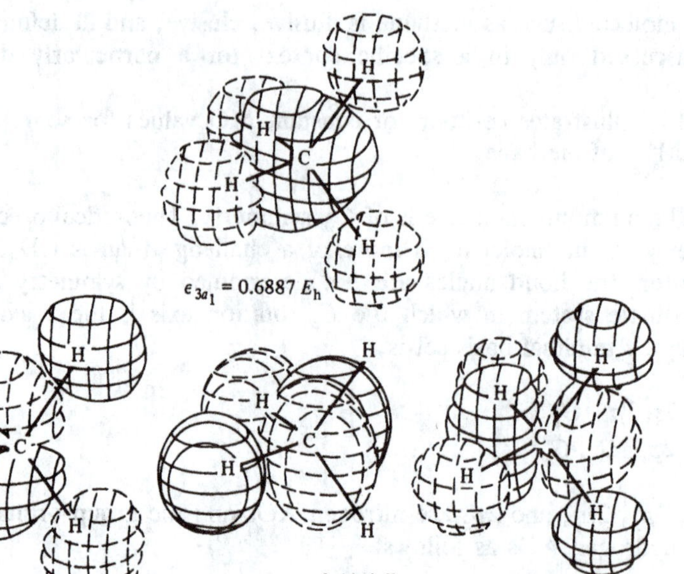

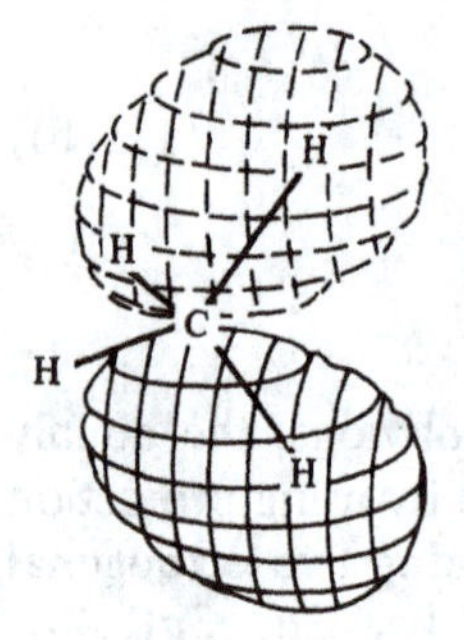

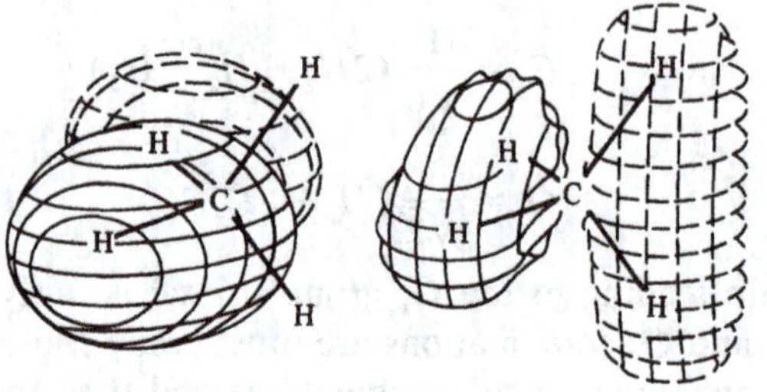

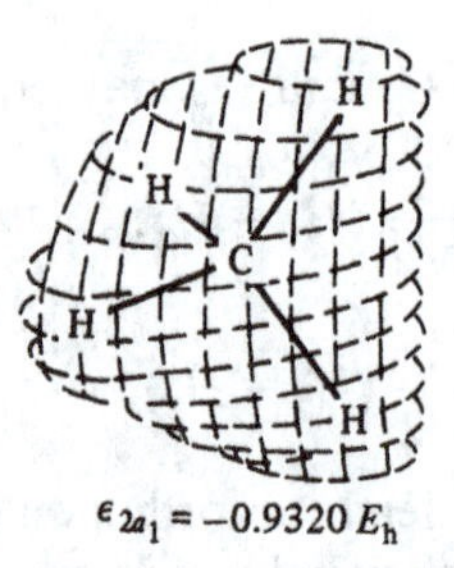

FIGURE 13-2
Contours of equal MO values for methane. (*Taken from W. L. Jorgensen and L. Salem*, The Organic Chemist's Book of Orbitals, *Academic, New York, 1973.*)

MO should be mainly nitrogen $2s$, with some nitrogen $2p_z$ and G_s. This MO and the two $1e$ MOs account for most of the N–H bonding pairs. The $3a_1$ MO should be mostly nitrogen $2p_z$ and represents, in part, the lone-pair or valence-shell nonbonding electrons as represented by the Lewis dot formula:

$$\begin{array}{c} \ddot{\text{N}} \\ \text{H}:\ddot{\text{N}}:\text{H} \\ \ddot{\text{H}} \end{array}$$

However, it should be borne in mind that the MOs defined in terms of irreducible representations of the point group do not generally lend themselves to descriptions of strictly localized bonding. Thus all five MOs play some role in an overall delocalized electron distribution of the molecule; our interpretation in terms of lone pairs, N–H bonding pairs, etc., does not correspond in a very precise way to the electron distributions represented by the MOs. Another way of saying this is that there are an infinite number of different ways of subdividing the total electron distribution of a molecule into subsets, and the Lewis model and the MO model do not do this in quite the same way. In a later section we will examine other ways of subdividing the total electron distribution such that localized bonding descriptions are facilitated.

When the ammonia calculation is done with a minimal basis set using STO-6G MOs and the experimental geometry (N–H bond length of 1.917 a_0 and an H–N–H bond angle of 106°47′), the electronic energy comes out to be $-55.294E_h$. The Hartree-Fock limit is estimated to be $-56.172E_h$. Also the dipole moment is predicted to be 2.54 D compared with the experimental value of 1.46 D. By contrast, a minimal basis set STO calculation by Palke and Lipscomb[6] (also using the experimental geometry) produces $E = -56.0052E_h$ and a dipole moment of 1.72 D. This latter calculation also predicts a charge of −0.464 on N and +0.155 on each H (based on a Mulliken population analysis).

When the ammonia basis set in the STO-6G calculation is extended by adding $2s$ and $2p$ AOs of hydrogen and $3d$ AOs of nitrogen, the energy is lowered slightly to $-55.578E_h$ and the dipole moment is also lowered slightly to 2.45 D. However, a rather dramatic improvement is made in the minimal basis set calculation by optimizing the geometry. This produces an energy of $-55.989E_h$ and a dipole moment of 1.90 D. The optimum geometry is predicted to be a N–H bond distance of 1.944 a_0 and an HNH bond angle of 104.46°. Thus the bond distance is predicted to be somewhat too long, and the bond angle is predicted to be somewhat too small. The MO coefficients and orbital energies obtained by this latter calculation are shown in Table 13-3. The N and C charges predicted by this calculation are almost exactly those reported in the calculation by Palke and Lipscomb.

One striking difference between the minimal basis experimental geometry calculation and the minimal basis geometry-optimized calculation reported

[6] W. E. Palke and W. N. Lipscomb, *J. Am. Chem. Soc.*, **88**:2384 (1966).

TABLE 13-3
Molecular orbitals of ammonia

AO	$1a_1$	$2a_1$	$1e_y$	$3a_1$
$1s$	0.9961	−0.2089		−0.0885
$2s$	0.1962	−0.7410		0.4759
$2p_z$	0.0039	0.1416		−0.8742
G_s	−0.0042	0.1573		−0.1356
$2p_y$			0.6468	
G_y			0.4680	
ϵ/E_h	−15.5130	−1.0929	−0.5660	−0.3625

Note: All calculations were made with the GAMESS program using an STO-6G minimal basis with geometry optimization.

above is in the relative energies of the $3a_1$ and $1e$ MOs. The former calculation produces $\epsilon(3a_1) = -0.588E_h$ and $\epsilon(1e) = -0.423E_h$, whereas the latter produces the opposite relative order: $\epsilon(3a_1) = -0.363E_h$ and $\epsilon(1e) = -0.566E_h$. Although it is difficult to interpret just what this means in terms of electronic structure, this occurrence is a clear demonstration of how apparently small changes in geometry affect the details of calculations. Certainly, the two calculations are at variance with respect to their predictions of the electronic structure of the ion resulting when ammonia loses its highest-energy electron.

If the geometry obtained from the minimal basis set is retained, but the basis is extended by adding $2s$ and $2p$ AOs of H and $3d$ AOs of N, the total energy comes out to be $-56.071E_h$—not too far from the estimated Hartree-Fock value.[7] Also, the dipole moment now comes out to be 1.36 D—much closer to the experimental value (but now too small). The rather marked decrease in the dipole moment is due largely to the inclusion of the $3d$ AO on nitrogen and the $2p$ AO on hydrogen; such orbitals permit more electron density to accumulate in the bonding regions and thus reduce the charge separation.

Figure 13-3 shows contours of constant MO values for some of the ammonia MOs.

WATER. The water molecule is of C_{2v} symmetry. The minimal basis set consists of H ($1s$) and O ($1s$, $2s$, $2p_x$, $2p_y$, and $2p_z$). If the coordinate system is chosen so that the principal axis is the z axis and the plane of the molecule is yz, these

[7] The effect of adding the $2s$ AO to hydrogen has virtually no effect on the energy and other properties, but a peculiarity of the GAMESS program makes it easier to include $2s$ and $2p$ together; these are given the same orbital exponent but differ in the expansion coefficients in the gaussian representation.

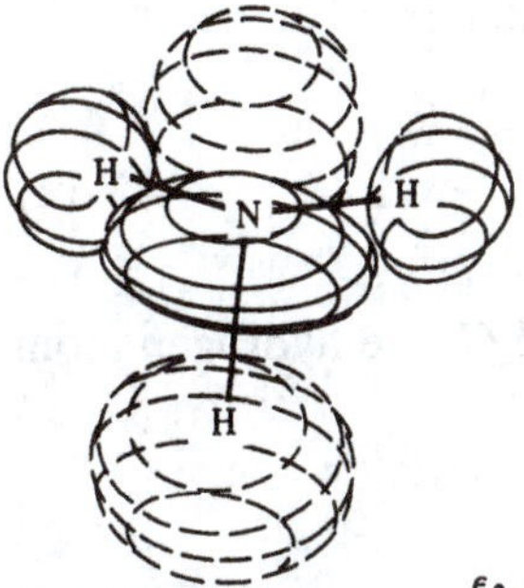

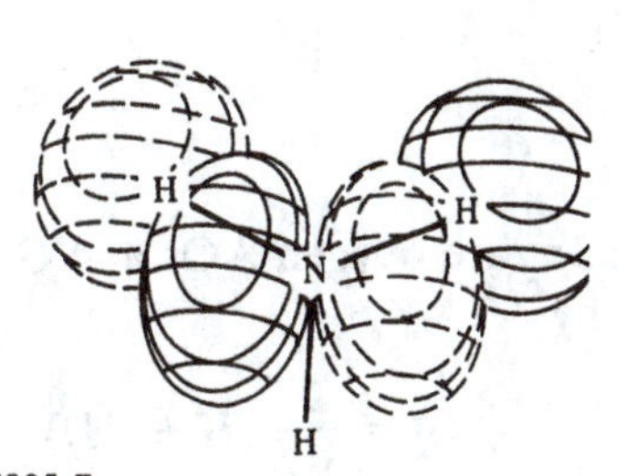

$\epsilon_{2e} = 0.6905\ E_h$

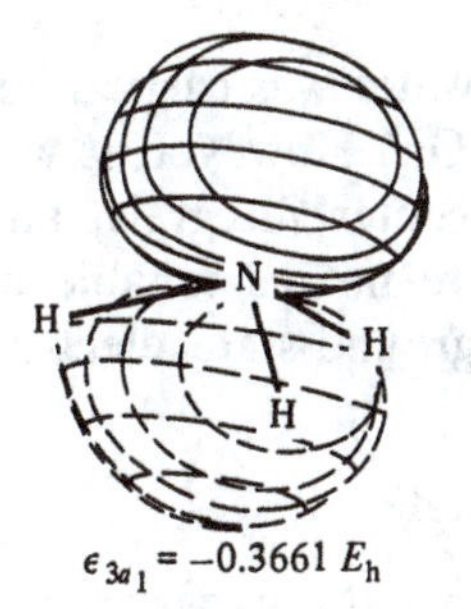

$\epsilon_{3a_1} = -0.3661\ E_h$

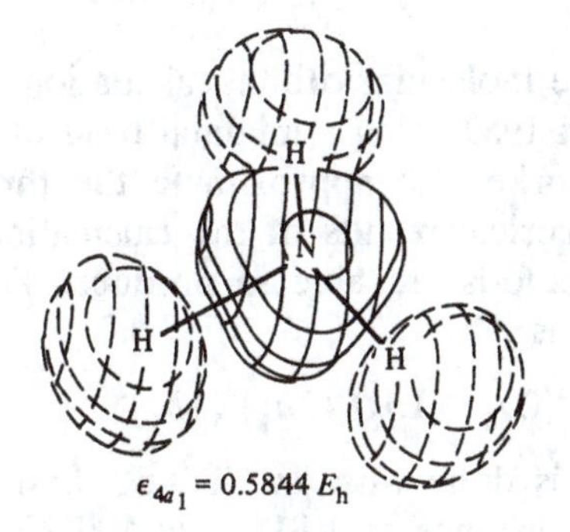

$\epsilon_{4a_1} = 0.5844\ E_h$

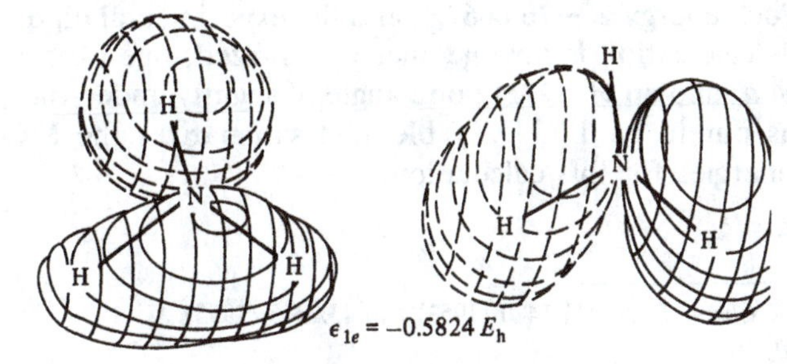

$\epsilon_{1e} = -0.5824\ E_h$

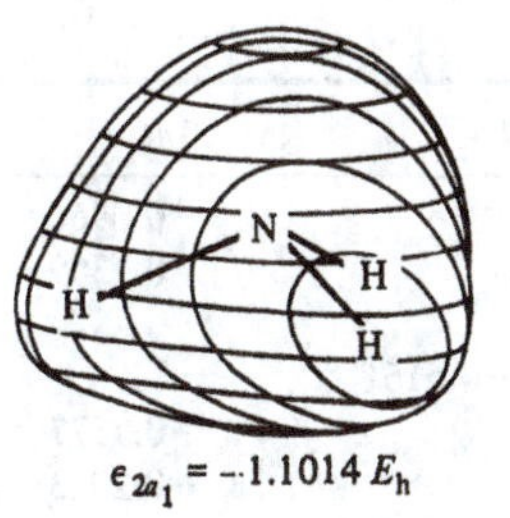

$\epsilon_{2a_1} = -1.1014\ E_h$

FIGURE 13-3
Contours of equal MO values for ammonia. (*Taken from W. L. Jorgensen and L. Salem*, The Organic Chemist's Book of Orbitals, *Academic, New York, 1973.*)

belong to the various irreducible representations as follows:

$$\begin{aligned} &A_1\text{: } 1s,\ 2s,\ 2p_z,\ G_s \\ &B_1\text{: } 2p_x \\ &B_2\text{: } 2p_y,\ G_y \end{aligned} \tag{13-12}$$

where $1s$, $2s$, $2p_y$, and $2p_z$ are oxygen AOs and G_s and G_y are hydrogen atom group orbitals defined by

$$\begin{aligned} G_s &= \frac{1}{\sqrt{2}}(H_1 + H_2) \\ G_y &= \frac{1}{\sqrt{2}}(H_1 - H_2) \end{aligned} \tag{13-13}$$

The first complete molecular orbital calculation on water was carried out by Ellison and Shull in 1953 using a minimal basis of STOs.[8] However, it was necessary for these workers to approximate the three-center integrals, and consequently, the numerical results of the calculation are not as reliable as those that modern methods are able to produce. The ground-state electron configuration of water is

$$(1a_1)^2(2a_1)^2(1b_2)^2(3a_1)^2(1b_1)^2 \qquad {}^1A_1 \text{ state}$$

When this calculation is done with an STO-6G basis using the experimental geometry (O–H bond distance of 1.811 a_0 and H–O–H bond angle of 105°), the total energy comes out to be $-75.6499E_h$, with a dipole moment of 1.95 D. The estimated Hartree-Fock energy is $-76.066E_h$, and the experimental dipole moment is 1.85 D. If this calculation is now geometry-optimized, one obtains an O–H distance of 1.864 a_0 and an H–O–H bond angle of slightly under 100°; the dipole moment turns out to be 1.75 D. Table 13-4 summarizes the MO coefficients and orbital energies for this calculation.

[8] F. O. Ellison and H. Shull, *J. Chem. Phys.*, **21**:1420 (1953) and **23**:2348 (1955).

TABLE 13-4
Molecular orbitals of water

AO	$1a_1$	$2a_1$	$1b_2$	$3a_1$	$1b_1$
$1s$	−0.9967	0.2225		−0.0966	
$2s$	−0.0155	−0.8473		0.5359	
$2p_x$					1.0000
$2p_y$			−0.6161		
$2p_z$	−0.0031	−0.1243		−0.7577	
G_s	0.0035	−0.1516		−0.2943	
G_y			0.4465		
ϵ/E_h	−20.5133	−1.2668	−0.5986	−0.4662	−0.3988

Note: All calculations were made with the GAMESS program using an STO-6G minimal basis with geometry optimization.

The $1a_1$ MO is, as would be expected, largely an oxygen $1s$ orbital. The $2a_1$ MO consists largely of hydrogen orbitals and an oxygen $2s$ AO. The $1b_2$ MO is entirely oxygen $2p_y$ and hydrogen AOs. The $3a_1$ MO and the $1b_1$ MO correspond to the two lone pairs. The $3a_1$ MO consists mainly of oxygen $2s$ and $2p_z$ and hydrogen AOs and has its maximum value in the $+z$ direction. The

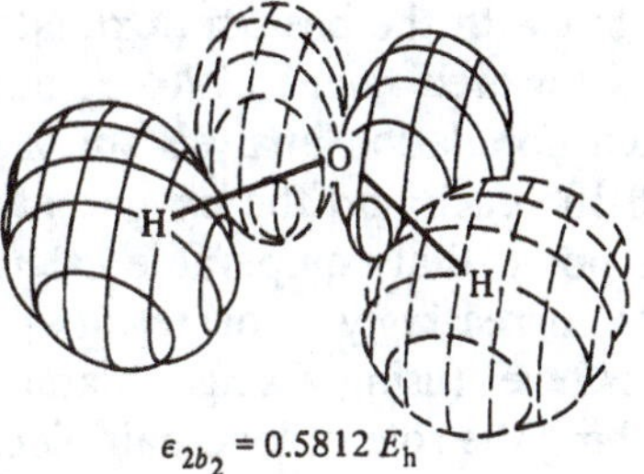

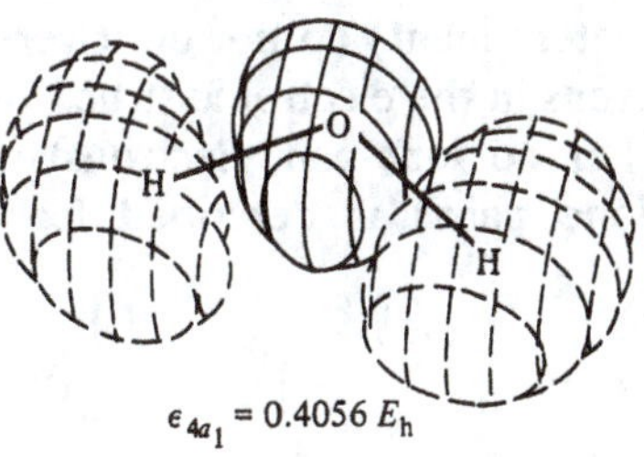

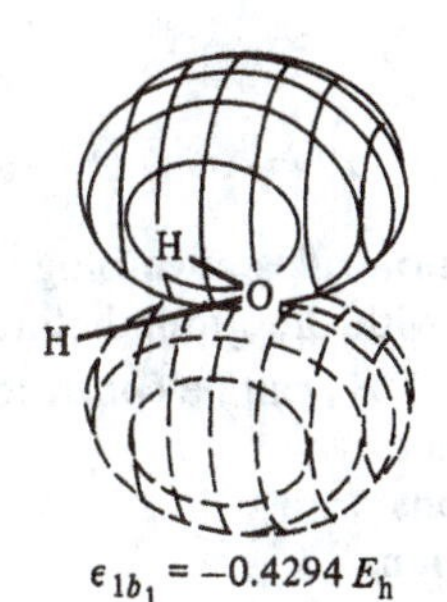

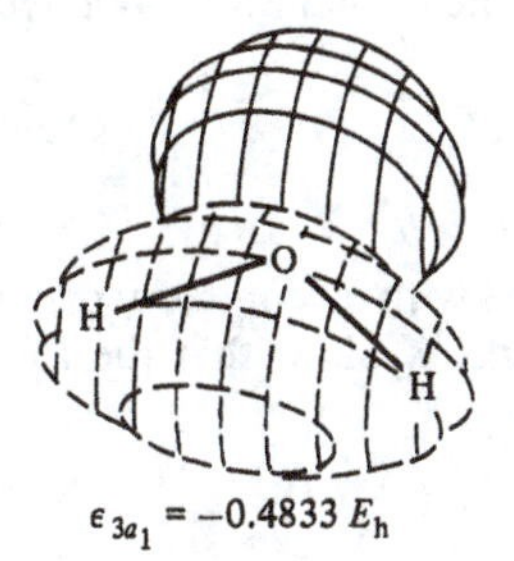

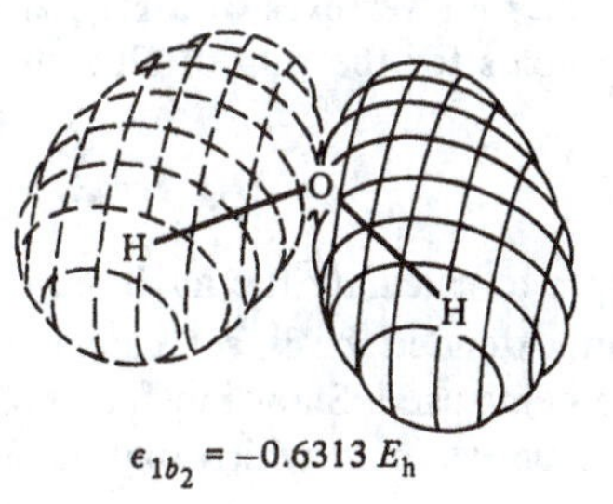

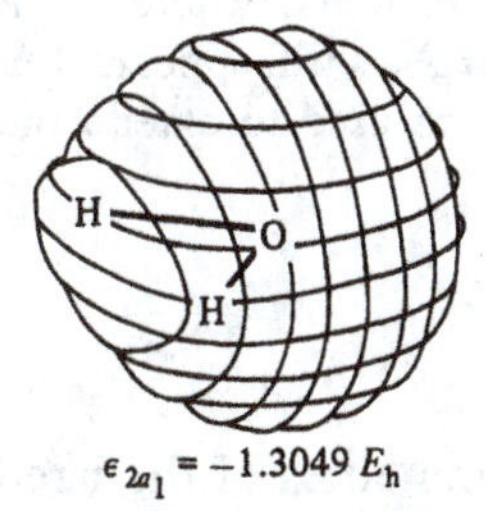

FIGURE 13-4
Contours of equal MO values for water. (*Taken from W: L. Jorgensen and L. Salem*, The Organic Chemist's Book for Orbitals, *Academic, New York, 1973.*)

$1b_1$ is entirely oxygen $2p_x$ and thus is directed at right angles to the plane of the molecule; the lone pairs thus have their maxima in the xz plane.

The appearance of the water MOs (excluding the $1a_1$ MO, which is essentially nonbonding) is depicted in the contour diagrams of Fig. 13-4. Recall that the contour lines of these diagrams represent surfaces along which the MO has a constant value.

If the water calculation is repeated using the geometry obtained from the minimal basis set but with the basis set extended ($2s$, $2p$ on H and $3d$ on O), the total energy is lowered to $-75.7765E_h$ and the dipole moment becomes 1.69 D. A Mulliken population analysis shows that the hydrogen atoms have net charges of +0.100 each and the oxygen atom has a charge of −0.200.

It should be noted that comparable calculations on methane, ammonia, and water lead to increasingly poor results—in that order. This is because calculations on this level (using a single Slater determinant) do not do a very good job of describing the role of lone-pair electrons in the electronic structure of molecules. Consequently, methane (which has no lone-pair electrons) is described best, and water (which has two lone pairs) is described least satisfactorily.

EXERCISES

13-1. What are the selection rules governing electronic state transitions in CH_4, NH_3, and H_2O, starting with the ground state in each case?

13-2. Show what states of NH_3 can be constructed from the virtual MOs obtained from the minimal basis set by
(*a*) Single excitations
(*b*) Double excitations
Show how these could be used in a CI calculation.

13-3. Let $\mathbf{r}_i$ $(i = 1, 2, 3, 4)$ represent four unit vectors with a common origin and pointing at the four vertexes of a regular tetrahedron. Show that the following relationship holds for the sum of all four vectors:

$$\mathbf{s} = \sum_{i=1}^{4} \mathbf{r}_i = 0$$

Then use $\mathbf{r}_i \cdot \mathbf{s}$ to calculate the angle between two vectors.

13-4. Consider an extended basis set calculation on ammonia, where $3d$ AOs of nitrogen are being used. Show in which MOs the $3d$ AOs will appear. Repeat for a calculation on water in which oxygen $3d$ AOs are used to extend the basis.

13-2 LOCALIZED MOLECULAR ORBITALS

When molecular orbitals are chosen to have the symmetries of the irreducible representations of the point groups of the molecules they represent, they are easy to set up and they make it possible to reduce the size of the secular determinants which must be solved.[9] But at the same time, the electron distributions implied by these MOs cannot generally be visualized in terms of the conventional formulas chemists use on a day-to-day basis. As pointed out in the preceding section, there are no individual MOs in the configurations of a molecule such as NH_3 which suggest three equivalent N–H bonds. This bonding is implicit in the total wavefunction (subject to certain arbitrary modes of interpretation), but it is not unequivocally delineated in terms of specific MOs. Consequently, it would appear to be desirable to develop alternative types of MOs which are confined to as small a space as possible, and at the same time, these MOs should be as far apart as possible. The more such MOs can be confined and separated from each other, the less they are influenced when modifications take place in more distant parts of the molecule. Thus, such localized MOs should exhibit *transferability*; e.g., the localized MO describing the C–H bond in methane should be very similar in other paraffinic hydrocarbons. Note that both localization and transferability are implicit in the ordinary language of descriptive chemistry—we learn these concepts in general chemistry and use them as conceptual tools throughout subsequent descriptive chemistry courses. Thus, it is highly desirable to develop quantitative quantum mechanical descriptions which overlap as much as possible with the commonly used qualitative descriptions—to the extent that significant overlaps are possible. This has a twofold effect: (1) to probe the legitimacy of the qualitative concepts and (2) to put these on a quantitative basis if they are indeed legitimate. As an example, the very successful VSEPR model of molecular structure is based on what is essentially a classical electrostatic picture of the minimization of repulsions between electron pairs. Is there a quantum mechanical analog of this model?

Using methane as an example, we will see that the nonequivalent MOs describing the ground state can be transformed to a new set of orbitals which are equivalent and whose concomitant electron distributions are localized in such a way as to suggest C–H bonds. Although the treatment can be carried out without recourse to any mathematical approximation, the procedure is conceptually clearer if one rather small approximation is made, namely, we assume that the $1a_1$ MO of CH_4 is purely $1s$ of carbon, so that the remaining ground-state MOs contain only $2s$, $2p_x$, $2p_y$, and $2p_z$ of carbon (as well as the four $1s$ AOs of hydrogen). These latter MOs are $2a_1$, $1t_x$, $1t_y$, and $1t_z$. The inverse of the transformation which produces group orbitals from hydrogen

[9] Note, however, that this is of no particular practical importance if programs such as GAMESS, GAUSSIAN 82, etc., are being used. Such programs generally generate the MOs in irreducible representation form, but not all identify individual MOs as to symmetry type; the latter must be done by inspection.

atom orbitals (see Sec. 13-1) will transform the MOs as follows:

$$[2a_1 \quad 1t_x \quad 1t_y \quad 1t_z]\begin{bmatrix} 1 & 1 & 1 & 1 \\ 1 & -1 & 1 & -1 \\ 1 & -1 & -1 & 1 \\ 1 & 1 & -1 & 1 \end{bmatrix} = [\Omega_1 \quad \Omega_2 \quad \Omega_3 \quad \Omega_4]$$

where
$$\Omega_1 = (2a_1 + 1t_x + 1t_y + 1t_z)$$
$$\Omega_2 = (2a_1 - 1t_x - 1t_y + 1t_z)$$
$$\Omega_3 = (2a_1 + 1t_x - 1t_y - 1t_z)$$
$$\Omega_4 = (2a_1 + 1t_x + 1t_y - 1t_z)$$

These orbitals span the reducible representation $A_1 \oplus T_2$; they transform under the group symmetry operations just as the hydrogen atom orbitals H_1, H_2, H_3, and H_4 do and, consequently, resemble vectors coincident with the C–H bond axes. We say the four orbitals Ω_1, Ω_2, Ω_3, and Ω_4 are *equivalent*; i.e., they transform into each other under the group symmetry operations and thus differ only in their arbitrary orientations in space. Such orbitals are called *equivalent orbitals* (EOs) and were first discussed in a systematic manner by Lennard-Jones and Hall.[10] As we will discuss shortly, the EOs are special cases of more general orbitals which maximize the sum of the orbital self-repulsions and thereby minimize interorbital repulsions; i.e., such orbitals represent maximum localization of electron bonding pairs.

In terms of EOs, the methane ground-state electron configuration may be written

$$(1a_1)^2(\Omega_1)^2(\Omega_2)^2(\Omega_3)^2(\Omega_4)^2$$

Unlike the original MOs, the EOs are not eigenfunctions of the Hartree-Fock operator, and hence, it is meaningless to associate orbital energies with them. Note, however, that the EOs represent the same *total* electron distributions as do the MOs used to form them; EOs and MOs merely subdivide the total electron distribution in different ways. Each of the EOs, although it suggests that portion of the total electron distribution associated with a single C–H bond, nevertheless is a five-center function. If we represent the normalized canonical MOs as follows:

$$2a_1 = N_s(2s + \lambda_s G_s) \qquad 1t_x = N_t(2p_x + \lambda_t G_x)$$
$$1t_y = N_t(2p_y + \lambda_t G_y) \qquad 1t_z = N_t(2p_z + \lambda_t G_z) \tag{13-14}$$

[10] J. E. Lennard-Jones, *Proc. Roy. Soc.* (*London*), **A198**:1, 14 (1949); G. G. Hall and J. E. Lennard-Jones, *Proc. Roy. Soc.* (*London*), **A202**:155 (1950). See also J. A. Pople, *Quart. Revs.*, **11**:273 (1957).

then each of the four EOs has the general form

$$\Omega_i = \tfrac{1}{2}[N_s(2s) + \sqrt{3}N_t(2p_i) + (N_s\lambda_s + 3N_t\lambda_t)H_i + (N_s\lambda_s - N_t\lambda_t)(H_j + H_k + H_l)] \quad i = 1, 2, 3, 4 \quad i \neq j \neq k \neq l \tag{13-15}$$

The quantities N_s and N_t are normalization constants, and λ_s and λ_t are variational parameters. The orbital $2p_i$ is a mixture of $2p_x$, $2p_y$, and $2p_z$ AOs which points at hydrogen atom i. The AO combinations are defined by

$$2p_1 = \frac{1}{\sqrt{3}}(2p_x + 2p_y + 2p_z) \qquad 2p_2 = \frac{1}{\sqrt{3}}(-2p_x - 2p_y + 2p_z)$$

$$2p_3 = \frac{1}{\sqrt{3}}(2p_x - 2p_y - 2p_z) \qquad 2p_4 = \frac{1}{\sqrt{3}}(-2p_x + 2p_y - 2p_z) \tag{13-16}$$

Coulson[11] carried out an MO calculation on CH_4 using a basis of Slater orbitals and a C–H distance of 2.0 a_0. The N and λ constants turned out to have the numerical values

$$N_s = 2.56 \qquad N_t = 1.697$$
$$\lambda_s = 0.121 \qquad \lambda_t = 0.589 \tag{13-17}$$

Thus, each EO of methane becomes

$$\Omega_i = \tfrac{1}{2}[2.56(2s) + 2.94(2p_i) + 3.31H_i - 0.69(H_j + H_k + H_l)] \tag{13-18}$$

Note that this is very nearly a two-center function; in each EO Ω_i, only the hydrogen atom AO H_i has a large numerical coefficient. The contribution of the other three hydrogen atoms represents a small nonlocalized "tail"; as we will see later, removal of this tail localizes the EO to a single C–H bond but destroys the orthogonality of the EO.

As was mentioned previously with the discussion of the Hartree-Fock method, a single-determinantal wavefunction of a closed-shell system is invariant under a unitary transformation among the orbitals. More generally, any single configurational wavefunction—including one which is multideterminantal—is invariant under unitary transformations among the doubly occupied orbitals. Consequently, it has long been a goal ot theoretical chemists to find unitary transformations which would *localize* the canonical orbitals obtained from Hartree-Fock calculations and thus produce orbitals more easily relatable to ordinary chemical descriptions of bonding pairs of electrons, lone pairs, etc. It is with this goal in mind that Edmiston and Ruedenberg[12] have

---

[11] C. A. Coulson, *Trans. Faraday Soc.*, **33**:388 (1937) and **38**:433 (1942).

[12] C. Edmiston and K. Ruedenberg, *Rev. Mod. Phys.*, **35**:457 (1963) and *J. Chem. Phys.*, **43**:S97 (1965). An extensive review, with applications to many specific molecular systems, is given by W. England, L. S. Salmon, and K. Ruedenberg in *Fortschritte der chemischen Forschung*, **23**:31–123 (1971).

developed an exact method for finding those molecular orbitals which maximize the sum of the orbital self-repulsion energies; these particular MOs are called *localized molecular orbitals* (LMOs). These LMOs are defined in such a way that only in the presence of a symmetry group can they acquire properties which make them, under certain conditions, the same as equivalent orbitals. In the Edmiston and Ruedenberg procedure, the proper unitary transformation is found by requiring the sum of all the exchange integrals K_{ij} to be minimized. The rationale for this procedure is as follows: To obtain a description of bonding in which maximum overlap with a classical electrostatic picture of bonding is obtained, one should eliminate (or at least minimize) nonclassical terms such as exchange. The Edmiston and Ruedenberg method may be briefly summarized as follows: First, define various sums:

1. The sum of the exchange integrals

$$X = \sum_{i \neq j} K_{ij}$$

2. The sum of the off-diagonal coulomb integrals

$$C = \sum_{i \neq j} J_{ij}$$

3. The sum of the diagonal coulomb integrals

$$D = \sum_{i} J_{ii}$$

Next, minimize the sum $X + C$ (and thus maximize D). This is done by means of an iterative procedure whereby successive rotations are carried out two orbitals at a time, with each rotation chosen so that there is the greatest possible mutual localization of a given pair. The rotations are continued until no further improvements are possible. The method does turn out to be quite time-consuming, but the results generally lead to LMOs which agree very well with qualitative chemical concepts and descriptions. An alternative method proposed by Boys[13] obtains LMOs by maximizing the product of the distances between the centroids of charge of the various orbitals. The Boys method is computationally superior in some respects to the Edmiston and Ruedenberg method and produces very similar results. However, as pointed out by England, Salmon, and Ruedenberg,[14] although canonical MOs are unique (apart from the degeneracy-induced arbitrary natures of some MOs), this is not necessarily true of LMOs since the localization sum may have several relative

---

[13] S. F. Boys, in P. O. Löwdin (ed.), *Quantum Theory of Atoms, Molecules, and the Solid State*, Academic, New York, 1966.

[14] England, Salmon, and Ruedenberg, op. cit. See also C. Edmiston and K. Ruedenberg, *Quantum Theory of Atoms, Molecules, and the Solid State*, Academic, New York, 1966, p. 263.

maxima representing different localizations. For example, the two sets of fluorine LMOs representing the six lone pairs of difluorine, F_2, may be rotated relative to each other without affecting the localization sum.

LMO calculations on simple hydrocarbons, including methane, have been carried out by Rothenberg.[15] In the case of methane, the contribution to the sum of the exchange integral portion of the energy was reduced from $0.9933E_h$ to $0.2957E_h$ upon localization. Rothenberg also carried out similar calculations on ethane and methanol and showed that the LMOs describing the C–H bond are approximately transferable from one of these molecules to the next: a result which is compatible with generally held qualitative concepts of bonding.

An interesting application of the Edmiston-Ruedenberg LMO procedure has been carried out on diborane, B_2H_6, by Switkes, Stevens, Lipscomb, and Newton.[16] This calculation produces three sets of equivalent orbitals from an original set of eight canonical orbitals: these are two boron core orbitals, four B–H terminal bond orbitals, and two B–H–B three-center orbitals. One interesting result of this calculation is that it reveals that the three-center bond has a population of 1.016 on the bridging hydrogen and 0.507 on each of the end boron atoms. The calculations also show that H atoms in B_2H_6 are very nearly neutral as opposed to a comparable hydrocarbon in which the hydrogens would be definitely positively charged.

Peters[17] has developed a means of generating LMOs directly without the necessity of obtaining the canonical MOs first and then transforming them. Peters's approach involves the use of a new secular determinant and the use of Rayleigh-Schrödinger perturbation theory. Calculations on the methane molecule using this approach produce some interesting results; foremost among these is that the carbon valence orbital is more nearly sp^2 hybridized than sp^3 hybridized.

As specific illustrations of how canonical MO and LMO descriptions differ, we will consider two of the molecules discussed in the previous section: ammonia and water. In each case we will consider transformations of the canonical MOs obtained from the minimal basis, geometry-optimized calculation to LMOs and examine each LMO in terms of its composition. All these transformations were carried out within the GAMESS program utilizing the Boys localization method.

Table 13-3 lists the canonical MOs obtained for ammonia when a minimal basis set is used and the geometry is optimized. Note that the occupied MOs are (in order of increasing energy) $1a_1$, $2a_1$, $3a_1$, and $1e$. The localized MOs

---

[15] S. Rothenberg, *J. Chem. Phys.*, **51**:3389 (1969).

[16] E. Switkes, R. M. Stevens, W. N. Lipscomb, and M. D. Newton, *J. Chem. Phys.*, **51**:2085 (1969).

[17] D. Peters, *J. Chem. Phys.*, **51**:1559, 1566 (1969) and *J. Chem. Soc.*, 2003, 2015, 4017 (1963); 2901, 2908, 2916 (1964); 3026 (1965); and 644, 652, 656 (1966)A.

become, in general, linear combinations of these. One of these LMOs is

$$\begin{aligned}\Omega_1 &= 0.997(1a_1) - 0.059(2a_1) - 0.046(3a_1)\\ &= 1.010(1s) - 0.046(2s) + 0.036(2p_z) - 0.007(H_1 + H_2 + H_3)\end{aligned} \tag{13-19}$$

This LMO, which spans the A_1 irreducible representation, is mostly $1a_1$ in character, which, in turn, is mostly $1s$ of nitrogen and $1s$ of the hydrogens. This LMO is a nitrogen atom core orbital and represents *inner-shell*, *nonbonding* electrons—to use the terminology common to descriptive chemistry. Three of the LMOs have forms in which the major differences are in algebraic signs of coefficients. One of these is

$$\begin{aligned}\Omega_2 &= -0.024(1a_1) - 0.546(2a_1) + 0.185(3a_1) + 0.724(1e_x) - 0.376(1e_y)\\ &= -0.074(1s) - 0.317(2s) - 0.243(2p_x) - 0.421(2p_y)\\ &\quad + 0.239(2p_z) - 0.092(H_1 + H_3) + 0.517H_2\end{aligned} \tag{13-20}$$

This LMO (and its two equivalents Ω_3 and Ω_4) span the $A_1 \oplus E$ reducible representation and represent the N–H bonds. Note that this LMO has significant contributions from all the basis AOs except the $1s$ and $2p_z$ of nitrogen; i.e., the bonding between nitrogen and hydrogen involves mostly $2s$ and $2p$ AOs of nitrogen overlapping with hydrogen $1s$ AOs. This is, of course, the standard description given at elementary levels. If the small, nonlocalized tail, i.e., $-0.092(H_1 + H_3)$ were not present, the above LMO would represent a bond strictly localized between the nitrogen atom and hydrogen atom 2. However, removal of this tail destroys orthogonality of the LMO with the other LMOs.

The fifth LMO is

$$\begin{aligned}\Omega_5 &= 0.062(1a_1) + 0.312(2a_1) + 0.946(3a_1)\\ &= -0.088(1s) + 0.687(2s) - 0.782(2p_z) - 0.079(H_1 + H_2 + H_3)\end{aligned} \tag{13-21}$$

This LMO spans the A_1 irreducible representation and contains no contributions from nitrogen $2p_x$ and $2p_y$ but does contain contributions from all other AOs (but very little from $1s$ of nitrogen). It also contains a small symmetrical tail from hydrogen $1s$ AOs. Consequently, this LMO is identified with the lone-pair electrons of ammonia.

Now consider the LMOs which arise from the canonical MOs of water displayed in Table 13-4. One of the LMOs is

$$\begin{aligned}\Omega_1 &= 0.996(1a_1) - 0.073(2a_1) + 0.047(3a_1)\\ &= -1.014(1s) + 0.072(2s) - 0.0300(2p_z) - 0.0008(H_1 + H_2)\end{aligned} \tag{13-22}$$

This LMO is of A_1 symmetry and clearly represents the inner-shell, nonbonding electrons (mostly $1s$ of oxygen); this would be called an *oxygen core orbital.*

Two of the LMOs are equivalent; one of these is

$$\begin{aligned}\Omega_2 &= -0.058(1a_1) - 0.421(2a_1) + 0.565(3a_1) - 0.707(1b_1) \\ &= -0.091(1s) + 0.660(2s) + 0.707(2p_x) - 0.376(2p_z) - 0.103(H_1 + H_2)\end{aligned} \tag{13-23}$$

These two LMOs span the $A_1 \oplus B_1$ reducible representation and are mostly $2s$ and $2p_x$ of oxygen; there is no contribution from $2p_y$ (this occurs in the $1b_2$ MO). Thus, these represent the lone pairs generally associated with the oxygen atom. Note that the lone-pair LMOs contain a symmetrical hydrogen atom tail. The remaining two LMOs are also equivalent; one of these is

$$\begin{aligned}\Omega_4 &= 0.022(1a_1) + 0.565(2a_1) + 0.707(1b_2) + 0.423(3a_1) \\ &= 0.063(1s) - 0.253(2s) - 0.436(2p_y) - 0.391(2p_z) + 0.526H_1 - 0.105H_2\end{aligned} \tag{13-24}$$

These two LMOs span the $A_1 \oplus B_2$ reducible representation and contain no $2p_x$ of nitrogen (found in the $1b_1$ MO); they are mostly $2s$, $2p_y$, and $2p_z$ of nitrogen. Consequently, these represent the two O–H bonds of water. Note once again that each O–H bond is not totally localized; each has a small tail from the other hydrogen.

EXERCISE

13-5. Where does the "lost" exchange energy go when one transforms a canonical MO wavefunction into LMO form?

13-3 HYBRID ORBITALS

Qualitative discussions of molecular geometries at elementary—and even advanced—levels are often carried out in terms of *hybrid orbitals*. Even at the first-year level, students are exposed to the device of "mixing" $2s$ and $2p$ AOs of carbon in various proportions to produce *sp* (digonal), sp^2 (trigonal), and sp^3 (tetrahedral) hybrid orbitals. These orbitals are then overlapped with $1s$ AOs of hydrogen to form C–H bonds (and other bonds) in a wide variety of compounds. In this section we will see how hybrid orbitals can be defined quantitatively and how these are related to accurate MO descriptions of chemical bonding.

Let us consider the construction of hybrid orbitals from $2s$ and $2p$ AOs centered on the same atom. A pair of these hybrid AOs is to point in the direction of two particular atoms A and B, as shown in Fig. 13-5. These two hybrids will have the general mathematical form

$$h_a = (2s) + \lambda_a(2p_a) \qquad h_b = (2s) + \lambda_b(2p_b) \tag{13-25}$$

where $2p_a$ and $2p_b$ are linear combinations of one or more $2p$ orbitals such that their maximum values lie in the directions of atoms A and B, respectively. The

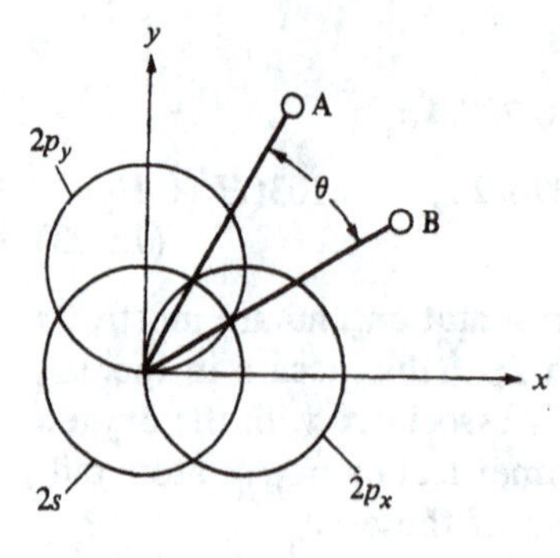

FIGURE 13-5
Diagram of $2s$ and $2p$ AOs before forming hybrid AOs pointing at atoms A and B. Only half of each p orbital is shown.

two hybrids will be orthogonal if the following relationship is satisfied:

$$\langle h_a | h_b \rangle = \langle (2s) + \lambda_a (2p_a) | (2s) + \lambda_b (2p_b) \rangle = 0 \tag{13-26}$$

Expanding the right-hand integral, we get

$$\langle 2s | 2s \rangle + \lambda_a \langle 2p_a | 2s \rangle + \lambda_b \langle 2s | 2p_b \rangle + \lambda_a \lambda_b \langle 2p_a | 2p_b \rangle = 0 \tag{13-27}$$

The second and third integrals vanish as a result of orthogonality of 2s and $2p$ AOs. If the $2s$ AO is normalized, Eq. (13-27) becomes

$$1 + \lambda_a \lambda_b \langle 2p_a | 2p_b \rangle = 0 \tag{13-28}$$

The integral in the latter equation is just the scalar product of two "vectors" forming an angle θ with each other. Thus we get

$$1 + \lambda_a \lambda_b \cos\theta = 0 \tag{13-29}$$

Alternatively,

$$\lambda_a \lambda_b = -\frac{1}{\cos\theta} = -\sec\theta \tag{13-30}$$

For nontrivial hybrids, λ_a and λ_b are greater than zero. It is then convenient to effect the substitution

$$\lambda_a = \sqrt{n_a} \qquad \lambda_b = \sqrt{n_b} \tag{13-31}$$

where n_a and n_b are real positive numbers. For equivalent hybrids it is necessary that $n_a = n_b = n$. Equation (13-30) then becomes

$$n = -\sec\theta \tag{13-32}$$

For $n = 3$ we obtain $\theta = 109°28'$. Then $\lambda_a = \lambda_b = \sqrt{3}$, and each hybrid is a tetrahedral, or sp^3, hybrid. In Table 13-5 are listed the equivalent hybrids for other integral values of n. Note that n is also the number of $2p$ AOs which are used to form the $2p_j$ AO. Also, the number of equivalent hybrids for a given integral n value is $n + 1$. For $n = 2$ we obtain trigonal, or sp^2, hybrids of the

TABLE 13-5
Normalized equivalent hybrid orbitals formed from primitive 2*s* and 2*p* atomic orbitals

Hybrid designation	n	θ	Normalized hybrid	Usual name	$n(n+1)$
sp^3	3	109°28′	$\frac{1}{2}(\phi_{2s} + \sqrt{3}\phi_{2p_j})$	Tetrahedral	12
sp^2	2	120°	$3^{-1/2}(\phi_{2s} + \sqrt{2}\phi_{2p_j})$	Trigonal	6
sp	1	180°	$2^{-1/2}(\phi_{2s} + \phi_{2p_j})$	Digonal	2
sp^n	n	$\sec^{-1}(-n)$	$(n+1)^{-1/2}(\phi_{2s} + \sqrt{n}\phi_{2p_j})$	—	$n(n+1)$
sp^∞	∞	90°	ϕ_{2p}	$2p$ AO	

general form

$$(\text{tr})_j = 2(s) + \sqrt{2}(2p_j) \tag{13-33}$$

where $\theta = 120°$.

Referring to the NH_3 molecule, we find that the experimental bond angle is 106°47′. We can construct three hybrids with such an interhybrid angle by letting $n = -\sec 106°47' = 3.463$. We obtain three hybrids, called $sp^{3.463}$ hybrids, having the form

$$h_j = \frac{1}{\sqrt{4.46}}[(2s) + 1.86(2p_j)] \tag{13-34}$$

Note that this implies a fourth hybrid, $sp^{1.611}$, which is not equivalent to the other three and which is oriented along the principal axis of the molecule. This hybrid could be used to describe the nitrogen lone pair. The above four hybrid AOs may be regarded as distorted tetrahedral hybrids; such a picture of bonding in ammonia is provided by the VSEPR model in which all four pairs of nitrogen electrons are originally placed into tetrahedral hybrid AOs but then the lone-pair repulsions "push" the other three pairs (which provide bonding to H atoms) closer together, thereby distorting the tetrahedral geometry.

Hybrid AOs formed from s and p orbitals must satisfy the relationship

$$\alpha + \beta + \gamma + \cdots = n(n-1) \tag{13-35}$$

where the hybrids are sp^α, sp^β, . . . and n is the total number of such hybrids. Thus, in the ammonia example, $n = 4$, so that $3(3.463) + 1.611 = 4(4-1) = 12$.

Since $n > 0$ requires that $\cos\theta < 0$, it is apparent that the limiting case of sp^n hybridization occurs when $\theta = 90°$. This means that molecules with bond angles less than 90° cannot be described in terms of sp^n hybrid AOs which point at each other, e.g., cyclopropane with C–C–C bond angles of 60°. The cyclopropane molecule must be described by using so-called *bent bonds*, i.e., bonds formed from hybrids whose interorbital angles are greater than the bond angles.

Referring to Coulson's numerical calculations on methane (see the preceding section), the ratio of $2p$ to $2s$ in the methane EOs is given by

$$\frac{2p}{2s} = \left(\frac{2.94}{2.56}\right)^2 = \frac{1.3}{1}$$

This means that insofar as the $N_s(2s) + \sqrt{3}(2p_i)$ part of Ω_i is concerned, it has $100(1/2.3) = 43$ percent $2s$ character and $100(1.3/2.3) = 57$ percent $2p$ character; a pure sp^3 hybrid would have 25 percent $2s$ character and 75 percent $2p$ character. One can "force" the $2s/2p$ part of Ω_i to be an sp^3 hybrid by redefining the EOs as follows:

$$\Omega_i' = \tfrac{1}{2}[k(2a_1) + 1t_x + 1t_y + 1t_z] \tag{13-36}$$

and so on for the other EOs. This leads to EOs of the general form

$$\Omega_i' = \tfrac{1}{2}[kN_s(2s) + \sqrt{3}N_t(2p_i) + (kN_s\lambda_s + 3N_t\lambda_t)H_i + (kN_s\lambda_s - N_t\lambda_t)(H_j + H_k + H_l)] \tag{13-37}$$

If the constant k is chosen as follows:

$$kN_s = N_t \tag{13-38}$$

then the ratio of $2p$ to $2s$ becomes

$$\frac{2p}{2s} = \left(\frac{\sqrt{3}}{1}\right)^2 = \frac{3}{1} \tag{13-39}$$

which is just the ratio required for sp^3 hybrid orbitals. Equation (13-37) would then be replaced by

$$\Omega_i' = \tfrac{1}{2}[1.70(2s) + 2.94(2p_i) + 3.21H_i - 0.79(H_j + H_k + H_l)] \tag{13-40}$$

However, this newly defined EO is not a good orbital. First, introduction of the constant k destroys the orthogonality of the EOs, since they are no longer related to the MOs by a unitary transformation. Second, the hybrids lead to a great increase in the amount of $2p$ in each EO, and this increases the energy obtainable from the wavefunction; this implies further that each EO is made less bonding, since an increase in $2p$ implies borrowing part of the virtual MOs. One should also note the LMO calculations by Peters on methane (see the preceding section) in which the s/p ratio in methane was found to be closer to sp^2 hybridization than sp^3 hybridization. Nevertheless, hybrid AOs are conceptually attractive for qualitative descriptions of chemical bonding, and it is unlikely that their use will disappear from elementary levels regardless of their quantitatively flawed basis.

EXERCISES

13-6. Using the H–O–H bond angle as 105°, find the hybrid AOs which could be used to describe the two O–H bonds of water and the two lone pairs.

13-7. The H–C–H bond angle in ethylene is 117°22′. What are the hybrid AOs needed to describe ethylene?

13-8. Metal-to-metal quadruple bonds are known in compounds such as $Cr_2(CO_3)_4(H_2O)_2$ and $[Tc_2Cl_3]^{2-}$. Show how to set up the following AO combinations to describe a $\sigma\pi^2\delta$ quadruple bond: $\sigma_g 3d$, $\pi_u 3d$, and $\delta_g 3d$. Let z represent the M–M bond axis.

13-4 THE ETHYLENE AND BENZENE MOLECULES

There have been a large number of theoretical studies of the ethylene and benzene molecules; both of these are examples of a class of hydrocarbons generally called *pi electron systems*. Ethylene may be regarded as the simplest example of a *linear* pi electron molecule, and benzene serves as an analogous model for *aromatic* pi electron molecules. In this section we will discuss some of the simpler *ab initio* calculations on these two molecules, and later we will discuss several types of simplified calculations—those generally known as *semiempirical* calculations.

ETHYLENE. We begin by illustrating a systematic method for constructing symmetry orbitals of ethylene (called *ethene* in the recommended IUPAC system of nomenclature). This 16-electron molecule belongs to the D_{2h} point group; the character table for this group and the corresponding coordinate system for ethylene are given in Fig. 13-6.[18] On the basis of a minimal basis set, the relevant AOs are as follows:

Hydrogen atoms: $1s$ AOs labeled H_1, H_2, H_3, and H_4
Carbon atom C_1: $1s_1$, $2s_1$, $2p_{x1}$, $2p_{y1}$, and $2p_{z1}$
Carbon atom C_2: $1s_2$, $2s_2$, $2p_{x2}$, $2p_{y2}$, and $2p_{z2}$

Since none of the above AOs is already a symmetry orbital, i.e., each belongs to some *reducible* representation of the D_{2h} point group, we first set up those linear combinations of AOs which belong to various irreducible representations. From these we can then set up the corresponding MOs. A simple, systematic way to do this is by use of Table 13-6, which shows how representative AOs transform under the symmetry operations of the group. The $2p_x$ and $2p_y$ AOs of carbon are chosen with their positive lobes along the positive x and y axes, respectively, but the $2p_z$ AOs have their positive lobes pointed in opposite directions along the C–C bond axis (the z axis). Such disposition of the $2p$ AOs is arbitrary and has no physical significance.

[18] The coordinate system used here is compatible with the D_{2h} character table given in app. III of F. A. Cotton, *Chemical Applications of Group Theory*, Wiley, New York, 1971. Some of the ethylene calculations found in the literature reverse the x and y axes; this has the effect of changing the irreducible representation symbols B_{1u} and B_{1g} to B_{3u} and B_{3g}, respectively.

D_{2h}	E	$C_2^{(x)}$	$C_2^{(y)}$	$C_2^{(z)}$	i	σ_{yz}	σ_{xz}	σ_{xy}	
A_g	1	1	1	1	1	1	1	1	x^2, y^2, z^2
A_u	1	1	1	1	−1	−1	−1	−1	
B_{1g}	1	−1	−1	1	1	−1	−1	1	R_z, xy
B_{1u}	1	−1	−1	1	−1	1	1	−1	z
B_{2g}	1	−1	1	−1	1	−1	1	−1	R_y, xz
B_{2u}	1	−1	1	−1	−1	1	−1	1	y
B_{3g}	1	1	−1	−1	1	1	−1	−1	R_x, yz
B_{3u}	1	1	−1	−1	−1	−1	1	1	x

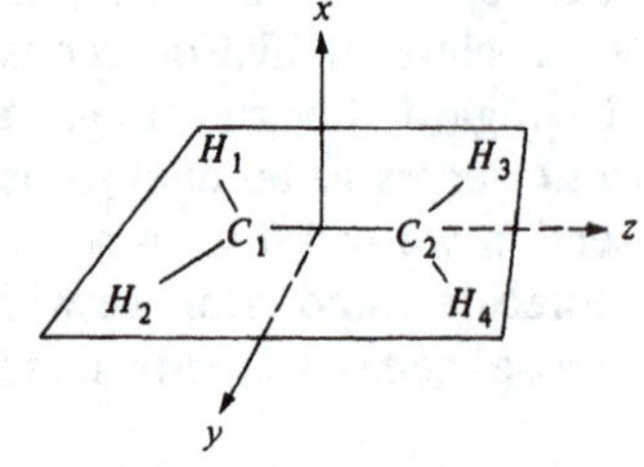

FIGURE 13-6

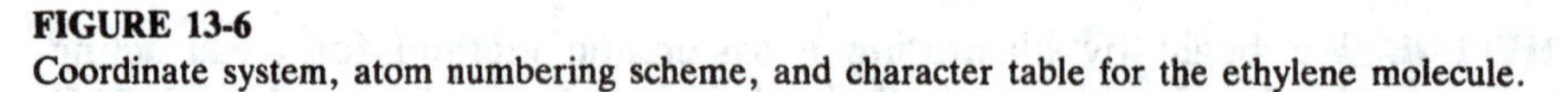
Coordinate system, atom numbering scheme, and character table for the ethylene molecule.

One finds that the hydrogen atom 1s orbitals form the following combinations (not normalized) which span irreducible representations:

$$
\begin{aligned}
G(A_g) &= H_1 + H_2 + H_3 + H_4 \\
G(B_{1u}) &= H_1 + H_2 - H_3 - H_4 \\
G(B_{2u}) &= H_1 - H_2 + H_3 - H_4 \\
G(B_{3g}) &= H_1 - H_2 - H_3 + H_4
\end{aligned}
\tag{13-41}
$$

TABLE 13-6
Behavior of minimal basis AOs of ethylene under the symmetry operations of the D_{2h} point group

$\hat{R}$	$\hat{R}H_1$	$\hat{R}ns_1$	$\hat{R}2p_{x1}$	$\hat{R}2p_{y1}$	$\hat{R}2p_{z1}$
$\hat{E}$	H_1	ns_1	$2p_{x1}$	$2p_{y1}$	$2p_{z1}$
$\hat{C}_2^{(x)}$	H_4	ns_2	$2p_{x2}$	$-2p_{y2}$	$2p_{z2}$
$\hat{C}_2^{(y)}$	H_3	ns_2	$-2p_{x2}$	$2p_{y2}$	$2p_{z2}$
$\hat{C}_2^{(z)}$	H_2	ns_1	$-2p_{x1}$	$-2p_{y1}$	$2p_{z1}$
$\hat{i}$	H_4	ns_2	$-2p_{x2}$	$-2p_{y2}$	$2p_{z2}$
$\hat{\sigma}_{yz}$	H_1	ns_1	$-2p_{x1}$	$2p_{y1}$	$2p_{z1}$
$\hat{\sigma}_{xz}$	H_2	ns_1	$2p_{x1}$	$-2p_{y1}$	$2p_{z1}$
$\hat{\sigma}_{xy}$	H_3	ns_2	$2p_{x2}$	$2p_{y2}$	$2p_{z2}$

Similarly, one obtains for the carbon AOs:

$$\begin{aligned}
1s(A_g) &= 1s_1 + 1s_2 & 2p_x(B_{2g}) &= 2p_{x1} - 2p_{x2} \\
1s(B_{1u}) &= 1s_1 - 1s_2 & 2p_y(B_{2u}) &= 2p_{y1} + 2p_{y2} \\
2s(A_g) &= 2s_1 + 2s_2 & 2p_y(B_{3g}) &= 2p_{y1} - 2p_{y2} \\
2s(B_{1u}) &= 2s_1 - 2s_2 & 2p_z(A_g) &= 2p_{z1} + 2p_{z2} \\
2p_x(B_{3u}) &= 2p_{x1} + 2p_{x2} & 2p_z(B_{1u}) &= 2p_{z1} - 2p_{z2}
\end{aligned} \tag{13-42}$$

None of these symmetry orbitals (group orbitals) has been normalized; such normalization is straightforward (but tedious) and requires overlap integrals between pairs of AOs.

The matrix representation of the Hartree-Fock operator in this basis could be factored by symmetry into two 4×4 matrices (one of A_g symmetry and one of B_{1u} symmetry), two 2×2 matrices (one of B_{2u} and one of B_{3g} symmetry), and two 1×1 matrices (one of B_{3u} symmetry and the other of B_{2g} symmetry). The general forms of the MOs are

$$\begin{aligned}
ja_g &= a_{1j}G(A_g) + a_{2j}1s(A_g) + a_{3j}2s(A_g) + a_{4j}2p_z(A_g) \qquad j = 1, 2, 3, 4 \\
kb_{1u} &= b_{1k}G(B_{1u}) + b_{2k}1s(B_{1u}) + b_{3k}2s(B_{1u}) + b_{4k}2p_z(B_{1u}) \qquad k = 1, 2, 3, 4 \\
mb_{2u} &= c_{1m}G(B_{2u}) + c_{2m}2p_y(B_{2u}) \qquad m = 1, 2 \\
nb_{3g} &= d_{1n}G(B_{3g}) + d_{2n}2p_y(B_{3g}) \qquad n = 1, 2 \\
b_{2g} &= 2p_x(B_{2g}) \qquad b_{3u} = 2p_x(B_{3u})
\end{aligned} \tag{13-43}$$

Although canonical MOs do not provide pictures of localized bonding, it is nevertheless possible to make fairly accurate guesses as to the electron configurations of many molecules by the use of rather elementary reasoning. For example, one would expect the two lowest MOs to represent the inner-shell nonbonding electrons ($1s$) of the carbon atoms (the carbon "core" orbitals). Thus the $1a_g$ and $1b_{1u}$ MOs would be expected to have large coefficients for the $1s(A_g)$ and $1s(B_{1u})$ symmetry orbitals, respectively, and small coefficients for the remaining symmetry orbitals. The remaining MOs would be expected to correspond in a very rough manner to the Lewis picture of the bonding electrons. Thus, the next MO should correspond to either a C–C single bond or a C–H bond. On the naive basis that a C–C bond energy is about 350 kJ $\cdot$ mol^{-1} (83 kcal $\cdot$ mol^{-1}) and a C–H bond is about 414 kJ $\cdot$ mol^{-1} (99 kcal $\cdot$ mol^{-1}), one would predict that the next-lowest MO corresponds to the latter. The four C–H bonds would be described by MOs having the symmetries of the hydrogen group orbitals, namely, $2a_g$, $2b_{1u}$, $1b_{2u}$, and $1b_{3g}$. Next we would have a $3a_g$ MO describing the C–C single bond. Finally, we have the remaining C–C bond: the so-called pi bond. We should expect this bond to be described by the highest energy MO, since this bond supposedly is the one which accounts for the principal chemical reactivity of ethylene. We should also expect this bond to be represented by the $1b_{3u}$ rather than the $1b_{2g}$, since the former has no nodes and the latter has one node. A usually quite reliable rule of thumb is that the energy of otherwise comparable orbitals increases with the number of nodes. Thus, the ground-state electron configuration is predicted to be:

$$(1a_g)^2(1b_{1u})^2(2a_g)^2(2b_{1u})^2(1b_{2u})^2(1b_{3g})^2(3a_g)^2(1b_{3u})^2 \quad {}^1A_g$$

Accurate calculations on ethylene indicate that there is only one minor discrepancy in this prediction: the energy of the $3a_g$ MO is actually slightly lower (less than $0.1E_h$) than that of the $1b_{3g}$ MO. The first accurate *ab initio* all-electron calculation on this molecule was carried out by Moskowitz and Harrison.[19] This calculation used gaussian orbitals to represent STOs; the most accurate calculation reported used the basis $(9s9p/3d)$. The molecular electronic energy obtained was $-77.950224E_h$. However, for purposes of detailed examination we will consider a simpler STO-6G calculation using a minimal basis set. This leads to a total energy of $-77.82576E_h$, admittedly not as good as obtained in the Moskowitz and Harrison calculation. This calculation also assumed the experimental geometry (C–C and C–H bond distances of 2.53 a_0 and 2.00 a_0, respectively, and an H–C–H bond angle of 117°22′), whereas Moskowitz and Harrison attempted to locate the equilibrium geometry by energy minimization. Table 13-7 lists the molecular orbital energies obtained by both methods of calculation; the difference between the two are typical of those obtained when calculations are done with different basis sets.

Of particular interest in the STO-6G calculation are the LMOs obtained using the Boys method. Table 13-8 lists the LMOs obtained in terms of the

[19] J. W. Moskowitz and M. C. Harrison, *J. Chem. Phys.*, **42**:1726 (1965); J. W. Moskowitz, *J. Chem. Phys.*, **43**:60 (1965). See also J. L. Whitten, *J. Chem. Phys.*, **44**:359 (1966).

TABLE 13-7
Molecular orbital energies of ethylene (ϵ/E_h)

MO	STO-6G*	$(9s9p/3s)^\dagger$
$1a_g$	−11.173	−11.249
$1b_{1u}$	−11.172	−11.248
$2a_g$	−0.984	−1.040
$2b_{1u}$	−0.759	−0.794
$1b_{2u}$	−0.614	−0.648
$3a_g$	−0.537	−0.569
$1b_{3g}$	−0.476	−0.510
$1b_{3u}$	−0.328	−0.366

* GAMESS calculation by the author using a minimal basis set.

† J. W. Moskowitz, *J. Chem. Phys.* **43**:60 (1965).

TABLE 13-8
LMOs of ethylene

AO	Core	C–H bond	C–C bond	$(CC_1 + CC_2)/2$	$(CC_1 - CC_2)/2$
$1s_1$	1.006	0.089	0.068	0.068	0.000
$2s_1$	−0.011	−0.346	−0.263	−0.263	0.000
$2p_{x1}$	0.000	0.000	−0.450	0.000	−0.450
$2p_{y1}$	0.000	−0.396	0.000	0.000	0.000
$2p_{z1}$	0.003	−0.253	0.293	0.293	0.000
$1s_2$	0.004	−0.014	0.068	0.068	0.000
$2s_2$	−0.016	0.060	−0.263	−0.263	0.000
$2p_{x2}$	0.000	0.000	−0.450	0.000	−0.450
$2p_{y2}$	0.000	0.0008	0.000	0.000	0.000
$2p_{z2}$	−0.018	0.052	−0.293	−0.293	0.000
H_1	−0.015	−0.531	0.038	0.038	0.000
H_2	−0.015	0.071	0.038	0.038	0.000
H_3	0.003	−0.048	0.038	0.038	0.000
H_4	0.003	0.037	0.038	0.038	0.000

Note: The figures were obtained through GAMESS calculation by the author using STO-6G with a minimal basis set. The entire calculation, including computation of the canonical MOs, required 40.82 s on a Digital VAX 8500 computer.

basic AOs of the basis; the original eight occupied MOs transform into two LMOs representing C cores, four LMOs representing C–H bonds, and two LMOs representing C–C bonds. Since LMOs within each one of these categories are equivalent, only one of each type is explicitly tabulated. However, the table also lists the sum and the difference of coefficients between the two LMOs describing the C–C bonds; these are labeled "$(CC_1 + CC_2)/2$" and "$(CC_1 - CC_2)/2$," respectively. We will discuss the significance of these shortly.

Several features of the LMOs are outstanding. First, note that the C core LMO is predominantly $1s$ of *one* of the carbon atoms; the second C core LMO (not tabulated) is of the same form, but it is the other $1s$ AO which predominates. Also note that this LMO does not contain $2p_x$ or $2p_y$ AOs. Also, this LMO contains some hydrogen $1s$ AOs—more from the H atoms bonded to the dominant C atom than to the other C atom. The C–H LMO contains no $2p_x$ AOs; this is because these AOs are at right angles to the plane of the C–H bonds. Note that these LMOs belong to the reducible representation $A_g \oplus B_{1u} \oplus B_{2u} \oplus B_{3g}$ (all irreducible representations except B_{3u}). In the simple description of a C–H bond in terms of hybridized AOs, each such bond is represented by an overlap of a carbon $2p^2$ (trigonal) hybrid and a hydrogen $1s$. The LMO description, although differing from this simple picture, nevertheless shows that the bond is largely $2s$, $2p_x$, and $2p_y$ of carbon (the components of trigonal hybrids), with the $1s$ orbital of one of the neighbor hydrogen atoms predominant. However, the LMO also contains small amounts of C $1s$ and the $1s$ orbitals of the other hydrogen atoms (the second neighbor hydrogen atom has a smaller coefficient than the first, and the others have coefficients which become smaller as the distance from the predominant hydrogen atom increases).

Perhaps the most prominent feature of the LMO description is that the sigma and pi bonds are not separated; they are described by a pair of equivalent LMOs. Note that these LMOs span the $A_g \oplus B_{3u}$ reducible representation; in the popular description of sigma and pi bonds as distinct, the sigma would be A_g and the pi bond would be B_{3u}. If one examines the second C–C LMO, one finds that it is identical with the one shown in Table 13-8 except that the coefficients of the $2p_x$ AOs have opposite signs. Thus, if one forms the sum and difference of these equivalent LMOs (and divides by two), there is obtained a separation of the sigma and pi bonds. Examination of the last two columns of Table 13-8 shows clearly that the sum represents mostly the overlap of $2s$ and $2p_z$ AOs of the two carbon atoms; this corresponds to the overlap of carbon sp^2 hybrids which describe this bond in the simple model. Furthermore, the difference is an overlap of $2p_x$ AOs of the two carbons and represents a pi bond in precisely the same way as in the simple hybrid AO model.

A Mulliken population analysis of the above calculation shows that the hydrogen atoms have a charge of +0.067 each and the carbon atoms have a charge of −0.133 each. This is qualitatively in accord with the normal expectation based on relative electronegativities.

BENZENE. The minimal basis set for benzene, C_6H_6, consists of 36 AOs: six $1s$ AOs of the hydrogen atoms, and six sets of $1s$, $2s$, and $2p$ AOs of the carbon atoms. Since no single AO spans an irreducible representation of the D_{6h} point group to which benzene belongs, it is necessary to form group orbitals of all of these in order to anticipate the forms of the MOs, as was done in the previous example of ethylene. Although the operations to do this are reasonably straighforward (but tedious), some awkward problems do arise with respect to the carbon $2p$ AOs which lie in the plane of the benzene ring (we will designate this the xy plane and let the z axis be coincident with the principal C_6 symmetry axis of benzene). If all the $2p_x$ and $2p_y$ AOs are oriented the same way (with respect to the coordinate system of the molecule), some will not transform into each other under all the group symmetry operations. Using the convenient abbreviations x_i for $2p_{xi}$ (i labels the carbon atoms 1, 2, . . . , 6), the general transformation may be written:

$$\hat{R}x_i = x_j \cos\theta - y_j \sin\theta = x'_j$$
$$\hat{R}y_i = x_j \sin\theta + y_j \cos\theta = y'_j \qquad (13\text{-}44)$$

where θ is the angle through which x_j must be rotated to form x'_j, etc. Note that a special case occurs when θ is a multiple of 90°: $\hat{R}x_i = \pm x_j$, etc. As a specific example, the $\hat{C}_6^+$ operation is of the general type. Figure 13-7 illustrates what happens when this operation is applied to the $2p_x$ and $2p_y$ AOs of carbon atom 1 when these are oriented relative to the chosen coordinate system and

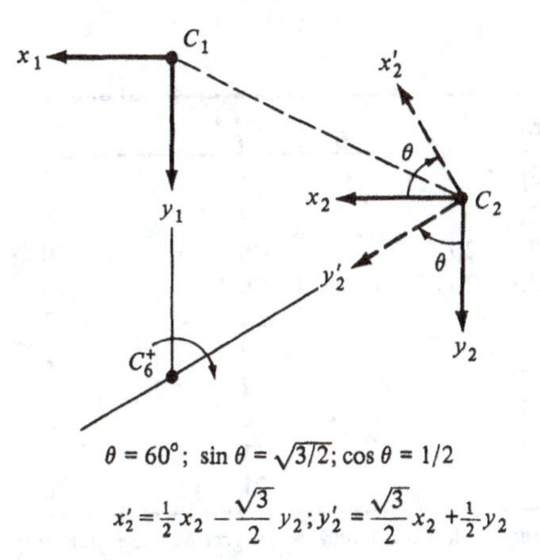

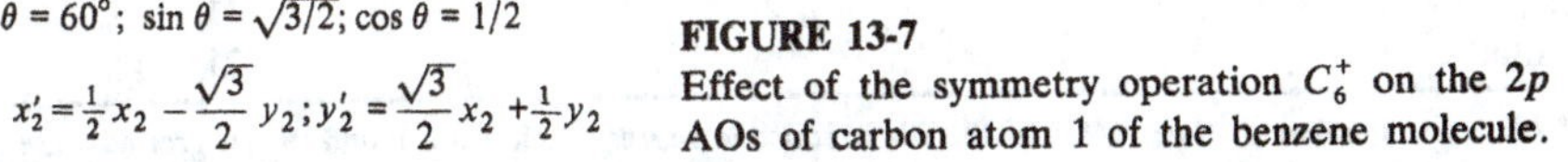

$\theta = 60°$; $\sin\theta = \sqrt{3}/2$; $\cos\theta = 1/2$
$x_2' = \frac{1}{2}x_2 - \frac{\sqrt{3}}{2}y_2$; $y_2' = \frac{\sqrt{3}}{2}x_2 + \frac{1}{2}y_2$

FIGURE 13-7
Effect of the symmetry operation C_6^+ on the $2p$ AOs of carbon atom 1 of the benzene molecule.

when the corresponding $2p$ AOs of carbon atom 2 are likewise oriented. Note that this operation does not transform $2p_{x1}$ and $2p_{y1}$ into $2p_{x2}$ and $2p_{y2}$, but rather into the following two linear combinations:

$$x_2' = \frac{1}{2}x_2 - \frac{\sqrt{3}}{2}y_2 \qquad y_2' = \frac{\sqrt{3}}{2}x_2 + \frac{1}{2}y_2 \tag{13-45}$$

An alternative approach is to orient each pair of $2p$ AOs differently vis-a-vis the x and y axes, but identically relative to the C atom to which they belong, e.g., as depicted in Fig. 13-8. In this setup each $2p$ AO is a linear combination of $2p_x$ and $2p_y$, and each will transform into an equivalent $2p$ AO under all symmetry operations of the group.

Table 13-9 lists all the AOs of the minimal basis set and the irreducible representations to which their group-orbital combinations belong. The table also shows the number of times that MOs of a given irreducible representation occur in the ground-state electron configuration and also the total number of MOs constructible from the group orbitals belonging to each irreducible

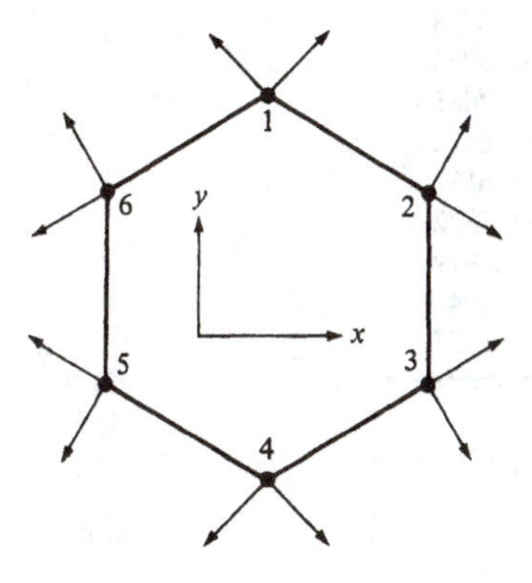

FIGURE 13-8
Six sets of $2p$ AOs of carbon arranged symmetrically about the benzene ring. Each set is a linear combination of $2p_x$ and $2p_y$ AOs.

TABLE 13-9
MOs of benzene

Irreducible representation i	Basis functions	N_i^g *	N_i^t †
A_{1g}	H_i, $1s_i$, $2s_i$, $2p_i'$ ‡	3	4
B_{1u}	H_i, $1s_i$, $2s_i$, $2p_i'$	2	4
E_{1u}	H_i, $1s_i$, $2s_i$, $2p_{xi}$, $2p_{yi}$	6	10
E_{2g}	H_i, $1s_i$, $2s_i$, $2p_{xi}$, $2p_{yi}$	6	10
A_{2u}	$2p_{zi}$	1	1
B_{2g}	$2p_{zi}$	0	1
E_{1g}	$2p_{zi}$	2	2
E_{2u}	$2p_{zi}$	0	2
B_{2u}	$2p_{xi}$, $2p_{yi}$	1	2
Total		21	36

* N_i^g is the number of times MOs of the ith irreducible representation are found in the ground-state electron configuration.

† N_i^t is the total number of MOs constructible from the basis AOs of the ith irreducible representation.

‡ $2p_i'$ refers to linear combinations of $2p_{xi}$ and $2p_{yi}$ AOs.

representation. Note that since benzene has a total of 42 electrons, it requires 21 MOs out of the total 36 to represent the ground state.

Table 13-10 lists the MOs and corresponding orbital energies obtained from a GAMESS calculation using a minimal basis set and the STO-6G method. The total molecular electronic energy obtained by this calculation was $-230.13105 E_h$. The calculation utilized the experimental geometry of a C–H bond distance of 2.04 a_0 and a C–C bond distance of 2.63 a_0. This table also

TABLE 13-10
Molecular orbital energies of benzene

Molecular orbital	Energy, ϵ/E_h *	Energy, ϵ/E_h †
$1a_{1g}$	−11.1804	−11.2434
$1e_{1u}$	−11.1802	−11.2429
$1e_{2g}$	−11.1795	−11.2416
$1b_{1u}$	−11.1792	−11.2411
$2a_{1g}$	−1.0945	−1.1487
$2e_{1u}$	−0.9595	−1.0138
$2e_{2g}$	−0.7715	−0.8230
$3a_{1g}$	−0.6669	−0.7081
$2b_{1u}$	−0.5989	−0.6427
$1b_{2u}$	−0.5572	−0.6187
$3e_{1u}$	−0.5379	−0.5872
$1a_{2u}$	−0.4606	−0.4979
$3e_{2g}$	−0.4370	−0.4946
$1e_{1g}$	−0.2846	−0.3337

* GAMESS calculation by the author, STO-6G, minimal basis set.

† W. C. Ermler and C. W. Kern, *J. Chem. Phys.*, **58**:3458 (1973), using a $(9s5p1d/4s1p)/[4s2p1d/2s1p]$ basis.

shows some of the results of a more extended calculation by Ermler and Kern[20] using a gaussian basis of ($9s5p1d/4p1s$) contracted to [$4s2p1d/2s1p$]. The latter calculations, which not only are of the double zeta type but also include polarization on both C and H as well, produce a total energy of $-230.74938E_h$; this is believed to be within $0.07 \pm 0.02E_h$ of the Hartree-Fock limit. Ermler and Kern also used a slightly different geometry from that of the GAMESS STO-6G calculation: a C–C distance of 2.637 a_0 and a C–H distance of 2.051 a_0. It should be noted that although the two calculations exhibit considerable differences in orbital energies, the order of the MOs is the same in both. Comparative Mulliken population analyses of the two calculations also show the following for net charges on the carbon and hydrogen atoms, respectively: GAMESS (−0.067 and +0.067), Ermler and Kern (−0.138 and +0.138). Thus the latter calculations show the hydrogen atoms to be somewhat more acidic than the former calculations indicate.

The four lowest-energy MOs ($1a_{1g}$, $1e_{1u}$, $1e_{2g}$, and $1b_{1u}$) belong principally to the carbon cores. The six electrons which are generally associated with the pi bonding of benzene in the simple treatments are those in the $1e_{1g}$ and $1a_{2u}$ MOs. These MOs are formed entirely from the $2p_z$ AOs of carbon. Note that these two sets of MOs are separated by a pair of doubly degenerate MOs, $3e_{2g}$, which are not part of the pi bonding, since they are made up of AOs other than carbon $2p_z$ AOs; these would be described as *sigma electrons* in the simple model. Since it is usually assumed that there is an energy separation between the pi electrons as a group and the sigma electrons as a group, it is clear that the MO picture of bonding in benzene is at variance with the elementary qualitative model generally employed.

The localized MO analysis of the benzene molecule is not as simple and straightforward as in the earlier cases discussed and, furthermore, is not unique; consequently, only a partial analysis will be presented here. When the Boys localization method is used, one finds that there are six LMOs which contain carbon $2p_z$ AOs; consequently, these encompass all the pi bonds. In Table 13-11 there are listed the coefficients of the carbon $2p_z$ AOs found in each of these LMOs. Note that the LMOs occur in pairs in which the absolute values of the coefficients for specific AOs are the same but the coefficients are opposite in sign.

It is apparent that each LMO must represent 1 pi electron and 1 sigma electron. The pi electron in each LMO is partly localized to a pair of adjacent carbon atoms, and the remainder is spread almost equally over the other remaining four carbon atoms. If we consider relative values of the AO coefficients (and ignore nonorthogonality of the AOs), the LMO pairs 1 and 2 place approximately $0.8e^-$ on carbon atoms 1 and 2, with $0.2e^-$ distributed

[20] W. C. Ermler and C. W. Kern, *J. Chem. Phys.*, **58**:3458 (1973).

TABLE 13-11
Pi electron LMOs of benzene

Atom	LMO 1	LMO 2	LMO 3	LMO 4	LMO 5	LMO 6
1	−0.402	0.402	0.130	−0.130	−0.136	0.136
2	−0.402	0.402	−0.136	0.136	0.130	−0.130
3	−0.136	0.136	−0.402	0.402	0.130	−0.130
4	0.130	−0.130	−0.402	0.402	−0.136	0.136
5	0.130	−0.130	−0.136	0.136	−0.402	0.402
6	−0.136	0.136	0.130	−0.130	−0.402	0.402

almost evenly over the remaining carbon atoms. LMO pairs 3 and 4 do likewise by concentrating the pi electron on atoms 3 and 4, and similarly LMO pairs 5 and 6 concentrate the pi electron on atoms 5 and 6. Thus, the composite picture of the pi bonding is that of complete delocalization of the six pi electrons. Each of these LMOs also describes one sigma electron; this sigma electron is distributed among both the carbon atoms and the hydrogen atoms.

England, Salmon and Ruedenberg[21] have demonstrated that there exist an infinite number of sets of LMOs for benzene with an equal degree of localization. One such set corresponds to a Kekulé structure of benzene and is just that described above. One of the other possibilities consists of LMOs in which the electrons are localized to *three* adjacent carbon atoms rather than two as in the Kekulé structures. Several other possibilities, intermediate to the two mentioned, are also illustrated by these authors.

A provocative paper by Cooper, Gerratt, and Raimondi[22] challenges the correctness of the delocalized pi bonding picture of benzene provided by the single-determinantal MO treatment. These authors carried out calculations utilizing a variation of the valence bond (VB) method which support the notion that the pi electrons in benzene are almost certainly localized and that the characteristic properties of such a system arise from the mode of spin coupling. In the simple VB treatment, the pi electrons of benzene are described by a linear combination of two wavefunctions, each of which represents one of the localized Kekulé structures of benzene. Cooper et al. show that if the AOs used as a basis are optimized (so that each spreads out a little toward its neighbors), then the two Kekulé structures alone give very nearly the same result. Thus the same final picture of benzene (six equivalent C–C bonds) can be arrived at either by delocalization (the MO method) or by localization (the modified VB method). A key factor here is the well-known fact that extended MO and extended VB methods can be shown to be exactly equivalent when carried out in certain ways. Specifically, if the single-determinantal MO

[21] W. England, L.S. Salmon, and K. Ruedenberg, op. cit., p. 31.

[22] D. L. Cooper, J. Gerratt, and M. Raimondi, *Nature*, 323:699 (1986).

calculation is extended by a full CI calculation, the end results are precisely the same, if the VB calculation with localized structures is extended with all possible ionic structures (there are 170 of these). The cogent question then is not which approach is right, but which gives the more rapid convergence and which produces the more convenient interpretation of chemical bonding.[23] Thus, although MO + full CI and VB + full ionic lead to the same end result, it may be that the latter method arrives at a chemically meaningful stage much earlier than the former does.

EXERCISES

13-9. Set up the MOs of cyclopentadiene (using a minimal basis set of STOs), classify them as to the irreducible representations involved, and guess what the ground-state electronic configuration might be. Is it possible to "see" where the double bonds are represented in the MO description? If you have access to a program such as GAMESS, you may wish to do the cyclopentadiene MO calculation using a basis such as STO-2G or STO-3G (these will run fairly fast and are reasonably reliable in geometric optimizations). GAMESS will also do the transformation to LMOs.

13-10. What would happen if you tried to transform the ground-state wavefunction of dihydrogen to an LMO description? Repeat for the two excited states obtainable from the $(1\sigma_g)(1\sigma_u)$ configuration.

13-11. Suppose you transformed the usual description of the singlet and triplet states of the helium atom (from the $1s2s$ configuration) such that the exchange energy was minimized. What would the "LMOs" look like, and what would they represent?

13-5 PSEUDO-POTENTIAL METHODS IN MO CALCULATIONS ON LARGE MOLECULES

Ab initio calculations on molecules become increasingly impractical as the number of electrons and internal coordinates increases. Since it has long been realized that many of the chemical and physical properties of molecules can be qualitatively accounted for by considering only the valence electrons of the constituent atoms, it appears reasonable to seek quantitative methods in which only the valence electrons are explicitly considered and the inner-shell electrons are included in some simpler way, e.g., as contributors to an "effective field" within which the valence electrons move. After all, the periodic properties of atoms can be rationalized in this way, and molecular behavior, in turn, can be satisfactorily described in terms of atoms. The question is: Can such a procedure be put on a rigorous quantum mechanical basis?

---

[23] See comments on this by R. McWeeney, *Nature*, **323**:666 (1986).

The general problem of treating atomic and molecular electronic structures in terms of two sets of electrons, inner-shell and valence electrons, has received considerable attention, and extensive literature on this topic now exists. We will discuss briefly one such treatment: the *pseudo-potential SCF method* developed by Ewig, Van Wazer, and others.[24] In this method the nonvalence electrons of the atoms are incorporated into a pseudo potential which not only represents the electrostatic effects of the nonvalence electrons but also reflects the effects of the Pauli exclusion principle. There are a number of ways of accomplishing this objective—none of them unique—but most are based on the original work of Phillips and Kleinman.[25] The Hartree-Fock equations can be rewritten in the form

$$(\hat{F} + \hat{V}^{\text{PP}})\chi_v^i = \epsilon_v^i \chi_v^i \tag{13-46}$$

where $\hat{V}^{\text{PP}}$ is the molecular Phillips-Kleinman pseudo-potential operator given by

$$\hat{V}^{\text{PP}} = \sum_i^{\text{val}} \sum_j^{\text{core}} \sum_k^{\text{nucl}} |\phi_{ck}^j\rangle(\epsilon_v^i - \epsilon_{ck}^j)\langle\phi_{ck}^j|\left(1 - \sum_{l \neq i}^{\text{val}} |\chi_v^l\rangle\langle\chi_v^l|\right)$$

and $\hat{F}$ is defined by

$$\hat{F} = \hat{F}' + V^{\text{mod}} \tag{13-47}$$

where $\hat{F}'$ is the Hartree-Fock operator for valence electrons and V^{mod} is a "model" potential representing the inner-shell electrons. The functions $\{\chi_v^i\}$ are pseudo orbitals, since they depend on pseudo potentials, and the functions $\{\phi_{ck}^i\}$ are core functions on nucleus k. The summations with indices i, j, and k refer to valence electrons, core electrons, and nuclei, respectively. The three main steps in solving the pseudo-potential SCF equations are as follows:

1. Conventional SCF calculations are carried out on the constituent atoms of the molecule, using the basis chosen for the ultimate molecular calculation.
2. Atomic pseudo-potential SCF calculations are used to produce model potentials for the nonvalence electrons. This determination of the model potentials represents the only parametrization in the procedure. In some cases, the model potentials may be represented by simple analytical forms.

---

[24] A reasonably detailed account is given by C. S. Ewig, R. Osman, and J. R. Van Wazer, *J. Chem. Phys.*, **66**:3557 (1977).

[25] J. C. Phillips and L. Kleinman, *Phys. Rev.*, **116**:287 (1959).

3. The above atomic model potentials are used along with the Phillips-Kleinman pseudo potential to form effective monoelectronic operators for use in the pseudo-potential SCF equations. The latter are then solved iteratively in the usual manner.

In one of their earlier papers, Ewig and Van Wazer[26] compared the results of a conventional minimal basis SCF calculation and extended basis set calculations on dibromine, Br_2, with a pseudo-potential SCF calculation in which the model potential was expressed by the simple relationship

$$V^{\text{mod}}(r) = \frac{N_c(1 - e^{-\alpha r})}{r} \tag{13-48}$$

where α is an adjustable parameter chosen to reproduce the sum of the s and p orbital energies. The two calculations were then compared with respect to the bond lengths, fundamental vibration frequencies, and ionization energies they predicted. Except for the case of the ionization energy, the results were in good agreement with near-Hartree-Fock calculations. Similar calculations were also reported on a number of dihalogen molecules, including a number of heteronuclear examples, such as FCl and ClBr.

Calculations by R. Ahlrichs[27] on ammonia, water, and other small molecules show that bond lengths and bond angles calculated by a particular pseudo-potential method are in close agreement with those obtained by conventional SCF calculations using exactly the same basis set.

Rothman, Bartell, Ewig, and Van Wazer[28] have applied the pseudo-potential SCF method to studies of various hypervalent compounds, e.g., the xenon fluorides XeF_2, XeF_4, XeF_5^+, and XeF_6. Experimental work on the last compound has revealed several puzzling features which hitherto have appeared difficult to interpret, but appear to be satisfactorily explained by the pseudo-potential SCF calculations.

The pseudo-potential method has also been extensively discussed by Szasz and McGinn,[29] and applications have been made to a number of atoms, ions, and diatomic molecules, e.g., Li, Na, K, Rb, Be^+, Al^{2+}, Cu, Zn^+, Li_2, Na_2, K_2, LiH, NaH, and KH.

Perhaps one of the best examples of the use of pseudo potentials is in the calculation of electronic structures of large molecules, particularly those of biological importance. In such cases, *ab initio* all-electron calculations are

[26] C. S. Ewig and J. R. Van Wazer, *J. Chem. Phys.*, **63**:4035 (1975).

[27] Reported by J. C. Berthelat and Ph. Durand in *Gazz. Chim. Ital.*, **108**:225 (1978).

[28] M. J. Rothman, L. S. Bartell, C. S. Ewig, and J. R. Van Wazer, *J. Chem. Phys.*, **73**:375 (1980).

[29] L. Szasz, *J. Chem. Phys.*, **49**:679 (1968) and L. Szasz and G. McGinn, *J. Chem. Phys.*, **45**:2898 (1966), **47**:3495 (1967), and **48**:2997 (1968).

generally out of the question. The following will provide a brief description of one such set of studies.

Examples of molecules of biological interest range all the way from the small amino acids such as glycine and alanine

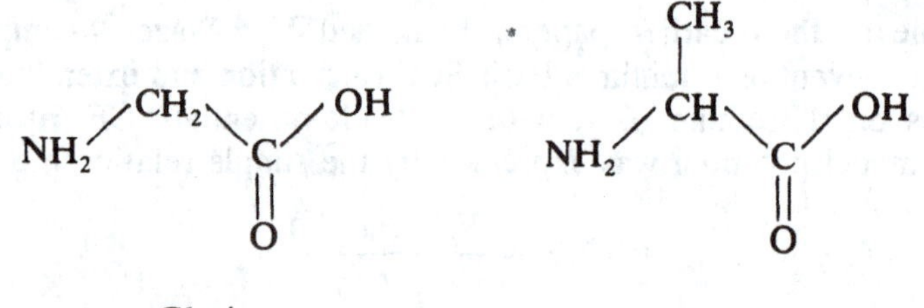

Glycine Alanine

to very large molecules such as RNA, DNA, and proteins. However, many of the molecules of greatest interest are of moderate size: e.g., histamine, epinephrine, and serotonin

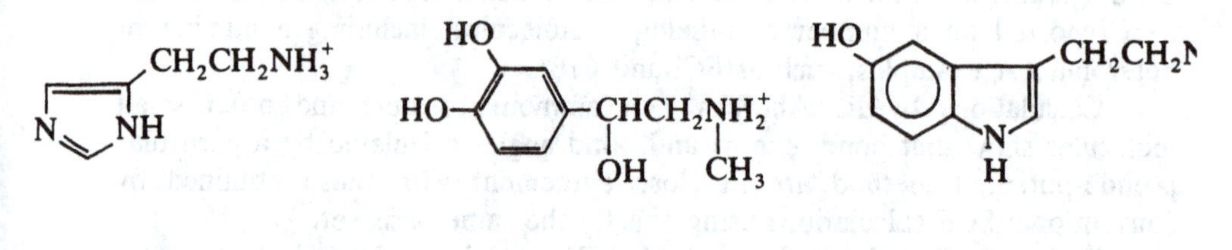

Histamine Epinephrine 5-Hydroxytryptamine (serotonin)

The basic question which arises in conjunction with calculations on such molecules is *what can one calculate*? Limiting ourselves to quantum chemical calculations of the molecular orbital variety, the answer is: We can calculate the *electronic structure* of the *isolated* molecule. The word "isolated" is emphasized, since it is obvious that biomolecules function in rather complex environments in which there may be strong solute-solvent interactions, electrostatic effects, etc. Unfortunately, the calculations we are capable of doing assume strictly isolated molecules, and we must either find ways to introduce these environmental effects after the fact or hope they aren't important enough to affect predictions and interpretations. Surprisingly, the latter situation often seems to be the case.

The next question is: What information can electronic structure calculations provide? The following are some examples of the information provided:

1. The calculations augment experimental data to help suggest mechanisms for biological processes.
2. The calculations permit the construction of semiempirical bases for predicting how new molecules work and for designing new bioactive structures.

The first step is to reduce the number of electrons which must be considered explicitly, with the rest to be included in a suitably chosen pseudo potential. We will use serotonin (5-hydroxytryptamine) as an illustration; this molecule ($C_{10}H_{12}N_2O$) has 25 atoms and a total of 94 electrons. If a minimal basis set is used, the 10 carbon, 2 nitrogen and 1 oxygen atoms require $1s$, $2s$, $2p_x$, $2p_y$, and $2p_z$ AOs each, or 65 AOs total. Adding $1s$ for each of the 12 hydrogens leads to a basis set of 77 AOs. Thus the calculations require repeated diagonalization of a 77×77 matrix [5929 matrix elements, each of which may consist of hundreds (and even thousands) of terms]. This produces 77 MOs, 47 of which will be doubly occupied in the ground state. This produces a mathematical problem of some complexity. Clearly, some simplification of the problem is needed if it is to be tractable for the computer.

One of the most-used simplifications is to include all the inner-shell, nonbonding electrons in the pseudo potential. Thus, for serotonin one would omit $1s$ electrons of the C, N, and O atoms and thereby reduce the number of electrons from 94 to 68.

To illustrate a conceptual framework for quantum chemical studies of biomolecules, let us consider the specific case of endogenous ligands (hormones, neurotransmitters) which are used in biological systems as "chemical messengers," i.e., molecules which bind to receptor sites on proteins which, for example, reside on nerve terminals. The binding of these ligands activates a receptor system, i.e., initiates a molecular process which, through some cascade of events, leads to a measurable physical response such as muscle contraction or gastric acid secretion. The two measurable events of general importance are those of binding and activation (also called *affinity* and *response* by physiologists). The role of quantum chemical calculations is to consider molecular models for the recognition (binding) and activation stages and to test these models on the measurable binding and response data.

There are three general types of endogenous ligands classified according to the manner in which they interact with receptors:

1. Agonists (positive response)
2. Partial agonists (weak response)
3. Antagonists (no response; may block the receptor)

Serotonin (also called *5-HT*) is of particular interest to physiologists, since it is an important neurotransmitter and since a large number of drugs (particularly hallucinogens) interact very readily with it. A neurotransmitter is generally a small molecule which enables a "message" to go across the synaptic gap (see Fig. 13-9). In this way neurons (nerve cells) on opposite sides of the synaptic gap can "communicate." One of the most publicized hallucinogenic molecules which interacts with 5-HT is lysergic acid diethylamide, better

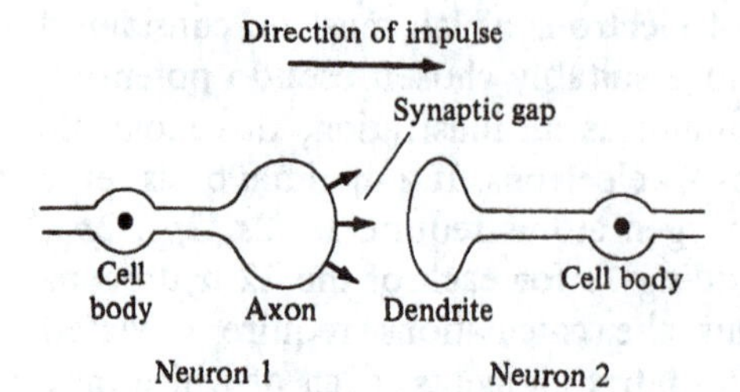

FIGURE 13-9
Schematic illustration of the structure and function of the synaptic gap. The nervous system is a network of neurons, and each neuron consists of a cell body, dendrites, and an axon. An impulse (message) travels from one cell body to another when the axon of one neuron releases a substance (the neurotransmitter) which crosses the synaptic gap to the dendrites of the adjoining neuron. This stimulates the second neuron to continue the message to its cell body. Enzymes clean up the expended neurotransmitters so that the synaptic gap is able to function over and over again.

known as *LSD*. This molecule has the chemical structure

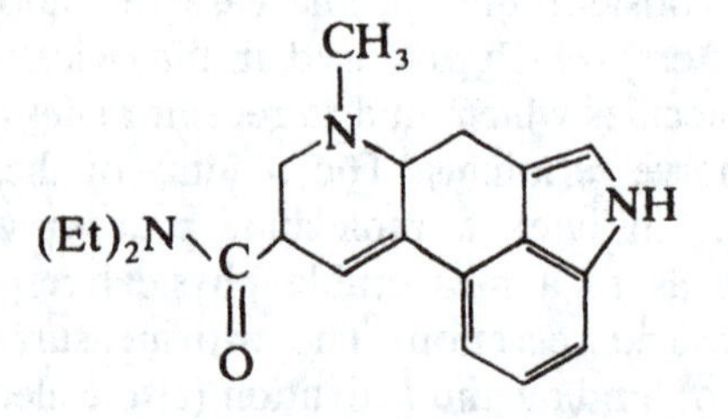

Lysergic acid diethylamide
(LSD)

The mechanism of how LSD interacts with 5-HT has been studied by a combination of experimental and theoretical methods, the latter being pseudopotential MO calculations by Sid Topiol et al.[30] The following is a summary of their approach.

At physiological pH, the ethylamine side chain of 5-HT is protonated; i.e., the side chain is $-CH_2CH_2NH_3^+$. The recognition of 5-HT as a ligand must be based on long-range properties of this molecule. The conjecture is that the electrostatic field produced by the $-CH_2CH_2NH_3^+$ is the primary element in the recognition stage. Experimental support for this comes from NMR studies of stacking complexes of 5-HT and ATP (adenosine triphosphate).[31] These studies show that when the side chain is removed from 5-HT, its affinity for ATP is significantly reduced—but not totally removed. Thus it is likely that other

[30] S. Topiol, J. W. Moskowitz, and C. F. Melius, *J. Chem. Phys.*, **68**:2364 (1978) and **70**:3008 (1979).

[31] T. Nogrady, P. D. Hrdina, and G. M. Ling, *Mol. Pharmacol.*, **8**:565 (1972).

factors also operate; most likely is the pi electron cloud over the indole ring of 5-HT. Consequently, Topiol and his coworkers used pseudo-potential MO methods to calculate long-range electrostatic properties of the 5-HT molecule. It should be noted that these properties appear totally inaccessible to direct experiment.

Once a wavefunction for 5-HT is obtained, the MO coefficients c_{ri} can be used to predict an electron charge distribution over the molecular framework (the map is done for the neutral molecule; once it is "recognized" by virtue of its charged side chain, it is bound and neutralized). In turn, this charge distribution permits the calculation of an interaction energy of this charge distribution with a positive point charge located outside the molecular framework. This can be repeated at any number of points to produce a map of the electrostatic potential of the molecule. For example, of particular interest in 5-HT is mapping of the electrostatic potential in a plane above and paralleling the indole ring portion of the molecule. When this is done for a plane located 0.16 nm above the indole ring, one finds an ellipsoidal double-minimum trough whose axis (a line joining the two minima) is as shown below (the locations of the two minima are indicated by small arrows pointing to the axis):

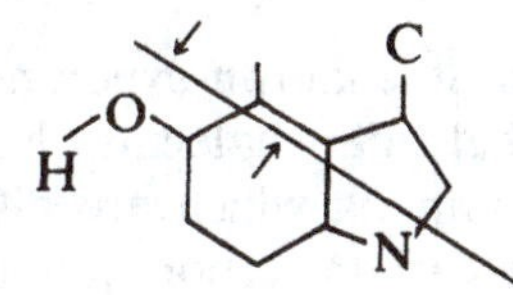

Somewhat surprisingly, a number of molecules which are known to bind to 5-HT (agonists) exhibit similar electrostatic potential maps.

Experiments show that 5-HT–imidazolium complexes are indeed electrostatic in nature. Furthermore, if one partitions the total interaction energy as calculated from the wavefunction according to a method proposed by Morokuma,[32] namely, as a sum of the electrostatic, polarization, exchange, and charge transfer terms, it is found that the first (electrostatic) term predominates.

Let us consider the use of an electrostatic potential map to help in the understanding of how the hallucinogenic drug LSD functions. The mechanism for the activity of this molecule is complex and not too well understood from the experimental standpoint, but it is generally agreed that it is an antagonist at both histamine and 5-HT receptor sites. The electrostatic map of LSD reveals an electrostatic potential minimum which closely corresponds to a minimum also found near the OH group of 5-HT. This LSD minimum is located just outside the carbon-carbon double bond found in the *N*-methyltetrahydro-

[32] K. Morokuma, *J. Chem. Phys.*, **55**:1236 (1971).

pyridine ring portion of the molecule:

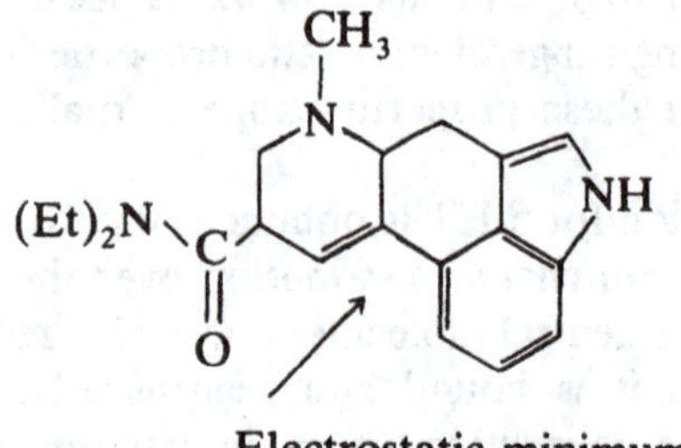

Such a similarity would be impossible (or at least difficult) to deduce on the basis of experimental evidence alone: the two molecules (LSD and 5-HT) simply do not appear to resemble each other sufficiently.

However, is this similarity of electrostatic potential regions of any real significance? One way to test this is to design a molecule just like LSD but without the double bond in the *N*-methyltetrahydropyridine ring. This has been done, and experiments show that the binding to 5-HT is dramatically reduced.[33]

As further corroboration, it is known experimentally that LSD also binds to histamine receptors and that both molecules have similar minima in their electrostatic potential maps, minima which are different from that in 5-HT. Consequently, removal of the carbon-carbon double bond in LSD should not affect the binding to histamine. Experiment shows that this is indeed the case.

In conclusion, such studies are in their infancy, but it appears clear that molecular orbital calculations are capable of giving considerable useful guidance to experimental chemists as they attempt to unravel some of the mysteries of biochemistry.

13-6 CALCULATIONS ON TRANSIENT AND EXPERIMENTALLY UNOBSERVED SPECIES

One of the major goals of chemists is to be able to carry out reliable calculations on the electronic and molecular structures of molecules, ions, radicals, etc., which are intrinsically unobservable by experiment or which, though theoretically observable, do not yield to extant state-of-the-art experimental methods. For example, one would like to be able to calculate

[33] H. Weinstein, R. Osman, J. P. Green, and S. Topiol, in P. Politzer and D. G. Truhlar (eds.), *Chemical Applications of Atomic and Molecular Electrostatic Potentials*, Plenum, New York, 1981, p. 309, and H. Weinstein, R. Osman, S. Topiol, and J. P. Green, *Ann. N.Y. Acad. Sci.*, **367**:434 (1981).

reliable energy surfaces which represent in detail how two molecules A and B interact to form a new molecule C (or two new molecules C and D). In many cases experimentalists have proposed possible mechanisms for such reactions, but in general, such mechanisms are incomplete and often lacking in true verifiability. In principle, it seems plausible that quantum calculations of sufficient sophistication could shed additional light on such mechanisms and perhaps even establish which of several alternatives is the "correct" one.

Another type of problem which appears inaccessible by direct experimentation is the establishment of the electronic states and equilibrium geometries of unusual molecules such as lithiated hydrocarbons. Although such compounds are of immense importance in chemistry, they generally exist as oligomers in solution and in the crystalline state and, consequently, structural information on the monomers has come primarily from calculations.[34] Furthermore, some lithium compounds appear to violate the octet rule. Recent studies at the single-determinantal SCF level have led to predictions of some truly remarkable structures for lithiated hydrocarbons which appear to violate conventional bonding rules for carbon. Some simple examples of these are the planar structures I and II of dilithiomethane and the bridged structure V of dilithioacetylene (see Fig. 13-10).

Previous studies by others at the restricted Hartree-Fock (RHF) level show that structures I–III are so close in energy that it is difficult to pick out the ground state.[35] However, the calculations make the unexpected prediction

[34] P. v.R. Schleyer, *Pure & Appl. Chem.*, **a56**:151 (1984).

[35] J. B. Collins et al., *J. Am. Chem. Soc.*, **98**:5419 (1976).

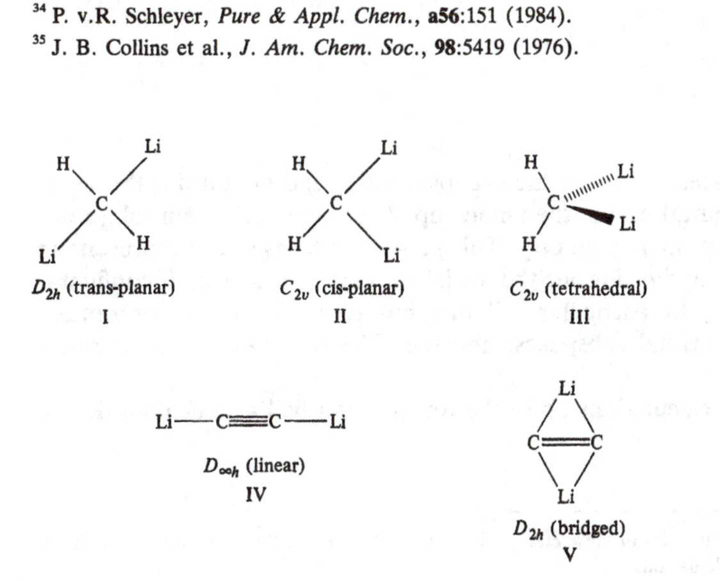

FIGURE 13-10
Stable structures of dilithiomethane (I, II, and III) and dilithioacetylene (IV and V).

that in each case there is a triplet state lower in energy than the corresponding singlet state. In the case of dilithioacetylene, many previous RHF calculations favor structure V as lower in energy than IV. However, exploratory calculations at the RHF level reveal that the relative energies of some of these simple lithiated hydrocarbons are strongly dependent on the basis used and that, therefore, this level of approach is generally unreliable. By contrast, full valence-electron, multiconfigurational self-consistent field (MCSCF) calculations (with full geometry optimization) on I–III strongly favor III as the ground state, with I being the highest in energy; in all cases the triplet state does appear to be slightly lower in energy than the corresponding singlet. Similarly, MCSCF calculations with very good basis sets and very large active configurational spaces suggest that it is the linear structure IV which is lower in energy.[36]

Two important reasons for questioning the adequacy of RHF calculations on organolithium are (1) the ionic nature of the C–Li bond and (2) the availability of vacant, low-lying $2s$ and $2p$ functions on lithium. These factors account for the remarkable bridging ability of lithium and also lead to electronic configurations which might be expected to contribute significantly to the total wavefunction. Since the MCSCF method is capable (in principle) of providing an optimal solution to such a situation, it is instructive to see how such calculations fare when applied to the problem. In particular, we will consider a special approach to the MCSCF method developed by Siegbahn and coworkers known as the *complete active space self-consistent field* (*CASSCF*) *method*.[37]

In the CASSCF procedure, the MO subspace is divided into three subspaces:

1. Inactive MOs
2. Active MOs
3. External MOs

The first two subspaces constitute the occupied MOs, and the third is the set of virtual MOs. Configurations are then made up of selected MOs from subgroups 2 and 3 which have proper symmetry. This procedure turns out to represent an outstanding way of optimizing all the variables inherent in a multiconfigurational wavefunction. In particular, all one has to do is select appropriate inactive and active orbital subspaces, and the CAS wavefunction is uniquely specified.

Møller-Plesset calculations up to the fourth order will also be considered.

[36] S. Su, Ph.D. thesis, University of New Hampshire, May 1988, and S. Su, F. L. Pilar, and R. P. Johnson, unpublished calculations.

[37] P. E. M. Siegbahn et al., *Phys. Scr.*, **21**:323 (1980), *Chem. Phys.*, **48**:157 (1980), and *J. Chem. Phys.*, **74**:2384 (1981).

The MCSCF calculations performed by Su and coworkers employed the following basis sets: 6-31G, 6-31G*, 6-31G**, and Dunning-Hay $(9s\,5p)/[3s\,2p]$ (the last basis will be abbreviated "D95V"). Note that the polarized, split-valence basis sets 6-31G* and 6-31G** are identical for the dilithioacetylene molecule, since no hydrogen atoms are present. A few calculations were also carried out with 6-311G and 6-311G* basis sets.[38]

While exploring the use of different basis sets at the Hartree-Fock level, the researchers made reasonably thorough searches for all possible minima on the potential-energy surfaces of the two molecules under study. These studies appear to verify that structures I–V represent the only such minima. Whereas some other workers have reported a bent version of V (C_{2v} symmetry),[39] the present calculations fail to substantiate the existence of an energy minimum for such a structure. The calculations also suggest that the potential-energy surfaces of these molecules are relatively flat around the various shallow minima; this suggests that it will be difficult to calculate relative values of such minima reliably.

Brief descriptions of the details of Su's calculations are given in the following.

DILITHIOMETHANE. All the calculations were carried out using the following general procedure for each particular basis set employed: first, an optimized geometry was obtained at the Hartree-Fock level, and this was used as the starting point for the first MCSCF-CASSCF calculation. The optimized geometry obtained from this calculation was then used as the starting point for a second MCSCF-CASSCF calculation, in which the size of the active space was made larger. Although actual bond lengths and bond angles did not change very much as the size of the wavefunction increased, this stratagem led to considerable saving of computational time. The electronic configurations for the various dilithiomethane species were as follows: the configurations shown explicitly are for the singlet state; those for the corresponding triplet states have one electron in the HOMO of the singlet and an additional electron in the MO to the left of the triplet-state term symbol.

Cis-planar: $(1a_1)^2(2a_1)^2(1b_2)^2(3a_1)^2(2b_2)^2(4a_1)^2(1b_1)^2 \quad {}^1A_1$; $(5a_1)^3B_1$

Trans-planar: $(1a_g)^2(1b_{1u})^2(2a_g)^2(3a_g)^2(1b_{2u})^2(2b_{1u})^2(1b_{3u})^2 \quad {}^1A_g$; $(4a_g)^3B_{3u}$

Tetrahedral: $(1a_1)^2(2a_1)^2(1b_2)^2(3a_1)^2(1b_1)^2(4a_1)^2(2b_2)^2 \quad {}^1A_1$; $(5a_1)^3B_2$

---

[38] Recall from Sec. 12-2 that 6-311G implies that the valence AOs are split into three parts, ϕ', ϕ'', and ϕ''', with ϕ' represented by three gaussians and ϕ'' and ϕ''' by one gaussian each.

[39] J. P. Ritchie, *Tetrahedron Lett.*, **23**:4999 (1982).

Table 13-12 lists the Hartree-Fock energies obtained for three different geometries of dilithiomethane ("tetrahedral" C_{2v}, cis-planar, and trans-planar) and for both singlet and triplet states. The singlet states are identified as "RHF" (restricted Hartree-Fock), and the triplet-state calculations are identified by "UHF" (unrestricted Hartree-Fock, where the MOs are not doubly occupied) and "ROHF" (restricted open-shell Hartree-Fock, where all MOs except the two highest are doubly occupied). Examination of the energies shows the following:

1. The Hartree-Fock energy of the tetrahedral (C_{2v}) geometry is always the lowest for both singlet and triplet states.
2. The Hartree-Fock energy of the trans-planar geometry is the highest for both singlet and triplet states.
3. For a given symmetry, the energy of the triplet state is lower than that of the singlet state with the simple exception of the calculation with the STO-6G basis.
4. As expected, the UHF calculation leads to a lower energy than does the corresponding ROHF calculation.

Except in the case of the trans-planar structure, singlet and triplet states of the same symmetry differ significantly in details of their geometries. In the case of the tetrahedral form, the C–Li bond lengths in the triplet state are about 0.01 nm longer than in the singlet, but the Li–Li separation in the former is shorter by almost 0.1 nm. Consequently, the H–C–Li bond angle in the triplet is about 10° larger than in the singlet, and the Li–C–Li bond angle is

TABLE 13-12
Hartree-Fock energies of dilithiomethane using various basis sets

	Energy/E_h			
Method	**Tetrahedral**	**Cis-planar**	**Trans-planar**	
RHF/STO-6G	−53.683524	−53.660360	−53.599967	Singlet
RHF/6-31G	−53.820860	−53.808334	−53.752104	Singlet
RHF/D95V	−53.825441	−53.818013	−53.760191	Singlet
RHF/6-31G*	−53.830163	−53.820650	−53.759979	Singlet
RHF/6-31G**	−53.834654	−53.825404	−53.767306	Singlet
UHF/STO-6G	−53.654154	−53.640889	−53.597938	Triplet
UHF/6-31G	−53.847344	−53.843632	−53.770301	Triplet
UHF/D95V	−53.853047	−53.850712	−53.766440	Triplet
UHF/6-31G*	−53.860654	−53.856875	−53.778837	Triplet
UHF/6-31G**	−53.864283	−53.860691	−53.784728	Triplet
ROHF/STO-6G	−53.701280	−53.686580	−53.593329	Triplet
ROHF/D95V	−53.851045	−53.848727	−53.774497	Triplet
ROHF/6-31G*	−53.857031	−53.853532	−53.776743	Triplet
ROHF/6-31G**	−53.860708	−53.857391	−53.781102	Triplet

smaller in the former by about 40 to 50°. Similar trends are exhibited in the cis-planar cases. In all cases except the STO-6G basis, the Mulliken net charges show the carbon atom to be negative and the lithium and hydrogen atoms to be positive.

MCSCF-CASSCF calculations were carried out using three different extended basis sets: 6-31G*, 6-31G**, and D95V. For all three types of symmetry of the singlet state, the complete active space consisted of three inactive orbitals and eight active orbitals; for the triplet state the active space consisted of three inactive orbitals and nine active orbitals. In each case the inactive orbitals were the core orbitals, those MOs lowest in energy. The results are summarized in Table 13-13. The calculations predict that for both singlet and triplet states, the lowest in energy is the tetrahedral geometry, with the trans-planar geometry being the highest. In all cases, for a given geometry, the triplet is lower in energy by a very small amount. Note that these MCSCF calculations support the RHF, UHF, and ROHF calculations for all basis sets but STO-6G.

Complete configuration-interaction calculations (the straightforward CI method) were also carried out on the singlet states in which *all* electrons—not just the valence electrons—were used to form all possible singly and doubly excited configurations. The results were qualitatively the same as for the MCSCF-CASSCF calculations.

DILITHIOACETYLENE. The electronic configurations for the two dilithioacetylene structures were as follows:

$$\textit{Linear}: (1\sigma_g)^2(1\sigma_u)^2(2\sigma_g)^2(2\sigma_u)^2(3\sigma_g)^2(3\sigma_u)^2(4\sigma_g)^2(1\pi_u)^4 \quad {}^1\Sigma_g^+$$

$$\textit{Bridged}: (1a_g)^2(1b_{2u})^2(2a_g)^2(1b_{1u})^2(3a_g)^2(2b_{2u})^2(2b_{1u})^2(4a_g)^2(1b_{3u})^2 \quad {}^1A_g$$

TABLE 13-13
MCSCF energies for dilithiomethane using various basis sets
All energies are in units of E_h

Method	Tetrahedral	Cis-planar	Trans-planar
	Singlet		
MCSCF/D95V	−53.927323	−53.920768	−53.819658
MCSCF/6-31G*	−53.930761	−53.921600	−53.825164
MCSCF/6-31G**	−53.934920	−53.926007	−53.832463
	Triplet		
MCSCF/D95V	−53.926215	−53.924066	−53.842219
MCSCF/6-31G*	−53.931691	−53.927742	−53.845785
MCSCF/6-31G**	−53.934978	−53.931207	−53.849954

With one minor exception, the energy order of the occupied MOs (listed above in increasing order of energy) is unaffected by the basis; the single exception is that the 6-31G* basis reverses the order of the two highest-occupied MOs in the case of the bridged structure.

As summarized in Table 13-14, the Hartree-Fock optimized energies for both the linear and the planar bridged forms of dilithioacetylene are dependent on the basis set. In particular, the addition of *d* functions plays an important role in stabilizing the bridged structure. It is of particular interest to note that when an STO-6G basis set is used, the bridged structure is favored over the linear structure by about $0.03E_h$ (about 82 $kJ \cdot mol^{-1}$), whereas for the 6-31G* and 6-311G* basis sets the bridged linear structure is favored by approximately 28 $kJ \cdot mol^{-1}$. However, the 6-31G, D95V, and 6-311G bases favor the linear structure by approximately 39, 27, and 18 $kJ \cdot mol^{-1}$, respectively. Clearly, no definite conclusions concerning the true ground state can be deduced from such inconsistent results.

Table 13-15 summarizes a number of MCSCF-CASSCF calculations on dilithioacetylene. Although the calculations do not provide a conclusive answer, the trend definitely appears to favor the linear structure as the lower in energy. In any event, the calculations strongly suggest that the last word is not yet in.

Tables 13-16 and 13-17 summarize Møller-Plesset calculations on linear and bridged dilithioacetylene. Note that calculations 1 through 5 and 11 through 13 favor the bridged structure, whereas calculations 6 through 10 favor the linear structure. It is difficult to draw any conclusions from this except that a definitive answer cannot be found in any of these calculations. Furthermore, since Møller-Plesset calculations, in common with all perturbation theory calculations, are not variational, it is difficult to compare any of these with any of the other calculations. In particular, perturbation methods often overestimate the correlation energy and can lead to total energies which are lower than the experimental values.

TABLE 13-14
Hartree-Fock energies for dilithioacetylene using various basis sets

	Linear $D_{\infty h}$		Planar bridged D_{2h}	
Basis	**Energy/E_h**	**ΔE^a**	**Energy/E_h**	**ΔE^a**
STO-6G	−90.284786	0.0	−90.316215	82^b
6-31G	−90.525607	−39	−90.510708	0.0
D95V^c	−90.539673	−27	−90.529369	0.0
6-311G	−90.555594	−18	−90.548812	0.0
6-31G*	−90.550953	0.0	−90.561614	28^b
6-311G*	−90.577568	0.0	−90.587950	27^b

^{a} $\Delta E = E_{\text{linear}} - E_{\text{bridged}}$ in $kJ \cdot mol^{-1}$.
^{b} The bridged form is lower.
^{c} D95V is the Dunning-Hay $(9s\,5p)/[3s\,2p]$ contracted basis.

TABLE 13-15
MCSCF energies for dilithioacetylene using various basis sets

Number of configurations	Energy/E_h		
	STO-6G	6-31G	6-31G*
Linear			
66	−90.367617	−90.598088	−90.615960
196	−90.372332	−90.601748	−90.619279
1,176	−90.392039	−90.618247	−90.650644
5,292	−90.427832	−90.653415	−90.669064
19,404	−90.432160	−90.664512	
Bridged			
66	−90.385739		−90.624062
196	−90.389202	−90.591434	−90.627029
1,176	−90.412267	−90.613190	−90.647245
5,292	−90.422174	−90.623972	−90.657453
19,404	−90.425363	−90.66	

TABLE 13-16
Møller-Plesset energies for dilithioacetylene (linear) for various basis sets and orders

Model	Energy/E_h				
	MP2[a]	MP3[a]	MP4DQ[a]	MP4SDQ[a]	MP4SDTQ[a]
1	−90.467245	−90.455779	−90.463619	−90.464600	−90.472238
2	−90.467867	−90.455904	−90.463869	−90.464847	−90.472622
3	−90.467780	−90.455948	−90.463972	−90.464966	−90.472720
4	−90.469213	−90.455987	−90.464429	−90.465424	−90.473578
5	−90.469579	−90.455878	−90.464446	−90.465438	−90.473726
6	−90.714941	−90.710303	−90.714243	−90.718961	−90.729998
7	−90.716196	−90.710305	−90.715327	−90.719086	−90.730236
8	−90.717460	−90.711608	−90.716632	−90.720400	−90.731566
9	−90.718728	−90.712153	−90.717422	−90.721318	−90.732829
10	−90.718793	−90.711977	−90.717326	−90.721263	−90.732882
11	−90.817547	−90.817413	−90.818307	−90.823507	−90.841890
12	−90.817648	−90.817586	−90.818459	−90.823633	−90.841958
13	−90.817395	−90.816509	−90.817584	−90.822979	−90.841966

Note: 1. STO-6G//MCSCF(66 CSFs); 2. STO-6G//MCSCF(196 CSFs); 3. STO-6G//MCSCF(1176 CSFs); 4. STO-6G//MCSCF(5292 CSFs); 5. STO-6G//MCSCF(19404 CSFs); 6. 6-31G//MCSCF(66 CSFs); 7. 6-31G//MCSCF(196 CSFs); 8. 6-31G//MCSCF(1176 CSFs); 9. 6-31G//MCSCF(5292 CSFs); 10. 6-31G//MCSCF(19404 CSFs); 11. 6-31G*//MCSCF(196 CSFs); 12. 6-31G*//MCSCF(1176 CSFs); 13. 6-31G*//MCSCF(5292 CSFs).

[a] MP*n* ($n = 2, 3, 4$) = Møller-Plesset *n*th order. The S, D, T, and Q represent addition of single, double, triple, and quadruple excitations; MP2 and MP3 include D only.

TABLE 13-17
Møller-Plesset energies for dilithioacetylene (bridged) for various basis sets and orders

	Energy/E_h				
Model	**MP2**[a]	**MP3**[a]	**MP4DQ**[a]	**MP4SDQ**[a]	**MP4SDTQ**[a]
1	−90.482873	−90.474074	−90.481409	−90.485066	−90.493219
2	−90.486556	−90.477576	−90.485061	−90.488453	−90.496212
3	−90.486334	−90.477583	−90.484954	−90.488358	−90.496113
4	−90.486312	−90.477497	−90.484885	−90.488332	−90.496162
5	−90.486488	−90.477571	−90.485003	−90.488456	−90.496308
6	−90.707642	−90.698047	−90.705017	−90.709181	−90.721030
7	−90.707557	−90.697762	−90.704818	−90.709009	−90.720929
8	−90.709531	−90.700095	−90.706996	−90.711168	−90.723015
9	−90.711046	−90.701598	−90.708500	−90.712701	−90.724604
10	−90.710872	−90.701297	−90.708253	−90.712467	−90.724411
11	−90.832879	−90.833521	−90.834449	−90.839239	−90.857385
12	−90.832647	−90.833446	−90.834302	−90.839059	−90.857123
13	−90.832275	−90.833011	−90.833863	−90.838645	−90.856776

Note: 1. STO-6G//MCSCF(66 CSFs); 2. STO-6G//MCSCF(196 CSFs); 3. STO-6G//MCSCF(1176 CSFs); 4. STO-6G//MCSCF(5292 CSFs); 5. STO-6G//MCSCF(19404 CSFs); 6. 6-31G//MCSCF(66 CSFs); 7. 6-31G//MCSCF(196 CSFs); 8. 6-31G//MCSCF(1176 CSFs); 9. 6-31G//MCSCF(5292 CSFs); 10. 6-31G//MCSCF(19404 CSFs); 11. 6-31G*//MCSCF(196 CSFs); 12. 6-31G*//MCSCF(1176 CSFs); 13. 6-31G*//MCSCF(5292 CSFs).

[a] MPn ($n = 2, 3, 4$) = Møller-Plesset nth order. The S, D, T, and Q represent addition of single, double, triple, and quadruple excitations; MP2 and MP3 include D only.

A recent paper by Jaworski et al. on dilithioacetylene claims that the bridged structure is definitely lower than the linear one.[40] These workers began with RHF calculations employing 6-31G* and 6-311G* basis sets and then augmented these with many-body perturbation theory (MBPT) and coupled-cluster theory (CCT) calculations; this approach is essentially the same as the Møller-Plesset calculations of Su et al., and consequently it is not surprising that the bridged structure is favored as lower in energy. In any event, it appears that the identity of the ground state of dilithioacetylene is still in some doubt.

In all the cases studied, one can note a tendency for the two lithium atoms of a dilithiohydrocarbon to be as close together as possible, consistent with other constraints. For example, quite apart from the question of which is truly the ground state, the bridged structure of dilithioacetylene, whose Li–Li distance is about 0.39 nm compared with about 0.51 nm in the linear structure,

[40] A. Jaworski, W. B. Person, L. Adamowicz, and R. J. Bartlett, *Int. J. Quantum Chem., Quantum Chem. Symp.*, **21**:613 (1987).

is unexpectedly stable. Similarly, the favored planar structure of dilithiomethane is the cis form, in which the two lithium atoms are adjacent. And finally, it should be noted that the Li–Li distance in the triplet state of the tetrahedral structure (0.25 nm) is significantly less than in the corresponding singlet state (0.34 nm). It is hard to resist the whimsical conclusion that a lithium atom behaves somewhat like half a carbon atom (which it is, electronically) and that two such atoms attempt to emulate a carbon atom by getting as close together as other constraints permit.

SUGGESTED READINGS

Čársky P., and M. Urban: Ab Initio *Calculations: Methods and Applications in Chemistry*, Springer-Verlag, New York, 1980.

Hehre, W. J., L. Radom, P. v.R. Schleyer, and J. A. Pople: Ab Initio *Molecular Orbital Theory*, Wiley-Interscience, New York, 1986.

Jorgensen, W. L., and L. Salem: *The Organic Chemist's Book of Orbitals*, Academic, New York, 1963.

Schaefer, H. F., III: *The Electronic Structure of Atoms and Molecules*, Addison-Wesley, Reading, Mass., 1972.

——— (ed.): *Modern Theoretical Chemistry*, vol. 3: *Methods of Electronic Structure Theory*, Plenum, New York, 1977.

Szabo, A., and N. S. Ostlund: *Modern Quantum Chemistry: Introduction to Advanced Structure Theory*, Macmillan, New York, 1982.

Szasz, L.: *Pseudopotential Theory of Atoms and Molecules*, Wiley-Interscience, New York, 1985.

CHAPTER 14

SEMIEMPIRICAL MOLECULAR ORBITAL METHODS I: PI ELECTRON SYSTEMS

With the exception of the pseudo-potential method of Sec. 13-5, all the molecular calculations discussed up to this point are of the type generally termed *ab initio*. In practice this means that one uses the nonrelativistic electronic hamiltonian within the Born-Oppenheimer approximation, and once the approximate wavefunction is chosen (along with a specific basis), no further significant approximations are made. In particular, all the electrons of the molecule are explicitly considered.

In this chapter and the next we will consider various versions of *semiempirical* calculations in which only some of the electrons (generally some of or all the valence electrons) are considered explicitly, with the contributions of the others being taken into account via various models. In addition, various numerical approximations are made—particularly with respect to integral evaluation—which serve to reduce drastically the computational demands of the problem.

14-1 PARTIAL ELECTRON-SHELL WAVEFUNCTIONS

All semiempirical methods used in the molecular orbital approximation deal explicitly with only part of the electron configuration of the molecule. In general, the electron configuration is divided into two parts: the *core* electrons and the *outer* electrons. Only the outer electrons are explicitly included in the electronic wavefunction; the core electrons are assumed to form a quasi-static shield about the nuclear framework of the molecule and, thus, to affect the outer electrons in some average way. The electronic hamiltonian for the outer electrons (N in number) is written

$$\hat{H} = \sum_{\mu=1}^{N} h'_\mu + \sum_{\mu<\nu}^{N} \frac{1}{r_{\mu\nu}} + \sum_{k<l} \frac{Z_k Z_l}{r_{kl}} \tag{14-1}$$

where h'_μ describes an electron moving in the field of the shielded nuclei, the so-called core of the molecule. The nuclear-repulsion term refers to repulsion among shielded nuclei so that Z_k and Z_l are the charges on the shielded nuclei.

Two different core-outer-electron separations are generally employed. With conjugated hydrocarbons (ethene, butadiene, benzene, etc.), the core electrons are the sigma electrons, and the outer electrons are the pi electrons. Thus the wavefunction contains pi electrons only, and the sigma electrons act only as part of the nuclear core. Also, the hydrogen atoms are totally ignored in this approach. The rationale behind this particular separation is that organic chemists have been highly successful in treating the chemistries of such species in terms of carbon pi electrons only. The present chapter is devoted to a discussion of such pi electron systems.

A more general type of separation—applicable to almost all types of molecules— is to choose the *inner-shell*, *nonbonding* electrons of the nuclei (other than hydrogen) as the core electrons. The wavefunction then takes explicit account of the remaining electrons, the valence-shell electrons. In practice, this means that the lowest-energy MOs—those consisting predominantly of 1s AOs—are not explicitly included in the MO calculation. In fact, these MOs are assumed to consist entirely of the 1s AOs of the constituent atoms with no admixture of any other AOs. Chapter 15 will discuss several versions of this more general approach.

The theoretical bases of core-outer-electron separations have been discussed by a number of writers.[1] Note that the general problem of core-outer-electron separability also enters into the pseudo-potential MO calculations discussed in Sec. 13-5.

14-2 THE HÜCKEL APPROXIMATION FOR CONJUGATED HYDROCARBONS

Empirical evidence suggests that the chemical behavior of conjugated hydrocarbons may be accounted for on the basis of the so-called pi electrons for

---

[1] H. Hellmann, *J. Chem. Phys.*, **3**:61 (1935); P. G. Lykos and R. G. Parr, *J. Chem. Phys.*, **24**:1166 (1956); P. G. Lykos, *Advan. Quantum Chem.*, **1**:117 (1964); W. H. E. Schwarz, *Theor. Chim. Acta*, **11**:30 (1968); S. Huzinaga and A. A. Cantu, *J. Chem. Phys.*, **55**:5543 (1971).

each pair of carbon atoms which are double-bonded in the classical formula. As early as 1931, E. Hückel[2] showed that it was possible to describe conjugated hydrocarbons such as benzene by use of a quantum mechanical model which considered only the pi electrons. Today, the model has been greatly extended and its basis clarified.

Consider a conjugated hydrocarbon containing m carbon atoms in a conjugated system; i.e., the carbon atoms form a chain which appears in the classical formula as an alternation of single and double bonds:

$$-C{=}C-C{=}C-C{=}C-$$

where hydrogen atoms have been omitted for clarity. In addition, the treatment which follows is also applicable to molecules in which the conjugated system is cyclic—or in which a number of cyclic and noncyclic conjugated chains are linked. For neutral molecules, the number of pi electrons is equal to the number of carbon atoms in the conjugated system; when the number of carbon atoms is odd, the neutral species is a radical. Thus, the allyl radical $CH_2{=}CH{-}CH_2\cdot$ is a conjugated system of three carbon atoms and three pi electrons, a fact made more apparent by examining the resonance structures

$$CH_2{=}CH{-}CH_2\cdot \leftrightarrow \cdot CH_2{-}CH{=}CH_2$$

The basis set for the pi electron wavefunction consists of $2p$ AOs of carbon which are perpendicular to the plane of the molecule. These are generally referred to as "$2p\pi$" AOs. The pi electron wavefunction for a molecule of $2N$ pi electrons is written as the single determinant

$$\psi_\pi = |\pi_1\,\bar{\pi}_1\,\pi_2\,\bar{\pi}_2 \cdots \pi_N\,\bar{\pi}_N| \tag{14-2}$$

where the $\{\pi_i\}$ are pi electron MOs given by

$$\pi_i = \sum_{r=1}^{m} \phi_r c_{ri} \tag{14-3}$$

The $\{\phi_r\}$ are the $2p\pi$ basis AOs, each centered on a different carbon atom of the conjugated chain. These AOs are treated as if they were strictly orthogonal; i.e., all overlap integrals $\langle \phi_r | \phi_s \rangle$ $(r \neq s)$ will be assumed to be zero.

Although the Hückel method was developed outside the framework of the Hartree-Fock SCF method, it is nevertheless instructive to recast it in the framework of the latter in order to highlight its limitations more easily. We can define the pi electron Hartree-Fock operator as

$$\hat{F}_\pi = h' + \sum_j (2J_j - K_j) = h' + V^{\text{eff}} \tag{14-4}$$

where h' describes a single electron moving in the field of the nuclei (shielded by the sigma electrons) and V^{eff} describes the repulsive potential of a single pi

[2] E. Hückel, *Z. Physik*, **70**:204 (1931) and **76**:628 (1932).

electron moving in an average field provided by all the other pi electrons. Proceeding as usual when solving Roothaan's equations we form the $m \times m$ secular determinant

$$\det|\mathbf{F}_\pi - \epsilon\Delta| = 0 \tag{14-5}$$

where $\mathbf{F}_\pi$ is the matrix representation of the pi electron Hartree-Fock operator in the $2p\pi$ basis. At this point there is a drastic departure from the procedure employed in an *ab initio* calculation; instead of specifying an explicit form for h' and evaluating matrix elements numerically, we use the following symbolic procedures: The matrix elements of $\mathbf{F}_\pi$ are defined as

$$\begin{aligned} F_{rr} &= \langle\phi_r|h' + V^{\text{eff}}|\phi_r\rangle = \langle\phi_r|h'|\phi_r\rangle + \langle\phi_r|V^{\text{eff}}|\phi_r\rangle = a'_r + J' = \alpha_r \\ F_{rs} &= \langle\phi_r|h' + V^{\text{eff}}|\phi_s\rangle = \langle\phi_r|h'|\phi_s\rangle + \langle\phi_r|V^{\text{eff}}|\phi_s\rangle = b'_{rs} + J'' \\ &= \beta_{rs} = F_{sr} \end{aligned} \tag{14-6}$$

The quantities α_r and β_{rs} are called the *coulombic* and *resonance integrals*, respectively.[3] Instead of assigning numerical values to these integrals, we define them as follows:

$$\begin{aligned} \alpha_r &= \alpha \qquad \text{for all } r \\ \beta_{rs} &= \begin{cases} \beta & \text{for nearest-neighbor atoms} \\ 0 & \text{otherwise} \end{cases} \end{aligned} \tag{14-7a}$$

The elements of the overlap matrix are assumed to be given by

$$\Delta_{rs} = \delta_{rs} \tag{14-7b}$$

In summary, overlap integrals are ignored, and all other integrals are either zero or equal to the numerically undefined quantities α or β. Thus the secular determinant [Eq. (14-5)] contains only three distinct terms: $\alpha - \epsilon$ for the diagonal terms and β's for nearest-neighbor, off-diagonal terms, with zeroes everywhere else. The m roots of the determinant, the pi MO energies have the general form

$$\epsilon_i = a' + J' + x_i(b' + J'') = \alpha + x_i\beta \tag{14-8}$$

where the $\{x_i\}$ are numerical coefficients characteristic of the molecule's pi electron system. It is often incorrectly stated that the Hückel method neglects electron repulsion; a more accurate statement is that the method relegates it to a minor implicit role.

The assumption stated in Eq. (14-7b) implies that the $2p\pi$ AO basis is orthonormal. Consequently, the normalization condition on each pi MO is of

---

[3] These names are far from appropriate, especially the former, since it invites comparison with previously defined coulombic integrals of the form $\langle\phi_r(\mu)\phi_s(\nu)|1/r_{\mu\nu}|\phi_r(\mu)\phi_s(\nu)\rangle$.

the simple form

$$\sum_{r}^{m} |c_{rj}|^2 = 1 \tag{14-9}$$

The total energy of the molecule (*sans* the nuclear-repulsion term, which is never appended) is

$$\langle \hat{H}_\pi \rangle = E_\pi = \sum_i n_i \epsilon_i - G \tag{14-10}$$

where n_i is the number of electrons in the ith MO ($n_i = 0, 1$, or 2 depending on whether the MO is unoccupied, singly occupied, or doubly occupied, respectively) and G is the electron-repulsion energy. This latter term is not elucidated, but it is tacitly assumed to depend only on the total number of pi electrons for a given conjugated hydrocarbon; consequently, it cancels out whenever one determines the energy difference between two electronic states.

The Hückel molecular orbital (HMO) approximation works best for a class of hydrocarbons known as *alternant hydrocarbons*. If the carbon atoms of which the conjugated molecule is composed can be divided into two sets such that members of one set are formally bonded only to members of the other set, the molecule is said to be alternant. It is readily seen that all linear or branched-chain conjugated molecules or even-numbered ring hydrocarbons are alternant. Combinations of these are also alternant. On the other hand, odd-numbered ring hydrocarbons (such as the cyclopentadienyl cation) are nonalternant. Figure 14-1 illustrates some examples of alternant and nonalternant hydrocarbons. The atoms of one set are represented by asterisks, and the atoms of the other set are unmarked. Note that there is no physical significance as to which set has the asterisks and which does not. The alternant property

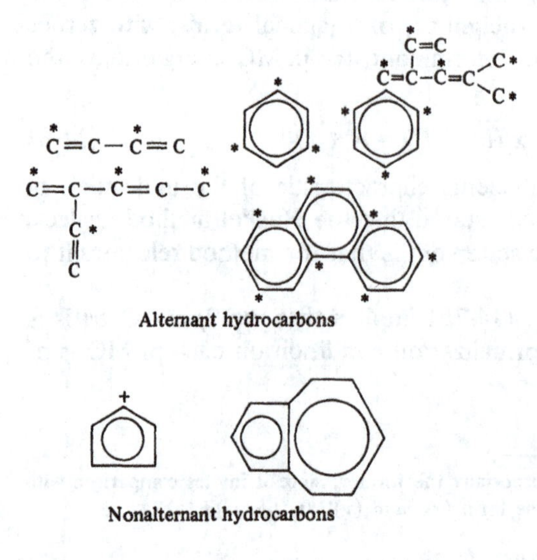

FIGURE 14-1
Examples of alternant and nonalternant hydrocarbons.

has certain topological implications which lead to several important theorems, and these theorems make the mathematical treatment of such systems very simple. The HMO approximation can also be applied to nonalternant hydrocarbons, but many of the mathematical simplifications are absent, and the results are generally less satisfactory.

According to Eq. (10-124), the sum of the orbital energies of a molecule with $2N$ electrons can be written in the form

$$2\sum_{i=1}^{N} \epsilon_1 = 2\,\mathrm{Tr}\,(\mathbf{RF}_{\pi}) \tag{14-11}$$

where $\mathbf{R}$ (the charge and bond-order matrix) is given by

$$\mathbf{R} = \mathbf{TT}^{\dagger} \tag{14-12}$$

with $\mathbf{T}$ being an $m \times N$ matrix of the eigenvectors of the N lowest MOs, that is,

$$\mathbf{T} = [\mathbf{C}_1 \quad \mathbf{C}_2 \quad \cdots \quad \mathbf{C}_N] \tag{14-13}$$

The matrix elements of $\mathbf{R}$ for a molecule of $2N$ electrons and $2N$ carbon atoms are

$$R_{rr} = \sum_{j=1}^{N} |c_{rj}|^2 \qquad R_{rs} = \sum_{j=1}^{N} c_{rj}c_{sj}^{*} \tag{14-14}$$

The diagonal elements are related to the electronic charges associated with the carbon atoms, and the off-diagonal elements are related to the order of the bond between atoms.[4] Specifically,

$$q_r = 2R_{rr} \qquad \text{and} \qquad p_{rs} = 2R_{rs} \tag{14-15}$$

where q_r is that part of the total pi electron population assigned to carbon atom r (in units of e^- per atom) and p_{rs} is the order of the pi bond between atoms r and s.[5] With Eqs. (14-14) and (14-15), Eq. (14-11) may be written

$$2\sum_{i=1}^{N} \epsilon_i = 2\sum_{r,s} R_{rs}F_{rs} = \sum_{r,s}\sum_{j} 2c_{rj}c_{sj}^{*}F_{rs}$$

$$= \sum_{r}\sum_{j} 2|c_{rj}|^2\alpha + 2\sum_{r<s}\sum_{j} 2c_{rj}c_{sj}^{*}\beta = \alpha\sum_{r} q_r + 2\beta\sum_{r<s} p_{rs} \tag{14-16}$$

A more general definition of charge densities and bond orders, applicable

---

[4] C. A. Coulson, *Proc. Roy. Soc.* (*London*), **A169**:419 (1939).

[5] A bond order of unity means that the bond consists of a pair of covalently linked electrons; an isolated double bond (as in ethene) has an order of 2 (1 for the sigma pair of electrons and 1 for the pi pair of electrons).

to all conjugated systems treated by the HMO approximation, is

$$q_r = \sum_i n_i |c_{ri}|^2 \qquad p_{rs} = \sum_i n_i c_{ri} c^*_{si} \tag{14-17}$$

where n_i is the number of pi electrons in the π_i MO ($n_i = 0$, 1, or 2).

Two of the most useful and important theorems applicable to alternant hydrocarbons are the following:

Theorem 14-1 The Coulson-Rushbrooke pairing theorem.[6] For every HMO energy $\alpha + x\beta$ in an alternant hydrocarbon there exists another energy $\alpha - x\beta$; i.e., the roots of the HMO secular determinant occur in pairs which are equal in magnitude and opposite in sign. Furthermore, the coefficients of paired MOs are the same except that the algebraic sign at every other atom is opposite.

The proof of this theorem is very simple and depends mainly on the assumptions of common α and β parameters, neglect of overlap, and the connectivity (topological) properties of the alternant system; the reader is referred to the original paper for the details.

A special case occurs when the alternant hydrocarbon contains an odd number of carbon atoms, e.g., the allyl radical. In such a case the pairing theorem still holds, but one obtains a root $x = 0$ ($\epsilon = \alpha$) for which pairing is trivial. The corresponding MO is called a *nonbonding MO*. The rth secular equation for such a root is

$$(c_{r0})^{(0)} + \sum_s c_{s0} = 0 \tag{14-18}$$

which states that the sum of the atom coefficients around any atom (with or without an asterisk) is zero for the nonbonding MO. An extensive and useful discussion of nonbonding MOs and their role in various types of HMO calculations has been given by Longuet-Higgins.[7]

Note that use of the pairing theorem enables one to use no more than half the roots of the secular determinant in order to deduce the entire set of MOs and MO energies.

Theorem 14-2 In the HMO approximation for alternant hydrocarbons the charge density on each atom of an N–pi electron, N–carbon atom molecule is unity, and the bond order between atoms belonging to the same set is zero.

The proof depends on the pairing theorem, the unitary nature of the matrix of HMO coefficients, the assumed orthogonality of the $2p\pi$ basis set, and the topological nature of the conjugated system.

---

[6] C. A. Coulson and G. S. Rushbrooke, *Proc. Cambridge Phil. Soc.*, **36**:193 (1940).

[7] H. C. Longuet-Higgins, *J. Chem. Phys.*, **18**:265, 275, 283 (1950).

For nonalternant hydrocarbons containing N pi electrons and N carbon atoms which are equivalent by symmetry (e.g., the cyclopentadienyl radical), the charge density is also equal to unity on each atom.

In solving the HMO secular determinant, it is convenient to use the substitution

$$\frac{\alpha-\epsilon}{\beta}=x \tag{14-19}$$

All the $\mathbf{F}_\pi$ matrix elements then become x for the diagonal terms, 1 for nearest-neighbor atoms, and 0 otherwise. In the case of linear conjugated molecules the HMO determinant is always of the general form:

$$D_n(x)=\begin{vmatrix} x & 1 & 0 & 0 & \cdots & 0 & 0 & 0 \\ 1 & x & 1 & 0 & \cdots & 0 & 0 & 0 \\ 0 & 1 & x & 1 & \cdots & 0 & 0 & 0 \\ \cdots & \cdots & \cdots & \cdots & \cdots & \cdots & \cdots & \cdots \\ 0 & 0 & 0 & 0 & \cdots & x & 1 & 0 \\ 0 & 0 & 0 & 0 & \cdots & 1 & x & 1 \\ 0 & 0 & 0 & 0 & \cdots & 0 & 1 & x \end{vmatrix}=0 \tag{14-20}$$

where n is the number of carbon atoms in the conjugated system and the determinant has n rows and n columns. Carrying out the first step in the expansion of $D_n(x)$ in terms of the cofactors of the first row, one obtains

$$D_n(x)=xD_{n-1}(x)-D_{n-2}(x)=0 \tag{14-21}$$

The above relationship is a recursion formula which is satisfied by mathematical functions known as *Chebyshev polynomials*.[8] The Chebyshev polynomials are defined formally by[9]

$$D_0(x)=1 \qquad D_1(x)=x \qquad D_n(x)=2^{1-n}\cos(n\cos^{-1}x) \tag{14-22}$$

If one carries out the substitution $x=2\cos\theta$, one obtains

$$D_n(x)=\frac{\sin(n+1)\theta}{\sin\theta} \tag{14-23}$$

Setting $D_n(\theta)=0$ for the case when $\sin\theta\neq 0$ leads to

$$\sin(n+1)\theta=0 \tag{14-24}$$

which is satisfied by

$$(n+1)\theta=j\pi \qquad \begin{matrix}\pi \text{ in radians} \\ j=1,2,\ldots,n\end{matrix} \tag{14-25}$$

[8] Also spelled "Tchebycheff," "Tschebycheff," "Tchebichef," etc. The present spelling is based on the Russian transliteration equivalents used by *Chemical Abstracts*.

[9] R. Courant and D. Hilbert, *Methods of Mathematical Physics*, vol. 1, Interscience, New York, 1953, pp. 88–90.

or, alternatively,

$$\theta = \frac{j\pi}{n+1} \qquad j = 1, 2, \ldots, n \tag{14-26}$$

The roots of the Chebyshev polynomials [and thus of the determinant $D_n(x)$] are then given by

$$x_j = 2\cos\theta = 2\cos\frac{j\pi}{n+1} \tag{14-27}$$

The jth MO then has the energy

$$\epsilon_j = \alpha - 2\beta\cos\frac{j\pi}{n+1} \qquad j = 1, 2, \ldots, n \tag{14-28}$$

The elements of the eigenvector $\mathbf{C}_j$ are given by

$$c_{rj} = \left(\frac{2}{n+1}\right)^{1/2} \sin\frac{rj\pi}{n+1} \tag{14-29}$$

For all cyclic hydrocarbons (alternant and nonalternant), the orbital energies and coefficients are given by

$$\left.\begin{aligned} \epsilon_j &= \alpha - 2\beta\cos\frac{2j\pi}{n} \\ c_{rj} &= \sqrt{n}\exp\left[\frac{2\pi i r(j-1)}{n}\right] \end{aligned}\right\} \quad j = 1, 2, \ldots, n \tag{14-30}$$

Cyclic hydrocarbons with an even number of carbon atoms will have $(n-2)/2$ doubly degenerate MOs; those with an odd number of carbon atoms will have $(n-1)/2$ doubly degenerate MOs. Thus, benzene ($n = 6$) will have two doubly degenerate pi MOs (from the HMO basis), and the cyclopentadienyl radical ($n = 5$) will also have two.[10]

14-3 HMO CALCULATIONS ON ETHENE AND 1,3-BUTADIENE

The ethene molecule, $H_2C{=}CH_2$, is the prototype of all conjugated systems. Since it is a two-electron system (in the pi electron approximation), its role in the chemistry of conjugated hydrocarbons is analogous to that of dihydrogen, H_2, in the chemistry of molecules in general. The pi electron wavefunction of ethene is written as

$$\psi_\pi = |\pi_1 \bar{\pi}_1| \tag{14-31}$$

[10] Referring back to Sec. 13-4, note that the *ab initio* treatment of benzene produced three pi MOs; two of these spanned the doubly degenerate, irreducible representation E_{1g}, and the remaining one was of A_{2u} symmetry.

where the MO is given by

$$\pi_1 = c_1\phi_1 + c_2\phi_2 \tag{14-32}$$

and where ϕ_1 and ϕ_2 are $2p\pi$ AOs centered on carbon atoms 1 and 2, respectively. Note that these $2p\pi$ AOs are perpendicular to the molecular plane formed by the two carbon atoms and the four hydrogen atoms (these are equivalent to the $2p_x$ AOs of Fig. 13-6). One should also note that writing the pi electron wavefunction in antisymmetrized form is superfluous, since the HMO method makes no use of this requirement. We do it, nevertheless, since later refinements of the approach will require explicit antisymmetrization.

The secular equations in their most primitive form are

$$c_1(\alpha - \epsilon) + c_2\beta = 0 \qquad c_1\beta + c_2(\alpha - \beta) = 0 \tag{14-33}$$

If we use the convenient substitution given by Eq. (14-19), these become

$$c_1x + c_2 = 0 \qquad c_1 + c_2x = 0 \tag{14-34}$$

The secular determinant is

$$D_2(x) = \begin{vmatrix} x & 1 \\ 1 & x \end{vmatrix} = 0 \tag{14-35}$$

which is equivalent to the algebraic equation: $x^2 - 1 = 0$. The solutions are $x_1 = -1$ and $x_2 = 1$. Thus the HMO energies of ethene are

$$\epsilon_1 = \alpha + \beta \qquad \epsilon_2 = \alpha - \beta \tag{14-36}$$

It is assumed that α and β are negative; thus $\epsilon_1 < \epsilon_2$. The higher root corresponds to a virtual MO, and the lower is used to describe the ground state.

The normalization conditions [Eq. (14-9)] for ethene are

$$c_{1j}^2 + c_{2j}^2 = 1 \tag{14-37}$$

For $j = 1$ (the ground state MO) this leads to

$$c_{11} = c_{21} = \frac{1}{\sqrt{2}} \tag{14-38}$$

and for $j = 2$ (the virtual MO) the corresponding values are

$$c_{12} = -c_{22} = \frac{1}{\sqrt{2}} \tag{14-39}$$

Thus the matrix $\mathbf{C}$ which diagonalizes the matrix representation of the pi electron Hartree-Fock operator is given by

$$\mathbf{C} = [\mathbf{C}_1 \quad \mathbf{C}_2] = \begin{bmatrix} c_{11} & c_{12} \\ c_{21} & c_{22} \end{bmatrix} = \frac{1}{\sqrt{2}} \begin{bmatrix} 1 & 1 \\ 1 & -1 \end{bmatrix} \tag{14-40}$$

Note that both the energies and the coefficients satisfy the pairing theorem.

The ground state MO, π_1, obtained by the HMO approximation corresponds to the $1b_{3u}$ MO obtained in the *ab initio* calculation described in Sec.

13-4. The total pi electron energy of ethylene is given by

$$E_\pi = 2\epsilon_1 = 2\alpha + 2\beta \tag{14-41}$$

This quantity represents the energy of an isolated double bond; we will use it later as a reference point for ascribing relative stabilities to other conjugated hydrocarbon systems.

The HMO approximation is insensitive to the actual geometry of a molecule, since it depends only on the connectivity of the carbon atoms composing the conjugated system. Thus, the method cannot distinguish between different conformations or diastereomers; e.g., the cis and trans diastereomers of 1,3-butadiene are treated exactly alike. In either case, the pi electron wavefunction is

$$\psi_\pi = |\pi_1\ \bar{\pi}_1\ \pi_2\ \bar{\pi}_2|$$

where each MO is a linear combination of the four $2p\pi$ AOs of the carbon atoms (numbered consecutively from one end of the carbon chain to the other). The corresponding HMO secular determinant becomes

$$D_4(x) = \begin{vmatrix} x & 1 & 0 & 0 \\ 1 & x & 1 & 0 \\ 0 & 1 & x & 1 \\ 0 & 0 & 1 & x \end{vmatrix} = 0 \tag{14-42a}$$

Solution of this determinant leads to the HMO energies:

$$\begin{aligned} \epsilon_1 &= \alpha + 1.6180\beta & \epsilon_2 &= \alpha + 0.6180\beta \\ \epsilon_3 &= \alpha - 0.6180\beta & \epsilon_4 &= \alpha - 1.6180\beta \end{aligned} \tag{14-42b}$$

The corresponding pi electron MOs are

$$\begin{aligned} \pi_1 &= 0.3718(\phi_1 + \phi_4) + 0.6015(\phi_2 + \phi_3) \\ \pi_2 &= 0.6015(\phi_1 - \phi_4) + 0.3718(\phi_2 - \phi_3) \\ \pi_3 &= 0.6015(\phi_1 + \phi_4) - 0.3718(\phi_2 + \phi_3) \\ \pi_4 &= 0.3718(\phi_1 - \phi_4) - 0.6015(\phi_2 - \phi_3) \end{aligned} \tag{14-43}$$

The total pi electron energy of 1,3-butadiene is given by

$$E_\pi = 2\epsilon_1 + 2\epsilon_2 = 4\alpha + 4.4720\beta \tag{14-44}$$

If one assumes that the α and β parameters in 1,3-butadiene have the same numerical values as in ethene, then one can compute the energy difference between having four pi electrons in two isolated double bonds (as in two ethene molecules) or in three conjugated C–C bonds (as in butadiene). This quantity is called the *quantum mechanical resonance energy* and is given by

$$\varepsilon = E_\pi(\text{butadiene}) - 2E_\pi(\text{ethene}) = 4\alpha + 4.4720\beta - 2(2\alpha + 2\beta) = 0.4720\beta \tag{14-45}$$

The resonance energy of butadiene is negative (since $\beta < 0$) and represents the stabilization due to having the four pi electrons spread out over the entire carbon framework as opposed to having them localized to a pair of separated bonds.

Figure 14-2 shows the MOs of 1,3-butadiene in a schematic representation based on their nodal properties. Especially noteworthy is the striking resemblance between the HMOs and the wavefunctions of a particle in a box. This can be shown to be more than a mere coincidence; if the one-electron HMO equations are solved approximately by a technique known as the *calculus of finite differences*, the resulting solutions are essentially equivalent to the particle-in-a-box solutions.[11] In fact, a number of workers, principally Platt,[12] have used the particle-in-a-box model explicitly to model electronic structures of a variety of conjugated hydrocarbons. This approach is often called the *free-electron molecular orbital (FEMO) model.*

Experiment shows that the trans diastereomer of 1,3-butadiene is the more stable. Using the fact that this molecule belongs to the C_{2h} point group and using the coordinate system shown in Fig. 14-3, one can assign the MOs of 1,3-butadiene to the irreducible representations of this group. One finds that the four HMOs span the reducible representation $2A_u \oplus 2B_g$ and, further-

[11] This has been discussed at some length by various authors, e.g., W. T. Simpson, *J. Chem. Phys.*, **17**:1218 (1949) and *J. Am. Chem. Soc.*, **73**:5361 (1951); A. A. Frost, *J. Chem. Phys.*, **32**:310 (1955); C. A. Coulson, *Proc Roy. Soc. (London)*, **66**:652 (1953); J. S. Griffith, *J. Chem. Phys.*, **21**:174 (1953); K. Ruedenberg, *J. Chem. Phys.*, **29**:1232 (1958).

[12] J. R. Platt, *J. Chem. Phys.*, **22**:1448 (1954).

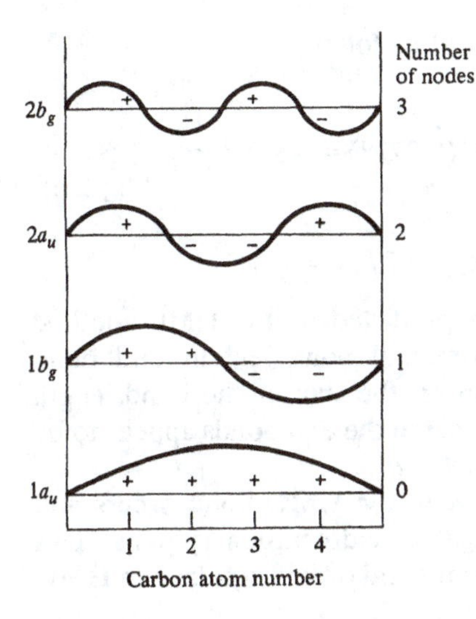

FIGURE 14-2
Schematic representation of the nodal properties of the HMO approximation to the pi electron MOs of butadiene. The algebraic signs correspond to the signs associated with the eigenvector components of a given MO. Note that the depiction of the MOs in this manner illustrates a correspondence with the wavefunctions of a particle in a box.

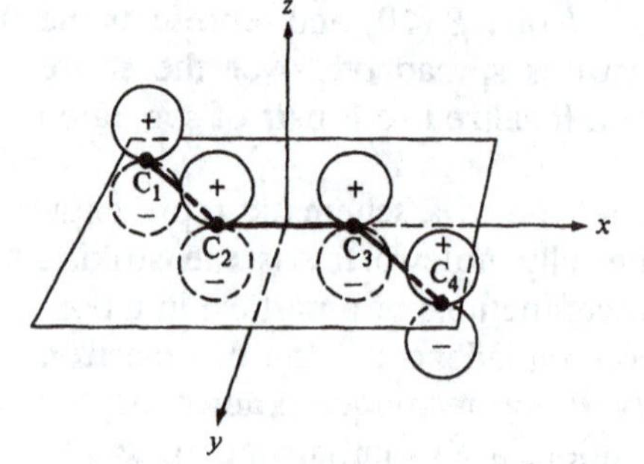

FIGURE 14-3
Coordinate system used to elucidate the symmetry operations of *trans*-1,3-butadiene.

more, π_1 and π_3 are of A_u symmetry and π_2 and π_4 are of B_g symmetry. Note that the members of a pair of MOs which are related by the Coulson-Rushbrooke pairing theorem span different irreducible representations.

The charge densities on atoms 1 and 2 are given by

$$\begin{aligned} q_1 = q_4 &= 2(0.3718)^2 + 2(0.6015)^2 = 1 \\ q_2 = q_3 &= 2(0.6015)^2 + 2(0.3718)^2 = 1 \end{aligned} \tag{14-46}$$

These results are consistent with Theorem 14-2. The present calculations can also be used to provide predictions of charge densities of ions of 1,3-butadiene. For example, the monopositive butadiene cation should have the following charge distributions:

$$\begin{aligned} q_1 = q_4 &= 2(0.3718)^2 + (0.6015)^2 = 0.64 \\ q_2 = q_3 &= 2(0.6015)^2 + (0.3718)^2 = 0.86 \end{aligned} \tag{14-47}$$

Thus it is predicted that the positive charge is concentrated on the terminal carbon atoms; such an ion would have positions 1 and 4 more susceptible to nucleophilic attack.

The butadiene bond orders are calculated as follows:

$$\begin{aligned} p_{12} &= 2c_{11}c_{21} + 2c_{12}c_{22} \\ &= 2(0.3718)(0.6015) + 2(0.6015)(0.3718) = 0.895 = p_{34} \\ p_{23} &= 2c_{21}c_{31} + 2c_{22}c_{23} \\ &= 2(0.6015)(0.6015) + 2(0.3718)(-0.3718) = 0.447 \end{aligned} \tag{14-48}$$

It is found empirically that the bond orders predicted by the HMO method often correlate reasonably well with experimental bond lengths and bond energies. Generally, the greater the bond order, the shorter the bond length and the greater the bond energy. For 1,3-butadiene the end bonds appear to be about 2.527 a_0 and the central bond about 2.903 a_0.

Figure 14-4 shows the relationship between the HMO bond orders and experimental bond lengths for a few conjugated hydrocarbon systems. This curve is often used to predict bond lengths from bond orders obtained in HMO calculations on other systems.

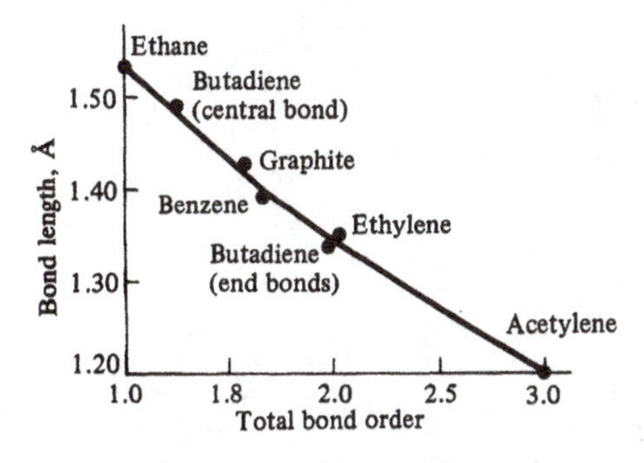

FIGURE 14-4
Correlation of bond length with bond order in the HMO approximation. (*From C. A. Coulson,* Valence, *Oxford University Press, Fair Lawn, N.J. 1961, p. 270. By permission.*)

EXERCISES

14-1. The two pi MOs resulting from the ethene HMO calculation (π and π^*) are of B_{3u} and B_{2g} symmetries, respectively, for the ground state and excited state. What spectroscopic states can one form from the electron configurations π^2, $\pi\pi^*$, and $(\pi^*)^2$?

14-2. Carry out the HMO treatment of the 1,3,5-hexatriene molecule. Determine orbital energies, normalized MOs, the resonance energy, and all nearest-neighbor bond orders. To what point group does this molecule belong (assuming an all-trans conformation)?

14-3. Carry out HMO calculations on the allyl radical, anion, and cation. Determine charge densities and bond orders for each of these species. What is the point group to which these belong?

14-4. Calculate the charge distributions and bond orders of the 1,3-butadiene anion. Compare these with the values obtained for the cation and neutral molecule. What do the results imply?

14-5. Classify the HMOs of 1,3-butadiene as to symmetry type for the cis diastereomer.

14-6. Verify that butadiene bond orders such as p_{13} (involving nonbonded atoms) are equal to zero.

14-4 HMO TREATMENT OF BENZENE

Although a full solution of the HMO equations for benzene can be obtained by application of Eq. (14-30), we will examine a method based on symmetry which can be extended to other pi electron systems, including those which are neither linear chains nor single rings.

This method is based on the symmetries of the MO coefficients with respect to two of the planes of symmetry of the carbon skeleton; these planes, labeled A and B, are illustrated in Fig. 14-5. The benzene HMO secular determinant is given by

$$\begin{vmatrix} x & 1 & 0 & 0 & 0 & 1 \\ 1 & x & 1 & 0 & 0 & 0 \\ 0 & 1 & x & 1 & 0 & 0 \\ 0 & 0 & 1 & x & 1 & 0 \\ 0 & 0 & 0 & 1 & x & 1 \\ 1 & 0 & 0 & 0 & 1 & x \end{vmatrix} = 0 \tag{14-49}$$

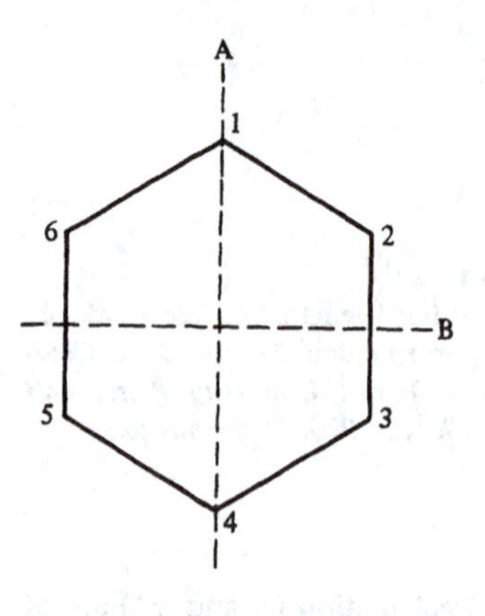

FIGURE 14-5
Atom numbering scheme for the HMO treatment of the benzene molecule. The dashed lines indicate two selected planes of symmetry of the molecule.

We begin by classifying relationships among the MO coefficients according to four symmetry classes, labeled A_sB_s, A_aB_a, A_sB_a, and A_aB_s, where, for example, "A_sB_a" denotes symmetric reflection in the plane A and antisymmetric reflection in the plane B. These symmetries do not make full use of the D_{6h} symmetry of benzene (they imply a lower symmetry D_{2h} instead), but they are sufficient to determine the coefficients. A systematic way to obtain these relationships is first to find all coefficient relationships for A_s, A_a, B_s, and B_a and then combine them. The results are

$$\begin{aligned} &A_sB_s\colon c_1 = c_4 && c_2 = c_6 = c_5 = c_3 \\ &A_aB_a\colon c_1 = -c_4 = 0 && c_2 = c_5 = -c_6 = -c_3 \\ &A_sB_a\colon c_1 = -c_4 && c_2 = c_6 = -c_5 = -c_3 \\ &A_aB_s\colon c_1 = c_4 = 0 && c_2 = -c_6 = -c_5 = c_3 \end{aligned} \tag{14-50}$$

Next, we take the first set of relationships and find in them the number, m, of distinct coefficient relationships, i.e., the smallest number of coefficients one must know in order to determine all the rest by symmetry. These relationships can then be used to factor out of the total determinant an $m \times m$ subdeterminant which is a factor of the total determinant. This is equivalent to constructing m secular equations in m different coefficients. For example, the A_sB_s relationships tell us that we need to know just *two* coefficients in order to fix the values of all six. This implies that the 6×6 determinant contains a 2×2 subdeterminant which is a factor of the total determinant and, thus, can be solved separately. Since the coefficients c_2 and c_6 are equal in this particular symmetry class, we can combine the second and sixth elements of the first row and write

$$x \quad 2 \quad 0 \quad 0 \quad 0 \quad 0$$

Similarly, since the coefficients c_2 and c_3 are also equal, the second row can be written

$$1 \quad x+1 \quad 0 \quad 0 \quad 0 \quad 0$$

At this point we have factored out a 2×2 determinant, namely,

$$\begin{vmatrix} x & 2 \\ 1 & x+1 \end{vmatrix} = 0 \tag{14-51}$$

which is equivalent to the algebraic equation $x^2 + x - 2 = 0$ with the solutions $x = -2$ and 1. Thus we have obtained two of the roots of the secular determinant with very little labor. In similar fashion we can find another 2×2 determinant and two 1×1 determinants. The factored 6×6 determinant is

$$\begin{array}{l} A_sB_s \\ \\ A_aB_a \\ A_sB_a \\ \\ A_aB_s \end{array} \begin{vmatrix} x & 2 & 0 & 0 & 0 & 0 \\ 1 & x+1 & 0 & 0 & 0 & 0 \\ 0 & 0 & x-1 & 0 & 0 & 0 \\ 0 & 0 & 0 & x & 2 & 0 \\ 0 & 0 & 0 & 1 & x-1 & 0 \\ 0 & 0 & 0 & 0 & 0 & x+1 \end{vmatrix} = 0 \tag{14-52}$$

Thus the six roots are seen to be -2, -1 (twice), 1 (twice), and 2. Note that there are two doubly degenerate MOs and also that the pairing theorem is satisfied. The MOs themselves are readily obtained by use of the coefficient symmetries in the appropriate secular equations and by normalization. For example, the secular equations corresponding to the A_sB_s subdeterminant are

$$c_1 x + 2c_2 = 0 \qquad c_1 + c_2(x+1) = 0 \tag{14-53}$$

Using the $x = -2$ root, these become

$$-2c_1 + 2c_2 = 0 \qquad c_1 - c_2 = 0 \tag{14-54}$$

This relationship tells us that c_1 and c_2 are equal for this particular MO; furthermore, the remaining coefficient relationships in A_sB_s tell us all six coefficients are now equal. Consequently, the normalization condition is

$$6c_i^2 = 1 \qquad \text{or} \qquad c_i = \frac{1}{\sqrt{6}} \qquad \text{for all } i \tag{14-55}$$

Similarly, using the $x = -1$ root in the same two secular equations produces another MO, namely,

$$\pi = \frac{1}{\sqrt{12}}(2\phi_1 - \phi_2 - \phi_3 + 2\phi_4 - \phi_5 - \phi_6) \tag{14-56}$$

Table 14-1 summarizes all the MOs and MO energies for the pi electrons of benzene. Also classified are all the MOs according to the irreducible representations of the D_{6h} point group. The ground-state pi electron configuration of benzene is seen to be

$$(a_{2u})^2(e_{1g})^4 \tag{14-57}$$

Recall that the all-electron *ab initio* calculation of Sec. 13-4 showed four sigma electrons in the doubly degenerate MO $3e_{2g}$, between the $1a_{2u}$ and $1e_{1g}$ pi MOs.

TABLE 14-1
HMO coefficients and MO energies of the benzene molecule

AO	$1a_{2u}$ A_sB_s $(1/\sqrt{6})$	$1e_{1g}$ A_sB_a $(1/\sqrt{12})$	$1e_{1g}$ A_aB_s $(1/2)$	$1e_{2u}$ A_aB_a $(1/2)$	$1e_{2u}$ A_sB_s $(1/\sqrt{12})$	$1b_{2g}$ A_sB_a $(1/\sqrt{6})$
1	1	2	0	0	2	1
2	1	1	1	1	−1	−1
3	1	−1	1	−1	−1	1
4	1	−2	0	0	2	−1
5	1	−1	−1	1	−1	1
6	1	1	−1	−1	−1	−1
x	−2.00	−1.00	−1.00	1.00	1.00	2.00

Note: The MOs are arranged symmetrically with respect to the pairing properties; e.g., $1a_{2u}$ and $1b_{2g}$ are paired, and $1e_{1g}$ and $1e_{2u}$ are paired. The factors in parentheses above each column are common factors to be included in each element of that column.

When applying the coefficient symmetry method to linear conjugated systems, one can always assume the highest possible symmetry for the system. For example, although 1,3-butadiene is actually C_{2h} (trans) or C_{2v} (cis), the correct HMO coefficients are obtained by writing the carbon skeleton as if it were linear, that is, C–C–C–C, and assuming a plane of symmetry bisecting the chain.

EXERCISES

14-7. Determine the butadiene HMOs and energies by applying the foregoing symmetry method. Note that you will obtain only two classes of coefficient relationships: $c_1 = c_4$, $c_2 = c_3$ (symmetric) and $c_1 = -c_4$, $c_2 = -c_3$ (antisymmetric). This will factor the original 4×4 determinant into two 2×2 determinants.

14-8. Carry out a complete HMO treatment of naphthalene using the symmetry method. Note that neither Eq. (14-21) nor (14-30) applies to this molecule. Determine orbital energies, normalized MOs, and all nearest-neighbor bond orders. Also show that all charge densities are unity.

14-9. Show that the benzyl radical nonbonding MO has $c_3 = -c_5 = c_7 = -c_2/2 = 1/\sqrt{7}$ and $c_2 = c_4 = c_6 = 0$. The $CH_2\cdot$ is numbered 1, and the C atom to which it is attached is numbered 2. All other atoms are numbered consecutively around the ring.

14-10. Calculate the charge densities of the atoms of the benzyl cation and benzyl anion and relate the results to the ortho-, para-, and meta-directive effects of functional groups attached to benzene.

14-11. Verify all the MO coefficients and orbital energies given for benzene in Table 14-1. Also compute charge densities for each carbon atom and the bond orders for nearest-neighbor carbons.

14-5 HMO TREATMENT OF HETERONUCLEAR CONJUGATED SYSTEMS

Molecules in which all the atoms and bonds cannot be considered equivalent, even to a rough approximation, pose some special problems in the HMO approximation. The basic modification necessary is to abandon the use of a common set of coulombic and resonance integrals for all atoms and bonds, respectively. One of the effects of this modification is to destroy the validity of the pairing theorem and of certain bond-order relationships which provide such a convenient framework for alternant hydrocarbons. From a practical standpoint, an even greater difficulty arises: what unique basis does one use to distinguish quantitatively among the different coulombic and resonance integral parameters used in the calculation? This is a difficult problem and probably has no single satisfactory solution. In this section we will examine one of the simpler approaches to this problem, using the pyridine molecule as a specific example. The atom numbering scheme for this molecule is shown in Fig. 14-6.

The pyridine molecule is quite similar to the benzene molecule, except that a nitrogen atom replaces a C–H group. The two molecules are also isoelectronic, and both are six–pi electron systems. However, the presence of the nitrogen heteroatom destroys much of the symmetry; i.e., pyridine has C_{2v} symmetry, whereas benzene has D_{6h} symmetry. Each MO of pyridine has the same general form as that of benzene, namely, a linear combination of six $2p\pi$ AOs, but now five of the AOs are identified with carbon atoms and one with a nitrogen atom. Of course, since there are no degenerate irreducible representations in the C_{2v} point group, the entire orbital pattern for the six pi electrons will be quite different in the case of pyridine: whereas benzene places two of the pi electrons into a nondegenerate MO and the other two in doubly degenerate MOs, pyridine has three pairs of pi electrons placed into three different nondegenerate MOs.

The minimum modifications we must make in the pyridine secular determinant to distinguish it from that of benzene is to modify the three matrix

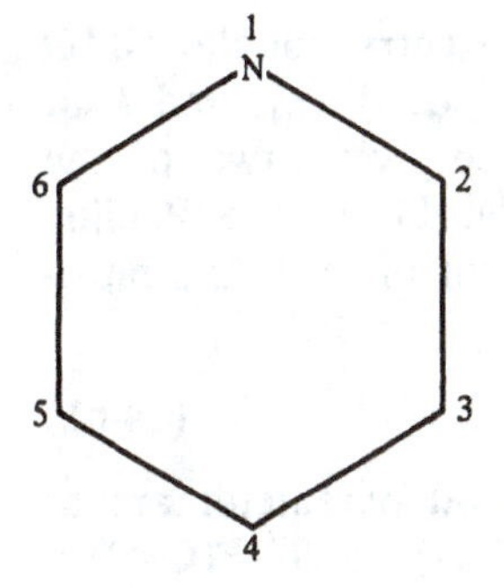

FIGURE 14-6
Atom numbering scheme for the HMO treatment of the pyridine molecule.

elements in which the $2p\pi$ AO of nitrogen appears: F_{11}, F_{12} (equal to F_{21}), and F_{16} (equal to F_{61}). The F_{11} matrix element is modified by using a very simple relationship for the coulombic integral used to describe the heteroatom:

$$\alpha_r = \alpha + h_r\beta \tag{14-58}$$

where α and β are reference integrals generally assumed to be the coulombic and resonance integrals describing the carbon atoms and bonds in the molecule under study. The quantity h_r, which distinguishes the heteroatom r from a reference carbon atom, is called the *electronegativity parameter*, since it can be thought of as differentiating among different atoms in terms of their electronegativity differences. The two off-diagonal matrix elements F_{12} and F_{16} describe equivalent C–N bonds. In general, the resonance integral for an r–s bond (in which one atom is not carbon) can be expressed in the simple relationship

$$\beta_{rs} = k_{rs}\beta \tag{14-59}$$

where k_{rs} is called the *bond parameter*, since it represents the difference between the r–s heterobond and a standard C–C bond. The particular forms used to denote the heteroatom-containing matrix elements allow one still to use the simple substitution formula Eq. (14-19) in simplifying the secular determinant. The diagonal elements now become $x + h_r$, the nearest-neighbor, off-diagonal elements become k_{rs}, and the remaining elements become zero as before. If r refers to a carbon atom, then $h_r = 0$; if both atoms of the r–s pair are carbon atoms, then $k_{rs} = 1$. The modifications needed for the case in which two heteroatoms are nearest neighbors are simple extensions of this procedure.

There are no generally satisfactory ways of choosing numerical values for the parameters h_r and k_{rs}. Perhaps the most often used are

$$\begin{aligned} h_r &= \chi_r - \chi_C \\ k_{rC} &= \frac{E(\text{C=}r) - E(\text{C–}r)}{E(\text{C=C}) - E(\text{C–C})} \end{aligned} \tag{14-60}$$

where the $\{\chi_r\}$ are Pauling electronegativities and the E's are empirical bond energies.

The above-described approach means that we can carry out the HMO treatment of pyridine using just four basic parameters: α, β, h_N, and k_{CN}. Since α and β are not given definite numerical values, only two of the parameters, h_N and k_{CN} need be evaluated explicitly. Using the Pauling electronegativity scale, the electronegativity parameter for the pyridine nitrogen atom is given by

$$h_N = \chi_N - \chi_C = 3.5 - 3.0 = 0.5 \tag{14-61}$$

The empirical bond energies used to calculate the bond parameter are as follows (all values in $\text{kJ} \cdot \text{mol}^{-1}$): $E(\text{C=N}) = 612$, $E(\text{C–N}) = 289$, $E(\text{C=C}) =$

612, and $E(\text{C–C}) = 345$. Thus the bond parameter becomes

$$k_{\text{CN}} = \frac{612 - 289}{612 - 345} = \frac{323}{267} = 1.21 \tag{14-62}$$

We will round off the latter value to 1.2. The secular determinant of pyridine then becomes

$$\begin{vmatrix} x+0.5 & 1.2 & 0 & 0 & 0 & 1.2 \\ 1.2 & x & 1 & 0 & 0 & 0 \\ 0 & 1 & x & 1 & 0 & 0 \\ 0 & 0 & 1 & x & 1 & 0 \\ 0 & 0 & 0 & 1 & x & 1 \\ 1.2 & 0 & 0 & 0 & 1 & x \end{vmatrix} = 0 \tag{14-63}$$

The symmetry-derived coefficient relationships (based on the plane which bisects the molecule and on the choice that carbon atoms 2, 3, 5, and 6 be equivalent) are determined as follows:

1. With respect to symmetric reflection in the plane, we have the relationships $c_2 = c_6$ and $c_3 = c_5$. In addition there are the trivial relationships $c_1 = c_1$ and $c_4 = c_4$.
2. With respect to antisymmetric reflection in the plane, we get first of all the relationships $c_1 = -c_1$ and $c_4 = -c_4$, which can be true only if $c_1 = c_4 = 0$. But if these two coefficients are zero and the four carbon atoms 2, 3, 5, and 6 are assumed equivalent (all have the same value of α), then two possibilities exist:
   a. $c_2 = c_3 = -c_5 = -c_6$
   b. $c_2 = -c_3 = c_5 = -c_6$

These relationships allow factorization of the 6×6 secular determinant into one 4×4 determinant and two 1×1 determinants. A useful check on the correctness of the roots of the full determinant is the fact that the sum of the $\{x_j\}$ must equal the negative of the sum of the electronegativity parameters. The second and third set of coefficient relationships produce $x = -1$ and $+1$, respectively, and thus produce the MO energies $\alpha + \beta$ and $\alpha - \beta$, respectively. The remaining 4×4 determinant is very tedious to solve algebraically, but, fortunately, there are now available a number of microcomputer programs which diagonalize HMO matrices very rapidly.[13] The results of the calculation are summarized in Table 14-2 along with an identification of the C_{2v} irreducible

[13] For this text, the program HUCKEL MO was used, written for the Apple II+ or IIe microcomputers by B. M. Peake; see B. M. Peake and R. Grauwmeijer, *J. Chem. Ed.*, **58**:692 (1981). Similar programs exist for the IBM PC microcomputer.

TABLE 14-2
HMOs of pyridine*

Atom	$1b_1$	$2b_1$	$1a_2$	$3b_1$	$2a_2$	$3a_2$
1	0.589	−0.483	0.000	0.506	0.000	−0.404
2	0.442	−0.149	−0.500	−0.313	−0.500	0.430
3	0.309	0.396	−0.500	−0.299	0.500	−0.398
4	0.269	0.639	0.000	0.607	0.000	0.388
5	0.309	0.396	0.500	−0.299	−0.500	−0.398
6	0.442	−0.149	0.500	−0.313	0.500	0.430
$-x$	2.230	1.239	1.000	−0.985	−1.000	−2.054

* Parameters used: $h_N = 0.5$, $k_{CN} = 1.2$.

representations to which the various MOs belong. Note that the pairing theorem no longer holds for the MO energies.

The charge densities and bond orders of pyridine derived from the above calculation are

$$q_1 = 1.16 \qquad q_2 = q_6 = 0.934 \qquad q_3 = q_5 = 1.004 \qquad q_4 = 0.961$$

$$p_{12} = p_{16} = 0.644 \qquad p_{23} = p_{56} = 0.655 \qquad p_{34} = p_{45} = 0.672 \tag{14-64}$$

The relative charge densities may be interpreted in terms of chemical reactivity; the fact that atoms 2 and 6 have a less negative charge than atoms 3 and 5 is compatible with the known tendency of pyridine to undergo electrophilic substitution at the latter two positions. The bond orders predict that the C–N bond should be longer than any of the C–C bonds, but, in fact, the C–N bond appears to be slightly shorter than either of the C–C bonds.[14]

An *ab initio* calculation on pyridine has been carried out by Clementi[15] using a contracted set of gaussian functions to build the basis set. The energies of the pi MOs were determined to be

$$\begin{aligned} \epsilon(1b_1) &= -0.622E_h \qquad (\alpha + 2.300\beta) \\ \epsilon(2b_1) &= -0.459E_h \qquad (\alpha + 1.239\beta) \\ \epsilon(1a_2) &= -0.447E_h \qquad (\alpha + 1.000\beta) \end{aligned} \tag{14-65}$$

The corresponding HMO values are given in parentheses. These results are in qualitative agreement with the HMO results. Clementi also reported a popula-

[14] L. E. Sutton (ed.), *Tables of Interatomic Distances and Configuration in Molecules and Ions*, special publication 11, The Chemical Society, London, 1958.

[15] E. Clementi, *J. Chem. Phys.*, **46**:4731 (1967).

tion analysis for pyridine; the atom charge densities were calculated to be

$$q_1 = 1.010 \qquad q_2 = 1.005 \qquad q_3 = 1.002 \qquad q_4 = 0.975 \qquad (14\text{-}66)$$

These calculations do not show the same relative orders for atoms 2 and 3, nor do they support the much larger charge density on nitrogen.

EXERCISES

14-12. Carry out an HMO treatment of the formaldehyde molecule using $h_O = 1$ and $k_{CO} = 1.1$. Compare the results obtained for the HMO treatment of ethene in Sec. 14-3. You should tabulate the orbital energies, the coefficients $\mathbf{C}_1$ and $\mathbf{C}_2$, and the density matrix $\mathbf{R}$. What do R_{11}, R_{22}, and R_{12} represent physically? One of the earliest *ab initio* calculations on formaldehyde employing a minimal basis of STOs is given by P. L. Goodfriend et al., *Rev. Mod. Phys.*, **32**:307 (1960).

14-13. Carry out an HMO treatment of the pyrrole molecule using the same parameters as for pyridine. Note that these are not the best parameter choices; the literature is replete with discussions on how pyridine and pyrrole should be parameterized, and no unique solutions seem forthcoming. An *ab initio* calculation similar to that on pyridine has been carried out on pyrrole by Clementi et al., *J. Chem. Phys.*, **46**:4725 (1967).

14-14. One can obtain "theoretical" values for the parameters α and β by setting the HMO values for two different MOs equal to the corresponding numerical values from *ab initio* calculations and solving for the two unknowns simultaneously. Try this for the pyridine calculation, using Clementi's *ab initio* MO energies. You should also repeat the calculation using different pairs of MOs and compare the results for consistency. This approach generally leads to fairly consistent values of α in the neighborhood of -8 to -9 eV, but values of β vary widely depending on just what system is used to calculate it. You might also wish to repeat this calculation using the ethene and benzene HMO calculations and the corresponding *ab initio* calculations reported in this and the previous chapters.

14-15. Those with access to a microcomputer Hückel program may wish to do a series of calculations on the following heteromolecules: imidazole, pyrazole, pyrimidine, purine, quinoline, isoquinoline, and carbazole. You may wish to experiment with using different parameter values, especially with respect to distinguishing between an NH group as on pyrrole (which contributes two pi electrons to the conjugated system) and an N atom as on pyridine (which contributes only one pi electron to the conjugated system).

14-6 THE PARISER-PARR-POPLE METHOD

As we have seen in the preceding sections, the HMO approximation has many serious deficiencies. Foremost among these is its lack of explicit treatment of electron repulsions and its total dependence on connectivities rather than actual geometries. Consequently, the method does not form a satisfactory basis for discussing excited states or molecular geometries. Some rectification of these deficiencies—but still within the pi electron basis—is met in approaches pioneered by Pariser and Parr[16] and by Pople.[17] Both approaches use the basic format of Roothaan's equations, and both make similar integral approximations and simplifications, but the former employs the device of configuration interaction and the latter uses self-consistency. The Pariser-Parr method may be succinctly described as HMO + CI, whereas the Pople approach is HMO + SCF. Nowadays, however, the two approaches are generally combined into what is known as the *Pariser-Parr-Pople*, or *PPP, method*. This method may be described as HMO + SCF with CI appended, if desired.

A major feature of the PPP method resides in the elucidation of the monoelectronic core hamiltonian h' introduced in Eq. (14-4). On the basis of earlier work of Goeppert-Mayer and Sklar,[18] the monoelectronic pi electron core hamiltonian is written

$$h'_\mu = -\frac{\nabla^2_\mu}{2} + \sum_r V^{n+}_{r\mu} \tag{14-67}$$

where the summation is over all the carbon atoms and where $V^{n+}_{r\mu}$ is the potential of the rth carbon atom, which has a charge of $n+$ due to loss of n pi electrons. (Usually $n = 1$, but other values arise when heteroatoms are involved.) Thus $\Sigma_r V^{n+}_{r\mu}$ is the potential seen by the μth pi electron as it moves in the field of the carbon framework and its sigma electron core. The potential $V^{n+}_{r\mu}$ may be written

$$V^{n+}_{r\mu} = V_{r\mu} - n\left\langle \phi_r(\nu) \left| \frac{1}{r_{\mu\nu}} \right| \phi_r(\nu) \right\rangle \tag{14-68}$$

where $V_{r\mu}$ is assumed to be a spherically symmetrical potential of a neutral carbon atom in the sp^2 valence state.[19] The integral in Eq. (14-68) is just the potential due to a single pi electron on carbon atom r. Perhaps the most important feature of the Goeppert-Mayer and Sklar core hamiltonian is the assumption that the potential $V^{n+}_{r\mu}$ is the potential-energy operator in the eigenvalue equation

$$\left(-\frac{\nabla^2_\mu}{2} + V^{n+}_{r\mu}\right)\phi_r(\mu) = \omega_r \phi_r(\mu) \tag{14-69}$$

where $-\omega_r$ is the ionization energy of carbon in its hypothetical sp^2 valence

---

[16] R. Pariser and R. G. Parr, *J. Chem. Phys.*, **21**:466, 767 (1953).

[17] J. A. Pople, *Trans. Faraday Soc.*, **49**:1475 (1953).

[18] M. Goeppert-Mayer and A. L. Sklar, *J. Chem. Phys.*, **6**:645 (1938).

[19] The sp^2 valence state is that hypothetical state a carbon atom would be in if it (with all its electrons) were removed from a conjugated hydrocarbon without undergoing any electron redistribution. Calculations of the energies of such valence states are discussed by W. Moffitt, *Rept. Progr. Phys.*, **17**:173 (1954); H. O. Pritchard and H. A. Skinner, *Chem. Rev.*, **55**:786 (1956); and J. Hinze and H. H. Jaffé, *J. Am. Chem. Soc.*, **84**:540 (1962).

state. This quantity is very close to the first ionization energy of the ^3P ground state of carbon: about 11.26 eV. Substituting Eq. (14-68) into (14-67) leads to

$$h'_\mu = -\frac{\nabla_\mu^2}{2} + \sum_{s \neq r} \left[V_{s\mu} - Z_s \left\langle \phi_s(\nu) \middle| \frac{1}{r_{\mu\nu}} \middle| \phi_s(\nu) \right\rangle \right] \tag{14-70}$$

The matrix element $h'_{rr} = \langle \phi_r(\mu) | h'_\mu | \phi_r(\mu) \rangle$ now becomes

$$h'_{rr} = \omega_r + \sum_{s \neq r} [\langle \phi_r(\mu) | V_{s\mu} | \phi_r(\mu) \rangle - Z_s \gamma_{rs}] \tag{14-71}$$

where γ_{rs} represents a two-center coulombic-repulsion integral of the type

$$\gamma_{rs} = \left\langle \phi_r(\mu)\phi_s(\nu) \middle| \frac{1}{r_{\mu\nu}} \middle| \phi_r(\mu)\phi_s(\nu) \right\rangle \tag{14-72}$$

and $\langle \phi_r(\mu) | V_{s\mu} | \phi_r(\mu) \rangle$ is called a *penetration integral*. This latter integral represents the potential energy due to interaction of the charge distribution $|\phi_r|^2$ and the neutral carbon atom s in its sp^2 valence state. Numerical values can be assigned to this integral, but it is usually omitted from the h'_{rr} term. Thus h'_{rr}, which was called a'_r in the HMO approximation [see Eq. (14-6)], now becomes

$$h'_{rr} = \omega_r - \sum_{s \neq r} Z_s \gamma_{rs} \tag{14-73}$$

The off-diagonal matrix element of h' is not elucidated; it is symbolized by

$$h'_{rs} = \beta_{rs} \tag{14-74}$$

Note that this quantity is analogous to b'_{rs} of Eq. (14-6) (and not to β_{rs}) of the HMO approximation.

The electron-repulsion part of the Hartree-Fock matrix $\mathbf{F}_\pi$ is given by

$$G_{rs} = \sum_{t,u} R_{tu}(2\langle rt|g|su \rangle - \langle rt|g|us \rangle) \tag{14-75}$$

[see Eq. (10-112)]. However, the PPP method employs a very important simplifying approximation which greatly reduces the number of integrals which must be used; all integrals in G_{rs} involving charge distributions such as $\phi_r^*(\mu)\phi_s(\mu)$ (for $r \neq s$) will be small and can be neglected. This simplification, known as the approximation of *zero-differential overlap* (*ZDO*), may be written formally as

$$\langle rt|g|su \rangle = \gamma_{rt}\delta_{rs}\delta_{tu} \tag{14-76}$$

This approximation reduces G_{rs} to one- and two-center integrals only. Using Eq. (14-76) in (14-75) leads to

$$G_{rr} = R_{rr}\gamma_{rr} + \sum_{t \neq r} 2R_{tt}\gamma_{rt} \qquad G_{rs} = -R_{rs}\gamma_{rs} \tag{14-77}$$

Combining the h' matrix elements of Eq. (14-73) and (14-74) with the above elements of **G** leads to

$$F_{rr} = \omega_r + R_{rr}\gamma_{rr} + \sum_{s \neq r} (2R_{ss} - Z_s)\gamma_{rs}$$
$$F_{rs} = \beta_{rs} - R_{rs}\gamma_{rs} \qquad (14\text{-}78)$$

Note that in the HMO approximation all F_{rr} were α and all F_{rs} were β or zero. The effect of these modifications is to put in some dependence on the actual geometry of the molecule, a feature which was lacking in the HMO method.

Owing to the many approximations made in arriving at the matrix-element expressions in Eq. (14-78), it is not satisfactory to evaluate the remaining quantities ω_r, β_{rs}, γ_{rr}, and γ_{rs} by direct integration. Rather, these quantities are chosen empirically or are chosen by means of certain mathematically simple formulas. The valence-state ionization energy ω_r cancels out in energy-difference calculations (if it is assumed to be the same for all carbon atoms), and hence, a numerical value for it is not needed in such cases. However, if a numerical value for ω_r is required, it is generally assigned a value of -11.2 eV for carbon atoms. For heteroatoms, the appropriate modification must be made.

Pariser and Parr suggested a simple exponential relationship for the β_{rs} parameter:

$$\beta_{rs} = a \exp(-bd_{rs}) \qquad (14\text{-}79)$$

where a and b are constants and d_{rs} is the distance between atoms r and s. The constants are evaluated for carbon atom systems as follows: The formula (14-79) is fitted to the experimental energy of the ${}^1A_{1g} \rightarrow {}^1B_{1u}$ transition of ethene ($d_{rs} = 2.55\ a_0$) and the experimental energy of the ${}^1A_{1g} \rightarrow {}^1B_{2u}$ transition of benzene ($d_{rs} = 2.63\ a_0$). This produces $a = -2517.5$ eV and $b = 2.649$ a_0^{-1}. An alternative formula proposed by Linderberg[20] relates the parameter β_{rs} to the corresponding overlap integral Δ_{rs}:

$$\beta_{rs} = \frac{1}{d_{rs}} \frac{\partial \Delta_{rs}}{\partial d_{rs}} \qquad (14\text{-}80)$$

This produces β_{rs} in units of E_h.

This integral γ_{rr} represents the energy required to transfer an electron from one neutral sp^2 carbon atom to another sp^2 carbon atom and, as shown by Pariser,[21] should be equal to the energy of the process

$$\text{C} + \text{C} \rightarrow \text{C}^+ + \text{C}^- \qquad (14\text{-}81)$$

The energy should be equal to $I - A$, where I and A are a valence-state ionization energy and electron affinity, respectively, of an sp^2 carbon atom.

---

[20] J. Linderberg, *Chem. Phys. Lett.*, **1**:39 (1967).

[21] R. Pariser, *J. Chem. Phys.*, **21**:568 (1953).

Mulliken[22] has estimated $I = 11.22$ eV and $A = 0.69$ eV, which leads to $\gamma_{11} = 10.53$ eV. This is the value generally adopted for alternant hydrocarbons. Paoloni[23] has suggested a simple formula which is useful for carbon atoms as well as heteroatoms, namely,

$$\gamma_{rr} = 3.29 Z_r^{\text{eff}} \tag{14-82}$$

where Z_r^{eff} is the effective charge on the atom r. For first-row atoms, Z_r^{eff} is given by Slater's rules for an atom of atomic number A_r as

$$Z_r^{\text{eff}} = A_r - 1.35 - 0.35(\sigma_r + 2R_{rr}) \tag{14-83}$$

where σ_r is the number of sigma electrons contributed by atom r to the molecular framework. For an sp^2 carbon atom, $\sigma_C = 3$ and $Z_C^{\text{eff}} = 3.25$. Thus $\gamma_{11} = (3.29)(3.25) = 10.69$ eV.

The original papers by Pariser and Parr introduced an extrapolation technique for evaluating the remaining integrals γ_{rs}. It was assumed that the two-center integral γ_{rs} was related to the 2 one-center integrals γ_{rr} and γ_{ss} by a quadratic dependence on the bond distance d_{rs}, given by

$$ad_{rs} + bd_{rs}^2 = \frac{\gamma_{rr} + \gamma_{ss}}{2} - \gamma_{rs} \tag{14-84}$$

The constants a and b [not the same constants as in Eq. (14-79)] are determined by calculating two values of γ_{rs} at 5.48 and 6.99 a_0 using the approximation of a uniformly charged sphere. For carbon atoms one obtains $a = -1.389$ eV/a_0 and $b = 6.036 \times 10^{-2}$ eV/a_0^2 when $\gamma_{rr} = \gamma_{ss} = 10.53$ eV.

Alternative methods for determining the γ_{rs} include the simple relationship

$$\gamma_{rs} = (d_{rs} + f_{rs})^n \tag{14-85}$$

where

$$f_{rs} = \frac{2}{\gamma_{rr} + \gamma_{ss}} \tag{14-86}$$

The exponent n is either chosen as -1 (Nishimoto and Mataga)[24] or as $-\frac{1}{2}$ (Ohno).[25]

In general, the PPP calculations are carried out iteratively, beginning with the HMO coefficients as the input approximation; self-consistency is reached when the output coefficients agree with the previous input to within some predetermined limit (generally to within about 10^{-5} or 10^{-6}).

---

[22] R. S. Mulliken, *J. Chem. Phys.*, **2**:782 (1934).

[23] L. Paoloni, *Nuovo Cimento*, **4**:410 (1956).

[24] K. Nishimoto and N. Mataga, *Z. Physik. Chem.* (*Frankfurt*), **12**:335 (1957) and **13**:140 (1957).

[25] K. Ohno, *Theoret. chim. Acta*, **2**:219 (1964).

14-7 PPP TREATMENTS OF ETHENE AND BUTADIENE

The pi electron MOs of ethene are determined solely by symmetry; therefore, the SCF calculations can be carried out in a single cycle. These MOs are

$$\pi_1 = \frac{1}{\sqrt{2}}(\phi_1 + \phi_2) \qquad 1b_{3u}$$
$$\pi_2 = \frac{1}{\sqrt{2}}(\phi_1 - \phi_2) \qquad 1b_{2g} \tag{14-87}$$

These are of exactly the same form as the *ab initio* MOs of the same symmetry discussed in Sec. 13-4, but the normalization coefficients are slightly different in the present context, since the two $2p\pi$ AOs are now assumed to be orthogonal. These two MOs can be used to form wavefunctions for four different electronic states:

1. The ground state (^1A_g); this is commonly designated N.
2. An excited state of the same symmetry as the ground state and formed by promotion of both electrons from the π_1 MO to the π_2 MO; this is designated Z.
3. A state of $^1B_{1u}$ symmetry formed by excitation of one electron from π_1 to π_2; this is called a V state.
4. The triplet counterpart ($^3B_{1u}$) to the V state; this is called a T state.

Explicit forms of these four wavefunctions are

$$\psi_N(^1A_g) = |\pi_1\ \bar{\pi}_1| \qquad \psi_Z(^1A_g) = |\pi_2\ \bar{\pi}_2|$$
$$\psi_V(^1B_{1u}) = \frac{1}{\sqrt{2}}(|\pi_1\ \bar{\pi}_2| - |\bar{\pi}_1\ \pi_2|) \qquad \psi_T(^3B_{1u}) = \frac{1}{\sqrt{2}}(|\pi_1\ \bar{\pi}_2| + |\bar{\pi}_1\ \pi_2|) \tag{14-88}$$

Note that all these wavefunctions are precisely analogous in form to the four helium atom states obtained from $1s$ and $2s$ AOs (see Secs. 7-5 and 9-4). One can make use of this fact to write down energy expressions for ethene which are based on those of helium but with π_1 and π_2 replacing $1s$ and $2s$, respectively. Of course, ethene would have an additional term for nuclear-nuclear repulsions, but these are generally not made explicit in the PPP method, and we will omit them in this discussion.

If we employ the SCF format for the PPP calculations, the coefficient matrix is given by

$$\mathbf{C} = [\mathbf{C}_1\ \ \mathbf{C}_2] = \frac{1}{\sqrt{2}}\begin{bmatrix} 1 & 1 \\ 1 & -1 \end{bmatrix} \tag{14-89}$$

The matrix **T**, which is needed to form the density matrix **R** of the ground state is given by

$$\mathbf{T} = \frac{1}{\sqrt{2}}\begin{bmatrix}1\\1\end{bmatrix} \tag{14-90}$$

Thus the density matrix **R** for the ground state of ethene is

$$\mathbf{R} = \mathbf{T}\mathbf{T}^{\dagger} = \begin{bmatrix}0.5 & 0.5\\0.5 & 0.5\end{bmatrix} \tag{14-91}$$

With Eq. (14-78) and the elements of the density matrix, the matrix elements of the Hartree-Fock matrix (in the PPP approximation) become

$$\begin{aligned} F_{11} &= F_{22} = \omega + \frac{\gamma_{11}}{2} \\ F_{12} &= F_{21} = \beta - \frac{\gamma_{12}}{2} \end{aligned} \tag{14-92}$$

where we have omitted subscripts on ω and β, since they are not needed. The secular determinant now becomes

$$\begin{vmatrix} \omega + \dfrac{\gamma_{11}}{2} - \epsilon & \beta - \dfrac{\gamma_{12}}{2} \\ \beta - \dfrac{\gamma_{12}}{2} & \omega + \dfrac{\gamma_{11}}{2} - \epsilon \end{vmatrix} = 0 \tag{14-93}$$

The two roots of this determinant are the pi electron MO energies

$$\epsilon_1 = \omega + \beta + \frac{\gamma_{11} - \gamma_{12}}{2} \qquad \epsilon_2 = \omega - \beta + \frac{\gamma_{11} + \gamma_{12}}{2} \tag{14-94}$$

The total electronic energy of the ground state is then given by

$$\begin{aligned} E_N &= 2\,\mathrm{Tr}\,(\mathbf{RF}) - \mathrm{Tr}\,(\mathbf{RG}) = 2\sum_i \epsilon_i - \sum_{r,s} R_{rs}G_{rs} \\ &= 2\epsilon_1 - 2(R_{11}G_{11} + R_{12}G_{12}) = 2(\omega + \beta) + \frac{\gamma_{11}}{2} - \frac{3\gamma_{12}}{2} \end{aligned} \tag{14-95}$$

The energies of the remaining states are evaluated by taking the appropriate expectation values and employing the integral approximations of the PPP method. In so doing it is well to recognize that the explicit form of the pi electron hamiltonian for ethene is

$$\hat{H}_\pi = h_1 + h_2 + \frac{1}{r_{12}} \tag{14-96}$$

Consequently, one can write the energy of the Z state as

$$E_Z = 2\langle \pi_2|h|\pi_2\rangle + \left\langle \pi_2(1)\pi_2(2)\left|\frac{1}{r_{12}}\right|\pi_2(1)\pi_2(2)\right\rangle = 2\epsilon_2^{(0)} + J_{22} \tag{14-97}$$

Since $\epsilon_2^{(0)} = \omega - \gamma_{12} - \beta$ in the PPP approximation [see Eqs. (14-73) and (10-29)] and $J_{22} = (\gamma_{11} + \gamma_{12})/2$, the energy of the Z state becomes

$$E_Z = 2(\omega - \beta) + \tfrac{1}{2}\gamma_{11} - \frac{3\gamma_{12}}{2} \tag{14-98}$$

If we employ the analogy with the helium excited $^1S_0$ state, the energy of the V state is given by

$$E_V = \epsilon_1^{(0)} + \epsilon_2^{(0)} + J_{12} + K_{12} \tag{14-99}$$

The PPP approximation reduces the various quantities to the following simple expressions:

$$\begin{aligned} \epsilon_1^{(0)} &= \omega - \gamma_{12} + \beta \qquad & \epsilon_2^{(0)} &= \omega - \gamma_{12} - \beta \\ J_{12} &= \frac{\gamma_{11} + \gamma_{12}}{2} \qquad & K_{12} &= \frac{\gamma_{11} - \gamma_{12}}{2} \end{aligned} \tag{14-100}$$

Thus the total energy of the V state is

$$E_V = 2\omega + \gamma_{11} - 2\gamma_{12} \tag{14-101}$$

Similarly, the energy of the triplet state is given by

$$E_T = \epsilon_1^{(0)} + \epsilon_2^{(0)} + J_{12} - K_{12} = 2\omega - \gamma_{12} \tag{14-102}$$

Subtracting Eq. (14-102) from (14-101) produces the energy of the singlet-triplet separation:

$$E_V - E_T = \gamma_{11} - \gamma_{12} \tag{14-103}$$

Using $\gamma_{11} = 10.53$ eV and $\gamma_{12} = 7.38$ eV (the standard PPP values) produces 3.15 eV as the predicted value of the singlet-triplet separation; the experimental value is 3.0 eV.

The energy of the ground state can be improved by a configuration-interaction calculation which utilizes a mixture of the N and Z states (both of ^1A_g symmetry). The CI wavefunction is

$$\psi_i(^1A_g) = C_{Ni}\psi_N + C_{Zi}\psi_Z \tag{14-104}$$

One of the roots ($i = 1$) is an improvement of the ground state; the other root ($i = 2$) is an improvement of the excited state. The energies of these two states are obtained by solving the secular determinant:

$$\begin{vmatrix} H_{NN} - E & H_{NZ} \\ H_{NZ} & H_{ZZ} - E \end{vmatrix} = 0 \tag{14-105}$$

The matrix elements are given by

$$\begin{aligned} H_{NN} &= \langle \psi_N | \hat{H}_\pi | \psi_N \rangle = E_N \\ H_{ZZ} &= \langle \psi_Z | \hat{H}_\pi | \psi_Z \rangle = E_Z \\ H_{NZ} &= H_{ZN} = \langle \psi_N | \hat{H}_\pi | \psi_Z \rangle \end{aligned} \tag{14-106}$$

Expressions for E_N and E_V have already been given in Eqs. (14-95) and (14-101), respectively. The remaining integral reduces to the simple expression

$$H_{NZ} = \left\langle \pi_1(1)\pi_1(2) \left| \frac{1}{r_{12}} \right| \pi_2(1)\pi_2(2) \right\rangle = \frac{\gamma_{11} - \gamma_{12}}{2} \tag{14-107}$$

Although solution of the secular determinant may at first appear to be an algebraic nightmare, it may be reduced to manageable form by employing some simple substitutions: subtract E_N from both diagonal elements and designate the off-diagonal element with the symbol C. The secular determinant then assumes the very simple form

$$\begin{vmatrix} -E & C \\ C & -4\beta - E \end{vmatrix} = 0 \tag{14-108}$$

which has the solutions

$$E = -\frac{4\beta \pm (16\beta^2 + 4C^2)^{1/2}}{2} = -2\beta \pm \left[4\beta^2 + \frac{(\gamma_{11} - \gamma_{12})^2}{4} \right]^{1/2} \tag{14-109}$$

The negative root reduces to -4β when the repulsion integrals are set equal to zero; this is just the HMO value for the $E_N - E_Z$ separation.

Pariser and Parr obtained an empirical value for the parameter β [one of the points used in Eq. (14-79)] by setting $E_V - E_N = 7.6$ eV; this is the experimental value for the ${}^1A_g \rightarrow {}^1B_{1u}$ transition of ethene. Using the value of E_N obtained via Eq. (14-109), they obtained $\beta = -2.92$ eV. If the Pople SCF value for E_N is used, the energy separation is

$$E_V - E_N = -2\beta + \frac{\gamma_{11} - \gamma_{12}}{2} \tag{14-110}$$

Using $\gamma_{11} = 10.53$ eV and $\gamma_{12} = 7.38$ eV leads to $\beta = -3.01$ eV—not too different from the Pariser-Parr value.

The SCF matrix elements for 1,3-butadiene (either *cis-* or *trans-*) in the PPP formulation are

$$\begin{aligned} &F_{11} = F_{22} = F_{33} = F_{44} = \omega + \gamma_{11} \qquad && F_{12} = F_{21} = \beta_{12} - R_{12}\gamma_{12} \\ &F_{14} = F_{41} = -R_{14}\gamma_{14} && F_{13} = F_{31} = F_{24} = F_{42} = 0 \\ &F_{23} = \beta_{23} - R_{23}\gamma_{23} \end{aligned} \tag{14-111}$$

The resulting 4×4 secular determinant is most conveniently solved by subtracting from each diagonal element the quantity ω, which is common to all. This has the effect of producing eigenvalues relative to ω. The values of β_{rs} are determined by use of formula (14-79), using the bond distance values appropriate for the molecule; values of the repulsion integrals are determined accordingly. The initial calculation assumes R_{rs} values which come from the HMO calculation on butadiene. The calculations are then carried out iteratively until the set of computed R_{rs} values agrees with the previously inputted set to within a previously agreed upon amount; e.g., the largest difference between successive R_{rs} values must be no larger than 1×10^{-6}. At this point one assumes self-consistency has been achieved.

Pople carried out the butadiene calculations assuming that all carbon-carbon bond distances were equal to 2.63 a_0 and that the C–C–C bond angle was 120°. This produces the following matrix of coefficients for the occupied MOs in the case of the trans diastereomer:

$$\mathbf{T} = \begin{bmatrix} 0.4246 & 0.5655 \\ 0.5655 & 0.4246 \\ 0.5655 & -0.4246 \\ 0.4246 & -0.5655 \end{bmatrix} \tag{14-112}$$

Thus, the charge and bond-order matrix **R** is given by

$$\mathbf{R} = \mathbf{T}\mathbf{T}^{\dagger} = \begin{bmatrix} 0.500 & 0.480 & 0.000 & -0.140 \\ 0.480 & 0.500 & 0.140 & 0.000 \\ 0.000 & 0.140 & 0.500 & 0.480 \\ -0.140 & 0.000 & 0.480 & 0.500 \end{bmatrix} \tag{14-113}$$

It should be noted that there are some significant differences between the SCF results and the HMO results, particularly with respect to the bond orders. For example:

$$p_{12} = 0.895 \text{ (HMO) or } 0.960 \text{ (SCF)} \qquad p_{23} = 0.447 \text{ (HMO) or } 0.280 \text{ (SCF)} \tag{14-114}$$

Since the pairing theorem is still valid in the PPP method, all atom charge densities still turn out to be unity (see, for example, the diagonal elements of **R**).

EXERCISES

14-16. Verify that the charge and bond-order matrix **R** of ethene is idempotent.

14-17. The occupied MOs of *cis*-1,3-butadiene, assuming all C–C distances are equal to 2.63 a_0 and the C–C–C bond angle is 120°, are given as follows:

$$\pi_1 = 0.4401(\phi_1 + \phi_4) + 0.5535(\phi_2 + \phi_3)$$
$$\pi_2 = 0.5535(\phi_1 - \phi_4) + 0.4401(\phi_2 - \phi_3)$$

Determine the charge and bond-order matrix and compute the charges and bond orders. Compare the results with those given in the text for the HMO calculations and for the trans diastereomer [see Eq. (14-114)].

14-18. Show by use of the pairing theorem that knowing the coefficients of the occupied MOs allows one to determine the coefficients of the unoccupied (antibonding) MOs. Verify this for the butadiene calculations given in this chapter.

14-19. Prove that the diagonal elements of a CI matrix involving singly excited configurations only (and utilizing the ZDO approximation) are given by

$$\langle \psi_j^k | \hat{H}_\pi - E_0 | \psi_j^k \rangle = \epsilon_k - \epsilon_j + K\mathbf{S}^\dagger(jk)\mathbf{\Gamma}\mathbf{S}^\dagger(jk) - \mathbf{S}^\dagger(jj)\mathbf{\Gamma}\mathbf{S}(kk)$$

where E_0 = ground-state energy
$\mathbf{\Gamma}$ = matrix of electron-repulsion integrals γ_{rs}
$\mathbf{S}(jk)$ = column matrix whose rth row is $c_{rj}c_{rk}$
$K = \text{constant} = 2(1-S) \qquad S = \begin{cases} 0 & \text{singlet states} \\ 1 & \text{triplet states} \end{cases}$

Also show that the corresponding off-diagonal elements are given by

$$\langle \psi_j^k | \hat{H}_\pi | \psi_i^l \rangle = K\mathbf{S}^\dagger(il)\mathbf{\Gamma}\mathbf{S}(jk) - \mathbf{S}^\dagger(ij)\mathbf{\Gamma}\mathbf{S}(kl)$$

In both the above expressions, i and j refer to MOs occupied in the ground state, and k and l refer to virtual MOs. ψ_j^k represents a wavefunction for a state arising by promotion of an electron from the jth MO of the ground state to the kth virtual MO. General formulas for matrix elements involving doubly excited configurations have been given by Čížek.[26]

14-8 ORTHOGONALIZED BASIS FUNCTIONS

Both the HMO method and its extension by Pariser, Parr, and Pople assume that the basis set of $2p\pi$ AOs is orthogonal, although neither method ever requires the explicit identification of a $2p\pi$ AO. Historically, this assumption of an orthogonal basis developed for purely practical reasons: sufficient mathematical simplification so that computations could be carried out with primitive devices such as mechanical calculators. Consequently, it is of interest to reexamine these methods in an effort to determine just what limitations such an assumption leads to. In particular, we will assume that the HMO and PPP methods use a basis which is strictly orthogonalized, and then we will transform this basis into its equivalent in a nonorthogonal basis of STOs. In this way we can determine what kind of a price one pays, if any, for the mathematical convenience of an orthogonal basis.

Suppose we have a basis set of AOs (the STO $2p\pi$'s of conjugated systems, for example) which is not assumed to be orthogonal, and in which the matrix of overlap integrals Δ_{rs} is represented by $\mathbf{\Delta}$. In principle, there are many different unitary transformations one could utilize to diagonalize this matrix of overlap integrals so as to define a new basis which is strictly orthogonal. However, not all these new bases would be convenient to use as an aid in the interpretation of calculations done with them; consequently, we must choose a unitary transformation with some desired properties in mind. The resulting orthogonalized AOs (which we will call *OAO*s from now on) should be chosen so that they give a small number of basic parameters which are invariant in the

[26] J. Čížek, *Theoret. chim. Acta*, **6**:292 (1966).

sense that they can be carried over from one molecular situation to another without appreciable change. Ordinary AOs possess this property only for atoms separated by very long distances; each such AO is localized about its own center, and each has a characteristic orbital energy. In order to achieve this at the distances typical of chemical bonding—where overlap cannot be neglected—*but be consistent with a minimum preservation of free AO character*, one can use the least-squares criterion[27]

$$\delta \sum_i \int |\bar{\phi}_i - \phi_i|^2 \, dx = 0 \tag{14-115}$$

where the overbar symbol indicates an OAO. A solution of the above equation is

$$\bar{\phi}_i = \sum_r \phi_r \Delta_{ri}^{-1/2} \tag{14-116}$$

where $\Delta_{ri}^{-1/2}$ is a matrix element of $\mathbf{\Delta}^{-1/2}$ and $\mathbf{\Delta}$ is the overlap matrix of the AOs $\{\phi_i\}$. The matrix form of Eq. (14-116) is

$$\bar{\boldsymbol{\phi}} = \boldsymbol{\phi}\mathbf{\Delta}^{-1/2} \tag{14-117}$$

Equation (14-117) represents a *symmetric orthonormalization* introduced into solid-state theory by Wannier[28] and into molecular electronic theory by Löwdin.[29] The formation of the OAO basis can be easily understood by considering the following process:

1. The ordinary AO basis satisfies $\mathbf{C}_i^\dagger \mathbf{\Delta} \mathbf{C}_j = \delta_{ij}$ and $\boldsymbol{\phi}^\dagger \boldsymbol{\phi} = \mathbf{\Delta}$.
2. The OAO basis is defined by $\bar{\mathbf{C}}_i = \mathbf{\Delta}^{1/2} \, C_i$ and leads to $\bar{\mathbf{C}}_i^\dagger \bar{\mathbf{C}}_j = \delta_{ij}$ and $\bar{\boldsymbol{\phi}}^\dagger \bar{\boldsymbol{\phi}} = \mathbf{1}$.

Expansion of Eq. (14-116) to first-order terms in Δ_{ji} (where j represents all atomic centers neighboring the ith atomic center) yields

$$\bar{\phi}_i = \phi_i - \frac{1}{2} \sum_j \phi_j \Delta_{ji}^{-1/2} \tag{14-118}$$

This means that each OAO has a negative cusp on each neighboring nucleus (this is what imparts the orthogonality). However, this cusp is small, so that the main effect of the transformation is to compress the original AO more tightly about its nucleus by canceling out its outer parts and renormalizing. This effect is illustrated in Fig. 14-7 for a pair of $1s$ AOs such as used in a calculation on dihydrogen. Note that each OAO is a linear combination of both nonorthogonal AOs; for example, $\overline{1s}_\mathrm{A}$ is mostly $1s_\mathrm{A}$, but with a small

[27] B. C. Carlson and J. H. Keller, *Phys. Rev.*, **105**:102 (1957).

[28] G. Wannier, *Phys. Rev.*, **52**:191 (1937).

[29] P. O. Löwdin, *J. Chem. Phys.*, **18**:365 (1950).

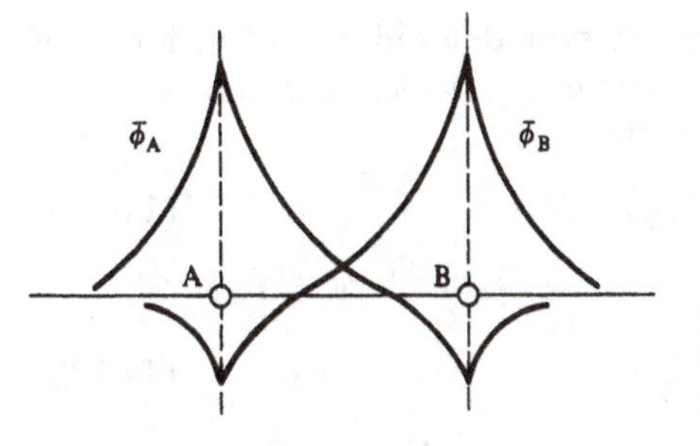

FIGURE 14-7
Orthogonalized 1*s* atomic orbitals for a diatomic molecule such as dihydrogen.

negative cusp contributed by $1s_B$. In actuality, each OAO is a *molecular orbital*, but it is strongly localized about a given atomic center much as a *localized* MO is strongly concentrated to a single interatomic region.

The transformation matrix $\mathbf{\Delta}^{-1/2}$ can be obtained as follows. We write

$$\mathbf{\Delta} = \mathbf{1} + \boldsymbol{\sigma} \tag{14-119}$$

where $\boldsymbol{\sigma}$ is the matrix of off-diagonal overlap integrals. One then applies the binomial expansion formula

$$(x + y)^n = \sum_{j=0} \binom{n}{j} x^{n-j} y^j \qquad y^2 < x^2 \tag{14-120}$$

The following steps clarify the process:

1. Expand $(x + y)^n$.
2. Set $n = 1$.
3. Replace n with $-n$.
4. Let $n = \frac{1}{2}$.

This leads to the expansion

$$\mathbf{\Delta}^{-1/2} = \mathbf{1} - \tfrac{1}{2}\boldsymbol{\sigma} + \tfrac{3}{8}\boldsymbol{\sigma}^2 - \tfrac{5}{16}\boldsymbol{\sigma}^3 + \cdots \tag{14-121}$$

A difficulty which sometimes arises with this method is that convergence may be slow unless all the overlap integrals Δ_{rs} are much smaller than unity; in fact, the series may even diverge in particular instances. An alternative approach is first to diagonalize $\mathbf{\Delta}$, that is, find the matrix $\mathbf{U}$ such that

$$\mathbf{U}^\dagger \mathbf{\Delta} \mathbf{U} = \boldsymbol{\lambda} \tag{14-122}$$

We then find the matrix $\boldsymbol{\lambda}^{1/2}$ by taking the positive square root of each element of $\boldsymbol{\lambda}$. The required transformation matrix is then given by

$$\mathbf{U}\boldsymbol{\lambda}^{-1/2}\mathbf{U}^\dagger = \mathbf{\Delta}^{-1/2} \tag{14-123}$$

where $\boldsymbol{\lambda}^{-1/2}$ is the inverse of $\boldsymbol{\lambda}^{1/2}$. As a simple illustration we will consider the HMO treatment of ethene, where the two basis functions are written $\bar{\phi}_1$ and $\bar{\phi}_2$

to denote that they are orthogonal. We will now determine explicit forms of these OAOs in terms of $2p\pi$ AOs ϕ_1 and ϕ_2, which are not assumed orthogonal and whose overlap integral is given by

$$\Delta = \langle \phi_1 | \phi_2 \rangle \tag{14-124}$$

The overlap matrix then is

$$\mathbf{\Delta} = \begin{bmatrix} 1 & \Delta \\ \Delta & 1 \end{bmatrix} \tag{14-125}$$

The eigenvalues of $\mathbf{\Delta}$ are readily found to be $1 + \Delta$ and $1 - \Delta$. From these we find the associated eigenvectors which form the transformatrix matrix $\mathbf{U}$, namely,

$$\mathbf{U} = \frac{1}{\sqrt{2}} \begin{bmatrix} 1 & 1 \\ 1 & -1 \end{bmatrix} \tag{14-126}$$

The matrix $\boldsymbol{\lambda}^{-1/2}$ is given by

$$\boldsymbol{\lambda}^{-1/2} = (\boldsymbol{\lambda}^{1/2})^{-1} = \begin{bmatrix} (1+\Delta)^{1/2} & 0 \\ 0 & (1-\Delta)^{1/2} \end{bmatrix}^{-1}$$

$$= (1-\Delta^2)^{-1/2} \begin{bmatrix} (1-\Delta)^{1/2} & 0 \\ 0 & (1+\Delta)^{1/2} \end{bmatrix} \tag{14-127}$$

Carrying out the transformation given in Eq. (14-123) produces the matrix needed to transform the nonorthogonal basis to the OAOs:

$$\mathbf{\Delta}^{-1/2} = \begin{pmatrix} a+b & a-b \\ a-b & a+b \end{pmatrix} \tag{14-128}$$

where

$$a = \frac{1}{2(1+\Delta)^{1/2}} \qquad b = \frac{1}{2(1-\Delta)^{1/2}} \tag{14-129}$$

The OAOs are then given by

$$\bar{\phi}_1 = (a+b)\phi_1 + (a-b)\phi_2 \qquad \bar{\phi}_2 = (a-b)\phi_1 + (a+b)\phi_2 \tag{14-130}$$

In the case of a molecule such as ethene where the MO coefficients are determined solely by symmetry, the charge densities on the atoms remain invariant under orthogonalization; thus $\bar{q}_1 = q_1 = \bar{q}_2 = q_2 = 1$. This is not the case, however, in the similar molecule formaldehyde, in which the coefficients no longer depend on symmetry alone (see Exercise 14-12 for an HMO calculation on formaldehyde). The HMO calculation using the auxiliary parameters $h_2 = 1.0$ (for the oxygen atom) and $k_{12} = 1.1$ (for the C–O bond) produces the following results:

$$\pi_1 = 0.541\bar{\phi}_1 + 0.841\bar{\phi}_2 \qquad \pi_2 = 0.841\bar{\phi}_1 - 0.541\bar{\phi}_2 \tag{14-131}$$

This produces $\bar{q}_1 = 2(0.541)^2 = 0.585$ and $\bar{q}_2 = 2(0.841)^2 = 1.415$. However, if we are to interpret the basis in terms of the assumed OAOs, these are charges not on a single atom but distributed over both atoms, since each OAO is a linear combination of AOs of both atoms. Thus, in order to obtain charge densities on single atoms, we must use the nonorthogonal AOs and take the overlap between these two into account. If we use the Mulliken population analysis method, the appropriate atom charge densities are readily obtained from the normalization condition for the nonorthogonal AOs. For the occupied MO this relationship is

$$c_{11}^2 + c_{21}^2 + 2\Delta c_{11}c_{21} = 1 \qquad (14\text{-}132)$$

The charge density on atom 1 is proportional to c_{11}^2 and to part of the overlap population $2\Delta c_{11}c_{21}$. Similarly, the charge density on atom 2 is proportional to c_{21}^2 and to the rest of the overlap population. As discussed in Sec. 12-7, the Mulliken assumption is to divide the overlap population equally between the two atoms. Thus the atom charge densities are

$$q_r = 2(c_{r1}^2 + \Delta c_{11}c_{21}) \qquad r = 1, 2 \qquad (14\text{-}133)$$

The factor of 2 is the total number of electrons in the MO. To evaluate the charge densities, one needs to have numerical values for the overlap integral Δ and the coefficients of the MO in the nonorthogonal basis. The overlap integral Δ is readily evaluated using formulas developed by Mulliken et al.[30] If we assume a carbon-oxygen bond distance of 2.33 a_0 and determine the atomic effective charges by Slater's rules, the overlap integral becomes equal to 0.2108. The coefficients are obtained from those of the OAOs by the back-transformation

$$\mathbf{C} = \mathbf{\Delta}^{-1/2}\bar{\mathbf{C}} = \begin{bmatrix} 1.0171 & -0.1081 \\ -0.1081 & 1.0171 \end{bmatrix}\begin{bmatrix} 0.541 & 0.841 \\ 0.841 & -0.541 \end{bmatrix}$$

$$= \begin{bmatrix} 0.459 & 0.797 \\ 0.797 & -0.459 \end{bmatrix} \qquad (14\text{-}134)$$

The charge densities are then given by

$$\begin{aligned} q_1 &= 2[(0.459)^2 + (0.211)(0.459)(0.797)] = 0.576 \\ q_2 &= 2[(0.797)^2 + (0.211)(0.459)(0.797)] = 1.424 \end{aligned} \qquad (14\text{-}135)$$

Since these values differ insignificantly from the OAO values, it appears as if the distinction between OAOs and their nonorthogonal counterparts does not need to be made when applying the HMO approximation.

[30] R. S. Mulliken, C. A. Rieke, D. Orloff, and H. Orloff, *J. Chem. Phys.*, **17**:1248 (1949).

SUGGESTED READINGS

Atkins, P. W.: *Molecular Quantum Mechanics*, 2d ed., Oxford University Press, New York, 1983

Coulson, C. A., and A. Streitwieser, Jr.: *Dictionary of π-Electron Calculations*, Freeman, San Francisco, 1965.

Daudel, R., R. Lefebvre, and C. Moser: *Quantum Chemistry: Methods and Applications* Interscience, New York, 1959.

Dias, J. R.: "Facile Calculations of the Characteristic Polynomial and π-Energy Levels of Molecules Using Chemical Graph Theory," *J. Chem. Ed.* **64**:213 (1987). This paper illustrates the application of chemical graph theory in the rapid calculation of the characteristic polynomial and $p\pi$ energy levels of many conjugated polyenes without solving the HMO secular determinant.

Higasi, K., H. Baba, and A. Rembaum: *Quantum Organic Chemistry*, Interscience, New York 1965.

McWeeny, R.: *Coulson's Valence*, 3d ed., Oxford University Press, New York, 1979.

Parr, R. G.: *Quantum Theory of Molecular Electronic Structure*, Benjamin, New York, 1963.

Peacock, T. E.: *Electronic Properties of Aromatic and Heterocyclic Molecules*, Academic, New York, 1965.

Pullman, B., and A. Pullman: *Les Théories électroniques de la chimie organique*, Masson et Cie Paris, 1952.

Roberts, J. D.: *Notes on Molecular Orbital Calculations*, Benjamin, New York, 1961.

Salem, L.: *The Molecular Orbital Theory of Conjugated Systems*, Benjamin, New York, 1966.

Streitwieser, A.: *Molecular Orbital Theory for Organic Chemists*, Wiley, New York, 1961.

CHAPTER 15

SEMIEMPIRICAL MOLECULAR ORBITAL METHODS II: ALL VALENCE-ELECTRON SYSTEMS

The preceding chapter discussed methods applicable to molecules in which the electrons could be divided into two groups: (1) pi electrons of nonhydrogen atoms and (2) all the other electrons. Thus the methods were restricted to that specific class of molecules generally called *conjugated hydrocarbons* and their derivatives. In the present chapter we will again divide the electrons into two groups: (1) the valence electrons of the constituent atoms and (2) the 1*s* core electrons of these atoms. Note that since the 1*s* electrons of hydrogen atoms are *valence* electrons, these belong to the former group. The present separation has the advantage of being capable of describing selected aspects of the electronic structures of virtually any type of molecule; i.e., it is not limited to a single type of molecular system.

15-1 THE EXTENDED HÜCKEL METHOD

The extended Hückel method (generally abbreviated *EHT* for "extended Hückel theory"), developed by Roald Hoffmann,[1] is much like the HMO method for conjugated hydrocarbons, but it can be applied to virtually any

[1] R. Hoffmann, *J. Chem. Phys.*, **39**:1397 (1963) and **40**:2245, 2474, 2480 (1964).

molecule, conjugated or not. The essential difference between the two methods resides in the particular partitioning of the electrons into two groups; whereas the HMO method separates pi electrons from all other electrons (the sigma core) and treats only the former in an explicit, albeit parametric, manner, the EHT method separates the valence-shell electrons from the inner-shell, nonbonding electrons and treats the former in an explicit, albeit parametric, manner. This means that the EHT basis set consists of the minimal basis set less the 1*s* AOs of all nonhydrogen atoms. Furthermore, the basis set is not assumed to be orthogonal, and all overlap integrals are computed analytically; this inclusion of overlap integrals imparts some dependence on geometry to the method. For example, for a molecule such as ethane (C_2H_6), the EHT basis set consists of $2s$, $2p_x$, $2p_y$, and $2p_z$ AOs of the two carbon atoms and the 1*s* AOs of the six hydrogen atoms. Explicitly, the basis AOs are Slater orbitals with the exponents chosen by use of Slater's rules. However, it must be mentioned that although overlap integrals are evaluated analytically, electron-repulsion integrals and other integrals appearing in the matrix elements are not; rather, they are incorporated into greatly simplified matrix elements whose numerical values are determined by rather simplistic formulas. Although the EHT method (like the HMO method) does not make use of self-consistency, it is nevertheless convenient to present the mathematical structure within the framework of Roothaan's solution to the Hartree-Fock SCF equations.[2] Thus, the method obtains orbital energies by solving the secular determinant

$$\det |\mathbf{F} - \epsilon \mathbf{\Delta}| = 0 \tag{15-1}$$

within the chosen basis. The diagonal elements, F_{rr}, are set equal to valence-state ionization energies of the atoms in question. For calculations on saturated and unsaturated hydrocarbons, Hoffmann used the numerical values obtained by Pritchard and Skinner for sp^3-hybridized carbon atoms:[3]

$$\begin{aligned} F_{rr} &= -11.4\text{ eV} \quad 2p \text{ AOs of C} \\ F_{rr} &= -21.4\text{ eV} \quad 2s \text{ AOs of C} \\ F_{rr} &= -13.6\text{ eV} \quad 1s \text{ AOs of H} \end{aligned} \tag{15-2}$$

Values for other atoms can be obtained from a paper by Stockis and Hoffmann.[4] The off-diagonal elements F_{rs} are given by

$$F_{rs} = 0.5K(F_{rr} + F_{ss})\Delta_{rs} \tag{15-3}$$

[2] G. Blyholder and C. A. Coulson, *Theoret. chim. Acta*, **10**:316 (1968), have demonstrated that the EHT equations can be obtained from Roothaan's equations by employing certain approximations for overlap charge distributions and electron-repulsion integrals.

[3] Methods for estimating these valence-state ionization energies are given by H. O. Pritchard and H. A. Skinner, *Chem. Rev.*, **55**:745 (1955) and J. Hinze and H. H. Jaffé, *J. Amer. Chem. Soc.*, **84**:540 (1962).

[4] A. Stockis and R. Hoffmann, *J. Am. Chem. Soc.*, **102**:2952 (1980).

where K is a constant to be chosen empirically. Thus, the off-diagonal elements are expressed as a sort of overlap-weighted average of diagonal elements. This form of relationship for the off-diagonal elements is known as the *Wolfsberg-Helmholtz approximation.*[5] Other approximations for the off-diagonal elements have also been used, but the above is the most popular. Generally, the parameter K is chosen to have a value between 1 and 3; Hoffmann used 1.75 in his earliest calculations on hydrocarbons.

The EHT method proceeds by finding the geometry which minimizes the sum of the orbital energies. Although this does not appear to be a valid approach, it leads to many successful applications. Theoretically, the most stable geometry of a molecule (in the Born-Oppenheimer approximation) is represented by the set of bond angles and bond distances which minimizes the total energy

$$E = \sum_i n_i \epsilon_i - G + V_{nn} \tag{15-4}$$

where n_i = occupation number (0, 1, or 2) of the ith MO
ϵ_i = energy of the ith MO
G = electron-repulsion energy
V_{nn} = nuclear-repulsion energy

Since the EHT method does not treat the electron-repulsion energy in a detailed fashion, one can use it to calculate only the sum of the orbital energies, and it is not immediately obvious how this can provide guidance for molecular geometries. At one time it was conjectured that the two terms G and V_{nn} approximately canceled each other so that calculating a sum of orbital energies was tantamount to calculating the total molecular energy, but an analysis by Ruedenberg[6] reveals that the justification is somewhat different. Ruedenberg showed that if one accepts an approximate relationship obtained by Politzer[7] for molecules at their equilibrium configurations, then the term $V_{nn} - G$ is not zero but is proportional to the sum of the orbital energies. Politzer's approximate relationship is

$$E = \tfrac{3}{7}(V_{ne} + 2V_{nn}) \tag{15-5}$$

where V_{ne} is the total nuclear-electronic potential energy (due to attractions between nuclei and electrons). If one employs the virial theorem along with Politzer's formula, one obtains

$$V = 2E = \tfrac{6}{7}(V_{ne} + 2V_{nn}) \tag{15-6}$$

The total potential energy V can also be written

$$V = V_{ne} + V_{nn} + G \tag{15-7}$$

---

[5] B. M. Wolfsberg and L. Helmholtz, *J. Chem. Phys.*, **20**:837 (1952).

[6] K. Ruedenberg, *J. Chem. Phys.*, **66**:375 (1977).

[7] P. Politzer, *J. Chem. Phys.*, **64**:4239 (1976).

Combination of the last two equations produces

$$V_{\text{ne}} - 5V_{\text{nn}} + 7G = 0 \tag{15-8}$$

Adding and subtracting $2V_{\text{nn}}$ in the above, one can solve for $V_{\text{nn}} - G$ to obtain

$$V_{\text{nn}} - G = \frac{V_{\text{nn}} + V_{\text{ne}}}{7} \tag{15-9}$$

Using the relationship (15-6) produces

$$V_{\text{nn}} - G = \frac{V}{6} = \frac{E}{3} \tag{15-10}$$

Using the exact Hartree-Fock expression for the total molecular energy [Eq. (15-4)] produces

$$E = \frac{3}{2} \sum_i n_i \epsilon_i \tag{15-11}$$

Robinson and Schaad[8] have developed another approximate relationship which preserves the general form of the Ruedenberg relationship but is valid for any geometric configuration—not necessarily the equilibrium configuration. This relationship is

$$E(Z, R) \approx \frac{1}{4} \sum_i n_i \epsilon_i \left(2Z, \frac{R}{2}\right) + V_{\text{nn}} \tag{15-12}$$

where the orbital energies have been scaled by doubling each nuclear charge and halving each internuclear distance in the molecule.

Only a few of the applications originally considered by Hoffmann will be mentioned here. Two of the more interesting are the barrier to internal rotation in ethane and the ring conformations of cyclohexane. In both cases Hoffmann's calculations predict the correct geometry of the more stable conformer. Values of the energy differences between conformers are qualitatively correct but quantitatively too high. In general, the method leads to an overemphasis of steric-repulsion effects. Nevertheless, the molecular energies are generally minimized at bond distances very close to their experimental equilibrium values. On the basis of these and many other calculations, Hoffmann concluded that the geometry of a molecule is perhaps its most easily predicted property.

EHT calculations on 1, 3-butadiene predict that the trans diastereomer is lower in energy than the cis form; this is the generally accepted conclusion. Calculations on ethene show the same order of MOs as indicated by the *ab initio* calculations summarized in Table 13-8. The *ab initio* calculations (in the GAMESS STO-6G version) obtained an energy of $-0.328E_{\text{h}}$ for the highest occupied MO (a pi MO), whereas the EHT value was somewhat lower: $-0.486E_{\text{h}}$.

---

[8] B. H. Robinson and L. J. Schaad, *J. Chem. Phys.*, **59**:6189 (1973).

The first EHT calculation on benzene was carried out before *ab initio* calculations on large molecules were generally available. An unexpected result was that the lowest bonding pi MO was below some of the sigma MOs; hitherto it was believed that all the pi MOs were higher in energy than any of the sigma MOs. A similar situation occurs in other conjugated hydrocarbons; e.g., Hoffmann found that the lowest pi MO of butadiene occurred below a sigma MO. As the *ab initio* results for benzene summarized in Table 13-11 clearly indicate, the EHT results appear to be correct. However, there is not complete agreement in detail between the EHT and *ab initio* calculations. For example, the *ab initio* calculations predict that the five highest-energy MOs (in increasing order of energy) are $1b_{2u}$, $3e_{1u}$, $1a_{2u}$, $3e_{2g}$, and $1e_{1g}$. This is in the order σ, σ, π, σ, and π. The corresponding EHT results are e_{1u}, a_{2u}, b_{2u}, e_{2g}, and e_{1g} with the order σ, π, σ, σ, and π. Thus, there is detailed agreement only for the two highest MOs. However, it is difficult to see what the significance of such discrepancies is; certainly there appear to be no experimental measurements which could be used to test which description is the "better." The EHT method and *ab initio* methods appear to be in basic agreement with respect to one very important assumption generally made in conjugated hydrocarbon chemistry: the highest occupied MO (HOMO) and the lowest unoccupied MO (LUMO) are invariably of the pi type. As we will see later, these two MOs generally suffice to facilitate some very important predictions concerning the chemical reactivity pathways for conjugated hydrocarbons.

The *ab initio* calculations on benzene (in the GAMESS STO-6G version) produce an energy of $-0.285E_{\rm h}$ for the HOMO, whereas the EHT result is somewhat lower: $-0.470E_{\rm h}$.

The EHT approach is not suitable for the description of excited states, since it does not take electron interactions into account in an explicit manner.

An important use of the EHT method is to provide an initial guess for the coefficients of the molecular orbitals being determined by an *ab initio* calculation or by other semiempirical methods to be discussed later in this chapter.

15-2 THE CNDO METHOD

In 1965, Pople and coworkers[9] introduced several self-consistent field models which are related to the extended Hückel method in much the same manner as the Pople-Pariser-Parr method is related to the HMO approximation. The simplest of these models is the *complete neglect of differential overlap* (*CNDO*) *method*. Nominally, the basis AOs used in the CNDO method are the same as in the EHT method: the valence-shell AOs of the nonhydrogenic atoms and

[9] J. A. Pople, D. P. Santry, and G. A. Segal, *J. Chem. Phys.*, **43**:S129 (1965); J. A. Pople and G. A. Segal, *J. Chem. Phys.*, **43**:S136 (1965) and **44**:3289 (1966).

the 1s AOs of all hydrogen atoms.[10] However, unlike in the EHT method, the basis is treated as if it were orthonormal. Perhaps the most important simplifying assumption of the CNDO method is that any integrals which contain differential overlap terms such as $\phi_\mu^*(1)\phi_\nu(1)$ will be zero whenever μ and ν refer to different AOs; it is this assumption which leads to the name "complete neglect of differential overlap." Consequently, electron-repulsion integrals of the form

$$\left\langle \mu(1)\lambda(2)\left|\frac{1}{r_{12}}\right|\nu(1)\sigma(2)\right\rangle = \langle \mu\lambda|g|\nu\sigma\rangle \tag{15-13}$$

will be nonzero only if $\mu = \nu$ and $\lambda = \sigma$. Thus, the only surviving electron-repulsion integrals are of the form[11]

$$\langle \mu\lambda|g|\mu\lambda\rangle = \gamma_{\mu\lambda} \tag{15-14}$$

Furthermore, in order to make the model invariant to a specific choice of atomic coordinate systems and to the type of hybridization assumed for an atom, the integrals $\gamma_{\mu\lambda}$ are assumed to depend only on the *atoms* to which the AOs ϕ_μ and ϕ_λ belong and *not* on the actual forms of these AOs. This means that the electron-repulsion integrals are of the general type γ_{AB}, where one electron is on atom A and the other is on atom B (A and B may be the same or different atoms). For example, if ϕ_1 and ϕ_2 are AOs of atom A and ϕ_3 and ϕ_4 are AOs of atom B, integrals such as γ_{13}, γ_{14}, γ_{23}, and γ_{24} are all equal to a common value of γ_{AB} regardless of the types of AOs that ϕ_1, ϕ_2, ϕ_3, and ϕ_4 actually are. Similarly, $\gamma_{11} = \gamma_{12} = \gamma_{22} = \gamma_{AA}$, and $\gamma_{33} = \gamma_{34} = \gamma_{44} = \gamma_{BB}$. Furthermore, each integral γ_{AB} is given by

$$\gamma_{AB} = \left\langle s_A(1)s_B(2)\left|\frac{1}{r_{12}}\right|s_A(1)s_B(2)\right\rangle = \langle s_A s_B|g|s_A s_B\rangle \tag{15-15}$$

where s_A and s_B are valence s functions chosen such that the integral γ_{AB} represents an *average* repulsion between an electron in one of the AOs of atom A and another electron in an AO of atom B. This relationship treats all electron repulsions in terms of spherically symmetrical charge distributions and, therefore, does not take into account directional properties of p orbitals and other higher orbitals.

Use of the CNDO approximation based on Eq. (15-14) leads to the

[10] Hitherto we have used lowercase roman subscripts r, s, . . . to denote basis AOs and lowercase Greek letters μ, ν, . . . to denote electron coordinates. However, Pople and coworkers employ the latter to designate basis AOs, and we will adopt their notation in order to avoid confusion between this text and the literature.

[11] Many authors, including Pople and coworkers, write these integrals as $(\mu\mu\,|\,\lambda\lambda)$; the left-hand part of the symbol refers to electron 1, and the right-hand part to electron 2.

following Hartree-Fock matrix elements in the Roothaan formulation:

$$F_{\mu\mu} = h_{\mu\mu} + R_{\mu\mu}\gamma_{\mu\mu} + 2\sum_{\sigma\neq\mu} R_{\sigma\sigma}\gamma_{\mu\sigma}$$
$$F_{\mu\nu} = h_{\mu\nu} - R_{\mu\nu}\gamma_{\mu\nu} \qquad (\mu \neq \nu) \tag{15-16}$$

Note that there is a similarity to the matrix elements of the PPP method [Eq. (14-78)]. Given Eq. (15-15), the diagonal matrix elements become

$$F_{\mu\mu} = h_{\mu\mu} - R_{\mu\mu}\gamma_{AA} + 2R_{AA}\gamma_{AA} + 2\sum_{B\neq A} R_{BB}\gamma_{AB} \tag{15-17}$$

where the AO ϕ_μ belongs to atom A, and

$$R_{BB} = \sum_{\nu}^{B} R_{\nu\nu} \tag{15-18}$$

(where ϕ_ν belongs to atom B) is half the total valence-electron density on atom B. The one-electron part of $F_{\mu\mu}$ is given by

$$h_{\mu\mu} = \left\langle \phi_\mu \left| -\frac{\nabla^2}{2} - V_A \right| \phi_\mu \right\rangle - \sum_{B\neq A} \langle \phi_\mu | V_B | \phi_\mu \rangle = U_{\mu\mu} - \sum_{B\neq A} \langle \phi_\mu | V_B | \phi_\mu \rangle \tag{15-19}$$

where V_A and V_B are potentials of atoms A and B, respectively (shielded by 1s electrons if A and B are nonhydrogenic), and $-\nabla^2/2 - V_A$ is a core monoelectronic hamiltonian of atom A. Thus, $U_{\mu\mu}$ is a valence-state ionization energy of atom A, a quantity analogous to the ω_r term used in the PPP method. The integral $\langle \phi_\mu | V_B | \phi_\mu \rangle$ is usually written V_{AB}; this parameter represents the interaction of a valence electron on atom A with the core of another atom B. This quantity is calculated by using s_A for the electron on atom A and a point charge for the core of atom B. Thus

$$V_{AB} = \left\langle s_A(1) \left| \frac{Z_B}{r_{1B}} \right| s_A(1) \right\rangle \tag{15-20}$$

where Z_B is the core charge of atom B. Note that, in general, $V_{AB} \neq V_{BA}$.

For the nondiagonal matrix elements $F_{\mu\nu}$ the one-electron part is written as follows when ϕ_μ and ϕ_ν belong to the same atom (say, A):

$$h_{\mu\nu} = U_{\mu\nu} - \sum_{B\neq A} \langle \phi_\mu | V_B | \phi_\nu \rangle \tag{15-21}$$

This matrix element is assumed to be zero when $\mu \neq \nu$; when $\mu = \nu$, this reduces to Eq. (15-19) and we obtain [using Eq. (15-20)]

$$h_{\mu\mu} = U_{\mu\mu} - \sum_{B\neq A} V_{AB} \tag{15-22}$$

When ϕ_μ and ϕ_ν belong to different atoms, we use the relationship

$$h_{\mu\nu} = \beta^0_{AB}\Delta_{\mu\nu} \tag{15-23}$$

where β^0_{AB} (called a *resonance* integral) is a parameter depending only on atoms A and B and $\Delta_{\mu\nu}$ is an overlap integral to be computed analytically from the explicit forms of the basis AOs. The resonance integral is generally determined empirically. The quantity does introduce some directional dependence owing to *p* orbitals and other higher orbitals, since it depends on the orientations and distances of the orbitals and also on a constant assigned to each type of bond (sigma, pi, etc.).

In summary, the matrix elements now become

$$\begin{aligned} F_{\mu\mu} &= U_{\mu\mu} + (2R_{AA} - R_{\mu\mu})\gamma_{AA} + \sum_{B\neq A} (2R_{BB}\gamma_{AB} - V_{AB}) \\ F_{\mu\nu} &= \beta^0_{\mu\nu}\Delta_{\mu\nu} - R_{\mu\nu}\gamma_{\mu\nu} \qquad \mu \neq \nu \end{aligned} \tag{15-24}$$

The expression for $F_{\mu\nu}$ is valid not only for ϕ_μ on atom A and ϕ_ν on atom B but also for both ϕ_μ and ϕ_ν on the same atom (say, A). In such a case $\Delta_{\mu\nu} = 0$, and γ_{AB} is replaced by γ_{AA}.

The total energy of the molecule in the CNDO approximation may be written

$$E = \sum_A E_A + \sum_{A<B} E_{AB}$$

where

$$\begin{aligned} E_A &= 2\sum_\mu^A R_{\mu\mu}U_{\mu\mu} + \sum_\mu^A \sum_\nu^A (2R_{\mu\mu}R_{\nu\nu} - R^2_{\mu\nu})\gamma_{AA} \\ E_{AB} &= 2\sum_\mu^A \sum_\nu^B (2R_{\mu\nu}\beta^0_{AB}\Delta_{\mu\nu} - R^2_{\mu\nu}\gamma_{AB}) \\ &\quad + \frac{Z_A Z_B}{r_{AB}} - 2R_{AA}V_{AB} - 2R_{BB}V_{BA} + 4R_{AA}R_{BB}\gamma_{AB} \end{aligned} \tag{15-25}$$

There are several versions of the CNDO approximation in use. Pople and coworkers originally introduced two versions: CNDO/1 and CNDO/2. The more popular of these, CNDO/2, proceeds as follows:

First, the electron-core repulsion integrals are approximated by the simple relationship

$$V_{AB} = Z_B\gamma_{AB} \tag{15-26}$$

Next, the parameter $U_{\mu\mu}$ is expressed in terms of valence-state ionization energies and electron affinities using the following relationships:

$$\begin{aligned} -I_\mu &= U_{\mu\mu} + (Z_A - \tfrac{1}{2})\gamma_{AA} \\ -A_\mu &= U_{\mu\mu} + Z_A\gamma_{AA} \end{aligned} \tag{15-27}$$

The average of these two quantities is given by

$$-\tfrac{1}{2}(I_\mu + A_\mu) = U_{\mu\mu} + (Z_A - \tfrac{1}{2})\gamma_{AA} \tag{15-28}$$

Thus the parameter $U_{\mu\mu}$ is given by

$$U_{\mu\mu} = -\tfrac{1}{2}(I_\mu + A_\mu) - (Z_A - \tfrac{1}{2})\gamma_{AA} \qquad (15\text{-}29)$$

The diagonal matrix element in Eq. (15-24) now becomes

$$F_{\mu\mu} = -\tfrac{1}{2}(I_\mu + A_\mu) + [(2R_{AA} - Z_A) - (R_{\mu\mu} - \tfrac{1}{2})]\gamma_{AA} + \sum_{B \neq A} (2R_{BB} - Z_B)\gamma_{AB} \qquad (15\text{-}30)$$

The first term in $F_{\mu\mu}$ is a fundamental electronegativity of the AO ϕ_μ and is closely related to the Mulliken scale. The remaining terms show how this electronegativity is modified by the actual molecular environment. In a molecule in which ϕ_μ contains one electron ($R_{\mu\mu} = \tfrac{1}{2}$) and all atoms have a net charge of zero ($2R_{AA} = Z_A$), the value of $F_{\mu\mu}$ is simply $-\tfrac{1}{2}(I_\mu + A_\mu)$. Note that this is similar to the HMO case in which $F_{\mu\mu} = \alpha_\mu + h_\mu \beta$ and the electronegativity parameter h_μ is related to the electronegativity difference between atom μ and a standard reference carbon atom.

Del Bene and Jaffé[12] have introduced a modification of the CNDO approximation (the *CNDO/S-CI method*) in which certain parameters are chosen empirically on the basis of CNDO calculations on a suitably chosen reference molecule, e.g., benzene, and then refined by a limited CI calculation. This approach was specifically designed to be useful in the interpretation of ultraviolet spectroscopic data.

A complete discussion of the CNDO method (and some of its related approximations) is given by Pople and Beveridge (see the suggested readings at the end of this chapter).

15-3 BETWEEN CNDO AND *AB INITIO*

One of the basic flaws in the CNDO methods is that electron-repulsion integrals are oversimplified; for example, all electron charges are assumed to be spherically symmetrical [see Eq. (15-15)]. Thus the directional nature of p orbitals enters in only through the one-electron resonance integrals [Eq. (15-23)], the sizes of which depend on orientations and distances of AOs and a constant designating each type of bond (sigma, pi, etc.). Also, the CNDO method does not adequately distinguish between the electron repulsions found in a molecule in which two orbitals have an electron each with parallel spin versus antiparallel spin. This inadequacy becomes important in situations analogous to distinguishing between the singlet and triplet states of helium arising from the electron configuration $1s2s$. Recall that these two states are distinguished in terms of the exchange integral between $1s$ and $2s$ AOs.

---

[12] J. del Bene and H. H. Jaffé, *J. Chem. Phys.*, **48**:1807, 4050 (1968). See also F. T. Marchese, C. J. Seliskar, and H. H. Jaffé, *J. Chem. Phys.*, **72**:4194, 4204 (1980).

The earliest modification introduce by Pople and coworkers is the *intermediate neglect of differential overlap* (*INDO*) *approximation*. The primary modification to the CNDO approximation is that one-center repulsion integrals between AOs on the same atom are not neglected, and thus some exchange integrals enter the expressions for the matrix elements. The INDO method is a substantial improvement over CNDO/2 in any problem in which electron spin distribution is important. However, the INDO approximation shares with CNDO the inadequate representation of electron repulsions involving AOs with directional properties.

The *neglect of diatomic differential overlap* (*NDDO*) *approximation* neglects differential overlap only when the AOs in question are on different atoms; thus dipole-dipole interactions are retained and are expressed in terms of integrals such as $\langle s_A(1)s_B(2)|g|p_A(1)p_B(2)\rangle$. Such integrals are calculated either from AOs or are determined empirically. In either event, the invariance conditions of the CNDO method must be satisfied. If we assume that the AOs μ and ν are on atom A and the AOs λ and σ are on atom B ($B \neq A$), the matrix elements for Roothaan's equations are given as follows (if needed, superscripts A or B assign a particular summation to atom A or B, respectively):

$$F_{\mu\mu} = U_{\mu\mu} + \sum_{B} V_{\mu\mu,B} + 2\sum_{\nu}^{A} R_{\nu\nu}(\langle \mu\nu|g|\mu\nu\rangle - \tfrac{1}{2}\langle \mu\mu|g|\nu\nu\rangle)$$

$$+2\sum_{B}\sum_{\lambda,\sigma}^{B} R_{\lambda\sigma}\langle \mu\lambda|g|\mu\sigma\rangle$$

$$F_{\mu\nu} = \sum_{B} U_{\mu\nu,B} + R_{\mu\nu}(3\langle \mu\mu|g|\nu\nu\rangle - \langle \mu\nu|g|\mu\nu\rangle) \tag{15-31}$$

$$+2\sum_{B}\sum_{\lambda,\sigma}^{B} R_{\lambda\sigma}\langle \mu\lambda|g|\nu\sigma\rangle$$

$$F_{\mu\lambda} = \beta_{\mu\lambda} - \sum_{\nu}^{A}\sum_{\sigma}^{B} R_{\nu\sigma}\langle \mu\lambda|g|\nu\sigma\rangle$$

The various NDDO matrix terms are identified as follows:

1. The $U_{\mu\mu}$ are one-center, one-electron energies and represent the sum of the kinetic energy of an electron in the AO ϕ_μ and its potential energy due to attraction of the core of atom A.
2. The one-center, two-electron repulsion integrals $\langle \mu\nu|g|\mu\nu\rangle$ are coulombic; the $\langle \mu\mu|g|\nu\nu\rangle$ integrals are exchange integrals.
3. The $\beta_{\mu\lambda}$ are two-center, one-electron core resonance integrals.
4. The $V_{\mu\mu,B}$ are two-center, one-electron integrals representing the attraction between an electron in $\phi_\mu^*(1)\phi_\nu(1)$ on atom A and the core of atom B.
5. The $\langle \mu\lambda|g|\nu\sigma\rangle$ are two-center, two-electron repulsion integrals.

Dewar and coworkers[13] have introduced a number of modified "NDO" approximations. The first of these was MINDO (modified INDO), and this appeared in several versions. The motivation for the modification was to remove several deficiencies of the INDO approximation, basically the manner in which one-electron repulsion integrals were to be evaluated. Whereas the original approach was to evaluate these integrals analytically, Dewar and colleagues evaluated them parametrically by fitting them to experimental data. In this way the INDO method becomes better-suited to calculations such as geometry optimizations. It is probably fair to say that MINDO represented a giant step toward encouraging chemists (organic chemists in particular) to use MO calculations in the interpretation of experimental data and as an aid to mechanistic studies.

Dewar and colleagues have also introduced a similar modification of NDDO, which they term *MNDO* (*modified neglect of diatomic overlap*). As in MINDO, various terms in the matrix elements are not evaluated analytically; rather they are determined either from experimental data or from semiempirical expressions which contain numerical parameters that can be adjusted to fit experimental data. The introduction of adjustable parameters is defended on twofold grounds: first, it may help to compensate for the fundamental deficiency of the single-determinantal MO method (i.e., neglect of electron correlation), and second, it may correct for certain specific assumptions of the NDDO scheme.

The MNDO method has been used to calculate heats of formation, molecular geometries, ionization energies and dipole moments and to provide guidance for a variety of experimental projects. For example, Angus and Johnson[14] used MNDO calculations to study the possibility of incorporating a 1, 2, 3-butatriene moiety into carbocyclic rings of various sizes. The calculations suggested that this moiety is readily bent and may remain intact in essentially any ring size. Consequently, these workers attempted the synthesis of 1, 2, 3-cyclononatriene and found this molecule to be stable under ordinary laboratory conditions.

Dewar and coworkers are also developing an alternative modification of the NDDO approximation which they call *AM1* (after *Austin Model 1*, for the University of Texas, Austin, where Dewar is located).[15] It is claimed that this new modification removes major weaknesses of MNDO, in particular MNDO's failure to reproduce hydrogen bonds.

MNDO, the most widely used Dewar program, is available on a number of computer systems, including the IBM PC microcomputer and compatibles.

---

[13] N. C. Baird and M. J. S. Dewar, *J. Chem. Phys.*, **50**:1262 (1969) and *J. Am. Chem. Soc.*, **97**:1285, 4787 (1975); M. J. S. Dewar and W. Thiel, *J. Am. Chem. Soc.*, **99**:4899, 4907 (1977).

[14] R. O. Angus and R. P. Johnson, *J. Org. Chem.*, **49**:2880 (1984).

[15] M. J. Dewar, E. G. Zoebisch, E. F. Healy, and J. J. P. Stewart, *J. Am. Chem. Soc.*, **107**:3902 (1985).

15-4 QUALITATIVE APPLICATIONS OF MO THEORY

The terminology and concepts of the Roothaan formulation of molecular Hartree-Fock SCF calculations are widely used by contemporary chemists in ways which make little or no use of calculated energies. One of the most important applications is in the study of the mechanisms of certain types of chemical reactions.[16] These applications have led to some valuable insights into questions which have puzzled chemists for many years; for example: Why are certain mechanisms, which appear inherently reasonable, not found? Also, why are certain thermodynamically spontaneous reactions so slow?

A reaction mechanism consists of the elementary steps—usually unimolecular or bimolecular—each of which must satisfy certain symmetry restrictions or rules. For example, each bimolecular step must obey the following rules:

1. As the two reacting species approach each other, electrons are transferred from the highest occupied MO (HOMO) of one reactant to the lowest unoccupied MO (LUMO) of the other. Other MOs may also be involved, but the roles of the HOMO and LUMO predominate. A partially filled MO can act either as a HOMO or LUMO.
2. The reacting species may be viewed as a larger pseudo molecule (transition-state complex). The HOMO and LUMO of the reactants must be components of the same irreducible representation of the point group to which the pseudo molecule belongs. This means that the HOMO and LUMO must have a large overlap.[17]
3. The orbital energy difference, $\epsilon_{LUMO} - \epsilon_{HOMO}$, must be less than about $0.22E_h$ (around 6 eV). However, if the LUMO and HOMO do not satisfy rule 2, other MOs in this energy range may take their place, provided these latter MOs do satisfy rule 2.[18]
4. If the HOMO and LUMO are both bonding MOs, then the HOMO must represent bonds to be broken, and the LUMO bonds to be formed. The opposite occurs if the HOMO and LUMO are both antibonding.

If no pair of MOs in the reacting species satisfies rules 2, 3, and 4, the reaction is said to be *symmetry-forbidden*.

[16] Simple accounts of this topic are given by R. G. Pearson, *Chem. Eng. News*, **48**:66 (1970) and *Accounts Chem. Research*, **4**:152 (1971).

[17] The HOMO and LUMO will appear combined in two new MOs of the pseudo molecule, hence the rationale behind this rule.

[18] It is an elementary mathematical fact that two MOs of the same symmetry but greatly different energies will not "mix" strongly; that is, their combinations will contain a large percentage of one MO and a small percentage of the other.

As a simple example of the role symmetry plays in reaction mechanisms, consider the well-studied hydrogen iodide reaction:

$$H_2(g) + I_2(g) \rightleftharpoons 2HI(g) \tag{15-32}$$

During the developmental days of chemical kinetics it was assumed that the reaction occurred as follows:

$$\begin{matrix} H\text{---}H \\ I\text{---}I \end{matrix} \rightleftharpoons \begin{matrix} H \cdots\cdots H \\ \vdots \;\times\; \vdots \\ I \cdots\cdots I \end{matrix} \longrightarrow \begin{matrix} H \\ | \\ I \end{matrix} \quad \begin{matrix} H \\ | \\ I \end{matrix} \tag{15-33}$$

This represents a broadside collision of H_2 and I_2. Let us consider the case where electrons flow from the σ_g HOMO of H_2 to the σ_u^* LUMO of I_2 (the asterisk is used to denote an *antibonding* MO). It is easy to show that such a step is symmetry-forbidden without bothering to specify the point group of the pseudo molecule and the irreducible representations to which the HOMO and LUMO would belong. As shown in Fig. 15-1, the HOMO σ_g and the LUMO σ_u^* have zero overlap in the complex and thus must be components of different irreducible representations in the point group of the complex.

Alternatively, consider electron flow from the π_g^* HOMO of I_2 to the σ_u^* LUMO of H_2. Figure 15-2 shows that these two MOs overlap and thus must be components of the same irreducible representation in the point group of the complex. Thus, these MOs lead to a symmetry-allowed mechanism. However, this mechanism implies that the I–I bond in the complex is strengthened (since electrons leave an antibonding MO), and consequently the step is *chemically forbidden*. Also, this step would have electrons flowing from a more electronegative element to a less electronegative element.

Another possible reaction step in the HI reaction is

$$I + H\text{–}H \longrightarrow IH + H \tag{15-34}$$

(the three atoms are collinear along the z axis), where the electron transfer is from σ_g of H_2 (the HOMO) to a $2p_z$ AO (the LUMO) of the iodine atom; this step is both symmetry-allowed and chemically allowed.

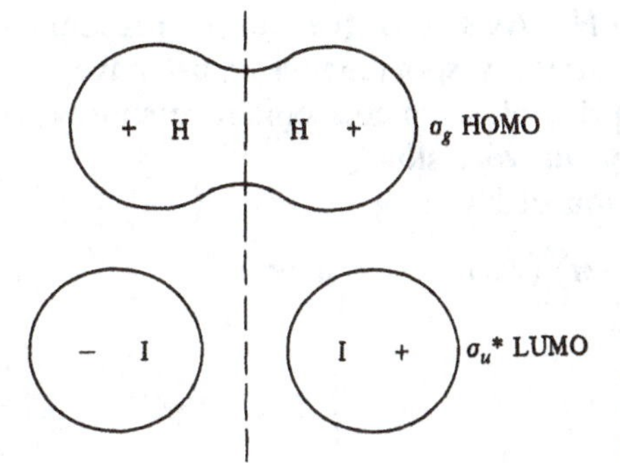

FIGURE 15-1
Symmetry-forbidden step in the reaction of H_2 and I_2 to form HI.

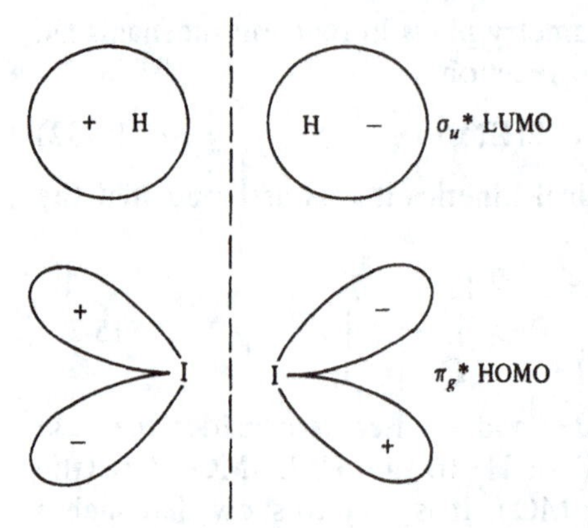

FIGURE 15-2
Symmetry-allowed, chemically forbidden step in the reaction of H_2 and I_2 to form HI.

As another simple example, consider the two reactions (at 298 K):[19]

$$\begin{aligned} 2\mathrm{NO}(g) &\rightleftharpoons N_2(g) + \mathrm{O}_2(g) \qquad \Delta G^0 = -173\ \mathrm{kJ} \\ 2\mathrm{NO}(g) + \mathrm{O}_2(g) &\rightleftharpoons 2\mathrm{NO}_2(g) \qquad \Delta G^0 = -69.7\ \mathrm{kJ} \end{aligned} \tag{15-35}$$

Although the first reaction is more thermodynamically favorable, it is much slower than the second reaction. The latter reaction is one of the critical steps in the production of photochemical smog; it occurs rapidly in the presence of sunlight and unburned hydrocarbons. Why are the rates of the two reactions so different? To answer this question let us examine the reverse of the first reaction. The electron configurations of dinitrogen and dioxygen in their ground states are as follows:

$$N_2: (1\sigma_g)^2(1\sigma_u)^2(2\sigma_g)^2(2\sigma_u)^2(3\sigma_g)^2(1\pi_u)^4 \qquad {}^1\Sigma_g^+ \text{ state}$$

$$O_2: (1\sigma_g)^2(1\sigma_u)^2(2\sigma_g)^2(2\sigma_u)^2(3\sigma_g)^2(1\pi_u)^4(1\pi_g)^2 \qquad {}^3\Sigma_g^- \text{ state}$$

If we assume that electron transfer is from the $1\pi_u$ HOMO of N_2 to the $1\pi_g$ LUMO of O_2 (this is a half-filled MO), Fig. 15-3 shows that the reaction is symmetry-forbidden. Also, if we consider the $1\pi_g$ of O_2 as the HOMO with electron transfer to the $1\pi_g$ LUMO of N_2, this reaction is chemically forbidden, since O_2 is more electronegative than N_2. As a consequence, the reaction between N_2 and O_2, although thermodynamically spontaneous, must have a high energy barrier, and according to the principle of microscopic reversibility, the forward reaction would be predicted to be very slow.

The ground-state electron configuration of NO is

$$(1\sigma)^2(2\sigma)^2(3\sigma)^2(4\sigma)^2(5\sigma)^2(1\pi)^4(2\pi) \qquad {}^2\Pi \text{ state}$$

[19] B. M. Fung, *J. Chem. Ed.*, **49**:26 (1972).

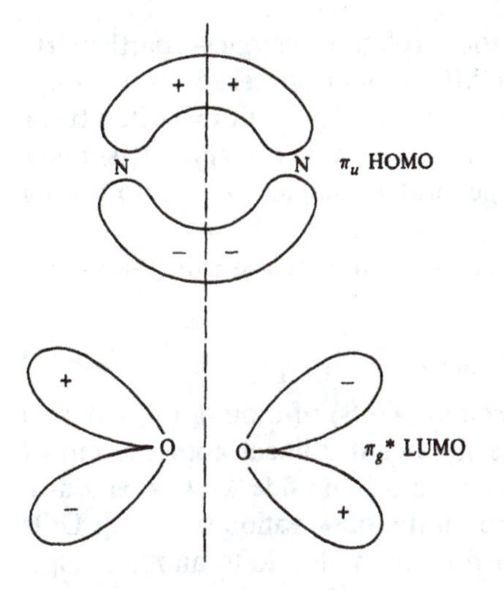

FIGURE 15-3
Symmetry-forbidden step in the reaction of N_2 and O_2 to form NO. The reverse process is also symmetry-forbidden.

The first step in the oxidation of NO is thought to be

$$NO(g) + O_2(g) \rightarrow NO_3(g) \tag{15-36}$$

If we consider electron transfer from the 2π HOMO of NO to the $1\pi_g$ LUMO of O_2 (see Fig. 15-4), we see the reaction can occur in a symmetry-allowed and chemically allowed manner.

Woodward and Hoffmann (see the suggested readings at the end of this chapter) have developed a systematic extended treatment of concerted chemical reactions by the use of methods similar to those discussed above.

It is often convenient and profitable to deduce at least the basic forms of molecular orbitals without resorting to actual calculations. A systematic ap-

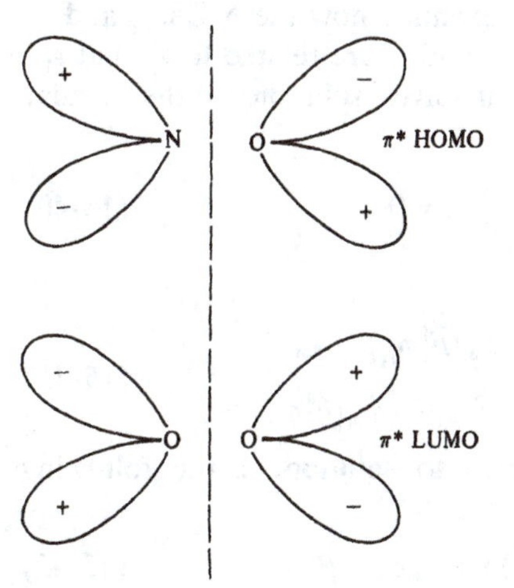

FIGURE 15-4
Symmetry-allowed step in the reaction of NO and O_2 to form NO_3. This is believed to be the first step in the oxidation of NO.

proach to "guessing" at such MOs and their relative energies—particularly useful with respect to the HOMO and LUMO—could be based on a group-theory procedure such as illustrated for ethene in Sec. 13-4, but this often turns out to be more awkward and time-consuming than necessary. A simpler approach can be based on some of the general principles of the following simple example.

Let us consider an MO of a molecule AB (not necessarily diatomic), which we write in the following form:

$$\chi_i = c_{\mathrm{A}i}\phi_{\mathrm{A}} + c_{\mathrm{B}i}\phi_{\mathrm{B}} \tag{15-37}$$

where ϕ_{A} and ϕ_{B} are, in general, *group orbitals* (GOs) of atom groups A and B, respectively. Each GO is, in general, a multicenter linear combination of AOs with the same form as these would have in a bona fide MO obtained in the usual manner. This formulation is based on the observation that such GOs are frequently approximately transferable from one molecule to another; e.g., a three-center GO used to describe methylene, CH_2, with C_{2v} symmetry appears in approximately the same form in other molecules containing the CH_2 group (ethene, cyclopropane, formaldehyde, etc.). Consequently, we assume that such GOs can be used to build up approximations to MOs of molecules in which the appropriate chemical groups are present.

It is convenient to assign an MO energy to each GO; we call this ϵ_{A} for the GO ϕ_{A} and ϵ_{B} for the GO ϕ_{B}. The MO given in Eq. (15-37) will then lead to the two possible combinations

$$\chi_1 = c_{\mathrm{A}1}\phi_{\mathrm{A}} + c_{\mathrm{B}1}\phi_{\mathrm{B}} \qquad \chi_2 = c_{\mathrm{A}2}\phi_{\mathrm{A}} + c_{\mathrm{B}2}\phi_{\mathrm{B}} \tag{15-38}$$

The corresponding "MO" energies are given by

$$\epsilon_i = \langle x_i|\hat{F}|x_i\rangle \qquad i = 1, 2 \tag{15-39}$$

where $\hat{F}$ is some unspecified effective one-electron operator which reduces to the Hartree-Fock operator when the MOs are generated strictly according to the Roothaan LCAO method. We will now examine how the MOs χ_1 and χ_2 are related to the GOs ϕ_{A} and ϕ_{B} and how ϵ_1 and ϵ_2 are related to ϵ_{A} and ϵ_{B}. The explicit formulation of the two MOs involves solution of the secular determinant

$$\begin{vmatrix} F_{\mathrm{AA}} - \lambda & F_{\mathrm{AB}} - \lambda\Delta \\ F_{\mathrm{BA}} - \lambda\Delta & F_{\mathrm{BB}} - \lambda \end{vmatrix} = 0 \tag{15-40}$$

where the matrix elements are defined by

$$\begin{aligned} \Delta = \langle \Phi_{\mathrm{A}}|\phi_{\mathrm{B}}\rangle \qquad & F_{\mathrm{AA}} = \langle \phi_{\mathrm{A}}|\hat{F}|\phi_{\mathrm{A}}\rangle = \epsilon_{\mathrm{A}} \\ F_{\mathrm{BB}} = \langle \phi_{\mathrm{B}}|\hat{F}|\phi_{\mathrm{B}}\rangle = \epsilon_{\mathrm{B}} \qquad & F_{\mathrm{AB}} = F_{\mathrm{BA}} = \langle \phi_{\mathrm{A}}|\hat{F}|\phi_{\mathrm{B}}\rangle = \beta \end{aligned} \tag{15-41}$$

Solution of this secular determinant reduces to solution of the following quadratic equation

$$(1 - \Delta^2)\lambda^2 + (2\beta\Delta - \epsilon_{\mathrm{A}} - \epsilon_{\mathrm{B}})\lambda + \epsilon_{\mathrm{A}}\epsilon_{\mathrm{B}} - \beta^2 = 0 \tag{15-42}$$

where λ represents the two possible roots ϵ_1 and ϵ_2. Although a general solution of this expression is straightforward in principle, the resulting expressions are unwieldy and difficult to interpret. Consequently, we will adopt a simpler and more informative approach: solution of the equation for two special extreme cases, from which the main features of the general solution are readily induced.

Let us consider first the case when the overlap integral Δ is zero. One circumstance in which this can exist is when ϕ_A and ϕ_B (which belong, in general, to reducible representations of the point group to which the molecule AB belongs) do not contain the same irreducible representations. In this case, ϕ_A and ϕ_B are of the wrong symmetry types to interact, and $\Delta = 0$. Also, the quantity β equals zero. This leads to the roots $\lambda = \epsilon_A$ and ϵ_B; that is, the two GOs do not mix. The coefficient values are the trivial relations $c_{A1} = 1$, $c_{B1} = 0$, $c_{A2} = 0$, and $c_{B2} = 1$.

Now consider the case when $\Delta \neq 0$. This case can be considered in terms of two subcases: (1) $\epsilon_A = \epsilon_B = \epsilon$ and (2) $\epsilon_A \neq \epsilon_B$. Considering the first subcase explicitly, one easily obtains

$$\epsilon_1 = \frac{\epsilon + \beta}{1 + \Delta} \qquad \epsilon_2 = \frac{\epsilon - \beta}{1 - \Delta} \tag{15-43}$$

where (assuming ϵ and β are negative) $\epsilon_1 < \epsilon_2$. This situation represents maximum mixing of the two GOs. As shown in Fig. 15-5, the two GOs combine to form two MOs: one of greatly increased energy (relative to ϵ) and the other of greatly decreased energy (relative to ϵ). The coefficients of the MOs are determined in the standard manner; these are

$$c_{A1} = c_{B1} = \frac{1}{\sqrt{2(1 + \Delta)}} \qquad c_{A2} = -c_{B2} = \frac{1}{\sqrt{2(1 - \Delta)}} \tag{15-44}$$

The general results of the second subcase are readily inferred without obtaining the explicit expressions for the results, since they are apparently intermediate

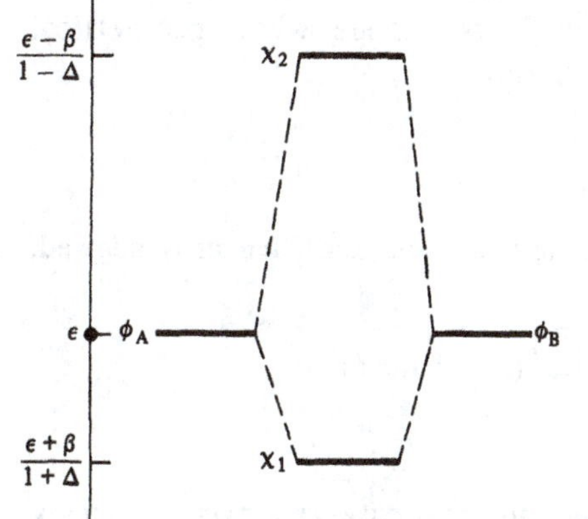

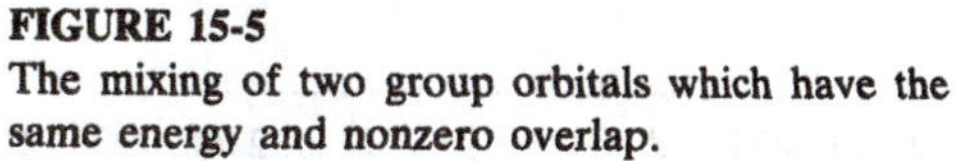
FIGURE 15-5
The mixing of two group orbitals which have the same energy and nonzero overlap.

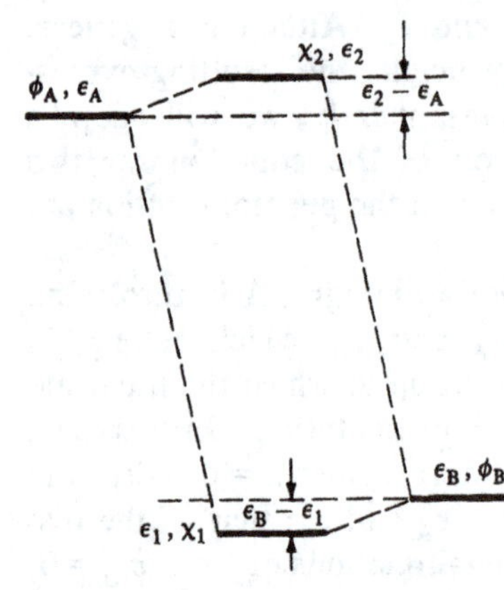

FIGURE 15-6
The restricted mixing of two group orbitals which have different energies and nonzero overlap.

to those of the two situations already considered: zero mixing of the GOs and maximum mixing of the GOs. In Fig. 15-6 we see the situation in which the energy difference $|\epsilon_A - \epsilon_B|$ is quite large. Note that one of the resulting MOs is only slightly lower in energy than one of the GOs and that the other MO is only slightly higher in energy than the other GO. In other words, the extent to which the two GOs mix decreases as their energy difference increases. Note that in the extreme case of very large GO energy differences, the overlap integral Δ tends to zero and, qualitatively, the situation resembles that in which $\Delta = 0$ owing to symmetry mismatch.

Since some of the same general conclusions can also be induced by the use of perturbation theory format, the approach is often called the *perturbational molecular orbital* (*PMO*) *method*. The approach is described in some detail by one of its main developers (see Dewar and Dougherty in the suggested readings at the end of this chapter).

There are a number of available detailed descriptions of the qualitative use of MOs: foremost of these are texts by Salem, by Clark, and by Gimarc (see the suggested readings at the end of this chapter). There is also a series of very important papers by A. D. Walsh[20] which discuss various aspects of qualitative MO theory dealing with how MOs change when geometrical changes are made in a given molecular structure.

EXERCISES

15-1. Verify that reaction (15-34) is both symmetry-allowed and chemically allowed.

15-2. Consider the reaction

$$2CH_2{=}CH_2 \longrightarrow \square \text{ (cyclobutane)}$$

[20] A. D. Walsh, *J. Chem. Soc.*, 2260, 2266, 2288, 2296, 2301, 2306, 2318, 2321, 2325, 2330 (1953).

Discuss the following mechanisms in which the two ethene molecules line up side by side in parallel planes:

(*a*) Both ethenes are in their ground states and the electron transfer is from a π_u HOMO of one molecule to a π_g (antibonding) LUMO of the other.

(*b*) One ethene is in the ground state and the other is in an excited state formed by promotion of one electron to a π_g MO. Electron transfer is from the π_g of the excited state to the π_g of the ground state.

15-3. Can the hydrogenation of ethene occur via the intermediate shown here:

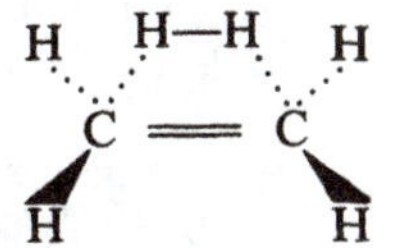

(The attacking H_2 is above the plane of the ethene molecule.)

in which electron transfer is from a σ_g HOMO of H_2 to a π_g LUMO (antibonding of ethene?

SUGGESTED READINGS

Clark, T.: *A Handbook of Computational Chemistry*, Wiley, New York, 1985.

Dewar, M. J. S., and R. C. Dougherty: *The PMO Theory of Organic Chemistry*, Plenum, New York, 1975.

Fukui, K.: *Theory of Orientation and Stereoselection*, Springer-Verlag, Berlin, 1975.

Gimarc, B. M.: *Molecular Structure and Bonding*, Academic, New York, 1979.

Pople, J. A., and D. L. Beveridge: *Approximate Molecular Orbital Theory*, McGraw-Hill, New York, 1970.

Salem, L.: *Electrons in Chemical Reactions: First Principles*, Wiley, New York, 1982.

Woodward, R. B., and R. Hoffmann: *The Conservation of Orbital Symmetry*, Verlag Chemie GmbH, Academic, Weinheim/Bergstr. 1971.

APPENDIX 1

VALUES OF PHYSICAL CONSTANTS

The values of most of the physical constants appearing in the following table are taken from E. R. Cohen and B. N. Taylor, *The 1986 Adjustment of the Fundamental Physical Constants: A Report of the CODATA Task Group on Fundamental Constants*, no. 63, Pergamon, November 1986.

The symbols used for SI units are as follows: m (meter), s (second), C (coulomb), kg (kilogram), J (joule), and K (degree Kelvin). An extensive discussion of SI units is given by M. A. Paul, *Chemistry*, **45**:14 (1972).

Constant	Symbol	Magnitude	SI unit
Speed of light in vacuum	c	2.99792458×10^{8}	$\mathrm{m \cdot s^{-1}}$
Elementary charge	e	$1.6021773 \times 10^{-19}$	C
Electron rest mass	m, m_e	9.109390×10^{-31}	kg
Proton rest mass	m_p	1.672623×10^{-27}	kg
Planck constant	h	6.626076×10^{-34}	$\mathrm{J \cdot s}$
Planck constant ($h/2\pi$)	$\hbar$	1.054577×10^{-34}	$\mathrm{J \cdot s}$
Rydberg constant	R_∞	1.09737312×10^{7}	$\mathrm{m^{-1}}$
Bohr radius	a_0	$5.2917725 \times 10^{-11}$	m
Boltzmann constant	k	1.380658×10^{-23}	$\mathrm{J \cdot K^{-1}}$
Debye unit of dipole moment	D	3.3358×10^{-30}	$\mathrm{C \cdot m}$

APPENDIX 2

ELECTROSTATIC AND ELECTROMAGNETIC UNITS

Coulomb's law for the interaction of two point charges q_1 and q_2 separated by a distance r_{12} (with a vacuum separating the charges) is given by

$$|\mathbf{F}| = \frac{Kq_1q_2}{r_{12}^2}$$

where **F** is the force between the charges and K is a constant depending on the units used for the charges. In the older gaussian units, an electrostatic unit of charge (*esu*, or statcoulomb) was defined in such a way that the force acting between two charges separated by 1 cm was equal to 1 dyne (10^{-5} N) if each charge was 1 esu. In this case the constant K was equal to unity and was dimensionless. In the SI system of units, charges are expressed in coulombs, distance in meters, and force in newtons. Thus K must be in units of $\mathrm{N \cdot m^2 \cdot C^{-2}}$. It is also customary to express K as the following:

$$K = \frac{1}{4\pi\epsilon_0}$$

where ϵ_0 is called the *permittivity of free space*. The permittivity enters into the dielectric constant κ, which is a ratio of the permittivity of a given medium (ϵ)

to the permittivity of a vacuum or free space (ϵ_0). To obtain correspondence between the older gaussian units and SI units, K has the numerical value of $8.98755 \times 10^9\ \mathrm{N \cdot m^2 \cdot C^{-2}}$.

When atomic units are used, the constant K is chosen as unity.

The SI unit of energy is the joule ($1\ \mathrm{J} = 1\ \mathrm{N \cdot m}$). Since the electronvolt is defined as the energy acquired by a charge e when accelerated through a potential of 1 volt (V), it is easy to see that 1 eV is equivalent to $1.6021917 \times 10^{-19}$ J, since $e = 1.6021917 \times 10^{-19}$ C and a joule is just a volt-coulomb.

APPENDIX 3

EVALUATION OF SOME SIMPLE HELIUM ATOM INTEGRALS

The simplest helium atom calculations described in Secs. 6-4 and 7-5 require evaluation of three basic types of integrals, namely,

1. $\langle 1s|\nabla^2|1s\rangle$ kinetic-energy integral
2. $\langle 1s|1/r|1s\rangle$ nucleus-electron attraction integral
3. $\langle 1s(1)1s(2)|1/r_{12}|1s(1)1s(2)\rangle$ electron-repulsion integral

The first two types of integrals are evaluated (after suitable manipulation to the proper form) by use of the standard integral

$$\int_0^\infty x^n e^{-ax}\,dx = \frac{n!}{a^{n+1}} \tag{A3-1}$$

In the following we assume $1s$ is given by the normalized hydrogenlike form $\sqrt{(\eta^3/\pi)}e^{-\eta r}$, where η is a variational (scaling) parameter ($\eta = Z =$ nuclear charge for unscaled functions).

The kinetic-energy integral may be expanded as follows

$$\langle 1s|\nabla^2|1s\rangle = \frac{\eta^3}{\pi}\left\langle e^{-\eta r}\left|\frac{1}{r^2}\frac{\partial}{\partial r}r^2\frac{\partial}{\partial r}\right|e^{-\eta r}\right\rangle \tag{A3-2}$$

where it must be remembered that the volume element of integration is $r^2 \sin\theta\, d\theta\, d\varphi\, dr$. Since the $1s$ orbital contains no angular factors, integration always contributes a factor of 4π, multiplying any one-electron integral in which it ($1s$) appears. Consequently, the explicit form of the kinetic-energy integral is

$$4\eta^3\int_0^\infty e^{-\eta r}\left(\frac{1}{r^2}\frac{d}{dr}r^2\frac{d}{dr}\right)e^{-\eta r}r^2\, dr \tag{A3-3}$$

Carrying out the differentiation indicated under the integral sign (differentiate only $e^{-\eta r}$, not the r^2 following it) one obtains

$$4\eta^3\int_0^\infty (\eta^2 r^2 - 2\eta r)e^{-2\eta r}\, dr \tag{A3-4}$$

Using Eq. (A3-1) shows that the kinetic-energy integral is equal to $-\eta^3$.

The nucleus-electron attraction integral becomes

$$4\eta^3\int_0^\infty re^{-2\eta r}\, dr \tag{A3-5}$$

Use of Eq. (A3-1) shows that this is equal to η.

The electron-repulsion integral is more difficult to evaluate. Written out in ordinary integral notation, this quantity is

$$J_{ii} = \int\int \frac{\phi_i^*(1)\phi_i(1)\phi_i^*(2)\phi_i(2)}{r_{12}}\, dv_1\, dv_2 = \langle ii|g|ii\rangle \tag{A3-6}$$

Note that Eq. (A3-6) is actually a six-dimensional integral; each of the integrals indicated involves integration over three components of the volume element of a single electron. One of the simplest methods for evaluating this integral is based on Lee's modification of an approach originally discussed by Margenau and Murphy.[1]

The volume element associated with the two-electron repulsion integral in terms of spherical polar coordinates is

$$r_1^2 \sin\theta_1\, d\theta_1\, d\varphi_1\, dr_1\, r_2^2 \sin\theta_2\, d\theta_2\, d\varphi_2\, dr_2 \tag{A3-7}$$

If the coordinates of the second particle are expressed relative to those of the first particle in terms of the interelectronic distance r_{12} and the angles χ and ω

---

[1] S. Lee, *J. Chem. Ed.*, **60**:935 (1983); also H. Margenau and G. M. Murphy, *The Mathematics of Physics and Chemistry*, Van Nostrand, Princeton, N.J., 1956, pp. 382–383.

as illustrated in Fig. A3-1, the volume element is transformed to

$$r_1^2 \sin\theta_1\, d\theta_1\, d\varphi_1\, dr_1\, r_{12}^2 \sin\omega\, d\omega\, d\chi\, dr_{12} \tag{A3-8}$$

The volume element may be transformed further. Using the law of cosines we obtain

$$r_2^2 = r_1^2 + r_{12}^2 - 2r_1 r_{12}\cos\omega \tag{A3-9}$$

If r_1 and r_{12} are held fixed, then

$$r_2\, dr_2 = r_1 r_{12} \sin\omega\, d\omega \tag{A3-10}$$

Substituting Eq. (A3-10) into Eq. (A3-8) for $\sin\omega\, d\omega$, the transformed volume element becomes

$$r_1 \sin\theta_1\, d\theta_1\, dr_1\, d\varphi_1\, r_2 r_{12}\, d\chi\, dr_2\, dr_{12} \tag{A3-11}$$

The integral in Eq. (A3-6) using the AO $\sqrt{(\eta^3/\pi)}e^{-\eta r}$ then becomes

$$J_{1s,\,1s} = \frac{\eta^6}{\pi^2}\int_0^{2\pi}\int_0^{2\pi}\int_0^{\pi}\int_{|r_1-r_{12}|}^{r_1+r_{12}}\int_0^{\infty}\int_0^{\infty} e^{-2\eta(r_1+r_2)} r_1 r_2\, dr_1\, dr_2\, dr_{12} \sin\theta_1\, d\theta_1\, d\varphi_1\, d\chi \tag{A3-12}$$

where we note that $1/r_{12}$ has been eliminated. Integrating over the three angles θ_1, φ_1, and χ, we get

$$J_{1s,\,1s} = 8\eta^6 \int_0^{\infty} dr_1 \int_0^{\infty} dr_{12} \int_{|r_1-r_{12}|}^{r_1+r_{12}} e^{-2\eta(r_1+r_2)} r_1 r_2\, dr_2 \tag{A3-13}$$

Because of the presence of the lower limit $|r_1 - r_{12}|$, the above integration must be done in two parts; i.e., the integral $J_{1s,\,1s}$ must be written as the sum of two integrals, I_1 and I_2, which are given by

$$I_1 = 8\eta^6 \int_0^{\infty} dr_{12} \int_{r_{12}}^{\infty} dr_1 \int_{r_1-r_{12}}^{r_1+r_{12}} e^{-2\eta(r_1+r_2)} r_1 r_2\, dr_2 \tag{A3-14a}$$

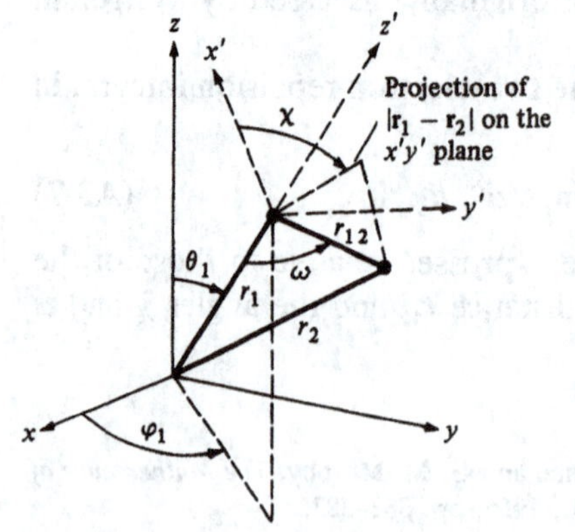

FIGURE A3-1
Coordinate system for one electron relative to another electron when both are interacting with a single nucleus. Electron 1 is described by the coordinates r_1, θ_1, and φ_1 relative to the x, y, and z axes. Electron 2 is described by the coordinates r_{12}, χ, and ω relative to the x', y', and z' axes. The z' axis is defined by the points (0, 0, 0) and (r_1, θ_1, φ_1) in the xyz system.

where $r_1 > r_{12}$, and

$$I_2 = 8\eta^6 \int_0^\infty dr_1 \int_{r_1}^\infty dr_{12} \int_{r_{12}-r_1}^{r_1+r_{12}} e^{-2\eta(r_1+r_2)} r_1 r_2 \, dr_2 \tag{A3-14b}$$

for the case in which $r_1 < r_{12}$. The first integral represents the case when the two electrons are close together, and the second represents the case when they are far apart. The integrations are straightforward and produce

$$I_1 = \frac{47}{216}\eta \qquad I_2 = \frac{11}{27}\eta \tag{A3-15}$$

Note that the first integral is smaller (indicating less repulsion) than the second, contrary to what one might at first expect. This is because the electron repulsion has a very long range. Adding these last two results produces

$$J_{1s,\,1s} = \frac{5}{8}\eta \tag{A3-16}$$

There are a number of alternative ways of evaluating electron-repulsion integrals. Some of these are more easily extended to multicenter integrals, for example, the *Fourier-transform method*. This method is illustrated for the case of the $J_{1s,\,1s}$ integral by Ornellas.[2] Other methods include the following:

1. Expansion of $1/r_{12}$ in terms of spherical harmonics; this makes use of the orthogonality of associated Legendre polynomials[3]
2. Use of physical arguments to develop an analogy with the electrostatic interaction energy of two spherically symmetrical charge distributions[4]

In general, the evaluation of electron-repulsion integrals—particularly over multicentered charge distributions—represents one of the most time-consuming aspects of computational quantum chemistry. Nevertheless, the efforts of a great many individuals have resulted in the development of efficient algorithms and techniques which have led to the availability of *ab initio* programs of many kinds. For example, representation of STOs as linear combinations of gaussian functions has greatly simplified many numerical evaluations.

---

[2] F. R. Ornellas, *J. Chem. Ed.*, **62**:378 (1985).

[3] H. Eyring, J. Walter, and G. E. Kimball, *Quantum Chemistry*, Wiley, New York, 1944, p. 103.

[4] L. Pauling and E. B. Wilson, Jr., *Introduction to Quantum Mechanics*, McGraw-Hill, New York, 1935, secs. 8-5 and 8-7; W. Kauzmann, *Quantum Chemistry*, Academic, New York, 1957, pp. 285–287.

Explicit forms for many of the molecular integrals based on STOs have been determined and compiled by Kotani et al.,[5] Matsen et al.,[6] and Preuss.[7]

EXERCISE

A3-1. Evaluate the kinetic-energy integral, nucleus-electron attraction integral, and electron-repulsion integral if $1s$ is expressed in the scaled gaussian form

$$1s = \left(\frac{2\eta}{\pi}\right)^{3/4} e^{-\eta r^2}$$

The final expression for the electron-repulsion integral is $2\sqrt{\eta/\pi}$. Formulas for the definite integrals involved can be found in standard tables of integrals, e.g., those found in *Handbook of Chemistry and Physics*, Chemical Rubber Publishing Co., Cleveland, Ohio.

---

[5] M. Kotani, A. Amemiya, E. Ishiguro, and T. Kimura, *Tables of Molecular Integrals*, Maruzen, Tokyo, 1963.

[6] J. Miller, J. M. Gerhausen, and F. A. Matsen, *Quantum Chemistry Integrals and Tables*, University of Texas Press, Austin, 1959.

[7] H. Preuss, *Integraltafeln zur Quantenchemie*, 4 vols., Springer-Verlag, Berlin, 1956.

APPENDIX 4

VECTOR AND OPERATOR ALGEBRA

A4-1 VECTOR ALGEBRA

Certain physical quantities such as energy, mass, speed, and distance are associated with numerical magnitudes having no relationship to direction. Other physical quantities such as linear and angular momentum, force, velocity, dipole moment, and displacement are inherently associated not only with a numerical magnitude but also with a specific direction (the direction is relative to an arbitrary fixed frame of reference). The former quantities are called *scalar* quantities, and the latter *vector* quantities. It is the algebra of the vector quantities we will consider here.

A *vector* is a mathematical construct defined as a *directed line segment*, whereas a *scalar* may be viewed as a *single point* on the real-number axis. A *vector quantity* will be defined as any physical quantity which can be mathematically represented as a vector. For the moment we will limit our discussion to the vectors of ordinary three-dimensional space. Extension of the treatment to n-dimensional spaces (where n may be infinite) will be made when the need arises.

A pair of vectors $\mathbf{V}_1$ and $\mathbf{V}_2$ can be multiplied in two distinct ways. One of these ways is called *scalar multiplication* and is expressed as follows:

$$\mathbf{V}_1 \cdot \mathbf{V}_2 = V_1 V_2 \cos\theta \qquad \text{(A4-1)}$$

where both vectors have the same origin and are separated by an angle θ. The product is called by several alternative names: *scalar product*, *dot product*, or *inner product*. The quantities V_1 and V_2 are the absolute values, or magnitudes, of the vectors $\mathbf{V}_1$ and $\mathbf{V}_2$, respectively, and are scalar quantities. The absolute values of a vector is called its *length* or *modulus* and is defined by

$$|\mathbf{V}| = V = \sqrt{\mathbf{V}\cdot\mathbf{V}} \tag{A4-2}$$

Since the scalar product is always a real number, Eq. (A4-1) may be viewed as a formula for establishing a correspondence between any pair of vectors and the set of real numbers. When the scalar product of two vectors is zero, i.e., when $\theta = 90°$, the two vectors are said to be *perpendicular* or, more commonly, *orthogonal*. Except for algebraic sign, the scalar product [Eq. (A4-1)] represents the modulus of the projection of one vector onto the other vector, which is then multiplied by the modulus of this other vector. Thus, when the two vectors are orthogonal, the projection of one onto the other is zero and neither vector has a component along the direction of the other.

Now consider three vectors $\mathbf{V}_1$, $\mathbf{V}_2$, and $\mathbf{V}_3$ which have different directions and are not coplanar, and no one of which can be transformed into one of the others by multiplication by a scalar. Note that this implies that two vectors differing in direction by integral multiples of π are to be excluded from the above. If we let $\mathbf{V}$ represent any arbitrary vector in three-dimensional space, it is always possible to express it as a linear combination of the three vectors $\mathbf{V}_1$, $\mathbf{V}_2$, and $\mathbf{V}_3$. This may be written

$$\mathbf{V} = c_1\mathbf{V}_1 + c_2\mathbf{V}_2 + c\mathbf{V}_3 \tag{A4-3}$$

where c_1, c_2, and c_3 are real numbers. The term $c_1\mathbf{V}_1$ is then seen to be a vector collinear with $\mathbf{V}_1$ and is the component of $\mathbf{V}$ in the direction of the vector $\mathbf{V}_1$. Vectors behaving as $\mathbf{V}_1$, $\mathbf{V}_2$, and $\mathbf{V}_3$ are said to be *linearly independent* and to form a *basis* for the representation of vectors in three-dimensional space. One of the most important characteristics of a set of linearly independent vectors is that if $\mathbf{V}$ is the zero vector, the only linear combination of basis vectors leading to $\mathbf{V} = \mathbf{0}$ is the one for which all the coefficients in the expansion (A4-3) are zero.

A commonly used set of linearly independent vectors in three-dimensional space is defined by the following relationships:

$$\begin{gathered} \mathbf{i}\cdot\mathbf{i} = \mathbf{j}\cdot\mathbf{j} = \mathbf{k}\cdot\mathbf{k} = 1 \\ \mathbf{i}\cdot\mathbf{j} = \mathbf{i}\cdot\mathbf{k} = \mathbf{k}\cdot\mathbf{j} = \mathbf{j}\cdot\mathbf{i} = \mathbf{k}\cdot\mathbf{i} = \mathbf{j}\cdot\mathbf{k} = 0 \end{gathered} \tag{A4-4}$$

Note that each of these vectors, **i**, **j**, and **k**, has a modulus of unity, and each is orthogonal to the others. This set of vectors constitutes a special case of linearly independent vectors called *orthonormal* vectors. In general, any set of vectors with unit modulus is said to be *normalized to unity* (or simply *normalized*). If the vectors are also orthogonal, the set is called *orthonormal*.

Since the vectors **i**, **j**, and **k** are mutually orthogonal, they may be viewed as lying along the x, y, and z axes, respectively, of three-dimensional space. Any arbitrary vector **V** [such as that in Eq. (A4-3)] may be expressed in terms of these orthonormal vectors as

$$\mathbf{V} = c_1\mathbf{i} + c_2\mathbf{j} + c_3\mathbf{k} \tag{A4-5}$$

Note: The coefficients $\{c_i\}$ will generally be different in the two cases illustrated by Eqs. (A4-3) and (A4-5). One very important practical advantage of using orthonormal vectors as a basis is that expressions for the scalar product of two vectors expressed in such a basis contain no cross products.

The second way of multiplying vectors is through the *vector product* (also called *cross product* or *outer product*), defined by

$$\mathbf{V}_1 \times \mathbf{V}_2 = V_1 V_2 \mathbf{C} \sin\theta \tag{A4-6}$$

where **C** is a unit vector perpendicular to the plane formed by $\mathbf{V}_1$ and $\mathbf{V}_2$. The direction of the vector **C** is given by the following convention: If the fingers of the right hand are curved so as to describe an arc from $\mathbf{V}_1$ to $\mathbf{V}_2$, the thumb points in the direction of **C**. This convention defines what is known as a *right-handed vector space*. It can be readily verified that the result of vector multiplication is to produce a new vector perpendicular to the plane of the multiplied vectors and having a modulus equal in value to the area of a parallelogram of sides V_1 and V_2 (with angle θ between these two sides).

The cross product of two vectors is conveniently represented in terms of an orthonormal basis by the determinantal form

$$\mathbf{V}_1 \times \mathbf{V}_2 = \begin{vmatrix} \mathbf{i} & \mathbf{j} & \mathbf{k} \\ V_{x1} & V_{y1} & V_{z1} \\ V_{x2} & V_{y2} & V_{z2} \end{vmatrix} \tag{A4-7}$$

where the scalar components of $\mathbf{V}_1$ in the x, y, and z directions are indicated by V_{x1}, V_{y1}, and V_{z1}, respectively, with similar designations for $\mathbf{V}_2$.

Since $\mathbf{V}_1 \times \mathbf{V}_2$ is itself a vector, it can be expressed in terms of the basis vectors **i**, **j**, and **k** as follows:

$$\mathbf{V}_1 \times \mathbf{V}_2 = \mathbf{U} = U_x\mathbf{i} + U_y\mathbf{j} + U_z\mathbf{k} \tag{A4-8}$$

The quantities U_x, U_y and U_z are the moduli of the x, y, and z components, respectively, of the product vector **U**. The numerical values of these moduli are given in terms of the cofactors (see Sec. A5-3) of **i**, **j**, and **k** when **U** is expressed in the determinantal form (A4-7). One finds

$$\begin{aligned} U_x &= V_{y1}V_{z2} - V_{y2}V_{z1} \\ U_y &= V_{x2}V_{z1} - V_{x1}V_{x2} \\ U_z &= V_{x1}V_{y2} - V_{x2}V_{y1} \end{aligned} \tag{A4-9}$$

where V_{y1} is the y component of vector $\mathbf{V}_1$, with similar relationships existing for the other terms.

EXERCISES

A4-1. A force sufficient to move a mass of 50 kg at an angle of 30° to the horizontal is applied to move a wagon a horizontal distance of 5 m. Interpret the work done as the scalar product of the force and displacement. What is the horizontal component of the force?

A4-2. Consider the two vectors $\mathbf{V}_1 = c_{a1}\mathbf{V}_a + c_{b1}\mathbf{V}_b + c_{c1}\mathbf{V}_c$ and $\mathbf{V}_2 = c_{a2}\mathbf{V}_a + c_{b2}\mathbf{V}_b + c_{c2}\mathbf{V}_c$, where $\mathbf{V}_a$, $\mathbf{V}_b$, and $\mathbf{V}_c$ form a basis for three-dimensional space. Evaluate the scalar product $\mathbf{V}_1 \cdot \mathbf{V}_2$ under each of the following cases:
(*a*) The basis is not orthonormal.
(*b*) The basis is orthonormal.

A4-3. Use the vectors given in the previous exercise to demonstrate that scalar multiplication is commutative, i.e., that the order in which the vectors are multiplied is irrelevant.

A4-4. Prove that the modulus of the vector multiplication $\mathbf{V}_1 \times \mathbf{V}_2$ is equal to the area of a parallelogram with sides V_1 and V_2 which form the angle θ.

A4-5. Prove that vector multiplication is not commutative (see Exercise A4-3). What happens to the vector product when the order of multiplication is reversed?

A4-6. Prove that the vector product of any vector with itself is the zero vector.

A4-7. Carry out all possible vector multiplications between pairs of the orthonormal vectors **i**, **j**, and **k**. What is the pattern illustrated by these multiplications?

A4-8. The vector definition of angular momentum of a single particle is $\mathbf{L} = \mathbf{r} \times \mathbf{p}$, where $\mathbf{r} = x\mathbf{i} + y\mathbf{j} + z\mathbf{k}$ and $\mathbf{p} = p_x\mathbf{i} + p_y\mathbf{j} + p_z\mathbf{k}$ ($p_x = m\,dx/dt$, etc.). Find the components of **L** in the equation $\mathbf{L} = L_x\mathbf{i} + L_y\mathbf{j} + L_z\mathbf{k}$.

A4-2 OPERATOR ALGEBRA

An *operator* may be defined as a shorthand notation for a set of well-defined mathematical operations to be carried out on a function (called the *operand*). For example, if $y = f(x)$, the symbol dy/dx may be interpreted in terms of the operator d/dx and the operand y. The operator d/dx is a symbol for the mathematical operation which will find the rate of change of the dependent variable y with respect to the independent variable x; that is, it tells us to take the first derivative of y with respect to x. Such operators are called *differential operators*. Other simple examples of operators are the square root operator $\sqrt{\ }$, which directs one to find a number such that its square is equal to the operand, and the operator $x+$, which tells one to add x to an operand. The simple vector product $\mathbf{i} \times \mathbf{V} = CV \sin\theta$ may be interpreted in terms of a vector operator $\mathbf{i}\times$ and a vector operand **V**. The operator $\mathbf{i}\times$ rotates the operand by 90° about the x axis and alters its magnitude by the factor $\sin\theta$. Note that the tip of the operand vector describes an arc parallel to the yz plane (see Exercise A4-7).

Several conventions are used in denoting quantities which are to be regarded as operators: script letters, circumflexes placed above the operator symbol, subscripts such as "op," etc. We will employ the second of these in this text and write "$\hat{A}$," "$\hat{B}$," etc., whenever we wish to emphasize that the symbols A, B, etc., are to represent general operators. However, in cases where the operator nature is obvious (as in d/dx) the circumflex will not be used. Also, certain cases arise in which the operator designation is irrelevant or not particularly useful, and we will not use the circumflex; e.g., the symbol x may be interpreted in expressions such as $xf(x)$ as an operator whose function is to multiply the operand $f(x)$ by x, but we will generally not write $\hat{x}$ in such cases unless for some reason the operator nature of x needs to be emphasized.

Consider a pair of arbitrary operators $\hat{A}$ and $\hat{B}$ which can operate on an arbitrary operand u. The operator expression $\hat{A}\hat{B}u$ is interpreted as follows:

$$\hat{A}\hat{B}u = \hat{A}(\hat{B}u) \tag{A4-10}$$

This expression is interpreted to mean "first, operate on the operand with the operator $\hat{B}$; then, operate on the result of this operation with the operator $\hat{A}$." In other words, operators are assumed (by convention) to operate successively, beginning with the operator immediately to the left of the operand. Successive operations of the same operator n times are written

$$\hat{A}\hat{A}\hat{A}\cdots\hat{A}u = \hat{A}^n u \tag{A4-11}$$

For example,

$$\left(\frac{d}{dx}\right)^2 y = \frac{d}{dx}\left(\frac{dy}{dx}\right) = \frac{d^2y}{dx^2} \tag{A4-12}$$

The *inverse* of an operator $\hat{A}$ is written $\hat{A}^{-1}$ and satisfies

$$\hat{A}\hat{A}^{-1}u = \hat{A}^{-1}\hat{A}u = u \tag{A4-13}$$

that is, $\hat{A}\hat{A}^{-1}$ is the *unit operator* (an operator which says "multiply the operand by unity"). It should be noted that some operators have no inverses. The multiplicative operator $\hat{0}$ (the *null operator*, which means "multiply the operand by zero") is a simple example of such an operator.

Two operators $\hat{A}$ and $\hat{B}$ are said to *commute* if the order in which they are applied to a given operand is immaterial, i.e., if the following relationship holds for any arbitrary operand

$$\hat{A}\hat{B}u = \hat{B}\hat{A}u \tag{A4-14}$$

It is convenient to introduce the *commutator* of two operators $\hat{A}$ and $\hat{B}$ by the relationship

$$[\hat{A}, \hat{B}] = \hat{A}\hat{B} - \hat{B}\hat{A} \tag{A4-15}$$

The commutator of the operators $\hat{A}$ and $\hat{B}$ is a single operator whose effect on an arbitrary operand is the same as

$$[\hat{A}, \hat{B}]u = (\hat{A}\hat{B} - \hat{B}\hat{A})u = \hat{A}\hat{B}u - \hat{B}\hat{A}u \tag{A4-16}$$

If the two operators commute, their commutator is the null operator, i.e., the operator which annihilates any arbitrary operand.

As an example of noncommuting operators, let us consider x and d/dx. The operation of $x(d/dx)$ on a arbitrary operand $u(x)$ is

$$x\left(\frac{d}{dx}\right)u = x\left(\frac{du}{dx}\right) \tag{A4-17}$$

Reversing the order of operation produces

$$\left(\frac{d}{dx}\right)xu = x\left(\frac{du}{dx}\right) + u\left(\frac{du}{dx}\right) = \left[x\left(\frac{d}{dx}\right) + 1\right]u \tag{A4-18}$$

The commutator of x and d/dx then is

$$\left[x, \frac{d}{dx}\right] = x\left(\frac{d}{dx}\right) - \left(\frac{d}{dx}\right)x = -1 \tag{A4-19}$$

The reader should verify that if the order of operations in the commutator is reversed, the commutator changes sign, i.e.,

$$\left[\frac{d}{dx}, x\right] = 1 \tag{A4-20}$$

An operator $\hat{A}$ is said to be *linear* if it satisfies

$$\hat{A}[f(x) + g(x)] = \hat{A}f(x) + \hat{A}g(x) \qquad \hat{A}[Cf(x)] = C\hat{A}f(x) \tag{A4-21}$$

where $f(x)$ and $g(x)$ are arbitrary functions and C is a scalar constant. The operators of quantum theory are linear and, in general, noncommutative. Many problems of quantum mechanics can be solved, at least in part, merely by establishing which relevant operators commute and which do not. A simple example of a nonlinear operator is $\sqrt{\ }$.

A vector operator of considerable importance in mathematical physics is the operator $\boldsymbol{\nabla}$ (called the *del*, or *nabla*, operator), whose cartesian coordinate representation is

$$\boldsymbol{\nabla} = \mathbf{i}\frac{\partial}{\partial x} + \mathbf{j}\frac{\partial}{\partial y} + \mathbf{k}\frac{\partial}{\partial z} \tag{A4-22}$$

When $\boldsymbol{\nabla}$ operates on a scalar quantity ϕ, the result is a vector called the *gradient of* ϕ. The gradient of ϕ is written in the alternative forms

$$\boldsymbol{\nabla}\phi = \text{grad}\ \phi \tag{A4-23}$$

As an example, if V represents a scalar potential, then one may associate with V a force $\mathbf{F}$ defined by

$$\mathbf{F} = -\boldsymbol{\nabla}V = -\text{grad}\ V \tag{A4-24}$$

Forces derived from potentials such as the above are called *conservative*. The vector $\boldsymbol{\nabla}V$ is perpendicular to the surface of constant V and is directed toward the steepest increase of V; its magnitude is equal to the slope of V in that direction. Thus the operator $\boldsymbol{\nabla}$ operating on V may be interpreted to mean "find the direction in which the force is the greatest and obtain its magnitude in that direction."

The scalar product of ∇ with itself leads to a scalar operator called the *laplacian operator*. The cartesian coordinate representation of this operator (read "del squared") is

$$\nabla \cdot \nabla = \nabla^2 = \frac{\partial^2}{\partial x^2} + \frac{\partial^2}{\partial y^2} + \frac{\partial^2}{\partial z^2} \tag{A4-25}$$

This operator appears in many differential equations of physics; in this text we encounter it in the hamiltonian operator of the Schrödinger equation.

The vector operator ∇ also operates on vector quantities by either scalar or vector multiplication, but these uses will not be encountered in this text.

EXERCISE

A4-9. Find the commutators of each of the following pairs of operators:
(*a*) d^2/dx^2, $x\,d/dx$ (*d*) d^2/dx^2, x^2
(*b*) d/dx, d^2/dx^2 (*e*) x^n, d^m/dx^m
(*c*) x, d/dy

APPENDIX 5

ELEMENTS OF MATRIX ALGEBRA

The manipulation of quantum mechanical equations and the proofs and general relationships of quantum theory frequently require some rather involved algebraic and geometric operations. Fortunately, most of these manipulations can be put into a very elegant and compact form by use of mathematical constructs called *matrices*. Also, it is the matrix form of quantum mechanics which lends itself most readily to computational procedures on modern high-speed electronic computers. This appendix concerns itself with those basic matrix rules and operations of greatest relevance to the topics of this text; once the basic definitions are committed to memory and some manipulative skill is acquired, it is the rare student who does not sense at least a slight elation in having acquired a simple but powerful mathematical tool. Although the subject matter included in this appendix is far from comprehensive, it should provide an adequate basis for an understanding of the formalism found in the current literature of quantum chemistry.

A5-1 ELEMENTARY PROPERTIES OF MATRICES

A matrix may be defined as an ordered array of elements subject to certain rules of operation. These elements, which may be real or complex, are arranged in rows and columns. An example of an arbitrary matrix **A** is the following:

$$\mathbf{A} = \begin{bmatrix} a_{11} & a_{12} & \cdots & a_{1n} \\ a_{21} & a_{22} & \cdots & a_{2n} \\ \cdots & \cdots & \cdots & \cdots \\ a_{m1} & a_{m2} & \cdots & a_{mn} \end{bmatrix} \tag{A5-1}$$

Matrix **A** is said to be an $m \times n$ matrix, since it has m rows and n columns. The general element of matrix **A** is written a_{ij} (or occasionally $\mathbf{A}_{ij}$), where the index i identifies the row and the index j identifies the column. When m and n are equal, we say **A** is a *square* matrix; when $n = 1$ and $m > 1$, we say **A** is a *column* matrix; when $n > 1$ and $m = 1$, we say **A** is a *row* matrix. There are also *rectangular* matrices (when $m \neq n \neq 1$), but these will not be encountered in this text.

Two matrices **A** and **B** are said to be equal if and only if all their corresponding elements are equal, that is,

$$\mathbf{A} = \mathbf{B} \qquad \text{if } a_{ij} = b_{ij} \text{ for all } i, j \tag{A5-2}$$

If every element a_{ij} of the matrix **A** is zero, then **A** is said to be a *null* matrix, represented by the symbol **0**, that is,

$$\mathbf{A} = \mathbf{0} \qquad \text{if } a_{ij} = 0 \text{ for all } i, j \tag{A5-3}$$

Another important matrix is the *unit* matrix, written **1** and defined by

$$\mathbf{A} = \mathbf{1} \qquad \text{if } a_{ij} = \delta_{ij} \tag{A5-4}$$

where δ_{ij} is the Kronecker delta. All the unit matrices we will use will be assumed to be square; thus all diagonal elements will be unity, and all off-diagonal elements will be zero. We will also encounter row and column matrices having one element of unity and all other elements zeros, but these will be referred to as *vectors* rather than as *unit matrices*.

If the matrix elements of **A** are given by

$$(\mathbf{A}_{ij}) = a_{ij}\delta_{ij} \tag{A5-5}$$

then **A** is called a *diagonal* matrix, i.e., all off-diagonal matrix elements are zero. A unit matrix is just a special case of a diagonal matrix in which all the diagonal elements are unity.

The sum of two matrices **A** and **B** (both $m \times n$) is a new matrix **C** whose elements c_{ij} are the sums of the corresponding elements of **A** and **B**, namely,

$$(\mathbf{A} + \mathbf{B})_{ij} = a_{ij} + b_{ij} = c_{ij} \tag{A5-6}$$

Perhaps the most distinctive way in which matrices differ from ordinary algebraic quantities is the way in which they are multiplied. For purposes of consistency we will interpret the product **AB** to mean "multiply the matrix **B** by the matrix **A**"; that is, matrix multiplication will always be from the left. This insistence on such a consistent interpretation is necessary, since, as we will see later, matrix multiplication is not generally commutative (see Exercise A5-1). Also, two matrices **A** and **B** can be multiplied only if the left matrix **A** has the same number of *columns* as the right matrix **B** has rows. This means that if **A** is $m \times q$, then **B** must be $q \times n$ and the product **C** will be an $m \times n$ matrix. In general, the element c_{ij} of **C** is obtained by taking the ith row of **A**, placing it alongside the jth column of **B**, multiplying each pair of adjoining elements, and adding all of these together. This may be written in algebraic form as

$$c_{ij} = \sum_k a_{ik} b_{kj} \tag{A5-7}$$

where the summation index k adds the q products arising from the "row by column" multiplications. Thus if **A** is 1×3 and **B** is 3×2, the matrix **C** is 1×2 and each of its elements is a sum of three binary products. From Eq. (A5-7) it is readily deduced that $\mathbf{AB} \neq \mathbf{BA}$; i.e., matrix multiplication is not generally commutative.

Multiplication of a matrix **A** by a constant λ implies that each element of **A** is multiplied by λ; that is, the elements of $\lambda\mathbf{A}$ are λa_{ij}.

The expression $\mathbf{A}^n\mathbf{B}$, where n is an integer, is interpreted as n successive multiplications of **B** by **A**. For example, $\mathbf{A}^2\mathbf{B} = \mathbf{A}(\mathbf{AB})$. Later we will consider the case in which n is negative. Section 14-8 also makes use of a situation in which $n = \pm 1/2$.

Whenever the matrix **A** obeys the relation

$$\mathbf{A}^2 = \mathbf{AA} = \mathbf{0} \tag{A5-8}$$

we say that **A** is a *nilpotent* matrix; that is, **A** annihilates itself. If, on the other hand,

$$\mathbf{A}^2 = \mathbf{A} \tag{A5-9}$$

we say **A** is an *idempotent* matrix. Note that idempotency of a matrix is analogous to the algebraic relationship $x^2 - x = 0$.

Although in ordinary algebra, the expression $ab = 0$ implies that either a, b, or both are zero, the same is not true for the matrix equation $\mathbf{AB} = \mathbf{0}$; that is, it is not necessarily true that either **A** or **B** is a null matrix. An example of this type is given in Exercise A5-3.

EXERCISES

A5-1. Verify the following relationships:

(*a*) $\begin{bmatrix} 1 & 2 \\ 0 & 3 \end{bmatrix}\begin{bmatrix} 3 & 4 \\ 9 & 1 \end{bmatrix} = \begin{bmatrix} 21 & 6 \\ 27 & 3 \end{bmatrix} \neq \begin{bmatrix} 3 & 4 \\ 9 & 1 \end{bmatrix}\begin{bmatrix} 1 & 2 \\ 0 & 3 \end{bmatrix} = \begin{bmatrix} 3 & 18 \\ 9 & 21 \end{bmatrix}$

(*b*) $\begin{bmatrix} 1 & 0 & 3 \end{bmatrix}\begin{bmatrix} 2 \\ 1 \\ 4 \end{bmatrix} = [14]$

(*c*) $\begin{bmatrix} 2 \\ 1 \\ 4 \end{bmatrix}\begin{bmatrix} 1 & 0 & 3 \end{bmatrix} = \begin{bmatrix} 2 & 0 & 6 \\ 1 & 0 & 3 \\ 4 & 0 & 12 \end{bmatrix}$

A5-2. Using Eq. (A5-7), show that if $\mathbf{ABC} = \mathbf{D}$, then

$$d_{ij} = \sum_k \sum_l a_{ik} b_{kl} c_{lj}$$

What are the restrictions on the rows and columns of **A**, **B**, and **C**? How many rows and columns, in general, does **D** have?

A5-3. Verify the following relationships:

$$(a)\ \begin{bmatrix} 0 & 1 \\ 0 & 0 \end{bmatrix}^2 = \mathbf{0} \qquad (b)\ \begin{bmatrix} 1 & 0 \\ 0 & 1 \end{bmatrix}^2 = \mathbf{1} \qquad (c)\ \begin{bmatrix} 1 & -2 & 4 \\ -2 & 3 & -5 \end{bmatrix}\begin{bmatrix} 2 & 4 \\ 3 & 6 \\ 1 & 2 \end{bmatrix} = \mathbf{0}$$

A5-4. Show that the family of real matrices **C** given by

$$\mathbf{C} = \begin{bmatrix} a & b \\ -b & a \end{bmatrix} = a\mathbf{1} + b\mathbf{i}$$

where

$$\mathbf{i} = \begin{bmatrix} 0 & 1 \\ -1 & 0 \end{bmatrix}$$

and a and b are real numbers, is isomorphic to the family of complex numbers.

A5-2 MATRICES AS TRANSFORMATION OPERATORS

A set of numbers $x_1, x_2, \ldots, x_n$ (real or complex) may be regarded as the components of a vector **X** in an n-dimensional space. It is convenient to write such a vector **X** as a column matrix in its n components, namely,

$$\mathbf{X} = \begin{bmatrix} x_1 \\ x_2 \\ \vdots \\ x_n \end{bmatrix} \tag{A5-10}$$

which is called a *column vector*. As an example drawn from a familiar area, an arbitrary vector in ordinary euclidean three-dimensional space, $\mathbf{r} = \mathbf{i}x + \mathbf{j}y + \mathbf{k}z$, would be represented by the column vector

$$\mathbf{r} = \begin{bmatrix} x \\ y \\ z \end{bmatrix} \tag{A5-11}$$

Just as in three-dimensional space, the components of a vector **X** in n-dimensional space can be used to form new vectors. Thus consider a new n-dimensional vector **Y** given by

$$\mathbf{Y} = \begin{bmatrix} y_1 \\ y_2 \\ \vdots \\ y_n \end{bmatrix} \tag{A5-12}$$

whose components are related to those of **X** by the n equations

$$\begin{aligned} y_1 &= a_{11}x_1 + a_{12}x_2 + \cdots + a_{1n}x_n \\ y_2 &= a_{21}x_1 + a_{22}x_2 + \cdots + a_{2n}x_n \\ &\cdots\cdots\cdots\cdots\cdots\cdots \\ y_n &= a_{n1}x_1 + a_{n2}x_2 + \cdots + a_{nn}x_n \end{aligned} \tag{A5-13}$$

If we define the $n \times n$ matrix **A** (whose elements are real or complex) by

$$\mathbf{A} = \begin{bmatrix} a_{11} & a_{12} & \cdots & a_{1n} \\ a_{21} & a_{22} & \cdots & a_{2n} \\ \cdots & \cdots & \cdots & \cdots \\ a_{n1} & a_{n2} & \cdots & a_{nn} \end{bmatrix} \tag{A5-14}$$

then the n equations in (A5-13) may be represented by the *single* matrix equation

$$\mathbf{AX} = \mathbf{Y} \tag{A5-15}$$

We may regard **A** as an operator which transforms the vector **X** into the vector **Y**. Since the matrix **AX** is a linear function of the components of **X**, we say the transformation by **A** is *linear*. In general, a transformation such as Eq. (A5-15) is linear if the following two conditions hold:

$$\begin{aligned} \mathbf{A}(\mathbf{X}_1 + \mathbf{X}_2) &= \mathbf{AX}_1 + \mathbf{AX}_2 = \mathbf{Y}_1 + \mathbf{Y}_2 \\ \mathbf{A}(\lambda\mathbf{X}) &= \lambda\mathbf{AX} \end{aligned} \tag{A5-16}$$

where $\mathbf{X}_1$, $\mathbf{X}_2$, and **X** are arbitrary column vectors and λ is a constant. In a linear transformation such as in Eq. (A5-16) the vectors $\mathbf{X}_1$ and $\mathbf{X}_2$ define the same plane as $\mathbf{Y}_1$ and $\mathbf{Y}_2$.

A5-3 DETERMINANTS

Most students are probably familiar with the use of determinants in the solution of n linearly independent equations for n unknowns. Two other uses of determinants occur in quantum chemistry:

1. To represent products of one-particle functions which are antisymmetric with respect to the full spatial and spin coordinates of a pair of indistinguishable fermions. The antisymmetric property is the result of a basic property of determinants: a change in sign when any two rows or columns are interchanged.
2. To aid in the solution of *secular equations* of the general form

$$\begin{aligned} c_1(G_{11} - \lambda) + c_2(G_{12} - \lambda\Delta_{12}) + \cdots + c_n(G_{1n} - \lambda\Delta_{1n}) &= 0 \\ c_1(G_{21} - \lambda\Delta_{21}) + c_2(G_{22} - \lambda) + \cdots + c_n(G_{2n} - \lambda\Delta_{2n}) &= 0 \\ \cdots\cdots\cdots\cdots\cdots\cdots\cdots\cdots\cdots & \\ c_1(G_{n1} - \lambda\Delta_{n1}) + c_2(G_{n2} - \lambda\Delta_{n2}) + \cdots + c_n(G_{nn} - \lambda) &= 0 \end{aligned} \tag{A5-17}$$

where the matrix elements G_{mn} are generally of the following specific types:

a. $G_{mn} = H_{mn}$, where $\hat{H}$ is the hamiltonian operator. The secular equations arise when one employs a trial wavefunction ψ which is a linear combination of basis functions and then invokes the variation principle requirement that $\partial\psi/\partial c_i = 0$ for all i.

b. $G_{mn} = F_{mn}$, where $\hat{F}$ is the Hartree-Fock SCF operator. The secular equations arise as a result of employing Roothaan's method for matrix solution of the Hartree-Fock SCF equations; the basis functions are then the AOs whose linear combinations constitute the molecular orbitals of the system.

In either case, the nontrivial solutions to the secular equations require that the determinant of the coefficients vanish:

$$\det(\mathbf{G} - \lambda\mathbf{\Delta}) = 0 \tag{A5-18}$$

It should be noted that an alternative (trivial) solution is that all the $\{c_i\}$ are equal to zero. The n roots (λ) of the secular equations are the eigenvalues of the relevant operator ($\hat{H}$ or $\hat{F}$). Also, corresponding to each root λ is a set of coefficients $\mathbf{C}_i$ (called an *eigenvector*). Each such eigenvector is a column matrix of n elements, and the collection of all n eigenvectors may be represented by the $n \times n$ matrix

$$\mathbf{C} = [\mathbf{C}_1 \quad \mathbf{C}_2 \quad \cdots \quad \mathbf{C}_n] \tag{A5-19}$$

The matrix $\mathbf{C}$ is, in general, a unitary matrix [see Eq. (A5-52)] which diagonalizes the matrix $\mathbf{G}$, namely,

$$\begin{aligned} \mathbf{GC} &= \mathbf{\Delta CX} \\ \mathbf{C}^\dagger\mathbf{GC} &= \mathbf{X} \end{aligned} \tag{A5-20}$$

where $\mathbf{X}$ is a diagonal matrix of eigenvalues.

It is useful to consider a determinant, $\det \mathbf{A}$, which is associated with a square matrix $\mathbf{A}$ such as given in Eq. (A5-14). Thus

$$\det \mathbf{A} = \begin{vmatrix} a_{11} & a_{12} & \cdots & a_{1n} \\ a_{21} & a_{22} & \cdots & a_{2n} \\ \cdots & \cdots & \cdots & \cdots \\ a_{n1} & a_{n2} & \cdots & a_{nn} \end{vmatrix} \tag{A5-21}$$

In our applications the elements of the determinant are numbers or functions (real or complex), and the determinant itself will also be a number or function whose value or form is determined by certain combinations of its elements. Whenever a determinant such as $\det \mathbf{A} = 0$, we say the matrix $\mathbf{A}$ is *singular*; otherwise, $\mathbf{A}$ is called *nonsingular*. We will now concern ourselves with singular matrices in which the elements are numbers arising from evaluation of integrals.

The numerical evaluation of a determinant involves expanding the determinant as a sum of products, each such product containing n terms. We will examine the procedure using the specific example of the 3×3 determinant

$$\det \mathbf{A} = \begin{vmatrix} a_{11} & a_{12} & a_{13} \\ a_{21} & a_{22} & a_{23} \\ a_{31} & a_{32} & a_{33} \end{vmatrix} \tag{A5-22}$$

This determinant may be expanded as a linear combination of 2×2 determinants as follows:

$$\det \mathbf{A} = a_{11}\begin{vmatrix} a_{22} & a_{23} \\ a_{32} & a_{33} \end{vmatrix} - a_{12}\begin{vmatrix} a_{21} & a_{23} \\ a_{31} & a_{33} \end{vmatrix} + a_{13}\begin{vmatrix} a_{21} & a_{22} \\ a_{31} & a_{32} \end{vmatrix} \tag{A5-23}$$

Note that the coefficient of each 2×2 determinant is one of the top-row elements, a_{ij}, of the original 3×3 determinant and that the algebraic signs of the terms alternate in the order $+, -, +$. In general, the sign of each term containing a_{1j} is given by $(-1)^{1+j}$. Furthermore, each 2×2 determinant multiplying a_{1j} is the determinant remaining when row 1 and column j are removed from the original determinant. Next, each 2×2 determinant is expanded in the same way:

$$\begin{aligned} \begin{vmatrix} a_{22} & a_{23} \\ a_{32} & a_{33} \end{vmatrix} &= a_{22}a_{33} - a_{23}a_{32} \\ \begin{vmatrix} a_{21} & a_{23} \\ a_{31} & a_{33} \end{vmatrix} &= a_{21}a_{33} - a_{23}a_{31} \\ \begin{vmatrix} a_{21} & a_{22} \\ a_{31} & a_{32} \end{vmatrix} &= a_{21}a_{32} - a_{22}a_{31} \end{aligned} \tag{A5-24}$$

Combining Eqs. (A5-23) and (A5-24) leads to the final result:

$$\det \mathbf{A} = a_{11}a_{22}a_{33} - a_{11}a_{23}a_{32} - a_{12}a_{21}a_{33} + a_{12}a_{23}a_{31} + a_{13}a_{21}a_{32} - a_{13}a_{22}a_{31} \tag{A5-25}$$

The method is readily generalized to determinants in which $n > 3$.

The entire determinantal expansion operation can be expressed in a formal mathematical manner. The 2×2 determinants appearing in Eq. (A5-23) are called *minors* of the elements which multiply them. For example, the first determinant is the minor of a_{11}, the second is the minor of a_{12}, and the third is the minor of a_{13}. In general, each element a_{ij} of an $n \times n$ determinant has a minor which is the $(n-1) \times (n-1)$ determinant resulting from removing the ith row and jth column from the original determinant. Symbolically, the minor of an element a_{ij} is written A_{ij}. Thus one writes the expansion in the form

$$\det \mathbf{A} = a_{11}A_{11} - a_{12}A_{12} + a_{13}A_{13} \tag{A5-26}$$

An alternative manner of expressing the same thing employs a construct called a *cofactor*. The cofactor of an element a_{ij} is written A^{ij} and is defined in terms of the minor A_{ij} as follows:

$$A^{ij} = (-1)^{i+j}A_{ij} \tag{A5-27}$$

When this form is used, Eq. (A5-26) is replaced with

$$\det \mathbf{A} = a_{11}A^{11} + a_{12}A^{12} + a_{13}A^{13} \tag{A5-28}$$

In actuality any row (or column) may be used to begin the expansion, but the top row is perhaps the most convenient. Thus, a general expression for the expansion of a determinant in terms of minors or cofactors is as follows:

$$\det \mathbf{A} = \sum_{j=1}^{n} a_{ij}(-1)^{i+j}A_{ij} = \sum_{j=1}^{n} a_{ij}A^{ij} \tag{A5-29}$$

These equations assume expansion in terms of the ith row; expressions for expansions in terms of the jth column are readily derived. Note that in general the minors (and cofactors) are themselves determinants. Thus, if one begins an expansion of an $n \times n$ determinant, the A_{ij} (and A^{ij}) will be $(n-1) \times (n-1)$ determinants. Each of these must then be expanded separately to produce new minors and cofactors which are $(n-2) \times (n-2)$ determinants, and the process is repeated until the determinant is fully reduced. In general, $n-1$ steps will be needed to reduce an $n \times n$ determinant completely.

Minors and cofactors will be useful in the context of certain matrix manipulations to be discussed in later sections.

A useful relationship (presented without proof) is that if $\mathbf{A}$ and $\mathbf{B}$ are two square matrices, such that $\mathbf{AB} = \mathbf{C}$, then

$$\det \mathbf{A} \det \mathbf{B} = \det \mathbf{C} \tag{A5-30}$$

Another useful relationship is illustrated in the following example: Suppose we have a 5×5 determinant of the form

$$\det \mathbf{A} = \begin{vmatrix} a_{11} & a_{12} & 0 & 0 & 0 \\ a_{21} & a_{22} & 0 & 0 & 0 \\ 0 & 0 & a_{33} & a_{34} & a_{35} \\ 0 & 0 & a_{43} & a_{44} & 45 \\ 0 & 0 & a_{53} & a_{54} & a_{55} \end{vmatrix} \tag{A5-31}$$

Such determinants (with blocks of zero elements) arise in practice when the basis functions for the matrix representation span different irreducible representations; i.e., a matrix element such as $\langle \phi_i | \hat{H} | \phi_j \rangle$ will be zero if ϕ_i and ϕ_j span different irreducible representations. In this case the above 5×5 determinant becomes a product of a 2×2 determinant and a 3×3 determinant as shown below:

$$\det \mathbf{A} = \begin{vmatrix} a_{11} & a_{12} \\ a_{21} & a_{22} \end{vmatrix} \begin{vmatrix} a_{33} & a_{34} & a_{35} \\ a_{43} & a_{44} & a_{45} \\ a_{53} & a_{54} & a_{55} \end{vmatrix} \tag{A5-32}$$

If the original determinant has a singular matrix, each determinantal factor can be equated to zero and solved separately.

It should be apparent that an $n \times n$ determinant such as Eq. (A5-21) is equivalent to a polynomial of the general form

$$a_1\lambda^n + a_2\lambda^{n-1} + \cdots + a_n\lambda + a_{n+1} = 0 \tag{A5-33}$$

EXERCISES

A5-5. Show that the sum of the products of the elements a_{ij} of row i and the cofactors of another row A^{kj} ($k \neq i$) is zero or, in general,

$$\sum_{j=1}^{n} a_{ij}A^{kj} = \delta_{kj} \det \mathbf{A}$$

A5-6. Let **A** be a 4×4 matrix with general elements given as in Eq. (A5-14). Perform a total expansion of the determinant of this matrix to obtain a sum of $4! = 24$ terms, each term being a product of four different matrix elements.

A5-7. Obtain the three roots of the following determinant:

$$\begin{vmatrix} 2-\lambda & 0 & 1-2\lambda \\ 1-2\lambda & 0 & 4-\lambda \\ 0 & 1-\lambda & 0 \end{vmatrix} = 0$$

Note: If one exchanges columns 2 and 3, the above determinant can be put into a form analogous to Eq. (A5-31). Thus, instead of having to solve a cubic equation in λ, one has to solve one quadratic equation and one first-degree equation.

A5-4 SOME SPECIAL MATRICES AND THEIR PROPERTIES

The *transpose* of a matrix **A** (not necessarily square) is defined by

$$(\tilde{\mathbf{A}})_{ij} = a_{ji} \tag{A5-34}$$

i.e., the elements of the transpose matrix $\tilde{\mathbf{A}}$ are those of the matrix **A** but with row and column indices transposed. For example, if

$$\mathbf{A} = \begin{bmatrix} a & b \\ c & d \end{bmatrix} \quad \text{then} \quad \tilde{\mathbf{A}} = \begin{bmatrix} a & c \\ b & d \end{bmatrix} \tag{A5-35}$$

The transpose of a row matrix is a column matrix and vice versa. Also, if $\mathbf{AB} = \mathbf{C}$, then one can readily prove that

$$\tilde{\mathbf{C}} = \widetilde{\mathbf{AB}} = \tilde{\mathbf{B}}\tilde{\mathbf{A}} \tag{A5-36}$$

The proof follows readily by the use of Eq. (A5-7).

The *adjugate* of the square matrix **A** is defined by

$$(\mathbf{A}_J)_{ij} = A^{ji} \tag{A5-37}$$

that is, the ij element of the adjugate matrix is the cofactor of the ji element of **A**. For example, if

$$\mathbf{A} = \begin{bmatrix} a & b \\ c & d \end{bmatrix} \quad \text{then} \quad \mathbf{A}_J = \begin{bmatrix} d & -b \\ -c & a \end{bmatrix} \tag{A5-38}$$

An important property of the adjugate matrix is revealed if we multiply a matrix **A** by its adjugate. The ik element of this product is

$$(\mathbf{A}_J\mathbf{A})_{ik} = \sum_j (\mathbf{A}_J)_{ij}a_{jk} = \sum_j A^{ji}a_{jk} \tag{A5-39}$$

Using the result of Exercise A5-5 allows us to write

$$(\mathbf{A}_J\mathbf{A})_{ik} = \delta_{ik} \det \mathbf{A} \tag{A5-40}$$

Since δ_{ik} is a general element of the unit matrix **1**, we can write

$$\mathbf{A}_J\mathbf{A} = \mathbf{1} \det \mathbf{A} \tag{A5-41}$$

The importance of this last relationship is that it provides a useful expression for the *inverse* of a matrix (which must be square). In general, the inverse of a (square) matrix **A** is defined by the relationship

$$\mathbf{A}^{-1}\mathbf{A} = \mathbf{A}\mathbf{A}^{-1} \tag{A5-42}$$

Provided that **A** *is nonsingular*, we may multiply Eq. (A5-41) from the right by $\mathbf{A}^{-1}$ and rearrange to

$$\mathbf{A}^{-1} = \frac{\mathbf{A}_J}{\det \mathbf{A}} \tag{A5-43}$$

Using Eq. (A5-37) we can write the *ij* element of the inverse of **A** as

$$(\mathbf{A}^{-1})_{ij} = \frac{A^{ji}}{\det \mathbf{A}} \tag{A5-44}$$

Note that if **A** is singular, then $\det \mathbf{A} = 0$ and **A** has no inverse.

The *complex conjugate* of a matrix **A** is formed by replacing each element a_{ij} by its complex conjugate, that is,

$$(\mathbf{A}^*)_{ij} = a_{ij}^* \tag{A5-45}$$

If all the elements of **A** are real, then $\mathbf{A} = \mathbf{A}^*$ and is called a *real matrix*.

The *adjoint* of a matrix **A** is defined as its *complex-conjugate transpose* and is written $\mathbf{A}^\dagger$. The general element of $\mathbf{A}^\dagger$ is given by

$$(\mathbf{A}^\dagger)_{ij} = a_{ji}^* = (\tilde{\mathbf{A}}^*)_{ij} \tag{A5-46}$$

Matrixes which are *self-adjoint* are said to be *hermitian*. Thus if

$$\mathbf{A} = \tilde{\mathbf{A}}^* = \mathbf{A}^\dagger \tag{A5-47}$$

then we say **A** is hermitian. Note that all hermitian matrices must be square.

A square matrix is said to be *symmetric* if it equals its transpose; i.e., if **A** is symmetric, then

$$\mathbf{A} = \tilde{\mathbf{A}} \tag{A5-48}$$

Note that any real symmetric matrix is hermitian.

A square matrix is called *orthogonal* if its transpose is equal to its inverse, i.e., if

$$\tilde{\mathbf{A}} = \mathbf{A}^{-1} \tag{A5-49}$$

The name *orthogonal* comes from the fact that the rows (or columns) of such a matrix behave like orthonormal row (or column) vectors. Since

$$\tilde{\mathbf{A}}\mathbf{A} = \mathbf{A}\tilde{\mathbf{A}} = \mathbf{1} \tag{A5-50}$$

then

$$\sum_k a_{ki}a_{kj} = \delta_{ij} \tag{A5-51}$$

If the adjoint of a square matrix **A** with complex elements is equal to its inverse, then **A** is said to be *unitary*, that is,

$$\tilde{\mathbf{A}}^* = \mathbf{A}^\dagger = \mathbf{A}^{-1} \tag{A5-52}$$

and

$$\mathbf{A}^\dagger\mathbf{A} = \mathbf{A}\mathbf{A}^\dagger = \mathbf{1} \tag{A5-53}$$

Note that a real orthogonal matrix is also a unitary matrix. From Eq. (A5-53) we obtain the relationship [analogous to Eq. (A5-51)]

$$\delta_{ij} = \sum_k u_{ki}^* u_{kj} = \sum_k u_{ik} u_{jk}^* \tag{A5-54}$$

where u_{ij} denotes a general element of a unitary matrix **U**. The foregoing shows that the rows (or columns) of a unitary matrix behave like complex orthonormal vectors.

An important and useful quantity associated with a square matrix **A** is the sum of the diagonal elements, called the *trace* (written "Tr **A**"). Thus

$$\mathrm{Tr}\,\mathbf{A} = \sum_{i=1}^{n} a_{ii} \tag{A5-55}$$

EXERCISES

A5-8. Find the inverse of the matrix **A** used in the examples of transpose and adjugate matrices [Eqs. (A5-35) and (A5-38), respectively]. Check, using Eq. (A5-42).

A5-9. Verify the following inverse matrix relationship:

$$\begin{bmatrix} 2 & 1 \\ 3 & 2 \end{bmatrix}^{-1} = \begin{bmatrix} 2 & -1 \\ -3 & 2 \end{bmatrix}$$

and demonstrate that Eq. (A5-42) is valid.

A5-10. Does the following matrix have an inverse? Explain.

$$\begin{bmatrix} 1 & 2 \\ 2 & 4 \end{bmatrix}$$

A5-11. Find the inverse of the matrix

$$\begin{bmatrix} 1 & 0 & -1 \\ 2 & -3 & 2 \\ 1 & 1 & 4 \end{bmatrix}$$

A5-12. Prove that if $\mathbf{AB} = \mathbf{C}$, where $\mathbf{A}$ and $\mathbf{B}$ are nonsingular matrices, then $\mathbf{C}^{-1} = \mathbf{B}^{-1}\mathbf{A}^{-1}$. *Hint*: Multiply both sides on the left by $\mathbf{B}^{-1}\mathbf{A}^{-1}$ and then on the right by $\mathbf{C}^{-1}$.

A5-13. Prove that if $\mathbf{AB} = \mathbf{C}$, then $\mathbf{B}^{\dagger}\mathbf{A}^{\dagger} = \mathbf{C}^{\dagger}$. Note the similarity to adjoint operators (Sec. 2-6), for example, $(\hat{F}\hat{G})^{\dagger} = \hat{G}^{\dagger}\hat{F}^{\dagger}$.

A5-14. Show that the trace of a product of two matrices, $\mathbf{A}$ and $\mathbf{B}$, is given by

$$\text{Tr}\,(\mathbf{AB}) = \sum_{i,j} a_{ij} b_{ji}$$

A5-15. Show that the trace of a product of matrices is invariant to a cyclic permutation of the matrices, for example,

$$\text{Tr}\,(\mathbf{ABCD}) = \text{Tr}\,(\mathbf{BCDA}) = \text{Tr}\,(\mathbf{CDAB}) = \text{Tr}\,(\mathbf{DABC})$$

A5-16. Show that the following matrices are unitary. Which ones are also orthogonal?

$$\begin{bmatrix} \cos\theta & \sin\theta \\ \sin\theta & -\cos\theta \end{bmatrix} \qquad \frac{1}{\sqrt{2}}\begin{bmatrix} 1 & 1 \\ 1 & -1 \end{bmatrix} \qquad \frac{1}{\sqrt{2}}\begin{pmatrix} 1 & i \\ 1 & -i \end{pmatrix}$$

$$\begin{bmatrix} 1/\sqrt{3} & 1/\sqrt{2} & 1/\sqrt{6} \\ -1/\sqrt{3} & 0 & 2/\sqrt{6} \\ 1/\sqrt{3} & -1/\sqrt{2} & -1/\sqrt{6} \end{bmatrix}$$

A5-5 LINEAR VECTORS SPACES

Any space which is closed under the operations of addition and multiplication by a scalar is called a *linear vector space* (LVS), and its elements are called *vectors*. We have greatest interest in a particular type of LVS which is called a *complete vector space*. Complete vector spaces are very useful in quantum theory, since they may be used to represent arbitrary vectors in the space of interest. The normal modes of a vibrating string form a complete set, since any arbitrary vibration satisfying the same boundary conditions as the normal modes can be represented as a linear combination of the normal modes. In general, a set of functions $\{\phi_i(t)\}$ is said to be complete if an arbitrary function $f(t)$ satisfying the same restrictive conditions as the $\{\phi_i(t)\}$ can be expanded as

$$f(t) = \sum_{k=1}^{\infty} c_k \phi_k(t) \tag{A5-56}$$

where the $\{c_k\}$ are constant coefficients (real or complex). Multiplying both sides of Eq. (A5-56) on the left by $\phi_m^*(t)$ and integrating over the configuration space, we obtain

$$\langle \phi_m(t) | f(t) \rangle = \sum_{k=1}^{\infty} c_k \langle \phi_m(t) | \phi_k(t) \rangle \tag{A5-57}$$

If the $\{\phi_i(t)\}$ are orthonormal, then $\langle \phi_k | \phi_m \rangle = \delta_{km}$ and Eq. (A5-57) becomes

$$c_m = \langle \phi_m(t) | f(t) \rangle \tag{A5-58}$$

The quantity c_m is called a *Fourier coefficient*. The functions $\{\phi_i(t)\}$ forming a complete space are commonly said to form a *basis* or *coordinate system* for the representation of functions in that space. One also says that the functions $\{\phi_i(t)\}$ *span* the space. In terms of simple three-dimensional space, the familiar unit vectors **i**, **j**, and **k** span the space. Alternatively, one can say that any vector in three-dimensional space can be represented by specifying the coordinates x, y, and z (assuming the vector's origin is at 0, 0, 0).

Another definition of completeness is as follows: A set of functions $\{\phi_i\}$ with norm 1 (that is, $\langle\phi_i|\phi_i\rangle = 1$) is complete if the only function (call this g) in the space which satisfies $\langle g|\phi_i\rangle = 0$ for all i is the zero function; in other words,

$$\langle g|\phi_i\rangle = 0 \qquad \text{for all } i \tag{A5-59}$$

implies $g = 0$.

To facilitate generalization to n-dimensional vector spaces, it is convenient to replace the three-dimensional space orthonormal unit vectors **i**, **j**, and **k** with $\mathbf{e}_1$, $\mathbf{e}_2$, and $\mathbf{e}_3$, respectively. These are written in column vector form as

$$\mathbf{e}_1 = \begin{bmatrix} 1 \\ 0 \\ 0 \end{bmatrix} \qquad \mathbf{e}_2 = \begin{bmatrix} 0 \\ 1 \\ 0 \end{bmatrix} \qquad \mathbf{e}_3 = \begin{bmatrix} 0 \\ 0 \\ 1 \end{bmatrix} \tag{A5-60}$$

Thus, any arbitrary vector in n-dimensional space may be written in the form

$$\mathbf{X} = \sum_{i=1}^{n} c_i \mathbf{e}_i \tag{A5-61}$$

Note that for the special case of three-dimensional space, $\mathbf{X} = c_1\mathbf{e}_1 + c_2\mathbf{e}_2 + c_3\mathbf{e}_3$, where c_1, c_2, and c_3 are x, y, and z, respectively (the vector **X** lies along the line whose ends are 0, 0, 0 and x, y, z).

The scalar product of 2 three-dimensional space vectors **U** and **V** given by

$$\mathbf{U} = \begin{pmatrix} u_1 \\ u_2 \\ u_3 \end{pmatrix} \qquad \mathbf{V} = \begin{pmatrix} v_1 \\ v_2 \\ v_3 \end{pmatrix} \tag{A5-62}$$

is written in matrix notation as

$$\mathbf{U}\cdot\mathbf{V} = \tilde{\mathbf{U}}\mathbf{V} = [u_1 \quad u_2 \quad u_3] \begin{bmatrix} v_1 \\ v_2 \\ v_3 \end{bmatrix} = \sum_{i=1}^{3} u_i v_i = u_1v_1 + u_2v_2 + u_3v_3 \tag{A5-63}$$

The length (or modulus) of a vector is given by

$$\sqrt{\mathbf{U}\cdot\mathbf{U}} = \sqrt{\tilde{\mathbf{U}}\mathbf{U}} = \sqrt{u_1^2 + u_2^2 + u_3^2} \tag{A5-64}$$

The results are readily generalized to n-dimensional space.

Let us now introduce the n-dimensional vectors

$$\mathbf{U} = \begin{bmatrix} u_1 \\ u_2 \\ \vdots \\ u_n \end{bmatrix} \qquad \mathbf{V} = \begin{bmatrix} v_1 \\ v_2 \\ \vdots \\ v_n \end{bmatrix}$$

and the n-dimensional basis vectors

$$\mathbf{e}_1 = \begin{bmatrix} 1 \\ 0 \\ \vdots \\ 0 \end{bmatrix}, \mathbf{e}_2 = \begin{bmatrix} 0 \\ 1 \\ \vdots \\ 0 \end{bmatrix}, \ldots, \mathbf{e}_n = \begin{bmatrix} 0 \\ 0 \\ \vdots \\ 1 \end{bmatrix} \tag{A5-65}$$

The basis vectors are readily shown to be orthonormal. The vector $\mathbf{U}$ may be written in terms of the basis set as

$$\mathbf{U} = u_1\mathbf{e}_1 + u_2\mathbf{e}_2 + \cdots + u_n\mathbf{e}_n \tag{A5-66}$$

Two important theorems pertaining to n-dimensional vector spaces are presented without proof as follows:

Theorem A5-1. The only vector which is orthogonal to every basis vector in a complete n-dimensional LVS is the zero vector.

Theorem A5-2. A necessary and sufficient condition for the linear independence of a set of vectors $\{\mathbf{V}_i\}$ is that $\det \mathbf{V} \neq 0$, where

$$\mathbf{V} = [\mathbf{V}_1 \quad \mathbf{V}_2 \quad \cdots \quad \mathbf{V}_n] = \begin{bmatrix} v_{11} & v_{12} & \cdots & v_{1n} \\ v_{21} & v_{22} & \cdots & v_{2n} \\ \cdots & \cdots & \cdots & \cdots \\ v_{n1} & v_{n2} & \cdots & v_{nn} \end{bmatrix} \tag{A5-67}$$

The first theorem is sometimes used as a criterion for the completeness of a set of vectors. If one can find no vector in the space other than the zero vector which is orthogonal to all the basis vectors, the set of functions (vectors) is complete. With respect to the second theorem, a set of vectors $\{\mathbf{V}_i\}$ is said to be linearly independent if the statement

$$c_1\mathbf{V}_1 = c_2\mathbf{V}_2 + \cdots + c_n\mathbf{V}_n = 0 \tag{A5-68}$$

implies that $c_i = 0$ for all $i = 1, 2, \ldots, n$.

In order to be able to discuss vector spaces whose elements are complex, it is convenient to introduce some new definitions. For example, in order to have all lengths (moduli) of vectors real, the scalar product of two vectors, $\mathbf{U}$ and $\mathbf{V}$, in a complex vector space is defined by

$$\mathbf{U} \cdot \mathbf{V} = \mathbf{U}^\dagger \mathbf{V} = \sum_{i=1}^{n} u_i^* v_i \tag{A5-69}$$

The length of any vector $\mathbf{U}$ is then given by

$$(\mathbf{U}^\dagger \mathbf{U})^{1/2} = \left(\sum_i u_i^* u_i\right)^{1/2} = \left(\sum_i |u_i|^2\right)^{1/2} \tag{A5-70}$$

When the elements of the vectors $\mathbf{U}$ and $\mathbf{V}$ are real, the above expressions reduce to the previous definition for real vectors. Note, however, that scalar multiplication of complex vectors is not generally commutative.

Any complete vector space in which the scalar product is defined analogously to Eq. (A5-69) is known as a *Hilbert space*. The wavefunctions of quantum mechanics may be regarded as vector quantities in a Hilbert space (generally of infinite dimensions). Thus if $U(t)$ and $V(t)$ are two functions in Hilbert space which satisfy

$$\int_0^1 |Q(t)|^2 \, dt < \infty \qquad (Q = U \text{ or } V) \tag{A5-71}$$

then the scalar product of $U(t)$ and $V(t)$ is given by

$$\int_0^1 U^*(t)V(t)\, dt = \langle U|V\rangle \tag{A5-72}$$

Note that the bra vector $\langle U|$ is symbolically analogous to $\mathbf{U}^\dagger$ and the ket vector $|V\rangle$ is symbolically analogous to $\mathbf{V}$. This analogy is often useful in the manipulation of quantum mechanical expressions containing integrals (matrix elements).

EXERCISES

A5-17. Prove that any vector orthogonal to the vectors $\mathbf{U}$ and $\mathbf{V}$ is also orthogonal to the linear combination $c_1\mathbf{U} + c_2\mathbf{V}$.

A5-18. Show that if $\mathbf{U}$ and $\mathbf{V}$ are any two vectors defined in the same space, the new vectors $\mathbf{U} - \mathbf{V}$ and $\mathbf{U} + \mathbf{V}$ are always orthogonal. Interpret this geometrically.

A5-19. Use the second theorem to determine the linear independence (or lack thereof) of the following two sets of vectors:
(*a*) The three vectors whose components are $(1, 2, 3)$, $(1, 3, 2)$, and $(2, 4, 6)$
(*b*) The three vectors whose components are $(1, 2, 3)$, $(1, 3, 2)$, and $(3, 2, 1)$

A5-6 MATRIX REPRESENTATION OF OPERATORS

The differential operators of quantum mechanics can be represented by matrices such that the algebraic properties of the operators are reflected in the representations.

In general, the eigenfunctions of a hermitian operator form a complete set of functions. Letting $\hat{A}$ represent a general hermitian operator having a discrete set of eigenfunctions (generally infinite in number), the corresponding eigenvalue equations may be written

$$\hat{A}\phi_i = \lambda_i\phi_i \qquad i = 1, 2, \ldots \tag{A5-73}$$

where the $\{\lambda_i\}$ are eigenvalues of the operator $\hat{A}$ and the $\{\phi_i\}$ are the corresponding eigenfunctions. Now let $\hat{F}$ and $\hat{G}$ be any two hermitian operators defined in the same domain as the operator $\hat{A}$, and define the integrals

$$F_{ij} = \langle \phi_i|\hat{F}|\phi_j\rangle \qquad G_{ij} = \langle \phi_i|\hat{G}|\phi_j\rangle \tag{A5-74}$$

which are called *matrix elements* of the operators $\hat{F}$ and $\hat{G}$, respectively, in the basis $\{\phi_i\}$. The operators $\hat{F}$ and $\hat{G}$ then have the *matrix representations*

$$\mathbf{F} = \begin{bmatrix} F_{11} & F_{12} & \cdots \\ F_{21} & F_{22} & \cdots \\ \cdots & \cdots & \cdots \end{bmatrix} \qquad \mathbf{G} = \begin{bmatrix} G_{11} & G_{12} & \cdots \\ G_{21} & G_{22} & \cdots \\ \cdots & \cdots & \cdots \end{bmatrix} \tag{A5-75}$$

In practice, one usually uses truncated basis sets so that the matrices **F** and **G** are finite. It is easy to demonstrate that both the operators and their matrix representations obey the same rules for addition and multiplication, namely,

$$\begin{aligned} (\mathbf{F}+\mathbf{G})_{ij} &= F_{ij} + G_{ij} \\ (\mathbf{FG})_{ij} &= \sum_k F_{ik} G_{kj} \end{aligned} \tag{A5-76}$$

Let us now consider the matrix eigenvalue equation

$$\mathbf{AC}_i = \lambda_i \mathbf{C}_i \tag{A5-77}$$

where λ_i is a numerical constant (real if $\hat{A}$ is hermitian) and $\mathbf{C}_i$ is a column vector. We call λ_i and $\mathbf{C}_i$ an eigenvalue and an eigenvector, respectively, belonging to the matrix **A**. Since $\lambda_i \mathbf{C}_i$ is collinear with $\mathbf{C}_i$, it is apparent that the operation of **A** on $\mathbf{C}_i$ is to produce a new vector having the same direction as $\mathbf{C}_i$ but, in general, a different length. If **A** is an $n \times n$ square matrix, Eq. (A5-77) is the matrix representation of the n simultaneous equations

$$\begin{aligned} a_{11}c_1 + a_{12}c_2 + \cdots + a_{1n}c_n &= \lambda_i c_1 \\ a_{21}c_1 + a_{22}c_2 + \cdots + a_{2n}c_n &= \lambda_i c_2 \\ \cdots\cdots\cdots\cdots\cdots\cdots\cdots\cdots \\ a_{n1}c_1 + a_{n2}c_2 + \cdots + a_{nn}c_n &= \lambda_i c_n \end{aligned} \tag{A5-78}$$

where we use a_{ij} instead of A_{ij} for the matrix elements of $\hat{A}$. Equation (A5-77) may also be written

$$(\mathbf{A} - \lambda_i \mathbf{1})\mathbf{C}_i = \mathbf{0} \tag{A5-79}$$

This corresponds to the n simultaneous equations

$$\begin{aligned} (a_{11} - \lambda_i)c_1 + a_{12}c_2 + \cdots + a_{1n}c_n &= 0 \\ a_{21}c_1 + (a_{22} - \lambda_i)c_2 + \cdots + a_{2n}c_n &= 0 \\ \cdots\cdots\cdots\cdots\cdots\cdots\cdots\cdots \\ a_{n1}c_1 + a_{n2}c_2 + \cdots + (a_{nn} - \lambda_i)c_n &= 0 \end{aligned} \tag{A5-80}$$

Equation (A5-79) is satisfied trivially if $\mathbf{C}_i = \mathbf{0}$; the nontrivial solutions require that the matrix $\mathbf{A} - \lambda_i \mathbf{1}$ be singular, that is, that

$$\det(\mathbf{A} - \lambda_i \mathbf{1}) = 0 \tag{A5-81}$$

In other words the determinant of the coefficients of the c_i must vanish.

Rewriting Eq. (A5-81) in explicit form, we obtain

$$\begin{vmatrix} a_{11}-\lambda_i & a_{12} & \cdots & a_{1n} \\ a_{21} & a_{22}-\lambda_i & \cdots & a_{2n} \\ \cdots & \cdots & \cdots & \cdots \\ a_{n1} & a_{n2} & \cdots & a_{nn}-\lambda_i \end{vmatrix} = 0 \tag{A5-82}$$

which is known as a *secular determinant*. Expansion of this determinant in terms of the cofactors will lead to an nth-order polynomial in λ_i of the general form

$$\lambda_i^n + \alpha_i \lambda_i^{n-1} + \alpha_2 \lambda_i^{n-2} + \cdots + \alpha_{n-1}\lambda_i + \alpha_n = 0 \tag{A5-83}$$

which is called the *characteristic equation* of the matrix **A**. The characteristic equation has n roots $\lambda_1, \lambda_2, \ldots, \lambda_n$ (not necessarily all distinct), which are called the *eigenvalue spectrum* of the matrix **A**. For each eigenvalue λ_i there exists an associated eigenvector $\mathbf{C}_i$ belonging to the matrix **A**. Consequently, we may replace Eq. (A5-77) with the more general equation

$$\mathbf{AC}_i = \lambda_i \mathbf{C}_i \qquad i = 1, 2, \ldots, n \tag{A5-84}$$

where

$$\mathbf{C}_i = \begin{bmatrix} c_{1i} \\ c_{2i} \\ \vdots \\ c_{n1} \end{bmatrix} \tag{A5-85}$$

The n simultaneous matrix equations implied by Eq. (A5-84) can be cast into the single compact matrix equation

$$\mathbf{AC} = \mathbf{C\Lambda} \tag{A5-86}$$

by defining the following matrices

$$\mathbf{C} = [\mathbf{C}_1 \quad \mathbf{C}_2 \quad \cdots \quad \mathbf{C}_n] \qquad \mathbf{\Lambda} = \begin{bmatrix} \lambda_1 & 0 & & \cdots & 0 \\ 0 & \lambda_2 & 0 & \cdots & 0 \\ \cdots & \cdots & \cdots & \cdots & \cdots \\ 0 & 0 & & \cdots & \lambda_n \end{bmatrix} \tag{A5-87}$$

If $\det \mathbf{C} \neq 0$, we can multiply Eq. (A5-86) from the left by $\mathbf{C}^{-1}$ to obtain

$$\mathbf{C}^{-1}\mathbf{AC} = \mathbf{\Lambda} \tag{A5-88}$$

We then say that the matrix **A** is diagonalized by a matrix of its eigenvectors. Finding the eigenvectors which diagonalize a given matrix is one of the most frequently encountered operations in computational quantum chemistry; a simple specific example of the process is given in the following discussion.

Let us consider the simple 2×2 matrix

$$\mathbf{A} = \begin{bmatrix} 7 & -3\sqrt{3} \\ -3\sqrt{3} & 13 \end{bmatrix} \tag{A5-89}$$

and find its eigenvalues and corresponding orthonormal eigenvectors. The eigenvalues are found first by solving the determinant

$$\det(\mathbf{A}-\lambda\mathbf{1}) = \begin{vmatrix} 7-\lambda & -3\sqrt{3} \\ -3\sqrt{3} & 13-\lambda \end{vmatrix} \tag{A5-90}$$

This determinant reduces to the quadratic equation

$$\lambda^2 - 20\lambda + 64 = 0 \tag{A5-91}$$

The roots are 4 and 16; these are the eigenvalues of **A**. The orthonormalized eigenvectors are found as follows: From $(\mathbf{A}-\lambda_i\mathbf{1})\mathbf{C}_i = 0$ ($i = 1$ or 2), we obtain

$$(7-\lambda_i)c_{1i} - 3\sqrt{3}c_{2i} = 0 \qquad \text{and} \qquad -3\sqrt{3}c_{1i} + (13-\lambda_i)c_{2i} = 0 \tag{A5-92}$$

The desired eigenvectors have the form

$$\mathbf{C}_i = \begin{bmatrix} c_{1i} \\ c_{2i} \end{bmatrix}$$

and (if orthonormal) must satisfy

$$c_{1i}^2 + c_{2i}^2 = 1 \tag{A5-93}$$

Letting $\lambda_i = 4$ when $i = 1$ and $\lambda_i = 16$ when $i = 2$, Eq. (A5-92) leads to

$$c_{21} = \frac{c_{11}}{\sqrt{3}} \tag{A5-94}$$

Substitution of this result into Eq. (A5-93) yields

$$c_{11} = \tfrac{1}{2}\sqrt{3} \qquad c_{21} = \tfrac{1}{2} \tag{A5-95}$$

which are the components of $\mathbf{C}_1$. Similarly, for $i = 2$ one obtains the components of $\mathbf{C}_2$, namely,

$$c_{12} = \tfrac{1}{2} \qquad c_{22} = -\tfrac{1}{2}\sqrt{3} \tag{A5-96}$$

The complete solution to the matrix equation may be summarized as follows:

$$\mathbf{C} = [\mathbf{C}_1 \quad \mathbf{C}_2] = \begin{bmatrix} \frac{1}{2}\sqrt{3} & \frac{1}{2} \\ \frac{1}{2} & -\frac{1}{2}\sqrt{3} \end{bmatrix} \qquad \Lambda = \begin{bmatrix} 4 & 0 \\ 0 & 16 \end{bmatrix} \tag{A5-97}$$

EXERCISES

A5-20. Prove the expressions given in Eq. (A5-76). *Hint*: The addition relationship follows at once by writing out $(\mathbf{F}+\mathbf{G})_{ij}$ in terms of bra-ket notation and carrying out suitable rearrangement. The second relationship is proved by also writing out $(\mathbf{FG})_{ij}$ in bra-ket form, expressing $\hat{G}\phi_j$ as a linear combination of orthonormal basis functions $\{\phi_i\}$ and carrying out suitable rearrangement and manipulation.

A5-21. Find the eigenvalues and normalized eigenvectors of the matrix

$$\mathbf{A} = \begin{bmatrix} 2 & 2 & 0 \\ 2 & 4 & 2 \\ 0 & 2 & 2 \end{bmatrix}$$

A5-7 SIMILARITY TRANSFORMATIONS

Let us consider a vector which can be expressed in two different coordinate systems $(x_1, x_2, \ldots, x_n)$ and $(x'_1, x'_2, \ldots, x'_n)$. Column matrix representations of this vector are

$$\mathbf{X} = \begin{bmatrix} x_1 \\ x_2 \\ \vdots \\ x_n \end{bmatrix} \quad \text{or} \quad \mathbf{X}' = \begin{bmatrix} x'_1 \\ x'_2 \\ \vdots \\ x'_n \end{bmatrix} \tag{A5-98}$$

The two coordinate systems are related by the matrix transformation equation

$$\mathbf{X} = \mathbf{T}\mathbf{X}' \tag{A5-99}$$

where $\mathbf{T}$ is an $n \times n$ matrix called the *transformation matrix*. Now consider the matrix transformation

$$\mathbf{A}\mathbf{X} = \mathbf{Y} \tag{A5-100}$$

where

$$\mathbf{Y} = \mathbf{T}\mathbf{Y} \tag{A5-101}$$

and $\mathbf{X}$ and $\mathbf{Y}$ are in the same coordinate system. We then assume that there exists a transformation of $\mathbf{X}'$ into $\mathbf{Y}'$ given by

$$\mathbf{A}'\mathbf{X}' = \mathbf{Y}' \tag{A5-102}$$

We now wish to find how the matrices $\mathbf{A}$ and $\mathbf{A}'$ are related. Using Eqs. (A5-99) and (A5-101), we can write Eq. (A5-100) as

$$\mathbf{A}\mathbf{T}\mathbf{X}' = \mathbf{T}\mathbf{Y}' \tag{A5-103}$$

If we assume that $\mathbf{T}$ is nonsingular, Eq. (A5-103) can be multiplied on the left by $\mathbf{T}^{-1}$ to obtain

$$\mathbf{T}^{-1}\mathbf{A}\mathbf{T}\mathbf{X}' = \mathbf{Y}' \tag{A5-104}$$

Comparison of Eqs. (A5-102) and (A5-104) shows that

$$\mathbf{T}^{-1}\mathbf{A}\mathbf{T} = \mathbf{A}' \tag{A5-105}$$

This latter transformation is known as a *similarity transformation*, and the matrices $\mathbf{A}$ and $\mathbf{A}'$ are said to be *similar*, or *equivalent*. Equation (A5-88) is also an example of a similarity transformation.

Following are three theorems (without proofs) concerning similarity transformations:

Theorem A5-3. Equivalent matrices have the same eigenvalues.

Theorem A5-4. The trace of a matrix is invariant under a similarity transformation.

Theorem A5-5. The determinant of a (square) matrix is invariant under a similarity transformation.

The following three theorems apply to hermitian matrices:

Theorem A5-6. The eigenvalues of a hermitian matrix ($\mathbf{A}^\dagger = \mathbf{A}$) are real.

Theorem A5-7. A transformation matrix $\mathbf{C}$ which diagonalizes a hermitian matrix ($\mathbf{C}^\dagger\mathbf{A}\mathbf{C} = \mathbf{\Lambda}$) is *unitary*. Note that a unitary matrix is a special case of an orthogonal matrix; thus, a unitary transformation is a special case of a similarity transformation.

Theorem A5-8. The eigenvectors of a nondegenerate hermitian matrix are orthogonal. In the event that the hermitian matrix is degenerate (i.e., has one or more repeated roots), one can always find a linear combination of the degenerate eigenvectors, e.g., by an orthogonalization procedure, which makes them mutually orthogonal.

We will now examine in more detail some of the important characteristics of unitary transformations. First we consider an important theorem:

Theorem A5-9. The scalar product of two vectors is invariant under a unitary transformation. This means that a unitary transformation preserves the lengths of the two vectors and the angle between them; i.e., the unitary transformation merely rotates the two vectors. Unitary transformations have the general property of transforming real linear operators into other real linear operators, and they leave invariant any algebraic equations which connect linear operators, bra vectors, or ket vectors.

It should be clear that Theorem A5-9 also applies to orthogonal transformations in real vector spaces.

The properties of unitary matrices are useful in the reduction of certain types of algebraic equations to simpler forms. For example, consider the equation

$$Q = \sum_i \sum_j a_{ij} x_i^* x_j \tag{A5-106}$$

We will see, through the use of a suitable unitary transformation, that this equation can be transformed to a quadratic form, i.e., a form containing no cross products. First, this equation may be regarded as the scalar product

$$\mathbf{Q} = \mathbf{X}^\dagger(\mathbf{A}\mathbf{X}) \tag{A5-107}$$

where

$$\mathbf{X} = \begin{bmatrix} x_1 \\ x_2 \\ \vdots \\ x_n \end{bmatrix} \qquad \mathbf{A} = \mathbf{A}^\dagger = \begin{bmatrix} a_{11} & a_{12} & \cdots & a_{1n} \\ a_{21} & a_{22} & \cdots & a_{2n} \\ a_{n1} & a_{n2} & \cdots & a_{nn} \end{bmatrix} \tag{A5-108}$$

Now let $\mathbf{U}$ be a unitary matrix which diagonalizes the hermitian matrix $\mathbf{A}$, namely,

$$\mathbf{U}^\dagger\mathbf{A}\mathbf{U} = \mathbf{D} \tag{A5-109}$$

where **D** is diagonal, and let the vector **X** be given by

$$\mathbf{X} = \mathbf{UY} \tag{A5-110}$$

where **Y** has the components $y_1, y_2, \ldots, y_n$. Then

$$\mathbf{X}^\dagger\mathbf{AX} = (\mathbf{UY})^\dagger\mathbf{AUY} = \mathbf{Y}^\dagger\mathbf{U}^\dagger\mathbf{AUY} = \mathbf{Y}^\dagger\mathbf{DY} = \sum_k \lambda_k y_k^* y_k = \sum_k \lambda_k |y_k|^2 \tag{A5-111}$$

where λ_k represents the diagonal elements of **D**.

As a specific example, consider the equation (in real space)

$$7x_1^2 + 13x_2^2 - 6\sqrt{3}x_1x_2 = 64 \tag{A5-112}$$

As seen in Fig. A5-1, this is the equation of an ellipse centered at the origin, with principal axes rotated counterclockwise by 30° from the normal coordinate axes x_1 and x_2. We will now find an equation for the same ellipse in terms of a new coordinate system y_1 and y_2 such that y_1 and y_2 coincide with the principal axes.

From Eq. (A5-112) we see that the matrix **A** is

$$\mathbf{A} = \begin{bmatrix} 7 & -3\sqrt{3} \\ -3\sqrt{3} & 13 \end{bmatrix} \quad \text{and that} \quad \mathbf{X} = \begin{bmatrix} x_1 \\ x_2 \end{bmatrix} \tag{A5-113}$$

The matrix **A** is just that given earlier in Eq. (A5-89) and used as an example of a matrix eigenvalue problem. The complete solution of this problem as needed for the present example is given by Eq. (A5-97). Thus we may write

$$\mathbf{U} = [\mathbf{C}_1 \quad \mathbf{C}_2] = \begin{bmatrix} \sqrt{3}/2 & 1/2 \\ 1/2 & -\sqrt{3}/2 \end{bmatrix} \tag{A5-114}$$

which is readily verified to be unitary (also orthogonal, since the space has real elements).

We now let **X** be given by

$$\mathbf{X} = \mathbf{UY} \tag{A5-115}$$

where

$$\mathbf{Y} = \begin{bmatrix} y_1 \\ y_2 \end{bmatrix} \tag{A5-116}$$

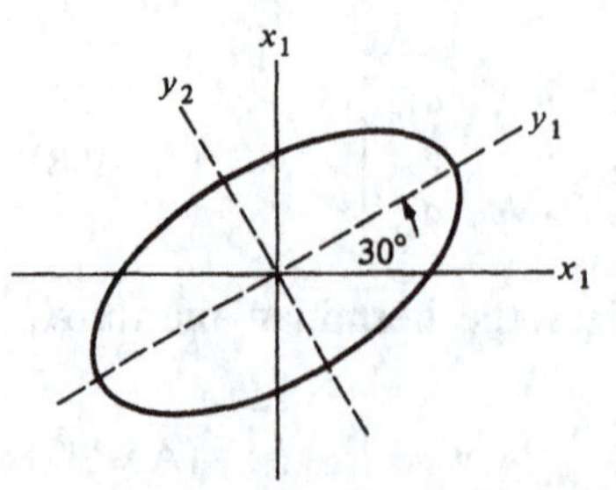

FIGURE A5-1
Graph of the ellipse $7x_1^2 + 13x_2^2 - 6\sqrt{3}x_1x_2 = 64$ or $y_1^2/16 + y_2^2/4 = 1$.

Equation (A5-115) means we have to carry out the transformations

$$x_1 = \frac{\sqrt{3}}{2} y_1 + \frac{y_2}{2} \quad \text{and} \quad x_2 = \frac{y_1}{2} - \frac{\sqrt{3}}{2} y_2 \tag{A5-117}$$

Substituting the expressions in Eq. (A5-117) into (A5-112), simplifying, and rearranging, we obtain

$$\frac{y_1^2}{16} + \frac{y_2^2}{4} = 1 \tag{A5-118}$$

which is the same ellipse as before, but now its equation is in the standard quadratic form.

In conclusion, a unitary transformation preserves normalization (lengths are preserved) and preserves linear independence (angles are preserved).

EXERCISES

A5-22. Show that a similarity transformation with a unitary matrix on a hermitian matrix produces another hermitian matrix.

A5-23. Show for the example of a unitary transformation on the equation of an ellipse that the unitary matrix can be generalized to

$$\begin{bmatrix} \cos\theta & \sin\theta \\ \sin\theta & -\cos\theta \end{bmatrix}$$

where θ is the angle between the principal axes and the coordinate axes.

A5-24. Prove that the determinant of a unitary matrix has an absolute value of unity. Recall that the determinant of a matrix is invariant under a similarity transformation, and consider the diagonal form of any unitary matrix. Suggest a physical interpretation of this result.

A5-25. The following equation illustrates the transformation of p orbitals in their real forms to their complex forms. Is this an example of a unitary transformation? Explain.

$$\frac{1}{\sqrt{2}}\begin{bmatrix} 1 & i & 0 \\ 1 & -i & 0 \\ 0 & 0 & \sqrt{2} \end{bmatrix}\begin{bmatrix} p_x \\ p_y \\ p_z \end{bmatrix} = \begin{bmatrix} p_1 \\ p_{-1} \\ p_0 \end{bmatrix}$$

A5-8 PROJECTION OPERATORS

The problem of diagonalizing a hermitian matrix, i.e., finding the basis set of orthonormal eigenvectors associated with the eigenvalues of the matrix, may be regarded as that of finding a set of *projection operators* $\{\mathbf{O}_i\}$ associated with the orthonormal basis set $\{\mathbf{e}_i\}$. Referring to Fig. A5-2, let us initially restrict ourselves to the ordinary euclidean space of three dimensions spanned by the orthonormal vectors $\mathbf{e}_1$, $\mathbf{e}_2$, and $\mathbf{e}_3$, whose matrix representations are given in Eq. (A5-60). Now let $\mathbf{X} = x_1\mathbf{e}_1 + x_2\mathbf{e}_2 + x_3\mathbf{e}_3$ represent an arbitrary vector in this space. From Fig. A5-2 it is readily established that

$$\mathbf{e}_1^\dagger\mathbf{X} = x_1 = X\cos\alpha \qquad \mathbf{e}_2^\dagger\mathbf{X} = x_2 = X\cos\beta \qquad \mathbf{e}_3^\dagger\mathbf{X} = x_3 = X\cos\gamma \tag{A5-119}$$

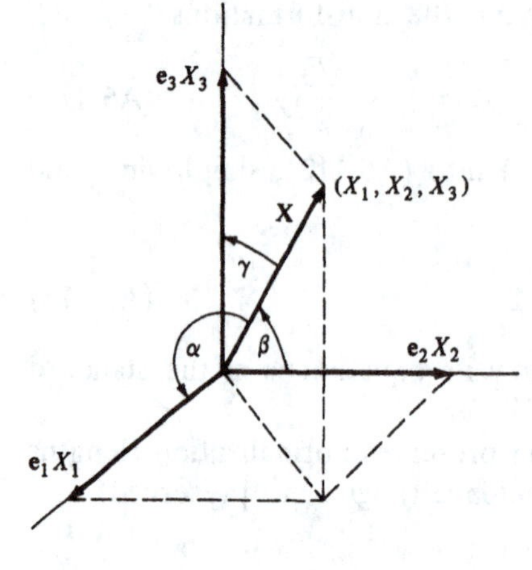

FIGURE A5-2
The projection of an arbitrary vector in three-dimensional space onto the axes of the unit vectors of the space.

where $X = \sqrt{\mathbf{X}^\dagger \mathbf{X}}$. It is evident that the scalar product of one of the basis vectors and an arbitrary vector in the space is just the projection of the modulus of the vector onto the axis of the basis vector. Now consider the product

$$\mathbf{e}_1(\mathbf{e}_1^\dagger \mathbf{X}) = x_1 \mathbf{e}_1 \tag{A5-120}$$

which is a vector of modulus x_1 and is collinear with $\mathbf{e}_1$. This vector $x_1\mathbf{e}_1$ may be regarded as a projection of $\mathbf{X}$ onto the axis of the basis vector $\mathbf{e}_1$.

If we define the matix representation of an operator $\hat{O}$ as

$$\mathbf{O}_i = \mathbf{e}_i \mathbf{e}_i^\dagger \qquad i = 1,\ 2,\ \text{or}\ 3 \tag{A5-121}$$

then Eq. (A5-120) may be written as

$$\mathbf{O}_1 \mathbf{X} = x_1 \mathbf{e}_1 \tag{A5-122}$$

We call the operator defined in Eq. (A5-121) a *projection operator*, since it has the property of projecting an arbitrary vector onto the axis of its associated basis vector. The projection operator $\mathbf{O}_i$ is sometimes written in Dirac notation as

$$\mathbf{O}_i = |e_i\rangle\langle e_i| \tag{A5-123}$$

Such a quantity is also known as a *dyad*. Note that a projection operator is defined in terms of a particular basis vector and that it projects onto the subspace of that basis vector. In the case we have considered, the subspace of the projection operator is one-dimensional.

The arbitrary vector $\mathbf{X}$ may be written

$$\mathbf{X} = \sum_{i=1}^{3} x_i \mathbf{e}_i = \sum_{i=1}^{3} \mathbf{O}_i \mathbf{X} \tag{A5-124}$$

which leads immediately to the important relationship

$$\sum_{i=1}^{3} \mathbf{O}_i = \mathbf{1} \tag{A5-125}$$

where **1** is the 3×3 unit matrix. Equations (A5-121) and (A5-125) may be written explicitly as

$$\mathbf{O}_1 = \begin{bmatrix} 1 \\ 0 \\ 0 \end{bmatrix} [1 \quad 0 \quad 0] = \begin{bmatrix} 1 & 0 & 0 \\ 0 & 0 & 0 \\ 0 & 0 & 0 \end{bmatrix} \qquad \mathbf{O}_2 = \begin{bmatrix} 0 \\ 1 \\ 0 \end{bmatrix} [0 \quad 1 \quad 0] = \begin{bmatrix} 0 & 0 & 0 \\ 0 & 1 & 0 \\ 0 & 0 & 0 \end{bmatrix}$$

$$\mathbf{O}_3 = \begin{bmatrix} 0 \\ 0 \\ 1 \end{bmatrix} [0 \quad 0 \quad 1] = \begin{bmatrix} 0 & 0 & 0 \\ 0 & 0 & 0 \\ 0 & 0 & 1 \end{bmatrix} \qquad \mathbf{O}_1 + \mathbf{O}_2 + \mathbf{O}_3 = \begin{bmatrix} 1 & 0 & 0 \\ 0 & 1 & 0 \\ 0 & 0 & 1 \end{bmatrix} = \mathbf{1} \tag{A5-126}$$

Finding the projection operators of a vector space is sometimes referred to as *resolution of the identity*. From the relationships already presented, it follows that

$$\mathbf{O}_i \mathbf{e}_i = \mathbf{e}_i \tag{A5-127}$$

that is, the projection operator projects the eigenvector associated with its own subspace onto itself. From this fact it follows that

$$\mathbf{O}_i^2 \mathbf{X} = \mathbf{O}_i(\mathbf{O}_i \mathbf{X}) = \mathbf{O}_i x_i \mathbf{e}_i = x_i \mathbf{e}_i \tag{A5-128}$$

so that we can write

$$\mathbf{O}_i^2 = \mathbf{O}_i \tag{A5-129}$$

This last relationship means that the projection operators are *idempotent*. It is also apparent from this that projection operators have only two distinct eigenvalues, namely, zero and unity. This latter conclusion follows from the fact that zero and unity are the only eigenvalues which satisfy Eq. (A5-129). It is useful to note that the idempotency of a matrix ($\mathbf{A}^2 - \mathbf{A} = \mathbf{0}$ with eigenvalues 0 and 1) is analogous to the algebraic relationship $x^2 - x = 0$, with solutions $x = 0$ and 1.

It is evident that the unit matrix itself is a projection operator which projects onto the entire space (the identity projection). It is also seen that the product of two different projection operators must be zero, provided the two projection operators are defined in terms of mutually orthogonal basis vectors.

Now let **A** be the matrix representation in three-dimensional space of a self-adjoint operator which is diagonalized by the orthonormal basis set $\mathbf{e}_1$, $\mathbf{e}_2$, and $\mathbf{e}_3$. Assuming that **A** is nondegenerate, we obtain

$$\mathbf{U}^\dagger \mathbf{A} \mathbf{U} = \begin{bmatrix} \lambda_1 & 0 & 0 \\ 0 & \lambda_2 & 0 \\ 0 & 0 & \lambda_3 \end{bmatrix} = \lambda_1 \mathbf{O}_1 + \lambda_2 \mathbf{O}_2 + \lambda_3 \mathbf{O}_3 = \sum_{i=1}^{3} \lambda_i \mathbf{O}_i \tag{A5-130}$$

This very important relationship is an equation expressing the operator $\hat{A}$ (in its diagonalized matrix representation) in terms of its eigenvalues and their respective subspaces. Provided the matrix is nondegenerate, the projected vectors $x_i \mathbf{e}_i$ obviously form an orthogonal basis set but are not, in general, normalized. It is apparent that if the arbitrary vector **X** (not one of the basis vectors) is normalized, its projections will not be normalized, since they lie along axes not collinear with **X**.

In the event that some of the eigenvalues of **A** are degenerate (as will be the case when the system contains a high degree of symmetry), the summation in Eq. (A5-130) must be replaced by a summation over distinct eigenvalues. Each g-fold degeneracy is now associated with a projection operator which projects onto a g-dimensional subspace. Such projections are not unique, and hence, one will be able to obtain an infinite number of them; however, no more than g linear combinations will be linearly independent, and these can always be made into g orthogonal combinations. As a simple example, consider a twofold degeneracy in three-dimensional space; i.e., the matrix **A** has the eigenvalues λ_1, λ_2, and λ_3 (which equals λ_2). The projection operators then are

$$\mathbf{O}_1 = \begin{bmatrix} 1 & 0 & 0 \\ 0 & 0 & 0 \\ 0 & 0 & 0 \end{bmatrix} \qquad \mathbf{O}_2 = \begin{bmatrix} 0 & 0 & 0 \\ 0 & 1 & 0 \\ 0 & 0 & 1 \end{bmatrix} \tag{A5-131}$$

where $\mathbf{O}_2$ projects onto a two-dimensional subspace (a plane). Out of all the vectors one can project onto this plane with $\mathbf{O}_2$, no more than two can be linearly independent, and from these no more than two orthogonal combinations can be formed. For instance, the sum and difference of any two linearly independent vectors of equal length form a pair of orthogonal vectors in the plane. The dimension of the subspace represented by a projection operator is given by its trace. Thus in the above example, $\mathrm{Tr}\,\mathbf{O}_1 = 1$ and $\mathrm{Tr}\,\mathbf{O}_2 = 2$.

Let us now consider a generalization of Eq. (A5-130) to an n-dimensional space, keeping in mind the modification necessary if **A** has degeneracies. We will now derive an equation for a projection operator which is useful in certain types of problems. Equation (A5-130) for the n-dimensional case can be written

$$\lambda_k \mathbf{O}_k = \mathbf{A} - \sum_{i \neq k}^{n'} \lambda_i \mathbf{O}_i \tag{A5-132}$$

where $n' \leq n$ is the number of distinct eigenvalues. Now using Eq. (A5-125) one can eliminate a given projection operator $\mathbf{O}_j$ $(j \neq k)$ to obtain the expression

$$\mathbf{O}_k = \frac{\mathbf{A} - \lambda_j \mathbf{1}}{\lambda_k - \lambda_j} + \frac{\sum_{i \neq j,k}^{n'} (\lambda_j - \lambda_i)\mathbf{O}_i}{\lambda_k - \lambda_j} \tag{A5-133}$$

Repeating this for all $\mathbf{O}_j$ $(j \neq k)$, one obtains $n' - 1$ different expressions for $\mathbf{O}_k$. Multiplying all these together, one obtains

$$\mathbf{O}_k^{n'-1} = \mathbf{O}_k = \prod_{j \neq k} \frac{\mathbf{A} - \lambda_j \mathbf{1}}{\lambda_k - \lambda_j} \tag{A5-134}$$

Only the product of the leading terms survives, since all other products contain factors such as $\mathbf{O}_i\mathbf{O}_j = \mathbf{0}$ $(i \neq j)$ or $(\mathbf{A} - \lambda_i \mathbf{1})\mathbf{O}_j = \mathbf{0}$ $(i = j)$. Thus usefulness of Eq. (A5-134) is apparent if we consider an arbitrary vector defined in the space spanned by the eigenvectors of the projection operator. Such a vector, say, **V**, can be expanded in terms of the basis set as

$$\mathbf{V} = \sum_i c_i \mathbf{e}_i \tag{A5-135}$$

The projection operator given in Eq. (A5-134) annihilates all components of $\mathbf{V}$ except $c_k \mathbf{e}_k$, a vector collinear with the kth eigenvector of the operator $\mathbf{A}$.

Equation (A5-134) is a useful form for the explicit construction of projection operators when the eigenvalues of $\mathbf{A}$ are known beforehand, e.g., when $\mathbf{A}$ refers to orbital and spin angular momenta of electrons.

Another useful relationship concerning projection operators (presented without proof) is seen in the fact that the expectation values of a projection operator $\hat{O}$ satisfy the inequality

$$0 \leq \langle \hat{O} \rangle \leq 1 \tag{A5-136}$$

This relationship can be used to prove Eq. (7-92) in Sec. 7-5. To carry out the proof, consider the two-electron transposition operator $\hat{P}_{12}$ whose function is as follows:

$$\hat{P}_{12} u(1) v(2) = v(1) u(2)$$

It is readily verified that the operator $\hat{P}_{12}$ is its own inverse and that it is unitary; thus it is also hermitian. We now prove the following:

$$-1 \leq \langle \hat{P}_{12} \rangle \leq 1 \tag{A5-137}$$

We consider the operator $\frac{1}{2}(1 - \hat{P}_{12})$ which can readily be seen to be idempotent, and thus is a projection operator. This fact leads immediately to the proof of (A5-137). Next, we consider an arbitrary two-electron function $\xi(1, 2)$. The expectation value of the operator $\hat{P}_{12}$ with respect to this function is

$$\langle \hat{P}_{12} \rangle = \langle \xi(1,2) | \hat{P}_{12} | \xi(1,2) \rangle = \langle \xi(1,2) | \xi(2,1) \rangle$$

Thus,

$$-1 \leq \langle \xi(1,2) | \xi(2,1) \rangle \leq 1 \tag{A5-138}$$

If $\xi(1,2)$ is normalized, we can write

$$-1 \leq \langle \xi(1,2) | \xi(2,1) \rangle \leq \langle \xi(1,2) | \xi(1,2) \rangle \tag{A5-139}$$

Now consider the special case

$$\xi(1,2) = u(1) v(2) \left(\frac{1}{r_{12}} \right)^{1/2}$$

We can now use Eq. (A5-139) to write

$$\left\langle u(1)v(2) \left| \frac{1}{r_{12}} \right| v(1)u(2) \right\rangle \leq \left\langle u(1)v(2) \left| \frac{1}{r_{12}} \right| u(1)v(2) \right\rangle$$

which is more simply written as

$$K_{uv} \leq J_{uv}$$

If we accept that K_{uv} is always positive (or zero), then the full inequality follows:

$$0 \leq K_{uv} \leq J_{uv}$$

This is Eq. (7-92) (where u, v are replaced with i, j).

EXERCISES

A5-26. Show that, in general, a projection operator has no inverse. Give a simple physical interpretation of this fact.

A5-27. Discuss the possibility of defining projection operators in terms of a basis set which is not orthogonal but is linearly independent.

SUGGESTED READINGS

Anderson, J. M.: *Mathematics for Quantum Chemistry*, Benjamin, New York, 1966.
Atkin, A. C.: *Determinants and Matrices*, Interscience, New York, 1956.
Hildebrand, F. B.: *Methods of Applied Mathematics*, 2d ed., Prentice-Hall, Englewood Cliffs, N.J., 1965.
Hollingsworth, C. A.: *Vectors, Matrices, and Group Theory for Scientists and Engineers*, McGraw-Hill, New York, 1967.
Margenau, H., and G. M. Murphy: *The Mathematics of Physics and Chemistry*, Van Nostrand, Princeton, N.J., 1956.
Schwartz., J. T.: *Introduction to Matrices and Vectors*, McGraw-Hill, New York, 1961.

APPENDIX 6

MOLECULAR SYMMETRY

According to a definition found in a standard dictionary, *symmetry* is a "correspondence in size, shape, and relative position of parts that are on opposite sides of a dividing line or median plane." Such symmetry is apparent in a molecule such as benzene; if a dividing line is passed through any pair of opposite atoms (or any midpoint of C–C bonds), the molecule is divided into two parts which are identical except for a sense of right and left. However, the above definition of symmetry is too restrictive, since benzene has other less obvious types of symmetry as well.

Most of the properties of a molecule depend on the atoms of which it is formed and the way in which they are joined. But some properties depend entirely or largely on the symmetry the molecule may possess.[1] In principle, full solution of the Schrödinger equation would produce all the properties of a molecule: those dependent on symmetry and those not dependent on it. However, it is possible to determine those properties depending only on symmetry without solving the Schrödinger equation; the purpose of this appendix is to show how to treat the symmetry elements of a molecule in a formal, systematic manner so that the consequences of these can be used to understand the molecule's chemical behavior.

[1] Symmetry is also important in atoms. For example, the $(2l + 1)$-fold degeneracy of the hydrogen atom's spherical harmonics is purely the result of spherical symmetry. Additional degeneracies arise owing to the central-field nature of the coulombic potential.

A6-1 THE SYMMETRY ELEMENTS OF MOLECULES

If an equilateral triangle in the plane of this paper is rotated 120°, 240°, or 360° (clockwise or counterclockwise) about an axis perpendicular to this paper, the appearance of the triangle is unchanged; were a person to turn out the lights and then either rotate or not rotate the triangle by some multiple of 120°, a second observer (once the lights were turned back on) would be unable to determine by how many multiples of 120° the triangle had been rotated—or if it had been rotated at all. Any operation which has the effect of sending a molecule (or any geometrical figure) into itself, i.e., into a position indistinguishable from the original, is called a *symmetry operation*. Such symmetry operations always leave some *point* of the figure invariant and are said to constitute a *point group*.[2] All the symmetry operations of the equilateral triangle leave its center unaffected.

The symmetry operations which send a molecule into itself are of five distinct types:

C_p: Rotation about an *axis of symmetry* by $2\pi/p$ radians. A molecule in which such a symmetry operation is possible is said to possess a p-fold axis of rotational symmetry. If the molecule possesses more than one axis of rotational symmetry, then the one with the highest value of p is called the *principal axis*.

σ: Reflection in a *plane of symmetry* (also called a *mirror plane*).

i: Inversion, i.e., "reflection" through a center of symmetry. In the case of a molecule, inversion means that the coordinates x_i, y_i, and z_i of each atom i are replaced by $-x_i$, $-y_i$, and $-z_i$. If this leads to an indistinguishable configuration, we say the molecule has a *center of symmetry*.

S_p: Rotation of $2\pi/p$ radians about an axis followed by reflection in a plane perpendicular to the axis of rotation. This operation is often called an *improper* rotation, and the molecule in which such an operation is possible is said to have a p-fold *alternating axis of symmetry*. Note that S_1 is equivalent to σ and S_2 is equivalent to i.

E: Identity operation. This operation means "do nothing." As we will see later, this operation is needed in order to define *inverse* operations.

All molecules possess the symmetry element E, and some may possess no other symmetry element, e.g., bromochlorofluoromethane. The ammonia molecule has one C_3 axis of symmetry, which passes through the nitrogen atom

---

[2] If symmetry operations which involve translation are involved, the point-group description must be replaced with a *space-group* description. Space groups will not be discussed in this text; they become necessary when discussing the type of symmetry found in crystalline solids.

and through the center of the equilateral triangle defined by the hydrogen atoms. In addition, there are three planes of symmetry bisecting the H–N–H bonds and containing the C_3 axis. Planes of symmetry containing the principal axis are called *vertical* planes of symmetry and are frequently labeled "σ_v." Planes of symmetry which are perpendicular to the principal axis are called *horizontal* planes of symmetry and are written "σ_h." There are also *diagonal* planes of symmetry (written "σ_d") which occur when a molecule has one p-fold axis and p twofold axes. These diagonal planes, p in number, then bisect the angles between successive twofold axes. The geometrical figure formed from the letter H by twisting one upright 90° in a vertical plane perpendicular to the plane of the paper has two σ_d planes. Each of the former uprights is contained by a σ_d plane. The allene molecule is similar to this and also contains two σ_d planes. Note that allene has an S_4 axis even though it has no C_4 axis. Although S_p always exists whenever C_p and a symmetry plane perpendicular to C_p exist, it is not necessary that either of these last two symmetry elements exist separately in order for S_p to exist.

The benzene molecule is an example of a molecule with a center of symmetry. If the coordinates x_i, y_i, and z_i of each atom are replaced by $-x_i$, $-y_i$, and $-z_i$, the molecular figure is unchanged. The staggered conformation of ethane has an S_6 operation: rotation by 60° about the C–C axis followed by reflection in the plane perpendicular to the C–C axis. Note that this molecule has neither a C_6 axis nor a mirror plane perpendicular to the S_6 rotation.

A6-2 THE CONCEPT OF GROUPS

The symmetry operations which send a molecule into itself constitute a point *group*, where "group" is used in its formal mathematical sense: a set of n elements $R_1, R_2, \ldots, R_n$ (where n may be infinite) which satisfy the following conditions with respect to a well-defined operation $\circ$ (symbolic multiplication):

1. $R_i \circ R_j = R_k$ for all i, j, and k. This condition is known as *closure* and means that all product elements must coincide with one of the n group elements.
2. $R_i \circ (R_j \circ R_k) = (R_i \circ R_j) \circ R_k$ for all i, j, and k. This condition means that the operation "$\circ$" is *associative*. This requirement is necessary in principle but is of little practical consequence as far as our needs are concerned.
3. One of the n elements (say, R_1), which we will denote by the special symbol E, must have the property $E \circ R_j = R_j \circ E$ for all j. The element E is called the *group identity* or *identity* element.
4. For every element R_j there must exist an *inverse* element R_j^{-1} (which is already a member of the set) such that $R_j \circ R_j^{-1} = R_j^{-1} \circ R_j = E$.

In some cases, a fifth condition is also met, namely,

5. $R_i \circ R_j = R_j \circ R_i$ for all i and j. This means that all the elements of the group *commute* under the operation "$\circ$,"

The set of n elements is said to constitute a *group of* n*th order*. Groups whose elements satisfy condition 5 for some given operation are called *commutative*, or *abelian*, groups. In general, we will deal with nonabelian groups. Note, however, that as a result of condition 4 every group must have within it an abelian subgroup; i.e., there must exist within the group a set of m elements ($m \leq n$) which themselves satisfy conditions 1 to 5.

The operation "$\circ$" requires some explanation. One can readily show that the real numbers form a group (abelian) if the operation "$\circ$" is identified either as ordinary addition or ordinary multiplication. Under addition, zero is the identity element, and each number has its own negative as its inverse. Under multiplication, unity is the identity element, and each number (zero excluded) has its own reciprocal as its inverse. Since zero has no multiplicative inverse, it must, of course, be excluded from membership in the group. When the elements of a group are operations (or their operator representations), then "$\circ$" is generally interpreted to mean "followed by"; that is, $R_1 \circ R_2$ means the "operation R_2 (on some arbitrary operand) followed by the operation R_1 (on the result of the first operation on the operand)." In such a case it is customary to omit the "multiplication" symbol and to write simply R_1R_2. Of primary concern in the following will be groups of finite order whose elements are the symmetry operations which represent the symmetry elements of molecules.

As a specific example of a molecular symmetry group, let us consider the ammonia molecule which is depicted in Fig. A6-1. The symmetry operations of this molecule are as follows:

C_3^+: Rotation about the z axis by 120° in the counterclockwise direction (when viewed as in Fig. A6-1)
C_3^-: Rotation about the z axis by 120° in the clockwise direction (when viewed as in Fig. A6-1)
σ_1: Reflection in the vertical plane containing hydrogen atom 1
σ_2: Reflection in the vertical plane containing hydrogen atom 2
σ_3: Reflection in the vertical plane containing hydrogen atom 3
E: The identity operation

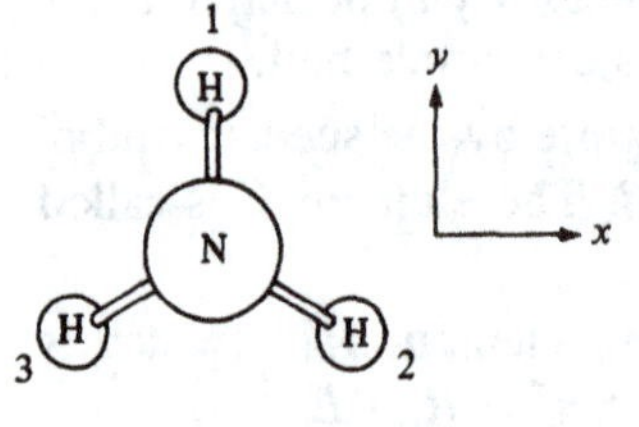

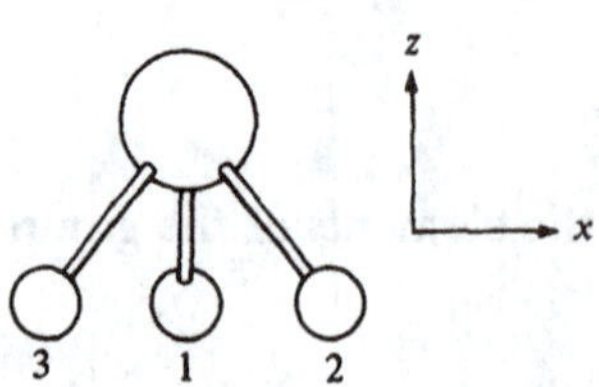

FIGURE A6-1
Cartesian coordinate system used to illustrate the symmetry operations of the ammonia molecule.

TABLE A6-1
Group multiplication table for the symmetry operations of the ammonia molecule

	E	C_3^+	C_3^-	σ_1	σ_2	σ_3
E	E	C_3^+	C_3^-	σ_1	σ_2	σ_3
C_3^+	C_3^+	C_3^-	E	σ_2	σ_3	σ_1
C_3^-	C_3^-	E	C_3^+	σ_3	σ_1	σ_2
σ_1	σ_1	σ_2	σ_3	E	C_3^-	C_3^+
σ_2	σ_2	σ_3	σ_1	C_3^+	E	C_3^-
σ_3	σ_3	σ_1	σ_2	C_3^-	C_3^+	E

Since there are six different symmetry operations, the group of operations is said to be of the *sixth order.*

We begin by constructing the group multiplication table. In so doing we adopt the convention that the symmetry operators rotate or reflect the axes, for example, $\hat{C}_3^+$ (the operator representation of the C_3^+ symmetry element) rotates the x and y axes by 120° in the counterclockwise direction and $\hat{\sigma}_1$ reflects the x and y axes in the plane containing the z axis and hydrogen atom 1. Carrying out all "multiplications" of the general form R_iR_j leads to the "products" in Table A6-1. Note that this is not an abelian group; for example, $\sigma_1\sigma_2 \neq \sigma_2\sigma_1$.

One of the benefits of constructing the group multiplication table is that symmetry elements which may have been overlooked in the original symmetry analysis may make their absence known. For example, had one not realized that the C_3^- operation, as well as the C_3^+ operation, was necessary to complete the group, one would have discovered the omission in forming a product such as $C_3^+C_3^+$ or $\sigma_1\sigma_2$. This illustrates the practical significance of the closure property of groups. Note also that there are symmetry operations such as rotation by 240° around the principal axis which are not separate symmetry operations, since they are equivalent to one of the basic six. In groups having a high degree of symmetry, it is often very difficult to identify all the symmetry elements by simple inspection; hence, the construction of a group multiplication table is a great aid in obtaining a complete description. For example, it is easy to see that the eclipsed conformation of ethane has two C_3 axes, three C_2 axes, one σ_h mirror plane, and three σ_v mirror planes, but it is not immediately obvious to the inexperienced that there are two S_3 improper rotation axes. However, the presence of these axes readily shows up when the group multiplication table is constructed.

EXERCISES

A6-1. Show by construction of a group multiplication table that the H_2O molecule has the four symmetry elements E, C_2, σ_{yz}, and σ_{xz}. Why are C_2^+ and C_2^- not needed? *Note*: Use the yz plane for the molecule, with the z axis bisecting the molecule.

A6-2. Find the inverses of all the symmetry operations of the ammonia and water molecules.

A6-3. Find all the elements of symmetry of ethane in both the staggered and eclipsed conformations.

A6-3 IRREDUCIBLE AND REDUCIBLE REPRESENTATIONS OF POINT GROUPS

Let us consider all functions $\{f_i\}$ that contain all the elements of symmetry of a given point group. If we operate on these functions with the symmetry operations (i.e., the operator representations of the symmetry elements), two different types of behavior are found. In one case we find

$$\hat{R}f_i = \pm f_i \tag{A6-1}$$

where $\hat{R}$ is any of the operators of the point group. In such a case we say the functions $\{f_i\}$ form bases for the *one-dimensional irreducible representations* of the group. The other type of behavior is

$$\hat{R}f_i = \sum_{k=1}^{g} f_k a_{ki} \tag{A6-2}$$

where the $\{a_{ki}\}$ are numerical coefficients. In this case the symmetry operation transforms f_i to a linear combination of functions which contains f_i itself. In this case the subset of functions $\{f_i\}$ $(i = 1, 2, \ldots, g)$ is such that $\hat{R}$, when operating on any one of them, always produces a linear combination of all of them. This type of behavior occurs whenever there is a subspace in which g directions cannot be uniquely defined, i.e., there is a g-fold degeneracy. We then say the subset $\{f_i\}$ $(i = 1, 2, \ldots, g)$ forms the basis of a g-*dimensional irreducible representation* of the group. By "irreducible," we imply that it is not possible to find a smaller set $\{f_i\}$ $(i = 1, 2, \ldots, m;\ m < g)$ within this subset to form a basis.

Functions $\{g_i\}$ which do not contain all the symmetry elements of the point group will behave as follows:

$$\hat{R}g_i = \sum_{k} f_k a_{ki} \tag{A6-3}$$

where the $\{f_k\}$ are bases for an irreducible representation. We then say g_i belongs to a *reducible* representation, since it can always be reduced to a sum of irreducible representations.

Some specific examples will make the above relationships clearer. First, consider the H_2O molecule set up in the coordinate system shown in Fig. A6-2, and consider the following functions: $1s$, $2s$, $2p_x$, $2p_y$, and $2p_z$ (AOs centered on the oxygen atom) and $1s_1$ and $1s_2$ (AOs centered on hydrogen atoms 1 and 2, respectively). The symmetry elements of the point group to which water belongs are E, C_2, σ_{yz}, and σ_{xz}. It is readily verified that

$$\hat{R}f_i = f_i \tag{A6-4}$$

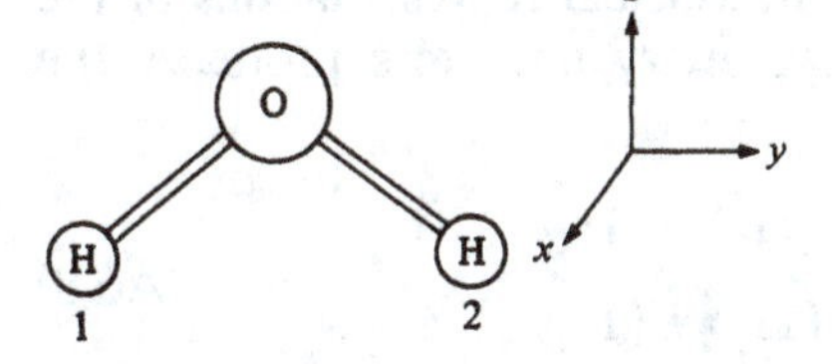

FIGURE A6-2
Cartesian coordinate system used to illustrate the symmetry operations of the water molecule.

whenever f_i is $1s$, $2s$, or $2p_z$; that is, these three functions are eigenfunctions of all the symmetry operators, and all have corresponding eigenvalues of +1. Thus we say that any one of these three functions ($1s$, $2s$, or $2p_z$) forms a one-dimensional irreducible representation of the point group of symmetry operations; we label the irreducible representation to which these belong as "Γ_1." Note that in a symmetry sense, the three functions are equivalent. Note also, that there will be many other functions which are equivalent to any of these three functions and they would also be members of the Γ_1 irreducible representation.

For the $2p_x$ function one obtains the following behavior:

$$\begin{aligned} \hat{E}(2p_x) &= (2p_x) & \hat{C}_2(2p_x) &= -(2p_x) \\ \hat{\sigma}_{zy}(2p_x) &= -(2p_x) & \hat{\sigma}_{zx}(2p_x) &= (2p_x) \end{aligned} \tag{A6-5}$$

Note that two of the symmetry operators have −1 eigenvalues and two have +1 eigenvalues. This different set of operator eigenvalues defines a second irreducible representation which we label "Γ_2."

For the $2p_y$ function we obtain a third set of eigenvalues

$$\begin{aligned} \hat{E}(2p_y) &= (2p_y) & \hat{C}_2(2p_y) &= -(2p_y) \\ \hat{\sigma}_{zy}(2p_y) &= (2p_y) & \hat{\sigma}_{zx}(2p_y) &= -(2p_y) \end{aligned} \tag{A6-6}$$

Thus, $2p_y$ forms the basis of a third irreducible representation which we label "Γ_3."

A fourth irreducible representation would be found by considering the function $3d_{xy}$ centered on the oxygen atom. One finds

$$\begin{aligned} \hat{E}(3d_{xy}) &= (3d_{xy}) & \hat{C}_2(3d_{xy}) &= (3d_{xy}) \\ \hat{\sigma}_{zy}(3d_{xy}) &= -(3d_{xy}) & \hat{\sigma}_{zx}(3d_{xy}) &= -(3d_{xy}) \end{aligned} \tag{A6-7}$$

We label this irreducible representation "Γ_4."

No matter what additional functions we might examine, no new irreducible representations will be found; i.e., the point group to which water belongs has only the four irreducible representations Γ_1, Γ_2, Γ_3, and Γ_4. In general, the total number of different irreducible representations equals the total number of *distinct* symmetry elements. Precisely what is meant by *distinct* will be illustrated later.

Now let us examine the functions $1s_1$ and $1s_2$. These functions do not contain the symmetry elements of the group, and thus they must be linear combinations of other functions which form irreducible representations of the group. Operating on $1s_1$ with the group symmetry operators produces the following results:

$$\hat{E}(1s_1) = (1s_1) \qquad \hat{C}_2(1s_1) = (1s_2)$$
$$\hat{\sigma}_{zy}(1s_1) = (1s_1) \qquad \hat{\sigma}_{zx}(1s_1) = (1s_2) \tag{A6-8}$$

Note that there are two cases in which $1s_1$ does not satisfy an eigenvalue equation. Consequently, $1s_1$ does not belong to any of the possible irreducible representations Γ_1, Γ_2, Γ_3, or Γ_4. It is apparent that $1s_2$ behaves similarly. However, if we define the two linear combinations

$$G_1 = \tfrac{1}{2}(1s_1 + 1s_2) \qquad G_2 = \tfrac{1}{2}(1s_1 - 1s_2) \tag{A6-9}$$

we see that G_1 belongs to Γ_1 and G_2 belongs to Γ_3. Thus, since

$$1s_1 = G_1 + G_2 \qquad \text{and} \qquad 1s_2 = G_1 - G_2 \tag{A6-10}$$

we see that $1s_1$ and $1s_2$ form bases for a *reducible* representation, one which can be reduced to a linear combination of the irreducible representations Γ_1 and Γ_3 [see Eq. (A6-3)].

All the preceding information is customarily summarized by means of a *character table* such as illustrated in Table A6-2. The symbol in the upper-left-hand corner of the table serves as the "name" or "label" of the particular point group; all molecules having only the E, C_2, σ_{zy}, and σ_{zx} symmetry elements belong to the C_{2v} point group. Other molecules belonging to this same point group include HCHO (formaldehyde), pyridine, pyrrole, and chlorobenzene. The numbers (±1 in this case) in the body of the table are called *characters*. In the special case of one-dimensional irreducible representations, these characters are also eigenvalues of the symmetry operators. The left-hand column contains the symbols for the irreducible representations. Rather than the Γ_1, Γ_2, Γ_3, and Γ_4 that we used, it is customary to employ the following:

A or *B* for one-dimensional irreducible representations
E or *T* for two- or three-dimensional irreducible representations, respectively

TABLE A6-2
Character table for any molecule belonging to the same symmetry point group as water

C_{2v}	E	C_2	σ_{zy}	σ_{zx}		
A_1 or Γ_1	1	1	1	1	z	x^2, y^2, z^2
A_2 or Γ_4	1	1	−1	−1	R_z	xy
B_1 or Γ_2	1	−1	−1	1	x, R_y	xz
B_2 or Γ_3	1	−1	1	−1	y, R_x	yz

A is used if the character of the main C_p element is $+1$; B is used if the character of the main C_p element is -1. Subscripts on the A, B, E, T symbols are used as follows:

$1 =$ symmetric under a σ_v operation
$2 =$ antisymmetric under a σ_v operation
g (gerade) $=$ symmetric under an i operation
u (ungerade) $=$ antisymmetric under an i operation

Single primes are used to denote a character of $+1$ for a σ_h operation; double primes denote a character of -1 for the same operation.

The right-hand columns contain sample functions which contain all of the symmetry elements of the group for the specific irreducible representation indicated: we say that such functions *span* that particular irreducible representation. For example, all *ns* AOs centered on O in H_2O span the A_1 irreducible representation. The first irreducible representation listed always has all characters equal to $+1$ and is called the *totally symmetric* irreducible representation; its symbol will be A, A_1, A_g, A_{1g}, etc., depending on the particular point group. The *np* AOs centered on oxygen in H_2O behave as their subscripts; for example $2p_z$ spans A_1 and $2p_x$ spans B_1. The *nd* AOs behave analogously; for example, $3d_{xy}$ spans A_2.

The functions R_x, R_y, and R_z are *rotational* functions. For example, R_x represents rotation about the x axis.

The C_{2v} point group is nondegenerate. Some additional features appear in the character table if some of the irreducible representations are degenerate. As an example, we consider the ammonia molecule which belongs to a point group called C_{3v}. A double degeneracy arises in this molecule (and in any molecule belonging to this same point group) because there is no way of uniquely defining a pair of orthogonal axes in the plane perpendicular to the principal (C_3) axis of rotation; any pair of orthogonal axes is as good as any other.

Using the coordinate system of Fig. A6-1 as a basis, let us consider using the AOs $2p_x$, $2p_y$, and $2p_z$ (centered on the nitrogen atom) as basis functions for the point group. We find that

$$\hat{R}(2p_z) = (2p_z) \tag{A6-11}$$

for all the operations. Thus $2p_z$ spans the totally symmetric irreducible representation which is labeled A_1 in the C_{3v} point group.

An important simplification arises if we note that the two operations C_3^+ and C_3^- are equivalent in the sense that they do the same thing; it is only our arbitrary notion of clockwise and counterclockwise that distinguishes them. Similarly, the operations σ_1, σ_2, and σ_3 are distinguished only by the arbitrary way in which we choose to label hydrogen atoms. Operations of this type are said to belong to the same *class*; thus the two C_3 operations belong to a single class, and the three σ_v operations belong to another class. Operations belonging to a given class are related by a similarity transformation. Thus, if $\hat{R}_k$ is any operator of the point group such that

$$\hat{R}_k^{-1}\hat{R}_j\hat{R}_k = \hat{R}_i \tag{A6-12}$$

then all the operators $\hat{R}_i$ generated in this way belong to the same class as $\hat{R}_j$.

As a consequence of the class property, we need consider only three distinct types of operations for the C_{3v} point group: E, C_3, and σ_v. It is these three distinct operations (or number of different classes) that determines the total number of irreducible representations in the group. In the C_{2v} point group there were *four* classes of symmetry operations and *four* different irreducible representations; in C_{3v} there are *three* classes of symmetry operations and *three* different irreducible representations. Using $\hat{C}_3^+$ as a specific example of a symmetry rotation (see Fig. A6-3), we find that

$$\begin{aligned} \hat{C}_3^+(2p_x) &= -\frac{1}{2}(2p_x) + \frac{\sqrt{3}}{2}(2p_y) = 2p_1 \\ \hat{C}_3^+(2p_y) &= \frac{\sqrt{3}}{2}(2p_x) + \frac{1}{2}(2p_y) = 2p_2 \end{aligned} \tag{A6-13}$$

This may be written in matrix form as

$$[2p_x \quad 2p_y]\begin{bmatrix} -\frac{1}{2} & \frac{\sqrt{3}}{2} \\ \frac{\sqrt{3}}{2} & \frac{1}{2} \end{bmatrix} = [2p_1 \quad 2p_2] \tag{A6-14}$$

This is an example of a unitary transformation. The 2×2 matrix in Eq. (A6-14) is the matrix representation of the $\hat{C}_3^+$ operation in the basis

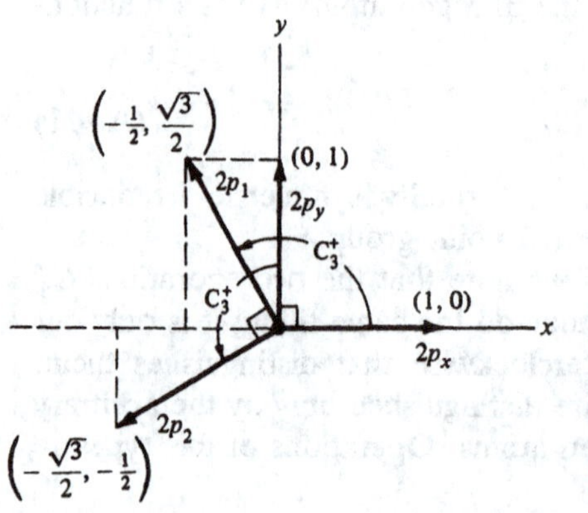

FIGURE A6-3
Transformations of $2p_x$ and $2p_y$ AOs of nitrogen in ammonia under the symmetry operation C_3^+ of the C_{3v} point group (see Fig. A6-1). The functions $2p_x$ and $2p_y$ are treated as vectors $2p_x\mathbf{i}$ and $2p_y\mathbf{j}$, respectively.

$\{2p_x, 2p_y\}$. Similarly, for the $\hat{C}_3^-$ operation one finds

$$[2p_x \quad 2p_y]\begin{bmatrix} \frac{1}{2} & \frac{\sqrt{3}}{2} \\ \frac{\sqrt{3}}{2} & -\frac{1}{2} \end{bmatrix} = [2p_1' \quad 2p_2'] \qquad \text{(A6-15)}$$

where

$$\begin{aligned} 2p_i' &= \frac{1}{2}(2p_x) + \frac{\sqrt{3}}{2}(2p_y) \\ 2p_2' &= \frac{\sqrt{3}}{2}(2p_x) - \frac{1}{2}(2p_y) \end{aligned} \qquad \text{(A6-16)}$$

Since the trace of a matrix is invariant under a similarity transformation (see Theorem A5-4), the traces of the matrix representation of $\hat{C}_3^+$ and $\hat{C}_3^-$ are the same (zero in each case), since both belong to the same class.

Using Fig. A6-4 as a guide, we can carry out a similar procedure for the three σ_v operations. One obtains

$$\boldsymbol{\sigma}_1 = \begin{bmatrix} -1 & 0 \\ 0 & 1 \end{bmatrix} \quad \boldsymbol{\sigma}_2 = \begin{bmatrix} \frac{1}{2} & -\frac{\sqrt{3}}{2} \\ -\frac{\sqrt{3}}{2} & -\frac{1}{2} \end{bmatrix} \quad \boldsymbol{\sigma}_3 = \begin{bmatrix} \frac{1}{2} & \frac{\sqrt{3}}{2} \\ \frac{\sqrt{3}}{2} & -\frac{1}{2} \end{bmatrix} \qquad \text{(A6-17)}$$

These operations, although they belong to a different class from C_3^+ and C_3^-, also happen to have a trace of zero.

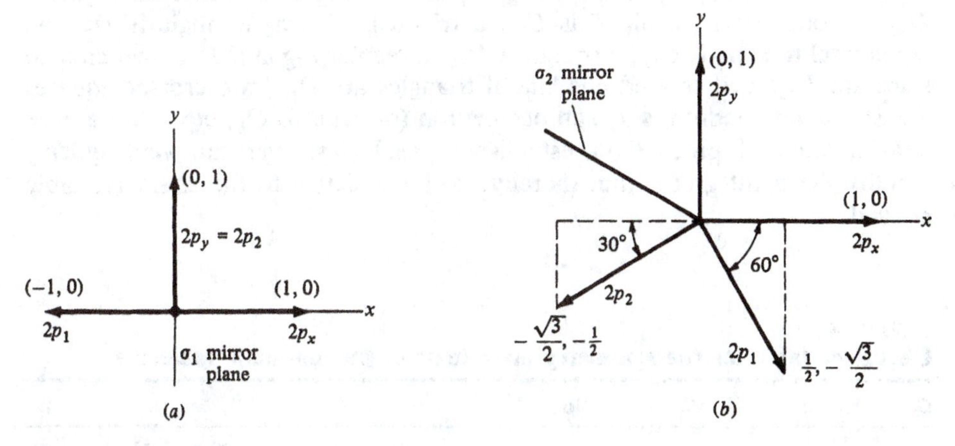

FIGURE A6-4
Transformation of $2p_x$ and $2p_y$ AOs of nitrogen in ammonia under (*a*) the $\hat{\sigma}_1$ operation and (*b*) the $\hat{\sigma}_2$ operation of the C_{3v} point group (see Fig. A6-1). The functions $2p_x$ and $2p_y$ are treated as vectors $2p_x\mathbf{i}$ and $2p_y\mathbf{j}$, respectively.

The matrix representation of the identity operation is always the unit matrix. Thus in the basis $\{2p_x, 2p_y)$

$$\mathbf{E} = \begin{bmatrix} 1 & 0 \\ 0 & 1 \end{bmatrix} \tag{A6-18}$$

We say that $\{2p_x, 2p_y\}$ form a basis for a doubly degenerate irreducible representation of the C_{3v} point group; alternatively, we say these functions span a doubly degenerate irreducible representation. If we had used any other basis functions to represent this irreducible representation, the actual matrices [Eqs. (A6-14, 15, 17, 18)] would generally have been different, but their traces would have been the same. Thus, for degenerate irreducible representations, the traces of the matrix representations of the symmetry operators—since they are basis-invariant—serve as the characters. Table A6-3 is a character table for the C_{3v} point group which summarizes all the symmetry behavior of the group. The symbol $2C_3$ means that there are two C_3 operations (C_3^+ and C_3^-) in this class of operations, and $3\sigma_v$ refers to the three operations (σ_1, σ_2, and σ_3) of this class.

The notation used in this table to label the point groups is called the *Schoenflies* notation. The interested reader should consult some of the more comprehensive texts listed in the suggested readings at the end of this appendix to obtain details of how these symbols are assigned. In practice, however, it is relatively easy to learn to recognize the point group to which a molecule belongs without having any familiarity with the formal rules. This is done by remembering some simple members of each point group and using analogy to place new molecules into the appropriate point group. With a little practice one can become quite proficient in correctly assigning even some of the trickier structures to the appropriate point group. For example, a cylindrical object is $D_{\infty h}$, a cone is $C_{\infty v}$, a bipod is C_{2v}, a tripod is C_{3v}, a rectangle is D_{2h}, an equilateral triangle is D_{3h}, a square is D_{4h}, a regular n-gon is D_{nh}, two crossed slabs are D_{2d}, two crossed equilateral triangles are D_{3d}, two crossed squares are D_{4d}, a tetrahedron is T_d, an octahedron (or cube) is O_h, etc. With a little imagination and practice in establishing analogies, one can very quickly identify the point group and, thereby, go immediately to the character table needed.

TABLE A6-3
Character table for the symmetry operations of the ammonia molecule

C_{3v}	E	$2C_3$	$3\sigma_v$		
A_1	1	1	1	z	x^2+y^2, z^2
A_2	1	1	−1	R_z	
E	2	−1	0	$(x, y), (R_x, R_y)$	$(x^2-y^2, xy), (xz, yz)$

Table A6-4 lists some commonly encountered point groups, along with simple examples of each.

Flow charts have been developed as aids to the rapid determination of point groups,[3] but these must be used with some caution, since some of the choices one needs to make are difficult to do correctly without using molecular models. This difficulty is especially evident in the case of the D_{nd} point groups.

In determining the symmetry elements of a molecule, it is helpful to assign arbitrary numbers to symmetrically indistinguishable atoms and to note that each symmetry operation represents a different way of labeling these atoms. Figure A6-5 illustrates the use of this technique for identifying the eight symmetry operations of the D_{2d} point group, using allene ($CH_2{=}C{=}CH_2$) as a specific example.

Before using any character table, it is important to find out what coordinate conventions the compiler has used. Otherwise, you and the compiler may not use the same symbols for some of the irreducible representations. Although some nonuniformity of coordinate conventions exists, most electronic spectroscopists use the following:

[3] R. L. Carter, *J. Chem. Ed.*, **45**:44 (1968); see the cautionary notes by D. Quane, *J. Chem. Ed.*, **53**:190 (1976) and J. H. Noggle, *J. Chem. Ed.*, **53**:190 (1976).

TABLE A6-4
Molecules belonging to some of the most common point groups*

Point group	Examples
$D_{\infty h}$	All linear molecules with a center of symmetry
$C_{\infty v}$	All linear molecules without a center of symmetry
C_{nv}	$n = 2$: H_2O, pyridine, pyrrole, cyclohexane (boat)
	$n = 3$: ammonia, PCl_3, CH_3Cl
D_{nh}	$n = 2$: ethene, naphthalene
	$n = 3$: eclipsed ethane
	$n = 4$: cyclobutadiene
	$n = 5$: cyclopentadienyl anion
	$n = 6$: benzene
D_{nd}	$n = 2$: allene
	$n = 3$: staggered ethane, cyclohexane (chair)
C_2	H_2O_2 (not planar)
C_s	Styrene (planar)
C_i	Staggered *anti*-ClFHC-CHFCl
T_d	Methane (tetrahedral)
O_h	SF_6 (octahedral)
C_{2h}	*trans*-1,2-Dichloroethene

* Character tables for each of these point groups are found in App. 7.

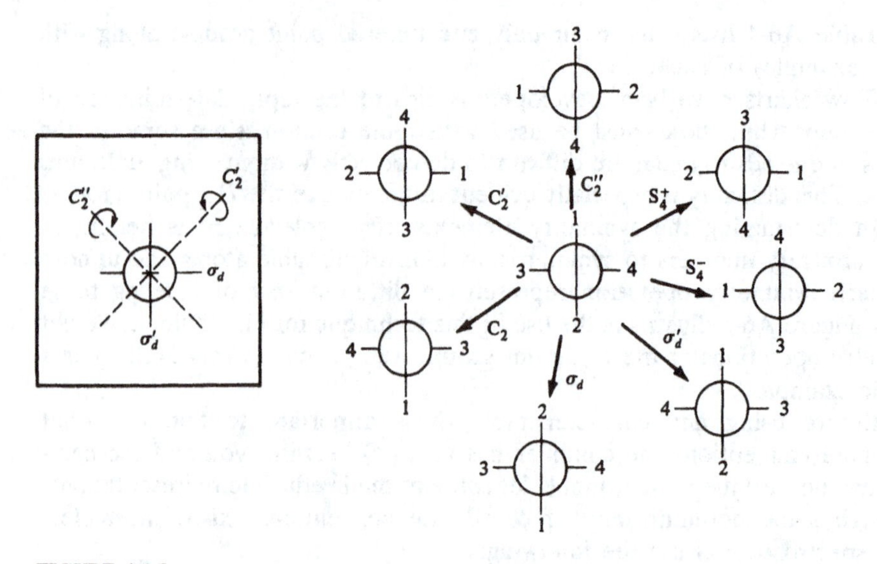

FIGURE A6-5
The symmetry operations of the D_{2d} point group using allene ($CH_2{=}C{=}CH_2$) as an example. The two C_2' elements are labeled C_2' and C_2'' in the illustration (see inset), the two σ_d are labeled σ_d and σ_d', and the two S_4 are labeled S_4^+ and S_4^-. Numbering the equivalent hydrogen atoms makes it easier to see how the various operations differ. Note that the molecule has $p = 2$, so that there are *one* C_p axis and p C_2' axes with p diagonal planes.

1. The origin is chosen as the center of mass of the molecule.
2. The z axis is chosen to coincide with the principal axis. If all the symmetry axes are of the same order, the one passing through the greatest number of atoms is the z axis. If this still leaves a choice, the axis passing through the greatest number of bonds is the z axis. Ethene, Fig. 13-6, is an example of this convention.
3. In planar molecules, the x axis is chosen perpendicular to the plane of the molecule unless convention 2 has led to choosing of the z axis as perpendicular to the plane of the molecule, e.g., as in benzene.

EXERCISES

A6-4. Use Eq. (A6-12) to demonstrate that σ_1, σ_2, and σ_3 belong to the same class in the C_{3v} point group.

A6-5. Determine the point groups to which each of the following belongs: H_2S, ferrocene, B_2H_6, isomers of dichlorobenzene, PCl_5, phenanthrene, S_8, HOCl, HCN, *cis*- and *trans*-1,4-butadiene, BF_3, triazole, methylacetylene, spiropentene, HD, $Fe(SCN)_6^{3-}$, CH_2Cl_2, $CHCl_3$, 1,5-dichloronaphthalene, the letters of the alphabet (uppercase), a football (with and without stitching), a cone, a tennis ball (include the seam), a tennis racket, a brick, scissors (ignore the screw and assume both handles are equivalent), a teacup, a horizontal piton (see Royal Robbins, *Basic Rockcraft*, La Siesta Press, Glendale, Calif., 1971), a hen's egg (not $D_{\infty h}$), and this book.

A6-6. The compound A_2H_8 (mythalane) has symmetry D_{2d}. Sketch a possible geometry for this molecule.

A6-4 GENERAL PROPERTIES OF IRREDUCIBLE REPRESENTATIONS

The characters of a point group have some important properties. One of these is that the sum of the squares of the characters of any given symmetry operation over all the irreducible representations satisfies

$$\sum_{\mu} [\chi^{(\mu)}(R)]^2 = \frac{g}{n_R} \tag{A6-19}$$

where $\chi^{(\mu)}(R)$ = character of the operation R for the μth irreducible representation
g = order of the group
n_R = number of operations in the same class as R

The characters of each irreducible representation satisfy

$$\sum_{R} [\chi^{(\mu)}(R)]^2 = g \tag{A6-20}$$

provided the summation is over all the operators in each class. The above shows that a set of characters associated with either a given operation (the columns of the table) or a given irreducible representation (the rows of the table) behave like vectors normalized to g/n_R or g, respectively. The characters also display "orthogonality," such that

$$\begin{aligned} \sum_{\mu} \chi^{(\mu)}(R_k)\chi^{(\mu)}(R_j) &= \frac{g\delta_{kj}}{n_R} \\ \sum_{R} \chi^{(\mu)}(R)\chi^{(\nu)}(R) &= g\delta_{\mu\nu} \end{aligned} \tag{A6-21}$$

In the case that the characters are complex, terms such as $\chi^{(\mu)}(R)\chi^{(\nu)}(R)$ must be replaced with $[\chi^{(\mu)}(R)]^*\chi^{(\nu)}(R)$, and $[\chi^{(\mu)}(R)]^2$ must be written $|\chi^{(\mu)}(R)|^2$.

It should be noted again that the number of classes of symmetry operations always equals the total number of different irreducible representations.

In general, it is easier to find reducible representations of a group than it is to find the irreducible representations. Fortunately, it is often not necessary to find irreducible representations explicitly; instead, it is often sufficient to determine how many times a given irreducible representation is found in a given reducible representation. The character of the matrix representation of the jth symmetry operation in a reducible representation can be written in the general form

$$\chi(R_j) = \sum_{\mu} a_\mu \chi^{(\mu)}(R_j) \tag{A6-22}$$

where a_μ is the number of times the μth irreducible representation is represented in a given reducible representation. This expression can be regarded as analogous to the expansion of an arbitrary vector $\chi(R_j)$ in the basis set $\{\chi^{(\mu)}(R_j)\}$. Multiplying Eq. (A6-22) from the left by $[\chi^{(\nu)}(R_j)]^*$ and summing over all operations j, we get

$$\sum_j [\chi^{(\nu)}(R_j)]^* \chi(R_j) = \sum_j \sum_\mu a_\mu [\chi^{(\nu)}(R_j)]^* \chi^{(\mu)}(R_j) = a_\mu \delta_{\mu\nu} n \tag{A6-23}$$

Solving for a_μ, we obtain

$$a_\mu = \frac{1}{n} \sum_j [\chi^{(\mu)}(R_j)]^* \chi(R_j) \tag{A6-24}$$

This is analogous to finding the Fourier coefficients of a vector expansion. As an example, suppose we wish to use the N–H bond orbitals h_1, h_2, and h_3 as a basis for a three-dimensional representation of the ammonia molecule (these bond orbitals may be regarded as hybrids of s and p AOs). The character table of the reducible representation (which we label Γ) is as follows:

C_{3v}	E	$2C_3$	$3\sigma_v$	
Γ	3	0	1	(h_1, h_2, h_3)

Using Eq. (A6-24) and the C_{3v} character table, we obtain

$$\begin{aligned} a(A_1) &= \tfrac{1}{6}[(3)(1) + 2(0)(1) + 3(1)(1)] = 1 \\ a(A_2) &= \tfrac{1}{6}[(3)(1) + 2(0)(1) + 3(1)(-1)] = 0 \\ a(E) &= \tfrac{1}{6}[(3)(2) + 2(0)(-1) + 3(1)(0)] = 1 \end{aligned} \tag{A6-25}$$

This means that the reducible representation Γ contains A_1 once, A_2 not at all, and E once. We then say the reducible representation Γ represents the direct sum

$$\Gamma = A_1 \oplus E \tag{A6-26}$$

It should be noted that the application of the above procedure presupposes the availability of the character table for the group in question. Also, the procedure is valid only when the group in question is of finite order. For groups of infinite order (e.g., $D_{\infty h}$ or $C_{\infty v}$), another procedure must be used.[4]

[4] D. P. Strommen and E. R. Lippincott, *J. Chem. Ed.*, **49**:3412 (1972).

Next we consider (without proof) a theorem by Wigner which is perhaps the most important single result of group theory:

Theorem A6-1. Every eigenfunction of the hamiltonian of a given system corresponds to an irreducible representation of the point group to which the system belongs; i.e., each eigenfunction spans a g-fold dimensional irreducible representation, where g is the degeneracy (≥ 1) of the state.

Sometimes it happens that two or more functions of different irreducible representations lead to the same value of the energy. This is called *accidental*, or *nonessential*, *degeneracy* and is the result of a special choice of constants in the system hamiltonian. This accidental degeneracy can be removed without changing the symmetry properties of the hamiltonian, i.e., by changing only the magnitude of some constant factor.

The hamiltonian operator of a system is invariant under all the symmetry operations of the group to which the system belongs and thus commutes with all these symmetry operations. Consequently, matrix elements of the hamiltonian between different irreducible representations must vanish (see Theorem 2-5). This fact is useful in the simplification and factorization of secular determinants arising in quantum mechanical problems. In fact, this enables one to separate a set of secular equations into several smaller and independent sets of equations, each set corresponding to some irreducible representation of the group.

A6-5 THE DIRECT-PRODUCT REPRESENTATION

Let R be a symmetry operation of a given point group, and let $\alpha_1, \alpha_2, \ldots, \alpha_\mu$ and $\beta_1, \beta_2, \ldots, \beta_\nu$ be two sets of functions (same or different) which form bases for two irreducible representations (same or different) of the group. If we let $\mathbf{R}_\alpha$ and $\mathbf{R}_\beta$ represent the matrix representations of the operation R in the bases $\{\alpha_i\}$ and $\{\beta_i\}$, respectively, we can write

$$\hat{R}\alpha_k = \sum_{i=1}^{\mu} \alpha_i(\mathbf{R}_\alpha)_{ik} \qquad \hat{R}\beta_l = \sum_{j=1}^{\nu} \beta_j(\mathbf{R}_\beta)_{jl} \tag{A6-27}$$

Now consider the $\mu\nu$ functions formed by all possible products of the form $\alpha_k\beta_l$ ($k = 1, 2, \ldots, \mu$; $l = 1, 2, \ldots, \nu$). These functions form the basis for a representation (generally reducible) of dimension $\mu\nu$. Letting $\mathbf{R}_{\alpha\beta}$ be the matrix representation of the operation R in the basis $\{\alpha_k\beta_l\}$, we can write [since $\hat{R}(\alpha_k\beta_l) = \hat{R}\alpha_k\hat{R}\beta_l$]

$$\hat{R}\alpha_k\beta_l = \sum_{i,j} \alpha_i\beta_j(\mathbf{R}_\alpha)_{ik}(\mathbf{R}_\beta)_{jl} = \sum_{i,j} \alpha_i\beta_j(\mathbf{R}_{\alpha\beta})_{ij,kl} \tag{A6-28}$$

where $(\mathbf{R}_{\alpha\beta})_{ij,kl}$ is the matrix element of the representation of $\hat{R}$ in the basis $\{\alpha_k\beta_l\}$. The matrix $\mathbf{R}_{\alpha\beta}$ is said to be a *direct product* (or Kronecker product) of the matrices $\mathbf{R}_\alpha$ and $\mathbf{R}_\beta$ and is defined formally by

$$(\mathbf{R}_{\alpha\beta})_{ij,kl} = (\mathbf{R}_\alpha \otimes \mathbf{R}_\beta)_{ij,kl} = (\mathbf{R}_\alpha)_{ik}(\mathbf{R}_\beta)_{jl} \tag{A6-29}$$

where the double index ij represents a row of the direct-product matrix and kl represents a column of this matrix. The character of the direct-product matrix is given by

$$\mathrm{Tr}\,(\mathbf{R}_\alpha \otimes \mathbf{R}_\beta) = \sum_{i,j} (\mathbf{R}_{\alpha\beta})_{ij,ij} = \sum_{i,j} (\mathbf{R}_\alpha)_{ii}(\mathbf{R}_\beta)_{jj} = \mathrm{Tr}\,\mathbf{R}_\alpha\ \mathrm{Tr}\,\mathbf{R}_\beta \tag{A6-30}$$

Thus the characters of the direct-product representation are just the products of the characters of the component irreducible representations.

As a simple example of direct-product representations, let us consider those we can construct from the irreducible representations of the C_{3v} point group. Using the C_{3v} character table and Eq. (A6-30), we find that the direct-product representation of A_1 and A_2 has a character table which is just the A_2 irreducible representation.

C_{3v}	E	$2C_3$	$3\sigma_v$
$A_1 \otimes A_2$	1	1	−1

Symbolically we write

$$A_1 \otimes A_2 = A_2 \tag{A6-31}$$

Similarly, we find

$$A_1 \otimes A_1 = A_1 \qquad A_2 \otimes A_2 = A_1 \tag{A6-32}$$

Equations (A6-31) and (A6-32) simply state that the direct product of symmetric and antisymmetric irreducible representations (with respect to σ_v) is antisymmetric and that the direct square of either a symmetric or an antisymmetric irreducible representation is symmetric. The concept of direct-product representations can be said to be a generalization of the multiplication of functions of even and odd parity.

The character table for the $E \otimes E$ product is

C_{3v}	E	$2C_3$	$3\sigma_v$
$E \otimes E$	4	1	0

which is a reducible representation. Applying Eq. (A6-24), one finds

$$a(A_1) = a(A_2) = a(E) = 1 \tag{A6-33}$$

which means that the $E \otimes E$ direct-product representation is reducible to the direct sum of A_1, A_2, and E, that is,

$$E \otimes E = A_1 \oplus A_2 \oplus E \tag{A6-34}$$

The characters of the direct-product representation satisfy

$$\chi^{(E\otimes E)}(R) = \chi^{(A_1)}(R) + \chi^{(A_2)}(R) + \chi^{(E)}(R) \tag{A6-35}$$

In general, the character of the matrix representation of the operation R in a direct-product representation formed from the irreducible representations Γ_1 and Γ_2 will satisfy

$$\chi^{(\Gamma_1\otimes\Gamma_2)}(R) = \sum_\mu a_\mu \chi^{(\mu)}(R) \tag{A6-36}$$

Table A6-5 is the direct-product multiplication table for the C_{3v} point group. One important feature of this table (common to all tables of direct products) is that the totally symmetric representation A_1 (or A_{1g}, etc.) is found only in the diagonal entries.

It is also useful to note that if Γ_A is the totally symmetric irreducible representation, then

$$\Gamma_A \otimes \Gamma_i = \Gamma_i \tag{A6-37}$$

For all nondegenerate irreducible representations,

$$\Gamma_i \otimes \Gamma_i = \Gamma_A \tag{A6-38}$$

For degenerate irreducible representations

$$\Gamma_i \otimes \Gamma_i = \Gamma_A + \text{others} \tag{A6-39}$$

Direct products for the $C_{\infty v}$ and $D_{\infty h}$ groups (of infinite order) are determined by the condition that the degeneracy of the product is the product of the degeneracies and from the trigonometric formula describing the infinite-fold rotation:

$$\chi^{(\mu)}\chi^{(\nu)} = \cos(\lambda_\mu \Phi)\cos(\lambda_\nu \Phi) = \cos[(\lambda_\mu + \lambda_\nu)\Phi] + \cos[(\lambda_\mu - \lambda_\nu)\Phi] \tag{A6-40}$$

where Φ represents an angle of rotation about the principal axis of the molecule. Table A6-6 gives the direct-product multiplication for the $C_{\infty v}$ group; that for the $D_{\infty h}$ group is easily obtained from the same table.

TABLE A6-5
Direct-product multiplication table for the C_{3v} point group

C_{3v}	A_1	A_2	E
A_1	A_1	A_2	E
A_2	A_2	A_1	E
E	E	E	$A_1 \oplus A_2 \oplus E$

TABLE A6-6
Direct-product multiplication table for some of the irreducible representations of the $C_{\infty v}$ point group

$C_{\infty v}$	A_1	A_2	E_1	E_2	E_3
A_1	A_1	A_2	E_1	E_2	E_3
A_2	A_2	A_1	E_1	E_2	E_3
E_1	E_1	E_1	(A_1, A_2, E_2)	(E_1, E_3)	(E_2, E_4)
E_2	E_2	E_2	(E_1, E_3)	(A_1, A_2, E_4)	(E_1, E_5)
E_3	E_3	E_3	(E_2, E_4)	(E_1, E_5)	(A_1, A_2, E_6)

Note: Direct products for $D_{\infty h}$ are obtained by adding the g and u subscripts and recalling $g \otimes g = u \otimes u = g$ and $u \otimes g = u$.

Direct-product representations are of particular usefulness in the evaluation of molecular integrals. The integral

$$I = \langle \psi_A | \hat{F} | \psi_B \rangle \tag{A6-41}$$

where ψ_A and ψ_B form basis functions for irreducible representations of the group and $\hat{F}$ is a hermitian operator, is nonzero only if it is invariant under all symmetry operations of the group. This is just a generalization of the fact that the integral

$$\int_{-a}^{a} f(x)g(x)\,dx \tag{A6-42}$$

is nonzero only if $f(x)$ and $g(x)$ are both even functions of x or both odd functions of x; that is, $f(x)$ and $g(x)$ must belong to the same irreducible representation of the C_i point group. Since the value of the integral can be thought of as representing an area, it is necessary that this area remain invariant under any symmetry operations of the group. This can happen only if the direct product of the irreducible representations to which $f(x)$ and $g(x)$ belong contains the A_1 irreducible representation (the totally symmetric representation). We can also say that the value of any integral representing a real, measurable quantity must be independent of the arbitrary labeling of the components of the system.

We then see that the integral in Eq. (A6-41) is nonzero only if the direct product $\Gamma_A \otimes \Gamma_F \otimes \Gamma_B$ contains the A_1 (or totally symmetric) irreducible representation. This will be true if the direct product $\Gamma_A \otimes \Gamma_B$ contains Γ_F. If $\hat{F}$ is the hamiltonian (always A_1, etc.), it is necessary that $\Gamma_A = \Gamma_B$ in order for the integral (A6-41) to be nonzero. This is just the conclusion reached in the last paragraph of Sec. A6-4 on the basis of Theorem 2-5.

In the determination of selection rules one considers integrals such as

$$I = \langle \psi_j | x | \psi_k \rangle \tag{A6-43}$$

It is apparent from the above discussion that such an integral vanishes unless $\Gamma_k \otimes \Gamma_j = \Gamma_x$; that is, the function x must transform in the same way as one of the irreducible representations contained in $\Gamma_k \otimes \Gamma_j$. If ψ_k is the ground state (usually A_1, etc.), then transitions $\psi_k \rightarrow \psi_j$ are allowed only if $\Gamma_j = \Gamma_x$; that is, x must transform under the group operations in the same way as the wavefunction of the upper state.

A6-6 SYMMETRY PROJECTION OPERATORS

It is often useful to take a function which, in general, is a direct sum of two or more irreducible representations (of a particular point group) and project functions from this which exhibit the symmetries associated with the component irreducible representations. A particular form of projection operator which is often adequate for this purpose is defined by

$$\hat{O}^{(\mu)} = \sum_R \chi^{(\mu)}(R)\hat{R} \tag{A6-44}$$

Using the C_{3v} point group as a specific example, one can form three projection operators of the above form: one for each of the three irreducible representations. These are

$$\begin{aligned} \hat{O}^{(A_1)} &= \hat{E} + \hat{C}_3^+ + \hat{C}_3^- + \hat{\sigma}_1 + \hat{\sigma}_2 + \hat{\sigma}_3 \\ \hat{O}^{(A_2)} &= \hat{E} + \hat{C}_3^+ + \hat{C}_3^- - \hat{\sigma}_1 - \hat{\sigma}_2 - \hat{\sigma}_3 \\ \hat{O}^{(E)} &= 2\hat{E} - \hat{C}_3^+ - \hat{C}_3^- \end{aligned} \tag{A6-45}$$

We will now consider the arbitrary function x^2 and show that the above projection operators will project from it two basis functions belonging to two of the irreducible representations of the C_{3v} point group. Using Eqs. (A6-14), (A6-15), and (A6-17), one can obtain the following when the various group operations are carried out on the function x^2:

$$\begin{aligned} \hat{C}_3^+ x^2 &= \left(-\frac{1}{2}x + \frac{\sqrt{3}}{2}y\right)^2 = \frac{1}{4}x^2 - \frac{\sqrt{3}}{2}xy + \frac{3}{4}y^2 \\ \hat{C}_3^- x^2 &= \left(-\frac{1}{2}x - \frac{\sqrt{3}}{2}y\right)^2 = \frac{1}{4}x^2 + \frac{\sqrt{3}}{2}xy + \frac{3}{4}y^2 \\ \hat{\sigma}_1 x^2 &= (-x)^2 = x^2 \\ \hat{\sigma}_2 x^2 &= \left(\frac{1}{2}x - \frac{\sqrt{3}}{2}y\right)^2 = \frac{1}{4}x^2 - \frac{\sqrt{3}}{2}xy + \frac{3}{4}y^2 \\ \hat{\sigma}_3 x^2 &= \left(\frac{1}{2}x + \frac{\sqrt{3}}{2}y\right)^2 = \frac{1}{4}x^2 + \frac{\sqrt{3}}{2}xy + \frac{3}{4}y^2 \\ \hat{E}x^2 &= x^2 \end{aligned} \tag{A6-46}$$

Now applying the projection operators [Eq. (A6-45)] to x^2, one obtains

$$\hat{O}^{(A_1)}x^2 = x^2 + y^2 \qquad \hat{O}^{(A_2)}x^2 = 0 \qquad \hat{O}^{(E)}x^2 = x^2 - y^2 \tag{A6-47}$$

where irrelevant multiplicative factors have been omitted. Looking back at the right-hand side of the C_{3v} character table (Table A6-3), we see that these are indeed basis functions of the given irreducible representations. Note that the function x^2 has no component belonging to the A_2 irreducible representation; thus we say that x^2 belongs to the $A_1 \oplus E$ reducible representation in the C_{3v} point group.

The 1s AOs of the hydrogen atoms of ammonia do not individually form bases for irreducible representations of the C_{3v} point group, and, therefore, each 1s AO must belong to a reducible representation. To determine what the irreducible representation components are, we use the projection-operator technique on one of the 1s AOs. For example, an A_1 component is obtained from H_1 (see Fig. A6-1) as follows:

$$\hat{O}^{(A_1)}H_1 = \tfrac{1}{6}(H_1 + H_2 + H_3 + H_1 + H_2 + H_3) = \tfrac{1}{3}(H_1 + H_2 + H_3) \tag{A6-48}$$

(the numerical factor $\frac{1}{3}$ is irrelevant and is often omitted). This combination is known as a *group orbital* (or *symmetry-adapted orbital*) and is given the symbol G_s (the subscript s denotes that the group orbital transforms like an s-type AO centered on N). Applying the A_2 projection operator to H_1 leads to

$$\hat{O}^{(A_2)}H_1 = 0 \tag{A6-49}$$

showing that the reducible representation does not contain an A_2 component. Applying the projection operator for the doubly degenerate irreducible representation (E) to the three hydrogen atom 1s AOs produces the following results:

$$\begin{aligned} \hat{O}^{(E)}H_1 &= 2H_1 - H_2 - H_3 \\ \hat{O}^{(E)}H_2 &= 2H_2 - H_1 - H_3 \\ \hat{O}^{(E)}H_3 &= 2H_3 - H_1 - H_2 \end{aligned} \tag{A6-50}$$

(where numerical coefficients have been reduced to the smallest numbers). Only two of these can be linearly independent. If these are also chosen to be orthogonal, one possibility is

$$\begin{aligned} \hat{O}^{(E)}H_1 &= 2H_1 - H_2 - H_3 \\ \hat{O}^{(E)}(H_1 - H_3) &= H_1 - H_3 \end{aligned} \tag{A6-51}$$

However, the choices in Eq. (A6-51) are not unique; in fact, there are an infinite number of equally acceptable combinations leading to a doubly degenerate set of functions. For example, the following two group orbitals are also of E symmetry and are orthogonal:

$$\hat{O}^{(E)}(H_1 + H_2) = H_1 + H_2 - 2H_3 \qquad \hat{O}^{(E)}(H_1 - H_2) = (H_1 - H_2) \tag{A6-52}$$

The first and second group orbitals in either Eq. (A6-50) or Eq. (A6-51) may be labeled G_x and G_y, respectively, since they behave much like a pair of $2p_x$ and $2p_y$ AOs centered on nitrogen.

In summary, the group orbitals may be written in normalized form as follows:

$$G_s = \frac{1}{\sqrt{3}}(H_1 + H_2 + H_3)$$

$$G_x = \frac{\sqrt{2}}{3}H_1 - \frac{1}{\sqrt{6}}(H_2 + H_3) \qquad \text{(A6-53)}$$

$$G_y = \frac{1}{\sqrt{2}}(H_2 - H_3)$$

Note that the above normalization assumes that H_1, H_2, and H_3 are themselves orthonormal. Consideration of their actual nonorthogonality leads to normalization factors involving the overlap integrals between two such functions. For example, the normalization factor for the G_s function becomes $[3(1+2\Delta)]^{-1/2}$, where Δ is the overlap integral between any two (normalized) hydrogen atom 1s AOs.

Equations (A6-53) can be written in the matrix form

$$\mathbf{HT} = \mathbf{G} \qquad \text{(A6-54)}$$

where

$$\mathbf{H} = [H_1 \quad H_2 \quad H_3] \qquad \mathbf{G} = [G_s \quad G_x \quad G_y] \qquad \mathbf{T} = \begin{bmatrix} \frac{1}{\sqrt{3}} & \sqrt{\frac{2}{3}} & 0 \\ \frac{1}{\sqrt{3}} & -\frac{1}{\sqrt{6}} & \frac{1}{\sqrt{2}} \\ \frac{1}{\sqrt{3}} & -\frac{1}{\sqrt{6}} & -\frac{1}{\sqrt{2}} \end{bmatrix}$$

(A6-55)

Note that the matrix **T** is unitary. If we let **G** be an arbitrary matrix of one A_1 function and a pair of E functions, the inverse transformation

$$\mathbf{H} = \mathbf{GT}^\dagger \qquad \text{(A6-56)}$$

defines a matrix **H** of functions forming a basis for the irreducible representation $A_1 \oplus E$. These functions can be regarded as *bond functions*, since they behave like vectors directed as much as possible along the directions of the N–H bonds.

A more complete discussion of projection operators and their use in chemistry has been given by Sannigrahi.[5]

---

[5] A. B. Sannigrahi, *J. Chem. Ed.*, **52**:307 (1975).

EXERCISES

A6-7. Analyze the function $(x + y + z)^2$ in the C_{2v} point group as to the irreducible representations it contains. Repeat for the same function in the C_{3v} point group.

A6-8. Using projection operators for the C_i point group, find the A_g and A_u components of the function $3x^2 - 6x^3 + x^4$. Do you need formal projection operators to do this?

A6-9. Demonstrate that the group orbitals of ammonia given in Eq. (A6-53) are orthogonal.

A6-10. Verify the normalization constant given for G_s in Eq. (A6-53) for the case of normalized but nonorthogonal 1*s* AOs for the H atom.

A6-11. Verify that the normalization factor for a group orbital $G = H_1 + H_2 + \cdots + H_n$ is given by

$$N = \frac{1}{\sqrt{n + n(n-1)\Delta}}$$

if it is assumed that $\langle H_i | H_j \rangle = \Delta$ for all $i \neq j$ and 1 for all $i = j$.

A6-12. Show that the normalization factor for the group orbital $H_1 + H_2 - H_3 - H_4$ is $(4 - 3\Delta)^{-1/2}$ if it is assumed that $\langle H_i | H_j \rangle = \Delta$ for all $i \neq j$ and 1 for all $i = j$.

A6-7 SYMMETRY AND MOLECULAR SPECTRA

The systematic use of symmetry is of great aid in the interpretation of infrared, Raman, and ultraviolet-visible spectroscopy. First, we will see how the vibrational modes of a molecule can be classified as to symmetry type and how the infrared and Raman activities of these modes can be predicted. Then we will see how selection rules are determined for molecular electronic transitions.

The symmetries of the normal vibrational modes of a molecule of N atoms are determined from a $3N$-dimensional reducible representation of the group to which the molecule belongs, using the cartesian coordinates (x, y, z) of each atom as a basis for the representation. This $3N$-dimensional reducible representation contains the irreducible representations which represent three translational motions, two (or three) rotational motions, and $3N - 5$ (or $3N - 6$) vibrational modes. The translational and rotational modes are readily identified; the remaining irreducible representations describe the fundamental vibrational modes. We illustrate the procedure using H_2O (C_{2v} point group) as an example.

When a symmetry operation $\hat{R}$ is performed on the coordinates of a particular atom of a molecule, two different results are possible:

$$\hat{R}\text{ (atom coordinates)} = \text{shift of position}$$
$$\hat{R}\text{ (atom coordinates)} = \text{no shift of position}$$

Using Fig. A6-2 as an example, we see that the two hydrogen atom and one oxygen atom coordinates (labeled as H_1, H_2, and O, respectively) behave as follows with respect to the $\hat{C}_2$ operation:

$$\hat{C}_2(\mathrm{H}_1) = \mathrm{H}_2 \qquad \hat{C}_2(\mathrm{H}_2) = \mathrm{H}_1 \qquad \hat{C}_2(\mathrm{O}) = \mathrm{O} \tag{A6-57}$$

Note that the H atom coordinates shift position (the two H atoms are interchanged), but the O atom remains unshifted. The characters of the $3N$-dimensional reducible representation we seek are just the traces of the matrices which represent what happens to the coordinates of the *unshifted* atoms. Thus the C_2 operation (which does not shift the O atom) has the following effect on the coordinates of the oxygen atom: $x \rightarrow -x$, $y \rightarrow -y$, and $z \rightarrow z$. This may be written in matrix form as

$$[x \quad y \quad z]\begin{bmatrix} -1 & 0 & 0 \\ 0 & -1 & 0 \\ 0 & 0 & 1 \end{bmatrix} = [-x \quad -y \quad z] \tag{A6-58}$$

where the 3×3 matrix (which is a representation of the C_2 operation) has a trace of -1; this is the character $\chi(C_2)$ in the $3N$-reducible representation we seek. Similarly, the effect of the operation $\hat{\sigma}_{zy}$ is $x \rightarrow -x$, $y \rightarrow y$, and $z \rightarrow z$ for *each* of the atoms and is represented by the matrix transformation

$$[x \quad y \quad z]\begin{bmatrix} -1 & 0 & 0 \\ 0 & 1 & 0 \\ 0 & 0 & 1 \end{bmatrix} = [-x \quad y \quad z] \tag{A6-59}$$

which leads to a trace of 1 and thus $\chi(\sigma_{zy}) = 1$ for each unshifted atom. Since there are *three* unshifted atoms, the character of the operation σ_{zy} is 3 for the reducible representation. In like manner, the operation σ_{zx} leaves only the O atom unshifted with $x \rightarrow x$, $y \rightarrow -y$, and $z \rightarrow z$. The matrix representation is

$$[x \quad y \quad z]\begin{bmatrix} 1 & 0 & 0 \\ 0 & -1 & 0 \\ 0 & 0 & 1 \end{bmatrix} = [x \quad -y \quad z] \tag{A6-60}$$

which leads to $\chi(\sigma_{zx}) = 1$.

In the case of the identity operation, it is readily seen that all atoms are unshifted, and since no coordinates change sign, each atom contributes 3 to the character; thus the total character is simply three times the total number of atoms, that is, $\chi(E) = 3N$ (9 in the example of the water molecule). Thus the character table of the $3N$-dimensional reducible representation becomes

C_{2v}	E	C_2	σ_{zy}	σ_{zx}
Γ_{3N}	9	−1	3	1

Use of Eq. (A6-24) shows

$$\Gamma_{3N} = 3A_1 \oplus A_2 \oplus 2B_1 \oplus 3B_2 \tag{A6-61}$$

Use of the C_{2v} character table shows that translations behave as x, y, and z do (centered along the principal axis) and thus are B_1, B_2, and A_1, respectively.

The rotations behave as R_x, R_y, and R_z do and thus are A_2, B_1, and B_2, respectively. Remaining are the $3(3) - 6 = 3$ vibrations which are A_1 (twice) and B_2 (once). The A_1 vibrational modes are the symmetric stretching mode and the bending mode; the asymmetric stretching mode is B_2. It should be noted that these three particular vibrational modes are not unique; the molecule does not necessarily vibrate in any of the pure normal modes, but these three modes constitute a complete basis for the description of how the molecule vibrates; i.e., any arbitrary vibrational motion of the water molecule can be expressed as some linear combination of the three normal modes.

Vibrational selection rules are derived not only from the condition that the transition-moment integral

$$P_{k \to l} = \langle \psi_{\text{final}} | \boldsymbol{\mu} | \psi_{\text{initial}} \rangle \tag{A6-62}$$

must be nonzero but also from the two additional conditions:

1. $\Delta v = +1$.
2. Only one vibrational change occurs during a transition.

Consequently (since $\boldsymbol{\mu}$ transforms as x, y, or z does), the only infrared-active vibrations will be A_1 (which transforms as z does) or B_2 (which transforms as y does). Note that no B_1 vibrations (which transform as x does) are present. Also, vibrations of A_2 symmetry are forbidden; these are absent in the specific case of water but are found in more complicated molecules of C_{2v} symmetry.

The Raman-active vibrations are those that belong to the same irreducible representations as the coordinate products x^2, y^2, z^2, xy, xz, or yz. Since x^2, y^2, and z^2 belong to A_1 and yz belong to B_2, both of these vibrational modes are also Raman-active.

If a molecule has a center of symmetry, then coordinate products and coordinates never appear in the same irreducible representation. This leads to the *mutual exclusion rule*: For all molecules with a center of symmetry, vibrational transitions which are infrared-allowed are Raman-forbidden, and vice versa. For example, H_2 ($D_{\infty h}$) has one vibrational frequency which is of A_{1g} symmetry; this is Raman-allowed (and infrared-forbidden), since it transforms as z^2 does. Ethene (D_{2h}) has the vibrational modes $3A_g$, A_u, $2B_{1g}$, B_{1u}, B_{2g}, $2B_{2u}$, and $2B_{3u}$. The infrared-active modes are B_{1u}, B_{2u}, and B_{3u} (z, y, x, respectively), and the Raman-active modes are A_g, B_{1g}, and B_{2g} (A_g is like x^2, y^2, or z^2, and the other two are like xy and xz, respectively). Note that the A_u vibration of ethene is neither infrared- nor Raman-active.

Selection rules for electronic transitions (ultraviolet and visible regions) are based on the transition-moment integral

$$P_{k \to l} = \langle \psi_l | \boldsymbol{\mu} | \psi_k \rangle \tag{A6-63}$$

where ψ_k and ψ_l are eigenfunctions of the electronic hamiltonian. If ψ_k is the ground-state wavefunction (generally belonging to the totally symmetric irreducible representation), then the above integral is nonzero only when ψ_l belongs to the same irreducible representation as one of the components of $\boldsymbol{\mu}$.

The quantum electronic states of molecules are generally designated by spectroscopic symbols based on the irreducible representations to which the states belong. The symbols have the general form

$$^{2S+1}\Gamma \tag{A6-64}$$

where $2S+1$ is the multiplicity of the state and Γ is the appropriate irreducible representation symbol. Thus, the ground state of H_2O (C_{2v} point group) is written $^1A_1$, and that of benzene (D_{6h} point group) is written $^1A_{1g}$. Just as in atoms, it is assumed that the spin selection rule $\Delta S = 0$ holds. The selection rules for the C_{2v} and D_{6h} groups are readily induced (assuming totally symmetric ground states) to be as follows:

$$\begin{aligned} C_{2v}&: \; {}^1A_1 \rightarrow {}^1A_1, {}^1B_1, {}^1B_2 \\ D_{6h}&: \; {}^1A_{1g} \rightarrow {}^1E_{1u}, {}^1A_{2u} \end{aligned} \tag{A6-65}$$

Angular momentum, so important in atoms, is of significance only in the case of linear molecules ($C_{\infty v}$ and $D_{\infty h}$). In nonlinear molecules neither $\hat{L}^2$ nor $\hat{L}_z$ commutes with $\hat{H}_{el}$, so that $\mathbf{L}^2$ and $\mathbf{L}_z$ are not constants of the motion. In linear molecules, $\hat{L}_z$ does commute with $\hat{H}_{el}$, so its eigenvalues ($\Lambda = 0, \pm 1, \pm 2, \ldots$) may be used to describe spectroscopic states. The spectroscopic symbols have the forms:

$$^{2S+1}\Lambda^{\pm}_{u,g} \; (D_{\infty h}) \qquad ^{2S+1}\Lambda^{\pm} \; (C_{\infty v}) \tag{A6-66}$$

where Λ is written $\Sigma, \Pi, \Delta, \Phi, \ldots$ when $\Lambda = 0, \pm 1, \pm 2, \pm 3, \ldots$, respectively. Thus, although Λ of linear molecules is analogous to M_L of atoms, it plays the same role in molecules as L does in atoms. The spectroscopic symbols correspond to the irreducible representation symbols as follows: A_1 is Σ^+, A_2 is Σ^-, and $\Pi, \Delta, \Phi, \ldots$ are $E_1, E_2, E_3, \ldots$, respectively. States with $\Lambda = 0$ are nondegenerate; all others are doubly degenerate (since Λ can have two different values). The $\pm$ superscript is used only when $\Lambda = 0$; it refers to the σ_v operation which changes the angle φ to $-\varphi$. A function ψ is called Σ^+ if $\Lambda = 0$ and $\hat{\sigma}_v \psi = \psi$; it is called Σ^- if $\hat{\sigma}_v \psi = -\psi$.

The ground electronic states of $C_{\infty v}$ molecules are usually $^1\Sigma^+$; those of $D_{\infty h}$ molecules are usually $^1\Sigma_g^+$. The selection rules from the ground states are

$$\begin{aligned} C_{\infty v}&: \; {}^1\Sigma^+ \rightarrow {}^1\Pi, {}^1\Sigma^+ \\ D_{\infty h}&: \; {}^1\Sigma_g^+ \rightarrow {}^1\Pi_u, {}^1\Sigma_u^+ \end{aligned} \tag{A6-67}$$

Alternatively, one may write

$$\begin{aligned} C_{\infty v}&: \; \Delta\Lambda = 0, \pm 1; \; + \rightarrow +; \; - \rightarrow -; \; + \not\rightarrow - \\ D_{\infty h}&: \; \Delta\Lambda = 0, \pm 1; \; g \rightarrow u; \; u \rightarrow g; \; + \rightarrow +; \; - \rightarrow -; \; + \not\rightarrow - \end{aligned} \tag{A6-68}$$

EXERCISES

A6-13. Determine the symmetries of the normal-mode vibrations of the formaldehyde molecule and classify them as to infrared and Raman activity. Repeat for the benzene molecule.

A6-14. Determine the electronic selection rules for the following molecules, assuming absorption from a totally symmetric ground state only: ammonia, naphthalene, *trans*-1,2-dichloroethene, and allene.

SUGGESTED READINGS

Altman, S. L.: *Induced Representations in Crystals and Molecules*, Academic, New York, 1977.

Cotton, F. A.: *Chemical Applications of Group Theory*, 2d ed., Wiley-Interscience, New York, 1971.

Flurry, R. L., Jr.: *Symmetry Groups: Theory and Chemical Applications*, Prentice-Hall, Englewood Cliffs, N.J., 1980.

Hall, L. H.: *Group Theory and Symmetry in Chemistry*, McGraw-Hill, New York, 1969.

Hargittai, I., and M. Hargittai, *Symmetry through the Eyes of a Chemist*, VCH Publishers, New York, 1987. This has to be the most delightful book on symmetry ever written! It surveys the entire field of chemistry from the point of view of symmetry—and does it with a wealth of interesting specific examples.

Jaffé, H. H., and M. Orchin: *Symmetry in Chemistry*, Wiley, New York, 1965.

McWeeny, R.: *Symmetry*, Macmillan, New York, 1963.

Orchin, M., and H. H. Jaffé: *Symmetry, Orbitals, and Spectra*, Wiley-Interscience, New York, 1970.

APPENDIX 7

CHARACTER TABLES FOR SOME COMMON MOLECULAR POINT GROUPS

C_1	E
A	1

C_s	E	σ_h		
A'	1	1	x, y, R_z	x^2, y^2, z^2, xy
A''	1	−1	z, R_x, R_y	yz, xz

C_i	E	i		
A_g	1	1	R_x, R_y, R_z	x^2, y^2, z^2 xy, xz, yz
A_u	1	−1	x, y, z	

C_2	E	C_2		
A	1	1	z, R_z	x^2, y^2, z^2, xy
B	1	−1	x, y, R_x, R_y	yz, xz

C_{2v}	E	C_2	$\sigma_v(xz)$	$\sigma_v'(yz)$		
A_1	1	1	1	1	z	x^2, y^2, z^2
A_2	1	1	−1	−1	R_z	xy
B_1	1	−1	1	−1	x, R_y	xz
B_2	1	−1	−1	1	y, R_x	yz

A more complete listing is given in F. A. Cotton, *Chemical Applications of Group Theory*, 2d ed., Wiley-Interscience, New York, 1971.

C_{3v}	E	$2C_3$	$3\sigma_v$		
A_1	1	1	1	z	x^2+y^2, z^2
A_2	1	1	−1	R_z	
E	2	−1	0	$(x, y)(R_x, R_y)$	$(x^2-y^2, xy)(xz, yz)$

C_{4v}	E	$2C_4$	C_2	$2\sigma_v$	$2\sigma_d$		
A_1	1	1	1	1	1	z	x^2+y^2, z^2
A_2	1	1	1	−1	−1	R_z	
B_1	1	−1	1	1	−1		x^2-y^2
B_2	1	−1	1	−1	1		xy
E	2	0	−2	0	0	$(x, y)(R_x, R_y)$	(xz, yz)

C_{2h}	E	C_2	i	σ_h		
A_g	1	1	1	1	R_z	x^2, y^2, z^2, xy
B_g	1	−1	1	−1	R_x, R_y	xz, yz
A_u	1	1	−1	−1	z	
B_u	1	−1	−1	1	x, y	

D_{2h}	E	$C_2(z)$	$C_2(y)$	$C_2(x)$	i	$\sigma(xy)$	$\sigma(xz)$	$\sigma(yz)$		
A_g	1	1	1	1	1	1	1	1		x^2, y^2, z^2
B_{1g}	1	1	−1	−1	1	1	−1	−1	R_z	xy
B_{2g}	1	−1	1	−1	1	−1	1	−1	R_y	xz
B_{3g}	1	−1	−1	1	1	−1	−1	1	R_x	yz
A_u	1	1	1	1	−1	−1	−1	−1		
B_{1u}	1	1	−1	−1	−1	−1	1	1	z	
B_{2u}	1	−1	1	−1	−1	1	−1	1	y	
B_{3u}	1	−1	−1	1	−1	1	1	−1	x	

D_{5h}	E	$2C_5$	$2C_5^2$	$5C_2$	σ_h	$2S_5$	$2S_5^3$	$5\sigma_v$		
A_1'	1	1	1	1	1	1	1	1		x^2+y^2, z^2
A_2'	1	1	1	−1	1	1	1	−1	R_z	
E_1'	2	2 cos 72°	2 cos 144°	0	2	2 cos 72°	2 cos 144°	0	(x, y)	
E_2'	2	2 cos 144°	2 cos 72°	0	2	2 cos 144°	2 cos 72°	0		(x^2-y^2, xy)
A_1''	1	1	1	1	−1	−1	−1	−1		
A_2''	1	1	1	−1	−1	−1	−1	1	z	
E_1''	2	2 cos 72°	2 cos 144°	0	−2	−2 cos 72°	−2 cos 144°	0	(R_x, R_y)	(xz, yz)
E_2''	2	2 cos 144°	2 cos 72°	0	−2	−2 cos 144°	−2 cos 72°	0		

D_{6h}	E	$2C_6$	$2C_3$	C_2	$3C_2'$	$3C_2''$	i	$2S_3$	$2S_6$	σ_h	$3\sigma_d$	$3\sigma_v$		
A_{1g}	1	1	1	1	1	1	1	1	1	1	1	1		x^2+y^2, z^2
A_{2g}	1	1	1	1	−1	−1	1	1	1	1	−1	−1	R_z	
B_{1g}	1	−1	1	−1	1	−1	1	−1	1	−1	1	−1		
B_{2g}	1	−1	1	−1	−1	1	1	−1	1	−1	−1	1		
E_{1g}	2	1	−1	−2	0	0	2	1	−1	−2	0	0	(R_x, R_y)	(xz, yz)
E_{2g}	2	−1	−1	2	0	0	2	−1	−1	2	0	0		(x^2-y^2, xy)
A_{1u}	1	1	1	1	1	1	−1	−1	−1	−1	−1	−1		
A_{2u}	1	1	1	1	−1	−1	−1	−1	−1	−1	1	1	z	
B_{1u}	1	−1	1	−1	1	−1	−1	1	−1	1	−1	1		
B_{2u}	1	−1	1	−1	−1	1	−1	1	−1	1	1	−1		
E_{1u}	2	1	−1	−2	0	0	−2	−1	1	2	0	0	(x, y)	
E_{2u}	2	−1	−1	2	0	0	−2	1	1	−2	0	0		

D_{2d}	E	$2S_4$	C_2	$2C_2'$	$2\sigma_d$		
A_1	1	1	1	1	1		x^2+y^2, z^2
A_2	1	1	1	−1	−1	R_z	
B_1	1	−1	1	1	−1		x^2-y^2
B_2	1	−1	1	−1	1	z	xy
E	2	0	−2	0	0	(x, y) (R_x, R_y)	(xz, yz)

D_{3d}	E	$2C_3$	$3C_2$	i	$2S_6$	$3\sigma_d$		
A_{1g}	1	1	1	1	1	1		x^2+y^2, z^2
A_{2g}	1	1	−1	1	1	−1	R_z	
E_g	2	−1	0	2	−1	0	(R_x, R_y)	(x^2-y^2, xy); (xz, yz)
A_{1u}	1	1	1	−1	−1	−1		
A_{2u}	1	1	−1	−1	−1	1	z	
E_u	2	−1	0	−2	1	0	(x, y)	

T_d	E	$8C_3$	$3C_2$	$6S_4$	$6\sigma_d$		
A_1	1	1	1	1	1		$x^2+y^2+z^2$
A_2	1	1	1	−1	−1		
E	2	−1	2	0	0		$(2z^2-x^2-y^2, x^2-y^2)$
T_1	3	0	−1	1	−1	(R_x, R_y, R_z)	
T_2	3	0	−1	−1	1	(x, y, z)	(xy, xz, yz)

O_h	E	$8C_3$	$6C_2$	$6C_4$	$3C_2(=C_4^2)$	i	$6S_4$	$8S_6$	$3\sigma_h$	$6\sigma_d$		
A_{1g}	1	1	1	1	1	1	1	1	1	1		$x^2+y^2+z^2$
A_{2g}	1	1	−1	−1	1	1	−1	1	1	−1		
E_g	2	−1	0	0	2	2	0	−1	2	0		$(2z^2-x^2-y^2,$ $x^2-y^2)$
T_{1g}	3	0	−1	1	−1	3	1	0	−1	−1	(R_x, R_y, R_z)	
T_{2g}	3	0	1	−1	−1	3	−1	0	−1	1		(xz, yz, xy)
A_{1u}	1	1	1	1	1	−1	−1	−1	−1	−1		
A_{2u}	1	1	−1	−1	1	−1	1	−1	−1	1		
E_u	2	−1	0	0	2	−2	0	1	−2	0		
T_{1u}	3	0	−1	1	−1	−3	−1	0	1	1	(x, y, z)	
T_{2u}	3	0	1	−1	−1	−3	1	0	1	−1		

$C_{\infty v}$	E	$2C_\infty^\Phi$	$\cdots$	$\infty\sigma_v$		
$A_1 \equiv \Sigma^+$	1	1	$\cdots$	1	z	x^2+y^2, z^2
$A_2 \equiv \Sigma^-$	1	1	$\cdots$	−1	R_z	
$E_1 \equiv \Pi$	2	$2\cos\Phi$	$\cdots$	0	$(x, y); (R_x, R_y)$	(xz, yz)
$E_2 \equiv \Delta$	2	$2\cos 2\Phi$	$\cdots$	0		(x^2-y^2, xy)
$E_3 \equiv \Phi$	2	$2\cos 3\Phi$	$\cdots$	0		
$\cdots$	$\cdots$	$\cdots$	$\cdots$	$\cdots$		

$D_{\infty h}$	E	$2C_\infty^\Phi$	$\cdots$	$\infty\sigma_v$	i	$2S_\infty^\Phi$	$\cdots$	∞C_2		
Σ_g^+	1	1	$\cdots$	1	1	1	$\cdots$	1		x^2+y^2, z^2
Σ_g^-	1	1	$\cdots$	−1	1	1	$\cdots$	−1	R_z	
Π_g	2	$2\cos\Phi$	$\cdots$	0	2	$-2\cos\Phi$	$\cdots$	0	(R_x, R_y)	(xz, yz)
Δ_g	2	$2\cos 2\Phi$	$\cdots$	0	2	$2\cos 2\Phi$	$\cdots$	0		(x^2-y^2, xy)
$\cdots$	$\cdots$	$\cdots$	$\cdots$	$\cdots$	$\cdots$	$\cdots$	$\cdots$	$\cdots$		
Σ_u^+	1	1	$\cdots$	1	−1	−1	$\cdots$	−1	z	
Σ_u^-	1	1	$\cdots$	−1	−1	−1	$\cdots$	1		
Π_u	2	$2\cos\Phi$	$\cdots$	0	−2	$2\cos\Phi$	$\cdots$	0	(x, y)	
Δ_u	2	$2\cos 2\Phi$	$\cdots$	0	−2	$-2\cos 2\Phi$	$\cdots$	0		
$\cdots$	$\cdots$	$\cdots$	$\cdots$	$\cdots$	$\cdots$	$\cdots$	$\cdots$	$\cdots$		

NAME INDEX

*Numbers in *italic* indicate tables or illustrations.

SUBJECT INDEX

*Numbers in *italic* indicate tables or illustrations.

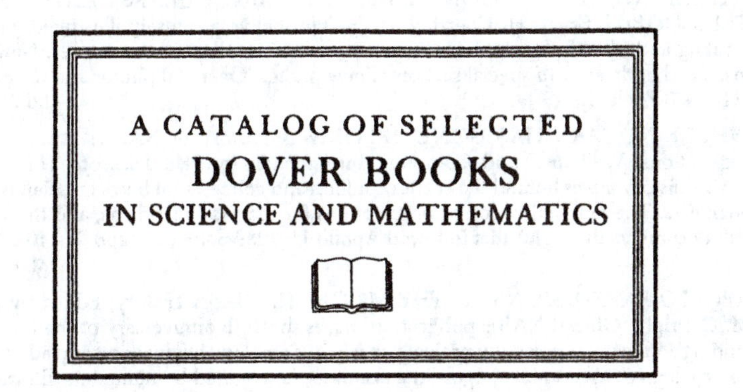
A CATALOG OF SELECTED
DOVER BOOKS
IN SCIENCE AND MATHEMATICS

Astronomy

CHARIOTS FOR APOLLO: The NASA History of Manned Lunar Spacecraft to 1969, Courtney G. Brooks, James M. Grimwood, and Loyd S. Swenson, Jr. This illustrated history by a trio of experts is the definitive reference on the Apollo spacecraft and lunar modules. It traces the vehicles' design, development, and operation in space. More than 100 photographs and illustrations. 576pp. 6 3/4 x 9 1/4. 0-486-46756-2

EXPLORING THE MOON THROUGH BINOCULARS AND SMALL TELESCOPES, Ernest H. Cherrington, Jr. Informative, profusely illustrated guide to locating and identifying craters, rills, seas, mountains, other lunar features. Newly revised and updated with special section of new photos. Over 100 photos and diagrams. 240pp. 8 1/4 x 11. 0-486-24491-1

WHERE NO MAN HAS GONE BEFORE: A History of NASA's Apollo Lunar Expeditions, William David Compton. Introduction by Paul Dickson. This official NASA history traces behind-the-scenes conflicts and cooperation between scientists and engineers. The first half concerns preparations for the Moon landings, and the second half documents the flights that followed Apollo 11. 1989 edition. 432pp. 7 x 10. 0-486-47888-2

APOLLO EXPEDITIONS TO THE MOON: The NASA History, Edited by Edgar M. Cortright. Official NASA publication marks the 40th anniversary of the first lunar landing and features essays by project participants recalling engineering and administrative challenges. Accessible, jargon-free accounts, highlighted by numerous illustrations. 336pp. 8 3/8 x 10 7/8. 0-486-47175-6

ON MARS: Exploration of the Red Planet, 1958-1978--The NASA History, Edward Clinton Ezell and Linda Neuman Ezell. NASA's official history chronicles the start of our explorations of our planetary neighbor. It recounts cooperation among government, industry, and academia, and it features dozens of photos from Viking cameras. 560pp. 6 3/4 x 9 1/4. 0-486-46757-0

ARISTARCHUS OF SAMOS: The Ancient Copernicus, Sir Thomas Heath. Heath's history of astronomy ranges from Homer and Hesiod to Aristarchus and includes quotes from numerous thinkers, compilers, and scholasticists from Thales and Anaximander through Pythagoras, Plato, Aristotle, and Heraclides. 34 figures. 448pp. 5 3/8 x 8 1/2. 0-486-43886-4

AN INTRODUCTION TO CELESTIAL MECHANICS, Forest Ray Moulton. Classic text still unsurpassed in presentation of fundamental principles. Covers rectilinear motion, central forces, problems of two and three bodies, much more. Includes over 200 problems, some with answers. 437pp. 5 3/8 x 8 1/2. 0-486-64687-4

BEYOND THE ATMOSPHERE: Early Years of Space Science, Homer E. Newell. This exciting survey is the work of a top NASA administrator who chronicles technological advances, the relationship of space science to general science, and the space program's social, political, and economic contexts. 528pp. 6 3/4 x 9 1/4. 0-486-47464-X

STAR LORE: Myths, Legends, and Facts, William Tyler Olcott. Captivating retellings of the origins and histories of ancient star groups include Pegasus, Ursa Major, Pleiades, signs of the zodiac, and other constellations. "Classic." – *Sky & Telescope.* 58 illustrations. 544pp. 5 3/8 x 8 1/2. 0-486-43581-4

A COMPLETE MANUAL OF AMATEUR ASTRONOMY: Tools and Techniques for Astronomical Observations, P. Clay Sherrod with Thomas L. Koed. Concise, highly readable book discusses the selection, set-up, and maintenance of a telescope; amateur studies of the sun; lunar topography and occultations; and more. 124 figures. 26 halftones. 37 tables. 335pp. 6 1/2 x 9 1/4. 0-486-42820-6

Browse over 9,000 books at www.doverpublications.com

Chemistry

MOLECULAR COLLISION THEORY, M. S. Child. This high-level monograph offers an analytical treatment of classical scattering by a central force, quantum scattering by a central force, elastic scattering phase shifts, and semi-classical elastic scattering. 1974 edition. 310pp. 5 3/8 x 8 1/2. 0-486-69437-2

HANDBOOK OF COMPUTATIONAL QUANTUM CHEMISTRY, David B. Cook. This comprehensive text provides upper-level undergraduates and graduate students with an accessible introduction to the implementation of quantum ideas in molecular modeling, exploring practical applications alongside theoretical explanations. 1998 edition. 832pp. 5 3/8 x 8 1/2. 0-486-44307-8

RADIOACTIVE SUBSTANCES, Marie Curie. The celebrated scientist's thesis, which directly preceded her 1903 Nobel Prize, discusses establishing atomic character of radioactivity; extraction from pitchblende of polonium and radium; isolation of pure radium chloride; more. 96pp. 5 3/8 x 8 1/2. 0-486-42550-9

CHEMICAL MAGIC, Leonard A. Ford. Classic guide provides intriguing entertainment while elucidating sound scientific principles, with more than 100 unusual stunts: cold fire, dust explosions, a nylon rope trick, a disappearing beaker, much more. 128pp. 5 3/8 x 8 1/2. 0-486-67628-5

ALCHEMY, E. J. Holmyard. Classic study by noted authority covers 2,000 years of alchemical history: religious, mystical overtones; apparatus; signs, symbols, and secret terms; advent of scientific method, much more. Illustrated. 320pp. 5 3/8 x 8 1/2. 0-486-26298-7

CHEMICAL KINETICS AND REACTION DYNAMICS, Paul L. Houston. This text teaches the principles underlying modern chemical kinetics in a clear, direct fashion, using several examples to enhance basic understanding. Solutions to selected problems. 2001 edition. 352pp. 8 3/8 x 11. 0-486-45334-0

PROBLEMS AND SOLUTIONS IN QUANTUM CHEMISTRY AND PHYSICS, Charles S. Johnson and Lee G. Pedersen. Unusually varied problems, with detailed solutions, cover of quantum mechanics, wave mechanics, angular momentum, molecular spectroscopy, scattering theory, more. 280 problems, plus 139 supplementary exercises. 430pp. 6 1/2 x 9 1/4. 0-486-65236-X

ELEMENTS OF CHEMISTRY, Antoine Lavoisier. Monumental classic by the founder of modern chemistry features first explicit statement of law of conservation of matter in chemical change, and more. Facsimile reprint of original (1790) Kerr translation. 539pp. 5 3/8 x 8 1/2. 0-486-64624-6

MAGNETISM AND TRANSITION METAL COMPLEXES, F. E. Mabbs and D. J. Machin. A detailed view of the calculation methods involved in the magnetic properties of transition metal complexes, this volume offers sufficient background for original work in the field. 1973 edition. 240pp. 5 3/8 x 8 1/2. 0-486-46284-6

GENERAL CHEMISTRY, Linus Pauling. Revised third edition of classic first-year text by Nobel laureate. Atomic and molecular structure, quantum mechanics, statistical mechanics, thermodynamics correlated with descriptive chemistry. Problems. 992pp. 5 3/8 x 8 1/2. 0-486-65622-5

ELECTROLYTE SOLUTIONS: Second Revised Edition, R. A. Robinson and R. H. Stokes. Classic text deals primarily with measurement, interpretation of conductance, chemical potential, and diffusion in electrolyte solutions. Detailed theoretical interpretations, plus extensive tables of thermodynamic and transport properties. 1970 edition. 590pp. 5 3/8 x 8 1/2. 0-486-42225-9

Browse over 9,000 books at www.doverpublications.com

Engineering

FUNDAMENTALS OF ASTRODYNAMICS, Roger R. Bate, Donald D. Mueller, and Jerry E. White. Teaching text developed by U.S. Air Force Academy develops the basic two-body and n-body equations of motion; orbit determination; classical orbital elements, coordinate transformations; differential correction; more. 1971 edition. 455pp. 5 3/8 x 8 1/2. 0-486-60061-0

INTRODUCTION TO CONTINUUM MECHANICS FOR ENGINEERS: Revised Edition, Ray M. Bowen. This self-contained text introduces classical continuum models within a modern framework. Its numerous exercises illustrate the governing principles, linearizations, and other approximations that constitute classical continuum models. 2007 edition. 320pp. 6 1/8 x 9 1/4. 0-486-47460-7

ENGINEERING MECHANICS FOR STRUCTURES, Louis L. Bucciarelli. This text explores the mechanics of solids and statics as well as the strength of materials and elasticity theory. Its many design exercises encourage creative initiative and systems thinking. 2009 edition. 320pp. 6 1/8 x 9 1/4. 0-486-46855-0

FEEDBACK CONTROL THEORY, John C. Doyle, Bruce A. Francis and Allen R. Tannenbaum. This excellent introduction to feedback control system design offers a theoretical approach that captures the essential issues and can be applied to a wide range of practical problems. 1992 edition. 224pp. 6 1/2 x 9 1/4. 0-486-46933-6

THE FORCES OF MATTER, Michael Faraday. These lectures by a famous inventor offer an easy-to-understand introduction to the interactions of the universe's physical forces. Six essays explore gravitation, cohesion, chemical affinity, heat, magnetism, and electricity. 1993 edition. 96pp. 5 3/8 x 8 1/2. 0-486-47482-8

DYNAMICS, Lawrence E. Goodman and William H. Warner. Beginning engineering text introduces calculus of vectors, particle motion, dynamics of particle systems and plane rigid bodies, technical applications in plane motions, and more. Exercises and answers in every chapter. 619pp. 5 3/8 x 8 1/2. 0-486-42006-X

ADAPTIVE FILTERING PREDICTION AND CONTROL, Graham C. Goodwin and Kwai Sang Sin. This unified survey focuses on linear discrete-time systems and explores natural extensions to nonlinear systems. It emphasizes discrete-time systems, summarizing theoretical and practical aspects of a large class of adaptive algorithms. 1984 edition. 560pp. 6 1/2 x 9 1/4. 0-486-46932-8

INDUCTANCE CALCULATIONS, Frederick W. Grover. This authoritative reference enables the design of virtually every type of inductor. It features a single simple formula for each type of inductor, together with tables containing essential numerical factors. 1946 edition. 304pp. 5 3/8 x 8 1/2. 0-486-47440-2

THERMODYNAMICS: Foundations and Applications, Elias P. Gyftopoulos and Gian Paolo Beretta. Designed by two MIT professors, this authoritative text discusses basic concepts and applications in detail, emphasizing generality, definitions, and logical consistency. More than 300 solved problems cover realistic energy systems and processes. 800pp. 6 1/8 x 9 1/4. 0-486-43932-1

THE FINITE ELEMENT METHOD: Linear Static and Dynamic Finite Element Analysis, Thomas J. R. Hughes. Text for students without in-depth mathematical training, this text includes a comprehensive presentation and analysis of algorithms of time-dependent phenomena plus beam, plate, and shell theories. Solution guide available upon request. 672pp. 6 1/2 x 9 1/4. 0-486-41181-8

Browse over 9,000 books at www.doverpublications.com

HELICOPTER THEORY, Wayne Johnson. Monumental engineering text covers vertical flight, forward flight, performance, mathematics of rotating systems, rotary wing dynamics and aerodynamics, aeroelasticity, stability and control, stall, noise, and more. 189 illustrations. 1980 edition. 1089pp. 5 5/8 x 8 1/4. 0-486-68230-7

MATHEMATICAL HANDBOOK FOR SCIENTISTS AND ENGINEERS: Definitions, Theorems, and Formulas for Reference and Review, Granino A. Korn and Theresa M. Korn. Convenient access to information from every area of mathematics: Fourier transforms, Z transforms, linear and nonlinear programming, calculus of variations, random-process theory, special functions, combinatorial analysis, game theory, much more. 1152pp. 5 3/8 x 8 1/2. 0-486-41147-8

A HEAT TRANSFER TEXTBOOK: Fourth Edition, John H. Lienhard V and John H. Lienhard IV. This introduction to heat and mass transfer for engineering students features worked examples and end-of-chapter exercises. Worked examples and end-of-chapter exercises appear throughout the book, along with well-drawn, illuminating figures. 768pp. 7 x 9 1/4. 0-486-47931-5

BASIC ELECTRICITY, U.S. Bureau of Naval Personnel. Originally a training course; best nontechnical coverage. Topics include batteries, circuits, conductors, AC and DC, inductance and capacitance, generators, motors, transformers, amplifiers, etc. Many questions with answers. 349 illustrations. 1969 edition. 448pp. 6 1/2 x 9 1/4. 0-486-20973-3

BASIC ELECTRONICS, U.S. Bureau of Naval Personnel. Clear, well-illustrated introduction to electronic equipment covers numerous essential topics: electron tubes, semiconductors, electronic power supplies, tuned circuits, amplifiers, receivers, ranging and navigation systems, computers, antennas, more. 560 illustrations. 567pp. 6 1/2 x 9 1/4. 0-486-21076-6

BASIC WING AND AIRFOIL THEORY, Alan Pope. This self-contained treatment by a pioneer in the study of wind effects covers flow functions, airfoil construction and pressure distribution, finite and monoplane wings, and many other subjects. 1951 edition. 320pp. 5 3/8 x 8 1/2. 0-486-47188-8

SYNTHETIC FUELS, Ronald F. Probstein and R. Edwin Hicks. This unified presentation examines the methods and processes for converting coal, oil, shale, tar sands, and various forms of biomass into liquid, gaseous, and clean solid fuels. 1982 edition. 512pp. 6 1/8 x 9 1/4. 0-486-44977-7

THEORY OF ELASTIC STABILITY, Stephen P. Timoshenko and James M. Gere. Written by world-renowned authorities on mechanics, this classic ranges from theoretical explanations of 2- and 3-D stress and strain to practical applications such as torsion, bending, and thermal stress. 1961 edition. 560pp. 5 3/8 x 8 1/2. 0-486-47207-8

PRINCIPLES OF DIGITAL COMMUNICATION AND CODING, Andrew J. Viterbi and Jim K. Omura. This classic by two digital communications experts is geared toward students of communications theory and to designers of channels, links, terminals, modems, or networks used to transmit and receive digital messages. 1979 edition. 576pp. 6 1/8 x 9 1/4. 0-486-46901-8

LINEAR SYSTEM THEORY: The State Space Approach, Lotfi A. Zadeh and Charles A. Desoer. Written by two pioneers in the field, this exploration of the state space approach focuses on problems of stability and control, plus connections between this approach and classical techniques. 1963 edition. 656pp. 6 1/8 x 9 1/4. 0-486-46663-9

Browse over 9,000 books at www.doverpublications.com

Mathematics–Bestsellers

HANDBOOK OF MATHEMATICAL FUNCTIONS: with Formulas, Graphs, and Mathematical Tables, Edited by Milton Abramowitz and Irene A. Stegun. A classic resource for working with special functions, standard trig, and exponential logarithmic definitions and extensions, it features 29 sets of tables, some to as high as 20 places. 1046pp. 8 x 10 1/2. 0-486-61272-4

ABSTRACT AND CONCRETE CATEGORIES: The Joy of Cats, Jiri Adamek, Horst Herrlich, and George E. Strecker. This up-to-date introductory treatment employs category theory to explore the theory of structures. Its unique approach stresses concrete categories and presents a systematic view of factorization structures. Numerous examples. 1990 edition, updated 2004. 528pp. 6 1/8 x 9 1/4. 0-486-46934-4

MATHEMATICS: Its Content, Methods and Meaning, A. D. Aleksandrov, A. N. Kolmogorov, and M. A. Lavrent'ev. Major survey offers comprehensive, coherent discussions of analytic geometry, algebra, differential equations, calculus of variations, functions of a complex variable, prime numbers, linear and non-Euclidean geometry, topology, functional analysis, more. 1963 edition. 1120pp. 5 3/8 x 8 1/2. 0-486-40916-3

INTRODUCTION TO VECTORS AND TENSORS: Second Edition--Two Volumes Bound as One, Ray M. Bowen and C.-C. Wang. Convenient single-volume compilation of two texts offers both introduction and in-depth survey. Geared toward engineering and science students rather than mathematicians, it focuses on physics and engineering applications. 1976 edition. 560pp. 6 1/2 x 9 1/4. 0-486-46914-X

AN INTRODUCTION TO ORTHOGONAL POLYNOMIALS, Theodore S. Chihara. Concise introduction covers general elementary theory, including the representation theorem and distribution functions, continued fractions and chain sequences, the recurrence formula, special functions, and some specific systems. 1978 edition. 272pp. 5 3/8 x 8 1/2. 0-486-47929-3

ADVANCED MATHEMATICS FOR ENGINEERS AND SCIENTISTS, Paul DuChateau. This primary text and supplemental reference focuses on linear algebra, calculus, and ordinary differential equations. Additional topics include partial differential equations and approximation methods. Includes solved problems. 1992 edition. 400pp. 7 1/2 x 9 1/4. 0-486-47930-7

PARTIAL DIFFERENTIAL EQUATIONS FOR SCIENTISTS AND ENGINEERS, Stanley J. Farlow. Practical text shows how to formulate and solve partial differential equations. Coverage of diffusion-type problems, hyperbolic-type problems, elliptic-type problems, numerical and approximate methods. Solution guide available upon request. 1982 edition. 414pp. 6 1/8 x 9 1/4. 0-486-67620-X

VARIATIONAL PRINCIPLES AND FREE-BOUNDARY PROBLEMS, Avner Friedman. Advanced graduate-level text examines variational methods in partial differential equations and illustrates their applications to free-boundary problems. Features detailed statements of standard theory of elliptic and parabolic operators. 1982 edition. 720pp. 6 1/8 x 9 1/4. 0-486-47853-X

LINEAR ANALYSIS AND REPRESENTATION THEORY, Steven A. Gaal. Unified treatment covers topics from the theory of operators and operator algebras on Hilbert spaces; integration and representation theory for topological groups; and the theory of Lie algebras, Lie groups, and transform groups. 1973 edition. 704pp. 6 1/8 x 9 1/4. 0-486-47851-3

Browse over 9,000 books at www.doverpublications.com

A SURVEY OF INDUSTRIAL MATHEMATICS, Charles R. MacCluer. Students learn how to solve problems they'll encounter in their professional lives with this concise single-volume treatment. It employs MATLAB and other strategies to explore typical industrial problems. 2000 edition. 384pp. 5 3/8 x 8 1/2. 0-486-47702-9

NUMBER SYSTEMS AND THE FOUNDATIONS OF ANALYSIS, Elliott Mendelson. Geared toward undergraduate and beginning graduate students, this study explores natural numbers, integers, rational numbers, real numbers, and complex numbers. Numerous exercises and appendixes supplement the text. 1973 edition. 368pp. 5 3/8 x 8 1/2. 0-486-45792-3

A FIRST LOOK AT NUMERICAL FUNCTIONAL ANALYSIS, W. W. Sawyer. Text by renowned educator shows how problems in numerical analysis lead to concepts of functional analysis. Topics include Banach and Hilbert spaces, contraction mappings, convergence, differentiation and integration, and Euclidean space. 1978 edition. 208pp. 5 3/8 x 8 1/2. 0-486-47882-3

FRACTALS, CHAOS, POWER LAWS: Minutes from an Infinite Paradise, Manfred Schroeder. A fascinating exploration of the connections between chaos theory, physics, biology, and mathematics, this book abounds in award-winning computer graphics, optical illusions, and games that clarify memorable insights into self-similarity. 1992 edition. 448pp. 6 1/8 x 9 1/4. 0-486-47204-3

SET THEORY AND THE CONTINUUM PROBLEM, Raymond M. Smullyan and Melvin Fitting. A lucid, elegant, and complete survey of set theory, this three-part treatment explores axiomatic set theory, the consistency of the continuum hypothesis, and forcing and independence results. 1996 edition. 336pp. 6 x 9. 0-486-47484-4

DYNAMICAL SYSTEMS, Shlomo Sternberg. A pioneer in the field of dynamical systems discusses one-dimensional dynamics, differential equations, random walks, iterated function systems, symbolic dynamics, and Markov chains. Supplementary materials include PowerPoint slides and MATLAB exercises. 2010 edition. 272pp. 6 1/8 x 9 1/4. 0-486-47705-3

ORDINARY DIFFERENTIAL EQUATIONS, Morris Tenenbaum and Harry Pollard. Skillfully organized introductory text examines origin of differential equations, then defines basic terms and outlines general solution of a differential equation. Explores integrating factors; dilution and accretion problems; Laplace Transforms; Newton's Interpolation Formulas, more. 818pp. 5 3/8 x 8 1/2. 0-486-64940-7

MATROID THEORY, D. J. A. Welsh. Text by a noted expert describes standard examples and investigation results, using elementary proofs to develop basic matroid properties before advancing to a more sophisticated treatment. Includes numerous exercises. 1976 edition. 448pp. 5 3/8 x 8 1/2. 0-486-47439-9

THE CONCEPT OF A RIEMANN SURFACE, Hermann Weyl. This classic on the general history of functions combines function theory and geometry, forming the basis of the modern approach to analysis, geometry, and topology. 1955 edition. 208pp. 5 3/8 x 8 1/2. 0-486-47004-0

THE LAPLACE TRANSFORM, David Vernon Widder. This volume focuses on the Laplace and Stieltjes transforms, offering a highly theoretical treatment. Topics include fundamental formulas, the moment problem, monotonic functions, and Tauberian theorems. 1941 edition. 416pp. 5 3/8 x 8 1/2. 0-486-47755-X

Browse over 9,000 books at www.doverpublications.com

Mathematics–Logic and Problem Solving

PERPLEXING PUZZLES AND TANTALIZING TEASERS, Martin Gardner. Ninety-three riddles, mazes, illusions, tricky questions, word and picture puzzles, and other challenges offer hours of entertainment for youngsters. Filled with rib-tickling drawings. Solutions. 224pp. 5 3/8 x 8 1/2. 0-486-25637-5

MY BEST MATHEMATICAL AND LOGIC PUZZLES, Martin Gardner. The noted expert selects 70 of his favorite "short" puzzles. Includes The Returning Explorer, The Mutilated Chessboard, Scrambled Box Tops, and dozens more. Complete solutions included. 96pp. 5 3/8 x 8 1/2. 0-486-28152-3

THE LADY OR THE TIGER?: and Other Logic Puzzles, Raymond M. Smullyan. Created by a renowned puzzle master, these whimsically themed challenges involve paradoxes about probability, time, and change; metapuzzles; and self-referentiality. Nineteen chapters advance in difficulty from relatively simple to highly complex. 1982 edition. 240pp. 5 3/8 x 8 1/2. 0-486-47027-X

SATAN, CANTOR AND INFINITY: Mind-Boggling Puzzles, Raymond M. Smullyan. A renowned mathematician tells stories of knights and knaves in an entertaining look at the logical precepts behind infinity, probability, time, and change. Requires a strong background in mathematics. Complete solutions. 288pp. 5 3/8 x 8 1/2. 0-486-47036-9

THE RED BOOK OF MATHEMATICAL PROBLEMS, Kenneth S. Williams and Kenneth Hardy. Handy compilation of 100 practice problems, hints and solutions indispensable for students preparing for the William Lowell Putnam and other mathematical competitions. Preface to the First Edition. Sources. 1988 edition. 192pp. 5 3/8 x 8 1/2. 0-486-69415-1

KING ARTHUR IN SEARCH OF HIS DOG AND OTHER CURIOUS PUZZLES, Raymond M. Smullyan. This fanciful, original collection for readers of all ages features arithmetic puzzles, logic problems related to crime detection, and logic and arithmetic puzzles involving King Arthur and his Dogs of the Round Table. 160pp. 5 3/8 x 8 1/2. 0-486-47435-6

UNDECIDABLE THEORIES: Studies in Logic and the Foundation of Mathematics, Alfred Tarski in collaboration with Andrzej Mostowski and Raphael M. Robinson. This well-known book by the famed logician consists of three treatises: "A General Method in Proofs of Undecidability," "Undecidability and Essential Undecidability in Mathematics," and "Undecidability of the Elementary Theory of Groups." 1953 edition. 112pp. 5 3/8 x 8 1/2. 0-486-47703-7

LOGIC FOR MATHEMATICIANS, J. Barkley Rosser. Examination of essential topics and theorems assumes no background in logic. "Undoubtedly a major addition to the literature of mathematical logic." – *Bulletin of the American Mathematical Society*. 1978 edition. 592pp. 6 1/8 x 9 1/4. 0-486-46898-4

INTRODUCTION TO PROOF IN ABSTRACT MATHEMATICS, Andrew Wohlgemuth. This undergraduate text teaches students what constitutes an acceptable proof, and it develops their ability to do proofs of routine problems as well as those requiring creative insights. 1990 edition. 384pp. 6 1/2 x 9 1/4. 0-486-47854-8

FIRST COURSE IN MATHEMATICAL LOGIC, Patrick Suppes and Shirley Hill. Rigorous introduction is simple enough in presentation and context for wide range of students. Symbolizing sentences; logical inference; truth and validity; truth tables; terms, predicates, universal quantifiers; universal specification and laws of identity; more. 288pp. 5 3/8 x 8 1/2. 0-486-42259-3

Browse over 9,000 books at www.doverpublications.com

Mathematics-Algebra and Calculus

VECTOR CALCULUS, Peter Baxandall and Hans Liebeck. This introductory text offers a rigorous, comprehensive treatment. Classical theorems of vector calculus are amply illustrated with figures, worked examples, physical applications, and exercises with hints and answers. 1986 edition. 560pp. 5 3/8 x 8 1/2. 0-486-46620-5

ADVANCED CALCULUS: An Introduction to Classical Analysis, Louis Brand. A course in analysis that focuses on the functions of a real variable, this text introduces the basic concepts in their simplest setting and illustrates its teachings with numerous examples, theorems, and proofs. 1955 edition. 592pp. 5 3/8 x 8 1/2. 0-486-44548-8

ADVANCED CALCULUS, Avner Friedman. Intended for students who have already completed a one-year course in elementary calculus, this two-part treatment advances from functions of one variable to those of several variables. Solutions. 1971 edition. 432pp. 5 3/8 x 8 1/2. 0-486-45795-8

METHODS OF MATHEMATICS APPLIED TO CALCULUS, PROBABILITY, AND STATISTICS, Richard W. Hamming. This 4-part treatment begins with algebra and analytic geometry and proceeds to an exploration of the calculus of algebraic functions and transcendental functions and applications. 1985 edition. Includes 310 figures and 18 tables. 880pp. 6 1/2 x 9 1/4. 0-486-43945-3

BASIC ALGEBRA I: Second Edition, Nathan Jacobson. A classic text and standard reference for a generation, this volume covers all undergraduate algebra topics, including groups, rings, modules, Galois theory, polynomials, linear algebra, and associative algebra. 1985 edition. 528pp. 6 1/8 x 9 1/4. 0-486-47189-6

BASIC ALGEBRA II: Second Edition, Nathan Jacobson. This classic text and standard reference comprises all subjects of a first-year graduate-level course, including in-depth coverage of groups and polynomials and extensive use of categories and functors. 1989 edition. 704pp. 6 1/8 x 9 1/4. 0-486-47187-X

CALCULUS: An Intuitive and Physical Approach (Second Edition), Morris Kline. Application-oriented introduction relates the subject as closely as possible to science with explorations of the derivative; differentiation and integration of the powers of x; theorems on differentiation, antidifferentiation; the chain rule; trigonometric functions; more. Examples. 1967 edition. 960pp. 6 1/2 x 9 1/4. 0-486-40453-6

ABSTRACT ALGEBRA AND SOLUTION BY RADICALS, John E. Maxfield and Margaret W. Maxfield. Accessible advanced undergraduate-level text starts with groups, rings, fields, and polynomials and advances to Galois theory, radicals and roots of unity, and solution by radicals. Numerous examples, illustrations, exercises, appendixes. 1971 edition. 224pp. 6 1/8 x 9 1/4. 0-486-47723-1

AN INTRODUCTION TO THE THEORY OF LINEAR SPACES, Georgi E. Shilov. Translated by Richard A. Silverman. Introductory treatment offers a clear exposition of algebra, geometry, and analysis as parts of an integrated whole rather than separate subjects. Numerous examples illustrate many different fields, and problems include hints or answers. 1961 edition. 320pp. 5 3/8 x 8 1/2. 0-486-63070-6

LINEAR ALGEBRA, Georgi E. Shilov. Covers determinants, linear spaces, systems of linear equations, linear functions of a vector argument, coordinate transformations, the canonical form of the matrix of a linear operator, bilinear and quadratic forms, and more. 387pp. 5 3/8 x 8 1/2. 0-486-63518-X

Browse over 9,000 books at www.doverpublications.com

Mathematics-Probability and Statistics

BASIC PROBABILITY THEORY, Robert B. Ash. This text emphasizes the probabilistic way of thinking, rather than measure-theoretic concepts. Geared toward advanced undergraduates and graduate students, it features solutions to some of the problems. 1970 edition. 352pp. 5 3/8 x 8 1/2. 0-486-46628-0

PRINCIPLES OF STATISTICS, M. G. Bulmer. Concise description of classical statistics, from basic dice probabilities to modern regression analysis. Equal stress on theory and applications. Moderate difficulty; only basic calculus required. Includes problems with answers. 252pp. 5 5/8 x 8 1/4. 0-486-63760-3

OUTLINE OF BASIC STATISTICS: Dictionary and Formulas, John E. Freund and Frank J. Williams. Handy guide includes a 70-page outline of essential statistical formulas covering grouped and ungrouped data, finite populations, probability, and more, plus over 1,000 clear, concise definitions of statistical terms. 1966 edition. 208pp. 5 3/8 x 8 1/2. 0-486-47769-X

GOOD THINKING: The Foundations of Probability and Its Applications, Irving J. Good. This in-depth treatment of probability theory by a famous British statistician explores Keynesian principles and surveys such topics as Bayesian rationality, corroboration, hypothesis testing, and mathematical tools for induction and simplicity. 1983 edition. 352pp. 5 3/8 x 8 1/2. 0-486-47438-0

INTRODUCTION TO PROBABILITY THEORY WITH CONTEMPORARY APPLICATIONS, Lester L. Helms. Extensive discussions and clear examples, written in plain language, expose students to the rules and methods of probability. Exercises foster problem-solving skills, and all problems feature step-by-step solutions. 1997 edition. 368pp. 6 1/2 x 9 1/4. 0-486-47418-6

CHANCE, LUCK, AND STATISTICS, Horace C. Levinson. In simple, non-technical language, this volume explores the fundamentals governing chance and applies them to sports, government, and business. "Clear and lively ... remarkably accurate." – *Scientific Monthly*. 384pp. 5 3/8 x 8 1/2. 0-486-41997-5

FIFTY CHALLENGING PROBLEMS IN PROBABILITY WITH SOLUTIONS, Frederick Mosteller. Remarkable puzzlers, graded in difficulty, illustrate elementary and advanced aspects of probability. These problems were selected for originality, general interest, or because they demonstrate valuable techniques. Also includes detailed solutions. 88pp. 5 3/8 x 8 1/2. 0-486-65355-2

EXPERIMENTAL STATISTICS, Mary Gibbons Natrella. A handbook for those seeking engineering information and quantitative data for designing, developing, constructing, and testing equipment. Covers the planning of experiments, the analyzing of extreme-value data; and more. 1966 edition. Index. Includes 52 figures and 76 tables. 560pp. 8 3/8 x 11. 0-486-43937-2

STOCHASTIC MODELING: Analysis and Simulation, Barry L. Nelson. Coherent introduction to techniques also offers a guide to the mathematical, numerical, and simulation tools of systems analysis. Includes formulation of models, analysis, and interpretation of results. 1995 edition. 336pp. 6 1/8 x 9 1/4. 0-486-47770-3

INTRODUCTION TO BIOSTATISTICS: Second Edition, Robert R. Sokal and F. James Rohlf. Suitable for undergraduates with a minimal background in mathematics, this introduction ranges from descriptive statistics to fundamental distributions and the testing of hypotheses. Includes numerous worked-out problems and examples. 1987 edition. 384pp. 6 1/8 x 9 1/4. 0-486-46961-1

Browse over 9,000 books at www.doverpublications.com

Mathematics–Geometry and Topology

PROBLEMS AND SOLUTIONS IN EUCLIDEAN GEOMETRY, M. N. Aref and William Wernick. Based on classical principles, this book is intended for a second course in Euclidean geometry and can be used as a refresher. More than 200 problems include hints and solutions. 1968 edition. 272pp. 5 3/8 x 8 1/2. 0-486-47720-7

TOPOLOGY OF 3-MANIFOLDS AND RELATED TOPICS, Edited by M. K. Fort, Jr. With a New Introduction by Daniel Silver. Summaries and full reports from a 1961 conference discuss decompositions and subsets of 3-space; n-manifolds; knot theory; the Poincaré conjecture; and periodic maps and isotopies. Familiarity with algebraic topology required. 1962 edition. 272pp. 6 1/8 x 9 1/4. 0-486-47753-3

POINT SET TOPOLOGY, Steven A. Gaal. Suitable for a complete course in topology, this text also functions as a self-contained treatment for independent study. Additional enrichment materials make it equally valuable as a reference. 1964 edition. 336pp. 5 3/8 x 8 1/2. 0-486-47222-1

INVITATION TO GEOMETRY, Z. A. Melzak. Intended for students of many different backgrounds with only a modest knowledge of mathematics, this text features self-contained chapters that can be adapted to several types of geometry courses. 1983 edition. 240pp. 5 3/8 x 8 1/2. 0-486-46626-4

TOPOLOGY AND GEOMETRY FOR PHYSICISTS, Charles Nash and Siddhartha Sen. Written by physicists for physics students, this text assumes no detailed background in topology or geometry. Topics include differential forms, homotopy, homology, cohomology, fiber bundles, connection and covariant derivatives, and Morse theory. 1983 edition. 320pp. 5 3/8 x 8 1/2. 0-486-47852-1

BEYOND GEOMETRY: Classic Papers from Riemann to Einstein, Edited with an Introduction and Notes by Peter Pesic. This is the only English-language collection of these 8 accessible essays. They trace seminal ideas about the foundations of geometry that led to Einstein's general theory of relativity. 224pp. 6 1/8 x 9 1/4. 0-486-45350-2

GEOMETRY FROM EUCLID TO KNOTS, Saul Stahl. This text provides a historical perspective on plane geometry and covers non-neutral Euclidean geometry, circles and regular polygons, projective geometry, symmetries, inversions, informal topology, and more. Includes 1,000 practice problems. Solutions available. 2003 edition. 480pp. 6 1/8 x 9 1/4. 0-486-47459-3

TOPOLOGICAL VECTOR SPACES, DISTRIBUTIONS AND KERNELS, François Trèves. Extending beyond the boundaries of Hilbert and Banach space theory, this text focuses on key aspects of functional analysis, particularly in regard to solving partial differential equations. 1967 edition. 592pp. 5 3/8 x 8 1/2. 0-486-45352-9

INTRODUCTION TO PROJECTIVE GEOMETRY, C. R. Wylie, Jr. This introductory volume offers strong reinforcement for its teachings, with detailed examples and numerous theorems, proofs, and exercises, plus complete answers to all odd-numbered end-of-chapter problems. 1970 edition. 576pp. 6 1/8 x 9 1/4. 0-486-46895-X

FOUNDATIONS OF GEOMETRY, C. R. Wylie, Jr. Geared toward students preparing to teach high school mathematics, this text explores the principles of Euclidean and non-Euclidean geometry and covers both generalities and specifics of the axiomatic method. 1964 edition. 352pp. 6 x 9. 0-486-47214-0

Browse over 9,000 books at www.doverpublications.com

Mathematics-History

THE WORKS OF ARCHIMEDES, Archimedes. Translated by Sir Thomas Heath. Complete works of ancient geometer feature such topics as the famous problems of the ratio of the areas of a cylinder and an inscribed sphere; the properties of conoids, spheroids, and spirals; more. 326pp. 5 3/8 x 8 1/2. 0-486-42084-1

THE HISTORICAL ROOTS OF ELEMENTARY MATHEMATICS, Lucas N. H. Bunt, Phillip S. Jones, and Jack D. Bedient. Exciting, hands-on approach to understanding fundamental underpinnings of modern arithmetic, algebra, geometry and number systems examines their origins in early Egyptian, Babylonian, and Greek sources. 336pp. 5 3/8 x 8 1/2. 0-486-25563-8

THE THIRTEEN BOOKS OF EUCLID'S ELEMENTS, Euclid. Contains complete English text of all 13 books of the Elements plus critical apparatus analyzing each definition, postulate, and proposition in great detail. Covers textual and linguistic matters; mathematical analyses of Euclid's ideas; classical, medieval, Renaissance and modern commentators; refutations, supports, extrapolations, reinterpretations and historical notes. 995 figures. Total of 1,425pp. All books 5 3/8 x 8 1/2.
Vol. I: 443pp. 0-486-60088-2
Vol. II: 464pp. 0-486-60089-0
Vol. III: 546pp. 0-486-60090-4

A HISTORY OF GREEK MATHEMATICS, Sir Thomas Heath. This authoritative two-volume set that covers the essentials of mathematics and features every landmark innovation and every important figure, including Euclid, Apollonius, and others. 5 3/8 x 8 1/2. Vol. I: 461pp. 0-486-24073-8
Vol. II: 597pp. 0-486-24074-6

A MANUAL OF GREEK MATHEMATICS, Sir Thomas L. Heath. This concise but thorough history encompasses the enduring contributions of the ancient Greek mathematicians whose works form the basis of most modern mathematics. Discusses Pythagorean arithmetic, Plato, Euclid, more. 1931 edition. 576pp. 5 3/8 x 8 1/2.
0-486-43231-9

CHINESE MATHEMATICS IN THE THIRTEENTH CENTURY, Ulrich Libbrecht. An exploration of the 13th-century mathematician Ch'in, this fascinating book combines what is known of the mathematician's life with a history of his only extant work, the Shu-shu chiu-chang. 1973 edition. 592pp. 5 3/8 x 8 1/2.
0-486-44619-0

PHILOSOPHY OF MATHEMATICS AND DEDUCTIVE STRUCTURE IN EUCLID'S ELEMENTS, Ian Mueller. This text provides an understanding of the classical Greek conception of mathematics as expressed in Euclid's Elements. It focuses on philosophical, foundational, and logical questions and features helpful appendixes. 400pp. 6 1/2 x 9 1/4. 0-486-45300-6

BEYOND GEOMETRY: Classic Papers from Riemann to Einstein, Edited with an Introduction and Notes by Peter Pesic. This is the only English-language collection of these 8 accessible essays. They trace seminal ideas about the foundations of geometry that led to Einstein's general theory of relativity. 224pp. 6 1/8 x 9 1/4. 0-486-45350-2

HISTORY OF MATHEMATICS, David E. Smith. Two-volume history – from Egyptian papyri and medieval maps to modern graphs and diagrams. Non-technical chronological survey with thousands of biographical notes, critical evaluations, and contemporary opinions on over 1,100 mathematicians. 5 3/8 x 8 1/2.
Vol. I: 618pp. 0-486-20429-4
Vol. II: 736pp. 0-486-20430-8

Browse over 9,000 books at www.doverpublications.com

Physics

THEORETICAL NUCLEAR PHYSICS, John M. Blatt and Victor F. Weisskopf. An uncommonly clear and cogent investigation and correlation of key aspects of theoretical nuclear physics by leading experts: the nucleus, nuclear forces, nuclear spectroscopy, two-, three- and four-body problems, nuclear reactions, beta-decay and nuclear shell structure. 896pp. 5 3/8 x 8 1/2. 0-486-66827-4

QUANTUM THEORY, David Bohm. This advanced undergraduate-level text presents the quantum theory in terms of qualitative and imaginative concepts, followed by specific applications worked out in mathematical detail. 655pp. 5 3/8 x 8 1/2. 0-486-65969-0

ATOMIC PHYSICS AND HUMAN KNOWLEDGE, Niels Bohr. Articles and speeches by the Nobel Prize–winning physicist, dating from 1934 to 1958, offer philosophical explorations of the relevance of atomic physics to many areas of human endeavor. 1961 edition. 112pp. 5 3/8 x 8 1/2. 0-486-47928-5

COSMOLOGY, Hermann Bondi. A co-developer of the steady-state theory explores his conception of the expanding universe. This historic book was among the first to present cosmology as a separate branch of physics. 1961 edition. 192pp. 5 3/8 x 8 1/2. 0-486-47483-6

LECTURES ON QUANTUM MECHANICS, Paul A. M. Dirac. Four concise, brilliant lectures on mathematical methods in quantum mechanics from Nobel Prize-winning quantum pioneer build on idea of visualizing quantum theory through the use of classical mechanics. 96pp. 5 3/8 x 8 1/2. 0-486-41713-1

THE PRINCIPLE OF RELATIVITY, Albert Einstein and Frances A. Davis. Eleven papers that forged the general and special theories of relativity include seven papers by Einstein, two by Lorentz, and one each by Minkowski and Weyl. 1923 edition. 240pp. 5 3/8 x 8 1/2. 0-486-60081-5

PHYSICS OF WAVES, William C. Elmore and Mark A. Heald. Ideal as a classroom text or for individual study, this unique one-volume overview of classical wave theory covers wave phenomena of acoustics, optics, electromagnetic radiations, and more. 477pp. 5 3/8 x 8 1/2. 0-486-64926-1

THERMODYNAMICS, Enrico Fermi. In this classic of modern science, the Nobel Laureate presents a clear treatment of systems, the First and Second Laws of Thermodynamics, entropy, thermodynamic potentials, and much more. Calculus required. 160pp. 5 3/8 x 8 1/2. 0-486-60361-X

QUANTUM THEORY OF MANY-PARTICLE SYSTEMS, Alexander L. Fetter and John Dirk Walecka. Self-contained treatment of nonrelativistic many-particle systems discusses both formalism and applications in terms of ground-state (zero-temperature) formalism, finite-temperature formalism, canonical transformations, and applications to physical systems. 1971 edition. 640pp. 5 3/8 x 8 1/2. 0-486-42827-3

QUANTUM MECHANICS AND PATH INTEGRALS: Emended Edition, Richard P. Feynman and Albert R. Hibbs. Emended by Daniel F. Styer. The Nobel Prize–winning physicist presents unique insights into his theory and its applications. Feynman starts with fundamentals and advances to the perturbation method, quantum electrodynamics, and statistical mechanics. 1965 edition, emended in 2005. 384pp. 6 1/8 x 9 1/4. 0-486-47722-3

Browse over 9,000 books at www.doverpublications.com

Physics

INTRODUCTION TO MODERN OPTICS, Grant R. Fowles. A complete basic undergraduate course in modern optics for students in physics, technology, and engineering. The first half deals with classical physical optics; the second, quantum nature of light. Solutions. 336pp. 5 3/8 x 8 1/2. 0-486-65957-7

THE QUANTUM THEORY OF RADIATION: Third Edition, W. Heitler. The first comprehensive treatment of quantum physics in any language, this classic introduction to basic theory remains highly recommended and widely used, both as a text and as a reference. 1954 edition. 464pp. 5 3/8 x 8 1/2. 0-486-64558-4

QUANTUM FIELD THEORY, Claude Itzykson and Jean-Bernard Zuber. This comprehensive text begins with the standard quantization of electrodynamics and perturbative renormalization, advancing to functional methods, relativistic bound states, broken symmetries, nonabelian gauge fields, and asymptotic behavior. 1980 edition. 752pp. 6 1/2 x 9 1/4. 0-486-44568-2

FOUNDATIONS OF POTENTIAL THERY, Oliver D. Kellogg. Introduction to fundamentals of potential functions covers the force of gravity, fields of force, potentials, harmonic functions, electric images and Green's function, sequences of harmonic functions, fundamental existence theorems, and much more. 400pp. 5 3/8 x 8 1/2. 0-486-60144-7

FUNDAMENTALS OF MATHEMATICAL PHYSICS, Edgar A. Kraut. Indispensable for students of modern physics, this text provides the necessary background in mathematics to study the concepts of electromagnetic theory and quantum mechanics. 1967 edition. 480pp. 6 1/2 x 9 1/4. 0-486-45809-1

GEOMETRY AND LIGHT: The Science of Invisibility, Ulf Leonhardt and Thomas Philbin. Suitable for advanced undergraduate and graduate students of engineering, physics, and mathematics and scientific researchers of all types, this is the first authoritative text on invisibility and the science behind it. More than 100 full-color illustrations, plus exercises with solutions. 2010 edition. 288pp. 7 x 9 1/4. 0-486-47693-6

QUANTUM MECHANICS: New Approaches to Selected Topics, Harry J. Lipkin. Acclaimed as "excellent" (*Nature*) and "very original and refreshing" (*Physics Today*), these studies examine the Mössbauer effect, many-body quantum mechanics, scattering theory, Feynman diagrams, and relativistic quantum mechanics. 1973 edition. 480pp. 5 3/8 x 8 1/2. 0-486-45893-8

THEORY OF HEAT, James Clerk Maxwell. This classic sets forth the fundamentals of thermodynamics and kinetic theory simply enough to be understood by beginners, yet with enough subtlety to appeal to more advanced readers, too. 352pp. 5 3/8 x 8 1/2. 0-486-41735-2

QUANTUM MECHANICS, Albert Messiah. Subjects include formalism and its interpretation, analysis of simple systems, symmetries and invariance, methods of approximation, elements of relativistic quantum mechanics, much more. "Strongly recommended." – *American Journal of Physics*. 1152pp. 5 3/8 x 8 1/2. 0-486-40924-4

RELATIVISTIC QUANTUM FIELDS, Charles Nash. This graduate-level text contains techniques for performing calculations in quantum field theory. It focuses chiefly on the dimensional method and the renormalization group methods. Additional topics include functional integration and differentiation. 1978 edition. 240pp. 5 3/8 x 8 1/2. 0-486-47752-5

Browse over 9,000 books at www.doverpublications.com

Physics

MATHEMATICAL TOOLS FOR PHYSICS, James Nearing. Encouraging students' development of intuition, this original work begins with a review of basic mathematics and advances to infinite series, complex algebra, differential equations, Fourier series, and more. 2010 edition. 496pp. 6 1/8 x 9 1/4. 0-486-48212-X

TREATISE ON THERMODYNAMICS, Max Planck. Great classic, still one of the best introductions to thermodynamics. Fundamentals, first and second principles of thermodynamics, applications to special states of equilibrium, more. Numerous worked examples. 1917 edition. 297pp. 5 3/8 x 8. 0-486-66371-X

AN INTRODUCTION TO RELATIVISTIC QUANTUM FIELD THEORY, Silvan S. Schweber. Complete, systematic, and self-contained, this text introduces modern quantum field theory. "Combines thorough knowledge with a high degree of didactic ability and a delightful style." – *Mathematical Reviews.* 1961 edition. 928pp. 5 3/8 x 8 1/2. 0-486-44228-4

THE ELECTROMAGNETIC FIELD, Albert Shadowitz. Comprehensive undergraduate text covers basics of electric and magnetic fields, building up to electromagnetic theory. Related topics include relativity theory. Over 900 problems, some with solutions. 1975 edition. 768pp. 5 5/8 x 8 1/4. 0-486-65660-8

THE PRINCIPLES OF STATISTICAL MECHANICS, Richard C. Tolman. Definitive treatise offers a concise exposition of classical statistical mechanics and a thorough elucidation of quantum statistical mechanics, plus applications of statistical mechanics to thermodynamic behavior. 1930 edition. 704pp. 5 5/8 x 8 1/4. 0-486-63896-0

INTRODUCTION TO THE PHYSICS OF FLUIDS AND SOLIDS, James S. Trefil. This interesting, informative survey by a well-known science author ranges from classical physics and geophysical topics, from the rings of Saturn and the rotation of the galaxy to underground nuclear tests. 1975 edition. 320pp. 5 3/8 x 8 1/2. 0-486-47437-2

STATISTICAL PHYSICS, Gregory H. Wannier. Classic text combines thermodynamics, statistical mechanics, and kinetic theory in one unified presentation. Topics include equilibrium statistics of special systems, kinetic theory, transport coefficients, and fluctuations. Problems with solutions. 1966 edition. 532pp. 5 3/8 x 8 1/2. 0-486-65401-X

SPACE, TIME, MATTER, Hermann Weyl. Excellent introduction probes deeply into Euclidean space, Riemann's space, Einstein's general relativity, gravitational waves and energy, and laws of conservation. "A classic of physics." – *British Journal for Philosophy and Science.* 330pp. 5 3/8 x 8 1/2. 0-486-60267-2

RANDOM VIBRATIONS: Theory and Practice, Paul H. Wirsching, Thomas L. Paez and Keith Ortiz. Comprehensive text and reference covers topics in probability, statistics, and random processes, plus methods for analyzing and controlling random vibrations. Suitable for graduate students and mechanical, structural, and aerospace engineers. 1995 edition. 464pp. 5 3/8 x 8 1/2. 0-486-45015-5

PHYSICS OF SHOCK WAVES AND HIGH-TEMPERATURE HYDRO DYNAMIC PHENOMENA, Ya B. Zel'dovich and Yu P. Raizer. Physical, chemical processes in gases at high temperatures are focus of outstanding text, which combines material from gas dynamics, shock-wave theory, thermodynamics and statistical physics, other fields. 284 illustrations. 1966–1967 edition. 944pp. 6 1/8 x 9 1/4. 0-486-42002-7

Browse over 9,000 books at www.doverpublications.com